Geotechnical, Geological and Earthquake Engineering

Volume 52

The book series entitled *Geotechnical, Geological and Earthquake Engineering* has been initiated to provide carefully selected and reviewed information from the most recent findings and observations in these engineering fields. Researchers as well as practitioners in these interdisciplinary fields will find valuable information in these book volumes, contributing to advancing the state-of-the-art and state-of-the-practice. This book series comprises monographs, edited volumes, handbooks as well as occasionally symposia and workshop proceedings volumes on the broad topics of geotechnical, geological and earthquake engineering. The topics covered are theoretical and applied soil mechanics, foundation engineering, geotechnical earthquake engineering, earthquake engineering, rock mechanics, engineering geology, engineering seismology, earthquake hazard, etc.

Prospective authors and/or editors should consult the **Series Editor Atilla Ansal** for more details. Any comments or suggestions for future volumes are welcomed.

EDITORS AND EDITORIAL BOARD MEMBERS IN THE GEOTECHNICAL, GEOLOGICAL AND EARTHQUAKE ENGINEERING BOOK SERIES:

More information about this series at https://link.springer.com/bookseries/6011

Lanmin Wang · Jian-Min Zhang ·
Rui Wang
Editors

Proceedings of the 4th International Conference on Performance Based Design in Earthquake Geotechnical Engineering (Beijing 2022)

Set 1

 Springer

Editors
Lanmin Wang
Lanzhou Institute of Seismology
China Earthquake Administration
Lanzhou, China

Jian-Min Zhang
Department of Hydraulic Engineering
Tsinghua University
Beijing, China

Rui Wang
Department of Hydraulic Engineering
Tsinghua University
Beijing, China

ISSN 1573-6059   ISSN 1872-4671   (electronic)
Geotechnical, Geological and Earthquake Engineering
ISBN 978-3-031-11897-5   ISBN 978-3-031-11898-2   (eBook)
https://doi.org/10.1007/978-3-031-11898-2

This Springer imprint is published by the registered company Springer Nature Switzerland AG
The registered company address is: Gewerbestrasse 11, 6330 Cham, Switzerland

Preface

The 4th International Conference on Performance-based Design in Earthquake Geotechnical Engineering (PBD-IV) will be held on July 15–17, 2022, in Beijing, China. The PBD-IV conference is organized under the auspices of the International Society of Soil Mechanics and Geotechnical Engineering—Technical Committee on Earthquake Geotechnical Engineering and Associated Problems (ISSMGE-TC203). The PBD-I, PBD-II, and PBD-III events in Japan (2009), Italy (2012), and Canada (2017), respectively, were highly successful events for the international earthquake geotechnical engineering community. The PBD events have been excellent companions to the International Conference on Earthquake Geotechnical Engineering (ICEGE) series that TC203 has held in Japan (1995), Portugal (1999), USA (2004), Greece (2007), Chile (2011), New Zealand (2015), and Italy (2019). The goal of PBD-IV is to provide an open forum for delegates to interact with their international colleagues and advance performance-based design research and practices for earthquake geotechnical engineering.

The proceedings of PBD-IV is the outcome of more than two years of concerted efforts by the conference organizing committee, scientific committee, and steering committee. The proceedings include 14 keynote lecture papers, 25 invited theme lecture papers, one TC203 Young Researcher Award lecture paper, and 187 accepted technical papers from 27 countries and regions. Each accepted paper in the conference proceedings was subject to review by credited peers. The final accepted technical papers are organized into five themes and six special sessions: (1) Performance Design and Seismic Hazard Assessment; (2) Ground Motions and Site Effects; (3) Foundations and Soil–Structure Interaction; (4) Slope Stability and Reinforcement; (5) Liquefaction and Testing; (6) S1: Liquefaction experiment and analysis projects (LEAP); (7) S2: Liquefaction Database; (8) S3: Embankment Dams; (9) S4: Earthquake Disaster Risk of Special Soil Sites and Engineering Seismic Design; (10) S5: Special Session on Soil Dynamic Properties at Micro-scale: From Small Strain Wave Propagation to Large Strain Liquefaction; and (11) S6: Underground Structures.

The Chinese Institution of Soil Mechanics and Geotechnical Engineering - China Civil Engineering Society and China Earthquake Administration provided tremendous support to the organization of the conference. The publication of the proceedings was financially supported by the National Natural Science Foundation of China (No. 52022046 and No. 52038005).

We would like to acknowledge the contributions of all authors and reviewers, as well as members of both Steering Committee and Scientific Committee for contributing their wisdom and influence to confer a successful PBD-IV conference in Beijing and attract the participation of many international colleagues in the hard time of COVID pandemic.

March 2022

Misko Cubrinovski
ISSMGE-TC203 Chair

Jian-Min Zhang
Conference Honorary Chair

Lanmin Wang
Conference Chair

Rui Wang
Conference Secretary

Organization

Honorary Chair

Jian-Min Zhang

Conference Chair

Lanmin Wang

Conference Secretary

Rui Wang

Steering Committee

Atilla Ansal
Charles W. W. Ng
George Gazetas
Izzat Idriss
Jian Xu
Jian-Min Zhang
Jonathan Bray
Kenji Ishihara
Kyriazis Pitilakis

Liam Finn
Lili Xie
Misko Cubrinovski
Ramon Verdugo
Ross Boulanger
Takaji Kokusho
Xianjing Kong
Xiaonan Gong
Yunmin Chen

Scientific Committee

Anastasios Anastasiadis
A. Murali Krishna
Alain Pecker
Amir Kaynia
Andrew Lees
António Araújo Correia
Antonio Morales Esteban

Arnaldo Mario Barchiesi
Carl Wesäll
Carolina Sigarán-Loría
Christos Vrettos
Deepankar Choudhury
Dharma Wijewickreme
Diego Alberto Cordero Carballo

Duhee Park
Eleni Stathopoulou
Ellen Rathje
Farzin Shahrokhi
Francesco Silvestri
Gang Wang
George Athanasopoulos
George Bouckovalas
Ioannis Anastasopoulos
Ivan Gratchev
Jan Laue
Jay Lee
Jean-François Semblat
Jin Man Kim
Jorgen Johansson
Juan Manuel Mayoral Villa
Jun Yang
Kemal Onder Cetin

Louis Ge
M. A. Klyachko
Masyhur Irsyam
Mitsu Okamura
Nicolas Lambert
Paulo Coelho
Sjoerd Van Ballegooy
Rubén Galindo Aires
Ryosuke Uzuoka
Sebastiano Foti
Seyed Mohsen Haeri
Siau Chen Chian
Stavroula Kontoe
Valérie Whenham
Waldemar Świdziński
Wei F. Lee
Zbigniew Bednarczyk

Organizing Committee

Ailan Che
Baitao Sun
Degao Zou
Gang Wang
Gang Zheng
Guoxing Chen
Hanlong Liu
Hongbo Xin
Hongru Zhang
Jie Cui
Jilin Qi
Maosong Huang
Ping Wang
Shengjun Shao

Wei. F. Lee
Wensheng Gao
Xianzhang Ling
Xiaojun Li
Xiaoming Yuan
Xiuli Du
Yanfeng Wen
Yangping Yao
Yanguo Zhou
Yong Zhou
Yufeng Gao
Yunsheng Yao
Zhijian Wu

Acknowledgements

Manuscript Reviewers
The editors are grateful to the following people who helped to review the manuscripts and hence assisted in improving the overall technical standard and presentation of the papers in these proceedings:

Amir Kaynia
Atilla Ansal
Brady cox
Dongyoup Kwak
Ellen Rathje
Emilio Bilotta
Haizuo Zhou
Jiangtao Wei
Jon Stewart
Kyriazis Pitilakis
Lanmin Wang
Mahdi Taiebat
Majid Manzari
Mark Stringer
Ming Yang

Rui Wang
Scott Brandenberg
Tetsuo Tobita
Ueda Kyohei
Wenjun Lu
Wu Yongxin
Xiao Wei
Xiaoqiang Gu
Xin Huang
Yang Zhao
Yanguo Zhou
Yifei Cui
Yu Yao
Yumeng Tao
Zitao Zhang

Contents

S6: Special Session on Underground Structures

Keynote Lectures

Performance-Based Seismic Assessment of Slope Systems

Jonathan D. Bray[1]([✉]) and Jorge Macedo[2]

[1] University of California, Berkeley, CA 94720-1710, USA
jonbray@berkeley.edu
[2] Georgia Institute of Technology, Atlanta, GA, USA

Abstract. Seismic slope displacement procedures are useful in the evaluation of earth embankments and natural slopes. The calculated seismic slope displacement provides an index of performance. Newmark-based sliding block models are typically employed. The manner in which the key components of the analysis are addressed largely determines the reliability of a particular procedure. The primary source of uncertainty in assessing the seismic performance of an earth slope is the input ground motion. Hence, sliding block procedures have advanced over the last two decades through the use of larger sets of ground motion records. Recent updates of the procedures developed by the authors are highlighted. The nonlinear fully coupled stick-slip sliding block model calculates reasonable seismic slope displacements. Displacements depend primarily on the earth structure's yield coefficient and the earthquake ground motion's spectral acceleration at the effective fundamental period of the sliding mass. Through their use, the sensitivity of the seismic slope displacements and their uncertainty to key input parameters can be investigated. These procedures can be implemented within a performance-based design framework to estimate the seismic slope displacement hazard, which is a more rational approach.

Keywords: Dam · Displacement · Earthquake · Hazard · Performance · Slope

1 Introduction

The failure of slope systems (e.g., earth dams, waste fills, natural slopes) during an earthquake can produce significant losses. Additionally, major damage without failure can have severe economic consequences. Accordingly, the seismic performance of earth structures and natural slopes requires evaluation. The assessment of the seismic performance of slope systems ranges from using straightforward pseudostatic procedures to advanced nonlinear effective stress finite element analyses. Performance should be evaluated through an assessment of the potential for seismically induced permanent displacement. Newmark (1965) sliding block analyses are typically utilized as part of the seismic evaluation of earth structures and natural slopes. They provide a preliminary assessment of an earth system's seismic performance. Aspects of these procedures are critiqued in this paper, and recently developed procedures for estimating earthquake-induced shear deformation in earth and waste structures and natural slopes are summarized and recommended for use in engineering practice.

© The Author(s), under exclusive license to Springer Nature Switzerland AG 2022
L. Wang et al. (Eds.): PBD-IV 2022, GGEE 52, pp. 3–23, 2022.
https://doi.org/10.1007/978-3-031-11898-2_1

2 Seismic Slope Stability Analysis

2.1 Critical Design Issues

Two critical design issues must be addressed when evaluating the seismic performance of an earth structure or slope:

1. First, most importantly, the engineer must investigate if there are materials in the structure or its foundation that will lose significant strength due to cyclic loading (e.g., soil liquefaction). If so, this should be the primary focus of the evaluation because a flow slide could result. The post-cyclic strength of materials that lose strength due to earthquake loading must be evaluated. The post-cyclic static slope stability factor of safety (*FS*) should be calculated. If it is near to or below one, a flow slide is possible. Mitigation measures or advanced analyses are warranted to address or to evaluate the flow slide and its consequences.
2. Second, if materials within or below the earth structure will not lose significant strength due to cyclic loading, the deformation of the earth structure or slope must be evaluated to assess if they jeopardize satisfactory performance of the system. The estimation of seismically induced slope displacement helps the engineer address this issue in combination with nonlinear effective stress finite element or finite different analyses when warranted. The calculation of seismic slope displacement using deformable sliding block analyses are the focus of this paper.

2.2 Shear-Induced Seismic Displacement

The calculated seismic slope displacement from a Newmark (1965)-type procedure, whether it is simplified or advanced, is an index of the potential seismic performance of the earth structure or slope. Seismic slope displacement estimates are approximate in nature due to the complexities of the dynamic response of the earth/waste materials involved and the variability of the earthquake ground motion, among other factors. However, when viewed as an index of potential seismic performance, the calculated seismic slope displacement can be used effectively in engineering practice to evaluate the seismic stability of earth structures and natural slopes.

A Newmark-type sliding block model captures that part of the seismically induced permanent displacement attributed to shear deformation (i.e., either rigid body slippage along a distinct failure surface or distributed shearing within the deformable sliding mass). Ground movement due to volumetric compression is not explicitly captured by Newmark models. The top of a slope can displace downward due to shear deformation or volumetric compression of the slope-forming materials. However, top of slope movements resulting from distributed shear straining within the sliding mass or stick-slip sliding along a failure surface are mechanistically different from top of slope movements that result from seismically induced volumetric compression of the materials forming the slope. Although a Newmark-type procedure may appear to capture the overall top of slope displacement for cases where seismic compression due to volumetric contraction of soil or waste is the dominant mechanism, this is merely because the seismic forces that produce large volumetric compression strains also often produce large calculated displacements in a Newmark method. This apparent correspondence should not imply that a sliding block model should be used to estimate seismic displacement due to volumetric strain. There are cases where the Newmark method does not capture the overall top of slope displacement, such as the seismic compression of compacted earth fills (e.g., Stewart et al. 2001). Shear-induced deformation and volumetric-induced deformation should be analyzed separately using procedures based on the sliding block model to estimate shear-induced displacement and using other procedures (e.g., Tokimatsu and Seed 1987) to estimate volumetric-induced displacement.

3   Components of a Seismic Slope Displacement Analysis

3.1   General

The critical components of a Newark-type sliding block analysis are: 1) the dynamic resistance of the structure, 2) the earthquake ground motion, 3) the dynamic response of the sliding mass, and 4) the permanent displacement calculational procedure. The dynamic resistance of the earth/waste structure or natural slope is a key component in the analysis. The system's yield coefficient defines its maximum dynamic resistance. The earthquake ground motion is the input to assessing the seismic demand on the system. The dynamic response of the potential sliding mass to the input earthquake ground motion should be considered because the sliding mass is rarely rigid. The Newmark calculational procedure should capture the coupled dynamic response and sliding resistance of the sliding mass during ground shaking. Other factors, such as topographic effects, can be important in some cases. In critiquing a seismic slope displacement procedure, one should consider how each procedure characterizes the slope's dynamic resistance, the earthquake ground motion, the dynamic response of the system to the ground motion, and the calculational procedure.

3.2   Dynamic Resistance

The slope's yield coefficient (k_y) represents its dynamic resistance. It depends primarily on the dynamic strength of the material along the critical sliding surface, the structure's geometry and weight, and the initial pore water pressures that determine the in situ effective stress within the system. The yield coefficient greatly influences the seismic slope displacement calculated by any Newmark-type sliding block model.

The primary issue in calculating k_y is estimating the dynamic strength of the critical strata within the slope. Several publications include extensive discussions of the dynamic strength of soil (e.g., Blake et al. 2002; Duncan and Wright 2005). The engineer should devote considerable attention and resources to developing realistic estimates of the dynamic strengths of key slope materials. Effective stress, drained strength parameters are appropriate for unsaturated or dilative cohesionless soil. Pore water pressure generation and post-cyclic residual shear strength are required to characterize saturated, contractive cohesionless soil. Newmark procedures should not be applied to cases involving soil that undergoes severe strength loss due to earthquake shaking (e.g., liquefaction) without considerable judgment.

For clay soil that does not liquefy, its dynamic peak undrained shear strength ($s_{u,dyn,peak}$) can be related to its static peak undrained shear strength ($s_{u,stat,peak}$) using adjustment factors (Chen et al. 2006) as:

$$s_{u,dyn,peak} = s_{u,stat,peak}(C_{rate})(C_{cyc})(C_{prog})(C_{def})$$ (1)

where C_{rate} = rate of loading factor, C_{cyc} = cyclic degradation factor, C_{prog} = progressive failure factor, and C_{def} = distributed shear deformation factor.

The shear strength of a plastic clay increases as the rate of loading increases (e.g., Seed and Chan 1966; Lacerda 1976; Biscontin and Pestana 2000, and Duncan and Wright 2005). The undrained shear strength of viscous clay materials can increase about 10% to 15% for each ten-fold increase in the strain rate. For example, Biscontin and Pestana (2000) found that $s_{u,dyn,peak}$ at earthquake rate of loadings was 1.3x larger than $s_{u,stat,peak}$ measured at conventional rates of loading in the vane shear test in a soft plastic clay. Rau (1998) found the shear strength mobilized in the first cycle of a rapid cyclic simple shear test on Young Bay Mud from Hamilton Air Force Base was up to 40% to 50% higher than that mobilized in a conventional static test performed at typical loading rates. Cyclic simple shear tests on Young Bay Mud at the San Francisco-Oakland Bay Bridge (Kammerer et al. 1999) indicated the strength mobilized in the first cycle of loading was about 40% greater than that mobilized in the monotonic test at the same level of strain but at the conventional strain-rate for monotonic shear tests. It depends on the clay and testing device, etc., but generally, the ratio of $s_{u,dyn,peak}/s_{u,stat,peak}$ in one cycle of loading at a strain rate representative of an earthquake loading relative to that for a conventional static test is on the order of 1.3 to 1.7.

With additional cycles of loading, the peak undrained shear strength of a plastic clay degrades (e.g., Seed and Chen 1966). This effect is captured with the cyclic degradation factor. For example, Rau (1998) found in her testing of Young Bay Mud that by the 15th load cycle, the cyclic shear strength was close to that obtained in the static tests. With an increasing number of cycles of loading its strength could reduce further. Shear strength reductions of 10% to 20% might be appropriate for large magnitude earthquakes with many cycles of loading. Due to cyclic degradation, as the number of cycles of loading increases, the clay's dynamic shear strength decreases, especially if the volumetric threshold strain of the material is exceeded, shear strains approach or exceed values that are half of its failure strain, and stress reversals occur.

The increased rate of loading increases the dynamic peak shear strength of a plastic clay while increasing the number of load cycles reduces its strength due to cyclic degradation. For example, Rumpelt and Sitar (1988) found Young Bay Mud's post-cyclic peak undrained shear strength ratio (s_u/σ'_{vo}) measured at slow strain-rates was about equal to its pre-seismic static strength of $s_u/\sigma'_{vo} = 0.35$. However, its s_u/σ'_{vo} was 0.55 for 2 cycles of rapid stress-controlled loading (an increase of 1.6), 0.44–0.48 for 12 load cycles (an increase of 1.3), and 0.41 for 22 load cycles (an increase of 1.2).

Additionally, a progressive failure factor less than one should be applied if the clay exhibits post-peak strain softening when it is likely that the dynamic peak shear strength will not be mobilized along the failure surface at the same time (Chen et al. 2006). A value of $C_{prog} = 0.9$ is often appropriate for moderately sensitive plastic clay. Additionally, deformations accumulate for stress cycles less than the dynamic peak shear strength due to the nonlinear elastoplastic response of soil (e.g., Makdisi and Seed 1978). A value of $C_{def} = 0.9$ is often appropriate to capture this effect.

Therefore, the dynamic peak shear strength of a plastic clay used in a sliding block analysis should depend on the combined effects of the rapid rate of earthquake loading and the equivalent number of significant cycles of loading, as well as the progressive failure and deformable sliding block effects. For example, if only one cycle of a near-fault, forward-directivity pulse motion occurs, the peak shear strength of Young Bay Mud might have a combined effect of $C_{rate} = 1.4$ and $C_{cyc} = 1.0$; whereas, if 30 cycles of loading are applied from a backwards-directivity long duration motion, the combined effect might be $C_{rate} = 1.4$ and $C_{cyc} = 0.8$. Combining the factors in Eq. 1, produces $s_{u,dyn,peak}/s_{u,stat,peak}$ ratios of $(1.4)(1.0)(0.9)(0.9) = 1.1$ and $(1.4)(0.8)(0.9)(0.9) = 0.9$ for the forward-directivity single pulse motion and for the backward-directivity multiple load cycles motion, respectively. As the shear strength of clay depends on the characteristics of the earthquake loading, one should use different shear strengths for the clay depending on the number of significant load cycles of the ground motion.

The use of a clay's dynamic peak shear strength would only be appropriate for a strain-hardening material or when limited seismic slope displacement is calculated. If moderate-to-large displacement is calculated for the case when the clay exhibits strain-softening, the dynamic shear strength used in the sliding block analysis needs to be compatible with the amount of shear strain induced in the clay. As the dynamic shear strength reduces as the clay is deformed beyond its peak shear strength, the resulting yield coefficient will reduce, and additional seismic slope displacement will be calculated. It is unconservative to use a constant k_y value based on peak strength when the soil exhibits strain-softening. If large displacement is calculated, the clay's residual shear strength is appropriate for calculating k_y. The residual strength of clay does not appear to be strain-rate-dependent (Biscontin and Pestana 2000).

Duncan (1996) found consistent estimates of a slope's static FS are calculated if a slope stability procedure that satisfies all three conditions of equilibrium is employed. Computer programs that utilize methods that satisfy full equilibrium, such as Spencer, Generalized Janbu, and Morgenstern and Price, should be used to calculate the static FS. These methods should also be used to calculate k_y, which is the horizontal seismic coefficient that results in a $FS = 1.0$ in a pseudostatic slope stability analysis.

Lastly, the potential sliding mass that has the lowest static FS may not be the most critical for dynamic analysis. A search should be made to find sliding surfaces that produce low k_y values as well. The most important parameter for identifying critical potential sliding masses for dynamic problems is k_y/k_{max}, where k_{max} is an estimate of the maximum seismic loading considering the dynamic response of the sliding mass.

3.3 Earthquake Ground Motion

An acceleration-time history provides a complete characterization of an earthquake ground motion. In a simplified description of a ground motion, its intensity, frequency content, and duration must be specified at a minimum. In this manner, a ground motion can be described in terms of parameters such as peak ground acceleration (PGA), mean period (T_m), and significant duration (D_{5-95}). It is overly simplistic to characterize an earthquake ground motion by just its PGA, because ground motions with identical

PGA values can vary significantly in terms of frequency content and duration, and most importantly, in terms of their effects on slope performance.

Spectral acceleration has been commonly employed in earthquake engineering to characterize an equivalent seismic loading on a structure from the earthquake ground motion. Bray and Travasarou (2007) found that the 5%-damped elastic spectral acceleration (S_a) at the degraded fundamental period of the potential sliding mass was the optimal ground motion intensity measure in terms of efficiency and sufficiency (i.e., it minimizes the variability in its correlation with seismic displacement, and it renders the relationship independent of other variables, respectively, Cornell and Luco 2001). An estimate of the initial fundamental period of the sliding mass (T_s) is required when using spectral acceleration; T_s is useful to characterize the dynamic response of a sliding mass. Additional benefits of using S_a are it can be estimated reliably with ground motion models and it is available at various return periods in ground motion hazard maps. Spectral acceleration captures the intensity and frequency content characteristics of an earthquake motion, but it fails to capture duration. Moment magnitude (M_w) can be added to capture the duration of strong shaking. Some Newmark-type models (e.g., Saygili and Rathje 2008, and Bray and Macedo 2019) also use peak ground velocity (*PGV*) to bring in frequency content or near-fault effects.

Ground motion characteristics vary systematically in different tectonic settings. Hence, it is important to use suites of a large number of ground motion records appropriate for the tectonic settings affecting the project. Thus, seismic slope displacement procedures should be developed for shallow crustal earthquakes along active plate margins, subduction zone interface and intraslab earthquakes, and stable continental earthquakes. As significant regional distinctions are identified (e.g., crustal attention in Japan vs. South America for subduction zone interface earthquakes), additional refinements may be justified. The exponential growth of the number of ground motion records in different regions of each tectonic setting is enabling researchers to examine these issues.

3.4 Dynamic Response and Seismic Displacement Calculation

The seismic slope displacement depends on the dynamic response of the potential sliding mass. With all other factors held constant, seismic displacement increases when the sliding mass is near resonance compared to that calculated for very stiff or very flexible slopes (e.g., Kramer and Smith 1997; Rathje and Bray 2000; Wartman et al. 2003). Many of the available seismic slope displacement procedures employ the original Newmark (1965) rigid sliding block assumption, which does not capture the dynamic response of the deformable sliding mass during earthquake shaking.

Seed and Martin (1966) introduced the concept of an equivalent acceleration to represent the seismic loading of a sliding earth mass. The horizontal equivalent acceleration (*HEA*)-time history when applied to a rigid sliding mass produces the same dynamic shear stresses along the sliding surface that is produced when a dynamic analysis of the deformable earth structure is performed. The calculation of the HEA-time history in a dynamic analysis that assumes no relative displacement occurs along the failure surface is decoupled from the rigid sliding block calculation that is performed using the HEA-time history to calculate the seismic slope displacement. Although an approximation, the decoupled approach provides a reasonable estimate of seismic displacement for many cases (e.g., Lin and Whitman 1982; Rathje and Bray 2000). However, it is not always reasonable, and it can lead to significant overestimation near resonance and some level of underestimation for cases where the structure has a large fundamental period or the ground motion is an intense near-fault motion. A nonlinear coupled stick-slip deformable sliding block model offers a more realistic representation of the dynamic response of an earth system by accounting for the deformability of the sliding mass and by considering the simultaneous occurrence of its nonlinear dynamic response and periodic sliding episodes (Fig. 1). Its validation with shaking table experiments provides confidence in its use (Wartman et al. 2003).

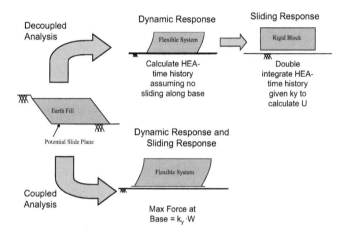

Fig. 1. Decoupled and fully coupled sliding block analysis (Bray 2007).

For seismic slope displacement methods that incorporate the seismic response of a deformable sliding block, the initial fundamental period of the sliding mass (T_s) of a relatively long sliding mass can be estimated as: $T_s = 4H/V_s'$, where H is the height of the sliding mass and V_s' is its equivalent shear wave velocity $= \Sigma[(V_{si})(m_i)]/\Sigma(m_i)$, where m_i is each differential mass i with shear wave velocity of V_{si} (Bray and Macedo 2021a, 2021b). For the case of a triangular-shaped sliding mass, $T_s = 2.6H/V_s'$ should be used (Ambraseys and Sarma 1967). The initial fundamental period of the sliding mass can be estimated approximately for other cases using a mass-weighted fundamental period (T_s') of rectangular slices of the sliding mass, as illustrated in Fig. 2. T_s' is calculated as the mass-weighted fundamental period of each incremental slice of the sliding mass, wherein the fundamental period of each rectangular slice of height H_i and shear wave velocity of V_{si}' is calculated as $T_{si} = 4H_i/V_{si}'$. The effective height of the entire slide mass (H') is calculated as $(T_s')(V_s')/4$, and the initial fundamental period of the sliding mass can be approximated as $T_s = 4H'/V_s'$. H' varies from $0.65\,H_v$ to $1.0\,H_v$, where H_v is the maximum height of a vertical line within the sliding mass (not the total height of the sliding mass from its base to its top). The use of T_s implicitly assumes the material below the sliding mass is rigid. Adjustments may be required if the base is not stiff relative to the potential sliding mass or if topographic effects are significant. As the method is based on 1D analysis, which may underestimate the seismic demand of shallow sliding at the top of 2D systems affected by topographic amplification, the input motion's intensity parameter should be amplified by 25% for moderately steep slopes and by 50% for steep slopes (Bray and Travasarou 2007). It may be amplified by 100% for localized sliding at the dam crest (Yu et al. 2012).

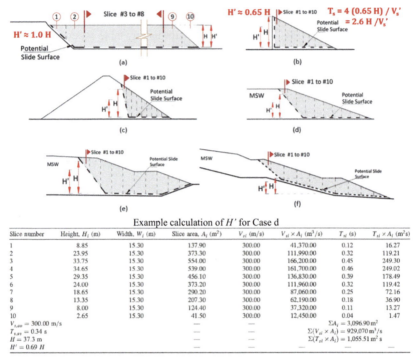

Example calculation of H' for Case d

Slice number	Height, H_i (m)	Width, W_i (m)	Slice area, A_i (m^2)	V_{si} (m/s)	$V_{si} \times A_i$ (m^3/s)	T_{si} (s)	$T_{si} \times A_i$ (m^{2}s)
1	8.85	15.30	137.90	300.00	41,370.00	0.12	16.27
2	23.95	15.30	373.30	300.00	111,990.00	0.32	119.21
3	33.75	15.30	554.00	300.00	166,200.00	0.45	249.30
4	34.65	15.30	539.00	300.00	161,700.00	0.46	249.02
5	29.35	15.30	456.10	300.00	136,830.00	0.39	178.49
6	24.00	15.30	373.20	300.00	111,960.00	0.32	119.42
7	18.65	15.30	290.20	300.00	87,060.00	0.25	72.16
8	13.35	15.30	207.30	300.00	62,190.00	0.18	36.90
9	8.00	15.30	124.40	300.00	37,320.00	0.11	13.27
10	2.65	15.30	41.50	300.00	12,450.00	0.04	1.47
$V_{s,av} = 300.00$ m/s			—	—	$\Sigma A_i = 3,096.90$ m^2		
$T_{s,av} = 0.34$ s			—	—	$\Sigma(V_{si} \times A_i) = 929,070$ m^3/s		
$H = 37.3$ m			—	—	$\Sigma(T_{si} \times A_i) = 1,055.51$ m^2 s		
$H' = 0.69$ H			—	—	—	—	—

Fig. 2. Estimating the initial fundamental period of potential sliding blocks using $T_S = 4\,H'/V_S'$ (Bray and Macedo 2021a, 2021b)

4 Selected Seismic Slope Displacement Procedures

4.1 General

The characteristics of the input ground motion are a key source of uncertainty in the estimate of seismic slope displacement. Thus, it is prudent to employ a comprehensive database of ground motion records for the tectonic setting of the governing earthquakes. Recently developed seismic slope procedures for shallow crustal earthquakes and for interface events and intraslab events in subduction earthquake zones are summarized in this section of the paper.

In these procedures, seismic slope displacement is modeled as a mixed random variable with a certain probability mass at zero displacement and a probability density for finite displacement values (Bray and Travasarou 2007). This approach allows the regression of seismic slope displacements (D) to not be controlled by meaningless values of calculated seismic displacement (i.e., $D < 0.5$ cm). The probability density function of seismic displacements is:

$$f_D(d) = \overline{P}\delta(d - d_0) + \left(1 - \overline{P}\right)\overline{f}_D(d) \qquad (2)$$

where $f_D(d)$ is the displacement probability density function; $\overline{P}$ is the probability mass at $D = d_0$; $\delta(d - d_0)$ is the Dirac delta function, and $\overline{f}_D(d)$ is the displacement probability density function for $D > d_0$. A mixed probability distribution has a finite probability at $D = d_0 = 0.5$ cm and a continuous probability density for $D > d_0$. The resulting model provides an equation for computing the probability of "zero" (i.e., negligible) displacement and an equation for computing the "nonzero" displacement.

The "zero" and "nonzero" displacement equations can be combined to calculate the probability of the seismic displacement exceeding a specified seismic slope displacement (d) for an earthquake scenario (i.e., $S_a(1.3T_s)$ and M_w) and slope properties (i.e., k_y and T_s). The probability of the seismic slope displacement (D) exceeding a specified displacement (d) is:

$$P(D > d) = [1 - P(D = 0)]P(D > d|D > 0) \qquad (3)$$

where $P(D = 0)$ is computed using the probability of "zero" displacement equations that follow, and the term $P(D > d|D > 0)$ is computed assuming that the estimated displacements are lognormally distributed as:

$$P(D > d|D > 0) = 1 - P(D \leq d|D > 0) = 1 - \Phi\left(\frac{Ln(d) - Ln\left(\hat{d}\right)}{\sigma}\right) \qquad (4)$$

where $Ln\left(\hat{d}\right)$ is calculated using the "non-zero" equations that follow, and σ is the standard deviation of the random error of the applicable equation.

4.2   Shallow Crustal Earthquakes

A total of 6711 ground motion records (with each record having 2 horizontal components) from shallow crustal earthquakes along active plate margins were employed in the Bray and Macedo (2019) update of the Bray and Travasarou (2007) procedure. Their study took advantage of the NGA-West2 empirical ground motion database (Bozorgnia et al. 2014). Each horizontal component of a ground motion recording was applied to the rigid base below the fully coupled, nonlinear, deformable stick-slip sliding block to calculate seismic displacement (Bray and Macedo 2019). The seismic displacement values calculated from the two horizontal components were averaged for ordinary ground motions, which are ground motions without near-fault forward-directivity pulses. The opposite polarity of the horizontal components, which represent an alternative excitation of the slope, were also used to compute an alternative average seismic displacement, and the maximum of the average seismic displacement value for each polarity was assigned to that ground motion record. For the near-fault forward-directivity pulse motions, the two recorded orthogonal horizontal components for each recording were rotated from $0°$ to $180°$ in $1°$ increments for each polarity to identify the component producing the maximum seismic displacement ($D100$) and median seismic displacement ($D50$). Nearly 3 million sliding block analyses were performed in the Bray and Macedo (2019) study.

In the near-fault region, the seismic slope displacement will be greatest for slopes oriented so their movement is in the fault-normal direction due to forward-directivity pulse motions. In this case, the $D100$ equations developed by Bray and Macedo (2019) should be used. If the slope is oriented so its movement is in the fault-parallel direction, the $D50$ equations are used. PGV is required for near-fault motions in combination with $S_a(1.3T_s)$, which is the 5%-damped spectral acceleration at the degraded period of the sliding mass estimated as $1.3T_s$. The resulting $D100$ equations are:

$$P(D100 = 0) = \left[1 + \exp(-10.787 - 8.717Ln(k_y) + 1.660Ln(PGV) + 3.150T_s \right.$$
$$\left. +7.560Ln(Sa(1.3T_s)))\right]^{-1} \text{when } T_s \leq 0.7\,\text{s} \tag{5a}$$

$$P(D100 = 0) = \left[1 + \exp(-12.771 - 9.979Ln(k_y) + 2.286Ln(PGV) - 4.965T_s \right.$$
$$\left. +4.817Ln(Sa(1.3T_s)))\right]^{-1} \text{ when } T_s > 0.7\,\text{s} \tag{5b}$$

$$Ln(D100) = c1 - 2.632Ln(k_y) - 0.278(Ln(k_y))^2 + 0.527Ln(k_y)Ln(Sa(1.3T_s))$$
$$+ 1.978Ln(Sa(1.3T_s)) - 0.233(Ln(Sa(1.3T_s)))^2 + c2T_s$$
$$+ c3(T_s)^2 + 0.01M_w + c4 * Ln(PGV) \pm \varepsilon \tag{6}$$

where $P(D100 = 0)$ is the probability of occurrence of "zero" seismic slope displacement (as a decimal number); $D100$ is the "nonzero" maximum component seismic displacement in cm; k_y is the yield coefficient; T_s is the fundamental period of the sliding mass in seconds; and $S_a(1.3T_s)$ the spectral acceleration at a period of $1.3T_s$ in the units of g of the design outcropping ground motion for the site conditions below the potential sliding mass (i.e. the value of $S_a(1.3T_s)$ for the earthquake ground motion at the elevation of the sliding surface if the potential sliding mass was removed); ε is a normally distributed random variable with zero mean and standard deviation $\sigma = 0.56$. When $60\,\text{cm/s} < PGV \leq 150\,\text{cm/s}$, $c1 = -6.951$, $c2 = 1.069$, $c3 = -0.498$, and $c4 = 1.547$ if $T_s \geq 0.10\,\text{s}$, and $c1 = -6.724$, $c2 = -2.744$, $c3 = 0.0$, and $c4 = 1.547$ if $T_s < 0.10\,\text{s}$. When $PGV > 150\,\text{cm/s}$, $c1 = 1.764$, $c2 = 1.069$, $c3 = -0.498$, and $c4 = -0.097$ if $T_s \geq 0.10\,\text{s}$, and $c1 = 1.991$, $c2 = -2.744$, $c3 = 0.0$, and $c4 = -0.097$ if $T_s < 0.10\,\text{s}$. The $D50$ equations are:

$$P(D50 = 0) = \left[1 + \exp(-14.930 - 10.383Ln(k_y) + 1.971Ln(PGV) + 3.763T_s \right.$$
$$\left. +8.812Ln(Sa(1.3T_s)))\right]^{-1} \text{ when } T_s \leq 0.7\,\text{s} \tag{7a}$$

$$P(D50 = 0) = \left[1 + \exp(-14.671 - 10.489Ln(k_y) + 2.222Ln(PGV) - 4.759T_s \right.$$
$$\left. +5.549Ln(Sa(1.3T_s)))\right]^{-1} \text{when} T_s > 0.7\,\text{s} \tag{7b}$$

$$Ln(D50) = c1 - 2.931Ln(k_y) - 0.319(Ln(k_y))^2 + 0.584Ln(k_y)Ln(Sa(1.3T_s))$$

$$+ 2.261Ln(Sa(1.3T_s)) - 0.241(Ln(Sa(1.3T_s)))^2 + c2T_s$$
$$+ c3(T_s)^2 + 0.05M_w + c4 * Ln(PGV) \pm \varepsilon \qquad (8)$$

where $P(D50 = 0)$ is the probability of occurrence of "zero" seismic slope displacement (as a decimal number); $D50$ is the "nonzero" median component seismic displacement in cm; ε is a normally distributed random variable with zero mean and standard deviation $\sigma = 0.54$. When 60 cm/s $< PGV \leq 150$ cm/s, $c1 = -7.718$, $c2 = 1.031$, $c3 = -0.480$, and $c4 = 1.458$ if $T_s \geq 0.10$ s, and $c1 = -7.497$, $c2 = -2.731$, $c3 = 0.0$, and $c4 = 1.458$ if $T_s < 0.10$ s. If $PGV > 150$ cm/s, $c1 = -0.369$, $c2 = 1.031$, $c3 = -0.480$, and $c4 = 0.025$ if $T_s \geq 0.10$ s, and $c1 = 2.480$, $c2 = -2.731$, $c3 = 0.0$, and $c4 = 0.025$ if $T_s < 0.10$ s.

Ordinary (non-pulse) motions produce these equations:

$$P(D = 0) = 1 - \Phi\left(-2.48 - 2.97Ln(k_y) - 0.12(Ln(k_y))^2 - 0.72T_sLn(k_y)\right.$$
$$\left. + 1.70T_s + 2.78Ln(S_a(1.3T_s))\right) \text{ when } T_s \leq 0.7 \text{ s} \qquad (9a)$$

$$P(D = 0) = 1 - \Phi\left(-3.42 - 4.93Ln(k_y) - 0.30(Ln(k_y))^2 - 0.35T_sLn(k_y)\right.$$
$$\left. -0.62T_s + 2.86Ln(S_a(1.3T_s))\right) \text{ when } T_s > 0.7 \text{ s} \qquad (9b)$$

$$Ln(D) = a1 - 2.482Ln(k_y) - 0.244(Ln(k_y))^2 + 0.344Ln(k_y)Ln(Sa(1.3T_s))$$
$$+ 2.649Ln(Sa(1.3T_s)) - 0.090(Ln(Sa(1.3T_s)))^2 + a2T_s$$
$$+ a3(T_s)^2 + 0.603M_w \pm \varepsilon_1 \qquad (10)$$

where $P(D = 0)$ is the probability of occurrence of "zero" seismic slope displacement (as a decimal number); Φ is the standard normal cumulative distribution function; D is the amount of "nonzero" seismic slope displacement in cm; k_y, T_s, $S_a(1.3T_s)$, and M_w are as defined previously, and ε_1 is a normally distributed random variable with zero mean and standard deviation $\sigma = 0.72$ In Eq. 10, $a1 = -5.981$, $a2 = 3.223$, and $a3 = -0.945$ for systems with $T_s \geq 0.10$ s, and $a1 = -4.684$, $a2 = -9.471$, and $a3 = 0.0$ for $T_s < 0.10$ s. The change in parameters at $T_s = 0.10$ s reduces the bias in the residuals for very stiff slopes. For the special case of the Newmark rigid-sliding block where $T_s = 0.0$ s, the "nonzero" D (cm) is estimated as:

$$Ln(D) = -4.684 - 2.482Ln(k_y) - 0.244(Ln(k_y))^2 + 0.344Ln(k_y)Ln(PGA)$$
$$+ 2.649Ln(PGA) - 0.090(Ln(PGA))^2 + 0.603M_w \pm \varepsilon \qquad (11)$$

where PGA is the peak ground acceleration in the units of g of the input base ground motion. If there are important topographic effects to capture for localized shallow sliding, the input PGA value should be adjusted as discussed previously (i.e., 1.3 PGA_{1D} for moderately steep slopes, 1.5 PGA_{1D} for steep slopes, or 2.0 PGA_{1D} for the dam crest). For long, shallow potential sliding masses, lateral incoherence of ground shaking reduces the input PGA value employed in the analysis (e.g., 0.65 PGA_{1D} for moderately steep slopes, Rathje and Bray 2001).

The "nonzero" seismic slope displacement equation for the entire ground motion database of ordinary and near-fault pulse motions can be used to calculate a seismic coefficient (k) consistent with a specified allowable calculated seismic slope displacement (D_a) for the general case when $PGV \leq 115$ cm/s (Bray and Macedo 2019). The owner and engineer should select D_a (cm) to achieve the desired performance level and the percent exceedance of this displacement threshold (e.g., median displacement estimate for $\varepsilon = 0$ or 16% exceedance displacement estimate for $\varepsilon = \sigma = 0.74$) considering the consequences of unsatisfactory performance at displacement levels greater than this threshold. The seismic demand is defined in terms of $S_a(1.3T_s)$ of the input ground motion for the outcropping site condition below the sliding mass and M_w of the governing earthquake event. If this value of k is used in a pseudostatic slope stability analysis and the calculated $FS \geq 1.0$, then the selected percentile estimate of the seismic displacement will be less than or equal to D_a. The minimum value of the acceptable FS should not be greater than 1.0, because FS varies nonlinearly as a function of the reliability of the system, and the procedure is calibrated to $FS \geq 1.0$.

The effects of specifying the allowable displacement as well as the level of the seismic demand in terms of $S_a(1.3T_s)$ on the value of k are illustrated in Fig. 3. Allowable displacement values of 5 cm, 15 cm, 30 cm, and 50 cm are used to illustrate the dependence of k on the selected level of D_a for a M_w 7.0 earthquake. Results are also provided at the 30 cm allowable displacement level for a lower magnitude event ($M_w = 6$). As expected, k increases systematically as the 5%-damped elastic spectral acceleration of the ground motion increases. Importantly, k also increases systematically as the allowable displacement value decreases. It also decreases as the earthquake magnitude decreases. The seismic coefficient varies systematically in a reasonable manner as the allowable displacement threshold and design ground shaking level vary.

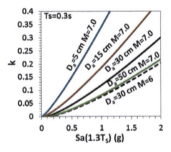

Fig. 3. Seismic coefficient as a function of the allowable displacement and seismic demand

4.3 Interface and Intraslab Subduction Zone Earthquakes

Macedo et al. (2022) recently updated the subduction zone interface earthquake seismic slope displacement procedure developed by Bray et al. (2018). They took advantage of the recently developed comprehensive NGA-Sub ground motion database (Bozorgnia and Stewart 2020) generated by the Pacific Earthquake Engineering Research (PEER)

Center. Macedo et al. (2022) utilized 6240 two-component horizontal ground motion recordings from 174 interface earthquakes with M_w from 4.8 to 9.1 to calculate seismic slope displacements with the Bray and Macedo (2019) coupled nonlinear sliding block model.

A robust seismic slope displacement developed using subduction zone intraslab earthquake ground motions did not exist. Given intraslab earthquake ground motion models differ from interface earthquake ground motion models, one might expect that the seismic slope displacement models for these two types of earthquakes to differ. Macedo et al. (2022) utilized 8299 two-component ground motion recordings from 200 intraslab earthquakes with M_w from 4.0 to 7.8 to calculate seismic slope displacements. They found there were significant biases in the residuals from the seismic slope displacements calculated using subduction zone seismic slope displacement models when comparing them with the displacements calculated using the intraslab earthquake records. Therefore, separate regressions were performed on the seismic slope displacements calculated using the interface and intraslab records.

As the two horizontal components of a ground motion record are highly correlated, the D value assigned to each two-component ground motion recording is the larger of the average displacement values calculated from the record's two polarities as was done for the ordinary ground motions in the Bray and Macedo (2019) study. This methodology mirrors what is typically done in engineering practice. Over 1.5 million and 1.8 million analyses were performed using the interface and intraslab records.

A logistic regression (Hosmer Jr., et al. 2013) is the basis of the model to estimate $P(D = 0)$ as a function of k_y, T_s, and $S_a(1.3T_s)$ with the resulting equation of:

$$ln\frac{P(D=0)}{1-P(D=0)} = c_1 + c_2 lnk_y + c_3(lnk_y)^2 + c_4T_slnk_y + c_5T_s + c_4T_slnSa(1.3T_s)$$

(12)

where c_1 to c_6 are coefficients provided in Table 1. The "nonzero" seismic displacement equation has a similar form to that used by Bray et al. (2018); however, the ground motion parameter PGV is included to minimize bias and reduce the residuals:

$$lnD = a_0 + a_1 lnk_y + a_2(lnk_y)^2 + a_3lnk_ylnSa(1.3T_s) + a_4lnSa(1.3T_s)$$
$$+ a_5(lnSa(1.5T_s))^2 + a_6T_s + a_7(T_s)^2 + a_6M + a_3lnPGV + \varepsilon$$

(13)

where a_0 to a_9 are model coefficients presented in Table 2 and ε is a Gaussian random variable with zero mean and standard deviation of $\sigma = 0.65$ for the interface event and $\sigma = 0.53$ for the interface. The values of the coefficients a_0, a_6, a_7, and a_9 are modeled as dependent on the values of T_s and PGV based on residual analyses. The residuals show negligible bias and no significant trends for the new seismic slope displacement models developed for interface and intraslab earthquakes.

The new Macedo et al. (2022) interface model produces seismic slope displacements fairly consistent with the Bray et al. (2018) interface model. However, the new Macedo et al. (2022) intraslab model produces significantly different results than the interface models (comparison with Bray et al. 2018 interface model is shown in Fig. 4). Most of the coefficients in the seismic slope displacement equations developed by Macedo et al. (2022) for subduction zone interface earthquakes and intraslab earthquakes differ significantly, which highlights the different scaling of seismic slope displacement for these different types of earthquakes. In addition, the standard deviation of the intraslab D model is smaller than that of the interface model, and both models have a standard deviation lower than $\sigma = 0.73$ of the Bray et al. (2018) model. The addition of PGV in the updated model lowers its standard deviation. Obviously, the uncertainty in estimating this additional ground motion parameter increases the uncertainty in estimating D when its uncertainty is included in a seismic slope displacement hazard estimate.

5 Seismic Slope Displacement Hazard

The slope displacement models discussed in previous sections can also be used in performance-based probabilistic assessments. The outcome of these assessments is a displacement hazard curve, which relates different displacement thresholds with their annual rate of exceedance. Displacement hazard curves are calculated as (Macedo et al. 2020):

$$\lambda_D(z) = \sum_{i=1}^{nk_y} \sum_{j=1}^{nT_s} \int_{M_{min}}^{M_{max}} \int_{IM} w_i w_j P\left(D > z | IM, M, k_y^i, T_s^j\right)$$
$$f(M|IM)\Delta\lambda(IM)d(IM)d(M) \tag{14}$$

Table 1. Coefficients for interface and intraslab earthquakes $P(D=0)$ equations

Coefficient	Interface	Intraslab
c_1	3.46 for $T_s < 0.6$; 3.57 for $T_s \geq 0.6$	5.22 for $T_s < 0.6$; 2.92 for $T_s \geq 0.6$
c_2	5.05 for $T_s < 0.6$; 9.39 for $T_s \geq 0.6$	6.55 for $T_s < 0.6$; 14.72 for $T_s \geq 0.6$
c_3	0.15 for $T_s < 0.6$; 0.55 for $T_s \geq 0.6$	0.43 for $T_s < 0.6$; 2.24 for $T_s \geq 0.6$
c_4	1.41 for $T_s < 0.6$; 1.64 for $T_s \geq 0.6$	4.73 for $T_s < 0.6$; 5.45 for $T_s \geq 0.6$
c_5	-1.08 for $T_s < 0.6$; 5.37 for $T_s \geq 0.6$	-0.87 for $T_s < 0.6$; 14.82 for $T_s \geq 0.6$
c_6	-5.13 for $T_s < 0.6$; -7.00 for $T_s \geq 0.6$	-6.50 for $T_s < 0.6$; -8.47 for $T_s \geq 0.6$

Table 2. Coefficients for interface and intraslab earthquakes 'nonzero" D equations

Coefficient	Interface	Intraslab
a_0	-5.62 for $T_s < 0.1$; -6.20 for $T_s \geq 0.1$	-5.91 for $T_s < 0.1$; -6.34 for $T_s \geq 0.1$
a_1	-3.26	-2.36
a_2	-0.36	-0.22
a_3	0.48	0.26
a_4	2.62	1.97

(continued)

Table 2. (*continued*)

Coefficient	Interface	Intraslab
a_5	-0.12	-0.02
a_6	-5.25 for $T_s < 0.1$; 2.06 for $T_s \geq 0.1$	-3.89 for $T_s < 0.1$; 2.31 for $T_s \geq 0.1$
a_7	0 for $T_s < 0.1$; -0.73 for $T_s \geq 0.1$	0 for $T_s < 0.1$; -0.90 for $T_s \geq 0.1$
a_8	0.19	0.38
a_9	0.62 for $PGV < 10$; 0.73 for $PGV \geq 10$	0.68 for $PGV < 30$; 0.72 for $PGV \geq 30$

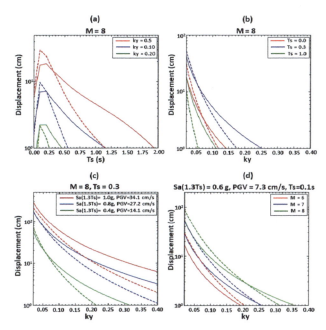

Fig. 4. Comparison of the Macedo et al. (2022) intraslab model (solid curves) with the Bray et al. (2018) interface model (dashed curves) with variations in k_y, T_s, $Sa(1.3T_s)$, PGV, and M.

where D represents the slope displacement, IM can be a scalar or a vector of ground motion intensity measures (i.e., $Sa(1.3T_s)$ or PGV), and $\lambda_D(z)$ in the mean annual rate slope displacement exceeding a given threshold z. $\Delta\lambda(IM)$ is the joint annual rate of occurrence of IM and $P(M|IM)$ is the conditional probability of M given IM, which can be estimated using from a probabilistic seismic hazard assessment (PSHA). nk_y and nT_s are the number of different k_y and T_s values considered for the slope system to account for the uncertainty in the slope properties. k_y^i and T_s^j are the i-th and j-th realizations of k_y and T_s with weighting factors w_i and w_j, respectively. $P\left(D > z|IM, M, k_y^i, T_s^j\right)$ is the conditionl probability of D exceeding z given the values of IM, M, k_y^i and T_s^j, which can be estimated as:

$$P(D > z) = \left(1 - P\left(D = 0|IM, M, k_y^i, T_s^j\right)\right)P(D > z|D > 0)$$

$$= \left(1 - P\left(D = 0|IM, M, k_y^i, T_s^j\right)\right)(1 - P(D > z|D > 0))$$

$$= \left(1 - P\left(D = 0|IM, M, k_y^i, T_s^j\right)\right)\left(1 - \Phi\left(\frac{lnz - ln\mu\left(IM, M, k_y^i, T_s^j\right)}{\sigma}\right)\right)$$

$$(15)$$

where Φ is the cumulative distribution function of the standard normal distribution and $\mu(IM, M, k_y^i, T_s^j)$ and σ are the median value and standard deviation of D, which can be estimated using slope displacement models given IM, M, T_s and k_y. $P(D = 0)$ can be estimated using equations that provide the probability of "zero" displacement. Equation 14 can be applied separately to different tectonic settings. For instance, when considering the subduction interface, subduction intraslab, and shallow crustal settings, three different annual rate of exceedance curves can be evaluates for each tectonic setting ($\lambda_D^{interface}$, $\lambda_D^{intraslab}$, and $\lambda_D^{crustal}$), which can be combined to estimate the total annual rate of exceedance λ_D^{total} as:

$$\lambda_D^{total} = \lambda_D^{interface} + \lambda_D^{intraslab} + \lambda_D^{crustal} \qquad (16)$$

Performance-based probabilistic assessments are amenable to incorporating uncertainties for ground motion models and slope properties. For example, Fig. 5 shows a typical assessment for the South American Andes where there is a contribution from multiple tectonic settings, i.e., shallow crustal, subduction interface, and subduction intraslab. First, a PSHA assessment is conducted, which requires different ground motion models (GMMs). Specifically, the results in Figs. 5a and 5b, consider the GMMs for $Sa(1.3T_s)$ and PGV from the NGASub project equally weighted in the case of subduction interface and subduction intraslab earthquake zones, whereas the GMMs from the NGAWest2 project equally weighted for shallow crustal settings. Of note, since the displacement models for subduction zones use two intensity measures ($Sa(1.3T_s)$ and PGV), their coefficient of correlation is required to estimate their join rate of occurrence, which is an input into Eq. 14. Engineers can use the coefficients of correlation in Baker and Bradley (2017) for shallow crustal settings and those in Macedo and Liu (2021) for subduction earthquake zones. Uncertainties in slope properties can also be incorporated. For instance, the results in Fig. 5, consider 9 realizations for T_s (0.25 to 0.41 s, best estimate of 0.33 s) and k_y (0.11 to 0.18, best estimate of 0.14). The weights for each realization are assigned as per Macedo et al. (2019).

Under these considerations, Figs. 5c, and 5d show the results of displacement hazard curves. Figure 5c shows the mean hazard curves, 5–95 percentiles, and individual realizations, whereas Fig. 5d shows the displacement hazard deagregation by tectonic settings showing that for this site in the South American Andes, intraslab seismic sources dominate followed by interface seismic sources, and the contribution from shallow crustal

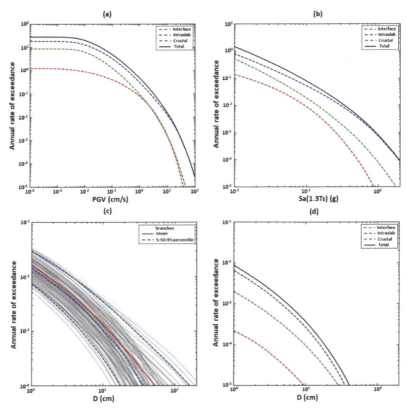

Fig. 5. Performance based assessment of a slope system with $T_s = 0.33$ and $k_y = 0.14$ in the South American Andes. (a) Deagregation of *PGV* hazard curves by tectonic mechanism, (b) Deagregation of $Sa(1.3T_s)$ hazard curves by tectonic mechanism, (c) realization of displacement hazard curves, (d) Deagregation of displacement hazard by tectonic mechanisms.

source is, comparatively, less important. The displacement hazard curves in Fig. 5d provide hazard-consistent estimates. There is no need to assume that the hazard level for intensity measures is consistent with that of displacements, which is an implicit assumption in assessments that dominate the state-of-practice. Using the displacement hazard curves in Fig. 5d, displacements for 475 and 2475 years return period are estimated as 8 cm and 27 cm, respectively. Macedo and Candia (2020) modified the procedures described in this section to estimate hazard-consistent seismic coefficients for pseudo-static analyses. Computational tools to use these methods have been implemented in Macedo et al. (2020) to facilitate their use in practice.

6   Conclusions

In evaluating seismic slope stability, the engineer must first investigate if there are materials in the system or its foundation that will lose significant strength due to cyclic loading. If there are materials that can lose significant strength, post-cyclic reduced strengths should be employed in a static slope stability analysis to calculate the post-cyclic *FS*. If it is low, this issue should be the primary focus of the evaluation, because a flow slide could occur. If materials will not lose significant strength due to cyclic loading, the deformation of the earth structure or slope should be evaluated to assess if they are sufficient to jeopardize satisfactory performance of the system.

A modified Newmark model with a deformable sliding mass provides useful insights for estimating seismic slope displacement due to shear deformation of the earth materials comprising earth dams and natural slopes for the latter case discussed above. The critical components of a Newark-type sliding block analysis are: 1) the dynamic resistance of the structure, 2) the earthquake ground motion, 3) the dynamic response of the potential sliding mass, and 4) the permanent displacement calculational procedure. Seismic slope displacement procedures should be evaluated in terms of how each procedure characterizes the slope's dynamic resistance, earthquake ground motion, dynamic response of the system, and calculational procedure.

The system's dynamic resistance is captured by its yield coefficient (k_y). This important system property depends greatly on the shear strength of the soil along the critical sliding surface. Assessment of the dynamic peak shear strength of a clay material requires consideration of the rate of loading, cyclic degradation, progressive failure, and distributed shear deformation effects. If the clay exhibits strain-softening and moderate-to-large displacements are calculated, the k_y value used in the sliding block analysis must be compatible with the reduction in clay shear strength with increasing displacement.

The primary source of uncertainty in assessing the seismic performance of an earth slope when there are not materials that can undergo severe strength loss is the input ground motion, so recent models have taken advantage of the wealth of strong motion records that have become available. The Bray and Macedo (2019) and Macedo et al. (2022) procedures are based on the results of nonlinear fully coupled stick-slip sliding block analyses using large databases of thousands of recorded ground motions. The model captures shear-induced displacement due to sliding on a distinct plane and distributed shear shearing within the slide mass.

The spectral acceleration at a degraded period of the potential sliding mass ($S_a(1.3T_s)$) is an optimal ground motion intensity measure. As it only captures the intensity and frequency content of the ground motion, M_w is added as a proxy to represent the important effect of duration. In some cases, the addition of *PGV* is necessary to minimize bias and reduce the scatter in the residuals. *PGV* is especially informative when applying the Bray and Macedo (2019) model to estimate seismic displacement for slopes with movement oriented in the fault-normal direction in the near-fault region. Forward-directivity velocity pulse motions tend to produce a large seismic slope displacement in the fault normal direction, captured by $D100$, which is systematically greater than the median component of motion ($D50$). When intense, pulse motions are likely, the $D100$ model should be used to estimate displacement.

The Bray and Macedo (2019) and Macedo et al. (2022) procedures use a mixed random variable formulation to separate the probability of "zero" displacement (i.e., $\leq$ 0.5 cm) occurring from the distribution of "nonzero" displacement, so that very low values of calculated displacement that are not of engineering interest do not bias the results. The calculation of the probability of "zero" displacement occurring provides a screening assessment of seismic performance. If the likelihood of negligible displacements occurring is not high, the "nonzero" displacement is estimated. The 16% to 84% exceedance seismic displacement range should be estimated as there is considerable uncertainty in the estimate of seismic slope displacement. This displacement range is approximately half to twice the median seismic displacement estimate.

These procedures provide estimates of seismic slope displacement that are generally consistent with documented cases of earth dam and solid-waste landfill performance for shallow crustal earthquakes and interface subduction zone earthquakes. Ongoing work is evaluating the Macedo et al. (2022) model for intraslab earthquakes. The proposed models can be implemented rigorously within a fully probabilistic framework for the evaluation of the seismic slope displacement hazard, or it may be used in a deterministic analysis. The estimated range of seismic displacement should be considered an index of the expected seismic performance of the earth slope.

The updated seismic slope displacement models are provided in the form of a spreadsheet at: http://www.ce.berkeley.edu/people/faculty/bray/research.

Acknowledgements. The Faculty Chair in Earthquake Engineering Excellence at UC Berkeley provided financial support to perform this research, which was supplemented by Georgia Tech. The PEER Center provided access to the NGA-West2 and NGA-Sub recordings.

References

Ambraseys, N.N., Sarma, S.K.: The response of earth dams to strong earthquakes. Geotechnique **17**, 181–213 (1967)

Baker, J.W., Bradley, B.A.: Intensity measure correlations observed in the NGA-West2 database, and dependence of correlations on rupture and site parameters. Earthq. Spectra **33**(1), 145–156 (2017)

Biscontin, G., Pestana, J.M.: Influence of peripheral velocity on undrained shear strength and deformability characteristics of a bentonite-kaolinite mixture. Geotech. Engrg. Report No. UCB/GT/99-19, Univ. of Calif. Berkeley, Revised December 2000

Blake, T.F., Hollingsworth, R.A., Stewart, J.P. (eds.): Recommended Procedures for Implementation of DMG Special Publication 117 Guidelines for Analyzing and Mitigating Landslide Hazards in California. Southern California Earthquake Center, June 2002

Bozorgnia, Y., Abrahamson, N.A., et al.: NGA-West2 research project. Earthq. Spectra **30**, 973–987 (2014). https://doi.org/10.1193/072113EQS209M

Bozorgnia, Y., Stewart, J.: Data resources for NGA-Subduction project. PEER Report 2020/02 (2020)

Bray, J.D.: Simplified seismic slope displacement procedures. In: Pitilakis, K.D. (eds.) Earthquake Geotechnical Engineering. Geotechnical, Geological and Earthquake Engineering, vol. 6., pp. 327–353. Springer, Dordrecht (2007). https://doi.org/10.1007/978-1-4020-5893-6_14

Bray, J.D., Rodriguez-Marek, A.: Characterization of forward-directivity ground motions in the near-fault region. SDEE **24**, 815–828 (2004)

Bray, J.D., Macedo, J.: Procedure for estimating shear-induced seismic slope displacement for shallow crustal earthquakes. J. Geotech. Geoenviron. Eng. **145**(12), 04019106 (2019)

Bray J.D., Macedo J.: Closure to: procedure for estimating shear-induced seismic slope displacement for shallow crustal earthquakes. J. Geotech. Geoenviron. Eng. (2021a). https://doi.org/10.1061/(ASCE)GT.1943-5606.0002143

Bray, J.D., Macedo, J.: Simplified seismic slope stability excel spreadsheets (2021b). https://www. ce.berkeley.edu/people/faculty/bray/research. Accessed 15 Dec 2021

Bray, J.D., Macedo, J., Travasarou, T.: Simplified procedure for estimating seismic slope displacements for subduction zone earthquakes. J. Geotech. Geoenviron. Eng. **144**(3), 04017124 (2018)

Bray, J.D., Travasarou, T.: Simplified procedure for estimating earthquake-induced deviatoric slope displacements. J. Geotech. Geoenviron. Eng. **133**(4), 381–392 (2007)

Chen, W.Y., Bray, J.D., Seed, R.B.: Shaking table model experiments to assess seismic slope deformation analysis procedures. In: Proceedings of the 8th US National Conference on Earthquake Engineering, EERI, Paper 1322 (2006)

Cornell, C., Luco, N.: Ground motion intensity measures for structural performance assessment at near-fault sites. In: Proceedings of the U.S.-Japan Joint Workshop and Third Grantees Meeting, U.S.-Japan Cooperative Research on Urban EQ. Disaster Mitigation, Seattle, Washington (2001)

Duncan, J.M.: State of the art: limit equilibrium and finite element analysis of slopes. J. Geotech. Eng. **122**(7), 577–596 (1996)

Duncan, J.M., Wright, S.G.: Soil Strength and Slope Stability. Wiley, Hoboken (2005)

Hosmer Jr., D.W., Lemeshow, S., Sturdivant, R.X.: Applied Logistic Regression, vol. 398. Wiley, Hoboken (2013)

Kammerer, A.M., Hunt, C., Riemer, M.: UC Berkeley geotechnical testing for the east bay crossing of the San Francisco-Oakland bridge, Geotech. Engrg. Report No. UCB/GT/99-18, Univ. of Calif. Berkeley, October 1999

Kramer, S.L., Smith, M.W.: Modified Newmark model for seismic displacements of compliant slopes. J. Geotech. Geoenviron Eng. **123**(7), 635–644 (1997)

Lacerda, W.J.: Sress relaxation and creep effects on the deformation of soils, Ph.D. thesis, Univ. of California, Berkeley (1976)

Macedo, J., Bray, J.D., Abrahamson, N., Travasarou, T.: Performance-based probabilistic seismic slope displacement procedure. Earthq. Spectra **34**(2), 673–695 (2018)

Macedo, J., Abrahamson, N., Bray, J.: Arias intensity conditional scaling ground-motion models for subduction zones. Bull. Seismol. Soc. Am. (BSSA) **109**(4), 1343–1357 (2019)

Macedo, J., Candia, G.: Performance-based assessment of the seismic pseudo-static coefficient used in slope stability analysis. Soil Dyn. Earthq. Eng. **133**, 106109 (2020)

Macedo, J., Candia, G., Lacour, M., Liu, C.: New developments for the performance-based assessment of seismically-induced slope displacements. Eng. Geol. J., **277**, 105786 (2020)

Macedo, J., Liu, C.: Ground-motion intensity measure correlations on interface and intraslab subduction zone earthquakes using the nga-sub database. Bull. Seismol. Soc. Am. **111**(3), 1529–1541 (2021)

Macedo, J., Bray, J.D., Liu, C.: Seismic slope displacement procedure for interface and intraslab subduction zone earthquakes. J. Geotech. Eng. (2022). Under review

Makdisi, F., Seed, H.B.: Simplified procedure for estimating dam and embankment earthquake-induced deformations. J. Geotech. Eng. **104**(7), 849–867 (1978)

Newmark, N.M.: Effects of earthquakes on dams and embankments. Geotechnique **15**(2), 139–160 (1965)

Rathje, E.M., Bray, J.D.: An examination of simplified earthquake induced displacement procedures for earth structures. Can. Geotech. J. **36**, 72–87 (1999)

Rathje, E.M., Bray, J.D.: Nonlinear coupled seismic sliding analysis of earth structures. J. Geotech. Geoenviron. Eng. **126**(11), 1002–1014 (2000)

Rathje, E.M., Bray, J.D.: One- and two-dimensional seismic analysis of solid-waste landfills. Can. Geotech. J. **384**, 850–862 (2001)

Rathje, E.M., Antonakos, G.: A unified model for predicting earthquake-induced sliding displacements of rigid and flexible slopes. Eng. Geol. **122**(1–2), 51–60 (2011)

Rau, G.A.: Evaluation of strength degradation in seismic loading of embankments on cohesive soils, Ph.D. thesis, Univ. of Calif., Berkeley, December 1998

Rumpelt, T.K., Sitar N.: Simple shear tests on bay mud from borehole GT-2 S-2: richmond sanitary landfill, & the effect of the rate of cyclic loading in simple shear tests on San Francisco Bay Mud, reports for EMCON ASSOC, 4 November and 13 December 1988

Saygili, G., Rathje, E.M.: Empirical predictive models for earthquake-induced sliding displacements of slopes. J. Geotech. Geoenviron. Eng. **134**(6), 790–803 (2008)

Seed, H.B., Chan, C.K.: Clay strength under earthquake loading condition. J. Soil Mech. Found. Div. **92**(SM 2) (1966)

Seed, H.B., Martin, G.R.: The seismic coefficient in earth dam design. J. Soil Mech. Found. Div. **92**(3), 25–58 (1966)

Song, J., Rodriguez-Marek, A.: Sliding displacement of flexible earth slopes subject to near-fault ground motions. J. Geotech. Geoenviron. Eng. **141**(3) (2014). https://doi.org/10.1061/(ASCE)GT.1943-5606.0001233

Stewart, J.P., Bray, J.D., McMahon, D. J., Smith, P.M., Kropp, A.L.: Seismic performance of hillside fills. J. Geotech. Geoenviron. Eng. **127**(11), 905–919 (2001)

Tokimatsu, K., Seed, H.B.: Evaluation of settlements in sands due to earthquake shaking. J. Geotech. Eng. **113**(8), 861–878 (1987)

Wang, M., Huang, D., Wang, G., Li D.,: SS-XGBoost: a machine learning framework for predicting newmark sliding displacements of slopes. J. Geotech. Geoenviron. Eng. **146**(9), 04020074 (2020)

Wartman, J., Bray, J.D., Seed, R.B.: Inclined plane studies of the newmark sliding block procedure. J. Geotech. Geoenviron. Eng. **129**(8), 673–684 (2003)

Yu, L., Kong, X., Xu, B.: Seismic response characteristics of earth and rockfill dams. In: 15th WCEE, Lisbon, Portugal, Paper No. 2563, September 2012

Recent Advances in Helical Piles for Dynamic and Seismic Applications

M. Hesham El Naggar$^{(\boxtimes)}$

Western University, London, ON N6A 5B9, Canada
naggar@uwo.ca

Abstract. Helical piles have become popular foundation option owing to their many advantages related to ease of installation and large load carrying capacity. They are typically manufactured of straight steel shafts fitted with one or more helices and are installed using mechanical torque. They can sustain static and dynamic loading and are increasingly used in applications that induce complex loading conditions on them. The behavior and design of single vertical helical piles subjected to static loading is well investigated. However, a few studies investigated the dynamic or seismic behavior of single helical piles and their group behavior. This paper presents recent advances in evaluating the axial and lateral capacity and performance of helical piles and their response to dynamic and seismic loads.

Keywords: Helical pile · Dynamic response · Seismic response · Field tests

1 Introduction

A helical pile (HP) consists of a central shaft with one or more pitched helical bearing plates affixed to it. It is typically manufactured of straight steel shaft fitted with one or more helices and is installed using mechanical torque. It could be installed to varying depths and at varying angle to suit the project needs and the available soil conditions. Helical piles are gaining wide popularity and are currently considered a preferred foundation option in a wide range of engineering projects to provide high compressive, uplift, and lateral resistance to static and dynamic loads. They are used to support buildings, bridges, telecommunication towers, power transmission lines and wind turbines (Elsharnouby and El Naggar, 2012a, 2012b; Elkasabgy and El Naggar 2013, 2015). More recently, they have become a preferred option in projects involving machine foundations and pipeline supports (Elkasabgy and El Naggar 2018). These applications involve various loading conditions ranging from the typical primarily compressive loading with minimal lateral loading (Alwalan and El Naggar, 2020a, 2020b and 2021), to more complex loading conditions with significant lateral loading and rocking moment accompanied by small compressive loading (Fahmy and El Naggar 2016a, 2016b; Elkasabgy and El Naggar 2019). It is necessary to have clear understanding of helical piles performance characteristics to ensure efficient and reliable design under these complex loading conditions.

© The Author(s), under exclusive license to Springer Nature Switzerland AG 2022
L. Wang et al. (Eds.): PBD-IV 2022, GGEE 52, pp. 24–49, 2022.
https://doi.org/10.1007/978-3-031-11898-2_2

Recent case histories proved that HPs exhibited outstanding performance during earthquakes (El-Sawy et al. 2019). In addition, Fayez et al. (2021, 2022) investigated the seismic response of HP groups in dry, dense sand deposits through full-scale shaking table tests. They reported that the pile group's lateral resistance comprised a flexural component and a rocking component, which was derived from the individual piles' axial resistance and was enhanced by the contribution of the helices. Orang et al. (2021) further studied the seismic response of HPs in liquefiable soil as a strategy to minimize the settlement of the shallow foundation employing large shaking table tests. Their results indicated that the HPs could reduce the post-liquefaction settlement of the supported foundation. Shahbazi et al. (2020) evaluated the damping characteristics of the soil-helical pile system consisting of single and grouped helical piles from large-scale shaking tests. The studies reviewed above collectively indicate the advantageous performance of helical piles under complex, dynamic and seismic loads. This paper provides the current state of knowledge related to the dynamic and seismic performance of helical piles.

2 Axial Capacity of Helical Piles

The compressive capacity of the helical pile is provided by a combination of shaft resistance and bearing resistance on the helices, i.e.

$$Q_c = Q_f + Q_b \tag{1}$$

where: Q_c is the ultimate compressive capacity; Q_f is the friction resistance along the shaft; and Q_b is the bearing resistance on the helical plates. The relative contributions of shaft and bearing resistances depend on the embedment depth, the soil layering and the pile geometry (i.e. shaft diameter, d, and helix diameter, D_h).

The ultimate compressive capacity of helical piles is predicted employing limit equilibrium methods, i.e., the static equilibrium of the pile at onset of soil failure around the pile. For helical piles with multiple helices, there are two possible failure mechanisms: individual bearing (IB) failure (Fig. 1a); or cylindrical shear (CS) failure (Fig. 1b) (Livneh and El Naggar 2008; Perko 2009; and Elsherbiny and El Naggar 2013). For helical piles installed in sand, the failure mechanism is primarily IB, while for piles installed in clay it depends on inter-helix spacing, S_r (CS for $S_r < 2D_h$; IB for $S_r > 3D_h$; and either for $2D_h < S_r < 3D_h$). Helix diameter is typically 2 to 3 times the shaft diameter: therefore, a helical pile with one helix would provide 4 to 9 times the bearing resistance of a conventional pile with same pile embedment, shaft diameter, and soil strength parameters. Consequently, the helical pile capacity is derived primarily from bearing resistance along helices, especially when they are placed within the strong layer(s) within the soil profile, or from a combination of shaft friction and end-bearing.

The pile capacity considering IB failure mechanism is simply calculated as the sum of individual helix bearing capacities (i.e., each helical plate is considered as an individual plate bearing on soil beneath it). The pile capacity considering CS failure mechanism is calculated as the sum of frictional resistance along the inter-helical soil cylinder and the bearing on the bottom helix. Tappenden (2007) evaluated the performance of these methods based on the results of 29 full-scale axial load tests. Similarly, the field study results indicated that there was no distinct transition from CS to IB behavior

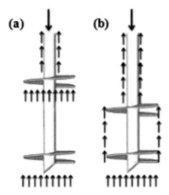

Fig. 1. Possible failure mechanisms of a multi-helix pile: a) individual bearing failure; and b) cylindrical shear failure

for multi-helix anchors in clay (Lutenegger 2009). Elsherbiny and El Naggar (2013) conducted a comprehensive study involving field testing and finite element modeling to evaluate the compressive capacity of helical piles installed in cohesive and cohesionless soil profiles. They reported that the predictions of theoretical equations for piles in cohesionless soil vary largely depending on the choice of bearing capacity factors and proper failure criteria. They also evaluated the interaction of closely spaced helices and proposed a bearing capacity reduction factor, R, and helix efficiency factor, E_H, to evaluate the compressive capacity of helical piles in cohesionless soil considering an industry acceptable ultimate load criterion corresponding to settlement equal to 5% of helix diameter, D_h, as discussed below.

2.1 Axial Capacity of Helical Piles in Sand

The axial capacity of helical piles installed in sand based on CS failure is given by:

$$Q_c = \gamma' H_2 A_2 N_q + \pi/2\, D_h \gamma' \left(H_2^2 - H_1^2 \right) K_s \tan\phi + P_s/2\, H_{eff}^2 \gamma' K_s \tan\delta \quad (2)$$

where: Q_c = ultimate compression capacity; γ' = effective unit weight of soil; K_s = coefficient of lateral earth pressure for piles in compression; ϕ = the soil peak internal friction angle; A_2 = surface area of the bottom helix; N_q = bearing capacity factor $N_q = e^{\pi \tan\phi} \tan^2(45° + \phi/2)$; H_{eff} = the effective shaft length, shaft length above top helix – helix diameter; H_1 = depth to top helix; H_2 = depth to bottom helix; P_s = pile shaft perimeter; and δ = pile-soil interface friction angle.

First term in Eq. 2 represents the bottom helix bearing capacity; second term represents the frictional resistance along the interface between the inter-helix soil cylinder and surrounding soil; and third term represents the pile shaft frictional resistance.

The compressive capacity based on IB is given by:

$$Q_c = \gamma' H_2 A_2 N_q + \gamma' H_1 A_1 Nq + P_s/2 H_{eff}^2 \gamma' K_s \tan\delta \quad (3)$$

where: A_1 = net surface area of the top helix = $\pi \left(D_h^2 - d^2 \right)/4$

The first term in Eq. 3 accounts for the bearing capacity of the bottom helix; the second term accounts for the bearing capacity of the top helix; and the third term represents the frictional resistance along the pile shaft and the surrounding soil.

Considerations of Pile Settlement and Helix Interaction. To limit the pile settlement to 5%D, especially for piles in dense sand, a reduction factor, R, is applied to the plate bearing capacity. The reduction factor is a function of the soil strength and helix diameter and is given by:

$$R = 2.255 - 0.0426\phi \left(\text{for } 45° \geq \phi \geq 30° \text{ with } 15° \geq \psi \geq 0° \text{ and } D_h \geq 500\,\text{mm}\right)$$
$$R = 1.0 \left(\text{for } \phi < 30° \text{ and } \psi = 0°\right) \tag{4}$$

The interaction between the helices is considered using a helix efficiency factor, E_H = bearing capacity of a multi-helix pile/sum of bearing capacities of individual helices, for different cohesionless soil strengths and helical spacing values. The helix efficiency factor applies to CS failure as the frictional shear resistance of the soil cylinder has to be transformed into bearing resistance onto the helix that confines the cylinder from the top, i.e., in essence CS failure is an end-bearing failure with E_H that accounts for interaction between the helices. Considering both R and E_H, the compressive capacity of helical piles installed in sand is given by:

$$Q_c = \gamma' H_2 A_2 N_q R + E_H \gamma' H_1 A_1 N_q R + P_s/2H_{eff}^2 \gamma' K_s \tan \delta \tag{5}$$

2.2 Axial Capacity of Helical Piles in Clay

The compressive capacity based on CS failure is given by:

$$Q_c = P_s H_{eff} \alpha C_{uf} + \pi D_{h1}(H_2 - H_1) C_{uc} S_f + C_{ub} N_c A_2 \tag{6}$$

where: C_{uf}, C_{uc}, C_{ub} = undrained shear strength along shaft, along soil cylinder, and below bottom helix, respectively; α = adhesion factor along shaft; S_f = inter-helix spacing factor for $S_r > 3$, a function of inter-helix spacing (Zhang 1999); A_2 = surface area of bottom helix; N_c = bearing capacity factor for cohesive soils; D_h = helix diameter; H_{eff} = effective shaft length = shaft length above top helix – helix diameter; H_1 = depth to top helix; H_2 = depth to bottom helix; P_s = perimeter of pile shaft. The first term in Eq. 6 provides the pile shaft adhesion resistance, the second term accounts for the resistance along the inter-helix soil cylinder and the third term represents the bearing capacity of the bottom helix.

The compressive capacity based on individual helix bearing is given by:

$$Q_c = P_s H_{eff} \alpha C_{uf} + C_{u1} N_{c1} A_1 + C_{u2} N_{c2} A_2 \tag{7}$$

where: C_{u1} and C_{u2} = shear strength below top and bottom helices, respectively; A_1 = net surface area of the top helix = $\pi(D_h^2 - d^2)/4$; *and* H_1 and H_2 = depth to top and bottom helices, respectively. In Eq. 7, the first term accounts for the adhesion resistance along the pile shaft, and the second and third terms account for the ultimate bearing capacity derived by the top and bottom helices, respectively. Elsherbiny and El Naggar (2013) suggested that the cylindrical shear failure mechanism is appropriate for piles with helices embedded into cohesive soils for helix spacing up to 3D, and the load transfer mechanism at failure comprises shear along the pile shaft, shear along the inter-helix soil cylinder, and end bearing on the bottom helix.

Axial Capacity of Helical Piles in Structured Clay. The capacity of helical piles in structured and sensitive clays is affected by its installation, which can cause disturbance to soil within the zone affected by the pile helices. This disturbance is attributed to the advancement of the pile shaft into the ground and the rotation of the helical plates, which weakens the soil at the pile interface (Mooney et al. 1985; Bradka 1997; Zhang 1999), and the disturbance caused by shaft advancement is even more severe. The soil disturbance can be ascribed to different phenomena: the helical plates disturb soil through spirally shaped cut with intervals equal to the pitch distance and the soil traversed by the helix is sheared, and the shaft displaces the soil laterally. Bagheri and El Naggar (2016) investigated different failure patterns for helical piles installed in structured clays utilizing the results of full-scale uplift and compression load tests data. They concluded that the helical piles' capacity is significantly affected by soil disturbance induced by penetration of pile shaft and proposed a methodology for evaluating the axial capacity of helical piles installed in cohesive soil accounting for soil disturbance.

3   Response of Helical Piles to Lateral Loads

The response of a laterally loaded helical pile can be predicted using three different approaches: (i) beam on Winkler foundation (BWF) method, which treats the soil as a series of disconnected springs, (ii) continuum approach, which considers the soil as a semi-infinite space, and (iii) numerical methods (e.g., finite element) that discretize the soil domain into elements. The BWF method simplifies the analysis of piles nonlinear behaviour with reasonable accuracy for practical applications using discrete nonlinear soil springs (Juirnarongrit and Ashford 2006; Brandenberg et al. 2013). The background of BWF method is discussed herein.

3.1   Beam on Winkler Foundation (BWF)

The soil is simulated as a series of unconnected linear-elastic on nonlinear distributed springs. It assumes that the soil deflection at a given point is proportionally related to the contact pressure at that point and is independent of the contact stresses at other points. The coefficient of subgrade reaction, k, is the ratio between the soil pressure, p, at any given point of the contact surface and the displacement, y, produced by the applied load at that point, i.e. $k = p/y$. The soil modulus of subgrade reaction, $E_s = p/y$ has units of (F/L2) and p is the soil reaction per unit pile length (F/L). E_s depends on the soil type, depth, and pile diameter, d, and is evaluated from field tests on piles. Vesić (1961) performed analyses of beams resting on an elastic, isotropic half-space medium and proposed a relationship between Es, and the material properties in the elastic continuum as:

$$E_s = \frac{0.65E}{(1 - \mu_s{}^2)} \left[\frac{Ed^4}{E_P I_P} \right]^{1/12} \tag{13}$$

where E = soil modulus of elasticity, μ_s = soil Poisson's ratio, and $E_P\, I_p$ is pile flexural rigidity. This model is suitable for medium to long piles.

p-y Curve Method

To account for soli nonlinearity, Matlock (1970) introduced the p-y curve method in which a series of nonlinear soil springs as shown in Fig. 2, known as p-y curve, are used to simulate the soil behaviour. The p-y curve is a force-deflection relationship that relates the soil resistance, p, to the pile deflection, y.

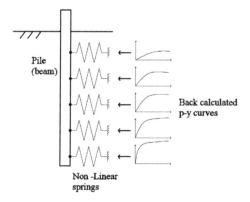

Fig. 2. Schematic of the p-y curve method

The p-y curves are established based on the results of lateral load tests conducted on full scale instrumented pile. The bending moment distribution along the pile shaft is calculated based on the pile curvature derived from the strain gage data. The soil reactions and pile deflections are determined by double integration and differentiation of the bending moment. Thus, variation of soil resistance with pile deflection, i.e. p-y curve, can be assessed at any given depth. Various p-y curves were obtained by curve fitting the results of testing piles embedded in clay and sand (McClelland and Focht 1958; Matlock 1970; Reese et al. 1975; Murchison and O'Neill 1984). The results of field tests verified the suitability of empirical p-y curves for predicting the piles lateral response, but the accuracy is sensitive to the implemented p-y curves. Proper selection of adequate p-y curves is crucial for improving accuracy using this method (Reese and Van Impe 2001). The behavior of laterally loaded pile is influenced by soil properties, pile type, cross-section shape and flexural stiffness, pile head fixity condition and type of loading. These important parameters are intrinsic characteristics of the developed p-y curves (Juirnarongrit and Ashford 2006).

p-y Curves of Helical Piles in Structured Clay: Elkasabgy and El Naggar (2019) conducted lateral load tests on full-scale helical piles installed in structured cohesive soil. They load tested five helical piles 6.0 and 9.0 m long with shaft diameter, d = 324 mm. The piles were either single or double-helix, with D_h = 610 mm and inter-helix spacing of 1.5 and 3.0 D_h. The piles were instrumented with strain gauges welded on the pile inner wall. The geometrical and material characteristics of the test piles are provided in Table 1. The testing program comprised two phases. In Phase I, the 6.0 m helical piles (SS11, SD11 and SD21) were tested 2 weeks after installation. In Phase II, the 9.0 m helical piles (LS12 and LD12) were tested 9 months after installation. The Phase I testing evaluated the pile behavior in disturbed soil disturbance due to pile installation, and Phase II testing evaluated the pile behaviour after soil regained most of its original strength. The load was applied using a 800 kN hydraulic jack.

The piles' load-displacement curves are displayed in Fig. 3. The piles response was generally nonlinear with an initial elastic region extending to a displacement of 1.5 to 2% d, and the final region (almost failure) initiated at displacement of 0.1d. The piles separated from the soil during the loading forming a gap behind the pile, which indicated that the soil has experienced plastic deformations.

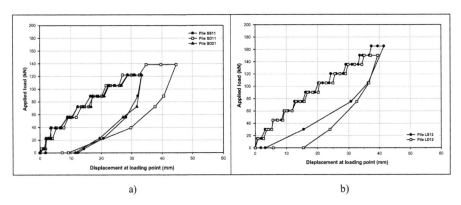

a)   b)

Fig. 3. Test results for: a) for 6.0 m piles (Phase I); b) 9.0 m piles (Phase II)

The bending moment profile was evaluated from measured strains. The bending moment profile exhibited a maximum value at a depth of about 5d below the ground elevation, and decreased to zero at the helices. Figure 4 displays the soil resistance along the pile shaft, corresponding to lateral displacement of pile head of 4.0, 6.25 and 12.5 mm. observed maximum soil resistance occurred at a depth of about 3d, then decreased rapidly with depth. The soil resistance for pile LD12 (Phase II) was higher than that for pile SD11 (Phase I). The soil p-y curves and its initial modulus of subgrade reaction were back-calculated from the strain measurements. It was found that soil disturbance during pile installation had a significant effect on the soil resistance and subgrade modulus, regardless of the number of helices and the value of the inter-helix spacing to helix diameter ratio. The effect of soil disturbance was introduced as a disturbance factor that can be incorporated in design. The p-y curves were numerically examined and a favorable agreement was achieved between the predicted and measured distributions of bending moments and deflections.

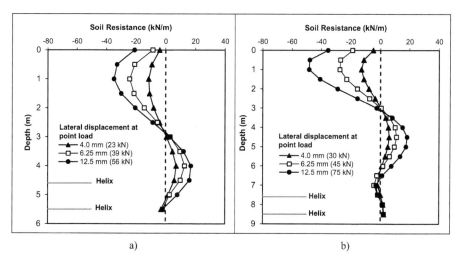

Fig. 4. Distribution of soil resistance for piles: a) SD11 and b) LD12

4 Dynamic Behavior of Helical Piles in Clay

Elkasabgy and El Naggar (2013, 2018) conducted full-scale dynamic load tests on helical and driven steel piles to evaluate their dynamic performance characteristics. The test piles were close-ended steel shafts with outer diameter of 324 mm. The piles were subjected to harmonic loading of increasing intensities and varying frequency. The dynamic properties of the subsurface soil were determined using the seismic cone penetration technique. A steel test body (59 machined circular steel plates) was added on top of the pile cap. The excitation force was produced by means of a mechanical oscillator mounted over the test body. The excitation force was harmonic with frequency between 3 and 60 Hz and was varied by altering the degree of eccentricity of rotating unbalanced masses with a maximum force of 23.5 kN. In addition, lateral free vibration tests were conducted on two piles. The vibration was monitored using two uniaxial piezoelectric accelerometers and one triaxial accelerometer, and the frequency was measured using a tachometer. The accelerometers were located on the test body such that two uniaxial accelerometers were mounted at equidistant positions from the foundation centre on the axis of symmetry. The triaxial accelerometer was mounted on the test body side, at the elevation of its CG. The displacement responses derived from the accelerometers measurements. The pile inner wall was instrumented with strain gauges to measure strain along the pile. Each level of gauges encompassed four half bridges allocated equidistantly from each other.

4.1 Vertical Vibration Tests

Phase I testing involved dynamic experiments conducted on 6.0 m helical and driven piles two weeks after installation. Phase II testing was conducted nine months after installation to allow the disturbed soil to regain some of its original stiffness and strength. Initially, low amplitude load was applied to avoid strong nonlinearity and the load frequency ranged from 3 to 60 Hz. The tests were conducted at five different load amplitudes (0.091, 0.12, 0.16, 0.18, and 0.21 kg.m) for the helical pile and three amplitudes (0.091, 0.16, and 0.21 kg.m) for the driven pile. The steady state dynamic response was measured over the frequency range for the assigned load amplitudes.

Figure 5 displays the test piles vertical response curves. The response varied with frequency and exhibited a resonant peak of 0.4 mm and 0.31 mm for the helical pile in Phases I and II and 0.32 mm for the driven pile in Phase I. As time passed (9 months after installation), the disturbed soil regained stiffness and the helical pile stiffness increased and its resonant frequencies increased by 16 to 26% and resonant amplitudes decreased by 22 to 55%. The driven pile response was close to that of helical pile in Phase I, which indicates that the dynamic response is dominated by the shaft resistance. The response curves demonstrated slight nonlinearity, especially for helical pile in Phase II where the measured resonant frequency shifted from 38 Hz for the lower excitation intensity to 35 Hz for the higher intensity.

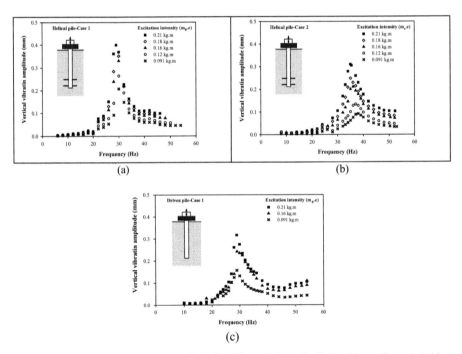

Fig. 5. Vertical response curves: a) helical pile (Phase I); b) helical pile (Phase II); and c) driven pile (Phase I)

4.2 Horizontal Vibration Tests

Free Vibration Test

The acceleration traces measured during the free vibration tests of piles LD1 (Phase II) and D1 (Phase I) are shown in Fig. 6. The piles natural frequencies were obtained from the period of the acceleration cycles and were found to be about 6.7 Hz and 3.0 Hz for piles LD1 and D1, respectively. The piles damping was determined from the decay in the acceleration amplitudes using the logarithmic decrement method. The equivalent viscous damping ratio, Ds, was varied between 4.5% and 6.0% for pile LD1 and 4.6% and 8.0% for pile D1.

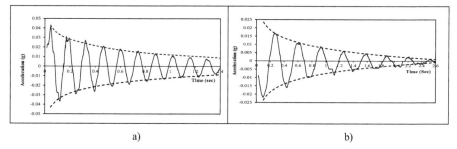

a)   b)

Fig. 6. Results of free vibration tests using test body 1; a) helical pile LD1-Phase II and b) driven pile D1-Phase I

Forced Horizontal Vibration

Figure 7 and Fig. 8 present the measured horizontal harmonic response curves for tested helical and driven piles. The response curves displayed two resonant peaks: The first dominated by the horizontal vibration mode and the second associated with the rocking vibration mode. The resonant frequencies decreased as the load amplitude increased due to moderate nonlinearity. The resonant frequencies of helical piles LS1 and LD1 tested in Phase II using test body 1 varied from 3.90 Hz to 3.80 Hz and 4.15 Hz to 3.75 Hz, respectively; and varied from 6.67 Hz to 6.42 Hz and from 6.95 Hz to 6.55 Hz, respectively, when tested with test body 2. The gap formed at the pile-soil interface at ground surface caused the observed nonlinear response. Their maximum horizontal amplitudes with test body 1 were 1.1 mm and 1.13 mm, and 1.78 mm and 1.87 with test body 2. The response of piles LS1 and LD1 was basically the same, indicating that the number of helices had no appreciable effect on the horizontal response. It should be noted that the depth of the upper helix of piles LS1 and LD1 was about 8.5 m (14D) and 6.7 m (11D) below the ground surface. The maximum amplitudes measured for piles LD1 and D1 (Phase I) using test body 1 were 0.85 mm and 1.17 mm, respectively. Figure 7 shows that the resonant frequencies for helical pile LD1 were approximately 3.5 Hz to 4.0 Hz, and were close to the resonant frequencies recorded for the driven pile D1. In addition, the maximum resonant amplitude for pile LD1 was close to the resonant amplitude for pile D1.

The measured resonant amplitudes and frequencies of pile D1 tested in Phase I are compared with those of helical piles LS1 and LD1 tested in Phase II to quantify the regain in soil stiffness. The resonant frequencies increased by 8 to 15% and resonant amplitudes decreased by 7 to 29% due to increase in stiffness. The soil stiffness was attributed to thixotropic hardening with time after installation.

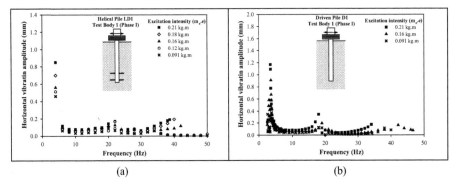

Fig. 7. Measured horizontal response curves for piles tested in Phase I using test body 1; a) helical pile-LD1 and b) driven pile-D1

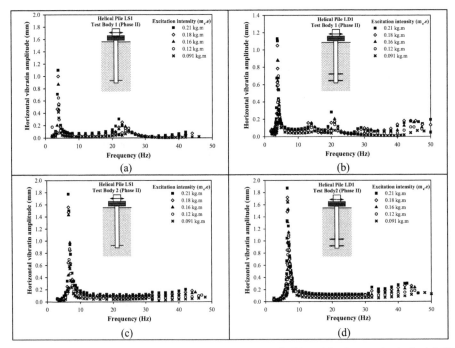

Fig. 8. Measured helical piles horizontal response curves in Phase II using test body 1; a) LS1 and b) LD1 and test body 2 c) LS1 and d) LD1

5 Dynamic Response of Helical Piles in Sand

Elshirbiny et al. (2017) conducted lateral dynamic load tests on a full-scale helical pile. The pile was 4.0 m long, shaft diameter d = 114.3 mm, wall thickness, t = 8.6 mm, and had a single helix with diameter, D = 254 mm. The pile head displacement rotation and acceleration were measured using linear string potentiometers, and unidirectional accelerometers. The pile was installed in soil pit 20 × 20 m and 9 m deep. The pit was filled with an uncontrolled granular soil fill. Two seismic cone penetration tests (SCPTs) were conducted to measure shear wave velocity. The shear wave velocity (V_s) measured was 900 m/s for the top 1.5 m and 300 m/s below that depth. The dynamic load was applied using a 320 kg eccentric mass shaker. The shaker was controlled by a variable frequency drive (VFD). It was attached to a 38 mm thick steel plate pinned to the test pile (SH2) while supported vertically on a roller supports, which were attached to reaction piles at the back such that the lateral load was transferred to SH2. The shaker was operated up to a speed of 316 rpm (i.e. 5 Hz) with force amplitude, f = 4.7 kn for SH2. The cyclic test setup involved two single acting hollow hydraulic jacks with capacity 60ton each situated against opposite sides of a steel plate. In order to apply two-way cyclic loading, one jack was pressurized and the other jack had zero pressure, and then reversed. The load was measured using two 900 kN hollow load cells with full-bridge strain gages.

Figure 9a presents average steady-state displacement response at different exciting frequency. The average displacement gradually increased as the load amplitude increased. The mean displacement was not zero especially at speeds ≥2.5 Hz. As the number of cycles increased at the same frequency, the pile response increased, and its stiffness decreased due to gap forming around the pile rather than changes in the soil stiffness. This was confirmed as the pile returned to its original position when the dynamic load was terminated. Five two-way load cycles were applied with gradually increasing amplitudes. The test was terminated when the maximum hydraulic jack stroke was reached. The cyclic load-displacement curve is shown in Fig. 9b. The cyclic loading reduced the pile stiffness.

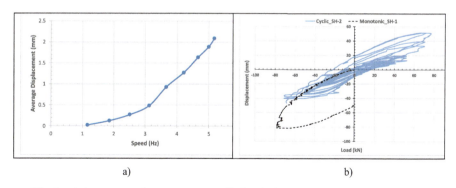

a)   b)

Fig. 9. a) Average steady-state response; b) Cyclic load-displacement curve for SH-2.

6 Seismic Loading of Helical Piles

6.1 Single Piles in Dry Sand

Elsawy et al. (2019a, 2019b) investigated the seismic response of helical piles embedded in dry sand. They conducted seismic shaking on ten full-scale helical and driven piles using the Large High Performance Outdoor Shake Table (LHPOST) at the University of California – San Diego. The effects of seismic load intensity and frequency on piles response were investigated. Nine steel helical piles and one steel driven pile were installed in well-graded, dense sand compacted to Dr = 100% inside a laminar shear box (6.7 m long, 3.0 m wide and 4.7 m high). The piles were instrumented with six or seven pairs of strain gauges placed at pre-determined elevations. The soil shaking was captured using 23 accelerometers placed within the soil bed. Pile heads were instrumented with accelerometers. Table 1 presents the piles properties.

Table 1. Tested Pile Information

Pile ID	Type (*Notation*)	Length (m)	Helix level (m)	Helix diameter (m)	Diameter/Wall thickness (mm)	Inertia $(m^4) \times 10^{-6}$	Yield Strength (MPa)	Mass (kg)
P1	Helical (*88C1HP*)	3.96	3.40	0.254	88/5	1.199	450	770
P2, P3	Helical (*88C1HP*)	3.66	– 3.15	0.254	88/5	1.199	450	750 780
P4	Helical (*88C2HP*)	3.66	– 2.55	0.203 0.254	88/5	1.199	450	750
P5	Driven (*88CDP*)	3.66	–	–	88/5	1.199	450	370
P6	Helical (*76S1HP*)	3.66	– 3.15	0.254	76/5	1.257	415	435

[*]C: circular shaft, S square shaft, 1HP & 2HP: single helix & double helix pile

The seismic loading tests consisted of two different shaking schemes: white noise records and ground motion records. The white noise contained a range of frequencies (0–40 Hz) and had a peak acceleration of 0.15 g. The ground motion records used were: Northridge (1994) and Takatori (1995) earthquakes. To observe both the linear and non-linear behavior of piles, each earthquake record was scaled to 75% and 50% resulting in a total of 6 earthquake records. Northridge record had a frequency range of 1.5 Hz to 5.0 Hz, while Takatori record had a range of 0.5 Hz to 1.5 Hz.

Figure 10 shows the Fourier spectra of pile P1 response obtained for strain gauge and accelerometer readings. Each Fourier spectrum displays two peaks corresponding to two different modes of motion, horizontal and rotational degrees of freedom.

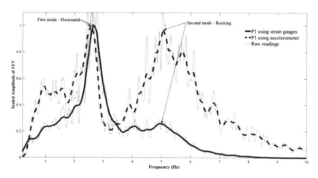

Fig. 10. Fourier spectra of Pile 1 response from strain gauges and accelerometer

Figure 11 compares the response of Piles 2, 5 and 6 to two ground motions with equal peak ground acceleration but different frequency content; 100% Northridge and 75% Takatori. All circular helical piles (Piles 1–4), had natural period ranging between 0.47 s–0.53 s, which was close to the predominant period of Takatori earthquake; therefore, all circular helical piles had a greater response during Takatori. On the other hand, the driven and the square helical piles had a period of 0.24 s and 0.29 s, respectively. This coincided with the peak of Northridge; therefore, they had higher response during Northridge earthquake.

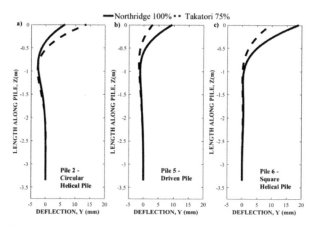

Fig. 11. Response to 100% Northridge & 75% Takatori a) Pile 2 b) Pile 5 c) Pile 6

The results showed no clear advantage for the shape of pile cross-section (i.e. square or circular). On the other hand, the response of the double-helix pile was less than that of the single-helix pile due to the negative moment developed close to the location of the second helix. It was also observed that the natural frequency of the driven pile was slightly higher than that of the helical piles. However, the response of the helical pile was very close to that of the driven pile, which illustrates the ability of helical piles to perform as good as conventional piles under seismic loading.

6.2   Helical Pile Groups in Dry Sand

Fayez et al. (2021 and 2022) conducted shake table tests on single and grouped helical piles and measured their responses to strong ground motions. They evaluated the
effects of earthquake characteristics (i.e., intensity and frequency content) on seismic
performance of the single and grouped HPs from the measured responses. In addition,
they discussed the performance characteristics of helical pile groups in terms of the
interaction between piles within a group and the contributions of vertical and lateral
stiffness of individual piles to the rocking stiffness and the overall capacity of the pile
group. Single piles as well as two groups of HPs with diameters 88 mm and 140 mm,
were subjected to strong ground motions. Each group consisted of four-piles with similar
diameters connected to a steel skid filled by sand to simulate a superstructure inertial
load. The 88 mm-diameter piles were loaded with 62 kN and the 140 mm-diameter piles
were loaded with 98 kN. Accelerometers were placed on the skid at mid-height to record
its acceleration. Figure 12 presents the shake table arrangement and Fig. 13 depicts the
distribution of piles within PG1 and PG2. Table 2 presents the piles dimensions and
material properties.

Fig. 12. Shake table tests of pile groups

Figure 14 compares the bending moment profiles for single piles and piles in a group
during Northridge Earthquake with 100% PGA (NOR-100) and Takatori Earthquake
with 75% PGA (TAK-75). Single piles experienced resonance during TAK earthquake
and exhibited significant response. Even though NOR-100 TAK-75 had the same PGA,
the maximum bending moment was more than twice during TAK-75. Moreover, PG1
experienced low bending moment in both earthquakes as it didn't experience resonance.
However, it exhibited a slightly larger bending moment during NOR-100 due to the
closeness of its natural frequency to the frequency content NOR-100 compared to TAK-
75. PG2 exhibited larger bending moment due to resonance during NOR-100. This
emphasizes the importance of considering the earthquake frequency content in seismic
design of helical pile foundations.

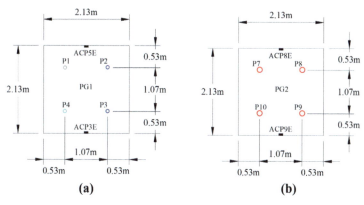

(a)   **(b)**

Fig. 13. Pile groups: (a) PG1 (b) PG2.

Table 2. Properties of test piles.

Pile Group	Pile	Type*	Total Length/Depth (m)	Helix level/Helix Diameter (m)	Diameter/Wall thickness (mm)	Yield strength (MPa)
PG1	P1	H	3.96/3.66	−3.51/0.254	88/5.3	448.2
	P2	H	3.66/3.35	−3.20/0.254		
	P3					
	P4	2H	3.66/3.35	−2.59/0.254		
				−3.20/0.203		
PG2	P7	H	4.22/3.35	−3.20/0.254	140/10.5	551.6
	P8					
	P9					
	P10					

* H (single helical pile), 2H (double helical pile)

Rocking Behaviour of Helical Pile Groups

The rocking resistance of the helical piles within a group is an important mechanism to resist the seismic loading. This rocking resistance is a function of the normal force developed in the piles during the seismic event. Figure 15 presents the variation of maximum normal forces with PGA. Figure 15a shows higher increase in normal force for larger diameter pile group (PG2) as the higher flexural rigidity of the piles promoted the rocking behaviour over lateral deflection. In addition, larger normal forces were observed during resonance (NOR earthquake) for both pile groups. On the other hand, Fig. 15b shows that the rate of increase in maximum normal force was lower than the increase in PGA for both pile groups. However, the rate of increase when resonance was experienced was higher. This indicates that rocking behaviour increased during resonance and consequently, the normal forces increased.

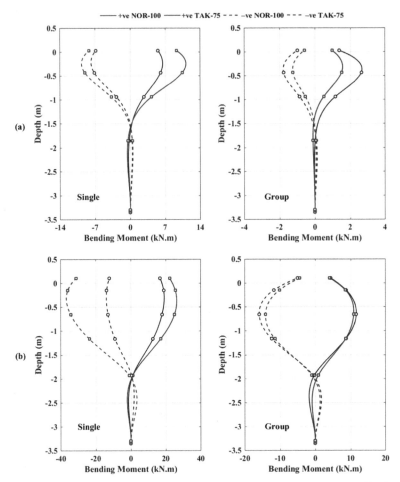

Fig. 14. Effect of earthquake frequency content on maximum bending moment profiles for single piles and fixed-head pile groups: P2 (a) and P7 (b).

6.3 Seismic Response of Helical Piles in Liquefiable Soil

Hussein and El Naggar (2021a, 2021b) investigated the seismic performance of helical piles in liquefiable soil employing three-dimensional nonlinear finite element models. The numerical models were validated using the published results from a comparative shaking table testing program of identical 4-pile groups installed in saturated and dry sand. The validated models were then used to analyze the seismic lateral and axial responses of helical piles to investigate the soil-helical pile-superstructure interaction.

The helical pile groups considered in the numerical models were installed in a soil profile comprising a surficial clay layer, loose sand layer and dense sand layer as shown in Fig. 16. The soil profile was dry in one set and saturated with water in the other set of shakings so that pore water pressure could develop during the shaking, and liquefaction could occur. The identical soil-pile group configuration in both sets of experiments

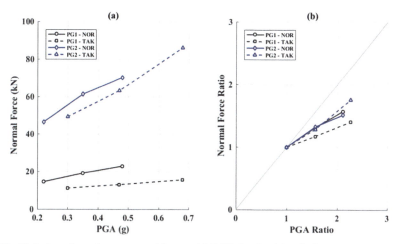

Fig. 15. Variation of maximum normal force with PGA for fixed-head pile groups: actual values (a) and normalized correlation (b).

allowed a direct comparison between the non-liquefiable and liquefiable tests. The soil bed was enclosed in a laminar box with dimensions of 3.2 m × 2.4 m × 3.1 m, which was situated on a 6 m × 6 m shaking table. The soil bed considered in the numerical modeling consisted of three layers: 0.3 m clay crust, 1.2 m loose sand with $D_r = 60\%$, and 1.6 m dense sand with $D_r = 90\%$. Table 3 describes the configuration of the analyzed helical pile group (HPG). The 1.95 m long HP shaft was divided into elements aligned with the adjacent soil elements. Figure 16 summarizes the layout of the soil elements, HP, cap, and superstructure, as well as elements types and material models.

The soil-helical piles-superstructure interaction models were developed using the OpenSees platform (Mazzoni et al. 2006). The test soil bed was discretized into 5148 elements. The soil elements were appropriately refined adjacent to the piles. The maximum elements' sizes in the longitudinal direction ($A_s = 3.6$ m) and transversal ($B_s = 2.4$ m) direction were 0.27 m and 0.34 m, and the largest element size along the depth ($d_s = 3.1$ m) was 0.3 m. The element size allowed a shear wave with maximum frequency to propagate efficiently considering shear wave velocity, $V_s = 100$ m/sec, at maximum frequency $F_{max} = V_s/(4H) = 73.5$ Hz (H = 0.34 m). The shaft was divided into elements aligned with the adjacent soil elements and rigid link elements were applied perpendicular to the shaft to fill the pile volume within the soil. The helix ($D_{helix} = 0.228$ m) was divided into elements of the same size as the surrounding soil elements, which facilitated the contact between HP and adjacent soil elements. The top and bottom helices were placed at a depth of 1.6 m and 1.9 m, respectively.

The soil was discretized using 8-node hexahedral linear isoperimetric elements (Brick u-p Element) (Yang and Elgamal 2008). The elastic-plastic pressure dependant constitutive model PressureDependMultiYield02 (PDMY02) simulated the nonlinear sand response. PressureIndependMultiYield elastic-plastic material model simulated the clay crust, in which the plasticity is considered in the deviatoric stress-strain response but does not account for the confining pressure variability (Elgamal et al. 2002). The

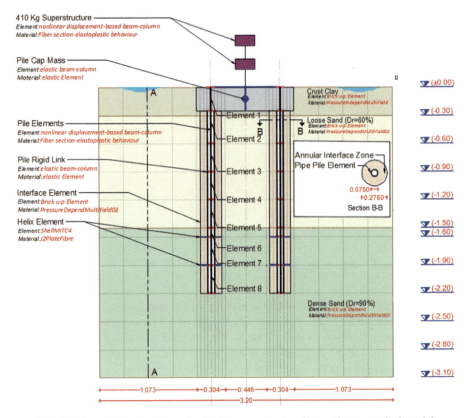

Fig. 16. Longitudinal section of soil-pile-superstructure interaction numerical model

Table 3. Helical pile group configuration, units in mm

Parameter	(mm)	Parameter	(mm)	Parameter	(mm)
Pile d	76	Helix Spacing	300	Pile Cap Thick	250
Shaft Wall Thickness	5	Pile Spacing	750	Helix Thick	10
Helix D_{helix}	228	Pile Cap width	1050		

interface element was 10 cm thick to simulate the large relative stiffness between the soil and helical piles. The measured soil parameters in the shake table tests were employed to correlate the material model geotechnical parameters. The particles' mean diameter and uniformity coefficient were used to calculate the maximum and minimum voids ratios (Sarkar et al. 2019), which were correlated to the low strain shear modulus at each depth (Das and Ramana 2011). The geotechnical parameters were adjusted through the calibration process to capture the proper soil element response.

Figure 17 compares the pile calculated lateral displacement of single helical pile (SHP) and double helical pile (DHP) with that for RC pile (RCP) of equivalent lateral stiffness. The lateral displacements of SHP and DHP were less than that for RCP in both saturated and dry tests. The higher axial stiffness of the HPs and the fixation provided by the helices increased the rocking and flexural resistance of the HPG. This is because the top helix offered significant resistance (and fixation), which reduced the rotation of the second helix. Thus, the contribution of the helices to lateral resistance is primarily due to passive resistance associated with their rotation (Elkasabgy and El Naggar 2019). Similarly, Al-Baghdadi et al. (2015) investigated the helix contribution to the lateral resistance and concluded that the contribution is important only for helical plates placed close to the surface. In addition, the flexural deformation was primarily concentrated within the liquefied layer.

The axial force was higher in the DHP shaft than the SHP at the top helix location; therefore, the axial force at the bottom helix decreased even more than the case for SHP. This response refers to the ability of the DHP to resist higher axial loads compared to the SHP. The shaft axial force below the bottom helix was marginal. These observations are further elaborated by monitoring the shaft friction and helix bearing forces. The shaft force within the loose sand decreased rapidly as the shaking progressed and diminished as liquefaction occurred. The static shaft force of the SHP was greater than that of the DHP, while its bearing force was less than the DHP. The end bearing forces for SHP and DHP increased to compensate for the reduction in the shaft forces. Thus, the bearing forces fluctuated as the shaft friction fluctuated, which was accompanied by pile. Cycles of tensile forces also occurred in both the leading and trailing DHP, which demonstrates that the complicated behaviour of HPs during the generation and dissipation of excess pore water pressure (EPWP). For both SHP and DHP, the shaft friction diminished as the shaking progressed and became zero (and even negative friction) at the end of motion due to the liquefaction of loose sand. On the other hand, the total bearing resistance increased, which compensated for the shaft friction and maintained the overall capacity of the pile. In addition, there was an appreciable difference in bearing forces of the leading and trailing piles due to the rocking of the pile group. Hussein and El Naggar (2022a) further extended this study to consider the behavior of prototype helical piles and reported similar findings.

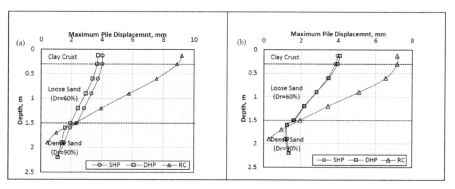

Fig. 17. Piles maximum lateral displacement for SHP, DHP and RCP during the 0.18 g Wolong motion: (a) saturated test; (b) dry test

Hussein and El Naggar (2021a, 2021b) concluded that the seismic lateral displacements of helical piles and associated soil were lower than those for RC piles. The helical pile groups exhibited rigid body movement in the saturated soil, especially DHPs, as the dense sand liquefied. In addition, the helical pile seismic settlements were much less than that of the RCP for both the saturated and dry soils. As liquefaction occurred, the HP shaft resistance diminished, and the pile settled. However, higher end-bearing force was mobilized, which compensated for the decrease in shaft force, which demonstrates the excellent performance of helical piles in maintaining their capacity during and after liquefaction and controlling the post-liquefaction settlement. Finally, the helical pile response in the saturated test was dominated by the rocking behaviour, while the flexural behaviour dominated the response in the dry tests.

6.4 Seismic Response of Helical Piles in Cohesive Soil

Hussein and El Naggar (2022b) evaluated the dynamic response of driven and helical piles in cohesive soil. They developed three-dimensional (3D) finite element models (FEMs) to analyze the response of driven piles based on the results of a full-scale field tests conducted by Fleming et al. (2016). These large-scale field tests examined the seismic response of driven piles in unimproved and improved soft clay.

Figure 18a presents a longitudinal section of the field test with the location and dimensions of the improved soil. The soil elements were simulated using 8-node hexahedral linear isoperimetric elements (Brick u-p Element). The element is used to simulate the dynamic response of the soil-fluid coupled material. The unimproved and improved clay materials were simulated using PressureIndependMultiYield elastic-plastic material model. In this model, the plasticity appears in the deviatoric stress-strain response only. For the volumetric response, it is independent of the deviatoric stress in the linear elastic response and is insensitive to the confining pressure variation. On the other hand, the sand layer was simulated using PressureDependMultiYield02, which is an elastic-plastic model that is sensitive to the change in the confining pressure. The FEM was validated by comparing its predictions with the measured pile response and bending moment profile during the field tests.

The validated FEM was employed to simulate the dynamic response of helical piles. Figure 18b presents the longitudinal section of the helical pile considered within the soil profile of the field test. Both single-helix and double-helix piles were considered in the analysis. The cross-section of the helical pile pipe cross-section was AISC-HSS12 × 0.5, and the helix was 762 mm in diameter and 25 mm in thickness. The bottom helix was placed at the interface of the soft clay-sand layers (i.e., depth 3.4 m below the ground), and the pile toe was embedded 150 mm in the sand layer. The helical pile was 5.35 m long, which was shorter than the driven pile by 1.75 m. The second helix was placed above the bottom helix with an inter-helix spacing of 1.5, 2.0, 2.5, and 3.0 times the helix diameter (D_{helix}).

The driven and helical piles' dynamic lateral responses were calculated and are compared in terms of the force-displacement response at the pile head and at the ground surface as shown in Fig. 19. Different analyses were conducted considering a SHP and DHP with helix spacing of 1.5, 2, 2.5 and $3D_{helix}$. Figure 19a shows that the lateral response of the DHP improved as the second helix was placed closer to the ground

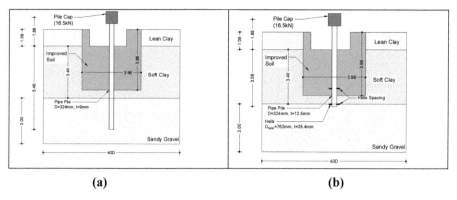

Fig. 18. Longitudinal section of the soil-pile profile: a) drive pile; b) helical pile

surface. The displacement of the DHP decreased from 9.6 cm to 8.7 cm as the inter-helix spacing increased from $1.5D_{helix}$ to $3D_{helix}$. This improvement in lateral response is attributed to the additional bearing resistance offered by the top helix as it was placed within the pile segment that experienced large deflection. This was observed from the axial forces calculated in the pile as will be discussed later. Figure 19b shows that the lateral displacement of the piles at the ground surface was much lower than at the head because the large free-standing segment of the pile caused most of the displacement, which is not affected much by the helices.

To further evaluate the dynamic lateral performance of driven and helical piles in cohesive soils, the undrained shear strength of the soil (Su) was varied increased to 50 kPa, 75 kPa, and 100 kPa. The maximum shear modulus was also increased with the same ratio of the undrained shear strength. Figures 19 (c, e, and g) present the lateral force-displacement responses of the driven and the helical piles at pile head, while Fig. 19 (d, f, and h) display the force-displacement responses at the ground level. As expected, the pile lateral displacements at both the pile and ground surface decrease the soil strength and stiffness increase. The driven pile lateral displacement decreased from 8.7 cm at Su = 25 kPa to 3.9 cm at Su = 100 kPa. Similarly, the SHP lateral displacement decreased as the soil strength increased, i.e., decreased from 10cm at Su = 25 kPa to 3.9 cm at Su = 100 kPa. The lateral response was enhanced further by using the DHP; for example, the maximum lateral displacement at the DHP head decreased from 8.7 cm at Su = 25 kPa to 3.6 cm at Su = 100 kPa (i.e., performed better than the driven pile).

Hussein and El Naggar (2022b) concluded that the lateral displacement of DHP was the same as or less than that of the driven pile for the same dynamic loading applied at the pile head. The lateral response of helical piles improved as the clay shear strength increased due to the additional passive resistance on the helices. They also concluded that during a seismic event, the helical pile would experience lower lateral deformation and bending moment compared to the driven pile due to the kinematic interaction. Furthermore, for piles fully embedded in the ground, the seismic performance of helical piles was better than that of the driven pile.

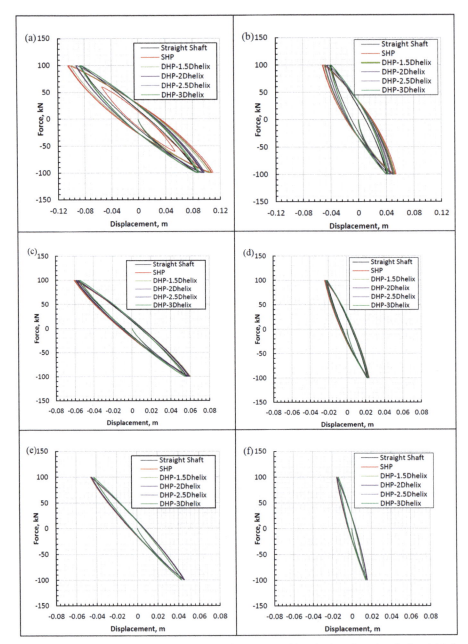

Fig. 19. Force-Displacement hysteretic loops in the test with cohesive clay: At pile head (a) Su = 25 kPa; (b) Su = 50 kPa; (c) Su = 75 kPa; (d) Su = 100 kPa, and at the ground level (e) Su = 25 kPa; (f) Su = 50 kPa; (g) Su = 75 kPa; (h) Su = 100 kPa

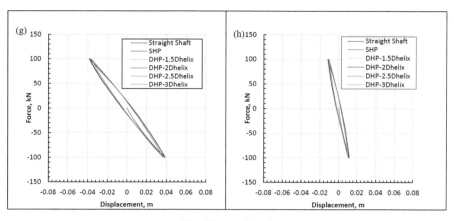

Fig. 19. continued

References

Alwalan, M.F., El Naggar, M.H.: Load-transfer mechanism of helical piles under compressive and impact loading. ASCE Int. J. Geomech. (2021). https://doi.org/10.1061/(ASCE)GM.1943-5622.0002037

Alwalan, M.F., El Naggar, M.H.: Analytical models of impact force-time response generated from high strain dynamic load test on driven and helical piles. Comput. Geotech. **128** (2020a). https://doi.org/10.1016/j.compgeo.2020.103834

Alwalan, M.F., El Naggar, M.H.: Finite element analysis of helical piles subjected to axial impact loading. Comput. Geotech. **123** (2020b). https://doi.org/10.1016/j.compgeo.2020.103597

Al-baghdadi, T.A., Brown, M.J., Knappett, J.A., Ishikura, R.: Modelling of laterally loaded screw piles with large helical plates in sand. In: Proceedings of the 3rd International Symposium on Frontiers in Offshore Geotechnics (Frontiers in Offshore Geotechnics III), Oslo, Norway, 10-12 June 2015, pp. 503–508 (2015)

Bagheri, F., El Naggar, M.H.: Effects of installation disturbance on behavior of multi-helix piles in structured clays. J. Deep Found. **9**(2), 80–91 (2016)

Bradka, T.D.: Vertical capacity of helical screw anchor piles. Master of Engineering Report, University of Alberta, Alberta, Canada (1997)

Brandenberg, S.J., Zhao, M., Boulanger, R.W., Wilson, D.W.: p-y plasticity model for nonlinear dynamic analysis of piles in liquefiable soil. J. Geotech. Geoenviron. Eng. **139**, 1262–1274 (2013)

Das, B., Ramana, G.: Principles of soil dynamics, 2nd International SI (edn.). Cengage learning, Boston (2011)

Elgamal, A., Yang, Z., Parra, E.: Computational modeling of cyclic mobility and post-liquefaction site response. Soil Dyn. Earthq. Eng. **22**, 259–271 (2002)

Elkasabgy, M., El Naggar, M.H.: Dynamic response of vertically loaded helical and driven steel piles. Can. Geotech. J. **50**(5), 521–535 (2013)

Elkasabgy, M., El Naggar, M.H.: Axial compressive response of large-capacity helical and driven steel piles in cohesive soil. Can. Geotech. J. **52**(2), 224–243 (2015)

Elkasabgy, M.A., El Naggar, M.H.: Lateral vibration of helical and driven steel piles installed in cohesive soils. ASCE J. Geotech. Geoenviron. Eng. **144**(9) (2018). https://doi.org/10.1061/(ASCE)GT.1943-5606.0001899

Elkasabgy, M., El Naggar, M.H.: Lateral performance and *p-y* curves for large-capacity helical piles installed in clayey glacial deposit. Geotech. Geoenviron. Eng. ASCE (2019). https://doi.org/10.1061/(ASCE)GT.1943-5606.0002063

El-Sawy, M.K., El Naggar, M.H., Cerato, A.B., Elgamal, A.: Data reduction and dynamic p-y curves of helical piles from large scale shake table tests. Geotech. Geoenviron. Eng. ASCE **145**(10), 04019075 (2019a)

El-Sawy, M.K., El Naggar, M.H., Cerato, A.B., Elgamal, A.: Seismic performance of helical piles in dry sand from large scale shake table tests. Geotechnique **69**(12), 1071–1085 (2019b). https://doi.org/10.1680/jgeot.18.P.001

Elsharnouby, M., El Naggar, M.H.: Field investigation of lateral monotonic and cyclic performance of reinforced helical pulldown micropiles. Can. Geotech. J. (2018a). https://doi.org/10.1139/cgj-2017-0330,Publishedon-lineJan23

ElSharnouby, M.M., El Naggar, M.H.: Axial monotonic and cyclic performance of fibre-reinforced polymer (FRP) – steel fibre–reinforced helical pulldown micropiles (FRP-RHPM). Can. Geotech. J. **49**(12), 1378–1392 (2012a)

ElSharnouby, M.M., El Naggar, M.H.: Field investigation of axial monotonic and cyclic performance of reinforced helical pulldown micropiles. Can. Geotech. J. **49**(5), 560–573 (2012b)

Elsherbiny, Z., El Naggar, M.H.: Axial compressive capacity of helical piles from field tests and numerical study. Can. Geotech. J. **50**(12), 1191–1203 (2013)

Fahmy, A., El Naggar, M.H.: Cyclic lateral performance of helical tapered piles in silty sand. J. Deep Found. **10**(3), 111–124 (2016a)

Fahmy, A., El Naggar, M.H.: Cyclic axial performance of helical tapered piles in sand. J. Deep Found. Inst. (2016b). https://doi.org/10.1080/19375247.2016.1211353

Fayez, A.F., El Naggar, M.H., Cerato, A.B., Elgamal, A.: Assessment of SSI effects on stiffness of single and grouped helical piles in dry sand from large shake table tests. Bull. Earthq. Eng. (2021). https://doi.org/10.1007/s10518-021-01241-7

Fayez, A., El Naggar, M.H., Cerato, A., Elgamal, A.: Seismic response of helical pile groups from shake table experiments. Soil Dyn. Earthq. Eng. **152** (2022). https://doi.org/10.1016/j.soildyn.2021.107008

Fleming, B.J., Sritharan, S., Miller, G.A., Muraleetharan, K.K.: Full-scale seismic testing of piles in improved and unimproved soft clay. Earthq. Spectra **32**(1), 239–265 (2016)

Hussein, A.F., El Naggar, M.H.: Seismic axial behaviour of pile groups in non-liquefiable and liquefiable soils. Soil Dyn. Earthq. Eng. **149** (2021a). https://doi.org/10.1016/j.soildyn.2021.106853

Hussein, A.F., El Naggar, M.H.: Seismic behaviour of piles in non-liquefiable and liquefiable soil. Bull. Earthq. Eng. (2021b). https://doi.org/10.1007/s10518-021-01244-4

Hussein, A.F., El Naggar, M.H.: Effect of model scale on helical piles response established from shake table tests. Soil Dyn. Earthq. Eng. **152** (2022a). https://doi.org/10.1016/j.soildyn.2021.107013

Hussein, A.F., El Naggar, M.H.: Seismic performance of driven and helical piles in cohesive soil. Acta Geotechnica (2022b, submitted)

Juirnarongrit, T., Ashford, S.A.: Soil-pile response to blast-induced lateral spreading. II: analysis and assessment of the *p–y* method. ASCE Geotech. Geoenviron. Eng. ASCE (2006). https://doi.org/10.1061/(ASCE)1090-0241(2006)132:2(163)

Livneh, B., El Naggar, M.H.: Axial load testing and numerical modeling of square shaft helical piles. Can. Geotech. J. **45**(8), 1142–1155 (2008)

Lutenegger, A.J.: Cylindrical Shear Or Plate Bearing? – Uplift behavior if multi-helix screw anchors in clay. In: International Foundation Congress and Equipment Expo, pp. 456–463 (2009)

Matlock, H.: Correlations for design of laterally loaded piles in soft clay. In: Proceedings of the 2nd Offshore Technology Conference, Houston, Texas, pp. 577–594 (1970)

Mazzoni, S., McKenna, F., Scott, M.H., Fenves, G.L.: OpenSees command language manual Pacific Earthquake Engineering Research (PEER) Center 264 (2006)

McClelland, B., Focht, J.A.: Soil modulus of laterally loaded piles. Trans. ASCE **123**, 1049–1063 (1958)

Mooney, J.S., Clemence, S.P., Adamczak, S.: Uplift capacity of helix anchors in clay and silt. In: ASCE Convention Conference Proceedings, New York, pp. 48–72. ASCE (1985)

Murchison, J., O'Neill, M.: Evaluation of p-y relationships in cohesionless soils: analysis and design of pile foundations. In: Proceedings of the Symposium in conjunction with ASCE National Convention, pp. 174–191 (1984)

Orang, M.J., Boushehri, R., Motamed, R., Prabhakaran, A., Elgamal, A.: Large-scale shake table experiment on the performance of helical piles in liquefiable soils. In: Proceedings of the 45th DFI Annual Conference on Deep Foundations, Deep Foundations Institute (2021)

Perko, H.A.: Helical Piles: A Practical Guide to Design and Installation. Wiley, New Jersey (2009)

Reese, L.C., Cox, W.R., Koop, F.D.: Field testing and analysis of laterally loaded piles in stiff clay. In: Proceedings of the 7th Offshore Technology Conference, Dallas, Texas, pp. 672–690 (1975)

Reese, L.C., Welch, R.C.: Lateral loading of deep foundations in stiff clay. J. Geotech. Eng. Div. ASCE **101**(GT7), 633–649 (1975)

Reese, L.C., Van Impe, W.F.: Single Piles and Pile Group Under Lateral Loading, 2nd edn. Balkema, Rotterdam (2001)

Sarkar, D., König, D., Goudarzy, M.: The influence of particle characteristics on the index void ratios in granular materials. Particuology **46**, 1–13 (2019)

Shahbazi, M., Cerato, A., El Naggar, M.H., Elgamal, A.: Evaluation of seismic soil-structure interaction of full-scale grouped helical piles in dense sand. ASCE Int. J. Geomech. **20**(12) (2020). https://doi.org/10.1061/(ASCE)GM.1943-5622.0001876

Tappenden, K.M.: Predicting the Axial Capacity of Screw Piles Installaed in Western Canadian Soils. The University of Alberta, Edmonton (2007)

Vesic, A.S.: Beam on elastic subgrade and the Winkler hypothesis. In: Proceedings of 5th International Conference on Soil Mechanics and Foundation Engineering, Paris, France, vol. 1, pp. 845–850 (1961)

Yang, Z., Lu, J., Elgamal, A.: OpenSees soil models and solid-fluid fully coupled ele-ments User's Manual Ver 1:27 (2008)

Zhang, D.J.Y.: Predicting capacity of helical screw piles in Alberta soils. M.E.Sc. thesis, University of Alberta, Edmonton, Alberta, Canada (1999)

Seismic Response of Offshore Wind Turbine Supported by Monopile and Caisson Foundations in Undrained Clay

Maosong Huang[1,2(✉)], He Cui[1,2], Zhenhao Shi[1,2], and Lei Liu[1,2]

[1] Key Laboratory of Geotechnical and Underground Engineering of Ministry of Education, Tongji University, Shanghai, China
mshuang@tongji.edu.cn
[2] Department of Geotechnical Engineering, Tongji University, Shanghai, China

Abstract. The seismic performance of offshore wind turbine (OWT) becomes increasingly critical as more wind farms are built/planned in sites subjected to active seismic events. Time-domain, continuum-based methods can provide comprehensive analyses of OWT during seismic shaking yet its application in engineering practice can be restricted by the complexity of soil constitutive models. This work presents a constitutive model for undrained clay that is simple yet replicates the essential soil behavior, including mechanical anisotropy, cyclic degradation of stiffness and strength, and path-dependent stiffness nonlinearity at small strains. The capacity of the soil model is assessed at various levels, by simulating soil element tests, the response of cyclically-loaded pile in centrifuge test, and the response of caisson-supported OWT in seismic centrifuge test. These assessments show that the seismic response of OWT can be reasonably represented by the proposed method. Based on the soil model, dynamic finite element analyses are performed to explore the seismic response of OWT supported by different foundation types: monopile and caisson. This investigation highlights: (1) the dynamic response of OWT foundation is governed by the kinematic-inertia interactions between soil, foundation, and superstructures; (2) the resonance of OWT structure system (including higher modes) can be a critical mechanism behind the developments of excessive foundation movements.

Keywords: Offshore wind turbine foundations · Monopile · Caisson · Seismic response · Constitutive relations

1 Introduction

The offshore wind sector represents an essential component for delivering large-scale, affordable, and clean energy [1]. Due to their harsh working environments, the foundation of offshore wind turbine (OWT) is required to sustain substantial cyclic lateral loading, e.g., wind and wave actions. Accordingly, the above aspect has been subjected to substantial investigations (e.g., see [2–4]). In the contrast, the seismic response of OWT foundations was traditionally considered secondary, largely due to the intrinsic

© The Author(s), under exclusive license to Springer Nature Switzerland AG 2022
L. Wang et al. (Eds.): PBD-IV 2022, GGEE 52, pp. 50–66, 2022.
https://doi.org/10.1007/978-3-031-11898-2_3

flexibility of OWT structures and the typical frequency contents of earthquakes [5]. This aspect, nevertheless, has begun to gain increasing attention from both academics and practitioners [6–12], due to the exponential expansion of offshore wind farms worldwide and, consequently, more frequent overlap between the active seismic regions and wind farms sites [6]. This paper has been focused on the seismic response of OWT supported by monopile and suction caisson, i.e., predominant OWT foundation types [3, 13–18].

Literature on the seismic response of OWT is not as extensive as that for the case of cyclic lateral response. Nevertheless, several trends have been gradually recognized. The OWT structures display inherent flexibility and their fundamental natural frequency typically is around 0.25 to 0.75 Hz [3, 6]. This range circumvents the frequency content of typical earthquakes that corresponds to the strongest excitation and makes the risk of earthquake-induced catastrophic collapse relatively low. Among 180 real earthquake records, De Risi et al. [6] show that only two lead to structural failure of OWT. Similar behavior is reported by other seismic fragility analyses [19, 20]. On the other hand, it should be emphasized that higher vibration modes can play important roles in affecting the seismic response of OWT [6, 10, 21, 22], because the corresponding resonance frequencies can coincide with those characterized by strong ground motions. The involvement of higher vibration modes can potentially shift the distributions and patterns of internal force demands [21, 22] and require special care in OWT structure design.

While the chance of the collapse of OWT during earthquake might be low, excessive deformation of OWT and exceeding the serviceability limit state (SLS) can be an important issue [6, 7, 10, 23]. For instance, Gao et al. [23] show that, in the liquefiable ground, OWT caisson foundation can develop considerable residual rotation (i.e., up to 40% of SLS) solely due to earthquakes. Kourkoulis et al. [7] and Anastasopoulos and Theofilou [10] show that, in undrained clay, OWT caisson and monopile foundation can rotate significantly during seismic shaking, however, it is transient and largely recoverable by the end of earthquake. An important factor that affects the deformation of OWT during earthquake is the coupling between seismic excitation and environmental actions. Kourkoulis et al. [7] and Anastasopoulos and Theofilou [10] show that the presence of wind and wave loads can significantly enlarge the accumulation of OWT foundation rotations during earthquake. Zheng et al. [24] and Yang et al. [25] show that the inclusion of wave actions can increase the seismic response of OWT structures. Another factor that can influence the OWT deformation is the frequency contents of seismic excitations. The foundation deformation and structure response can both be amplified when earthquake displays more intense motion at the natural frequency of OWT structure system [6, 26].

The seismic behavior of OWT is complex and governed by many factors. Among them, accurate modeling of soil-structure interaction (SSI) has been shown essential for predicting the performance of OWT during earthquake. The latter aspect has been considered by two main approaches: Winkler spring-based models [6, 25, 27] and continuum-based models [7, 10, 26]. The former method is attractive from the perspective of simplicity. Nevertheless, its reliability highly depends on the employed soil reaction curves, which can be empirical and limited to certain applications. Moreover, Winkler spring-based models often cannot accurately consider the near-field ground shaking that is affected by the presence of foundation.

Time-domain, continuum-based approaches (e.g., finite element analysis (FEA)) can naturally account for the seismic wave propagations in the ground and the kinematic-inertia interactions between soil, foundation, and superstructures. Despite these advantages, the investigations using continuum analysis are limited, compared to the counterparts of Winker spring-based methods. This at least partially results from the complexity of the method, which might be to a large extent attributed to the complexity of the constitutive models for soils. The goal of this paper is to show the possibility of delivering realistic and efficient FEA of the seismic response of OWT based on the constitutive models that are simple yet captures essential features of soil behavior. In the following, one of such models for undrained clay will be described, followed by the assessment of its capacity to represent cyclic soil-structure interactions against cyclic lateral loading and seismic shaking centrifuge tests. Lastly, dynamic FEA based on the soil model is performed to explore the seismic response of OWT supported by different foundation types: monopile and caisson.

2   Total Stress-Based Constitutive Model for Undrained Clay

In many built and planned OWT farms, fine-grained soils constitute the primary seabed [13, 17, 28]. Given the low permeability of the clayey soils, undrained conditions are often prevailing during earthquake. This study idealizes undrained clay as a single-phase material governed by total stresses. The formulation of the total stress model will be briefly described in this section, followed by examples illustrating its capacity to represent the behavior of cyclically-loaded undrained clay.

2.1 Model Formulation

The employed total stress soil model (hereafter referred to as AUC-Clay-Small) can capture the salient features of undrained clay, including mechanical anisotropy, cyclic degradation of stiffness and strength, and path-dependent nonlinear variation of stiffness at small strains. More detailed description of the constitutive model can be found from the authors' previous publications [29–31].

The AUC-Clay-Small model employs a Mises type yield surface (see $f = 0$ in Fig. 1):

$$f = \sqrt{\frac{3}{2}}\|S_{ij} - \alpha_{ij}\| - A = 0 \tag{1}$$

where S_{ij} indicates the deviatoric stress, α_{ij} denotes the backstress controlling the translation of the yield surface, while the constant A indicates the size of the yield surface.

The evolution of the backstress is controlled by a nonlinear kinematic hardening law proposed by Huang et al. [29] that accounts for the anisotropy of undrained clay and its cyclic softening:

$$\dot{\alpha}_{ij} = \langle \lambda \rangle \left[\frac{2}{3} \rho_e \rho_s C \frac{\partial f}{\partial \sigma_{ij}} - \rho_e \gamma (\alpha_{ij} - \beta_{ij}) \right] \tag{2}$$

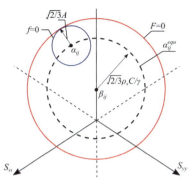

Fig. 1. Schematics illustrating yield surface $f = 0$ and anisotropic strength surface $F = 0$.

In Eq. (2), λ indicates the plastic multiplier, while C and γ are material constants. The tensor β_{ij} in Eq. (2) accounts for the anisotropic shear strength in that it implies a strength surface that is centered around β_{ij} (see $F = 0$ in Fig. 1). The variables ρ_e and ρ_s are included to consider the cyclic degradation of soil stiffness and strength (see Huang et al. [29] for detailed discussion).

The elastic response of undrained clay is described by an intergranular strain elastic model [30]:

$$G = \rho^{\chi} G_0 + (1 - \rho^{\chi}) G_0 + \begin{cases} \rho^{\chi} |l|^n (G_e - G_0) & \text{if} \quad l > 0 \\ 0 & \text{if} \quad l \leq 0 \end{cases} \tag{3}$$

where G indicates the elastic shear modulus, while G_0 and G_e indicate the value of G at very small strains (e.g., less than 1×10^{-5}) and relatively large strains (e.g., greater than 1×10^{-2}), respectively. The transition from G_0 to G_e depends on both the magnitude of shear strains and the rotation of current loading path from the previous history, as observed from experiments [32, 33]. This objective is realized by employing the intergranular strain ζ_{ij} as a state variable for memorizing deformation history. In general, the modulus G computed by Eq. (3) will decrease from G_0 to G_e during monotonic loading, attain enhanced stiffness upon strain path rotation, and regain the highest stiffness, i.e., G_0, upon full stress reversal. More detailed discussion of the intergranular strain elastic model can be found from [30, 31].

2.2   Simulation of Soil Element Tests

Figure 2 compares the measured and simulated response of Cloverdale clays from undrained triaxial cyclic shear tests [31, 34, 35]. It is seen that the soil model can well represent the cyclic softening observed in the tests (i.e., enlarged and rotated stress-strain hysteresis loops). Moreover, the model can reasonably capture the soil response under cyclic loading with different amplitudes. This feature is further illustrated in Fig. 3, which depicts the observed and computed residual strains (i.e., those attained at the end of each cycle) versus the number of cycles.

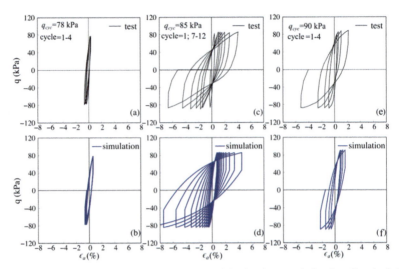

Fig. 2. Measured and simulated response of Cloverdale clay from undrained cyclic triaxial tests: (a)–(b) cyclic shear stress q_{cyc} = 78 kPa; (c)–(d) q_{cyc} = 85 kPa; (e)–(f) q_{cyc} = 90 kPa.

3   Analyses of Cyclically-Loaded Pile in Centrifuge Test

The application of the total stress model in cyclic soil-structure interaction is demonstrated by simulating a series of centrifuge tests where pile foundations are subjected to cyclic lateral loading [36, 37]. The soils used in the tests are Malaysia kaolin clays. The variation of shear modulus and damping ratio of the soils at small strains, measured by using resonant column tests, are given in Fig. 4. This figure also includes the corresponding model simulations [31].

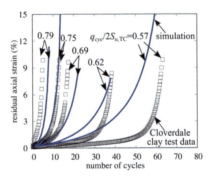

Fig. 3. Measured and computed residual strain versus number of cycles.

Figure 5 presents the load-displacement relationships at pile head measured from centrifuge tests and computed by FEA based on the AUC-Clay-Small model (detailed discussion on the FEA mesh and boundary conditions can be found from [29]). The graphs (a) and (b) show the response of the semi-rigid pile with a free head subjected

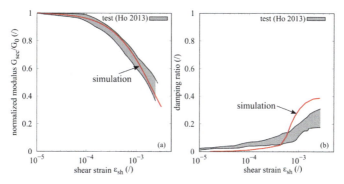

Fig. 4. Small-strain characteristics of Malaysia kaolin clay: (a) shear modulus; (b) damping.

to small and large magnitudes of cyclic pile head displacements, respectively, while the graphs (c) and (d) show the counterpart for the rigid pile with a fixed head. It is seen that the measured hysteresis can be reasonably represented by the FEM simulations, as well as the gradual drop of pile head reactions at peak displacement during cyclic loading (i.e., the softening of pile foundation stiffness).

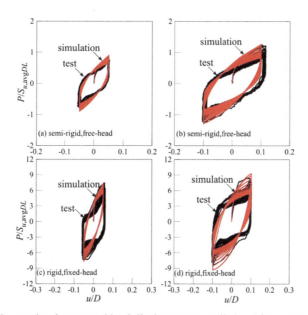

Fig. 5. Measured and computed load-displacements at pile head from centrifuge test.

4 Analyses of Caisson Supported OWT in Seismic Centrifuge Test

The total stress soil model described in Sect. 2 is incorporated into time-domain, dynamic finite element analysis (FEA) and is applied to simulate a seismic centrifuge test of OWT.

The test is conducted at TJL-150 centrifuge facility at Tongji University. A laminar box (550 mm in height, 500 mm in length, and 450 mm in width) is used to contain the soils (see Fig. 6(a)). The earthquake loading is applied under a centrifuge acceleration of 50 g. The model OWT is simplified as a one-degree-of-freedom system, where an aluminum tube (inner and outer diameters are 0.9 m and 1.0 m, respectively, in prototype size) and a concentrated mass (5.47×10^3 kg, in prototype size) are used to represent the turbine tower and the rotor-nacelle head, respectively. The OWT foundation is a caisson with a diameter of 4.2 m and an embedment depth of 4.2 m (prototype size). The test is instrumented by accelerometers (A1 to A4 in Fig. 6(a)) and LVDTs (L1 and L2 Fig. 6(a)). The soils used in the centrifuge test are the normally consolidated Malaysia kaolin clays discussed in the previous section.

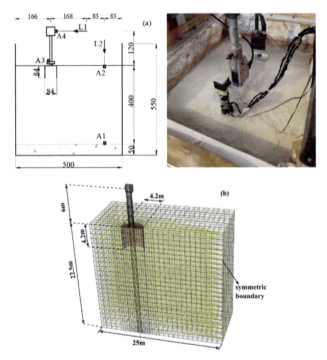

Fig. 6. Layout of (a) seismic centrifuge test and (b) dynamic FEA.

Figure 6(b) shows the mesh used in the FEA. To simulate the laminar box, the degrees of freedom at the same elevations along the opposite sides of the model perpendicular to the excitation direction are tied together (i.e., symmetric boundary). The degrees of freedom at the nodes along the other two sides are fixed along the normal direction. At the base of the FEA model, all degrees of freedom are fixed except the one along the direction of earthquake excitation, for which the acceleration input in the centrifuge test (see Fig. 7) is prescribed.

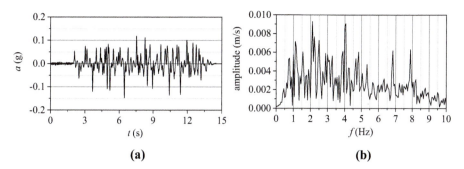

Fig. 7. Input acceleration in centrifuge test: (a) time history; (b) spectrum.

Figures 8, 9 and 10 show the measured acceleration at the ground surface (A2 in Fig. 6(a)), caisson foundation near mudline (A3 in Fig. 6(a)), and the OWT tower header (A4 in Fig. 6(a)), respectively. The comparison between Fig. 8(a) and 9(a) shows that the ground motion is attenuated during upward propagation, in particular, high frequency contents. This might be attributed to the low shear strength of normally consolidated clays near ground surface. Figure 9 shows that the acceleration of the caisson foundation is de-amplified relative to ground motion. In contrast to the foundation response, Fig. 10 shows that the OWT tower head exhibits considerable amplification in relative to that of the ground surface. The FEA computations are also included in Figs. 7, 8 and 9, which can reasonably represent the above general trends observed from the experiments. It is noted that the FEA can reasonably capture the frequencies of local peak response of the OWT header (see Fig. 10(c) and (d)), but tends to over-estimate the overall amplitudes.

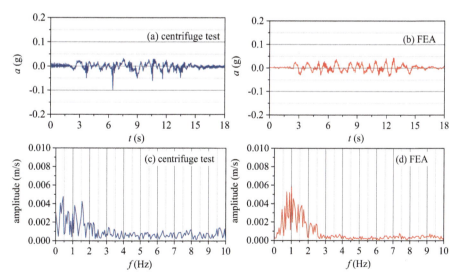

Fig. 8. Measured and computed acceleration response at ground surface: (a)–(b) time history; (c)–(d) spectrum.

This discrepancy might suggest that the adopted soil parameters (i.e., determined solely based on the cyclic loading centrifuge test discussed in Sect. 3) can over-estimate the stiffness and strength of the soils used in the seismic centrifuge test.

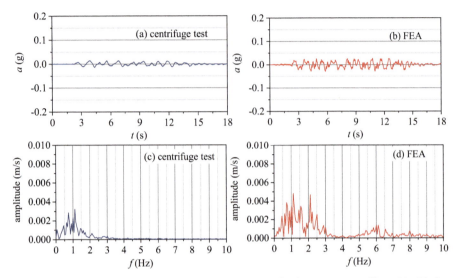

Fig. 9. Measured and computed acceleration response of caisson near mudline: (a)–(b) time history; (c)–(d) spectrum.

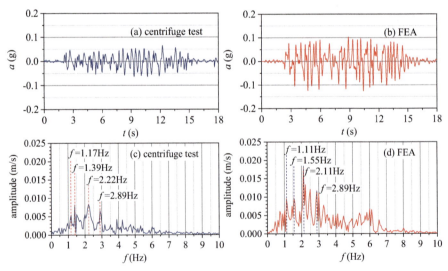

Fig. 10. Measured and computed acceleration response of OWT tower header: (a)–(b) time history; (c)–(d) spectrum.

Figure 11 presents the measured and computed ground settlement and lateral movement of OWT header during seismic shaking. The FEA can reflect the gradual development of surface settlement and OWT displacement observed from the centrifuge test, but under-estimate the overall magnitude. This mismatch is consistent with those regarding the acceleration response (e.g., Fig. 10) and, as discussed above, might be explained by that the actual soil properties in the centrifuge test are weaker than those implied by the soil parameters adopted in the FEA.

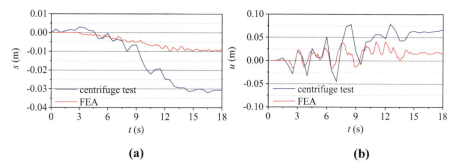

(a)   (b)

Fig. 11. Measured and computed time history of (a) ground surface settlement and (b) OWT header lateral movement (prototype scale).

5   Seismic Response of OWT Supported by Different Foundation Types

Previous sections show the capacity of FEA based on the simple soil model to represent the performance of OWT during earthquake loading. In this section, this numerical tool is used to explore the influences of foundation types on the seismic response of OWT. For this purpose, monopile and caisson foundation are considered (see Fig. 12). The superstructure is a simplified representation of typical 5 MW OWT installed in China. The tower is modeled as a steel hollow cylinder (outer and inner diameters are 6 m and 5.84 m, respectively; the density is 7.8×10^3 kg/m^3), while the rotor-nacelle head is simplified as a concentrated mass (267.8 Tons). The sizes of the caisson foundation are chosen in accordance with those adopted in an OWT farm in China. The sizes of the monopile are adjusted to match the lateral stiffness of the caisson foundation near the mudline in preliminary static elastic analysis. Infinite elements are prescribed along the lateral sides of the FEA mesh to minimize wave reflections. The acceleration time history prescribed at the bottom of the model is given in Fig. 13. The soils adopt the parameters used in the above centrifuge test but follow a different profile of shear modulus G_0 = 7 + 4.35z MPa/m and undrained shear strength S_u = 20 + 2z kPa/m, respectively. The gradient of these profiles is close to in situ measurements in Shanghai area. To reduce the computational cost, the proposed total stress model is approximated by a built-in kinematic hardening model in Abaqus, in conjunction with a user-defined field for cyclic softening of stiffness and strength. By doing so, the mechanical anisotropy of undrained clay and the path-dependent non-linearity at small strains is neglected. Figures 14 and 15 suggest that such a simplification (denoted by AUC-Clay) does not lead to significant alteration of the general patterns of computed response.

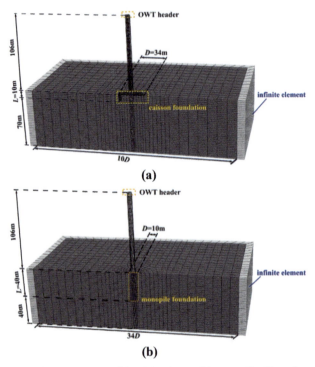

Fig. 12. FEA model of OWT supported by: (a) caisson; (b) monopile. Note that only half of the model is depicted to show the location of the foundations.

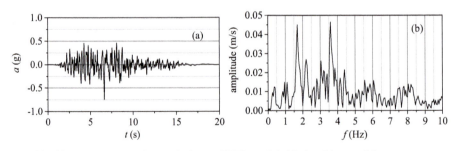

Fig. 13. Input acceleration at the base of FEA model: (a) time history; (b) spectrum.

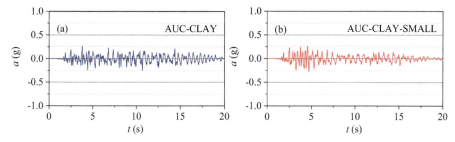

Fig. 14. Acceleration response of monopile at mudline computed by two soil models: (a) AUC-Clay; (b) AUC-Clay-Small.

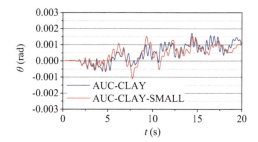

Fig. 15. Rotation of monopile at mudline computed by using AUC-Clay and AUC-Clay-Small model.

Figures 16 and 17 show the acceleration response of the ground surface (at locations relatively far from the foundation) and the two types of foundation at the mudline, respectively. In general, the presence of foundation de-amplifies the ground motion, similar to that observed from the centrifuge test. The comparison between Fig. 16(b), 17(b) and 17(d) shows that the above filtering is more significant for high-frequency contents, indicating that it is caused by the kinematic interaction between foundation and ground, i.e., the high stiffness of the foundation in relative to the ground makes it cannot follow the high-frequency vibration of the ground [38]. Moreover, the monopile has stronger filtering effects than the caisson foundation (see Fig. 17(b) and (d)), suggesting that the former foundation is stiffer. Figure 18 shows the ratio between the displacement amplitude of foundation and ground surface under different excitation frequencies. It is seen that, despite the above filtering effects, the response of foundation is amplified under certain frequencies.

Figure 19 shows the acceleration response of the OWT header. For the caisson-supported OWT, local peak amplifications are seen at 0.2 Hz and 2.05 Hz, close to the range of the first and second natural frequency of typical OWT structure system [6], indicating these local peaks are caused by the resonance. Moreover, Fig. 19(b) shows that the resonance at the second mode leads to a greater OWT response than that of the first mode, thus suggesting that sufficient attention should be paid to higher modes of OWT structure during seismic analysis. Similar trends are also observed for the monopile-supported OWT (i.e., Fig. 19(c) and (d)), except that the natural frequencies tend to be greater. This is consistent with the above-mentioned fact that the monopile is stiffer than the caisson. Comparing Figs. 18 and 19 shows that, under excitation frequencies close to the second natural frequency of OWT structure, the foundation response is amplified relative to the reference free-field conditions, i.e., the inertial interaction between foundation and superstructures [38].

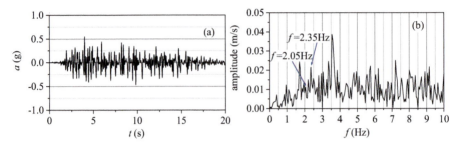

Fig. 16. Free-field ground surface acceleration: (a) time history; (b) spectrum.

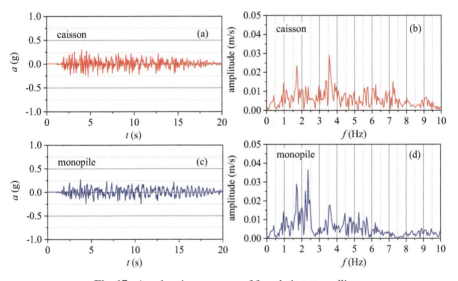

Fig. 17. Acceleration response of foundation at mudline.

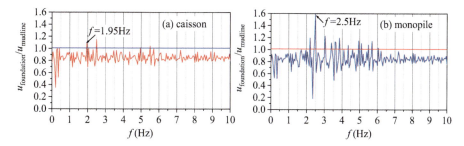

Fig. 18. Ratio between displacement amplitude of foundation and ground surface.

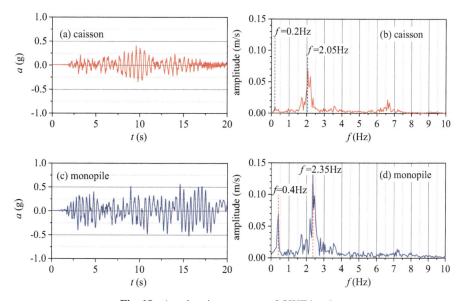

Fig. 19. Acceleration response of OWT header.

Figure 20(a) shows the time history of the ground settlement and foundation settlement during seismic shaking. The caisson foundation settles following that of its surrounding soils. In contrast, the monopile settles less than the ground. Figure 20(b) shows the time history of foundation rotation. It should be emphasized that, despite that the monopile displays greater stiffness than the caisson (i.e., larger natural frequency in Fig. 19), it develops considerably greater residual rotation than the caisson. This response can be attributed to the relationships between the frequency contents of the ground motions and the natural frequency of the OWT system. The comparison between Fig. 16 and 19 shows that the ground shaking at the second natural frequency of the monopile-supported OWT ($f = 2.35$ Hz) is larger than that of the caisson-supported OWT ($f = 2.05$ Hz). Therefore, avoiding resonance of OWT system including higher modes can be important for preventing excessive deformation of OWT foundations.

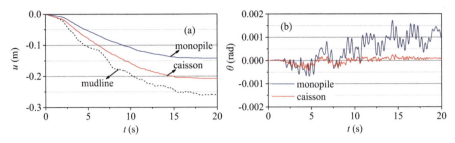

Fig. 20. Time history of (a) foundation and ground settlements and (b) foundation rotation.

6 Conclusions

The seismic performance of offshore wind turbine (OWT) becomes increasingly critical as wind farms expand to seismically active areas. Time-domain, continuum-based methods (e.g., finite element analyses, FEA) can provide comprehensive analyses of the performance of OWT subjected to earthquake loading, yet their practical applicability can be restricted by the complexity of the soil constitutive models. This work illustrates the possibility of delivering realistic and efficient continuum analysis based on constitutive models that are simple but capture the essential features of soil behavior. For this purpose, undrained clay is idealized as single-phase material governed by total stresses. Non-linear kinematic hardening is used to reflect mechanical anisotropy and the cyclic degradation of soil stiffness and strength, while intergranular-strain based elastic model is employed to reflect path-dependent stiffness non-linearity at small strains. The soil model is verified at different levels, by simulating soil element tests, the response of cyclically-loaded pile foundation in centrifuge test, and the response of caisson-supported OWT in seismic centrifuge test. Lastly, the soil model is applied to analyze the influences of foundation types, i.e., monopile and caisson, on the seismic performance of OWT. The main conclusions that can be drawn from this study include:

- The proposed continuum method can reasonably represent foundation load-displacement hysteresis and stiffness degradation during cyclic loading, as well as the response of OWT during seismic loading.
- The kinematic-inertial interactions between soils, foundations, and superstructures can govern the response of OWT foundation during seismic loading.
- Higher vibration modes can be important for the dynamic response of OWT and the accumulation of foundation deformation.
- Foundations displaying high stiffness during static loading might, otherwise, lead to a more intense dynamic response of OWT and larger foundation deformation, depending on the frequency contents of seismic shaking.
- Avoiding resonance of OWT structure system (including higher modes) can be important for preventing excessive foundation deformation.

References

1. GWEC, Global Wind Energy Council: Global wind report, 2021 annual market update. Brussels, Belgium (2021)
2. Byrne, B.W., et al.: Cyclic laterally loaded medium scale field pile testing for the PISA project. In: 4th International Symposium on Frontiers in Offshore Geotechnics, pp. 1323–1332 (2020)
3. Lombardi, D., Bhattacharya, S., Muir Wood, D.: Dynamic soil–structure interaction of monopile supported wind turbines in cohesive soil. Soil Dyn. Earthq. Eng. **49**, 165–180 (2013)
4. Achmus, M., Kuo, Y.-S., Abdel-Rahman, K.: Behavior of monopile foundations under cyclic lateral load. Comput. Geotech. **36**(5), 725–735 (2009)
5. Bhattacharya, S., Goda, K.: Use of offshore wind farms to increase seismic resilience of nuclear power plants. Soil Dyn. Earthq. Eng. **80**, 65–68 (2016)
6. De Risi, R., Bhattacharya, S., Goda, K.: Seismic performance assessment of monopile-supported offshore wind turbines using unscaled natural earthquake records. Soil Dyn. Earthq. Eng. **109**, 154–172 (2018)
7. Kourkoulis, R., Lekkakis, P., Gelagoti, F., Kaynia, A.: Suction caisson foundations for offshore wind turbines subjected to wave and earthquake loading: effect of soil–foundation interface. Géotechnique **64**(3), 171–185 (2014)
8. DNV, Det Norske Veritas: Design of offshore wind turbine structure. Offshore Standard DNV-OS-J101 (2004)
9. IEC, International Electrotechnical Commission: IEC 61400-3 wind turbines Part 3: design requirements for offshore wind turbines. International Electrotechnical Commission, Geneva, Switzerland (2009)
10. Anastasopoulos, I., Theofilou, M.: Hybrid foundation for offshore wind turbines: environmental and seismic loading. Soil Dyn. Earthq. Eng. **80**, 192–209 (2016)
11. Herrera, J., Aznárez, J.J., Padrón, L.A., Maeso, O.: Observations on the influence of soil profile on the seismic kinematic bending moments of offshore wind turbine monopiles. Procedia Eng. **199**, 3230–3235 (2017)
12. Ju, S.-H., Huang, Y.-C.: Analyses of offshore wind turbine structures with soil-structure interaction under earthquakes. Ocean Eng. **187**, 106–190 (2019)
13. Byrne, B.W., et al.: PISA design model for monopiles for offshore wind turbines: application to a stiff glacial clay till. Géotechnique **70**(11), 1030–1047 (2020)
14. Kallehave, D., Byrne, B.W., LeBlanc Thilsted, C., Mikkelsen, K.K.: Optimization of monopiles for offshore wind turbines. Philos. Trans. R. Soc. A Math. Phys. Eng. Sci. **373**(2035), 20140100 (2015)
15. Seidel, M.: Substructures for offshore wind turbines-current trends and developments. Festschrift Peter Schaumann, 363–368 (2014)
16. Ding, H., Liu, Y., Zhang, P., Le, C.: Model tests on the bearing capacity of wide-shallow composite bucket foundations for offshore wind turbines in clay. Ocean Eng. **103**, 114–122 (2015)
17. Liu, B., et al.: Design considerations of suction caisson foundations for offshore wind turbines in Southern China. Appl. Ocean Res. **104**, 1023–1058 (2020)
18. Wang, X., Zeng, X., Li, J., Yang, X., Wang, H.: A review on recent advancements of substructures for offshore wind turbines. Energy Convers. Manag. **158**, 103–119 (2018)
19. Kim, D.H., Lee, S.G., Lee, I.K.: Seismic fragility analysis of 5 MW offshore wind turbine. Renew. Energy **65**, 250–256 (2016)
20. Asareh, M.-A., Schonberg, W., Volz, J.: Fragility analysis of a 5-MW NREL wind turbine considering aero-elastic and seismic interaction using finite element method. Finite Elem. Anal. Des. **120**, 57–67 (2016)

21. Prowell, I., Elgamal, A., Lu, J.: Modeling the influence of soil structure interaction on the seismic response of a 5 MW wind turbine. In: 5th International Conference on Recent Advances in Geotechnical Earthquake Engineering and Soil Dynamics, No. 5.09a, pp. 1–8 (2020)
22. Yang, Y., Li, C., Bashir, M., Wang, J., Yang, C.: Investigation on the sensitivity of flexible foundation models of an offshore wind turbine under earthquake loadings. Eng. Struct. **183**, 756–769 (2019)
23. Gao, B., Ye, G., Zhang, Q., Xie, Y., Yan, B.: Numerical simulation of suction bucket foundation response located in liquefiable sand under earthquakes. Ocean Eng. **235**, 1093–1094 (2021)
24. Zheng, X.Y., Li, H., Rong, W., Li, W.: Joint earthquake and wave action on the monopile wind turbine foundation: an experimental study. Mar. Struct. **44**, 125–141 (2015)
25. Yang, Y., Bashir, M., Li, C., Michailides, C., Wang, J.: Mitigation of coupled wind-wave-earthquake responses of a 10 MW fixed-bottom offshore wind turbine. Renew. Energy **157**, 1171–1184 (2020)
26. Zhang, J., Cheng, W., Cheng, X., Wang, P., Wang, T.: Seismic responses analysis of suction bucket foundation for offshore wind turbine in clays. Ocean Eng. **232**, 109–159 (2021)
27. Sapountzakis, E.J., Dikaros, I.C., Kampitsis, A.E., Koroneou, A.D.: Nonlinear response of wind turbines under wind and seismic excitations with soil–structure interaction. J. Comput. Nonlinear Dyn. **10**(4), 041007 (2015)
28. Houlsby, G., Kelly, R., Huxtable, J., Byrne, B.: Field trials of suction caissons in clay for offshore wind turbine foundations. Géotechnique **55**(4), 287–296 (2005)
29. Huang, M., Liu, L., Shi, Z., Li, S.: Modeling of laterally cyclic loaded monopile foundation by anisotropic undrained clay model. Ocean Eng. **228**, 108915 (2021)
30. Shi, Z., Huang, M.: Intergranular-strain elastic model for recent stress history effects on clay. Comput. Geotech. **118**, 103316 (2020)
31. Shi, Z., Liu, L., Huang, M., Shen, K., Wang B.: Simulation of cyclic laterally-loaded piles in undrained clays accounting for soil small strain characteristics. Ocean Eng.. (under review)
32. Atkinson, J., Richardson, D., Stallebrass, S.: Effect of recent stress history on the stiffness of overconsolidated soil. Géotechnique **40**(4), 531–540 (1990)
33. Finno, R.J., Kim, T.: Effects of stress path rotation angle on small strain responses. J. Geotech. Geoenviron. Eng. **138**(4), 526–534 (2012)
34. Zergoun, M.: Effective stress response of clay to undrained cyclic loading. Ph.D. thesis, University of British Columbia (1991)
35. Zergoun, M., Vaid, Y.: Effective stress response of clay to undrained cyclic loading. Can. Geotech. J. **31**(5), 714–727 (1994)
36. Li, S.: Total-loading EMSD method and the behavior of laterally loaded single pile in clay. Ph.D. thesis, Tongji University (2019). (in Chinese)
37. Li, S., Yu, J., Huang, M., Leung, C.: Application of T-EMSD based p-y curves in the three-dimensional analysis of laterally loaded pile in undrained clay. Ocean Eng. **206**, 107256 (2020)
38. Gazetas, G.: Seismic response of end-bearing single piles. Soil Dyn. Earthq. Eng. **3**(2), 82–93 (1984)

Development of the Earthquake Geotechnical Engineering (EGE) in ISSMGE

Kenji Ishihara[✉]

Chuo University, Tokyo, Japan
`kenji-ishihara@e-mail.jp`

After the original studies at the University of California led by H.B. Seed and I.M. Idriss, the study of EGE was introduced to ISSMFE in 1985 as one of the activities of a technical committee (TC) of ISSMFE. Since then, it has grown to an area of importance through the activities of TC-4 (1985 – 2009) and TC-203 (2009-present).

Factor of Safety

$$F = \frac{\text{Capacity}}{\text{Demand}} = \frac{\text{Resistance of soils}}{\text{Earthquake induced force}}$$

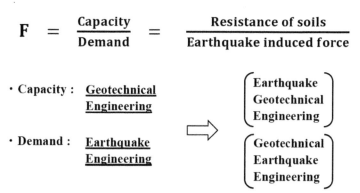

• Capacity : Geotechnical Engineering ⟹ (Earthquake Geotechnical Engineering)

• Demand : Earthquake Engineering ⟹ (Geotechnical Earthquake Engineering)

Fig. 1. Geotechnical engineering versus earthquake engineering

The development of the EGE can be considered roughly as a challenge to explore "capacity" versus "demand", as illustrated in Fig. 1. There has been steadily increased demand coming mainly from the advances of earthquake engineering. Due to proliferation of the network of strong motion recorders, the magnitude of recorded acceleration in recent earthquakes has substantially increased to a level as large as 1000 gal (1 g) as illustrated in Fig. 2.

L. Wang et al. (Eds.): PBD-IV 2022, GGEE 52, pp. 67–73, 2022.
https://doi.org/10.1007/978-3-031-11898-2_4

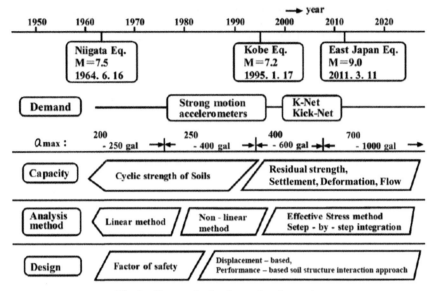

Fig. 2. Changes in the demand and capacity

In response to the growing demand year after year, researchers and engineers in the geotechnical engineering discipline have been requested to take initiative and to cope with this increasing demand. This involved extensive endeavor on the capacity side as illustrated in Fig. 1. Detailed studies on the deformation characteristics, cyclic strength, residual strength and estimate of allowable displacements, etc. have been targets of studies in the geotechnical engineering. In the course of these efforts, the methodologies to evaluate the ground behavior has changed from the simple factor of safety approach, to linear and non-linear, and further to the analysis based on the effective stress, as illustrated in Fig. 2. In accordance to the general trends as above, the norm or protocol of design has changed from simple evaluation of factor of safety to displacement-based or performance-based approach as illustrated in Fig. 2.

The development of the earthquake geotechnical engineering has been enhanced through the quadrennial International Conference on Earthquake Engineering (ICEGE) as shown in Fig. 3. The names of the chair or Secretary General (SG) of the conferences and the chairperson in change of the operation of the TC for the 4 year interval period are shown in Fig. 3.

In the meanwhile, another conference series was inaugurated in 2009, that is, the International Conference on Performance-based Design in Earthquake Geotechnical Engineering (ICPBD) which is intended to focus more on the problems associated with practice in EGE. This conference series has been held in the mid-term year in the 4 year period of the ICEGE as shown in Fig. 3. The venues and names of the chairperson in charge of the conference are also shown in Fig. 3.

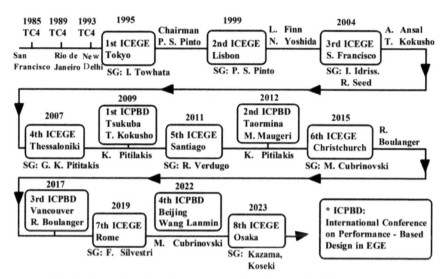

Fig. 3. International conferences on earthquake geotechnical engineering

In addition, a lecture series named "Ishihara Lecture" was inaugurated in 2003. The list of the lecturers is shown below.

1st: Finn W.D.L. 2003, at the 3rd ICEGE, San Francisco, USA
  "An overview of the behavior of pile foundations in liquefiable and non-liquefiable soils during earthquake excitation"

2nd: Idriss I.M. (paper with Boulanger, R.) 2007, at the 4th ICEGE, Thessaloniki, Greece
  "SPT-and CPT-based relationships for the residual shear strength of liquefied soils"

3rd: Dobry R. (paper with Abdoun, T.) 2011, at the 5th ICEGE, Santiago, Chile
  "An investigation into why liquefaction charts work:
  A necessary step toward integrating the states of the work:
  A necessary step toward integrating the states of art and practice"

4th: Gazetas G. 2013, at the 18th International Conference on Soil Mechanics and Geotechnical Engineering, Paris, France
"Soil-foundation structure systems beyond conventional seismic failure thresholds"

5th: Kokusho T. 2015, at the 6th ICEGE in Christchurch, New Zealand
"Liquefaction research by laboratory tests versus in-situ behavior"

6th: Bray J.D. (paper with Macedo J.) 2017, at the 19th International Conference on Soil Mechanics and Geotechnical Engineering, Seoul, Korea
"Simplified procedure for estimating liquefaction-induced building settlement"

7th: Towhata I. 2019, at the 7th ICEGE, Rome, Italy
"Summarizing Geotechnical activities after the 2011 Tohoku Earthquake of Japan"

8th: Cubrinovski M. (paper with Ntritsos N.) 2022, at the 4th ICPBD, Beijing, China
"Holistic evaluation of liquefaction problems"

1   Fundamental Laws for Deformation of Granular Soils

While the basic laws of deformation had been established for cohesive soils and incorporated into the framework such as the Cam-clay model and critical-state soil mechanics, studies in a similar vein were lacking and left behind for cohesionless soils such as silt, sand, and gravel. Since the issues of sand liquefaction were recognized as an important problem after the occurrence of the Niigata earthquake in 1964, needs were instigated for studying the basic features of deformation mechanism for the cohesionless granular soils particularly under dynamic and cyclic loading conditions. The trend in the main stream of development in soil dynamics is briefly illustrated in Fig. 4, as contrasted to the advances in soil mechanics which are mainly associated with the static loading condition. It is to be noted that, while the soil mechanics has been developed by considering the soil mass as a continuum, the same thought has also been adopted as the basis for developing theorems of deformation for granular soils. The constitutive laws of deformation for the granular soils have been known to be composed of four basic laws, that is, 1) Friction law, 2) Dilatancy law, 3) Flow rule, and 4) Harding rule.

1.   Friction Law

In the early period of studies in late 1960s at the University of Tokyo, a set of data was obtained on sand using the triaxial test apparatus 1). In the classical theory of plasticity developed for the deformation of steel or concrete, the yielding is assumed to

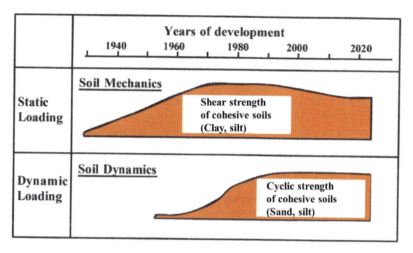

Fig. 4. Development of soil dynamics in comparison to that of soil mechanics

take place merely with increase in shear stress. In contrast, the irrecoverable permanent deformation in the granular soils was shown to occur, only with increase in the stress ratio, as illustrated in Fig. 5, leading eventually to the failure criterion known as Coulomb law of friction. The results of the tests are considered as the basis of seminal importance to substantiate validity of the "friction rule" for granular soils.

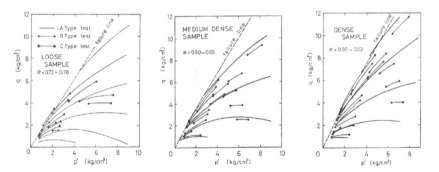

Fig. 5. Yield loci for granular soils [1]

2. Dilatancy Law

As well-known, significant volume change can occur easily in the granular soils not only by compressive stress but more predominantly by the application of shear stress. The dilatancy as above has been investigated thoroughly and formulated in various fashions by several investigators in the early period of development in soil mechanics. The most widely used formula is the relation, as follows, which was derived from the energy balance during the stress application.

$$\frac{\Delta\varepsilon x + \Delta\varepsilon z}{\sqrt{\Delta\gamma 2xz + \left(\frac{\Delta\varepsilon z - \Delta\varepsilon x}{2}\right)2}} = M - \frac{\sigma z - \sigma x}{\sigma z + \sigma x} \tag{1}$$

where $\Delta\varepsilon_x$, $\Delta\varepsilon_z$ and $\Delta\gamma_{xz}$ are increments of plastic strain component, and $\Delta\sigma_x$ and $\Delta\sigma_z$ are stress increment-components, as illustrated in Fig. 6. It is to be noted that when the current shear stress ratio is smaller than M, the volume change is positive, that is, volume decrease takes place. In contrast, if the current shear stress ratio is larger than M, the increase in volume takes place.

3. Non-coincidence of the principal axes of stresses and plastic strain increments.

In executing the stress and strain analysis in the two-dimensional plane strain conditions, what is tacitly assumed in the classical theory of metal plasticity is what is called "flow rule" which is formulated as below.

$$\frac{\tau zx}{(\sigma z - \sigma x)/2} = \frac{\Delta\gamma zx}{\Delta\varepsilon z - \Delta\varepsilon x/2} \tag{2}$$

As illustrated in Fig. 6, Eq. (2) implies that the vector formed by the two components of plastic strain increments should be directed in the same direction as the vector formed by the two components of current state of stresses. In this context, the flow rule is alternatively called "coincidence of principal axes".

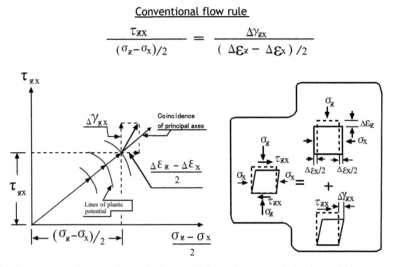

Fig. 6. Non-coincidence of principal axes of shear stresses and plastic strain increments.

However, as a result of comprehensive laboratory tests, it has become clear that the rule as above does not hold valid for the granular materials such as sand and gravel which are deposited under gravity. The laboratory tests using the triaxial torsion tests have shown the results as typically demonstrated in Fig. 7. Pending detailed description here, a new method for constructing the flow rule was proposed by Gutierrez et al. (2). 2) It is to be noted that the effects of non-coaxiality could exert some influence particularly when dealing with the sand performance subjected to cyclic stresses involving continuous rotation of principal stress axis such as those under wave-induced or traffic induced loads. Detail account is given in Ref. 3.

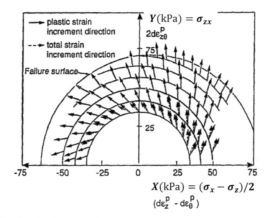

Fig. 7. Plastic strain development at various stress states [2].

4. Plastic stress-strain relation

In addition, to conduct the effective stress analysis in 2D condition, what is called "hardening rule" needs to be specified. Various relations have been used, but the hyperbolic type relation is used most commonly. Details of the hardening function are omitted here.

References

1. Tatsuoka, F., Ishihara, K.: Yielding of sand in triaxial compression. Soils Found. **14**(2), 63–76 (1974)
2. Gutierrez, M., Ishihara, K., Towhata, I.: Flow theory for sand during rotaion of principal stress direction. Soils Found. **31**(4), 121–132 (1991)
3. Ishihara, K.: Soil response in cyclic loading induced by earthquakes, traffic and waves. In: Proceedings of the 7th Asion Regional Conference on Soil Mechanics and Foundation Engineering, Haifa, Israel, vol. 2, pp. 42–66 (1983)

Transient Loading Effects on Pore Pressure Generation and the Response of Liquefiable Soils

Steven L. Kramer[1]([⊠]) and Samuel S. Sideras[2]

[1] 132E More Hall, University of Washington, Seattle, WA 98195-2700, USA
kramer@uw.edu
[2] Shannon & Wilson, 3990 Collins Way, Suite 100, Lake Oswego, OR 97035, USA
sam.sideras@shanwil.com

Abstract. Like ground motions, the loading applied to potentially liquefiable soils in earthquakes is complex and unique. Conventional procedures for characterizing liquefaction hazards represent seismic loading in simplified manners that do not account for characteristics that can influence pore pressure generation and potential ground deformation. This paper presents the results of cyclic direct simple shear tests performed with transient, irregular loading derived from recorded earthquake ground motions. Data from these tests show the influence of the order in which individual pulses of shear stress occur on generated pore pressure and provide insight into limitations of evolutionary intensity measures based on time integration to provide efficient predictions of liquefaction. The form of a new intensity measure for triggering of liquefaction is proposed and calibrated on the basis of available data. The response of cyclic simple shear specimens subjected to transient loading superimposed upon static shear stresses is also illustrated and comments on existing procedures for estimation of the adjustment factor, K_α, are provided.

Keywords: Liquefaction · Pore pressure · Loading · Relative density

1 Introduction

Soil liquefaction has caused significant damage to the natural and built environment in many historical earthquakes and has been a topic of intense geotechnical engineering research for nearly 60 years. Over that period of time, many studies have taken place to understand the mechanical behavior of liquefiable soils and to develop useful and practical procedures to assess liquefaction hazards. Most of the research on soil liquefaction has been oriented toward evaluation of soil resistance to the triggering of liquefaction and, more recently, to prediction of its consequences. Both triggering and consequences, however, are strongly influenced by the loading imposed on liquefiable soils and less attention has historically been paid to this aspect of the problem. This paper focuses on characterization of loading and its effects on liquefiable soils.

2 Ground Motions and Liquefaction Loading

The seismic loading imposed on a liquefiable soil deposit is related to the characteristics of the ground motions that propagate through that deposit. Ground motions are unique and the response of a liquefiable soil deposit can be sensitive to many details of their specific characteristics. A 60-s ground motion recorded at 200 Hz will have 12,000 acceleration values each of which provides potentially important information about the motion's ability to generate pore pressure and produce permanent strain in an element of a liquefiable soil deposit. Ground motions are typically characterized by a relatively small number of intensity measures (*IMs*) that are intended to reflect their amplitudes, frequency contents, and durations since each of those factors can influence the seismic response of physical systems (structures and soil deposits) of interest. The degree to which these characteristics influence response, damage, and loss, however, is problem-specific. Thus, characterization of ground motions for the purpose of estimating liquefaction potential and the consequences of its triggering becomes important. The *IM* historically used for characterizing earthquake loading imposed on liquefaction-susceptible soils is peak ground acceleration, due largely to its close relationship to peak shear stress at the relatively shallow depths at which liquefaction has most commonly been observed in the field. Early experimental research showed, however, that pore pressure generation also depends on the number of cycles of loading applied to the soil in laboratory tests. To eliminate the need for a parameter related to number of loading cycles in addition to peak ground acceleration in the characterization of loading, other intensity measures have been proposed for use in liquefaction triggering relationships. Most of these alternative *IMs* have been evolutionary in nature, i.e., they have consisted of measures that are integrated with time over the duration of a ground motion. Such intensity measures, such as Arias intensity [16], cumulative absolute velocity and variations thereof [19], and dissipated energy [10, 18, 27], increase with time and reflect the amplitude, frequency content, and duration of a ground motion.

Early research on liquefaction made use of laboratory testing for evaluation of liquefaction resistance, which was interpreted in terms of cyclic shear stresses. Cyclic triaxial tests were commonly used, although cyclic direct simple shear testing is becoming more common now due to its closer correlation to the type of deformation usually experienced by liquefiable soils subjected to vertically propagating shear waves. Cyclic testing has nearly always subjected soil specimens to uniform harmonic loading – a series of harmonic loading cycles of constant shear stress amplitude applied at constant frequency. These tests have provided great insight into the effects of factors such as soil density, initial effective stress, shear stress amplitude, number of loading cycles, etc. that strongly affect the resistance of a soil to triggering of liquefaction.

The type of transient loading produced by actual earthquake shaking, however, is very different than that used in typical laboratory testing programs. The shear stress histories imposed on soils by earthquakes are highly non-stationary, meaning that their amplitudes and frequency contents change over the duration of shaking. The number of loading cycles is difficult to predict for a given earthquake but is expected to increase with increasing ground motion duration. Since ground motion duration is known to increase with increasing earthquake magnitude, the concept of "equivalent" loading cycles was developed [29]. This concept held that the pore pressure generated by a transient loading history would be equivalent to that generated by a certain number of uniform harmonic loading cycles at a reference amplitude and required definition of both the reference loading amplitude and the corresponding number of loading cycles.

Characterization of ground motions for liquefaction becomes even more complex when the effects of sloping ground are considered. In such cases, the cyclic shear stress imposed by earthquake shaking is superimposed on constant static shear stresses with the result that the soil is subjected to asymmetric loading that influences pore pressure generation and causes permanent strain to develop in a preferential direction.

3 Pore Pressure Response to Cyclic Loading

The contractive tendency of soils at relatively low shear strain levels cause pore pressures to increase in saturated soils subjected to cyclic loading. This tendency is at the heart of the soil liquefaction problem and establishing the rate at which pore pressures are generated is key to the prediction of liquefaction potential. The level of contraction that occurs in liquefiable soils is related to the shear strain experienced by the soil, and undrained cyclic testing of soils has established the close relationship between pore pressure and cyclic strain amplitude [8]. Pore pressure generation can then be viewed as a largely kinematic process, i.e., one that depends on the relative movement of soil grains with respect to each other. Prediction of that relative movement, i.e., of shear strain levels, is very difficult. As a result, pore pressure generation, and the liquefaction it can produce, are generally related to the cyclic stresses induced in the soil by earthquake shaking. Laboratory tests are usually performed under stress-controlled conditions.

3.1 Uniform Harmonic Loading

Laboratory tests on liquefiable soils have historically been performed using uniform (constant-amplitude) harmonic loading, typically at a constant frequency of 1 Hz. Plots of cyclic shear stress, usually normalized by initial effective stress, σ'_{v0}, in the form of a cyclic stress ratio, CSR, versus the number of loading cycles required to trigger liquefaction are often referred to as cyclic strength curves. Figure 1 shows cyclic strength curves for reconstituted samples of a clean sand at three relative densities. The curves show that the cyclic stress ratio required to trigger liquefaction decreases with increasing number of loading cycles and increases with increasing relative density, behaviors that are intuitive, well demonstrated, and widely accepted. The relationship between the cyclic resistance and the number of loading cycles is particularly important as it indicates that liquefaction resistance, for a given soil of a given density, is influenced by both amplitude and number of loading cycles. The rate of pore pressure generation in constant-amplitude tests has been shown to follow a repeatable pattern (Fig. 2) of starting rapidly, slowing down, and then speeding up as initial liquefaction (pore pressure ratio, $r_u = 100\%$) is approached. The first phase likely involves some degree of particle reorientation as tenuous contacts are broken, the second involves incremental generation of pore pressure with a more stable particle structure, and the third involves particle reorientation during dilation as the effective stress path approaches and crosses the phase transformation line. As the soil dilates, the particles reorient themselves into a structure that efficiently resists shear stress in the applied direction; upon stress reversal, however, that structure is inefficient for resisting shear in the new direction, which increases its tendency to contract and generate pore pressure until it begins to dilate in the new direction.

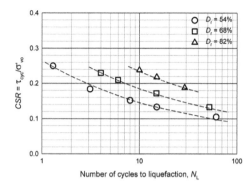
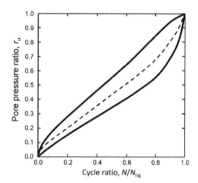

Fig. 1. Variation of number of cycles required to trigger liquefaction of Monterey No. 0 sand with amplitude of applied loading for soils of different relative density (after [8]).

Fig. 2. Variation of pore pressure ratio with cycle ratio for constant-amplitude harmonic loading (after [8]).

In the presence of a static shear stress, τ_{static}, as would exist in the field under sloping ground or in the vicinity of structures, the loading that drives pore pressure generation has both static and cyclic components. This situation is relatively easily modeled in laboratory tests by applying uniform harmonic loading to a specimen to which a static shear stress has been applied; the result is a harmonic loading history in which the same number of loading cycles is applied but of shear stress amplitudes that are larger in one direction than the other. Such tests have shown that the interaction of static and cyclic stresses is complex and that the presence of static shear stresses can increase or decrease triggering resistance depending on soil density. This behavior is accounted for in triggering analyses by a static shear stress adjustment factor,

$$K_\alpha = \frac{CRR_\alpha}{CRR_{\alpha=0}} \tag{1}$$

where $\alpha = \tau_{static}/\sigma'_{vo}$, which the cyclic resistance for level ground conditions ($\alpha = 0$) is multiplied by to account for the effect of the static shear stress. Values of $K_\alpha > 1.0$ increase liquefaction resistance and values less than 1.0 decrease it. Figure 3 shows four proposed K_α relationships, each of which indicate that static shear stresses decrease the liquefaction resistance of loose sands and increase it in dense sands. With the transient loading applied by actual earthquakes, however, the existence of a static shear stress causes a reduction in the number of loading cycles that result in reversals of shear stress in addition to a reduction in the amplitudes of the reversed cycles.

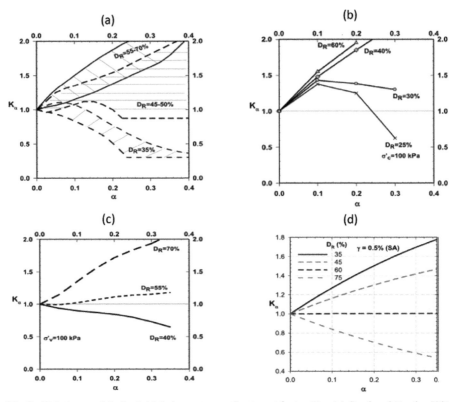

Fig. 3. Existing models for initial shear stress adjustment factor, Ka: (a) Seed and Harder [30]; (b) Vaid et al., [35]; (c) Boulanger [4]; and (d) Cetin and Bilge [6].

3.2 Transient Loading

Several investigators have performed cyclic loading tests with irregular (variable amplitude) loading. Ishihara and Yasuda [14, 15] performed cyclic triaxial tests using loading histories scaled in proportion to accelerograms measured in the basement of the Kawagishi-cho apartment building in the 1964 Niigata earthquake. The tests showed irregular generation of pore pressure with the largest increases being associated with the highest axial stress pulses, and tests performed with reversed polarity showed significant differences in pore pressure development, presumably due to differences in the loading histories relative to the compression/extension stress states inherent in cyclic triaxial testing. Tatsuoka and Silver [33] performed cyclic simple shear tests with uniform and irregular loading and focused on the development of strain following initial liquefaction. Wang and Kavazanjian [36] performed cyclic triaxial tests with the same numbers of smaller and larger amplitude cycles but reversed in order and found that the order of loading influenced the rate and magnitude of generated pore pressures. More recent investigations [2, 3, 22, 26] have largely focused on the potential of energy-based intensity measures for prediction of liquefaction potential. Transient loading tests with static shear stresses are not available in the literature.

4   Intensity Measures for Liquefaction

The triggering of liquefaction is influenced by ground motion characteristics and hence by the ability of ground motion intensity measures to represent the characteristics that are most strongly influence pore pressure generation. Ideally, a ground motion intensity measure would have a unique relationship to pore pressure generation. Because such an intensity measure does not exist, an intensity measure that best predicts pore pressure is sought. The notion of what constitutes "best" is usually expressed in terms of the "efficiency" [31], "sufficiency," and "predictability" of the intensity measure. Efficiency is a measure of the uncertainty with which a measure of response can be estimated for a given value of the intensity measure; an efficient *IM* for liquefaction would be one that the generation of pore pressure is closely related to, i.e., one for which the uncertainty in pore pressure given *IM* ($\sigma_{r_u|IM}$) would be low. Sufficiency is a measure of the completeness with which an *IM* characterizes the ground motion; a sufficient *IM* for liquefaction would be one for which no additional information about the motion would improve its correlation to pore pressure generation. Finally, predictability is a measure of how precisely the value of an *IM* can be predicted for a particular earthquake rupture scenario; a predictable *IM* would have a low standard error, i.e., $\sigma_{\ln IM}$, in a predictive ground motion model.

The laboratory data described in the preceding section clearly shows that an *IM* based solely on the peak amplitude is not sufficient to characterize loading on a liquefiable soil. Since pore pressure builds up incrementally from cycle to cycle, the potential for triggering of liquefaction depends on both the amplitude and number of cycles of loading. Under the transient loading of an actual earthquake, the relationship between number of cycles and ground motion amplitude is complicated. The approach that underlies most current liquefaction triggering procedures is based, explicitly or implicitly, on representation of a transient loading history by a number of equivalent uniform loading cycles. This approach is usually implemented in the form of a magnitude scaling factor that can be used to adjust the loading to account for duration effects.

4.1   Number of Loading Cycles

Various cycle-counting procedures have been proposed to identify the amplitude and number of cycles of uniform harmonic loading that are equivalent, in terms of pore pressure generation potential, to a given non-stationary shear stress history. Hancock and Bommer [12] provide an extensive review of procedures for computing numbers of equivalent loading cycles for structural and geotechnical purposes.

The number of loading cycles is important for problems, such as metal fatigue, where damage accumulates with each cycle of loading. In such problems, a given level of damage can be caused by a small number of large loading cycles or by a larger number of smaller loading cycles. Liquefiable soils, in which excess pore pressure builds incrementally in response to successive cycles of loading, are an example of materials in which "damage," in this case pore pressure generation, evolves over time. Seismic loading of liquefiable soils in actual earthquakes involves highly non-stationary loading cycles with a small number of high-amplitude cycles preceded and followed by many more cycles of lower amplitude; each of these cycles can potentially contribute to pore pressure generation but do so in a complex manner. As such, a defined number of loading cycles at an effective peak amplitude that considers the relative contributions of the various cycles is needed to properly characterize the pore pressure generation potential of the loading history.

The most basic procedures for characterizing numbers of equivalent cycles [24, 25] are based on the assumption that damage accumulates linearly, i.e., that the damage produced by each of a series of uniform amplitude loading cycles produces the same increment of damage. Miner's derivation was focused on high-cycle fatigue of aluminum alloys used in the aircraft industry and inherently assumed material linearity in which cumulative absorbed work, i.e., dissipated energy, is proportional to number of loading cycles. Under this assumption, the damage, D, done by n_i cycles of loading of amplitude S_i can then be expressed as

$$D = \sum \frac{n_i(S_i)}{N_i(S_i)} \tag{2}$$

where N_i is the number of uniform loading cycles at amplitude S_i required to cause failure. With this criterion, failure occurs when $D = 1.0$. The assumption of strict linearity can be relaxed, however, by assuming that the damage produced by each cycle is a power law function of its amplitude, i.e.

$$N(S) = aS^b \tag{3}$$

where a and and b are constants that reflect the intercept and slope of a log-log plot of S vs. N. Then, substituting Eq. (2) into Eq. (1), the damage caused by a number of uniform equivalent cycles, N_{eq}, of amplitude S_{ref} can be set equal to the damage produced by the set of variable amplitude cycles

$$\frac{N_{eq}(S_{ref})}{aS_{ref}^b} = \sum \frac{n_i(S_i)}{aS_i^b} \tag{4}$$

from which the equivalent number of cycles at the reference stress can be expressed as

$$N_{eq}(S_{ref}) = S_{ref}^b \sum \frac{n_i(S_i)}{S_i^b} \tag{5}$$

Since the intercept constant, a, cancels out in Eq. (3), the slope constant, b, which describes the rate at which the amplitude required to cause failure decreases with increasing number of loading cycles, can be seen to strongly influence the number of equivalent cycles.

The concept of a number of equivalent cycles was originally applied to liquefaction problems by Seed et al. [29] using a modified version of the Palmgren-Miner cumulative damage hypothesis. The reference stress was generally taken as a fraction of the peak stress; for liquefaction purposes, this fraction has historically been taken to be 0.65. This approach allows the number of equivalent cycles to be obtained from experimental data that use uniform harmonic loading to establish the relationship between load amplitude and number of cycles to failure. Cycle counting schemes based on Palmgren-Miner damage theory treat all loading cycles equally regardless of the order in which they occur. In addition, Seed et al. [29] considered only loading cycles of amplitude greater than 30% of the peak amplitude to be capable of generating significant pore pressure.

Liu et al. [23] used the vector sum of two orthogonal horizontal acceleration histories and individual peak amplitude weighting factors based on both laboratory and field data to define numbers of equivalent cycles. Using a large database of recorded ground motions along with consideration of source, path, and site effects, predictive models for numbers of equivalent cycles based on laboratory, field, and lab/field data were developed. The models predict number of equivalent cycles as functions of magnitude, distance, and site condition (soil or rock) while accounting for near-fault effects and show that N_{eq} increases with both magnitude and distance.

Recognizing that softening due to pore pressure generation prior to the triggering of liquefaction produces significantly nonlinear behavior, Green and Terri [11] proposed a cycle-counting scheme based on dissipated energy. In this approach, the dissipated energy of the transient and uniform motions are equated. The number of equivalent cycles was taken as the ratio of the cumulative dissipated energy to the energy dissipated in one cycle of loading at the reference stress, i.e., as

$$N_{eq}(S_{ref}) = \frac{\sum \omega_i}{[\omega(S_{ref})]_{1 \text{ cycle}}} \tag{6}$$

where ω_i is the energy dissipated in the ith cycle of loading and $[\omega(S_{ref})]_{1 \text{ cycle}}$ is the energy dissipated in one cycle of loading at $S = S_{ref}$. The energy-based N_{eq} increases with distance in a manner similar to that of the Liu et al. [23] model at distances less than about 75 km beyond which it remains nearly constant as the diminishing ground motion amplitude balances the effect of increasing duration with distance. Notably, N_{eq} also decreases with depth for typical soil profiles.

Each of these cycle counting procedures accounts for the relative sizes of the individual pulses that comprise a transient loading history and each assigns a weight to each pulse's expected contribution to pore pressure generation. The weighting functions are nonlinear and some include a threshold, generally relative to the peak amplitude of the entire history, below which no contribution to pore pressure is assumed. In all cases, the order or sequence in which the pulses arrive at a given element of soil is not considered – each pulse is considered to contribute according to its amplitude independent of the preceding loading history.

4.2 Magnitude Scaling Factor

The concept of equivalent cycles was implemented into liquefaction potential evaluations by means of a magnitude scaling factor, *MSF*. Recognizing that numbers of equivalent cycles increased with increasing ground motion duration, and that ground motion duration was most strongly influenced by earthquake magnitude, Seed et al. [29] related N_{eq} to magnitude with a magnitude of 7.5 being assumed to produce 15 equivalent cycles of loading. In the simplified procedure for evaluation of liquefaction potential, the magnitude scaling factor is applied to the peak acceleration for the entire ground motion; as a result, the effective *IM* is *PGA/MSF*. The magnitude scaling factor can be defined as

$$MSF = \frac{CRR_M}{CRR_{M=7.5}} = \left(\frac{N_{M=7.5}}{N_M}\right)^b \tag{7}$$

Using a value of $b = 0.34$ as being representative of clean sands from laboratory tests and a relationship between magnitude and number of equivalent cycles (Fig. 4), Idriss [13] proposed the relationship

$$MSF = 6.9\exp(-M/4) - 0.058 \le 1.8 \tag{8}$$

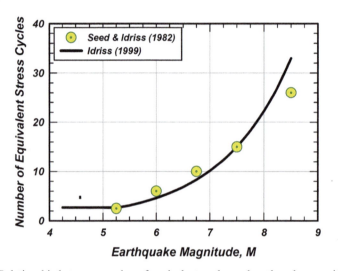

Fig. 4. Relationship between number of equivalent cycles and earthquake magnitude [5].

Boulanger and Idriss [5], noted that the slopes of laboratory cyclic resistance curves, represented by the *b*-parameter in Eqs. 2, 4, and 6, increased with increasing soil density [17] and proposed that

$$MSF = 1 + (MSF_{max} - 1)\left[8.64\exp(-M/4) - 1.325\right] \tag{9}$$

where MSF_{max} is limited to being less than or equal to 1.8 for sand and 1.09 for clay and plastic silt and $MSF_{max} = 1.09 + (q_{c1Ncs}/180)^3 \le 2.2$ for CPT-based and $MSF_{max} = 1.09 + (N_{1,60cs}/31.5)^2 \le 2.2$ for SPT-based evaluations. Finding that the number of equivalent cycles for a given event depended on factors in addition to magnitude, Liu et al. [23] related a *CRR* scaling factor to number of equivalent cycles as

$$\ln CRR_{SF} = 1.3 - 0.41\ln N_{eq} \tag{10}$$

The CRR_{SF} term can be used in a manner equivalent to *MSF* in the simplified method for evaluation of liquefaction potential. The variation of *MSF* and CRR_{SF} with magnitude from several triggering models is illustrated in Fig. 5.

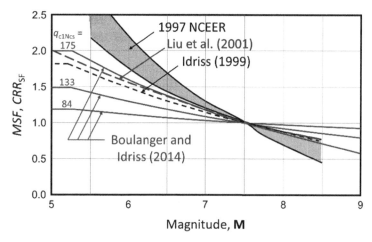

Fig. 5. Comparison of magnitude scaling factors.

5   Transient Loading Laboratory Testing Program

In order to explore alternative, and more efficient, intensity measures for evaluation of liquefaction potential under loading conditions more representative of actual earthquakes, a series of cyclic direct simple shear tests was performed on wet-pluviated samples of Nevada sand [20, 21, 32]. Two types of tests were performed: (a) tests with harmonic, constant-frequency loading of variable amplitude, and (b) transient loading derived from recorded earthquake motions.

5.1   Variable Amplitude Harmonic Loading

Figure 6 shows the results of a series of cyclic simple shear tests on specimens loaded with nine cycles of constant amplitude ($CSR = 0.075$) and one (either the 2nd, 4th, 6th, 8th, or 10th loading cycles) with twice that amplitude ($CSR = 0.15$). The applied loading produced relatively low pore pressure ratios so that phase transformation behavior was not exhibited. The pore pressure ratios after the first cycle of loading, which was identical in all tests, decreased exponentially with increasing specimen relative density. To reduce the effect of these differences, Fig. 6(c) shows the pore pressure ratios normalized by their values after the first cycle of loading (which was identical in all tests). The results show that (a) the increment of pore pressure response prior to the large cycle was generally consistent for all specimens (and consistent with the behavior shown in Fig. 2), (b) the large cycle produced a significantly larger pore pressure increment than the smaller cycles that preceded it, (c) very little additional pore pressure was generated by the lower-amplitude cycles that followed the large cycle even though cycles of the same amplitude had produced significant pore pressures before the large cycle, and (d) the final pore pressures were influenced by the order of the loading cycles. These results suggest that the generation of pore pressure is influenced by both the order of the loading cycles and the maximum level of loading that a specimen has been subjected to.

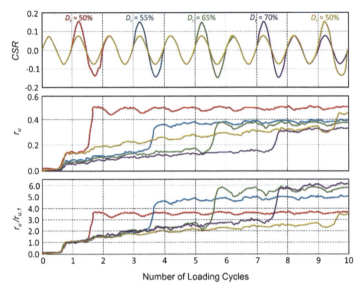

Fig. 6. Responses of cyclic simple shear specimens subjected to variable amplitude harmonic loading.

Another variable amplitude harmonic test was performed by subjecting a torsional shear test specimen to 40 cycles of modulated harmonic loading of smoothly increasing and decreasing amplitude as shown in Fig. 7. Examination of the figure shows that approximately 80% of the generated pore pressure developed in the first 20 cycles when the shear stress amplitude increased from one cycle to the next; the last 20 cycles, each of which had an amplitude that corresponded to a cycle in the first 20 cycles, produced much less pore pressure and the last 14 cycles, which had amplitudes less than 85% of the peak prior amplitude, produced essentially no additional pore pressure.

5.2   Transient Loading – Level-Ground Conditions

Actual earthquake ground motions are transient in nature and induce transient, highly irregular loading histories in potentially liquefiable soils. Transient ground motions are generally characterized by spectral content, either in terms of Fourier spectra or, more commonly, response spectra. Such characterizations typically do not capture ground motion duration well and do not capture the temporal variation of ground motion amplitude at all. For purposes of liquefaction hazard evaluation, peak acceleration and earthquake magnitude are also independent of temporal amplitude variation. Figure 8 shows the results of two cyclic torsional shear tests using the same loading history but reversed in time (i.e., run forward and backward). In both cases, pore pressures built up until the highest cyclic stress was reached and then remained essentially constant after that time. Cyclic shear strain amplitudes, however, were not noticeably influenced by peak past loading level. Peak and final pore pressures and cyclic strains were approximately 20% greater for the forward loading case.

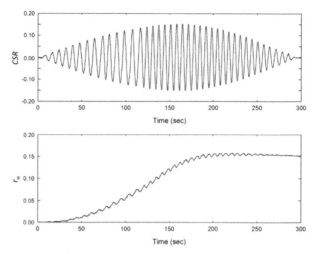

Fig. 7. Responses of cyclic simple shear specimens subjected to modulated harmonic loading (data courtesy of Fujen Ho, personal communication).

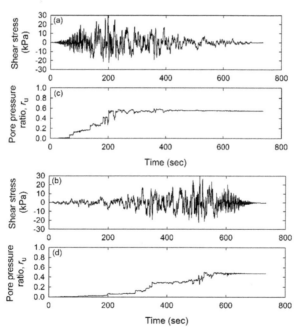

Fig. 8. Response of cyclic torsional shear specimens to the same loading history: (a) shear stress – forward loading, (b) pore pressure ratio – forward loading, (c) shear stress – reverse loading, (d) pore pressure ratio – reverse loading (data courtesy of Fujen Ho, personal communication).

To further investigate response to transient loading and to determine whether alternative intensity measures would be more suitable for liquefaction hazard evaluation, a cyclic simple shear testing program involving many transient loading histories was undertaken. The loading histories were obtained by a process intended to emphasize ground motion characteristics that would distinguish between the efficiencies of several evolutionary intensity measures. This involved identifying motions with combinations of amplitude, frequency content, duration, and phasing that differed significantly relative to each other. Figure 9(a) shows a scatter plot of two IMs computed from some 7000 recorded outcrop motions ($V_{S30} > 360$ m/sec) that span a wide range of magnitudes, distances, styles of faulting, basin depths, etc. [1]. There is a clear correlation between the two IMs but there are also motions that vary widely for one of the IMs with constant values of the other IM. Figure 9(b) shows examples of the four motions highlighted in Fig. 9(a). Two of the motions have the same value of PGA/MSF but very different values of CAV_5 [19] and two have the same value of CAV_5 but very different values of PGA/MSF. Tests using shear stress histories based on motions with the same value of PGA/MSF but high and low values of CAV_5 that produced similar pore pressures would indicate that PGA/MSF was an efficient predictor of pore pressure generation (and hence liquefaction). An initial set of 15 motions developed to distinguish the relative efficiencies of PGA/MSF, Arias intensity, and cumulative absolute velocity is shown in Fig. 10. A supplemental set of six additional motions in which energy built up at different rates is shown in Fig. 11.

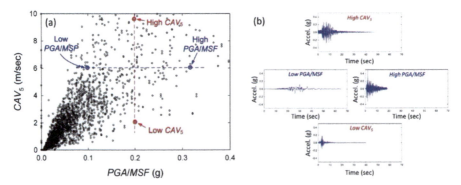

Fig. 9. Relationship between PGA/MSF and $CAV5$ for 7000 recorded outcrop motions highlighting motions with same PGA/MSF (red) and same $CAV5$ (blue) [1].

The previously described motions were converted to shear stress histories by applying them at the base of an equivalent linear model of a 6-m-thick profile of medium dense ($D_r = 55\%$) sand underlain by 1 m of cemented sand over a rigid base and extracting the shear stress history at a depth of 5 m; this profile corresponded to that of a prototype soil profile used in parallel centrifuge tests. The shear stress histories were applied to cyclic simple shear specimens prepared at relative densities ranging from 37–89% at 1/5 their actual rate to allow for accurate pore pressure measurement. Several of the tests were not used in subsequent analyses due to low (<0.90) B-values or apparent problems with the testing system. Because the loading system was not able to reproduce the input loading

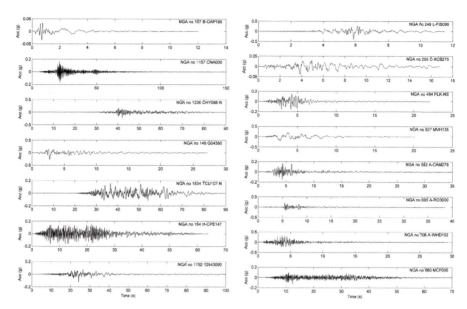

Fig. 10. Acceleration histories for 15 motions used to develop transient loading histories.

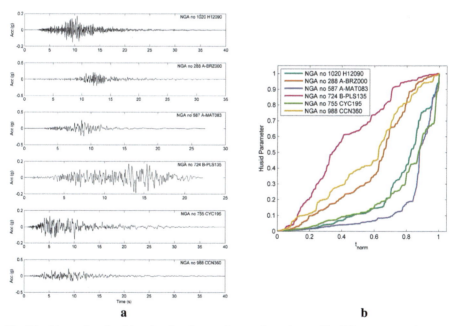

a   **b**

Fig. 11. (a) acceleration histories for six supplemental motions with different rates of energy buildup and (b) Husid plots illustrating rates of energy buildup.

histories exactly, the measured shear stress histories were converted to input acceleration histories using a shear stress-based equivalent linear analysis program (Sideras 2019) developed for the project.

An example of the measured response is shown in Fig. 12. The measured pore pressure increases relatively quickly while the shear stress amplitude is increasing but stalls (e.g., from 138–142 s) when cyclic stresses are lower than the peak past cyclic stress. Shear strains are small until the pore pressure ratio is nearly 1.0 and then become large; dilation-induced stiffening is clearly seen in the stress-strain and stress path curves.

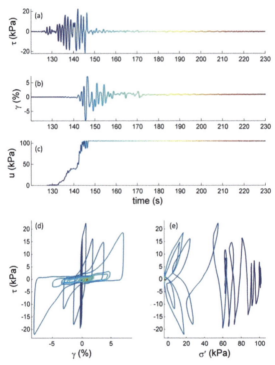

Fig. 12. Measured response of cyclic simple shear specimen subjected to transient loading history: (a) applied shear stress, (b) shear strain, (c) excess pore pressure, (d) stress–strain behavior, and (c) effective stress path.

The results of the transient loading tests were used to evaluate the relative efficiencies of four evolutionary liquefaction-related intensity measures. In order to do so, the measured pore pressure histories were modified by removing temporary reductions in pore pressure due to unloading and dilation; modified pore pressures were held constant during periods of measured pore pressure reduction. The evolutionary intensity measures examined were $PGA_M = PGA/MSF$, I_a, CAV, and CAV_5. Two versions of PGA/MSF were used – one in which MSF was obtained by correlation to number of loading cycles (Liu et al. 2001) and one in which MSF was also related to relative density through laboratory cyclic strength curves [5]. These IMs are referred to subsequently as

$PGA_{M,L}$ and $PGA_{M,B}$, respectively. At a particular relative density, the modified pore pressure can be plotted as a function of the evolving intensity measure. The rate of pore pressure generation would increase with increasing intensity measure and decrease with increasing relative density. Because the pore pressure ratio is bounded by values of 0.0 and 1.0, the test data was interpreted in terms of the *IM* values from $r_u = 0.05$ up to the point of triggering (taken as $r_u = 0.99$). The relationship between the three variables $(IM, D_r, \text{and } r_u)$ were found to be described well by the function

$$IM = \frac{1 + \alpha_1}{1 + \alpha_1 \cdot r_u^{\alpha_2}} \cdot \beta_1 \exp(\beta_2 D_r) \tag{11}$$

where α_1, α_2, β_1, and β_2 are coefficients obtained by regression for each evolutionary intensity measure. When the coefficients were determined based on the transient cyclic simple shear test results over the range of $0.05 < r_u < 0.99$, the standard deviations of the *IM* residuals were as indicated in Table 1. These values show considerable differences in the abilities of the different *IM*s to track pore pressure development up to and at the point of triggering.

Table 1. Coefficients and uncertainties for *IM* models

IM	α_1	α_2	β_1	β_2	$\sigma_{\ln IM}$
$PGA_{M,L}$	0.2367	−1.228	0.07713	0.01441	0.31
$PGA_{M,B}$	0.3137	−1.082	0.1199	0.01404	0.24
I_a	0.0676	−2.658	0.06688	0.02624	0.82
CAV	0.372	−1.322	0.8443	0.01537	0.58
CAV_5	0.3167	−1.412	0.7972	0.01597	0.60

Discussion. The results of the transient loading tests indicate that PGA_M-based intensity measures are significantly more closely related to pore pressure generation during transient loading than the integral evolutionary intensity measures. Several factors appear to contribute to this finding:

1. The PGA_M-based intensity measures are based on the relative amplitudes of individual loading pulses without regard to the frequencies at which they occur.
2. Integral *IM*s are strongly influenced by frequencies and can increase significantly due to low frequency components of ground motions that would not produce significant shear strain (hence, pore pressure generation) in liquefiable soil layers.
3. The PGA_M-based intensity measures are nonlinearly related to the amplitudes of individual pulses in a way that emphasizes larger pulses and discounts (or ignores) the effects of small pulses that contribute little to pore pressure generation.
4. Integral *IM*s consider all pulses in proportion to their amplitudes and can have their values inflated by low-amplitude cycles that do not generate pore pressure.

5. PGA_M is more efficient at predicting the onset of phase transformation behavior after which pore pressures tend to increase more rapidly.

The relative manners in which PGA_M-based and integral evolutionary intensity measures account for irregular loading can be illustrated by considering the initial set of variable amplitude harmonic tests shown in Fig. 13. Assuming the tests represent an element of soil at the same depth, shear stresses and CSR values would be proportional to accelerations in a simplified model analysis. Considering the tests with large pulses in the 4th, 6th, and 8th cycles, which had similar relative densities, comparison of the shapes of the recorded pore pressure responses (Fig. 13b) and the $PGA_{M,B}$ (Fig. 13f) and CAV (Fig. 13g) show how much more closely $PGA_{M,B}$ correlates to pore pressure generation over the entire course of loading than CAV. As an integral parameter, CAV continues to increase from cycle to cycle with a relatively small jump during the large cycle. $PGA_{M,B}$, on the other hand, increases significantly during each large cycle and, as does the pore pressure, remains essentially constant thereafter.

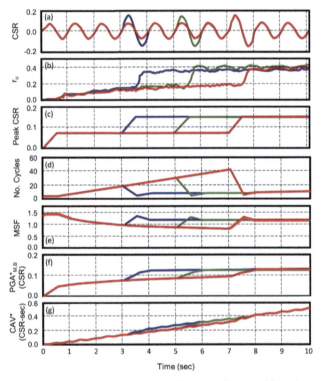

Fig. 13. Response histories of three cyclic simple shear specimens subjected to variable amplitude harmonic loading: (a) applied cyclic stress ratio, (b) generated pore pressure ratio, (c) peak cyclic stress ratio, (d) number of equivalent loading cycles, (e) magnitude scaling factor, (f) PGA^*_{MB} (computed as the quotient of (c) and (e)), and (g) CAV^* (computed using CSR rather than acceleration).

5.3   Transient Loading – Sloping Ground Conditions

In order to better understand the response of liquefiable soils to transient loading under conditions that commonly exist in the field, another series of cyclic simple shear tests were performed with transient loading superimposed on constant, static shear stresses (deLaveaga 2016). The tests were performed at the Norwegian Geotechnical Institute using two transient loading histories with very different temporal characteristics (Fig. 14). The Palm Springs motion (MVH 135) was from the 1986 N. Palm Springs (**M**6.1) earthquake; this motion had a relatively short duration and was characterized by a few low-frequency pulses of acceleration that occurred early in the record. The Landers record (MCF 000) was from the 1992 Landers (**M**7.3) earthquake and had a long duration with many high-frequency pulses of similar amplitude.

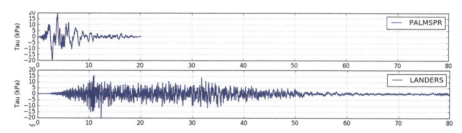

Fig. 14. Accelerograms for Palm Springs MVH 135 and Landers MCF 000 motions scaled to peak shear stresses of 20 kPa.

Tests with both uniform harmonic and transient loading were applied to the soil specimens and the transient loading tests were applied with nominal static stress ratios of 0.00, 0.05, 0.10, 0.15, and 0.20. Test specimens of Nevada sand were reconstituted by wet pluviation to relative densities ranging from approximately 35% to 70%. Tests were performed at 5% of the rate of the actual recorded earthquake motion. The responses of specimens subjected to the Palm Springs and Landers loading histories with different

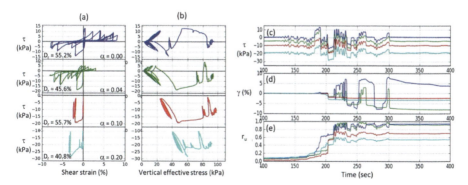

Fig. 15. Response of Nevada sand specimens to Palm Springs loading histories with constant *CSR* but different static shear stress levels: (a) stress-strain curves, (b) stress paths, (c) shear stress, (d) shear strain, and (e) pore pressure ratio.

static shear stress levels are shown in Figs. 15 and 16, respectively. Accounting for differences in relative densities, the specimens exhibited behavior generally consistent with that indicated in prior studies based on uniform harmonic loading (Fig. 3). The specimens with relative densities in the range of 55–62% generated lower levels of excess pore pressure with increasing static shear stress levels, thus indicating K_α values greater than 1.0 and increasing with α.

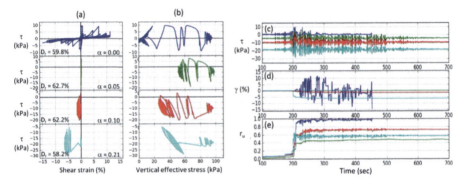

Fig. 16. Response of Nevada sand specimens to Landers loading histories with similar *CSR* but different static shear stress levels: (a) stress-strain curves, (b) stress paths, (c) shear stress, (d) shear strain, and (e) pore pressure ratio.

6   An Improved Intensity Measure for Triggering of Liquefaction

The variable amplitude and transient loading tests showed consistently that the generation of pore pressure was influenced by the temporal variation of ground motion amplitude, which is not accounted for in the intensity measures currently used in the evaluation of liquefaction potential. An intensity measure that does account for temporal variation as well as the amplitude and number of loading cycles would be beneficial in (a) the interpretation of liquefaction case histories in areas with nearby ground motion recordings, (b) the prediction of liquefaction for certain rupture scenarios in which repeatable temporal variations are expected (e.g., near-fault, forward and backward directivity cases), and (c) cases where ground motion hazards are defined in terms of ground motion histories (e.g., in the case of physics-based modeling).

Defining half-cycles as the periods between successive zero-crossings, the evolutionary form of $PGA_{M,B}$ at the jth cycle of loading can be expressed as

$$PGA_{M,B,j} = \frac{PGA_j}{MSF_j} = PGA_j \left(\frac{N_j}{N_{ref}} \right)^b \tag{12}$$

where PGA_j is the peak absolute acceleration as of the jth cycle, MSF_j is the magnitude scaling factor as of the jth cycle, N_{ref} is the reference number of cycles (generally taken as 15), N_j is the number of equivalent cycles as of the jth cycle, and b is an exponent that describes the slope of a cyclic strength curve. Assuming a linear relationship between damage and number of loading cycles,

$$N_j = \sum_{i=1}^{j} N_i = \sum_{i=1}^{j} \frac{1}{2} \left(\frac{pk_i}{0.65PGA_j} \right)^{1/b} \tag{13}$$

where pk_i is the absolute value of the amplitude of the ith local peak. This allows $PGA_{M,B}$ at the jth cycle of loading to be written as

$$PGA_{M,B,j} = \frac{1}{0.65N_{ref}^{b}} \left(\frac{1}{2} \sum_{i=1}^{j} pk_i^{1/b} \right)^{b} = C \left(\sum_{i=1}^{j} pk_i^{1/b} \right)^{b} \tag{14}$$

where C groups the constant terms for given values of N_{ref} and b. To allow more flexibility in modeling laboratory results, the two exponents on the right side of Eq. (14) can be freed from their reciprocal relationship. To account for the observed order of cycles effect, in which cycles of increased amplitude produce relatively high increments of pore pressure and cycles of amplitude lower than the prior peak (absolute) amplitude between zero crossings (i.e., the PPA) produce small increments of pore pressure, a peak ratio factor, r, that amplifies the effects of peaks that define new PGA values and reduces the effects of peaks that are preceded by a higher peak. This allowed a liquefaction triggering intensity measure to be defined as

$$IM_L = \left[\sum_{i=1}^{j} \left(pk_i \cdot \frac{2}{1 + \left(\frac{pk_i}{PPA} \right)^{\theta_3}} \right)^{\theta_2} \right]^{\theta_1} \tag{15}$$

The peak ratio factor ranges from zero to 2 and takes on values less than 1.0 for peaks preceded by a higher peak (i.e., $pk_i < PPA$), greater than 1.0 for peaks that exceed all prior peaks ($pk_i > PPA$), and equal to 1.0 for peaks equal to the prior peak ($pk_i = PPA$). The coefficient, θ_3, controls the rate at which r changes with pk_i/PPA. Examination of Equation (c) shows that $\theta_1 = b$ and $\theta_2 = 1/b$ in the Boulanger and Idriss formulation; that formulation would produce $b = 0.236$ for $(N_1)_{60,cs} = 15$ blows/ft. The values of the coefficients, θ_1–θ_3, were determined through a process of constrained optimization with the constraints set to avoid convergence to coefficient values that differed greatly from those based on laboratory data (Boulanger and Idriss 2014). The coefficients θ_1 and θ_2 were constrained to lie within ranges of 0.15 to 0.55 and $0.7/\theta_1$ to $1.4/\theta_1$, respectively; θ_3 was constrained to being between -5.0 and zero. Using an objective function defined as the sum-of-squared-errors in $\ln IM_L$, values of $\theta_1 = 0.298$, $\theta_2 = 2.53$, and $\theta_3 = -2.91$ were obtained. The uncertainty in IM_L was characterized by $\sigma_{\ln IM_L} = 0.16$, which is significantly lower than the corresponding values for the other evolutionary IMs listed in Table 1.

Figure 17 shows how the peak ratio factor discounts the contributions of pulses preceded by larger pulses and amplifies the contributions of those that represent new peak loadings. Note that Seed et al. (1975) assumed that accelerations less than 30%

of the *PGA* do not contribute to N_{eq}; Sassa and Yamazaki (2017) did not count shear stresses less than 60% of the peak shear stress toward their effective number of stress cycles. Examination of transient test data showed that pore pressure ratios at the first exceedance of 30% of the peak shear stress were generally less than 0.1 but could be as high as 0.3 in some tests. The peak ratio factor shown in Fig. 17 provides a smooth and continuous variation with the evolving *PPA* as opposed to peak values for the entire motion.

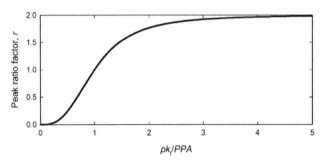

Fig. 17. Variation of peak cycle weighting factor applied to individual cycles in proposed intensity measure.

7 Toward an Improved K_{α} Adjustment Factor

The effect of static shear stress on the triggering of liquefaction under transient loading can be quantified by a K_{α} term (Eq. 1) but care must be taken in the definition of that term due to the transient nature of actual earthquake motions. For uniform harmonic loading, the *CRR* terms can be defined using the peak shear stress, which is the same for all loading cycles. For transient loading, however, the amplitudes of individual loading pulses vary over the duration of the loading, which complicates definition of K_{α}. In this study, the evolving magnitude-corrected cyclic stress ratio could be computed at the point at which liquefaction was observed to have been triggered, i.e., as

$$CRR_{\alpha,state} = CSR_M(t_{state}) = \frac{CSR(t_{state})}{MSF(t_{state})} = \frac{[\tau_{max}(t_{state}) - \tau_{static}]/\sigma'_{vo}}{f[N_{eq}(t_{state})]} \qquad (16)$$

where t_{state} is the time of at which some state of interest (typically the triggering of liquefaction) is reached and $f[N_{eq}(t_{state})]$ is the *CRR* scaling factor (analogous to *MSF*) at time, t_{state}, obtained using a cycle counting-based procedure applied to the cyclic portion of the loading history (i.e., to $\tau_{max}(t) - \tau_{static}$). The transient load testing program for sloping ground conditions had a relatively small number of tests in which initial liquefaction was actually reached and those tests were all conducted at α-values less than about 0.05. In order to gain insight into the effects of static shear stress over a broader range of conditions relevant to practice, the finding of close correlation between intensity measures at the point of phase transformation ($r_u \approx 0.6$) and $r_u \approx 1.0$ for level-ground transient loading [32] was used to define $CRR_{\alpha,0.6}$ (i.e., CRR_{α} for the state of $r_u = 0.6$). This roughly doubled the number of data points and extended the range of α out to 0.21. A smooth function quadratic in both α and D_r was then fit to the data as illustrated in Fig. 18. This function can then be used to define K_α for the state of interest, i.e.,

$$K_{\alpha,0.6} = \frac{CSR_{\alpha,0.6}}{CSR_{\alpha=0,0.6}} \qquad (17)$$

for which curves at different relative densities are shown in Fig. 19.

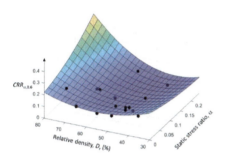

Fig. 18. Smooth function fit to cyclic resistance ratios corresponding to the state of $r_u = 0.60$.

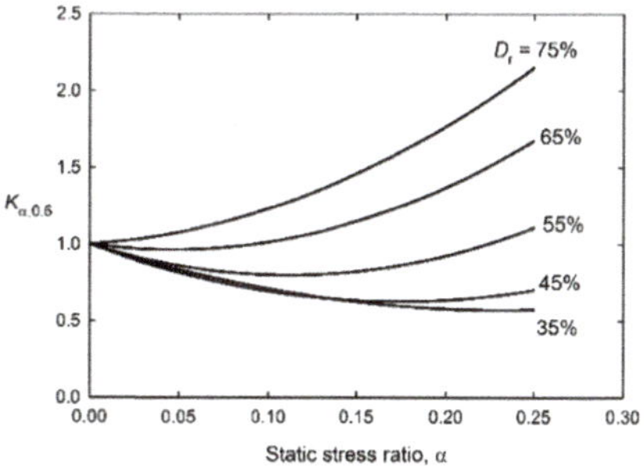

Fig. 19. Curves of $K_{\alpha,0.6}$ from transient loading tests with Palm Springs motion for different soil densities.

The $K_{\alpha,0.6}$ curves of Fig. 19, while not a substitute for K_α, provide insight into how improved models might be developed. The curves can be seen to be generally consistent with those in Fig. 3. They show $K_\alpha < 1.0$ for loose sands and $K_\alpha > 1.0$ for medium dense to dense sands. The shapes of the curves, however, differ from those shown in Fig. 3 in that they are all concave upward, i.e., all show increasing K_α at large values of α for all densities. This behavior appears to be reasonable given the differences in loading between the uniform loading histories used to develop the relationships shown in Fig. 3 and the transient loading upon which the curves shown in Fig. 19 are based. As the static shear stress is increased in a series of uniform harmonic loading cycles, the amplitudes of the stress reversals decreases but the number of stress reversals remains unchanged until dropping to zero instantaneously when α exceeds CSR. Under transient loading, however, both the amplitude and the number of stress reversals decrease with increasing static shear stress. As a result, the effects of static shear stress should appear earlier and more intensely for transient loading than for uniform loading.

8 Summary and Conclusions

Improved understanding of the mechanics of liquefiable soil behavior has led to the development of improved tools for analyzing the response of liquefiable soil profiles in recent years. Advanced numerical models employing well-documented, validated constitutive models are capable of representing important behaviors of liquefiable soils up to and, increasingly, after the triggering of liquefaction. These tools allow engineers to visualize the mechanisms by which pore pressures generate, redistribute, and dissipate, and by which resulting deformations develop. The insight and understanding gained from these analyses can be invaluable and their use in practice will continue to increase.

At the same time, the depositional complexities of many soil profiles prevent subsurface characterization so complete that all behaviors can be anticipated or quantified even in the most capable of numerical models. Both the mechanical and hydraulic response of a liquefiable soil profile can be sensitive to the presence of thin layers or lateral inhomogeneities that are not captured by typical subsurface investigations. As a result, the use of empirical procedures for evaluating liquefaction potential and its consequences will remain an important part of geotechnical earthquake engineering practice. Empirical procedures have evolved and improved as understanding of liquefiable soil behavior has developed and as more case histories have been documented in large and small earthquakes around the world.

Empirical models for the triggering of liquefaction require characterization of demand and capacity, i.e., of the loading imposed on liquefiable soils and the resistance of those soils to the generation of high pore pressures. The great majority of liquefaction research to date has focused on liquefaction resistance with less attention being paid to loading. The laboratory tests that have elucidated important behaviors that cannot be obtained from available case histories have represented earthquake loading by harmonic shear stresses of constant amplitude, and empirical models are based on the concept of representing the transient, irregular nature of actual earthquake loading by a number of equivalent uniform harmonic stress cycles. This was due, in large part to the common use of pneumatic loading systems driven by function generators available at the time. With the advanced control systems of modern laboratory testing equipment now available, however, that restriction has been overcome and it is possible to apply loading more representative of actual earthquake loading and to understand how liquefiable soils respond differently to such loading.

This paper presents some of the results of cyclic simple shear tests in which loading histories representative of actual earthquake loading are applied. The intent of the tests was to help identify optimal intensity measures for evaluation of liquefaction hazards, and to develop data that could be used by constitutive modelers to validate or improve their models. Tests were performed to simulate level-ground conditions with a large number of loading histories based on different recorded ground motions, and to simulate sloping ground conditions with two motions. The results of the tests have led to several significant conclusions.

1. The generation of excess pore pressure in liquefiable soils is more closely related to the number and amplitude of the individual stress pulses that make up a loading history than to the duration of that history. Thus, the potential for time-integrated intensity measures such as Arias intensity or cumulative absolute velocity to be efficient predictors of liquefaction triggering appears to be limited.

2. The manner in which pore pressures develop under transient loading is influenced by the order of the stress pulses imposed on the soil. Thus, commonly used intensity measures and cycle-counting procedures, which treat all pulses equally based solely on their amplitudes, are limited in their ability to represent pore pressure generation.

3. The incremental pore pressure generated by an individual stress pulse is strongly influenced by the amplitude of that pulse relative to the past peak amplitude. If the pulse is preceded by a larger pulse, it will generate a significantly smaller increment of pore pressure than if it had not. This leads to the irregular generation of pore pressures under earthquake loading in which large increments of pore pressure are associated with stress pulses that exceed all preceding pulses followed by nearly constant pore pressures between those pulses.

4. The transient testing performed in this investigation supports the concept of a new intensity measure, IM_L, that explicitly considers the order of cycles in a ground motion. It therefore offers the potential to better distinguish between motions of short and long duration, motions of forward and backward directivity, and motions with different source rupture characteristics. In the context of empirical procedures, such an intensity measure can be used with recorded ground motions in the vicinity of liquefaction case histories to better characterize the loading imposed on the soil in those case histories.

5. The presence of static shear stresses, as occur under sloping ground conditions or in the vicinity of structures, appears to influence pore pressure generation and liquefaction triggering in a manner generally consistent with that exhibited in tests involving uniform harmonic loading. However, the relationship and interaction between transient shear stress histories and static shear stresses is significantly different than for uniform harmonic loading, particularly when static stresses are high relative to cyclic stresses, a condition for which little test data is currently available.

6. Additional testing with transient loading is needed. The investigations described in this paper were limited and exploratory and used only uniformly-graded clean sand. Additional testing is needed with a broader range of soils (different grain size distributions, fines contents, fines plasticities) and broader ranges of relative densities and loading amplitudes. Tests that extend beyond the point of triggering, particularly tests with static shear stresses, will also be useful to better understand the development of large strains and to identify intensity measures that predict those strains efficiently and sufficiently. Such intensity measures can be expected to improve empirical models for prediction of the consequences of liquefaction.

Acknowledgements. The research described in this paper was supported by the National Science Foundation (Award 0936408). The efforts of Masters students Stephanie Abegg and Katie deLaveaga with analytical and experimental work is greatly appreciated; Ms. deLaveaga's work at NGI was supported by a Valle Foundation fellowship from the University of Washington. The use of NGI testing equipment and the assistance of NGI personnel, particularly Dr. Brian Carlton and Dr. Amir Kaynia, is also greatly appreciated.

References

1. Abegg, S.: Identification of optimal evolutionary intensity measures for evaluation of liquefaction hazards, Masters thesis, University of Washington, 185 p. (2010)

2. Azeiteiro, R.N., Coelho, P.A.L.F., Taborda, D.M.G., Grazina J.C.: Dissipated energy in undrained cyclic triaxial tests. In: Proceedings of the 6th International Conference on Earthquake Geotechnical Engineering, Christchurch, New Zealand, 1–4, paper no. 220 (2015)
3. Azeiteiro, R.N., Coelho, P.A.L.F., Taborda, D.M.G., Grazina, J.C.: Energy-based evaluation of liquefaction potential under-non-uniform cyclic loading. Soil Dyn. Eq. Eng. **92**, 650–665 (2017)
4. Boulanger, R.W.: Relating Kα to relative state parameter index. J. Geotech. Geoenviron. Eng. ASCE **129**(8), 770–773 (2003)
5. Boulanger, R.W., Idriss, I.M.: CPT and SPT based liquefaction triggering procedures. Rep. No. UCD/CGM-14/01, Center for Geotech. Modeling, U.C. Davis, Davis, California, 138 p. (2014)
6. Çetin, K., Bilge, H.: Stress scaling factors for seismic soil liquefaction engineering problems: a performance-based approach. In: Ansal, A., Sakr, M. (eds.) Perspectives on Earthquake Geotechnical Engineering. Geotechnical, Geological and Earthquake Engineering, vol. 37, pp. 113–139. Springer, Cham (2015). https://doi.org/10.1007/978-3-319-10786-8_5
7. deLaveaga, K.M.: Slope effects on liquefaction potential and pore pressure generation in earthquake loading, Masters thesis, University of Washington, 113 p. (2016)
8. De Alba, P., Seed, H.B., Chan, C.K.: Sand liquefaction in large-scale simple shear tests. J. Geotech. Eng. Div. ASCE **102**(GT9), 909–927 (1976)
9. Dobry, R., Ladd, R., Yokel, F., Chung, R., Powell, D.: Prediction of pore water pressure buildup and liquefaction of sands during earthquakes by the cyclic strain method. NBS Building Science Series 138, Nat. Bur. Stand., U.S. Department of Commerce (1982)
10. Green, R.A., Mitchell, J.K., Polito, C.P.: An energy-based excess pore pressure generation model for cohesionless soils. In: Smith, D.W., Carter, J.P. (eds.) Proceedings the John Booker Memorial Symposium – Developments in Theoretical Geomechanics, pp. 383–390. A.A. Balkema, Rotterdam (2000)
11. Green, R.A., Terri, G.A.: Number of equivalent cycles concept for liquefaction evaluations – revisited. J. Geotech. Geoenviron. Eng. **131**(4), 477–488 (2005)
12. Hancock, J., Bommer, J.J.: The effective number of cycles of earthquake ground motion. Earthq. Eng. Struct. Dyn. **34**, 637–664 (2005)
13. Idriss, I.M.: An update to the Seed-Idriss simplified procedure for evaluating liquefaction potential. In: Proceedings, TRB Workshop on New Approaches to Liquefaction, Pub. No. FHWARD-99-165, Fed. Hwy. Admin. (1999)
14. Ishihara, K., Yasuda, S.: Sand liquefaction under random earthquake loading conditions. In: Proceedings of the 5th World Conference on Earthquake Engineering (1973)
15. Ishihara, K., Yasuda, S.: Sand liquefaction in hollow cylinder torsional under irregular excitation. Soils Found. **15**(1), 45–59 (1975)
16. Kayen, R.E., Mitchell, J.K.: Assessment of liquefaction potential during earthquakes by Arias intensity. J. Geotech. Geoenviron. Eng. ASCE **123**(12), 1162–1174 (1997)
17. Kishida, T., Tsai, C.-C.: Seismic demand of the liquefaction potential with equivalent number of cycles for probabilistic seismic hazard analysis. J. Geotech. Geoenviron. Eng. ASCE **140**(3), 1–14 (2014)
18. Kokusho, T.: Liquefaction potential evaluations: energy-based method versus stress-based method. Can. Geotech. J. **50**, 1088–1089 (2013)
19. Kramer, S.L., Mitchell, R.A.: Ground motion intensity measures for liquefaction hazard evaluation. Earthq. Spectra **22**(2), 413–438 (2006). https://doi.org/10.1193/1.2194970
20. Kwan, W.S.: Laboratory investigation into evaluation of sand liquefaction under transient loadings. Ph.D. dissertation, University of Texas, Austin, TX (2015)
21. Kwan, W.S., Sideras, S.S., Kramer, S.L., El Mohtar, C.: Experimental database of cyclic simple shear tests under transient loadings. Earthq. Spectra **33**(3), 1219–1239 (2017). https://doi.org/10.1193/093016eqs167dp

22. Liang, B.L., Figueroa, J.L., Saada, A.S.: Liquefaction under random loading: unit energy approach. J. Geotech. Eng. **121**(11), 776–781 (1995)
23. Liu, A.H., Stewart, J.P., Abrahamson, N.A., Moriwaki, Y.: Equivalent number of uniform stress cycles for soil liquefaction analysis. J. Geotech. Geoenviron. Eng. ASCE **127**(12), 1017–1026 (2001)
24. Miner, M.A.: Cumulative damage in fatigue. Trans. ASME **67**, A159–A164 (1945)
25. Palmgren, A.: Die lebensdauer von kugella geru. ZVDI **68**(14), 339–341 (1924)
26. Pan, K., Yang, Z.X.: Effects of initial static shear on cyclic resistance and pore pressure generation of saturated sand. Acta Geotech. **13**(2), 473–487 (2018)
27. Polito, C., Green, R.A., Dillon, E., Sohn, C.: Effect of load shape on relationship between dissipated energy and residual excess pore pressure generation in cyclic triaxial tests. Can. Geotech. J. **50**(11), 18–28 (2013)
28. Sassa, S., Yamazaki, J.: Simplified liquefaction predictions and assessment method considering waveforms and durations of earthquakes. J. Geotech. Geoenviron. Eng. **143**(2), 04016091 (2017)
29. Seed, H.B., Idriss, I.M., Makdisi, F., Banerjee, N.: Representation of irregular stress time histories by equivalent uniform stress series in liquefaction analysis. Earthq. Eng. Res. Cen. Rep. EERC 75-29, University of California, Berkeley (1975)
30. Seed, R.B., Harder Jr., L.F.: SPT-based analysis of cyclic pore pressure generation and undrained residual shear strength. In: H. Bolton Seed Memorial Symposium Proceedings (1990)
31. Shome, N., Cornell, C.A.: Probabilistic seismic demand analysis of nonlinear structures. Report RMS-35, Dept. Civil Eng., Stanford Univ., 320 p. (1999)
32. Sideras, S.S.: Evolutionary intensity measures for more accurate and informative evaluation of liquefaction triggering, Ph.D. dissertation, University of Washington, 717 p. (2019)
33. Tatsuoka, F., Silver, M.L.: Undrained stress-strain behavior of sand under irregular loading. Soils Found. **21**(1), 51–66 (1981)
34. Vaid, Y.P., Stedman, J.D., Sivathayalan, S.: Confining stress and static shear effects in cyclic liquefaction. Can. Geotech. J. **38**(3), 580–591 (2001). https://doi.org/10.1139/t00-120
35. Wang, J.N., Kavazanjian, E.: Pore pressure development during non-uniform cyclic loading. Soils Found. **29**(2), 1–14 (1989)

Performance-Based Design for Earthquake-Induced Liquefaction: Application to Offshore Energy Structures

Haoyuan Liu[1] and Amir M. Kaynia[2,3](✉)

[1] Norwegian Geotechnical Institute (NGI), Sognsveien 72, 0806 Oslo, Norway
[2] Norconsult AS, Vestfjordgaten 4, N-1338 Sandvika, Norway
[3] Norwegian University of Science and Technology (NTNU), 7491 Trondheim, Norway
amir.kaynia@ntnu.no

Abstract. Liquefaction has been a major challenge to design of structures founded on loose silt and sand in moderate and specially highly seismic regions. While assessment of liquefaction susceptibility and potential have been largely based on empirical methods, the design of structures on liquefiable soil requires reliable numerical tools and clear performance criteria. In this paper, solutions are provided based on the well-established SANISAND model and its more recent extension, SANISAND-MSu, implemented in the open-source finite element platform OpenSEEs. Applications are presented for structures commonly encountered in offshore energy sector such as conventional subsea facilities on mudmats and offshore wind turbines founded on large-diameter monopiles. The impact of pore-water pressure, and ultimately liquefaction, on the offshore structures is assessed by performing both quasi-static cyclic loading and earthquake shaking. The general behavior of these offshore structures during liquefaction are presented from a numerical modelling perspective. The simulation results indicate that the response of these structures is considerably affected by structural features and environmental loading conditions. The results presented in this work motivates the use of SANISAND-MSu model in enhanced 3D finite element modelling in offshore structural dynamic analyses.

Keywords: Earthquake · Pore-water pressure · Numerical modelling · Constitutive model · Dynamic

1 Introduction

The proportion of the renewable energy in the overall energy consumption has kept increasing in recent years. Offshore energy as an important component of the renewable energy is in tune with the trend and experiencing rapid growth in energy share. The global offshore wind capacity cumulated to about 22 GW in 2019 and is expected to reach about 100 GW in 2030 [1]. In the longer term, European Union aims to make offshore energy the main electricity source by 2040. Offshore wind will be a key element in the renewable energy considerations in the years to come [2].

© The Author(s), under exclusive license to Springer Nature Switzerland AG 2022
L. Wang et al. (Eds.): PBD-IV 2022, GGEE 52, pp. 100–119, 2022.
https://doi.org/10.1007/978-3-031-11898-2_6

The offshore energy-related structures have been constructed mostly in the areas with low seismicity. The research performed in the relevant subject span from experimental studies [3–8] to numerical investigation [9–12]. The growth of the construction of structures related to offshore energy in seismic regions requires a comprehensive understanding of the impact of liquefaction on these structures [13].

The main challenge is that the soil behaviour is in practice undrained or partially drained during load cycles in most dynamic excitations including earthquakes [14]. The accumulation of strain and pore-water pressure in the soil domain during the dynamic loading can change the soil's stiffness and strength significantly, which in turn alters the dynamic structural response [15]. Besides, during earthquake shaking, the combined effect from the environmental load may lead to a distinct behaviour of offshore structures.

Significant efforts have recently been devoted to the study of the dynamic response of offshore structures using three-dimensional Finite Element (3D FE) analysis enhanced by advanced implicit constitutive models (here 'implicit' is used to refer to models that calculate soil stress-strain response in incremental sub-septs). For instance, 3D FE analysis of monopile dynamic behaviour is investigated by Corciulo et al. [16] using multi-surface model [17], and by Kementzetzidis et al. [18] and Esfeh and Kaynia [14] using bounding surface SANISAND04 model (where SANISAND04 is used here refers to the original simple anisotropic sand model developed by Dafalias and Manzari in the 2004 [19]). Recently, Esfeh et al. [20] numerically studied the seismic response of subsea structures on caissons and mudmats with SANISAND04 as the constitutive model in FLAC3D.

Accurate implicit 3D FE analysis of structural dynamic response rely on the simulation ability of the adopted constitutive models. Proper capturing of soil cyclic behaviour at element level is required to evaluate the performance of constitutive models at both qualitative and quantitative levels. Recently, Liu et al. [21] adopted the concept of 'memory surface' [22] to realistically reflect sand fabric evolution (and effects) on sand cyclic behavior, based on Dafalias and Manzari's SANISAND04 model framework (denoted by SANISAND-MS). The model presents considerable improved accuracy compared with many other models when simulating drained ratcheting behaviour of sand. Later, Liu et al. [23] improved SANISAND-MS on aspects of hardening rule and flow rule for better simulation of undrained cyclic behaviour of sand. For this reason, the model is referred to as SANISAND-MSu in this work. SANISAND-MSu model can predict with good accuracy the cycle-by-cycle pore water pressure evolution and accumulation in the soil pre-dilative straining process. Besides, SANISAND-MSu predicts realistic strain accumulation in cyclic-mobility regime. The model has been successfully implemented in the open-source finite element platform OpenSEEs and shows good prediction abilities in simulating monopile response with the presence of pore-water pressure in quasi-static cyclic loading events [24].

Due to its promising performance at both element and finite element levels, SANISAND-MSu model is employed in this study to further investigate the dynamic response of different structures used in offshore energy. For completeness, a conceptual introduction to the SANISAND-MSu model is presented together with the model performance in reproducing element-level cyclic response. Then simulation results for quasistatic cyclic response of a typical offshore monopile are presented using SANISAND-MSu model in 3D FE analysis with emphasis on the pore-water pressure effects. In the end, dynamic responses of two types of structures in offshore energy are studied. The selected structures include wind turbines on an extra-large 'XL' monopile and subsea manifolds on mudmat foundation. The focus of the dynamic analyses is: (1) the combined effect of earthquake shaking and wind load on XL monopile, and (2) the effect of mass distribution on the earthquake response of mudmat.

2 SANISAND-MSu Model

SANISAND-MSu model was developed mainly for the simulation of undrained cyclic behaviour of sand with the objective of accurate simulation of pore pressure generation and cyclic mobility. Some key features of the SANISAND-MSu model are summarized in this section.

SANISAND-MSu model adopts bounding surface plasticity theory and is built based on the SANISAND04 model proposed by Dafalias and Manzari [18]. Sand fabric is taken into consideration in SANISAND-MSu model through the so-called 'memory-surface'. Fabric evolution takes place during both plastic contraction and dilation. Compared with SANISAND04, within which sand fabric only evolves and activates during the post-dilative stage, SANISAND-MSu simulates the effects of soil fabric in the whole plastic deformation stage. Such an improvement allows realistic simulation of sand cyclic behaviour under both drained and undrained conditions [21, 23]. The model follows critical soil state theory. Soil contractive/dilative behaviour is determined through the well-established 'state-parameter' concept [25]. Such features allow the model to simulate the behaviour of a given sand over wide density range with a single set of model parameters.

SANISAND-MSu model includes four loci. Figure 1 illustrates the model loci in the deviatoric stress ratio plane:

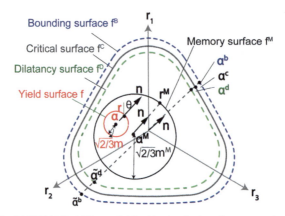

Fig. 1. SANISAND-MSu model loci in the deviatoric stress ratio plane.

1. A circular yield surface (f) that defines a pure elastic range;
2. An Argyris-shaped bounding surface (f^B) that encloses the admissible soil state. The bounding surface is associated with critical state through state parameter;
3. An Argyris-shaped dilatancy surface (f^D) that distinguishes contractive state from dilative soil state. The dilatancy surface is defined through state parameter;
4. A circular memory surface (f^M) that evolves during plastic deformation and is related to the evolution of stress-induced anisotropy.

The memory surface tracks the evolution of soil fabric during the plastic straining. SANISAND-MSu model [23] improves the original SANISAND-MS model [21] on two aspects, namely, proper simulation of sand fabric evolution history effects and stress-ratio effects.

The SANISAND-MSu model uses kinematic hardening rule and adopts non-associated flow rule. Soil stiffness is determined through distances between (a) yield surface and memory surface and (b) yield surface and bounding surface. Besides, the stress ratio effects are introduced in the hardening coefficient. Such modifications allow SANISAND-MSu to accurately predict the cycle-by-cycle pore-water pressure accumulation. Fabric evolution history and lode angle effects are introduced in the dilatancy coefficient to predict realistic strain accumulation in the cyclic mobility regime.

2.1 Stress-Strain and Pore Pressure Accumulation

The performance of SANISAND-MSu model is investigated here by comparing the results from the undrained cyclic triaxial tests [26] with model simulation results. The selected sand is Karlsruhe fine sand, with the following soil index properties: maximum void ratio $e_{max} = 1.054$, $e_{min} = 0.677$, soil uniformity coefficient $C_u = \frac{D_{60}}{D_{10}} = 1.5$ and medium particle diameter $D_{50} = 0.14$ mm. SANISAND-MSu parameters are calibrated against drained monotonic, undrained monotonic and undrained cyclic triaxial tests as summarized in [23]. The parameters are summarized in Table 1.

Table 1. SANISAND-MSu model parameters for Karlsruhe fine sand: calibration against triaxial test results from [26].

Elasticity		Critical state					Plastic modulus		
G_0	ν	M_c	c	λ_c	e_0	ξ	h_0	c_h	n^b
95	0.05	1.35	0.81	0.55	1.035	0.36	7.6	0.97	1.2
Yield	Dilatancy					Memory surface			
m	A_0	n^d	β_1	β_2	k	μ_0	ζ	w_1	w_2
0.01	0.74	1.79	4	3.2	2	65	0.0003	2.5	1.5

Figure 2 compares the experimental results (Fig. 2a–b) with the simulation results from both SANISAND-MSu model (Fig. 2c–d) and SANISAND04 model (Fig. 2e–f). The test conditions are: relative density $D_r = 80\%$, initial confining pressure $p = 200$ kPa, cyclic amplitude $q^{ampl} = 50$ kPa.

In general, both SANISAND-MSu and SANISAND04 models capture the main features of the undrained cyclic triaxial behaviour of sand – i.e., reduction of mean effective stress p with loading cycles in the pre-dilative regime and cyclic mobility behaviour (repetitive increasing and decreasing mean effective stress butterfly loop with cycles). Note that SANISAND04 clearly overestimate the reduction of p for each loading cycle while SANISAND-MSu predicts more accurately the cycle-by-cycle pore-water pressure evolution. This is also demonstrated by Fig. 3 which shows the pore water pressure ratio r_u (defined as $1 - \sigma_v/\sigma_{v0}$, where σ_v is the vertical effective stress and σ_{v0} is the initial vertical effective stress prior to shearing) against number of loading cycles N. SANISAND-MSu model predicts 76 loading cycles to trigger the initial liquefaction (i.e., the first time p approaches 0), which agrees well with the experimental result (indicating 74 loading cycles to trigger initial liquefaction). However, SANISAND04 model suggests only 5 cycles to trigger initial liquefaction.

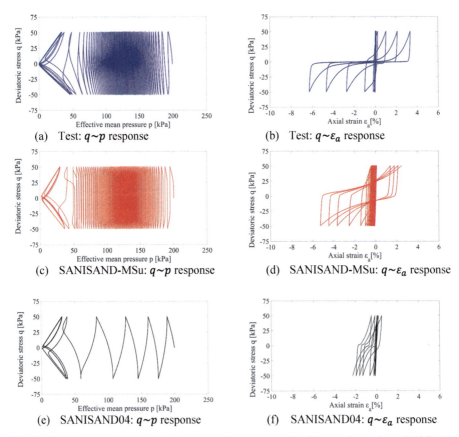

Fig. 2. Undrained cyclic triaxial tests: comparison among: (a)–(b) experimental result [26], (c)–(d) SANISAND-MSu simulation result and (e)–(f) SANISAND04 simulation result.

Another important improvement of SANISAND-MSu is that it can predict realistic strain accumulation in the cyclic mobility regime. As indicated in Fig. 2b, under stress-controlled undrained triaxial condition, axial strain accumulate in both positive and negative directions. The phenomenon has been properly captured by SANISAND-MSu (Fig. 2d). However, SANISAND04 model suggests an accumulation of axial strain only in the negative direction (Fig. 2f).

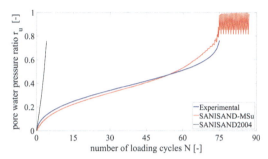

Fig. 3. Evolution of pore-water pressure ratio against number of cycles: comparison between experimental result [26], SANISAND-MS simulation and SANISAND04 simulation results.

2.2 Stiffness Reduction and Damping

While the ability for prediction of the pore-pressure generation is an important feature of a constitutive model, it is also important to have knowledge about the stiffness variation and hysteretic damping during cyclic loading. These two characteristics steer the dynamic response of the soil domain and soil-structure dynamics. A common method to assess these features is through the variation of the stiffness reduction G/G_{max} and damping with shear strain γ, as exemplified by Fig. 4a and Fig. 4b, respectively. The shear modulus G is the peak-trough soil secant shear modulus during a stress cycle and G_{max} is the initial shear modulus. Realistic reproduction of the variation of soil stiffness and damping is thus an important aspect to evaluate a constitutive model.

The performance of SANISAND-MSu in terms of shear modulus reduction (Fig. 4a) is investigated by comparing the model simulation results with two empirical results from the literature [27, 28] for different relative densities D_r in the range 40%–80%. The comparison suggests that while the trend is captured properly, SANISAND-MSu predicts greater shear stiffness for the same shear strain level.

The material damping ratio is defined as the ratio between the dissipated energy and the maximum potential energy during a load cycle. Figure 4b shows the damping ratio predicted by SANISAND-MSu for the first stress cycle but sheared to different shear strain levels. Different relative densities $D_r = 40\%$, 60% and 80% are selected. The simulation results are compared with the empirical data by Darendeli [28] and Masing rule. Masing rule generally yields large damping ratios at medium to large shear strain levels – as indicated by the comparison between Masing rule result and the rest results. SANISAND-MSu model agrees better with Darendeli's empirical results, which

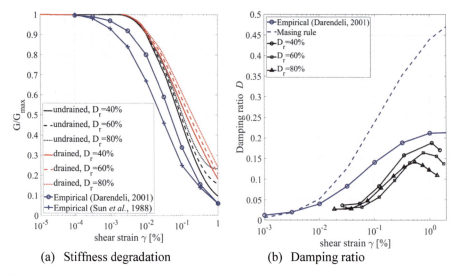

Fig. 4. SANISAND-MSu model simulation results for (a) stiffness reduction and (b) damping ratio as function of shear strain. The simulations are performed under load controlled undrained cyclic DSS condition, with $\sigma_{v0} = 200$ kPa.

gives confidence in using SANISAND-MSu model in earthquake analyses. Besides, SANISAND-MSu predicts a reduction of the damping ratio at relatively high strain levels – larger than typically 0.5% in this case. Undrained sand exhibits such performance after entering the cyclic mobility or soil dilative behavior at larger strain levels, as also observed in lab tests results [29].

3 Application of SANISAND-MSu in Offshore Energy Design

This section discusses representative results for the earthquake response of offshore energy structures founded on liquefaction-prone ground. Implicit 3D Finite Element modelling is performed using the open-source FE platform OpenSEES with the SANISAND-MSu as the user-defined material. Monopile response subject to quasi-static cyclic loading is first presented in Sect. 3.1 with focus on the effect of pore-water pressure. Section 3.2 is devoted to the dynamic response of two types of offshore energy-related structures, namely, wind turbines founded on monopiles and manifolds resting on mudmats. Karlsruhe fine sand, as described in Sect. 2, was selected for the simulation purposes with the model parameters listed in Table 1.

3.1 Monopile Response Under Quasi-Static Cyclic Loading

Effects of pore-water pressure on the cyclic response of monopile was studied using the 3D FE model presented in Fig. 5. The monopile has the following geometry features: pile embedded length $L_{pile} = 25$ m, pile outer diameter $D_{pile} = 5$ m (hence, aspect ratio $L_{pile}/D_{pile}=5$, which is comparable to a monopile supporting 5-MW offshore wind turbine) and pile wall thickness $= 10$ cm. The soil profile is assumed to have a uniform relative density $D_r = 80\%$

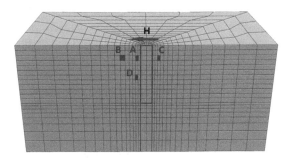

Fig. 5. FE model domain for quasi-static analysis.

The monopile is assumed to be subject to symmetrical sinusoidal load cycles that are applied at the pile head (soil surface level), which means the load eccentricity is zero. Such a loading does not represent a realistic condition for monopiles of wind turbines. However, the objective is to better understand the response of large monopiles when their cyclic motions generate pore pressure in the soil.

To highlight the effects of pore-water pressure, two different simulations are performed: (1) monopile cyclic response under fully drained condition – which is realized by setting large permeability: 10^{10} m/s; (2) monopile cyclic response under fully undrained condition, in which case the permeability is set to 10^{-10} m/s. The number of loading cycles is limited to 10 which is a realistic representation of moderate to strong earthquake loading [31]. Detailed FE model settings can be found in [24].

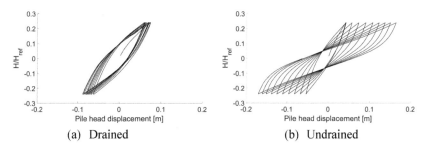

(a) Drained   (b) Undrained

Fig. 6. Simulated pile force-displacement response for: (a) drained and (b) undrained sand.

Figure 6 compares the pile head load-displacement results obtained for the two drainage conditions. The normalized load ratio H/H_{ref} is selected to represent the load level, with H_{ref} being the horizontal load applied at the pile head to induce a displacement $0.1D_{pile}$ at the pile head for drained condition. Figure 6a shows that for the first loading cycle, $N = 1$, a non-closed load-displacement loop is formed due to the soil inhomogeneity induced by the initial virgin loading (when the initial loading up to H reaches its maximum value for the first time). Pile head displacement decreases from 0.07 m after the initial virgin loading to about 0.05 m after the first re-loading at $H = H_{max}$ (i.e., $N = 2$). For the following loading cycles, the pile head displacement at $H = H_{max}$ keeps increasing slowly. For the undrained simulation (Fig. 6b), rapid increase of pile displacement or progressive softening pile response, is induced with loading cycles.

The distinct behaviour captured in the two simulations can be attributed to different mechanisms, including:

(1) the generation and accumulation of the pore-water pressure in undrained case progressively reduces soil stiffness and thus leads to larger strain in the surrounding soils at large N. The surrounding soil in general shows stiffer behaviour in drained simulation.

(2) soil fabric evolves differently under the two drainage conditions. Under undrained condition, the rapid reduction of the mean effective stress accelerates the soil entering the dilative phase before the load increment reversal. Upon repeated unloading and reloading, the state of soil element likely alters from dilative to contractive to dilative. Such a process largely erases the load-induced fabric. In other words, sand fabric stiffening effects in the undrained simulation is smaller than that in drained simulation in the scope of current simulation.

Figure 7 presents the soil resistance against pile displacement response (that is, cyclic $p - y$ response) simulated using SANISAND-MSu enhanced implicit 3D FE. Two different representative depths say $z = 2.5 D_{pile}$ and $0.3 D_{pile}$, are studied.

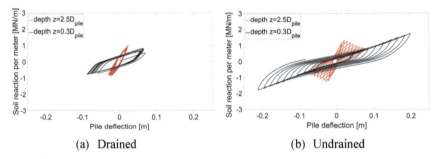

(a) Drained   (b) Undrained

Fig. 7. Soil resistance against pile displacement: comparison between (a) drained and (b) undrained simulation results using SANISAND-MSu model.

For the drained simulation, (Fig. 7a), stable cyclic $p - y$ loops are obtained after the second loading cycle for both depths. For the undrained case, shown in Fig. 7b, the soil reaction first increases then decease at deep soil layer ($z = 2.5 D_{pile}$). However, at shallow depth ($z = 0.3 D_{pile}$), the soil reaction keeps increasing with loading cycles, which might be the direct result of easy soil dilatancy being triggering under undrained condition when initial stress level is low.

Figure 8 presents the stress paths ($q \sim p$ response, Fig. 8a and Fig. 8c) and stress-strain responses ($\tau \sim \gamma$ response, Fig. 8b and Fig. 8d) for the two selected soil elements

A and C as indicated in Fig. 5. Different colors represent for the lode angle defined within 0°–60° as indicated by the color bar. Soil element A lies on the fore side (the left-hand side of the monopile in Fig. 5 of the monopile. It enters cyclic mobility regime after only one loading cycle, as shown in Fig. 8a, with shear strain γ, presented in see Fig. 8b, evolving mostly at moments lode angle equals to 0° (corresponding to triaxial compression) or 60° (triaxial extension). For element C that lies on the rear side (i.e., the right-hand side of the pile in Fig. 5) of the pile, more loading cycles are required to trigger the initial liquefaction (Fig. 8c), despite of the same initial mean effective stress. The accumulated strain is much smaller in element C compared with that in element A. Besides, the lode angle that corresponds to the rapid γ evolution (Fig. 8d) differs from that as indicated in Fig. 8b.

Understanding the local soil behaviour and pore-pressure generation are believed to be the key in the estimation of pile tilting due to cyclic loading and during earthquake shaking combined with environmental loading.

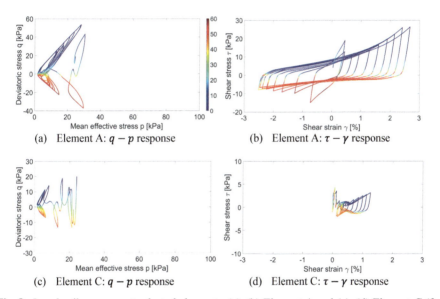

Fig. 8. Local soil response at selected elements: (a)–(b) Element A and (c)–(d) Element C (for element locations, see Fig. 5).

3.2 Dynamic Response of Offshore Structures

In this section, the earthquake responses of structures founded on mudmat and monopiles in liquefiable soil are investigated through 3D nonlinear SANISAND-MSu enhanced dynamic analyses. In total, four simulation cases are included: two for a mudmat foundation for a subsea manifold and two for a monopile foundation used for an offshore wind turbine. The recorded earthquake, Kobe-L [14, 31], was used for the simulation purpose as indicated in Fig. 9. Uniform sand profile with relative density $D_r = 50\%$ and permeability $k_{soil} = 10^{-5}$ m/s was used in the analyses in this section. The earthquake shaking was applied to the base of the models. The tied boundary condition was employed to link the two side boundaries as successfully employed in [14]. The earthquake response is studied through pore-water pressure evolution as well as the foundation motions.

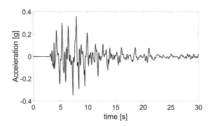

Fig. 9. Acceleration time history for horizontal component of Kobe-L earthquake.

Dynamic Response of Wind Turbine on XL Monopile

The so-called 'XL monopile' (monopiles with diameter exceeding 7 m) has become a practical offshore foundation solution in response to the demand for economic construction of larger and more efficient wind turbines. In the present study, a 3D FE model with monopile diameter equal to 9 m was constructed as illustrated in Fig. 10.

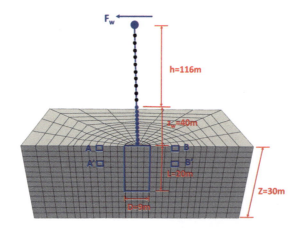

Fig. 10. FE model domain with wind turbine on XL monopile

A monopile with an embedded length $L = 20$ m, which corresponds to an aspect ratio $\frac{L}{D_{pile}} = 2.22$, has been adopted for the analyses. This aspect ratio is currently considered the limit for plausible monopile foundations. The other parameters of the structure are the wall thickness t_{pile} of monopile = 10 cm (giving $\frac{t_{pile}}{D_{pile}} = 0.011$), pile length above soil surface (transition piece) = 40 m (in sea water), the tower length = 116 m, and mass density of tower material = 7850 kg/m³. In addition, a lumped mass of 700,000 kg is placed at the top of the tower to represent the masses of rotor and nacelle and a mass of 500,000 kg was distributed in 23 m above the sea level to represent the mass of the transition piece and added hydrodynamic mass. A Rayleigh damping ratio of 2% was adopted at the frequencies 0.35 Hz and 1.75 Hz. A relatively coarse mesh was used for the soil domain without performing mesh sensitivity checks. The same mesh size was shown to be satisfactory for high-level assessments of the response in [14]. For simulations involving wind load, a static lateral load of 700 kN is applied at the hub.

The distributions of the pore-water pressure ratios at the end of the earthquake shaking ($t = 30$ s) are summarized in Fig. 11. The SANISAND-MSu simulation result (Fig. 11 b) is compared with that using SANISAND04 model (Fig. 11a) to highlight the influence of the adopted constitutive model on the pore-water pressure evolution in the soil domain during. The loading condition is the combined earthquake shaking (horizontal motion applied at the soil base) and a constant wind load applied at the hub as indicated in Fig. 10.

Both simulations clearly demonstrate soil liquefaction on the fore side of the monopile (the left hand-side of the monopile as indicated in Fig. 10) with the SANISAND04 simulation indicating a significantly larger liquefied soil zone. None of the simulations indicate any noteworthy pore-water pressures inside the monopiles, as also observed and discussed in [14]. On the rear side of the monopile (i.e., right-hand side of the model), only a limited liquefaction zone is observed away from the monopile in SANISAND-MSu simulation case. However, SANISAND04 simulation results suggests no significant difference between the two sides which could be because free-field liquefaction is more prevalent when using SaniSAND04. Negative pore-water pressure ratios (r_u) are detected in shallow soil layers and below soil tip on the fore side of the monopile in both simulation cases. For SANISAND-MSu simulation, negative r_u is also detected on the rear side of the monopile in the shallow layers.

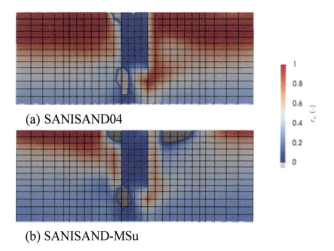

(a) SANISAND04

(b) SANISAND-MSu

Fig. 11. Distribution of pore-water pressure ratio. Simulations using (a) SANISAND04 and (b) SANISAND-MSu. Loading case: combined earthquake shaking and wind load.

Figure 12 presents the time-history of the pore-water pressure ratio at selected elements A′ and B′ (ref. Fig. 10). For both elements, SANISAND-MSu simulations indicate smaller peak r_u values compared with SANISAND04. The different pore-water distribution in Fig. 11 and the time evolutions in Fig. 12 are the direct consequences of the different hardening laws in the two models. As explained earlier, SANISAND-MSu model accounts for the load-induced sand fabric stiffening in the pre-dilative soil response and thus delays the occurrence of initial liquefaction.

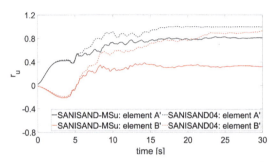

Fig. 12. Time history of pore-water pressure ratio at selected elements A′ and B′ (ref. Fig. 10) using both SANISAND04 and SANISAND-MSu for combined earthquake and wind load.

Figure 13 summarizes the SANISAND-MSu simulation results of r_u evolutions at the four representative soil elements (indicated in Fig. 10), under combined earthquake shaking and wind load. On the fore side of the monopile (element A and A′), relatively large r_u values are obtained for both shallow soil (element A) and deeper soil (element A′). An opposite conclusion is drawn on the rear side of the monopile: initial liquefaction is triggered in neither element with the element at shallower layer showing smaller r_u compared with the element at deeper layer.

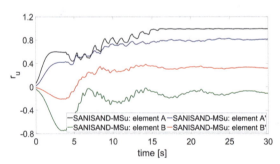

Fig. 13. Time history of pore-water pressure ratio at selected elements A, B, A′ and B′ in Fig. 10 using SANISAND-MSu for combined earthquake shaking and wind load.

As mentioned above, the accumulation of pore-water pressure alters the soil stiffness and affects the foundation behavior. This conclusion can be confirmed by the plots in Fig. 14, which display the time histories of the hub horizontal displacement (Fig. 14a) and the pile head rotation (Fig. 14b). As expected, SANISAND04 simulation generates larger hub displacement U_x and larger pile head rotation θ_r. The observed differences in rotations can be explained by the same principles described in Sect. 2.1.

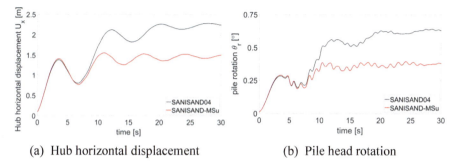

(a) Hub horizontal displacement   (b) Pile head rotation

Fig. 14. Comparison of time histories of (a) hub horizontal displacement and (b) pile head rotation using SANISAND04 and SANISAND-MSu models for combined earthquake and wind load.

These results have important design implications. Following the pile rotation limit specified in DNV-OS-J101 [30] for the normal drivetrain operation, SANISAND-MSu predicts rotations within the acceptable limit while SANISAND04 indicates unacceptable performance.

For reference, the distribution of r_u under only earthquake shaking is presented in Fig. 15. Pore-water pressure ratio is nearly symmetrical on both sides of the monopile. No negative r_u is detected which is different from the case with wind load and tower inertia as observed in Fig. 11b. Therefore, the simultaneous loading from earthquake and wind lead to a more critical condition regarding liquefaction.

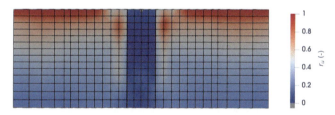

Fig. 15. Distribution of pore-water pressure ratio for earthquake shaking only.

The effect of wind load on the vertical displacement and rotation of the monopile are presented in Fig. 16. When subjected to solely earthquake shaking, the accumulated vertical displacement (Fig. 16a) after the shaking is nearly zero; this is significantly smaller than the displacement in the presence of the wind load (9.5 cm in this case). However, for the two loading conditions, the amplification of the vertical displacement during shaking are comparable. Similar conclusion can be drawn from the pile rotation response under the two loading conditions as indicated in Fig. 16b; namely, there is nearly no permanent pile head rotation accumulated after the shaking for pure earthquake loading, while for combined wind load and earthquake shaking the permanent tilt of the monopile reaches nearly 0.4° which is very close to the normal operational limit set by DNV-OS-J101 (0.5°) [32].

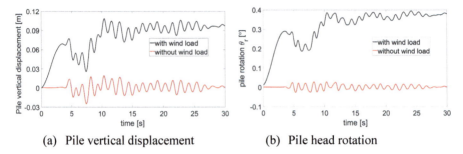

(a) Pile vertical displacement   (b) Pile head rotation

Fig. 16. Pile response subject to earthquake shaking with and without wind load: (a) vertical displacement and (b) rotation.

Dynamic Response of Manifold on Mudmat

Figure 17 illustrates the three-dimensional models used for the simulation of the response of a manifold on a mudmat. The mudmat has a dimension (width × length × height) = 15 m × 15 m × 5 m; however, due to symmetry of the simulated case, only half of the width, that is, 7.5 m, is modelled), the same as that adopted in [20]. The dimensions of the soil domain are indicated in the figure.

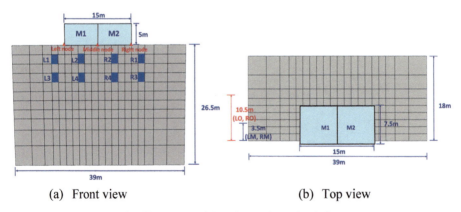

(a) Front view   (b) Top view

Fig. 17. FE model domain: mudmat simulation.

An earlier study by Esfeh et al. [20] has indicated generally small rotations for typical symmetrical manifolds during earthquake shaking and in presence of liquefaction. The objective of the analyses in this section is to study the effect of manifold's unsymmetrical configuration in a simple model. To this end, two simulations were performed: a) mudmat with symmetric manifold, and b) mudmat with unsymmetrical mass distribution of manifold (denoted as symmetric case and unsymmetrical case in the following). For the symmetric case, the masses for parts $M1$ and $M2$ (See Fig. 17) are both 60, 000 kg (and therefore a total mass of 120,000 kg). For the unsymmetrical case, $M1$ has a mass of 80,000 kg, while $M2$ has a mass of 40,000 kg (same total weight as the symmetric case). The same Rayleigh damping parameters were used in this model.

The distribution of the pore-water pressure ratio at times $t = 20$ s and 30 s are presented in Fig. 18. In general, simulations of both symmetric and unsymmetrical cases indicate that liquefaction can be triggered at shallow soil layers, with pore-water pressure ratios in the elements beneath the mudmat tending to be smaller than those in the outer elements.

At $t = 20$ s, different r_u distributions are achieved for the two simulations. For the symmetric case, r_u distribute almost symmetrically. While for the unsymmetrical simulation case, large pore-water pressure ratio (larger than 0.8) is triggered in elements not only at shallow soil layer, but also in the area located slightly deeper and outside $M2$ (the part with smaller mass). A possible reason can be the different initial mean effective stresses at different zones if the manifold mass is distributed unsymmetrically. More specifically, on the side with smaller mass, the initial mean effective stress is smaller than the other side; therefore, the soil elements on the lighter side tend to accumulate more pore-water pressure under the same shaking and inertial load of the manifold. At $t = 30$ s, the differences in r_u distribution for two simulation cases become negligible.

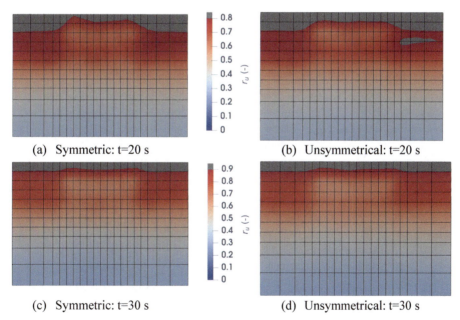

(a) Symmetric: t=20 s   (b) Unsymmetrical: t=20 s

(c) Symmetric: t=30 s   (d) Unsymmetrical: t=30 s

Fig. 18. Distribution of pore-water pressure ratios at time t = 20 s and t = 30 s. for symmetric and unsymmetrical cases (refer to figure texts for cases).

It is also instructive to compare the time histories of r_u for both the symmetric case (Fig. 19a) and the unsymmetrical case (Fig. 19b) at different representative elements $L1$–$L4$ and $R1$–$R4$ as indicated in Fig. 17a.

For the symmetric case (Fig. 19a), elements lying at shallow soil layer ($L1$, $L2$, $R1$ and $R2$) enter the liquefaction state at the end of the shaking. However, elements that lie outside the mudmat edge (elements L_1 and R_1) enter liquefaction much earlier than the elements within the mudmat edge (elements L_2 and R_2). For elements at deeper soil layer ($L3$, $L4$, $R3$ and $R4$), liquefaction is triggered.

For unsymmetrical case (Fig. 19b), soil elements at relatively deeper depth ($L3$, $L4$, $R3$ and $R4$) show almost same trend as in the symmetric case. At shallow layer, element $R1$ shows difference in r_u evolution when compared with $L1$ response – the element located outside $M1$ edge, that is, element $L1$ enter liquefaction earlier than the element located outside $M2$ edge (element $R1$).

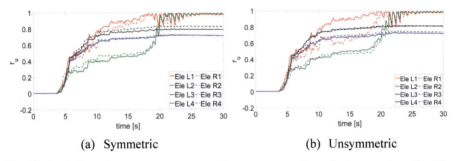

(a) Symmetric	(b) Unsymmetric

Fig. 19. Evolution of pore-water pressure ratio at representative soil elements (see Fig. 17), simulation cases of: (a) symmetric manifold, and (b) unsymmetrical manifold.

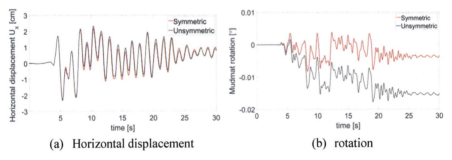

(a) Horizontal displacement	(b) rotation

Fig. 20. Time history of horizontal displacement and rotation of mudmat during the Kobe-L earthquake shaking for symmetric and unsymmetrical manifolds.

Finally, Fig. 20 compares the responses of the mudmats for the symmetric and unsymmetrical cases. As expected, for the horizontal displacement (Fig. 20a), no significant differences are observed for the two simulation cases. For the rotation, on the other hand (Fig. 20b), the unsymmetrical case experiences considerably larger mudmat rotation compared with the symmetric case. This is an important consideration in design of heavy subsea facilities if a simple (so-called stick) model of the manifold is adopted for the global response of the system. These results indicate that one should properly capture the unsymmetrical stress distribution under the mudmat in the analyses.

4 Conclusion

The features and performance of SANISAND-MSu model in reproducing the soil element response has been presented in this paper. The model is shown to be reasonable in reflecting the soil undrained cyclic behavior, in particular, in terms of cycle-by-cycle pore water pressure accumulation and progressive strain accumulation in the cyclic mobility regime. Subsequently, SANISAND-MSu model is implemented into OpenSEEs for analysis of two types of offshore energy structure.

Pore-water pressure effects on monopile lateral behaviour is first studied under quasi-static cyclic loading. The simulation results indicate progressive decreasing of pile stiffness due to increasing of the pore-water pressure in the surrounding soil. The simulated cyclic $p - y$ response is also presented. Local soil elements responses are extracted. The $q \sim p \sim \theta$ (mean effective stress ~ deviatoric stress ~ Lode angle) and $\tau \sim \gamma \sim \theta$ (shear stress ~ shear strain ~ Lode angle) relationships of the selected elements suggests that for soil elements located at the symmetry plane, shear strain is mostly generated when the mean effective stress is small and the Lode angle enters 'triaxial compression' and 'triaxial extension' states.

Dynamic response of an XL monopile (pile diameter of 9 m with aspect ratio $\frac{L_{pile}}{D_{pile}} = 2.22$) has been studied. The horizontal component of Kobe-L earthquake shaking is applied at the base of the soil. The simulation results indicate that soil liquefaction mainly occur at shallow soil layers and outside of the monopile. For soils inside the monopile, negligible increase in pore-water pressure can be observed. For the loading case with only earthquake shaking, very small accumulated pile displacement (or pile head rotation) is achieved. While if the earthquake shaking is combined with the static load wind applied at turbine hub (which is the most probable load case in reality), the loading can significantly increase the pile head rotation. Besides, in the combined wind load and earthquake shaking, the soil elements located in the fore side of the monopile are more likely to liquefy.

Effect of mass distribution on the dynamic response of a manifold over a mudmat is also studied. Soil elements beneath the mudmat but inside the foundation edge can might need longer shaking to enter liquefaction. For the unsymmetrical mass distribution, during earthquake shaking, soil elements located beneath the lighter side and outside the mudmat edge show slightly higher pore-water pressure ratios. However, such a difference progressively vanishes after the strong shaking. With the same total weight, the unsymmetrical weight distribution leads to a larger foundation rotation compared with the case where the weight is distributed symmetrically.

The simulations in this work aim to give general impression on the cyclic and dynamic response of two typical offshore structures using SANISAND-MSu enhanced 3D FE analysis approach. Detailed calibration of model parameters and more rigorous consideration of the model dynamic properties need to be undertaken together with more refined FE mesh, are required to achieve more accurate results in actual design cases. Regardless, sensitivity analyses are highly encouraged considering the uncertainties in the soil parameters and the computational models.

References

1. Europe Wind: Offshore wind in Europe: Key trends and statistics 2019 (2020)
2. IEA: Offshore wind outlook 2019. Technical report (2019)
3. Bransby, M., Randolph, M.: Combined loading of skirted foundations. Géotechnique **48**(5), 637–655 (1998)
4. Byrne, B.W., Houlsby, G.T.: Experimental investigations of the response of suction caissons to transient combined loading. J. Geotech. Geoenviron. Eng. **130**(3), 240–253 (2004)
5. Byrne, B., Houlsby, G.: Assessing novel foundation options for offshore wind turbines. In: World Maritime Technology Conference, London (2006)
6. LeBlanc, C., Houlsby, G.T., Byrne, B.W.: Response of stiff piles in sand to long-term cyclic lateral loading. Géotechnique **60**(2), 79–90 (2010)
7. Wang, X., Yang, X., Zeng, X.: Seismic centrifuge modelling of suction bucket foundation for offshore wind turbine. Renew. Energy **114**, 1013–1022 (2017)
8. Richards, I., Bransby, M., Byrne, B., Gaudin, C., Houlsby, G.: Effect of stress level on response of model monopile to cyclic lateral loading in sand. J. Geotech. Geoenviron. Eng. **147**(3), 04021002 (2021)
9. Cuéllar, P., Mira, P., Pastor, M., Merodo, J.A.F., Baeßler, M., Rücker, W.: A numerical model for the transient analysis of offshore foundations under cyclic loading. Comput. Geotech. **59**, 75–86 (2014)
10. Tasiopoulou, P., Chaloulos, Y., Gerolymos, N., Giannakou, A., Chacko, J.: Cyclic lateral response of OWT bucket foundations in sand: 3D coupled effective stress analysis with Ta-Ger model. Soils Found. **61**(2), 371–385 (2021)
11. Liu, H.Y., Kementzetzidis, E., Abell, J.A., Pisanò, F.: From cyclic sand ratcheting to tilt accumulation of offshore monopiles: 3D FE modelling using SANISAND-MS. Géotechnique **72**(9), 753–768 (2022)
12. Chaloulos, Y.K., Tsiapas, Y.Z., Bouckovalas, G.D.: Seismic analysis of a model tension leg supported wind turbine under seabed liquefaction. Ocean Eng. **238**, 109706 (2021)
13. Kaynia, A.M.: Seismic considerations in design of offshore wind turbines. Soil Dyn. Earthq. Eng. **124**, 399–407 (2019)
14. Esfeh, P.K., Kaynia, A.M.: Earthquake response of monopiles and caissons for offshore wind turbines founded in liquefiable soil. Soil Dyn. Earthq. Eng. **136**, 106213 (2020)
15. Jostad, H.P., Dahl, B.M., Page, A., Sivasithamparam, N., Sturm, H.: Evaluation of soil models for improved design of offshore wind turbine foundations in dense sand. Géotechnique **70**(8), 682–699 (2020)
16. Corciulo, S., Zanoli, O., Pisanò, F.: Transient response of offshore wind turbines on monopiles in sand: role of cyclic hydro–mechanical soil behaviour. Comput. Geotech. **83**, 221–238 (2017)
17. Yang, Z., Elgamal, A.: Multi-surface cyclic plasticity sand model with lode angle effect. Geotech. Geol. Eng. **26**(3), 335–348 (2008)
18. Kementzetzidis, E., Corciulo, S., Versteijlen, W.G., Pisano, F.: Geotechnical aspects of offshore wind turbine dynamics from 3D non-linear soil-structure simulations. Soil Dyn. Earthq. Eng. **120**, 181–199 (2019)
19. Dafalias, Y.F., Manzari, M.T.: Simple plasticity sand model accounting for fabric change effects. J. Eng. Mech. **130**(6), 622–634 (2004)

20. Esfeh, P.K., Govoni, L., Kaynia, A.M.: Seismic response of subsea structures on caissons and mudmats due to liquefaction. Mar. Struct. **78**, 102972 (2021)
21. Liu, H.Y., Abell, J.A., Diambra, A., Pisanò, F.: Modelling the cyclic ratcheting of sands through memory-enhanced bounding surface plasticity. Géotechnique **69**(9), 783–800 (2019)
22. Corti, R., Diambra, A., Muir Wood, D., Escribano, D.E., Nash, D.F.: Memory surface hardening model for granular soils under repeated loading conditions. J. Eng. Mech., 04016102 (2016)
23. Liu, H.Y., Diambra, A., Abell, J.A., Pisanò, F.: Memory-enhanced plasticity modeling of sand behavior under undrained cyclic loading. J. Geotech. Geoenviron. Eng. **146**(11), 04020122 (2020)
24. Liu, H.Y., Kaynia, A.M.: Characteristics of cyclic undrained model SANISAND-MSu and their effects on response of monopiles for offshore wind structures. Géotechnique, 1–16 (2021)
25. Been, K., Jefferies, M.G.: A state parameter for sands. Géotechnique **35**(2), 99–112 (1985)
26. Wichtmann, T., Triantafyllidis, T.: An experimental database for the development, calibration and verification of constitutive models for sand with focus to cyclic loading: Part I — tests with monotonic loading and stress cycles. Acta Geotech. **11**(4), 739–761 (2016)
27. Sun, J.I., Golesorkhi, R., Seed, H.B.: Dynamic moduli and damping ratios for cohesive soils. Earthquake Engineering Research Center, University of California Berkeley (1988)
28. Darendeli, M.B.: Development of a new family of normalized modulus reduction and material damping curves. The University of Texas at Austin (2001)
29. Blaker, Ø., Andersen, K.H.: Cyclic properties of dense to very dense silica sand. Soils Found. **59**(4), 982–1000 (2019)
30. Ramirez, J., et al.: Site response in a layered liquefiable deposit: evaluation of different numerical tools and methodologies with centrifuge experimental results. J. Geotech. Geoenviron. Eng. **144**(10), 04018073 (2018)
31. Kramer, S.L.: Geotechnical earthquake engineering. Prentice-Hall Civil Engineering and Engineering Mechanics Series (1996)
32. DNV: Design of offshore wind turbine structures, vol. DNV-OS-J101. Det Norske Veritas, Høvik, Norway (2016)

Factors Affecting Liquefaction Resistance and Assessment by Pore Pressure Model

Mitsu Okamura$^{(\boxtimes)}$ (iD)

Ehime University, Matsuyama, Ehime 790-8522, Japan
okamura@cee.ehime-u.ac.jp

Abstract. Assessment of the liquefaction resistance of clean sand still involves considerable uncertainties, which are a current research topic in the field of soil liquefaction. Factors to be considered include shaking history, overconsolidation, degree of saturation and partial drainage. The effects of these factors on liquefaction resistance have been studied in the laboratory and empirical relationships are derived. This paper describes the development of pore pressure generation model similar to that of Martin et al. [18] but based on stress-controlled triaxial tests. The effects of various factors on the pore pressure generation and liquefaction resistance of clean sand are explained using the unique index of volumetric strain. The model is verified through comparisons with the results of laboratory tests. It is confirmed that the plastic volumetric strain accumulated in sand either by drained or undrained loading dominates the increase in liquefaction resistance of pre-sheared, overconsolidated and unsaturated sand. The model provides a better understanding of the physical processes leading to liquefaction of saturated and unsaturated sand with and without stress histories.

Keywords: Excess pore pressure · Liquefaction resistance · Sand · Volumetric strain · Degree of saturation · Stress history

1 Introduction

Resistance of sand to liquefaction during undrained cyclic shearing has been investigated for several decades. Significant progress has been made in understanding the fundamental mechanism and influence of various factors on liquefaction resistance, including soil density, grain fabric, testing apparatus, stress anisotropy, all listed by Seed [23] (cited in Finn [5]). Although considerable research efforts have been devoted to the topic, assessment of liquefaction resistance of clean sand still involves considerable uncertainties and the effects of influential factors are a current topic in the field of soil liquefaction. Factors include loading history (shaking history and over consolidation), degree of saturation and partial drainage. The effects of these factors on liquefaction resistance have been examined in several studies, mostly in the laboratory.

With regard to the effect of shaking history, it is recognized that sand in seismically active areas has a higher resistance to liquefaction than that in calm areas. Small pre-shearing events during earthquakes are believed to make the sand more resistant to liquefaction (Dobry and Abdoun [2], and Towhata et al. [31]). Observations based on field evidences have been corroborated by laboratory test results. It has been reported that cyclic pre-shearing enhanced the liquefaction resistance of sand, first by Finn et al. [4], followed by numerous researchers including Singh et al. [26], Tokimatsu and Hosaka [29], and more recently Kiyota et al. [15], Goto and Towhata [7], Wu et al. [36], Toyota and Takada [32], Nelson and Okamura [19] and Wu and Kiyota [37]. For example, Tokimatsu et al. [30] conducted cyclic triaxial tests on medium dense and dense Niigata sand and found that drained pre-shearing of up to ten thousand cycles tripled the liquefaction resistance, even though the increase in the relative density was only up to several per cent. Results of similar tests in the literature consistently indicate that the liquefaction resistance increases with the number of cycles and cyclic stress ratio of the pre-shearing. However, none of the testing parameters alone quantified the improvement of the liquefaction resistance. Finn et al. [4] reported that pre-shearing with shear strain amplitude larger than 0.5% decreased rather than increased liquefaction resistance. This contradictory results of the effect of pre-shearing add further complexity. Further studies sought to define the separation between beneficial and deleterious pre-shearing (Ishihara and Okada [12], Suzuki and Toki [27], Nelson and Okamura [19]).

The loading history (i.e., over-consolidation) has substantial effects on liquefaction resistance. The effect of the overconsolidation ratio (OCR) and the value of the lateral pressure coefficient at rest (K_0) were separately investigated. The OCR effect was demonstrated experimentally by Seed and Peacock [24] and confirmed later by Ishibashi and Sherif [9]. Experimental data has been further accumulated (Ishihara and Takasu [13], Kokusho et al. [16], Tatsuoka et al. [28] and Toyota and Takada [32]) and an empirical equation of liquefaction resistance was proposed as a power function of OCR (Kokusho [17]).

The liquefaction resistance of unsaturated sand has been studied in the laboratory. In early research, the degree of saturation of the tested specimens was generally close to 100%, because the primary objective in the studies was to establish a standard for laboratory cyclic shear tests, and it was therefore necessary to avoid undesirable unsaturated conditions that would result in overestimation of the liquefaction resistance. Thereafter, unsaturated sand with a lower degree of saturation, down to approximately 70%, has been tested by several researchers. The results were expressed in the relationship between the degree of saturation and the liquefaction resistance of unsaturated sand normalized with respect to that of fully saturated sand (Huang et al. [8], Yoshimi et al. [34], Yasuda et al. [33], Ishihara et al. [14] and Goto and Shamoto [6]). Extensive data accumulated conclusively indicates that the liquefaction resistance ratio increases with decreasing degree of saturation. However, the liquefaction resistance ratios were considerably different for different sands tested under different conditions, indicating that the degree of saturation is not the only parameter affecting the normalized liquefaction resistance ratio of unsaturated sand.

Parameters such as the number of cycles and cyclic stress ratio of pre-shearing, over consolidation ratio (OCR), and degree of saturation are all influential; however, none of these are sufficient to explain the nature of sand behavior. In recent years, attempts have been made to explain the effects of those factors through a unified single parameter of volumetric strain. For unsaturated sand, volumetric strain estimated according to the increase in excess pore pressure has been found to be a dominant factor in liquefaction resistance (e.g., Okamura and Soga [20] and Okamura and Noguchi [21]). The liquefaction resistance ratio uniquely correlates with the volumetric strain during undrained cyclic shearing. Regarding pre-shearing effects, volumetric strain accumulated by small shakings is also a factor that can explain the enhancement of liquefaction resistance during a subsequent undrained shaking event (Okamura et al. [22] and Nelson and Okamura [19]).

The principal mechanics of pore pressure buildup and triggering liquefaction have emerged from a finding made early in the history of liquefaction study by Martin et al. [18]. They related the volumetric densification of sand, when subjected to cyclic shear loading in drained condition, to the increase in pore pressure that occurs when the same cyclic loading is applied to the saturated sand in undrained condition. The plastic volumetric strain due to the densification of the sand skeleton is cancelled by the elastic volumetric strain to maintain the sand volume constant, and the elastic strain can be correlated with the increase in pore pressure. The validity of this model showing the relationship between densification and pore pressure buildup was verified by several laboratory studies in the 1970s and 1980s. Martin et al. [18] and Finn [5] proposed a pore pressure generation model based on volumetric strain evolution under constant cyclic strain tests using simple shear apparatus. Their model successfully explained the effects of stress and strain history as well as the degree of saturation on the excess pore pressure generation in undrained cyclic shearing. Although the basic concept is simple and clear, its practical application is difficult because many experimental parameters are involved in the model, some of which are determined by strain-controlled simple shear tests instead of the stress-controlled tests commonly used in practice. Changes in horizontal earth pressure in the simple shear apparatus add further complexity to the model. Nevertheless, it has been established that volumetric plastic strain characteristics in conjunction with elastic stiffness are the properties determining the excess pore pressure not only for freshly deposited and reconstituted sand but also for sand with stress and strain history and for unsaturated sand.

This paper describes a pore pressure generation model based on cyclic stress ratio that is similar to that of Martin et al. [18] but based on a stress-controlled triaxial test. Attempts are made to explain the effect of various factors on the pore pressure generation and liquefaction resistance of clean sand using the index of volumetric strain. The model is verified through a comparison with the results of laboratory tests. Note that effect of partial drainage is being investigated in ongoing research and has not been included in this paper but will be presented elsewhere in the near future.

2   Stress-Based Pore Pressure Model

2.1   Pore Pressure Model

The basic mechanics of densification and pore pressure generation in the undrained cyclic shearing were conceptually demonstrated by Martin et al. [18]. Volumetric strain (densification) of sand consists of recoverable strain stored by elastic deformation at grain contacts (ε_{ve}) and irrecoverable strain due to slippage at the contacts (ε_{vp}). ε_{vp} is directly associated with ε_{ve} ($= -\varepsilon_{vp}$) and is converted into a change in effective stress if a constant volume condition is imposed. This is explained in a conceptual and rheology model in Fig. 1. Slippage at soil grain contacts occurs when subjected to cyclic shearing, which is the cause of plastic volumetric strain, ε_{vp},—indicated as the distance between points A and B in Fig. 1(a) and measured by a plastic slider in Fig. 1(b). When the undrained condition is imposed, ε_{vp} has to be compensated by the same amount of negative volumetric strain,

$$\varepsilon_{ve} + \varepsilon_{vp} = 0 \tag{1}$$

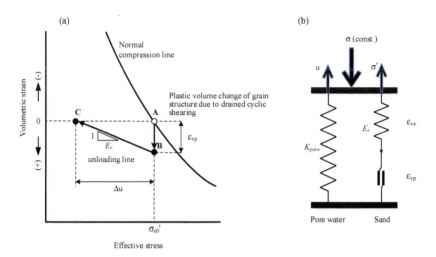

Fig. 1. Densification in undrained cyclic shearing and generation of pore pressure.

In other words, the soil grain structure rebounds to the extent required to maintain the volume constant—indicated from points B to C on the rebound line. With the bulk modulus of water, K_w, being sufficiently high to ignore the volume change of water, change in pore pressure can be expressed using the elastic (rebound) modulus of sand, E_r, as.

$$-\Delta_\varepsilon = -\Delta_{\sigma'} = -\varepsilon_{ev}E_r. \tag{2}$$

Silver and Seed [25] performed cyclic simple shear tests and found that there is a unique relationship between ε_{vp} and cyclic shear strain amplitude for a particular density of a sand at a given number of cycles irrespective of the effective confining stress, indicating that volumetric strain is uniquely related to cyclic shear strain. Martin et al. [18] and Finn [5] modeled the evolution of ε_{vp} as a function of number of cycles and shear strain amplitude based on drained cyclic simple shear tests where specimens were subjected to cycles of constant shear strain amplitude.

To estimate volume changes during the stress-controlled cyclic shear tests commonly conducted in engineering practice and for the field liquefaction assessment for a specified earthquake acceleration time history, the shear stress has to be converted to corresponding shear strain in order to apply the strain-based model. This can be done using shear modulus; however, shear modulus is expressed by non-linear functions of shear strain and effective confining pressure. For this reason, the model of Martin et al. [18] and Finn [5] has many parameters to be determined experimentally. It is clear the volumetric change is more closely related to cyclic shear strain than to cyclic shear stress and that the strain-based ε_{vp} model is appealing. However, the advantage of the model that it does not use the effective confining stress is debased when introducing shear modulus to convert shear stress into shear strain. Moreover, the confining pressure effects that were found to be negligible by Silver and Seed [25] and Youd [35] have a substantial influence on the volumetric strain behavior of sand tested in a simple shear apparatus (Duke et al. [3]). Nevertheless, the model provided a clear basis for better understanding the physical processes of progressive pore water pressure increase during undrained cyclic shearing, leading to liquefaction.

2.2 Densification Due to Constant Stress Amplitude Cycles

Although shear strain is more clearly correlated with plastic volumetric strain than is shear stress, a stress-based model has an apparent advantage over the shear strain-based model, namely that in liquefaction triggering assessment, acceleration and thus shear stresses are directly used to obtain volumetric strain in the stress-based model. In addition, simple shear testing used by Martin et al. suffers possible influences of horizontal stress change. In order to avoid further complication associated with change in horizontal stress, triaxial testing is employed to build a stress-based model in the present study.

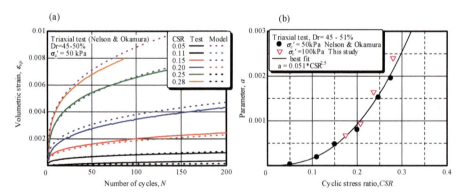

Fig. 2. Plastic volumetric strain. (a) Evolution of volumetric strain of loose Toyoura sand in drained cyclic shearing together with best fitted curves. (b) Variation of parameter a with CSR.

The sand used for all the tests conducted and described in this study to develop and verify the pore pressure model was medium dense Toyoura sand. The specific gravity and minimum and maximum void ratio of the sand are $G_s = 2.64$, $e_{min} = 0.609$, and $e_{max} = 0.973$, respectively. Specimens were air-pluviated medium dense Toyoura sand in a range of relative density Dr $= 45 - 51\%$, confined isotropically at effective stresses of 50 kPa and 100 kPa. Since all the specimens were prepared using the same sand and method, the effects of sand type, relative densities and sand fabric are not discussed in this paper.

Figure 2(a) indicates the evolution of volumetric strain with the number of uniform shear stress cycles in drained triaxial tests, where the volumetric strain at the end of each cycle is plotted. The rate of increase in ε_{vp} is higher for a higher cyclic stress ratio (CSR), however, the curves are similar. All the curves are well approximated with a function of the form

$$\varepsilon_{vp} = a \times N^b \tag{3}$$

where a and b are parameters and $b = 0.31$ gives a good fit to the experimental data over the range up to 200 cycles. The parameter a is plotted in Fig. 2(b) and can be well expressed by the following function, irrespective of the confining pressures:

$$a = c \times \mathrm{CSR}^d \tag{4}$$

A set of three parameters can be used to completely defines the volume change behavior under drained cyclic shearing of constant CSR: $(b, c, d) = (0.31, 0.051, 2.5)$. It should be noted that there is a limiting shear strain below which no pore water pressure develops regardless of the number of loading cycles. This value is of the order of 0.01% (Dobry et al. [1]). Conversely, as shown by Eqs. (2) and (3), plastic volumetric strain occurs for any cyclic stress ratio. However, strain in the range corresponding to a very small CSR, less than approximately 0.05 for the particular soil, may not generate significant excess pore pressure leading to soil liquefaction, and is not considered in this study.

Duke et al. [3] reported that cyclic volumetric strain decreased with increasing confining pressure, based on simple shear tests at a constant shear strain amplitude. This was more significant for a higher confining pressure range (>100 kPa). Possible reasons for the results in Fig. 2(b), which does not show a dependency on confining stress, are that the stress range tested in this study was relatively low (<100 kPa) and narrow (50–100 kPa), and a triaxial test apparatus was used. Further research is apparently needed to investigate the stress level dependency of volumetric strain on stress level.

2.3 ε_{vp} Due to Random Stress Cycles

Cycles of constant shear stress amplitude, τ_d, are applied to specimens in most common laboratory liquefaction tests in which the cyclic shear stress ratio defined as $\mathrm{CSR} = \tau_d/\sigma'_{c0}$ is constant throughout the test, and σ'_{c0} denotes the initial effective confining pressure. However, the actual cyclic stress ratio, $\mathrm{CSR}_a = \tau_d/\sigma'_c$, where σ'_c is effective stress at a given time, is not constant but increases due to the generation of excess pore pressure, and thus decreases the effective stress. This means that in order to estimate plastic volumetric strain for sand in undrained cyclic shearing from the ε_{vp} model (Eqs. (3) and (4)), the value of CSR_a in each cycle should be considered even for a constant CSR test. Variation in actual cyclic stress ratio from cycle to cycle is also the case when random shear stress caused by earthquake acceleration is considered.

It is clear that sand that has already been subjected to cycles of shearing and therefore has accumulated ε_{vp} shows a less contractive response to cyclic shearing than that of freshly deposited sand without any accumulated ε_{vp}. A basic assumption similar to that used by Martin et al. [18] is employed in this study: plastic volumetric strain increment for a cycle of shearing, $\Delta\varepsilon_{vp}$, depends on CSR_a and plastic volumetric strain accumulated since the deposition of the sand, ε_{vp}. For sand accumulating ε_{vp}, the plastic volumetric strain increment, $\Delta\varepsilon_{vp}$, for a cycle of CSR_a can be obtained by finding the equivalent number, N_e, for the ε_{vp} on the CSR_a curve, and then reading $\varepsilon_{vp} + \Delta\varepsilon_{vp}$ corresponding to $N = N_e + 1$, as shown in Fig. 3. Freshly deposited sand with $\varepsilon_{vp} = 0$ has the highest potential to densify and $\Delta\varepsilon_{vp}$ decreases as the accumulated plastic strain increases.

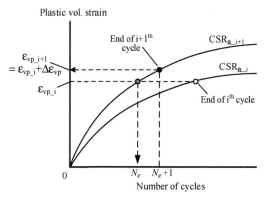

Fig. 3. Dependency of volumetric strain increment ($\Delta\varepsilon_{vp}$) on accumulated plastic strain (ε_{vp}).

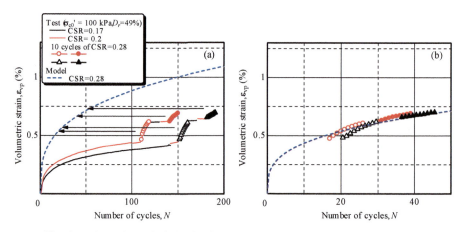

Fig. 4. Volumetric strain behavior for tests with constant CSR and varying CSR.

Figure 4 shows results of tests conducted to verify this assumption. Medium dense specimens were subjected to an initial confining pressure of 100 kPa. Two specimens were subjected drained cyclic shearing with CSR of 0.17 and 0.2 until an ε_{vp} value of approximately 0.4% was attained, followed by an additional 10 cycles of CSR $= 0.29$, 0.2 or 0.17. As indicated in Fig. 4(b), sections of additional 10 cycles of CSR $= 0.29$ (indicated by circles and triangles in the figure), when horizontally shifted leftward, falls on the model curve of the corresponding CSR_a, showing validity of the assumption. The effect of ε_{vp} accumulated in sand after deposition on volumetric densification characteristics is hereafter in this paper referred to as the ε_{vp} effect.

2.4   Elastic Bulk Modulus

The elastic (rebound) modulus of sand, E_r, was observed in triaxial tests. Specimens were isotopically consolidated at $\sigma' = 100\,kPa$, followed by unloading to 30 kPa and reloading back to 100 kPa, before and after drained cyclic shearing. A typical response is shown in Fig. 5. Drained cyclic shearing, which caused 0.6% plastic volumetric strain, did not alter the elastic modulus. Although the rebound curves were non-linear, a linear approximation was adopted for simplicity in this study. The curve was reasonably approximated by lines of $E_r = 24.0$ MPa and 13.5 MPa for the ranges from 100 to 50 kPa and 50 kPa to 25 kPa, respectively, which have been used in subsequent sections to evaluate excess pore pressures. It is reasonable to use these E_r values for ranges of higher effective stress, from 100 to 50 kPa for specimens with $\sigma'_{c0} = 100\,kPa$ and from 50 to 25 kPa for those with $\sigma'_{c0} = 50\,kPa$, in order to estimate pore pressure behavior in liquefaction tests. This is because, as cyclic shearing proceeds and the effective stress decreases, CSR_a will be dominant in the estimation of Δu increment after the excess pore pressure ratio becomes about 50% and higher. E_r plays more important role in the early stages of the liquefaction test.

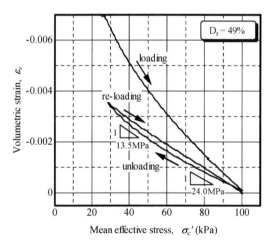

Fig. 5. Elastic rebound during unloading.

3 Excess Pore Pressure Buildup in Undrained Cyclic Shearing

Triaxial liquefaction tests were conducted on medium dense Toyoura sand. All specimens were fully saturated with the Skempton's B value higher than 0.95 and confined at σ'_{c0} = 50 kPa. Figure 6 shows typical results of the observed excess pore pressures ratio (EPPR) during cyclic shearing. This was also simulated using the pore pressure model as shown in the figure. The excess pore pressure ratio increment for a single cycle in the triaxial test is large for the first cycle, decreased gradually with the number of cycles and increased sharply after the excess pore pressure ratio (EPPR) reached approximately 0.6. Finally, the EPPR jumped up to 100% shortly after the effective stress path reached the phase transformation line (Ishihara et al. [10]). It can be seen that the overall trend of EPPR evolution for the test observations and the pore pressure model estimation agree quite well. The rapid increase in estimated excess pore pressure after the EPPR exceeded 0.6 is also evident. The plastic strain increment ($\Delta\varepsilon_{vp}$) and actual cyclic stress ratio (CSR_a) in each cycle estimated from the model are shown in Fig. 7. $\Delta\varepsilon_{vp}$ and CSR_a increase particularly after EPPR exceeded approximately 0.6. This is attributed to the sharp increase in excess pore pressure.

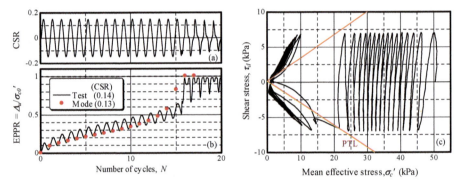

Fig. 6. Liquefaction test results on medium dense Toyoura sand ($D_r = 46\%$).

The liquefaction strength curve for the sand is shown in Fig. 8. Open circles show the CSR versus the number of cycles to reach the liquefaction condition (EPPR = 0.95) for four liquefaction tests. The solid line corresponds to the liquefaction resistance curve estimated with the pore pressure model, which predicts the test results quite well.

In the tests, the actual cyclic stress ratio increased with an increase in excess pore pressure and with the number of cycles. The filled circles are the plots of CSR_a and the remaining number of cycles to liquefaction from that point. For instance, point A in the figure corresponds to a specimen before shearing, which liquefied in $N_l = 16$ cycles of uniform CSR of 0.135. Filled circles indicate the relationship between CSR_a in each cycle and the remaining number of cycles needed to liquefy the specimen, $N_l - N$. At eleventh cycle (N = 11), the excess pore pressure was approximately 20 kPa and CSR_a was 0.23(= $0.135 \times 50/(50 - 20)$) as indicated by the point B. This indicates that the specimen at this time withstood 5 cycles of CSR = 0.23, which shows higher resistance

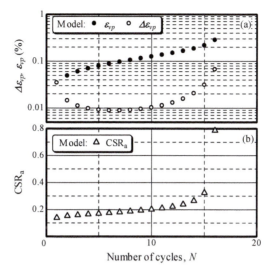

Fig. 7. Plastic volumetric strain and actual cyclic stress ratio estimated the model.

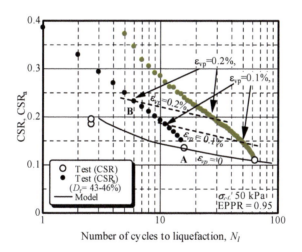

Fig. 8. Relationship between CSR and number of cycles to liquefy.

to liquefaction than that of freshly deposited medium dense Toyoura sand specimen consolidated at 29 kPa. The difference between CSR_a and CSR for a given number of cycles in this figure represents an increment in resistance to liquefaction which was gained due to cyclic shearing. During cyclic shearing, even in an undrained condition, sand gradually accumulates ε_{vp} and decreases its potential for yielding further volumetric strain. The estimated ε_{vp} using observed excess pore pressure and Eq. (1) are shown in Fig. 8. Equi-strain conditions are depicted with broken lines. The specimen is estimated to have accumulated ε_{vp} of 0.2% at point B.

4 Influence of Factors on Excess Pore Pressure Generation and Liquefaction Resistance

4.1 Small Pre-shearing

In order to study the effects of pre-shearing, a series of triaxial tests was conducted by Nelson and Okamura [19]. Stress controlled cyclic shearing was applied to specimens in drained condition first, and undrained cyclic shearing was then applied to observe the excess pore pressure generation and resistance to liquefaction. For the pre-shearing, the cyclic stress ratio and number of cycles were systematically varied between tests to achieve the target volumetric strains of $\varepsilon_{vp\_ps} = 0.1\%$, 0.3% or 0.8%. The cyclic stress ratios applied in the pre-shearing (CSR_{ps}) were determined so that the shear strain double amplitude in all cycles did not exceed 0.35%, which is believed to be lower than the threshold shear strain beyond which pre-shearing becomes deleterious. More detailed test conditions and results can be found in Nelson and Okamura [19]. Typical responses to the undrained cyclic shearing of specimens with different target volumetric strains ($\varepsilon_{vp\_ps} = 0$, 0.1%, and 0.3%) are shown in Fig. 9. For the same CSR, specimens accumulating higher ε_{vp_ps} required more cycles to liquefy compared to that without pre-shearing. CSR and number of cycles, N_l, to reach the excess pore pressure ratio (EPPR) of 95% for all the tests are shown in Fig. 10. The pre-shearing enhances the liquefaction resistance significantly, although the increase in the relative density due to pre-shearing is marginal. A pre-shearing volumetric strain of 0.8% corresponds to only a 3.9% increase in the relative density. The data points corresponding to the same ε_{vp_ps} lie almost on the same line irrespective of the CSR and number of cycles in the antecedent pre-shearing. The cyclic stress ratio to cause liquefaction for a given number of cycles increases with increasing ε_{vp_ps}, confirming that the volumetric strain in pre-shearing dominates the pre-shearing effects on the liquefaction resistance.

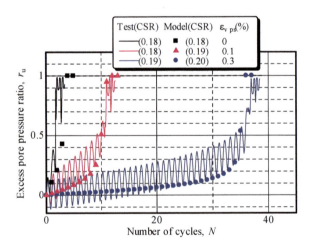

Fig. 9. Excess pore pressure ratio for different volumetric strain by pre-shearing.

The pore pressure model was used to predict the pore pressure response for the pre-sheared sand. The primary effect of pre-shearing is to reduce the volume strain potential. The accumulated volumetric strain during the pre-shearing, ε_{vp_ps}, was reflected in the model calculation by introducing ε_{vp_ps} as the initial volumetric strain. Typical excess pore pressure responses are depicted in Fig. 9 for the three specimens, which accumulated ε_{vp_ps} of either 0, 0.1% or 0.3% and were subjected to undrained cyclic shearing with a similar CSR. The model simulated the observed excess pore pressure generation quite well. It is of interest that excess pore pressure in the tests increased more or less linearly with the number of cycles for pre-sheared sand, while for the virgin sand, the pressure increase in the first cycles was large and slowed down in the following cycles. This pattern of excess pore pressure response is well represented by the model.

The solid lines in Fig. 10 show the liquefaction resistance curves for pre-sheared sand with different ε_{vp_ps} estimated with the pore pressure model. The model successfully predicted the liquefaction curves for pre-sheared medium dense Toyoura sand. It should be noted that the line for $\varepsilon_{vp\_ps} = 0.1\%$ in this figure almost coincide with that for $\varepsilon_v = 0.1\%$ in Fig. 8. Sand subjected to undrained cyclic shearing accumulates plastic volumetric strain in accordance with generated excess pore pressure as shown in Fig. 1 and Eq. (2) and the pre-shearing volumetric strain has the same effect in reducing the potential for further densification.

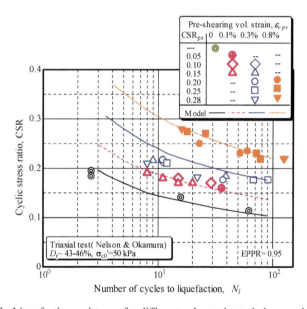

Fig. 10. Liquefaction resistance for different volumetric strain by pre-shearing.

4.2 Degree of Saturation

Unsaturated sand can change its volume even if undrained condition is imposed. The volume change of an unsaturated sand mass in undrained condition is equal to that of its void, which is comprised of not only uncompressible water but a water-air mixture. The volumetric constraint shown by Eq. (1) for fully saturated sand in undrained condition can be extended for unsaturated sand as,

$$\varepsilon_v = \varepsilon_{vp} - \varepsilon_{ve} \tag{5}$$

where ε_v denotes the volumetric strain of the sand mass, which arises from compression of air existing as a part of the pore fluid in the sand. It is reasonable to assume that the pore air pressure is the same as the pore water pressure for sand at relatively high degree of saturation, say 90% or higher, where the air bubbles exist separately in a form of insular saturation. The difference between the air bubble pressure and the surrounding water pressure given by the Young-Laplace equation is, for a spherical bubble,

$$p_{air} - p_{water} = 2T_s/r \tag{6}$$

where p_{air} is the pressure inside the air bubble, p_{water} is pressure in the surrounding water, T_s is the surface tension, and r is the bubble radius. Using a surface tension for the air–water interface of 0.075 N/m, the pressure difference for a bubble of radius 0.5 mm is $p_{air} - p_{warer} = 0.6$ kPa, which is small compared to the pore pressure discussed and can be considered negligible. The value of ε_v that arises from excess pore pressure Δu is given as,

$$\varepsilon_v = \Delta u \cdot n/K_{pf} \tag{7}$$

where K_{pf} and n are the bulk modulus of pore fluid and the porosity, respectively. From Eqs. (2), (5) and (7),

$$\Delta u = \frac{K_{pf} E_r}{K_{pf} + nE_r} \varepsilon_{vp} \tag{8}$$

Ignoring the volume change of water and assuming the Δu in each cycle of shearing is relatively small, K_{pf} can be expressed using Boyle's law for an ideal gas as

$$K_{pf} = p/(1 - S_r) \tag{9}$$

where S_r is the degree of saturation, and p is absolute pore pressure (Okamura and Soga [20]). Figure 11(a) depicts the volumetric strain-effective stress plane, on which paths are indicated for fully saturated sand in undrained and drained shearing, as well as for unsaturated sand in undrained shearing. A cycle of a given actual stress ratio CSR_a yields plastic strain $\Delta\varepsilon_{vp}$ indicated from points A to B, which is converted to the pore pressure increase Δu (B to C) if the sand is fully saturated and in undrained condition as described in Fig. 1. For unsaturated sand, the generated pore pressure and resultant volumetric strain can be defined as the intersection point D of two lines (Eqs. (2) and (7)). Figure 11(b) presents the results of undrained cyclic tests for three different degree of saturation, $S_r = 100$, 98 and 95.5%, subjected to CSR in a narrow range of 0.18–0.21 (Okamura and Soga [20]). Plastic strains derived from the model are also indicated by crosses. The numbers shown in this figure correspond to the number of cycles of shearing. In comparing the points at the end of the first cycles labeled "1", the CSR_a

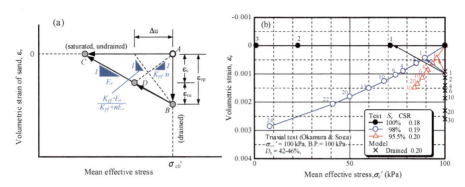

Fig. 11. Volumetric strain in undrained cyclic shearing and generation of pore pressure for saturated and unsaturated sand.

of the tests are approximately the same, and the excess pore pressure decreases with deceasing S_r. Δu was reduced by more than 80% by decreasing S_r from 100% to 95.5%.

Figure 12 shows the effect of the degree of saturation on the relationship between cyclic stress ratio and the number of cycles required to liquefy medium dense Toyoura sand obtained from triaxial tests (Okamura and Soga 2006). Both the initial effective confining pressure and the back pressure were 100 kPa and the degree of saturation was set to either 100%, 98% or 96%. The tests were also simulated using the pore pressure model and results are indicated by solid lines. The bulk modulus of the fluid was set at $K_{pf} = 11.0$ MPa and 5.1 MPa for $S_r = 98\%$ and 96%, respectively. The test results show that lowering the degree of saturation by 4% almost doubled the liquefaction resistance and the model simulated the liquefaction resistance curves quite well.

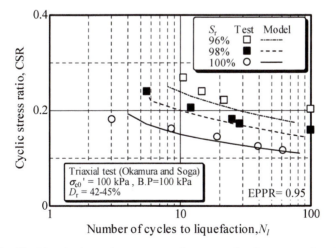

Fig. 12. Liquefaction resistance curves for saturated and unsaturated sand.

4.3 Overconsolidation

Results of undrained cyclic triaxial tests on normally consolidated and overconsolidated medium dense Toyoura sand derived from the literature are shown in Fig. 13. Liquefaction resistance of the overconsolidated sand reported by Tatsuoka et al. [28] and Toyota and Takada [32] are quite consistent, which increased linearly with logarithm of OCR. These tests are simulated using the pore pressure model.

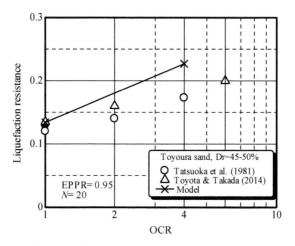

Fig. 13. Effect of OCR on liquefaction resistance.

Figures 14 and 15 are results of a triaxial test conducted in this study. Figure 14 shows relationship between volumetric strain and the effective stress. The specimen was normally consolidated at 100 kPa (point A), further consolidated to 400 kPa (C) and unloaded to 100 kPa (D). Elastic moduli measured at A and D with overconsolidation ratios OCR of 1 and 4, respectively, confirms that Er does not change practically. Duke et al. [3] reported that the modulus increased with increasing OCR for a sand with the overburden stress of 100 kPa, and this was not observed for the sand with lower stresses. The appropriate values of Er used in the simulation by the model is 24.0 MPa for OCR = 4.

The sand accumulated plastic volumetric strain $\varepsilon_{vp} = 0.0035$ in the overconsolidation process from A to D. In the model, overconsolidated sand is treated in the same way as the pre-sheared sand in the model calculation that the sand accumulated ε_{vp} in the overconsolidation process. Simulated liquefaction resistance using the model is given in Fig. 13, which overestimates the test results. Following the loading–unloading cycle, the specimen was subjected to drained cyclic loading from point F in Fig. 14. A solid line in Fig. 15 indicates observed evolution of volumetric strain. ε_{vp} reached 0.002 at 100 cycles of uniform CSR of 0.18, which is smaller than that for normally consolidated sand without accumulated volumetric strain ($\varepsilon_{vp} = 0$), as shown in the broken line in the figure. It is expected that two curves overlap if the curve for OCR = 4 are shifted $\varepsilon_{vp} = 0.0035$ vertically and equivalent number of Ne horizontally. As explained in Sect. 2.3, the equivalent number corresponding to $\varepsilon_{vp} = 0.0035$ for CSR = 0.18 is Ne = 179.

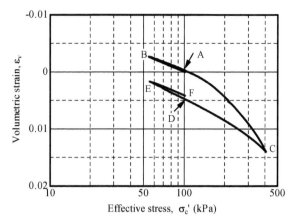

Fig. 14. Volumetric strain due to isotropic loading–unloading cycles and subsequent cyclic shearing (Toyoura sand, Dr = 49%).

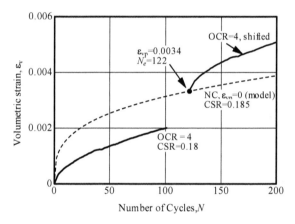

Fig. 15. Volumetric strain due to cyclic shearing on overconsoliaded sand.

The shifted curve, however, does not fall on the broken line, which indicates that the plastic volumetric strains accumulated during isotropic loading-unloading cycles and cyclic shearing have different effects on densification of the sand in subsequent cyclic shearing.

This might be explained as follows. Plastic volumetric strain induced by cyclic shearing is caused predominantly by sliding at sand particle contacts. In contrast, sand is carrying normal stress at the summits of the highest asperities of the particle surface, and as the consolidation pressure is increased, significant number of contact asperities may crash and deform plastically. Area of contacts increases resulting in densifying the sand. This plastic volumetric strain, however, does not reduce the potential of volumetric strain for subsequent cyclic shearing that much. Figures 14 and 15 are results of a triaxial test conducted in this study. Figure 14 shows relationship between volumetric strain and the effective stress. The specimen was normally consolidated at 100 kPa (point A), further consolidated to 400 kPa (C) and unloaded to 100 kPa (D). Elastic moduli measured at A and D with overconsolidation ratios OCR of 1 and 4, respectively, confirms that E_r does not change practically. Duke et al. (2008) reported that the modulus increased with increasing OCR for a sand with the overburden stress of 100 kPa, and this was not observed for the sand with lower stresses. The appropriate values of E_r used in the simulation by the model is 24.0 MPa for OCR = 4.

The sand accumulated plastic volumetric strain $\varepsilon_{vp} = 0.0035$ in the overconsolidation process from A to D. In the model, overconsolidated sand is treated in the same way as the pre-sheared sand in the model calculation that the sand accumulated ε_{vp} in the overconsolidation process. Simulated liquefaction resistance using the model is given in Fig. 13, which overestimates the test results. Following the loading-unloading cycle, the specimen was subjected to drained cyclic loading from point F in Fig. 14. A solid line in Fig. 15 indicates observed evolution of volumetric strain. ε_{vp} reached 0.002 at 100 cycles of uniform CSR of 0.18, which is smaller than that for normally consolidated sand without accumulated volumetric strain ($\varepsilon_{vp} = 0$), as shown in the broken line in the figure. It is expected that two curves overlap if the curve for OCR = 4 are shifted $\varepsilon_{vp} = 0.0035$ vertically and equivalent number of N_e horizontally. As explained in Sect. 2.3, the equivalent number corresponding to $\varepsilon_{vp} = 0.0035$ for CSR = 0.18 is $N_e = 179$. The shifted curve, however, does not fall on the broken line, which indicates that the plastic volumetric strains accumulated during isotropic loading-unloading cycles and cyclic shearing have different effects on densification of the sand in subsequent cyclic shearing.

This might be explained as follows. Plastic volumetric strain induced by cyclic shearing is caused predominantly by sliding at sand particle contacts. In contrast, sand is carrying normal stress at the summits of the highest asperities of the particle surface, and as the consolidation pressure is increased, significant number of contact asperities may crash and deform plastically. Area of contacts increases resulting in densifying the sand. This plastic volumetric strain, however, does not reduce the potential of volumetric strain for subsequent cyclic shearing that much.

5   Conclusions

This study developed a pore pressure generation model based on stress-controlled tri-axial tests. The model follows a basic assumption in Martin et al. [18] that a unique relationship exists between volumetric strains in drained tests and excess pore pressure in undrained tests. The plastic volumetric strain developed during undrained cyclic shar-ing is absorbed by elastic rebound in the soil skeleton due to the reduction in effective stress, preserving the condition of constant volume. The plastic volumetric strain in drained tests was modeled based on constant CSR triaxial tests. This cyclic stress-based model has an apparent advantage over the shear strain-based model of Martin et al. in two aspects. In the assessment of liquefaction triggering conditions, acceleration and thus shear stress is directly used to obtain volumetric strain in the stress-based model, while strains must be converted into stresses using shear modulus in the strain-based model. Simple shear testing used in the model of Martin et al. experiences possible influences from complicated variations in horizontal stress, whereas triaxial testing is free from this issue.

The effects of various factors on the pore pressure generation and liquefaction resistance of clean sand are explained using the unique index of volumetric strain through triaxial tests and simulation using the proposed model. The major findings are summarized as follows.

- Progressive pore pressure buildup during undrained cyclic shearing in saturated or unsaturated sand with and without stress histories was investigated with triaxial tests and the pore pressure model. There was a general trend of excess pore pressure buildup observed in triaxial tests in sand without pre-shearing and loading history, with the rate of pore pressure generation decelerating in the early stages and accelerating in the later stages with the number of cycles. This is attributed to the ε_{vp} accumulation and the increase in actual cyclic stress ratio (CSR_a). The accumulation of ε_{vp}, which decreases the potential for volumetric strain, has the beneficial effect of enhancing resistance to the generation of excess pore pressure (ε_{vp} effect), whereas the increase in actual cyclic stress ratio due to the generation of excess pore pressure has a detrimental effect (CSR_a effect).
- The excess pore pressure observed in triaxial tests is the result of the combined effects of accumulated ε_{vp} and CSR_a. Comparisons of triaxial specimens at the same excess pore pressure, and thus the same ε_{ve}, with specimens having the same ε_{vp} due to pre-shearing makes it possible to better understand the individual effects of ε_{vp} and CSR_a.
- A small pre-shearing has a beneficial effect in enhancing resistance to pore pressure generation and liquefaction resistance of sand. This can be explained by the ε_{vp} effect accumulated prior to the undrained cyclic shearing. Plastic volumetric strain accumulated in sand due to undrained cyclic shearing and that due to pre-shearing have the same effect in reducing the potential for further densification.
- Unsaturated sand changes its volume during cyclic shearing even if undrained condition is imposed. Volumetric compatibility and the bulk modulus of pore fluid (water-air mixture) are considered in the pore pressure model. The model simulates pore pressure generation and liquefaction resistance curves of triaxial tests reasonably well.
- Overconsoidated sand has a higher resistance to liquefaction than normally consolidated sand. This is partly explained by the plastic volumetric strain caused in the loading-unloading process. However, effects of plastic volumetric strain caused by overconsolidation on reducing the potential may be somewhat different from those caused by cyclic shearing.

The proposed pore pressure model simulated the buildup of pore pressure during cyclic shearing. The ability of the model to predict porewater pressures in pre-sheared, overconsolidated and unsaturated sands is confirmed. The model has considerable benefit in improving the understanding of the effects of different factors on liquefaction resistance.

References

1. Dobry, R., Powell, D.J., Yokel, F.Y., Ladd, R.S.: Liquefaction potential of saturated sand: the stiffness method. In: Proceedings of the 7th WCEE, vol. 3, pp. 25–32 (1980)

2. Dobry, R., Abdoun, T.: Recent findings on liquefaction triggering in clean and silty sands during earthquakes. J. Geotech. Geoenviron. Eng. **143**(10), 0401707 (2017)
3. Duku, P.M., Stewart, J.P., Whang, D.H., Yee, E.: Volumetric strains of clean sands subject to cyclic loads. J. Geotech. Geoenviron. Eng. **134**(8), 1073–1085 (2008)
4. Finn, W.D.L., Bransby, P., Pickering, D.: Effect of strain history on liquefaction of sand. J. Soil Mech. Found. Div. **96**(SM6), 1917–1934 (1970)
5. Finn, W.D.L.: Liquefaction potential: developments since 1976. In: Proceedings of the International Conference on Recent Advances in Geotechnical Earthquake Engineering and Soil Dynamics, vol. II, pp. 655–681 (1981)
6. Goto, S., Shamoto, Y.: Estimation method for the liquefaction strength of unsaturated sandy soils (Part II). In: Proceedings of the 37th Japan National Conference on Geotechnical Engineering, pp. 1987–1988 (2002). (in Japanese)
7. Goto, S., Towhata, I.: Acceleration of aging effect of drained cyclic pre-shearing and high temperature consolidation on liquefaction resistance of sandy soils. Geotech. Eng. J. JGS **9**(4), 707–719 (2014). (in Japanese)
8. Huang, Y., Tsuchiya, H., Ishihara, K.: Estimation of partial saturation effect on liquefaction resistance of sand using P-wave velocity. In: Proceedings of JGS Symposium, vol. 113, pp. 431–434 (1999). (in Japanese)
9. Ishibashi, I., Sherif, M.A.: Soil liquefaction by torsional simple shear device. J. Geotech. Eng. Div. **100**(GT8), 871–888 (1974)
10. Ishihara, K., Tatsuoka, F., Yasuda, S.: Undrained deformation and liquefaction of sand under cyclic stresses. Soils Found. **15**(1), 29–44 (1975)
11. Ishihara, K., Iwamoto, S., Yasuda, S., Takatsu, H.: Liquefaction of anisotropically consolidated sand. In: Proceedings of the IXth ICSMFE, vol. 2, pp. 261–264 (1977)
12. Ishihara, K., Okada, S.: Effects of stress history on cyclic behavior of sand. Soils Found. **18**(4), 31–45 (1978)
13. Ishihara, K., Takatsu, H.: Effects of overconsolidation and K0 condition on the liquefaction characteristics of sands. Soils Found. **19**(4), 59–68 (1979)
14. Ishihara, K., Tsuchiya, H., Huang, Y., Kamada, K.: Recent studies on liquefaction resistance of sand – effect of saturation. In: Proceedings of the 4th International Conference on Recent Advances in Geotechnical Earthquake Engineering and Soil Dynamics, pp. 1–7 (2001)
15. Kiyota, T., Koseki, J., Sato, T., Kuwano, R.: Aging effects on small strain shear moduli and liquefaction properties of the in situ frozen and reconstituted sandy soils. Soils Found. **49**(2), 259–274 (2009)
16. Kokusho, T., Yoshida, Y., Nishi, K., Esashi, Y.: Evaluation of seismic stability of dense sand layer (Part 1) dynamic strength characteristics of dense sand. CRIEPI report 383025, Central Research Institute of Electric Power Industry (1983)
17. Kokusho, T.: Innovative Earthquake Soil Dynamics. CRC Press, Boca Raton (2017)
18. Martin, G.R., Finn, W.D.L., Seed, H.B.: Fundamentals of liquefaction under cyclic loading. J. Geotech. Eng. Div. **101**(5), 423–438 (1975)
19. Nelson, F., Okamura, M.: Influence of strain histories on liquefaction resistance of sand. Soils Found. **59**(5), 1481–1495 (2019)
20. Okamura, M., Soga, Y.: Effect on liquefaction resistance of volumetric strain of pore fluid. Soils Found. **46**(5), 703–708 (2006)
21. Okamura, M., Noguchi, K.: Liquefaction resistance of unsaturated non-plastic silt. Soils Found. **49**(2), 221–229 (2009)
22. Okamura, M., Nelson, F., Watanabe, S.: Pre-shaking effects on volumetric strain and cyclic strength of sand and comparison to unsaturated soils. Int. J. Soil Dyn. Earthq. Eng. **124**, 307–316 (2018)
23. Seed, H.B.: Evaluation of Soil liquefaction effects of level ground during earthquakes. Presented at the ASCE Annual Convention and Exposition, Philadelphia (1976)

24. Seed, H.B., Peacock, W.H.: Test procedure for measuring soil liquefaction characteristics. J. Soil Mech. Found. Div. **97**(SM8), 1099–1119 (1971)

25. Silver, M.L., Seed, H.B.: Volume changes in sands during cyclic loading. J. Soil Mech. Found. Div. **97**(9), 1171–1182 (1971)

26. Singh, S., Seed, H.B., Chan, C.K.: Undisturbed sampling of saturated sands by freezing. J. Geotech. Eng. Div. **108**(2), 247–264 (1982)

27. Suzuki, T., Toki, S.: Effects of preshearing on liquefaction characteristics of saturated sand subjected to cyclic loading. Soils Found. **24**(2), 16–28 (1984)

28. Tatsuoka, F., Kato, H., Kimura, M., Pradhan, T.B.S.: Liquefaction strength of sands subjected to sustained pressure. Soils Found. **28**(1), 119–131 (1988)

29. Tokimatsu, K., Hosaka, Y.: Effects of sample disturbance on dynamic properties of Sand. Soils Found. **26**(1), 53–64 (1986)

30. Tokimatsu, K., Yamazaki, T., Yoshimi, Y.: Soil liquefaction evaluations by elastic shear moduli. Soils Found. **26**(1), 25–35 (1986)

31. Towhata, I., Taguchi, Y., Hayashida, T., Goto, S., Shintaku, Y., Hamada, Y., Aoyama, S.: Liquefaction perspective of soil ageing. Geotechnique **67**(6), 467–478 (2017)

32. Toyota, H., Takada, S.: Variations of liquefaction strength induced by monotonic and cyclic loading histories. J. Geotech. Geoenviron. Eng. **143**(4), 04016120 (2017)

33. Yasuda, S., Kobayashi, T., Fukushima, Y., Kohari, M., Simazakl, T.: Effect of degree of saturation on the liquefaction strength of Masa. In: Proceedings of the 34th Japan National Conference on Geotechnical Engineering, pp. 2071–2072 (1999). (in Japanese)

34. Yoshimi, Y., Tanaka, K., Tokimatsu, K.: Liquefaction resistance of a partially saturated sand. Soils Found. **29**(3), 157–162 (1989)

35. Youd, T.L.: Compaction of sands by repeated shear straining. J. Soil Mech. and Found. Div. **98**(7), 709–725 (1972)

36. Wu, C., Kiyota, T., Katagiri, T.: Effects of drained and undrained cyclic loading history on small strain shear moduli and liquefaction resistance of medium dense Toyoura sand. J. JSCE A1 **72**(4), 482–488 (2016). (in Japanese)

37. Wu, C., Kiyota, T.: Effects of specimen density and initial cyclic loading history on correlation between shear wave velocity and liquefaction resistance of Toyoura sand. Soils Found. **59**(6), 2324–2330 (2019)

How Important is Site Conditions Detailing and Vulnerability Modeling in Seismic Hazard and Risk Assessment at Urban Scale?

Kyriazis Pitilakis$^{(\boxtimes)}$ [iD], Evi Riga [iD], and Stefania Apostolaki [iD]

Aristotle University of Thessaloniki, P.O.B. 424, 54124 Thessaloniki, Greece
kpitilak@civil.auth.gr

Abstract. This work investigates the effect of the level of detailing of site conditions as well as the selection of the fragility and vulnerability models on large scale seismic risk assessment. For this we consider the application of selected components of the recent European Seismic Hazard (ESHM20) and Risk (ESRM20) Models and we focus on Thessaloniki, Greece, a city which is very well documented in terms of local site conditions and exposure. Seismic risk results are compared in terms of expected damages and economic losses for a seismic hazard with a 475-year return period. The results indicate that the level of site conditions modelling does not significantly affect the estimated aggregate damages and economic losses at city scale, however significant discrepancies may occur at local scale. On the other hand, the selection of the vulnerability model for the building stock may considerably affect the intensity and the spatial distribution of damages, resulting in a considerable differentiation in the economic losses estimate.

Keywords: ESHM20 · ESRM20 · Site modelling · Site amplification · Fragility and vulnerability models · Seismic risk model

1 Introduction

In large-scale seismic hazard and risk studies the level of detailing of site conditions is a real challenge both from scientific and practical point. Site amplification of seismic ground motion due to the local site conditions and other induced phenomena, like liquefaction, may affect considerably the seismic hazard and the final risk assessment in terms of physical, human and immaterial losses. Regarding seismic hazard assessment at urban scale, and in the absence of detailed microzonation studies, which is a very common situation, site effects are necessarily estimated using simple parameters like $V_{s,30}$, i.e., the time-averaged shear wave velocity in the upper 30 m of the soil deposits. Again, in large scale applications, $V_{s,30}$ values are often inferred from proxies such as the topographic slope and geology, or from simplified site categorization based on seismic codes site characterization and mapping. The risk assessment, on the other hand, for a given large scale urban building exposure, is conducted using fragility and vulnerability models, which, due to the complexity and diversification of the building typologies, are usually of generic nature, while they are also tightly related to the intensity measures

© The Author(s), under exclusive license to Springer Nature Switzerland AG 2022
L. Wang et al. (Eds.): PBD-IV 2022, GGEE 52, pp. 140–161, 2022.
https://doi.org/10.1007/978-3-031-11898-2_8

used, which in turn, are explicitly related to the results of the seismic hazard. It is evident that the uncertainties involved in all stages of this complex process are important and may affect the final risk assessment, with consequent high impact on the frame of strategic decision making.

To this regard, the scope of this work is to investigate on one hand the effects of the method applied for site characterization in the seismic hazard assessment and on the other hand the appropriateness of the selection of the vulnerability functions in urban large scale seismic risk assessment. For this we consider the application of selected components of the recent European Seismic Hazard (ESHM20, Danciu et al. 2021) and Risk (ESRM20, Crowley et al. 2021) Models and we focus on Thessaloniki, Greece, a city which is very well documented in terms of local site conditions, exposure and vulnerability curves specific for the main building typologies. Thessaloniki, located in Northern Greece, is the second largest city in Greece, with over 1 million inhabitants in its metropolitan area. The area studied herein, includes 16 municipalities and covers an area of 126 km^2 (Fig. 1).

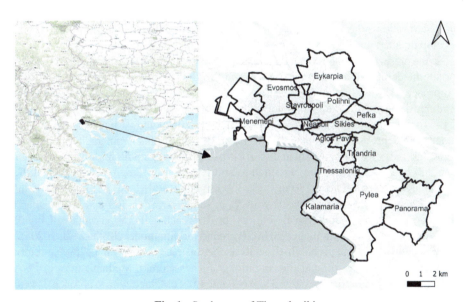

Fig. 1. Study area of Thessaloniki

To investigate the impact of the method applied for site modelling in seismic hazard assessment and of the selection of the vulnerability functions on the seismic risk assessment of Thessaloniki, we compare seismic risk results obtained using two different approaches for site conditions modelling for the seismic hazard combined with two different fragility and vulnerability models. Regarding site modelling for seismic hazard assessment, in the first approach, which may be considered as more rigorous but also more demanding, appropriate site models in terms of $V_{s,30}$ are directly used in the ESHM20 hazard logic tree, while in the second approach, which is more simplified and efficient in practice, the study area is classified into different site classes based on a code-oriented classification scheme (Riga et al. 2022). With respect to fragility and vulnerability modelling, we apply two sets of analytical generic fragility and vulnerability models, recommended for large scale applications, i.e., the ESRM20 models (Romão et al. 2021), derived from equivalent single degree-of-freedom oscillators and the models by Kappos et al. (2006, 2010), derived from 2D numerical models of structures representative for Greece and Southern Europe in general.

Seismic risk results are presented in terms of expected damages and economic losses for a seismic hazard with a 475-year return period. Both hazard and risk assessment calculations are undertaken with the OpenQuake Engine (Pagani et al. 2014; Silva et al. 2013).

2   Site Modelling and Seismic Hazard Assessment

Thessaloniki is one of the best-documented urban areas in terms of local site conditions. In the framework of its microzonation study (Anastasiadis et al. 2001), a significant amount of laboratory and in situ geophysical and geotechnical surveys have been performed to validate the dynamic properties and geometry of the main soil formations and develop detailed geotechnical maps, 1-D profiles and 2-D cross sections.

In the present work, following Riga et al. (2022), we applied two different approaches for site modelling to assess the seismic hazard in the study area. In the first approach (Approach 1), we adopted a detailed site model developed for Thessaloniki using the available data from the microzonation study. The $V_{s,30}$ values of this model are shown in Fig. 2 for a grid consisting of 142 points, with a horizontal and vertical spacing of 0.01 geographic degrees (approximately 1×1 km). Based on this detailed site model, the north-east part of the city is situated on rock-like formations, with $V_{s,30}$ values greater than 800 m/s, and is therefore classified as soil type A according to EC8. $V_{s,30}$ at the coastal area ranges between 180 m/s and 360 m/s (EC8 soil class C/S2), while the in-between region consists of deposits with $V_{s,30}$ values between 360 and 800 m/s, and is categorized as EC8 soil type B (Fig. 2). In Approach 1 the $V_{s,30}$ values of Fig. 2 were directly used as input in the ESHM20 ground motion model (Kotha et al. 2020) to estimate with a classical probabilistic analysis the seismic hazard with a 475-year return period for the specific soil and site conditions of Thessaloniki. Median values of PGA and spectral accelerations, S_a, at 0.3 s, 0.6 s and 1.0 s obtained with Approach 1 were stored for all grid points. The specific ground motion parameters were selected, as they are needed as intensity measures by the applied fragility and vulnerability models, as described in detail in Sect. 4. Although Approach 1 is a very rigorous method, it is also very demanding, as it entails the existence of a site model for the whole study area as well as advanced knowledge and skills on seismic hazard assessment by the user (Riga et al. 2022).

In the second Approach (Approach 2), we used the detailed site model to classify the study area into appropriate site classes following the classification scheme by Pitilakis et al. (2020), which uses as classification parameters the approximate depth to seismic bedrock, the equivalent shear wave velocity $V_{s,H}$ (equal to $V_{s,30}$ for soil deposits with depth greater than 30 m, otherwise equal to the mean shear wave velocity up to the seismic bedrock depth) and the fundamental period of the site T_0. The classification of

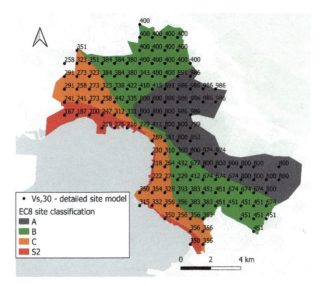

Fig. 2. Spatial distribution of $V_{s,30}$ (m/s) in the study area based on the detailed site model developed from the microzonation study of Thessaloniki, applied in Approach 1 for seismic hazard assessment. The classification of the study area according to EC8 (CEN 2004) based on the detailed site model is also included in the figure.

the study area based on the Pitilakis et al. (2020) scheme is shown in Fig. 3. Compared to the EC8 classification shown in Fig. 2, the Pitilakis et al. (2020) classification differs mainly in the coastal area, as it includes more site classes, and in the spatial extent of zones (B1: shallow very stiff soil deposits and B2: Intermediate depth stiff soil deposits whose mechanical properties increase with depth) and C (C1: deep stiff soil deposits and C3: deep soil deposits of medium stiffness) due to the difference in the definitions of the respective soil classes in the two site classification schemes. For the application of Approach 2 for seismic hazard assessment, we evaluated for the soil classes of Fig. 3 the period-dependent site amplification factors according to Pitilakis et al. (2020) as the ratio between the elastic response spectrum for each soil class and the elastic response spectrum for soil class A which represents rock or rock-like site conditions. We should highlight that the amplification factors by Pitilakis et al. (2020) are intensity-dependent, i.e., they decrease for increasing levels of ground shaking, taking in this way soil nonlinearity into account. The resulting amplification factors for PGA and S_a at 1.0 s, $S_a(1.0$ s) are shown in Fig. 4. We observe that for the specific ground shaking levels, the amplification factors for $S_a(1.0$ s) at the part of the coastal area classified as class D (deep soil deposits consisting of soft to medium stiffness clays and/or loose sandy to sandy-silt formations with substantial fines percentage) - X (special soils requiring site-specific evaluations), and in Kalamaria region (see Fig. 1), classified as soil class C3 (deep soil deposits, consisting of medium dense sand and gravel and/or medium stiffness clay), the amplification factors for $S_a(1.0$ s) range between 2.1 and 2.9, and are significantly higher than the respective factors for PGA, as well as the amplification factors of EC8 (Riga et al. 2022).

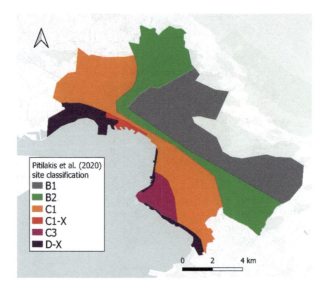

Fig. 3. Site classification of the study area based on the classification scheme by Pitilakis et al. (2020), applied in Approach 2 for seismic hazard assessment.

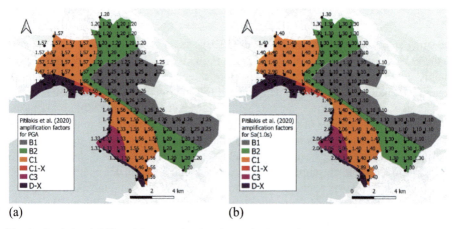

Fig. 4. Spatial variability of the amplification factors in the study area according to Pitilakis et al. (2020) for mean return period of 475 years and for (a) PGA and (b) $S_a(1.0 \text{ s})$.

The amplification factors estimated for PGA and spectral accelerations, S_a, at 0.3 s, 0.6 s and 1.0 s were then multiplied with the respective median ESHM20 hazard output for rock site conditions with a 475-year return period, publicly available through the hazard platform of EFEHR (http://hazard.efehr.org). Compared to Approach 1, Approach 2 is more simplified and easier to be applied, as it entails a site classification of the study area based on a commonly applied site classification scheme and the existence of a seismic hazard assessment for rock-site conditions for the study area, both of which can be quite easily obtained (Riga et al. 2022).

Median values of PGA and spectral accelerations, S_a, at 0.3 s, 0.6 s and 1.0 s obtained with both approaches were stored for all grid points. The spatial distribution of PGA and S_a (1.0 s) at the ground surface obtained with Approaches 1 and 2 is compared in Fig. 5 and Fig. 6 respectively. Regarding PGA, the more detailed Approach 1 results in PGA values ranging between 0.27 g at the rock-like formation and 0.45 g at the coastal area, significantly higher than the respective PGA values obtained with Approach 2, where a maximum PGA equal to 0.38 g is observed (Fig. 5). This trend is reversed for S_a(1.0 s) is some areas(Fig. 6), especially at the part of the coastal area classified as soil class D-X, and in Kalamaria region, classified as soil class C3, where the Pitilakis et al. (2020) amplification factors for S_a(1.0 s) are quite high (between 2.1 and 2.9), resulting in an overestimation of S_a(1.0 s) with Approach 2 compared to Approach 1. This is expected to have a significant impact on the estimated damages at these regions, since one of the two adopted vulnerability models uses S_a(1.0 s) as intensity measure for a significant number of buildings.

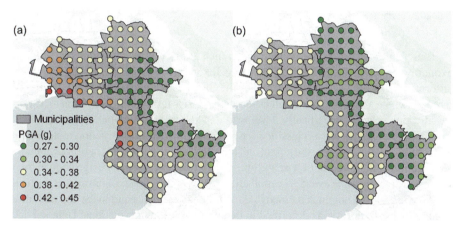

Fig. 5. Spatial distribution of PGA at the ground surface for seismic hazard with a 475-year return period obtained with (a) Approach 1 (detailed site model) and (b) Approach 2 (Pitilakis et al. 2020 classification system).

3   Exposure Model

For the development of the exposure model for the study area we used the data from the 2011 Population – Housing and Building Census (ELSTAT 2011). The exposure model contains the residential buildings of the study area which are constructed either from reinforced concrete or masonry, as they constitute the vast majority of the residential buildings (99.7%) in the study area. The exposure model therefore contains 75,169 residential buildings that are either concrete, or unreinforced masonry (brick or stone).

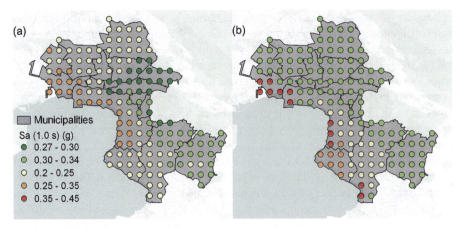

Fig. 6. Spatial distribution of $S_a(1.0\ s)$ at the ground surface for seismic hazard with a 475-year return period obtained with (a) Approach 1 (detailed site model) and (b) Approach 2 (Pitilakis et al. 2020 classification system).

These buildings were classified into different building classes following two different taxonomy schemes, the GED4ALL Building Taxonomy (Silva et al., 2022), which is a uniform classification system developed by the Global Earthquake Model (GEM) aiming to be applicable at a global scale and is adopted in ESRM20 and the Kappos et al. (2006) classification scheme, specifically developed to address the Greek building stock. Both classification schemes use as classification attributes the main construction material, the lateral load resisting system, the height and the ductility level, which is herein assumed to be a function of the construction period and, thus, respective seismic design code in force at the time of the seismic event. The main differences of the two taxonomy schemes lie in the way the building height is defined in the different typologies and in the consideration or not of the soft-storeys, as described in detail in the following sections. Detailed data for the type of use, main construction material, number of storeys and the period of construction of the buildings are provided by the Hellenic Statistical Authority ELSTAT (2011) for each census sector of Thessaloniki. The census sector is the geographic unit adopted by ELSTAT for the census and contains an average of 600 residences. The lack of data for the lateral load resisting system has led to unavoidable assumptions based on the feedback from the SERA European Building Exposure Workshop questionnaire (https://sites.google.com/eucentre.it/sera-exposure-workshop/questionnaire).

3.1 Exposure Model Following the GED4ALL Building Taxonomy Scheme

Using the GED4ALL Building Taxonomy (Silva et al., 2022), the residential buildings of the study area were classified according to the main construction material, the lateral load resisting system, the number of storeys (i.e. height), the ductility level, which is herein assumed to be a function of the construction period and, thus, respective seismic design code in force at the time of the seismic event, and lateral load coefficient used at the time of the design. The symbolization of each attribute is provided in Table 1. Level 1 is the first level of detail required to describe an attribute, whereas Level 2 provides additional detail on the Level 1 attributes. Building height is defined in terms of exact number of storeys. The existence of irregularly infilled frames with soft storey, which is a common practice in Greece, is not taken into account in this scheme, which is aimed to be applicable at a global scale.

The 75,169 residential buildings of the exposure model were finally classified in 107 building typologies. The vast majority (94.98%) of them are reinforce concrete (CR), while 59.7% have been designed with low level (DUCL or CDL) or even no seismic code (DNO or CDN). Figure 7 shows the distribution of the examined residential buildings based on (a) the main construction material and lateral load resisting system, (b) the seismic code level and (c) the height expressed in terms of number of storeys.

Table 1. Values of attributes of GED4ALL Building Taxonomy (Silva et al., 2022) used to describe the building stock of Thessaloniki.

Attribute	Element code	Level 1 value	Element code	Level 2 value
Material	CR	Concrete, reinforced		
	MUR	Masonry, unreinforced	CL99	Fired clay unit, unknown type
			STDRE	Stone
Lateral load-resisting system (LLRS)	LFM	Moment frame		
	LFINF	Infilled frame		
	LWAL	Walls and frames where the walls, due to their substantial lengths, resist the vast majority of the lateral load		
	LDUAL	Moment frames and shear walls acting together to resist seismic effects		
Ductility level – seismic code level	DNO or CDN	Non-ductile (Period of construction: before 1959)		
	DUCL or CDL	Ductile, low (Period of construction: 1960–1985)		
	DUCM or CDM	Ductile, medium (Period of construction: 1986–1995)		
	DUCH or CDH	Ductile, high (Period of construction: 1996-present)		
Height	H	Exact number of storeys above ground		
Lateral force coefficient	Number expressed in %	The value of the lateral force coefficient, i.e. the fraction of the weight that was specified as the design lateral force in the seismic design code (Applied to reinforced concrete moment and infilled frames only)		

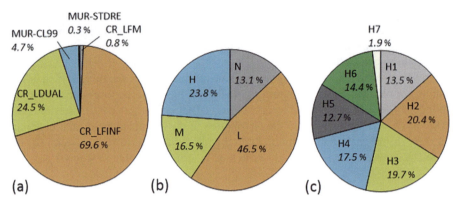

Fig. 7. Classification of the building taxonomies in the study area based on a) material and lateral load resisting system (LLRS), b) code level and c) height according to the GED4ALL Building Taxonomy scheme of Table 1.

3.2 Exposure Model According to Kappos et al. (2006)

Using the taxonomy scheme defined by Kappos et al. (2006), which has been specifically developed to address the Greek building stock, the residential buildings of the study area were classified according to the main construction material, the lateral load resisting

Table 2. Values of attributes of Kappos et al. (2006) taxonomy scheme used to describe the building stock of Thessaloniki.

Type	Structural system	Height (number of storeys)	Seismic design level
RC1	Concrete moment frames	(L)ow-rise (1–3) (M)id-rise (4–7) (H)igh-rise (8+)	(N)o/pre code (L)ow code (M)edium code (H)igh code
RC3	Concrete moment frames with unreinforced masonry infill walls		
	RC3.1 Regularly infilled frames		
	RC3.2 Irregularly infilled frames (pilotis)		
RC4	RC dual systems (RC frames and walls)		
	RC4.2 Regularly infilled dual systems		
	RC4.2 Irregularly infilled dual systems (pilotis)		
MBr	Unfeinforced masonry of bricks	12 (1–2)	
		3+ (3 or more)	
MSt	Unfeinforced masonry of stone	12 (1–2)	
		3+ (3 or more)	

system, the height category, and the relevant seismic code level which is linked with the period of construction. The symbolization of each attribute is provided in Table 2. Regarding building height, this scheme provides three height categories defined in terms of ranges of number of storeys, while the existence of a soft storey (commonly referred to as pilotis in Greece) is considered.

The 75,169 residential buildings of the exposure model were classified in 64 building typologies. Most of the buildings (75.6%) are regularly infilled reinforced concrete buildings (CR) whereas 18.45% are irregularly reinforced as they have at least one soft-storey. Figure 8 shows the distribution of these buildings based on (a) the main construction material and lateral load resisting system, (b) the seismic code level and (c) the height category.

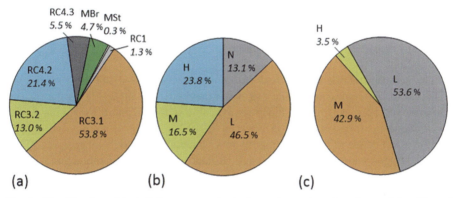

Fig. 8. Classification of the building taxonomies in the study area based on a) material and lateral load resisting system (LLRS), b) code level and c) height category according to the Kappos et al. (2006) taxonomy scheme of Table 2.

4 Vulnerability

Fragility models are a critical component of any seismic risk assessment methodology as they describe the probability of exceeding a damage state conditional to a ground shaking intensity (Silva et al. 2019). They may be developed from analyses of appropriate mechanical models, statistical analysis of damage data, expert judgment or with a combination of the above-mentioned approaches (Kappos et al. 2006). Vulnerability models, i.e., the probability of loss ratio conditional on a set of ground shaking intensities can be produced by the fragility functions using a damage-to-loss model, which expresses the relation between a damage state and the corresponding fraction of loss.

For large scale seismic risk analyses, e.g., at urban, national, continental or even global scale, generic fragility and vulnerability models are usually applied, covering the most common building typologies, such as the model by Martins and Silva (2021), which has been applied at the global seismic risk model developed by of the Global Earthquake Model (GEM) Foundation. These generic models are inevitably developed mainly analytically due to the lack of sufficient empirical damage data to cover all possible building typologies and all seismic intensities in different regions of the world (e.g., Martins and Silva 2021; Romão et al. 2021). For simplicity reasons, their development is usually based on analyses of equivalent single-degree-of-freedom (SDOF) oscillators. Generic models may also be obtained with the combination of analytical methodologies and empirical data, such as the models by Kappos et al. (2006, 2010), which have been derived from analyses of 2D numerical models, representative again for specific building typologies, combined with statistical data from past earthquakes. Finally, for cases of buildings with strategic interest, such as hospitals or schools, building-specific fragility functions can be developed from nonlinear analysis of detailed 3D numerical models, which may be updated through field measurements (Karapetrou et al. 2016; Fotopoulou et al. 2022).

The selection of the fragility/vulnerability model is one of the most significant sources of uncertainties in seismic risk assessment (Riga et al. 2017), as the dispersion observed between fragility functions available in the literature for similar building typologies can be very high (Silva et al. 2019). In the present study we investigate the effect of the selection of fragility and vulnerability models by applying a) the ESRM20 fragility and vulnerability models (Romão et al. 2021) in combination with the GED4ALL exposure and b) the Kappos et al. (2006, 2010) fragility and vulnerability models, in combination with the Kappos et al. (2006) exposure model.

The ESRM20 models have been derived for simplified equivalent single degree-of-freedom oscillators and their use is recommended for large scale applications (e.g., the European building stock). These functions use as intensity measures PGA and spectral accelerations, S_a (0.3 s), S_a(0.6 s) and S_a(1.0 s) depending on building typology. More specifically, for each building typology the final intensity measure is selected as that with the lowest lognormal dispersion in the fragility function, given that it is related to efficiency (Crowley et al. 2021). Following this approach, for the building typologies in the exposure model of Thessaloniki (Sect. 3.1), S_a(1.0 s) is the most commonly adopted intensity measure, as is shown in Fig. 9. Probabilities of exceedance are provided for four damage states, i.e., slight, moderate, extensive and complete, with the thresholds between the different damage states expressed in terms of yield and ultimate displacement in the capacity curves.

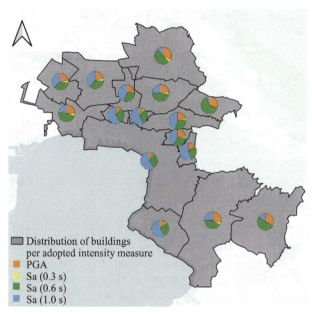

Fig. 9. Distribution of the residential buildings in the exposure model of Thessaloniki per adopted intensity measure in the Romão et al. (2021) vulnerability models

The Kappos et al. (2006, 2010) fragility functions are based on a hybrid approach, which is a combination of statistical data with appropriately processed results from inelastic analyses of advances 2D numerical models. However, in this application we used only the analytical component of the fragility models (G. Panagopoulos, personal communication), as the hybrid functions are not available for all building types in the exposure model. Fragility functions by Kappos et al. (2006, 2010) use PGA as an intensity measure for all building types, a selection which has often been questioned for flexible buildings. Probabilities of exceedance are provided for five damage states, DS1 (slight), DS2 (moderate), DS3 (substantial to heavy), DS4 (very heavy) and DS5 (collapse), with the thresholds between the different damage states expressed in terms of loss indices for RC structures and in terms of yield and ultimate displacement for masonry structures.

We should highlight that there is no direct compatibility between the damage states which share the same damage state label (e.g. slight, moderate etc.) in the two above-mentioned vulnerability models, as the definition of the damage states is not the same. However, we can try to make a rough correspondence between the damage states of the two models by comparing the values of the loss indices (ratio of repair cost to replacement cost) used to relate the distribution of physical damages to the probability of damage repair cost (Table 3). Based on Table 3, slight, moderate, extensive and complete damage in ESRM20 can be roughly matched to DS2, DS3, DS4 and DS5 damage states by Kappos et al. (2006, 2010) respectively. This correspondence will be used in the following to compare the damages estimated with the two fragility models.

Table 3. Damage states and loss indices for the ESRM20 and Kappos et al. (2006, 2010) vulnerability models

ESRM20		Kappos et al. (2006, 2010)		
Damage state	Loss index	Damage state	Range of loss index	Central loss index
		DS1	0.005	0.005
Slight	0.05	DS2	0.05	0.05
Moderate	0.15	DS3	0.2	0.2
Extensive	0.6	DS4	0.45	0.45
Complete	1.0	DS5	0.80	0.80

The fragility curves of the two models cannot be directly compared due to the differences in (a) the way the building classes are defined in the two taxonomy schemes, (b) the adopted intensity measures, (c) the definition of the damage states. Their performance will be compared through the comparison of the estimated damages and mainly, economic losses in the study area. However, indicative fragility and vulnerability curves of the two models for the most common building typologies in the exposure model with the GED4ALL and the Kappos et al. (2006) taxonomies are plotted in Fig. 10 and Fig. 11 respectively.

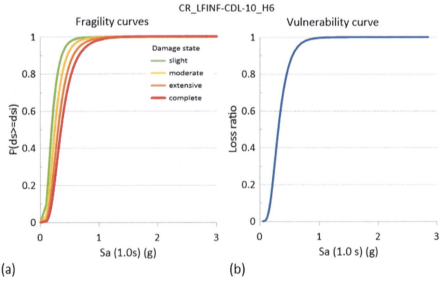

Fig. 10. Indicative fragility curves (a) and vulnerability curve (b) for the most common building typology in Thessaloniki for the exposure model with the GED4ALL taxonomy, i.e., reinforced concrete (CR) - infilled frame (LFINF) - low ductility (CDL-10) - 6 storey (H6) buildings.

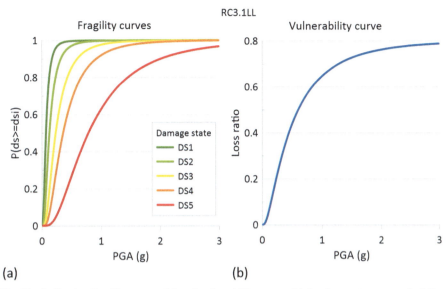

Fig. 11. Indicative fragility curves (a) and vulnerability curve (b) for the most common building typology in Thessaloniki for the exposure model with the Kappos et al. (2006) taxonomy, i.e., reinforced concrete - infilled frame (RC3.1) – low height (L) – low ductility (L).

5 Physical Damages

Expected physical damages have been derived through scenario-based damage analyses with precomputed ground motion fields in OpenQuake Engine (Pagani et al. 2014; Silva et al. 2013) where the seismic hazard estimated with Approach 1 and 2 (Sect. 2) and the exposure models with the GED4ALL and the Kappos et al. (2006) taxonomies (Sect. 3) and the respective fragility models (Sect. 4) are combined to estimate the probabilities of exceedance for all the damage states, resulting in a total number of four analyses.

Figure 12 shows the distribution of the aggregate damage for the whole study area for the two fragility models (ESRM20 and Kappos et al. 2006, 2010) combined with the two hazard cases (Approach 1 and Approach 2). To examine the effect of the selection of the fragility model, expected damages estimated with the same site modelling approach (1 or 2) and with different fragility models are compared. In this regard, as mentioned in the previous section, the two fragility models differ on the definition of the damage states, however a rough correspondence between the two models based on the loss indices can be the following: slight, moderate, extensive and complete damage in ESRM20 can be matched to DS2, DS3, DS4 and DS5 damage states by Kappos et al. (2006, 2010) respectively (Table 3). Based on this assumption, 80–85% of the buildings in the study area are estimated to show no or slight damage using the ESRM20 fragility model depending on site modelling approach, while using the Kappos et al. (2006, 2010) model this percentage is approximately 50% (DS0+DS1+DS2). Also, according to ESRM20, around 5% of the total number of buildings, i.e., 3,730 buildings, are expected to collapse, while using the Kappos et al. (2006, 2010) model these numbers are doubled, thus leading to more devastating scenarios of about 10–11% of the buildings to be expected to collapse. At this point we should highlight that in this application we used the purely analytical functions by Kappos et al. (2006, 2010), which often tend to overestimate damages due to some conservative assumptions used in the modelling of the structures. The use of the hybrid functions would probably lead to less severe damages, but this would require adequate damage data from past events for a lot of typologies and for different intensities. In any case, given the uncertainties involved in the development of the different fragility models, the discrepancies between the two models may be considered to be within reasonable limits.

Regarding the site modelling method, this does not seem to significantly affect the distribution of the aggregate damage for the whole study area. This, however, may not be the case when moving to smaller-scale level of analysis, as shown in Riga et al. (2022).

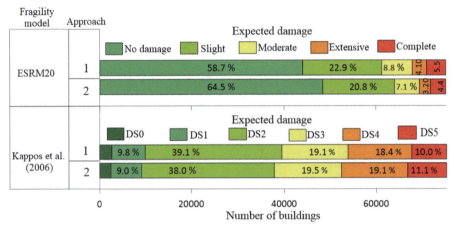

Fig. 12. Aggregate damage distribution of the residential buildings of Thessaloniki, using the fragility model by Romão et al. (2021) and the fragility model by Kappos et al. (2006, 2010) for a seismic hazard with a 475-year return period, estimated using Approach 1 (detailed site model) and Approach 2 (Pitilakis et al. 2020).

Indeed, when moving to municipality level the effect of site modelling method in the estimated seismic damage is becoming more obvious (Fig. 13). For instance, using the ESRM20 fragility model there are significant discrepancies at the south-east coastal area (Kalamaria region, see Fig. 1), where Approach 2 with the amplification factors by Pitilakis et al. (2020) results in significantly higher percentage of complete damage (15.5%, Fig. 13b) compared to Approach 1 with the detailed site model (7.5%, Fig. 13a). This is justified by the increased hazard values in terms of $S_a(1.0 \text{ s})$ (Fig. 6) in this area with Approach 2, due to the high amplification values for $S_a(1.0 \text{ s})$ in soil types C3 and D. It is reminded that $S_a(1.0 \text{ s})$ is used as intensity measure by the ESRM20 fragility model for the majority of the buildings in the area (Fig. 9). On the contrary, for the fragility model by Kappos et al. (2006, 2010), this percentage is higher for Approach 1 (12%, Fig. 13c) compared to Approach 2 (9%, Fig. 13d), as the only intensity measure used in this model, i.e. PGA, is higher for Approach 1 in Kalamaria (Fig. 5). However, the overall discrepancies of the two site modelling methods are smoothed down in this case, making the simplified Approach 2 equally efficient with the more demanding Approach 1. Regarding the effect of the adopted fragility model, the Kappos et al. (2006, 2010) model leads to significantly higher percentages of extensive and complete damages of the buildings in all municipalities compared to the ESRM20 fragility model, as was the case for the whole study area.

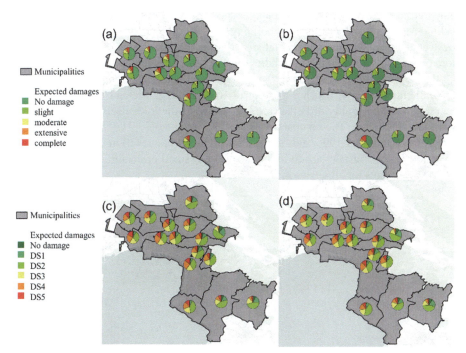

Fig. 13. Spatial distribution of the expected damages of the residential buildings of Thessaloniki using (a) the Romão et al. (2021) fragility model and Approach 1, (b) the Romão et al. (2021) fragility model and Approach 2, (c) the Kappos et al. (2006, 2010) fragility model and Approach 1 and (d) the Kappos et al. (2006, 2010) fragility model and Approach 2.

Similar observations can be made when moving to even smaller scale, as is shown in Fig. 14, which illustrates the percentages of the buildings which are expected to be in the complete (for ESRM20)/DS5 (for Kappos et al. 2006, 2010) damage state at census sector level. It is reminded that all census sectors have more or less the same number of residences. This damage state was selected due to its devastating nature and to its comparability in the two fragility models. Using the ESRM20 fragility model, the effect of the site modelling at the Kalamaria region discussed above is confirmed (Fig. 14a and Fig. 14b). In addition, the results for the municipality of Thessaloniki (see Fig. 1), i.e., the center of the broader metropolitan area, are of great interest. The municipality of Thessaloniki has a total of 20,680 residential buildings, representing 27.5% of the residential buildings of the whole study area. 74.5% of the buildings in the municipality of Thessaloniki are designed with low code level or even no seismic code at all. The results of the seismic risk analyses show higher percentages of complete damage in the coastal area than in the rest of the municipality in all four analyses (Fig. 14). The highest percentages are observed in the southern part of the coastal area for the case of ESRM20 fragility model and Approach 2 (around 40%), where the soft soil D and the accompanying amplification factors of Pitilakis et al. (2020) result in higher $S_a(1.0$ s) values (see Fig. 6b) and, thus, in higher damages than Approach 1. The Kappos et al. (2006, 2010) fragility model adopts the PGA values which are higher in this area compared to the rest of the municipality, especially with site modelling Approach 1, leading to a significant percentage of buildings in DS5, around 20–30%. We should note, however, that the sectors in the coastal area, despite their large area in some cases, have more or less the same number of buildings as the rest of the sectors in the municipality, and thus do not affect significantly the aggregate probability of complete damage for the whole municipality, which comes up to 16% (see Fig. 13c and Fig. 13d). On the contrary, the municipality of Pefka (see Fig. 1) shows the lowest seismic risk in terms of structural damage. The comparatively new building stock (71.3% of the buildings were constructed after 1996, thus designed with high code level) in combination with the relatively low seismic hazard in the area lead to small probabilities of complete or extensive damages for all analyses.

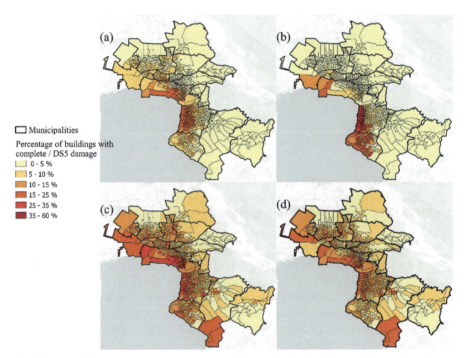

Fig. 14. Spatial distribution of the percentage of the residential buildings with complete damage using (a) the Romão et al. (2021) fragility model and Approach 1 (b) the Romão et al. (2021) fragility model and Approach 2, (c) the Kappos et al. (2006, 2010) fragility model and Approach 1 and (d) the Kappos et al. (2006, 2010) fragility model and Approach 2.

6   Economic Losses

Expected economic losses have been derived through scenario-based risk analyses with precomputed ground motion fields in OpenQuake Engine (Pagani et al. 2014; Silva et al. 2013) where the seismic hazard estimated with Approach 1 and 2 (Sect. 2) and the exposure models with the GED4ALL and the Kappos et al. (2006) taxonomies (Sect. 3) and the respective fragility models (Sect. 4) are combined to estimate the economic losses and loss ratios, resulting in a total number of four analyses. For all analyses we assumed a replacement cost of 1000 €/m^2, which is the average cost proposed by ESRM20 for the study area.

The aggregate economic losses for the study area, as well as the total loss ratio, estimated as the ratio between the total replacement cost and the total repair cost, resulting from the four analyses are included in Table 4, while Fig. 15 illustrates the spatial distribution of loss ratio at the study area at census sector level.

Table 4. Expected aggregate economic losses and loss ratios for the residential buildings of Thessaloniki for seismic hazard with a 475-year return period

Vulnerability model	Seismic hazard	Aggregate economic losses	Aggregate loss ratio
ESRM20	Approach 1 - Detailed site model	4.4 billion €	0.12
	Approach 2 - Pitilakis et al. (2020)	3.8 billion €	0.11
Kappos et al. (2006, 2010)	Approach 1 - Detailed site model	8.8 billion €	0.25
	Approach 2 - Pitilakis et al. (2020)	8.2 billion €	0.24

It is obvious that the Kappos et al. (2006, 2010) vulnerability model leads to increased, almost double, aggregate losses compared to ESRM20. Similarly to what was observed for the aggregate damages, the site modelling method does not affect the aggregate economic losses and the aggregate loss ratio. However, significant discrepancies, show up when moving to census sector level (Fig. 15). For instance, in the southern part of the coastal area of the municipality of Thessaloniki (see Fig. 1) which has relatively old building stock, the ESRM20 vulnerability model leads to substantially increased loss ratio than the rest of the area, around 30% using Approach 1 and around 45% using Approach 2 (Fig. 15a and Fig. 15b). This is in line with the spatial distribution of $S_a(1.0 \text{ s})$ values shown in Fig. 6, which is the main intensity measure adopted by ESRM20 for the buildings of this area. On the contrary, using the Kappos et al. (2006) vulnerability model, the estimated loss ratio ranges between 35% and 45% (Fig. 15c and Fig. 15d) with higher values observed in the case of the Approach 1, due to higher PGA values than the ones resulting from Approach 2 in this area. In the municipality of Pefka, which is characterized by lower seismic hazard, the seismic risk in terms of loss ratio

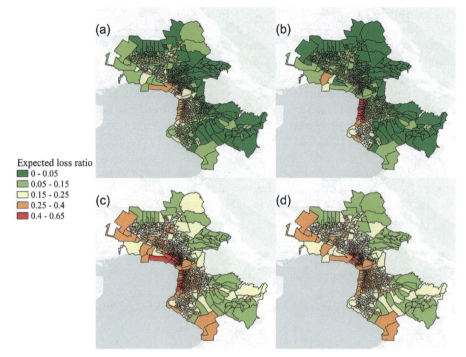

Fig. 15. Spatial distribution of the expected loss ratio of the residential buildings of Thessaloniki using (a) the Romão et al. (2021) fragility model and Approach 1, (b) the Romão et al. (2021) fragility model and Approach 2, (c) the Kappos et al. (2006) fragility model and Approach 1 and (d) the Kappos et al. (2006) fragility model and Approach 2

is lower than for the rest of the study area. However, even in this case, the vulnerability model by Kappos et al. (2006, 2010) leads to slightly higher loss ratio (5–15%) than ESRM20 (less than 5%).

These results are in accordance with the probabilities of complete damage/DS5 shown in Fig. 14. It is reminded that in both vulnerability models, the complete damage/DS5 is mainly affecting the economic losses, as it is the damage state with the highest loss index (ratio of repair cost to replacement cost).

7   Discussion

The large number of uncertainties involved in the methodological chain of seismic risk assessment (Riga et al., 2017) inevitably makes the ability of a seismic risk model to accurately predict future losses a real challenge. In this work we investigated the effect of two of the most important sources of uncertainties, i.e., the level of precision of the site modelling method and the selection of the fragility and vulnerability models, on the seismic risk assessment of the residential buildings of Thessaloniki, the second largest city of Greece. In this regard, we considered two different approaches for site effects modelling and two different vulnerability models, including in both cases the respective components of the recent European Seismic Hazard (ESHM20, Danciu et al. 2021) and Risk (ESRM20, Crowley et al. 2021) Models, which are expected to be applied widely for large scale seismic risk application in Europe in the near future.

Regarding site modelling for seismic hazard assessment, in Approach 1 we used a detailed $V_{s,30}$ model for the study area based on the microzonation study of Thessaloniki as input in the ESHM20 hazard logic tree, while in Approach 2 we classified the study area into different site classes based on the classification scheme by Pitilakis et al. (2020) and amplifies the ESHM20 hazard for rock using the Pitilakis et al. (2020) amplification factors. Approach 1 is considered as more rigorous compared to Approach 2 which is more simplified. The site modelling certainly affects the seismic hazard output, with Approach 1 resulting in significantly higher PGA values for a return period of 475 years at the coastal area compared to Approach 2. On the contrary, the high amplification factors for $S_a(1.0 \text{ s})$ by Pitilakis et al. (2020) for the soil types in the coastal area of the centre of Thessaloniki and in Kalamaria, result in an overestimation of $S_a(1.0 \text{ s})$ with respect to Approach 1. These discrepancies are expected to be propagated in the estimated damages and losses, depending of course on the intensity measures used in the adopted vulnerability models. Indeed, at small scale, i.e., census sector level, significant differences in the estimated damages and losses were observed in these particular areas, with contradictory trends for the two different vulnerability models, due to the different adopted intensity measures. With the ESRM20 vulnerability model, the higher $S_a(1.0 \text{ s})$ values with Approach 2 result in increased risk in the coastal area and Kalamaria, while with the Kappos et al. (2006, 2010) vulnerability model, the higher PGA values of Approach 1 result in higher risk. However, the observed discrepancies are smoothed down when moving to larger scale, encouraging the use of simplified methods for site modelling at large scale seismic risk analyses at regions lacking detailed information on site parameters. These results are in accordance with the observations made by Riga et al. (2022).

Regarding the effect of the selection of fragility/vulnerability model, this parameter was found to have a much more significant impact on the estimated risk, both at small scale and large scale. With the models by Kappos et al. (2006, 2010), the estimated number of buildings in DS5 (collapse), as well as the estimated economic losses for the whole study area, are twice the number of the respective estimated with the ESRM20 models regardless of adopted site modelling method. These discrepancies are important, but yet within reasonable limits, considering the very high dispersion observed between fragility functions in the literature for similar building typologies. Part of the discrepancies could be attributed to the differences in the definition of the building typologies and damage thresholds, the adopted intensity measures and the way the functions have been developed, while they are also influenced by the hazard and site effect model adopted in the analysis and the evaluation of the IM. The use of the hybrid functions by Kappos et al. (2006, 2010) instead of the analytical ones adopted in this study would probably lead to less severe damages, but this would require adequate damage data from past events for a lot of typologies and for different intensities.

In overall, the main conclusions with respect to physical damages can be summarized as follows: (a) the highest probabilities of complete damage/DS5 are observed in the central part of the study area, and specifically at the coastal area of the historical center of the town, (b) the Kappos et al. (2006, 2010) fragility model overestimates the expected damages compared to ESRM20 (Romão et al. 2021), (c) the site modelling method may affect significantly the results at smaller scale analysis, but at large scale more simplified methods, such as Approach 2, seem to be able to provide reliable results, (d) the fragility models and the intensity measure adopted by the vulnerability model has a significant impact on the results. The same observations can be made for the expected economic losses and loss ratios.

Finally, it is noted that in this study we used only the seismic hazard with a 475-year return period (10% probability of exceedance in 50 years), which is widely adopted by seismic codes for the seismic design of structures. Considering lower or higher return periods, as is often the case for seismic risk analyses, or probabilistic seismic risk assessment may lead to different conclusions.

Acknowledgements. We are grateful to Dr. Georgios Panagopoulos and Professor Andreas Kappos for providing data on the fragility and vulnerability curves by Kappos et al. (2006, 2010), Laurentiu Danciu for his support on ESHM20 and Helen Crowley for her support on ESRM20.

Data and Resources. OpenQuake Engine is available for download at https://www.globalquakemodel.org/oq-get-started.

The main datasets and OpenQuake input files of ESHM20 and ESRM20 are online available at https://gitlab.seismo.ethz.ch/efehr.

The results of the ESHM20 are open to access and download at hazard.efehr.org, whereas those of the ESRM20 are distributed by risk.efehr.org.

References

Anastasiadis, A., Raptakis, D., Pitilakis, K.: Thessaloniki's detailed microzoning: subsurface structure as basis for site response analysis. Pure Appl. Geophys. **158**(12), 2597–2633 (2001)

Crowley, H., et al.: European seismic risk model (ESRM20). EFEHR Technical report 002 V1.0.0 (2021). https://doi.org/10.7414/EUC-EFEHR-TR002-ESRM20

Danciu, L., et al.: The 2020 update of the European seismic hazard model: model overview. EFEHR Technical report 001, v1.0.0 (2021). https://doi.org/10.12686/a15

ELSTAT (2011): Population-housing census, Hellenic statistical authority (2011)

Fotopoulou, S., Karafagka, S., Petridis, C., Manakou, M., Riga, E., Pitilakis, K.: Vulnerability assessment of school buildings: generic versus building-specific fragility curves (2022). (under review)

Kappos, A., Panagopoulos, G., Panagiotopoulos, C., Penelis, G.: A hybrid method for the vulnerability assessment of R/C and URM buildings. Bull. Earthq. Eng. **4**(4), 391–413 (2006). https://doi.org/10.1007/s10518-006-9023-0

Kappos, A.J., Panagopoulos, G.: Fragility curves for reinforced concrete buildings in Greece. Struct. Infrastruct. Eng. **6**(1–2), 39–53 (2010)

Karapetrou, S., Manakou, M., Bindi, D., Petrovic, B., Pitilakis, K.: "Time-building specific" seismic vulnerability assessment of a hospital RC building using field monitoring data. Eng. Struct. **112**, 114–132 (2016). https://doi.org/10.1016/j.engstruct.2016.01.009

Kotha, S.R., Weatherill, G., Bindi, D., Cotton, F.: A regionally-adaptable ground-motion model for shallow crustal earthquakes in Europe. Bull. Earthq. Eng. **18**(9), 4091–4125 (2020). https://doi.org/10.1007/s10518-020-00869-1

Martins, L., Silva, V.: Development of a fragility and vulnerability model for global seismic risk analyses. Bull. Earthq. Eng. **19**, 6719–6745 (2021)

Pagani, M., et al.: OpenQuake Engine: an open hazard (and risk) software for the global earthquake model. Seismol. Res. Lett. **85**(3), 692–702 (2014)

Pitilakis, K., Riga, E., Anastasiadis, A.: Towards the revision of EC8: proposal for an alternative site classification scheme and associated intensity-dependent amplification factors. In 17th World Conference on Earthquake Engineering, Sendai, Japan (2020)

Riga, E., Karatzetzou, A., Mara, A., Pitilakis, K: Studying the uncertainties in the seismic risk assessment at urban scale applying the capacity spectrum method: the case of Thessaloniki. Soil Dyn. Earthq. Eng. **92**, 9–24 (2017)

Riga, E., Apostolaki, S., Karatzetzou, A., Danciu, L., Pitilakis, K: The role of site modelling in seismic hazard and risk assessment at urban scale. The case of Thessaloniki, Greece. Italian J. Geosci. **141**(2), 1–18 (2022). https://doi.org/10.3301/IJG.2022.16. (Under review)

Romão, X., et al.: European building vulnerability data repository. Zenodo (2021). https://doi.org/10.5281/zenodo.4062410

Silva, V., Brzev, S., Scawthorn, C., Yepes, C., Dabbeek, J., Crowley, H.: A building classification system for multi-hazard risk assessment. Int. J. Disaster Risk Sci. **13**, 161–177 (2022). https://doi.org/10.1007/s13753-022-00400-x

Silva, V., Crowley, H., Pagani, M., Monelli, D., Pinho, R.: Development of the OpenQuake engine, the global earthquake model's open-source software for seismic risk assessment. Nat. Hazards **72**(3), 1409–1427 (2013). https://doi.org/10.1007/s11069-013-0618-x

Silva, V., et al.: Current challenges and future trends in analytical fragility and vulnerability modeling. Earthq. Spectra **35**(4), 1927–1952 (2019)

The Influence of Soil-Foundation-Structure Interaction on the Seismic Performance of Masonry Buildings

Francesco Silvestri$^{(\boxtimes)}$ (iD), Filomena de Silva(iD), Fulvio Parisi(iD),
and Annachiara Piro(iD)

University of Naples Federico II, Naples, Italy
francesco.silvestri@unina.it

Abstract. Most of the damage and of the casualties induced even by the most recent strong-motion earthquakes which stroke Central and Southern Italy can be attributed to the extreme seismic vulnerability of the ordinary residential buildings. Both in small villages and in mid-size towns, these latter are mainly constituted by two- to four-story masonry structures built without anti-seismic criteria, with direct foundations corresponding to an in-depth extension of the loadbearing walls or to an underground level. For such structures, especially when founded on soft soils, soil-foundation-structure interaction can significantly affect the seismic performance; on the other hand, its influence must be handled with methods which should be as simple and straightforward as possible, in order to be cost-effective and accessible by practitioners. The contribution wishes to summarize the studies carried out in the last years at University of Napoli Federico II, based on parametric numerical analyses on complete soil-foundation-structure models reproducing the most recurrent building configurations combined with different subsoil conditions. The analyses provided calibration criteria for: i) predicting the elongation of the fundamental period of the structure, ii) defining and optimizing fragility functions for different damage mechanisms accounting for soil-foundation-structure interaction. The effectiveness of these simplified tools was validated against well-documented case studies at the scale of single instrumented buildings or of extended areas, with building properties and subsoil conditions comparable to those adopted in the parametric analyses.

Keywords: Soil-foundation-structure interaction · Masonry building · Replacement oscillator · Fragility function

1 Introduction

According to the World Bank [1], the number of natural disasters in the last years has increased both in magnitude and frequency. Certainly, earthquakes are among the natural events with the greatest impact on the world economy and on the loss of lives. Such evidence makes the seismic protection of the built heritage to be urgent, in order to limit damages and consequent human and economic losses.

L. Wang et al. (Eds.): PBD-IV 2022, GGEE 52, pp. 162–194, 2022.
https://doi.org/10.1007/978-3-031-11898-2_9

The seismic risk in Europe is maximum in the South-East Mediterranean countries, due to both the significant seismic hazard, as shown by the European Seismic Hazard Map [2] in Fig. 1a, and the high structural vulnerability [3], as remarked after the most recent earthquakes (see Fig. 1b) that struck Italy [4, 5] or Greece [6].

The high vulnerability is due to the fact that most of the existing structures are made of unreinforced masonry (URM) (see Fig. 1a) [3, 7] and were built before the emanation of seismic codes. The consequent lack of structural elements to withstand horizontal actions makes the load-bearing walls to be subjected to significant out-of-plane (OOP) lateral actions, frequently resulting in local collapse mechanisms.

Moreover, numerous URM buildings are founded on shallow soft covers which amplify the seismic motion and affect the structural response through soil-foundation-structure interaction (SFSI) mechanisms [8]. Amplification of the seismic waves propagating through a soft soil deposit may be further affected by the kinematic interaction between the embedded masonry foundation and the surrounding soil, which modifies the foundation input motion with respect to the free-field conditions. Simultaneously, the structure transfers to its base inertial forces and moments which induce foundation swaying and rocking motions. These latter affect the structural response in terms of displacements and accelerations, as well as by increasing the period, T^*, and damping, ξ^*, due to the additional energy dissipated by wave radiation and soil hysteresis.

Nevertheless, few research studies have investigated the effects of SFSI on the seismic response of URM structures, like as towers [9–13], fortresses [14, 15], masonry bridges [16] and school buildings [17]. Even fewer are the cases in which fragility curves were computed accounting for SFSI effects [18, 19].

All the above-mentioned studies were developed for specific case studies or very peculiar structures, rather than ordinary residential buildings which this paper focuses on. Thus, after a synthetic overview of a hierarchy of approaches available for the analysis of SFSI (Sect. 2), this study was addressed to the following main objectives:

(i) to evaluate the effects of soil deformability on the dynamic response of typical URM residential buildings (Sect. 3), by updating traditional simplified approaches according to the results of advanced numerical analyses (Sect. 4);
(ii) to account for SFSI effects in order to define fragility curves relevant to the OOP damage mechanisms of masonry walls (Sect. 5).

The main methodological advances developed with reference to objectives (i) and (ii) were assessed and validated through applications to well-documented case studies at territorial scale.

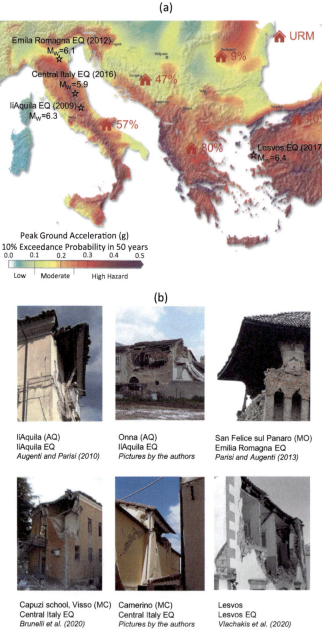

Fig. 1. (a) SHARE seismic hazard map of peak ground acceleration with a probability of exceedance equal to 10% in 50 years [2] and percentage of URM buildings in some European countries [3], (b) masonry buildings damaged by out-of-plane mechanisms in recent seismic events.

2 Approaches for the SFSI Analysis: An Overview

Figure 2 shows a hierarchy of models with different degrees of refinement for soil and structure adopted in the literature, to account for SFSI in seismic performance assessments. The structural models can be sorted according to an increasing complexity level as follows:

- a single-degree-of-freedom (SDOF) oscillator with mass m, flexural stiffness k, and damping ratio ξ, which is characterized by a single vibration mode and, consequently, by a single natural period (Fig. 2a and Fig. 2d);
- a multi-degree-of-freedom (MDOF) system with n lumped masses m_i, stiffness matrix **K**, and damping matrix **C**, which is characterized by n vibration modes (Fig. 2b and Fig. 2e);
- a continuum model with mass density ρ, shear modulus G, Poisson's ratio ν, and a given shape and size, which is characterized by infinite vibration modes and can be discretized by a numerical technique, such as the finite element method or finite difference method (Fig. 2c and Fig. 2f).

On the other hand, the presence of the soil can be simulated through:

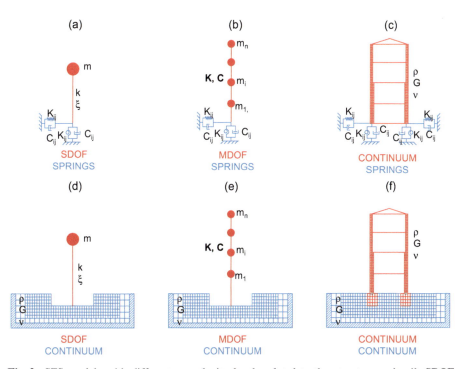

Fig. 2. SFS models with different complexity levels related to the structure and soil: SDOF oscillator, MDOF system, continuum structural model on springs and dashpots (a, b, c), and on continuum soil model (d, e, f).

- a combination of springs and dashpots with stiffness K_{ij} and damping coefficients C_{ij}, related to each translational and rotational component of the foundation motion (Fig. 2a-b-c);
- a continuum model with mass density ρ, shear modulus G, Poisson's ratio v, characterized by suitable in-depth and lateral extensions as well as by reflecting or absorbing boundaries (Fig. 2d-e-f).

In the simplest approaches, the spring stiffness and the constant of the dashpots simulating the soil compliance to the foundation motion are calibrated through the impedance functions [20]. They are the sum of a real part representing the dynamic stiffness and an imaginary part accounting for the damping:

$$\bar{K}_{ij} = k_{ij}(a_0)K_{ij} + \mathrm{i}\,\omega c_{ij}\,(a_0)C_{ij} \tag{1}$$

In Eq. (1):

- i is the imaginary unit;
- the subscripts i, j indicate that $\bar{K}_{ij}$ links the component i of the vector of the loads transmitted by the foundation into the soil to the component j of the displacement vector;
- the low-frequency stiffness, K_{ij}, and the dashpot coefficient, C_{ij}, depend on the soil shear modulus, G, and Poisson's ratio, v, as well as on a characteristic dimension of the foundation, r;
- the dynamic coefficients, $k_{ij}(a_0)$ and $c_{ij}(a_0)$, depend on the vibration frequency, ω, the characteristic dimension of the foundation, r, and the soil shear wave velocity, V_S, through the dimensionless frequency factor, $a_0 = \omega r / V_S$.

Unless experimentally measured from records on existing structures during free and forced vibration tests [21, 22], the impedances are typically calibrated through analytical expressions referred to rigid massless foundations, more or less embedded in the soil. The latter is generally assumed to be an elastic homogeneous half-space [20, 23], an elastic stratum placed on a half-space (e.g. [20]) or a layered soil profile (e.g. [24]).

As usual in soil dynamics, the variation of the impedance under moderate to strong motions due to non-linear and dissipative soil behavior can be taken into account through the equivalent-linear approach [25]. This is the main limitation of the impedance functions, which can be overcome through macro-element approaches, reproducing the overall soil-foundation behavior through a single constitutive relationship, capable of describing the non-linear behavior until failure [26, 27].

The simple oscillator with compliant base (Fig. 2a) is the system more extensively adopted as a reference to derive simplified approaches to calculate the SFS dynamic response parameters (T^*, ξ^*) since the pioneering work by Veletsos and Meek [28] to the most recent analytical developments [29] and adaptations to non-trivial soil-foundation-structure systems (e.g. [30] and [31]).

In terms of analytical procedures, the study of interaction effects through each of the models reported in Fig. 2 can be classified as:

– uncoupled approaches, in which the system is analyzed by decoupling the 'kinematic' from the 'inertial' interaction with the so-called "substructure method";
– coupled approaches, in which all the effects of the interaction can be evaluated simultaneously, performing dynamic analyses on a model including soil, foundation and structure.

In the first kind of procedures, a dynamic analysis is performed on a subsoil model including the foundation stiffness but neglecting the structural mass. The resulting 'foundation input motion' (FIM) is used as dynamic load at the base of a complementary structural model, in which the soil-foundation system is replaced by a set of springs and dashpots. If the foundation is shallow, the kinematic interaction is negligible, hence the FIM is almost coincident with the '*free field* motion' derived from a conventional seismic response analysis. This approach is typically applied to the models shown in Fig. 2a, b, c.

Conversely, the coupled procedures jointly analyze the structure and the soil, with this latter modeled as a continuum (see Fig. 2 d, e, f). In such cases, a rigorous calibration of all the parameters involved in the simulation is necessary; otherwise, the results may not be the most realistic. On the other hand, in the uncoupled approach, an accurate definition of equivalent properties is required to consider material nonlinearity and the actual geometry of the single elements of the SFS system (i.e. structure with distributed mass, embedded and/or flexible foundation, soil inhomogeneity and/or irregular morphology).

Some of these aspects represent a significant limitation in seismic performance assessment of URM structures, usually characterized by bearing wall embedments or underground stories which cannot be assumed rigid, due to the material nature and deterioration caused by aging. It follows that the more refined coupled procedures should be in principle adopted for most URM structures, but the corresponding experimental and analytical effort can be justified only for high-value historical buildings. For ordinary residential structures, however, numerical simulations in which the SFS system is regarded as a coupled model can support the calibration and validation of simpler and more sustainable approaches, as shown in the following sections.

3   Soil – Foundation – Structure Systems Analyzed

The reference soil-foundation-structure models analyzed in this study are summarized in Fig. 3 [30]. The structural geometry reproduces the transverse section of a masonry building for residential use with single-span floors (Fig. 3a). The width, b, and inter-story height are constant and respectively equal to 8 m and 4 m; the total height, h, varies considering 2, 3 and 4 above-ground stories, which correspond to aspect ratios h/b equal to 1, 1.5 and 2. Such structural configuration is recurrent in constructions located in the Euro-Mediterranean region [32, 33].

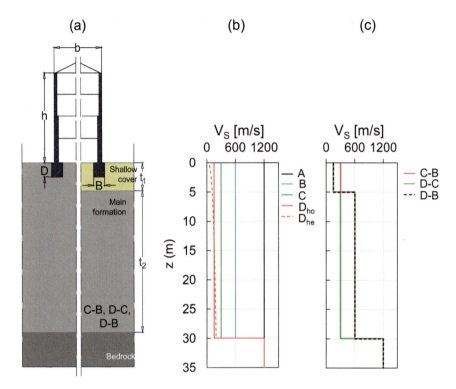

Fig. 3. (a) SFS model; profiles of shear wave velocity for (b) homogeneous and (c) layered soil.

The structure consists of two load-bearing masonry walls connected each other by mixed steel-tile floor systems, pinned to the walls and composed of steel I-beams, clay tiles and poor filling material. The pitched roof is made of timber elements. The thickness of the walls reduces along the building height, leading to a fairly homogeneous distribution of vertical stresses from the ground floor to the top. As typical for masonry residential buildings [33], the lowly embedded foundations were assumed to be made of the same material of the above-ground structure, with a width, B, and a depth, D, equal to 2.0 m and 2.5 m, respectively.

The building schemes were settled on:

- four homogeneous subsoil models, with lithology and properties representative of the Eurocode-conforming ground types A, B, C and D [34];
- three-layered subsoil profiles, D-B, D-C and C-B, constituted by a shallow cover with a thickness $t_1 = 5$ m overlying a main formation as thick as $t_2 = 25$ m.

A stiff bedrock pertaining to ground type A was systematically assumed at a depth of 30 m. The stiffness and strength parameters of the subsoil model belonging to the class D were assumed as either constant (homogeneous profile, D_{ho}) or variable with depth (heterogeneous profile, D_{he}). The profiles of shear wave velocity, V_S, for each model are shown in Fig. 3b–c.

Table 1 and Table 2 summarize the material properties adopted in the linear and nonlinear analyses, respectively. Soil and masonry were modeled as a continuum with mass density, ρ, bulk modulus, K, shear modulus, G, and Poisson's ratio, v. Floors and roof were modeled as equivalent beam elements made of a homogenized material.

Table 1. Material properties adopted for linear analyses.

Material	V_S (m/s)	ρ (kg/m^3)	K (MPa)	G (MPa)	v
Soil A – bedrock	1200	2200	4224	3170	0.20
Soil B – gravel	600	2000	1200	720	0.25
Soil C – dense sand	300	1800	351	162	0.30
Soil D – loose sand	150	1600	108	36	0.35
TSM – tuff stone masonry	–	1600	–	360	0.49
Steel-tile floor	–	1750	–	12500	0.20
Timber roof	–	300	–	542	0.20

Table 2. Material properties adopted for nonlinear analyses.

Material	V_S (m/s)	ρ (kg/m^3)	K (MPa)	G (MPa)	v	c_u (kPa)	ϕ (°)	c (MPa)	σ_c (MPa)	σ_t (MPa)	D_0 %
Soil A - bedrock	1200	2200	4224	3170	0.20	–	–	–	–	–	2
Soil C - dense sand	300	1800	351	162	0.30	–	35	–	–	–	2
Soil D_{ho} - o.c. clay	150	1600	1788	36	0.49	100	–	–	–	–	2
Soil D_{he} - n.c. clay	100 191	1600	915 2970	18 60	0.49	10 70	–	–	–	–	2
RSM Rubble stone masonry	–	1900	14500	290	0.49	–	27	0.45	1.5	0.15	5
CBM Clay brick masonry	–	1600	25000	500	0.49	–	36	0.87	3.5	0.35	5
Steel-tile floor	–	1350	30000	12500	0.20	–	–	–	–	–	–
Timber roof	–	300	1300	542	0.20	–	–	–	–	–	–

The soil density and Poisson's ratio were realistically assumed as respectively increasing and decreasing with V_S, and representative of soft rock (A), gravel (B), dense sand (C) and loose sand (D). The properties of the tuff stone masonry (TSM) considered for the linear analyses were defined from the experimental results collected by [35].

In nonlinear analyses, a limit shear strength was introduced for soil types C and D through a Mohr-Coulomb criterion with a friction angle, ϕ, equal to 35° for the former, and a Tresca criterion for the latter, which this time was assumed as a fine-grained soil characterized by an undrained strength, c_u. The homogeneous and heterogeneous soil profiles, D_{ho} and D_{he}, were defined as representative, respectively, of a lightly overconsolidated and a normally consolidated clay with a plasticity index $IP = 30\%$ for the homogeneous (D_{ho}) and $IP = 20\%$ for the heterogeneous (D_{he}) profile. The variations with depth of shear stiffness at small strains and undrained strength in the heterogeneous soil profile follow the model adopted by Capatti et al. [36]. Due to the light overconsolidation, for soil type D_{ho} the undrained strength was set constant with depth and higher than that of the heterogeneous soil profile.

For both soft soil profiles, a pre-failure hysteretic behavior was modelled. The strain-dependent variation of normalized shear modulus, G/G_0, and of the damping ratio, D, was described for the soil type C by the standard curves reported by Seed and Idriss [37], while for soil profiles D by those suggested by Vucetic and Dobry [38] for the relevant plasticity indexes. The standard curves were implemented in the numerical model by fitting them through 'sigmoidal' functions.

The energy dissipation at very small strains, for nonlinear analyses, was simulated through a Rayleigh approach, with a minimum damping ratio, D_0, equal to 2% and 5%, for soil and structure, respectively. The control frequency was calibrated on the fundamental frequency of the input motion joint to that of the *free-field* soil or the fixed-base structure, as detailed by Piro [39].

Finally, in nonlinear analyses two different types of structural material were considered, namely, rubble stone masonry (RSM) and clay brick masonry (CBM). The masonry was modeled as an equivalent homogeneous material adopting an elastic-perfectly plastic constitutive model with the Mohr-Coulomb failure criterion, to limit the computational work. The elastic parameters listed in Table 2 were set equal to the median values reported by the Italian Building Code Commentary [40] for existing masonry buildings. The value of the friction angle, ϕ, was set based on the friction coefficient, depending on the ratio between the compression stress, σ_0, at the base of the above-ground structure under the static load and the uniaxial compression strength, σ_c, equal to the value reported by [40]. The cohesion, c, was back-calculated from ϕ and σ_c on the basis of the Mohr-Coulomb criterion. A tensile cut-off, σ_t, equal to $0.1\sigma_c$ was also implemented in the model.

Table 3 summarizes all the combinations of soil and structure models with the corresponding analyses represented by different color shadings. It should be underlined that, while the linear analyses were focused on the masonry type characterized by average mechanical properties (TSM) and extended to all subsoil profiles considered, the nonlinear simulations were addressed to assessing the SFSI effects by comparing each other the extreme combinations (i.e. softest vs. stiffest soils, as well as most vs. less deformable masonry types).

4   Prediction of the Fundamental Frequency of the SFS System

4.1   Coupled Approach

The SFS systems described in Sect. 3 were analyzed as continuum coupled models (see Fig. 2f) by means of the 2D finite difference code FLAC ver. 7.0 [41]. As an example, Fig. 4 shows the model adopted for the SFS systems on a homogeneous soil profile. The

Table 3. Summary of the dynamic analyses performed (linear: light blue, nonlinear: red).

Soil model	Squat structure (h/b=1)			Slender structure (h/b=1.5)			Very slender structure (h/b=2)		
	TSM	RSM	CBM	TSM	RSM	CBM	TSM	RSM	CBM
A	blue	red	red	blue			blue	red	red
B	blue			blue			blue		
C	blue	red		blue			blue	red	red
D_{ho}	blue	red		blue			blue	red	red
D_{he}	blue	red		blue			blue	red	red
D-B	blue			blue			blue		
D-C	blue			blue			blue		
C-B	blue			blue			blue		

subsoil domain is 50 m wide, 30 m deep and includes the top of the bedrock through a finite layer with a thickness of 10 m.

The infinite extension of bedrock in depth was simulated by dashpots attached to the bottom nodes and oriented along the normal and shear directions. To minimize the model size, *free-field* boundary conditions were imposed to the vertical sides of the soil volume, simulating an ideal horizontally layered soil profile connected to the main-grid domain through viscous dashpots. The soil was discretized into a mesh of quadrilateral elements, the size of which was defined to satisfy the criterion by [42] for accurately reproducing the shear wave propagation up to a frequency of 25 Hz. In the proximity of the structure, the size of the quadrilateral elements was reduced in order to approximate the dimensions (length and height) of a single masonry brick. Floor and roof systems were modeled through one-dimensional (1D) beam elements with 1 m-wide homogenized cross-section and with pinned connections to load-bearing walls. The input motions were applied as a shear stress time-history at the bottom of the bedrock layer.

The fundamental frequency of each model was firstly computed at strain levels well below structural and geotechnical failure states. Thus, a linear visco-elastic behavior of the materials was assumed, with the properties listed in Table 1 and a very low Rayleigh damping ratio equal to 0.1%. The latter was properly minimized in order to isolate the effect of the radiation damping.

Being impossible to perform modal analyses through FLAC software, the procedure developed by [12] was used to detect the fundamental frequency of each SFS system. The base of the model was subjected to a low-amplitude random input motion with a frequency content ranging between 1 and 25 Hz and a duration of 10 s, after which the free vibration of the SFS system was numerically monitored over 20 s.

The fundamental frequency, f^*, of each SFS system was back-figured from the peaks of the Fourier spectra of the displacement time histories recorded during the free vibration at the control points shown in Fig. 4. The plots in Fig. 5a-b show the dynamic response (in terms of displacement amplitudes, U_i, normalized with respect to the roof maximum

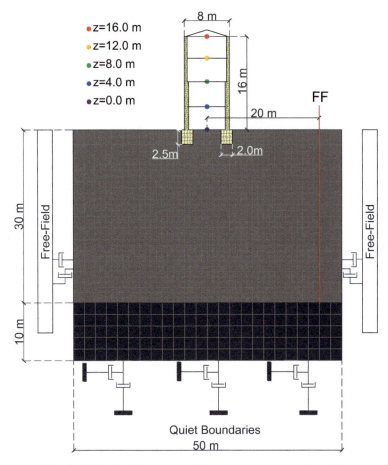

Fig. 4. 2D finite difference mesh for a homogeneous soil profile.

value, U_{TOPmax}) of two-story ($h/b = 1$) and four-story ($h/b = 2$) structures, respectively, laying on homogeneous soil A and layered soil D-B. The SFS fundamental frequency f^* is highlighted by spectral peaks at all elevations, whereas dashed lines indicate the *free-field* soil natural frequencies, denoted as f_{soil}. The fundamental frequency of the fixed-base (FB) structural model, f_0, was assumed to be coincident with that of the SFS system characterized by homogeneous soil type A (upper plots).

A non-negligible reduction of the fundamental frequency of the shortest structure ($h/b = 1$) on the layered soil D-B is observed with respect to the fixed-base value, i.e. 4.18 Hz vs 5.01 Hz. Conversely, the frequency of the tallest structure (Fig. 5b) was found to be much less affected by the soil deformability (f^* from 2.02 Hz to 1.94 Hz), highlighting that the dynamic response of the masonry building is increasingly influenced by SFSI with the increase of the structure-soil stiffness ratio [13, 30].

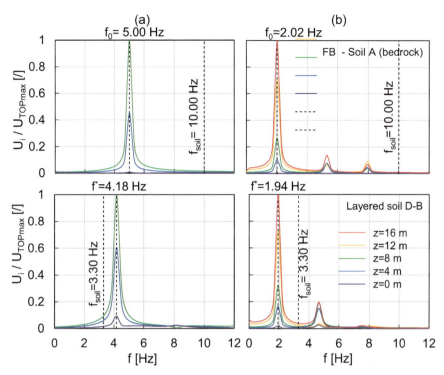

Fig. 5. Frequency response at different structural elevations for (a) two-story ($h/b = 1$) and (b) four-story ($h/b = 2$) SFS systems on rock outcrop and layered soil D-B.

4.2 Simplified Approaches

Veletsos and Meek [28] firstly proposed a closed-form solution to evaluate the fundamental frequency, f^*, of a SFS system (Fig. 6a) based on the ideal scheme of a compliant-base SDOF as that drawn in Fig. 6b.

The dynamic response of the compliant-base SDOF is in turn assumed as equivalent to that of the 'replacement oscillator' (Fig. 6c), i. e. a fixed-base SDOF with the same frequency, f^*, and damping ratio, ξ^*. The total flexibility of this latter to dynamic loadings can be taken as the sum of the flexibilities of each SFS component, as follows:

$$\frac{1}{k^*(f^*)} = \frac{1}{k} + \frac{1}{k_u(f^*)K_u} + \frac{h^2}{k_\theta(f^*)K_\theta} \tag{2}$$

The second and the third denominators on the right side of Eq. 2 are the real parts of the translational and the rotational impedances equivalent to those of the building foundation. By replacing Eq. 2 in the well-known expression of the fundamental frequency of the SDOF, that of the SFS system can be easily calculated as:

$$f^* = \frac{1}{2\pi}\sqrt{\frac{k^*}{m}} \tag{3}$$

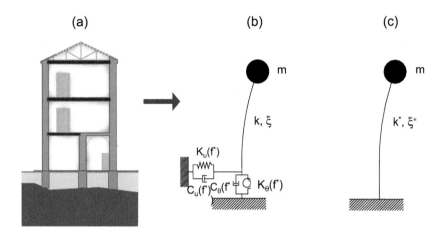

Fig. 6. Definition of replacement oscillator for SFS system: (a) typical transverse cross-section of a URM building; (b) compliant-base SDOF system; (c) replacement oscillator.

The same authors proposed a set of curves representing the dependence of f^* normalized with respect to the fixed base frequency, f^*/f_0, on the soil-structure relative stiffness, σ, defined as:

$$\sigma = \frac{V_S}{f_0 h} \tag{4}$$

The latter parameter is hard to define for URM buildings with irregular geometry above and under the ground level, as well as with flexible foundations placed on layered soil. To overcome such limitation, Piro et al. [30] proposed to calculate σ based on an equivalent shear wave velocity, $V_{S,eq}$, resulting from the weighted contributions - through appropriately calibrated coefficients - of the stiffness and the mass of the SFS components falling in the volume underlying the building significantly affected by the inertial interaction mechanism (see Fig. 7b). For a given aspect ratio, h/b, the equivalent stiffness ratio, $\sigma_{eq} = V_{S,eq}/hf_0$, leads to the corresponding value of the frequency reduction factor, f^*/f_0, by referring to the same curves suggested by Veletsos and Meek [28], as shown in Fig. 7c.

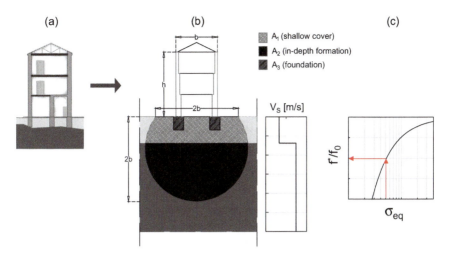

Fig. 7. Approach for the estimation of the frequency reduction factor, f^*/f_0: (a) typical transverse cross-section of a URM building; (b) soil volume affected by inertial interaction mechanism; (c) frequency reduction factor, f^*/f_0, versus equivalent soil-structure stiffness ratio, σ_{eq}.

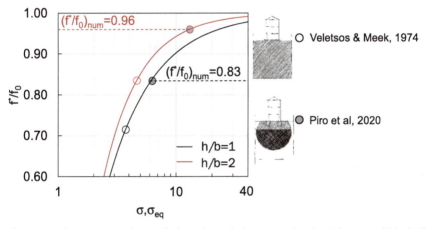

Fig. 8. Comparison between the predictions through the approaches by Veletsos and Meek [28] and Piro et al. [30] for two and four-story SFS systems on layered soil profile D-B.

As an example, Fig. 8 compares the frequency reduction induced by the soil compliance predicted by the traditional formulation of σ [28] (hollow circles) and by σ_{eq} [30] (full circles) for the SFS systems characterized by $h/b = 1$ and $h/b = 2$ on the layered soil profile D-B (see Fig. 5). It is apparent that the traditional formulation leads to a higher reduction of the fundamental frequency because σ is calibrated on the V_S value of the upper softer layer. Moreover, it can be checked that the frequency reduction factors, $(f^*/f_0)_{num}$, shown by the horizontal dashed lines, resulting from the numerical analyses as the ratios between those reported in the lower and upper plots in Fig. 5, are in a perfect agreement with those predicted through the procedure based on σ_{eq}.

4.3 Urban-Scale Application to the City of Matera

The simplified approach above described was applied to seven buildings located in the historical city of Matera, located in Southern Italy and well-known for its peculiar 'Sassi' caves. The two main geological formations (Fig. 9a–b) are the Altamura limestone and Gravina calcarenite which outcrop in the North-West and South-East areas of the urban center. The latter is covered by the Sub-Apennine clays, with thickness varying from a few meters, near the Sassi area, to 40–50 m inwards. Down-hole and seismic refraction tests revealed a shear wave velocity increasing with depth from 146 m/s to 450 m/s in the Sub-Apennine clays and ranging between 394 m/s and 1185 m/s in the Gravina calcarenite depending on its degree of cementation, while it is almost constant (around 950 m/s) in the Altamura limestone.

Figure 9c reports the cross-sections of seven soil-foundation-structure systems analyzed through the simplified approaches based on the replacement oscillator to estimate the fundamental frequency. The soil stratigraphy and the shear wave velocity profiles were inferred from available investigations. In lack of direct measurements, the unit weight was set equal to typical values of 17.50 kN/m^3, 19.50 kN/m^3 and 27.50 kN/m^3 respectively for the clay, calcarenite and limestone formations.

The geometries of the selected buildings (height, width and number of stories) were obtained from the survey by Gallipoli et al. [43], while empirical correlations were applied to estimate the thickness of the bearing walls, made of calcarenite.

An enlargement of 0.15 m at each side of the bearing wall was considered to determine the foundation width, while the embedment was set equal to 1.5 m. In lack of specific survey, the properties of walls with irregular texture provided by the Italian Building Code Commentary [40] were assigned to the masonry, referring to the most recurrent typology in the historical center.

The fundamental frequency of all the seven buildings was measured through Horizontal/Vertical Spectral Ratios (HVSR) of white noise records during the CLARA project [44]. Even though the number of stories is the same for the buildings 1-2-5 in Fig. 9c, the measured value, f_{exp}^*, reduces with the thickness of the clay layer, highlighting a first evidence of the effects of subsoil conditions and of SFS interaction.

Such influence is confirmed by the application of the simplified approaches to estimate f^*, the results of which are compared to f_0 in Table 4. The latter was estimated through a correlation between the experimental frequency and the height of masonry buildings directly founded on calcarenite outcrops, obtained by Gallipoli et al. [43].

On average, the traditional formulation overestimates the frequency reduction, while a significantly better agreement was found between the proposed procedure and the experimental data, as shown by the low values of the percentage error ε, computed as:

$$\varepsilon = \left(\frac{f_{exp}^* - f^*}{f_{exp}^*}\right) \times 100 \tag{5}$$

Note that the only exceptions are represented by buildings 5 and 7, which is likely to be due to their hollow geometrical plan.

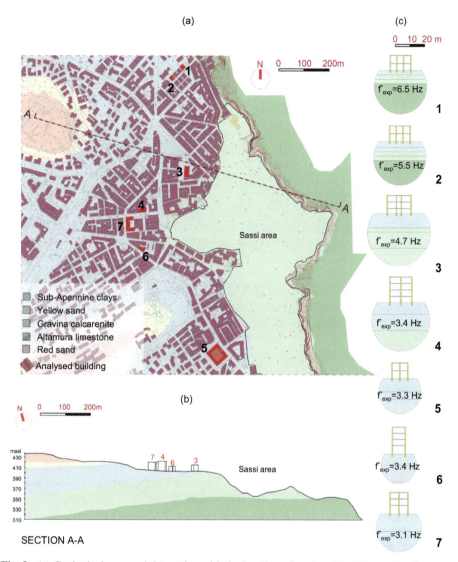

Fig. 9. (a) Geological map and (b) section with the location of analyzed buildings; (c) reference schemes of structure and subsoil for the application of the approach by Piro et al. [30].

5 Nonlinear Performance of SFS Systems

5.1 Reference Input Motions

The influence of SFS interaction on the activation of limit states for both out-of-plane mechanisms of masonry walls and plastic straining in the soil-foundation system was investigated using nonlinear time history analyses.

Table 4. Comparison between experimental and analytical frequencies calculated with the traditional and proposed approaches.

Building ID			Veletsos and Meek [28]				Piro et al. [30]			
	f_0 (Hz)	f^*_{exp} (Hz)	V_S (m/s)	σ (−)	f^* (Hz)	$\varepsilon(\%)$	$V_{S,eq}$ (m/s)	σ_{eq} (−)	f^* (Hz)	$\varepsilon(\%)$
1	7.0	6.5	146	2.0	3.1	53.0	561	7.7	6.1	6.5
2	6.7	5.5	146	2.0	2.9	47.0	411	5.6	5.5	0.0
3	5.9	4.7	203	2.8	3.6	23.8	730	10.0	5.4	−14.4
4	4.1	3.4	239	3.3	3.0	12.0	554	7.6	3.8	−10.0
5	6.5	3.3	190	2.6	3.8	-13.6	198	2.7	3.9	−17.2
6	3.9	3.4	239	3.3	2.9	15.0	519	7.1	3.6	−5.0
7	4.9	3.1	239	3.3	3.3	−9.0	414	5.7	4.1	−33.0

The so-called 'cloud method' [45] was adopted by selecting a set of 15 reference input motions from the SIMBAD database [46]. The 15 ground motions were selected according to the following criteria [45]:

– to consider a wide range of spectral accelerations, i.e. $S_a(T^*) = 0.01$ g–3.00 g;
– to avoid records of the same seismic events;
– to select ground motions recorded on stiff outcropping formations, being site effects properly accounted for in the coupled SFS analysis.

In detail, twelve ground motions were recorded on soils classified as type A, whereas the others were recorded on type B with equivalent shear wave velocity, V_{S30}, higher than 500 m/s. Each input motion was applied to the base of the 16 coupled SFS models subjected to nonlinear time history analyses, i.e. those corresponding to the red-hatched cells in Table 3.

The high variability of the selected input motions is shown in Fig. 10a through the scatter plot of peak ground acceleration (PGA) versus moment magnitude (M_W) and epicentral distance (R). As a matter of fact, the acceleration spectra reported in Fig. 10b are characterized by a significant variation of spectral shapes and amplitudes, these latter by an order of magnitude. The green dashed-dotted lines correspond to the soil fundamental periods, with reference to a linear *free-field* seismic response. The red and black dashed lines identify the ranges of fundamental periods numerically evaluated for brick and rubble masonry buildings, respectively, relevant to the structural models with either $h/b = 1$ or $h/b = 2$ laying on different subsoil profiles (see Table 3).

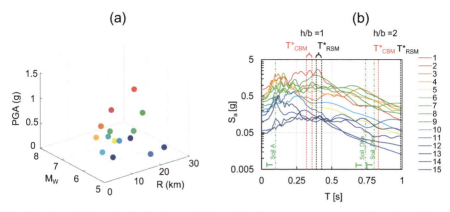

Fig. 10. (a) PGA–M_W–R distribution and (b) acceleration response spectra of the selected input motions.

5.2 Performance Assessment of SFS Components

Soil amplification in *free-field* conditions was firstly investigated in terms of spectral ratios, $S_{a,s}/S_{a,b}$, between surface and bedrock motions. Response spectra were obtained from acceleration time histories predicted at the foundation level ($z = -2.5$ m) and at a distance of 20 m from the structural model axis, which was assumed to be representative of *free-field* conditions (see Fig. 4). The comparison between spectral ratios computed at the foundation level for the soil profiles D_{ho} and D_{he} is shown in Fig. 11 for the ground motions #1, 8 and 14, which were respectively characterized by PGA values equal to 1.15 g, 0.40 g and 0.06 g. Just like in Fig. 10b, the vertical dashed and dashed-dotted lines in Fig. 11 correspond to structural and soil periods, respectively.

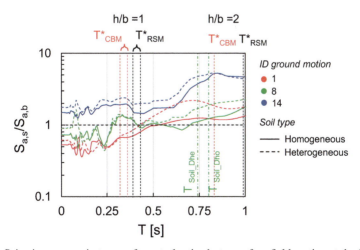

Fig. 11. Seismic response in terms of spectral ratios between free-field motion at the foundation level ($S_{a,s}$) and bedrock motion ($S_{a,b}$) for homogeneous and heterogeneous D soil profiles.

Under the strongest motion #1, significant reductions of *PGA* and spectral amplitudes (resulting in $S_{a,s}/S_{a,b} < 1$) can be observed in the period range of the two-story buildings ($h/b = 1$) located on both soft soil profiles. Conversely, the spectral amplitudes of the weakest motion #14 are found to be amplified (resulting in $S_{a,s}/S_{a,b} > 1$), especially in the case of heterogeneous soil profile. On the other hand, in the period range of four-story buildings ($h/b = 2$), spectral amplitudes of all the three records are amplified, with the fundamental period of the tallest CBM structure so close to that of the soils to induce double resonance.

The maximum settlement (w) and tilting rotation (θ) were assumed as engineering demand parameters, *EDPs*, for the assessment of foundation performance, as illustrated in Fig. 12. Time histories of the settlements (w_1 and w_2) of the opposite corners of each footing were recorded during the analysis and θ was calculated as their difference ($w_1 - w_2$) divided by the foundation width (B).

(a)   (b)

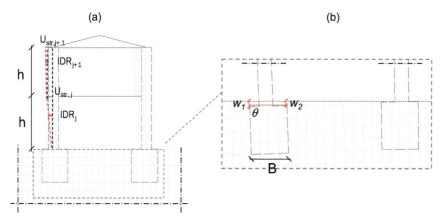

Fig. 12. Definition of (a) *MIDR* and (b) foundation settlement (w) and rotation (θ).

Figure 13 shows the values of the engineering demand parameters computed for clay brick masonry buildings with $h/b = 1$ (left column) and $h/b = 2$ (right column) founded on soil types D_{ho} and D_{he} (red and blue circles, respectively), plotted versus *PGA* at the bedrock. The lower stiffness of the heterogeneous soil profile close to the surface leads to higher foundation settlements (Fig. 13a–b) than those resulting for the homogeneous soil profile. The values of w calculated for structures with $h/b = 2$ are visibly larger than those associated with $h/b = 1$, due to more pronounced rocking induced in the foundation soil by the heavier and taller four-story buildings. In any case, the values of w predicted under even the highest *PGA* values are significantly below the conventional threshold levels adopted in engineering practice for loadbearing masonry walls, which are typically around 2.5 cm [47].

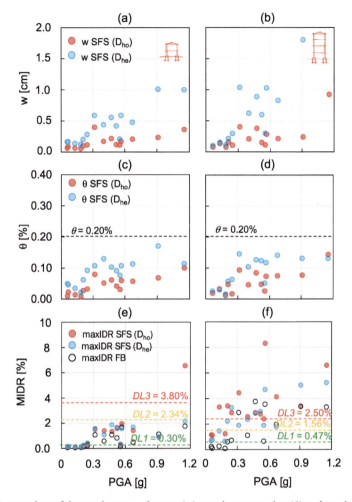

Fig. 13. Scatter plots of the maximum settlement (w), maximum rotation (θ) and maximum inter-story drift ratio (*MIDR*) versus *PGA*, produced by the selected input motions for two-story (a, c, e) and four-story (b, d, f) clay brick masonry structures.

Figure 13c–d show that the values of θ predicted for the homogeneous soil are again lower than those relevant to the heterogeneous profile, but the differences are significantly smaller than those observed on settlements. In all cases, even the maximum θ-values are below the typical thresholds of 0.2% (i.e. 1/500) for infills and 0.5% (i.e. 1/200) for loadbearing walls [48] adopted in engineering practice.

The engineering demand parameter representative of the structural out-of-plane response was the maximum inter-story drift ratio, *MIDR*, i.e. the peak relative displacement at the j-th floor divided by the inter-story height (h_j):

$$MIDR = \max\left(\frac{u_{str,j+1} - u_{str,j}}{h_j}\right)_{j=1,N} \quad (6)$$

where $u_{str,j}$ and $u_{str,j+1}$ are the horizontal displacements at the j-th and j-th $+ 1$ floors, respectively. The values of u_{str} were obtained as the total displacements minus the rigid base motions corresponding to the foundation rotation and translation.

Figure 13e–f show the *MIDR* values resulting from the nonlinear analyses on fixed-base (black circles) and compliant-base (red and blue circles) SFS models. The *MIDR* resulting from compliant-base models on both soft soil profiles are similar to each other, with the highest values pertaining to the tall structures on homogeneous soil profiles subjected to high *PGA* levels. Moreover, *MIDR* values are in some cases comparable to those predicted by the fixed-base models, implying less significant SFSI effects.

The data points are compared to three different damage level thresholds (DLs):

- DL1: formation of tensile cracks at the toe of the wall, due to the attainment of tensile strength of masonry;
- DL2: activation of the rocking mechanism;
- DL3: near-collapse limit state due to overturning.

The latter two were referred to the ultimate limit state in which the wall collapse is caused by overturning under out-of-plane motion, which takes place when the inter-story drift ratio is equal to [49]:

$$IDR_u = \frac{s}{2h} \quad (7)$$

being s the masonry wall thickness. In the case of rubble stone masonry, IDR_u is reduced by 35% [50] in order to account for nonlinear effects an d possible loss of masonry integrity. In the present study, values of $0.25 IDR_u$ The comparison between the structural demand (*MIDR*) and the capacity (damage thresholds) show that the vulnerability of the four-story structures is greater than that of the two-story buildings. This is not only due to the increase of *MIDR* with the building slenderness, but also to the lower inter-story drift ratio capacity related to the reduced masonry thickness at the highest floors. Such a difference is enhanced in case of compliant base models, due to the proximity of the soil and structural fundamental periods, inducing double resonance (see Fig. 11). Both effects make only the taller structural models to overcome the thresholds associated with the most severe DLs.

5.3 Fragility Curves

Once defined the *MIDR* as *EDP* for the structural performance, the probability P of overcoming one of the above described damage thresholds, DL$_i$, for a given seismic intensity measure, *IM*, of the input motion, was estimated as follows:

$$P[MIDR > MIDR_i | IM] = \Phi\left(\frac{\ln(IM/\eta_{DLi})}{\beta_{DL}}\right) \quad (8)$$

where Φ is the cumulative normal distribution function, while η_{DLi} and β_{DL} are the median and standard deviation of the lognormal distribution of the IM value causing the attainment of $MIDR_i$. The 'cloud method' [45] assumes a linear relationship between $MIDR$ and IM in the log-log scale, so that η_{DLi} was estimated as the abscissa of the intersection between the i-th damage threshold and the linear regression model that fits the $(MIDR, IM)$ data points resulting from the analyses.

Among the several peak, spectral and integral synthetic parameters of the input motion which can be correlated to the seismic damage, in this study only three were selected as intensity measures which can reasonably satisfy the requirements of efficiency, sufficiency and hazard computability. They are the peak ground acceleration, PGA, and velocity, PGV, and the Housner intensity, I_H, i.e. the integral of spectral velocity $S_v(T)$:

$$I_H = \int_{T_1}^{T_2} S_v(T)dT \qquad (9)$$

evaluated in the period range $[T_1, T_2]$ equal to $[0.1\text{ s}, 2.0\text{ s}]$.

As an example, Fig. 14 shows the correlations of the $MIDR$ with PGA, I_H and PGV for the clay brick masonry buildings founded on subsoil profiles A (black circles), D_{ho} (red circles) and D_{he} (blue circles). The median values η_{DL1}, η_{DL2} and η_{DL3} are highlighted in Fig. 14b for the tallest structure on soil type A and $IM = PGA$.

The plots in Fig. 14a reveal that the rate of increase of $MIDR$ with each one of the selected IMs is almost independent of the soil model for the shortest structure. The same rate significantly reduces in case of the tallest structure on soft soil D_{ho} and D_{he} with respect to that founded on soil type A. In other words, while the performance of slender masonry buildings on soft soils results more significantly affected by soil-foundation-structure interaction, their vulnerability appears less sensitive to an increase of the seismic loading. This is likely due to a higher amount of energy dissipated through foundation damping associated to rocking motion [39].

The efficiency of the selected IMs was checked by comparing them in terms of standard deviation ($\sigma_{EDP|IM}$) and coefficient of determination (R^2) of the regression model. The best-fit lines computed for all SFS models returned $\sigma_{EDP|IM}$ ranging between 0.27 and 0.79, and R^2 varying between 0.38 and 0.94. The ranking of IMs according to decreasing values of $\sigma_{EDP|IM}$ and increasing values of R^2 revealed that PGV is the most efficient parameter, followed by I_H and PGA. This outcome points out that the commonly adopted PGA is not the best option for predicting structural damage of masonry buildings accounting for SFS interaction.

Figure 15 compares the fragility curves at each DL for two-story (dark lines) and four-story (light lines) clay brick masonry buildings, as computed for the most efficient IMs (i.e. I_H and PGV).

The fragility functions for fixed-base models (solid lines) are always shifted towards higher IM levels with respect to those associated with compliant base models, implying that neglecting site effects and soil-foundation-structure interaction on soft soils leads to a significant underestimation of the probability of damage.

The heterogeneity in soft soil profiles (dotted lines) reduces the structural fragility, except for the shortest structure at DL1. As a matter of fact, settlements and rotations

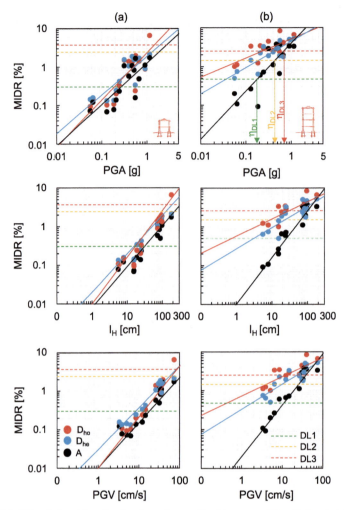

Fig. 14. *EDP–IM* relationships for fixed- and compliant-base models of clay brick masonry buildings with (a) $h/b = 1$ and (b) $h/b = 2$.

of the foundation on the heterogeneous soil are larger than in case of homogeneous profile (see Fig. 13), leading to a higher dissipation of seismic energy and a consequent reduction of the *MIDR* (see the log–log plots in Fig. 14). For example, if the activation of the rocking mechanism of a two-story structure is considered ($h/b = 1$, DL2), the median value of *PGV* is reduced from 81 cm/s to 49 cm/s (i.e. by almost 50%) in the case of homogenous soft soil and to 59 cm/s in the case of heterogeneous profile (Fig. 15b).

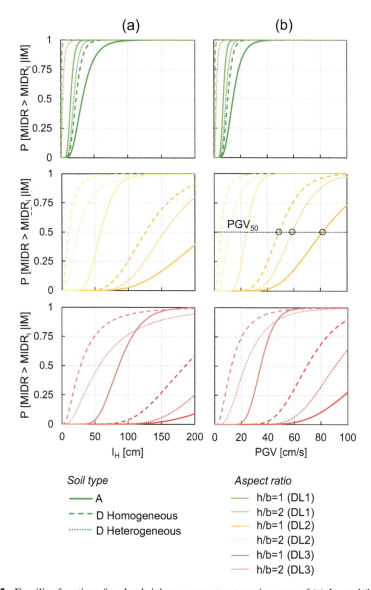

Fig. 15. Fragility functions for clay brick masonry structures in terms of (a) I_H and (b) PGV.

Whatever the soil conditions, the significant shift to the left of the thin curves ($h/b = 2$) with respect to the thick ones ($h/b = 1$) once again shows the higher vulnerability of the slender structures with respect to OOP damage mechanisms of the masonry walls.

The histograms in Fig. 16 summarize, for each damage level, the reduction factor, R_F, of the 16th, 50th and 84th percentiles of the IMs relevant to RSM and CBM buildings founded on soft soils, with respect to fixed-base conditions. For a given percentile x, such a reduction factor is defined as follows:

$$R_F = \frac{IM_{xD}}{IM_{xA}} \tag{10}$$

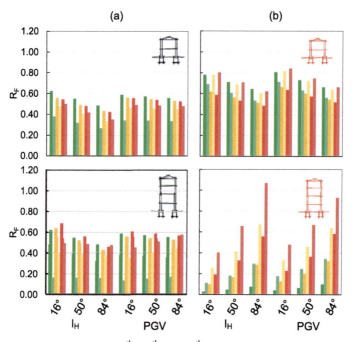

Fig. 16. Reduction factors R_F for 16th, 50th and 84th percentiles of I_H and PGV in case of (a) rubble stone and (b) clay brick masonry structures with $h/b = 1$ (upper plots) and $h/b = 2$ (lower plots)

Similarly to the factor proposed by Petridis and Pitilakis [51], R_F quantifies the combined influence of site amplification and SFSI on the OOP seismic vulnerability of masonry structures, by measuring the distance between the fragility functions. A R_F value lower or higher than unity implies that the combination of seismic amplification and SFSI is detrimental or beneficial, respectively.

It can be noted that:

- in most cases, R_F is significantly lower than unity, i.e. both site amplification and SFSI lead to detrimental effects on the OOP behavior of the structure;
- for two-story structures, R_F increases with the stiffness of the masonry type (upper plots); the opposite occurs for the four-story structures (lower plots) because the proximity between the soil and structural fundamental periods (see Fig. 11) leads to double resonance phenomena (except for DL3 and heterogeneous soil profile);
- for a given masonry type, at the lower damage levels R_F on average decreases (i.e. the detrimental effects of soil deformability are more significant) with the building height;

– at any DL, the distance between the functions relevant to fixed- and compliant-base structures – on average – decreases for D_{ho} and increases for D_{he}, with the exception of the stiffer clay brick structures, for which the effects of soil inhomogeneity with depth are less detrimental on the OOP fragility at higher damage levels.

5.4 Urban-Scale Application to the Village of Onna

Aisa et al. [52] reported a damage analysis of Onna, a hamlet close to L'Aquila city (Central Italy), in the middle of the Aterno river valley, which was severely damaged by a M_W 6.1 earthquake on the 6th of April 2009 (Fig. 17a). The structural typology most widespread in the village corresponds to two- or three-storey buildings with rubble stone loadbearing masonry walls and mixed steel-tile floor systems. Hence, they were considered as pertaining to vulnerability class B (Fig. 17b) according to the European Macroseismic Scale EMS-98 [53], i. e. masonry structures with irregular texture and efficient connections or with regular texture and inefficient connections. The same authors also observed that masonry disgregation and OOP mechanisms were the most common failure modes (as exemplified by the picture in Fig. 1b). The above factors suggested to consider this case history in order to validate the fragility curves developed in this study.

Figure 17a shows the shakemap in terms of macroseismic intensity and *PGA* contours of the mainshock, as reported by National Institute of Geophysics and Volcanology. In lack of any record of the seismic motion in the village area, in order to infer the intensity measures characterizing the site during the mainshock, the bedrock motion at Onna was assimilated to that recorded by the closest seismic station, AQG. The assumption is reasonable because both the seismic station and the village fall within the surface projection of the fault plane. In such conditions and up to a Joyner and Boore source-to-site distance equal to 4 km, ground motion prediction equations are flat (e.g. Bindi et al. [54]), hence there is no need to scale the recorded signal.

Since the weathered and fractured rock underlying the AQG station is far from being a stiff rock outcrop, the horizontal components of the recorded motion were deconvoluted to the bedrock by Evangelista et al.[55], then projected along the fault parallel (FP) and fault normal (FN) directions. *PGA* resulted equal to 0.31 g along both directions, while I_H was equal to 91.7 cm and 83.7 cm along FP and FN, respectively, due to the impulsive and directivity effects influencing the velocity spectrum.

Such values were used to predict the damage through the fragility curves of the SFS systems. To this aim, the shear wave velocity profiles measured in the alluvial coarse-grained deposit for the seismic microzonation emergency study [56] were firstly examined. Three geophysical surveys, i.e. two refraction microtremors tests (ReMi1 and ReMi2 in Fig. 17b) and a MASW test (see purple, blue and light blue circles) were performed. Figure 17c reports the comparison between the measured shear wave velocity profiles and that adopted in this study for ground type C (see Sect. 3), which represents the most suitable soil category for the site of Onna.

Figure 18a shows the comparison between the fragility curves computed in this study for the same kind of masonry buildings (rubble stone, $h/b = 1$) founded on homogeneous soil types A, C and D, in terms of *PGA* (as the most commonly used *IM*) and one of optimal *IM* for this case, i.e. Housner Intensity computed between 0.1 s and 2.0 s.

(a)

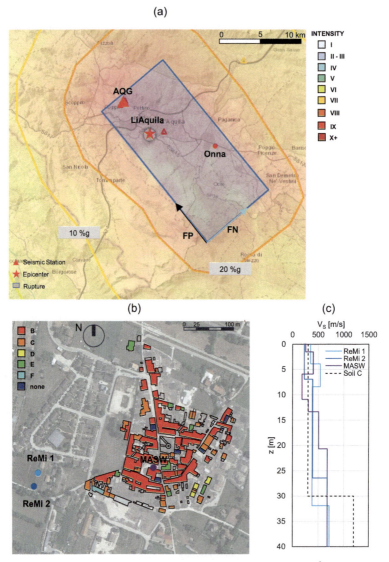

Fig. 17. (a) Shakemap representative of the M_W 6.1 seismic event on 6th April 2009 (http://shakemap.ingv.it/shake4/viewLeaflet.html?eventid=1895389); (b) distribution of vulnerability classes according to EMS-98 [53] with location of geophysical surveys; (c) shear wave velocity profiles.

Since the RSM short buildings had a fundamental frequency close to that of the C soil profile (2.50 Hz), it can be observed, for DL1, that the fragility functions for soil type C (dotted lines) are shifted further towards lower *IM* levels with respect to those related to the D soil profile (dashed lines), leading to higher probability of damage. On the other hand, at high levels of I_H the probability of exceeding DL2 and DL3 in ground

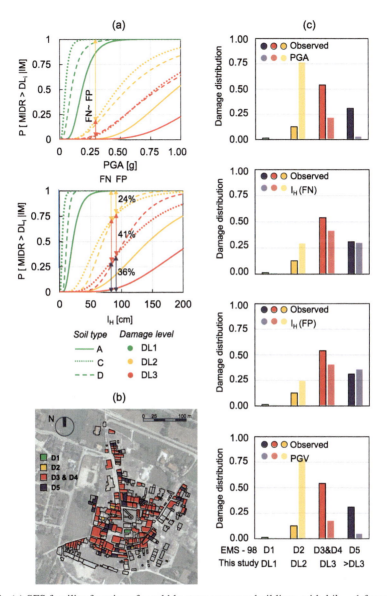

Fig. 18. (a) SFS fragility functions for rubble stone masonry buildings with $h/b = 1$ founded on ground types A, C, D; (b) territorial damage distribution at Onna for vulnerability class B; (c) comparison between statistical distributions of observed and predicted damage.

type C appears reduced. The vertical arrows in Fig. 18a show the values in terms of *PGA* and I_H which correspond to the reference bedrock motion inferred at Onna.

Figure 18b shows the territorial distribution of the observed damage as classified by [53] for vulnerability class B, by referring to the EMS-98 scale, which identifies the following six damage levels:

- D0: no damage;
- D1: development of few cracks in several walls;
- D2: significant cracks in many walls and collapse of plaster;
- D3: development of extensive and extended crack in many walls;
- D4: significative collapse of walls, or partial structural collapse of roofs;
- D5: destruction of the structure.

In order to compare such damage levels with the thresholds defined in the fragility study (see Sect. 5.3), D1 and D2 were considered equivalent to DL1 and DL2, respectively, while D3 and D4 were merged and assimilated to DL3.

The histograms in Fig. 18c show the comparison between the statistical distributions of observed vs. predicted damage, the latter derived by computing the difference between the probabilities of exceeding DL_i and DL_{i+1}. It can be observed that the use of either *PGA* or *PGV* as *IM* can lead to significant overestimation of damage at DL2, while the opposite occurs for DL3 or higher. Instead, a much better agreement with the observed distribution was found at all damage levels using the spectral intensity I_H as intensity measure.

6 Conclusions and Perspectives

This paper was dedicated to report the main methodological outcomes of a long-term comprehensive study, which was developed by the authors with the purpose of calibrating up-to-date straightforward tools to account for soil-foundation-structure interaction in assessing the seismic safety of the most widespread typologies of residential masonry buildings in the most hazardous countries in the Mediterranean area. Although not exhaustive of all the possible geometries, material properties and damage mechanisms, the parametric studies sampled a range of representative combinations extended enough to highlight the role of several factors on the dynamic response and seismic damage, such as the building slenderness, the soil-structure stiffness ratio and the degree of inhomogeneity in the subsoil profile.

The procedure enabling to predict the elongation of the fundamental period of a 'replacement oscillator' was proved to be effectively extended to SFS systems for which some critical factors may be overlooked by assuming simplified hypotheses. The case study of Matera provided a significant validation test based on several measurements of the soil-building fundamental frequency; however, the proposed method deserves further improvements to account for irregular structural geometries. Once soil nonlinearity and overall system damping are appropriately considered, the method might be fruitfully adopted for an evaluation of the seismic inertial loads acting on structures and foundations, by using free-field response spectra derived from seismic response analyses, at both local and territorial scale.

On the other hand, the formulation of fragility curves for typical soil-foundation-structure systems as a function of synthetic intensity measures of the reference bedrock motion can ideally support the simulation of damage scenarios at a territorial scale, by a 'convolution' of shakemaps through site and building classification databases. The consistency between the statistical distributions of damage predicted and observed in the village of Onna seems an encouraging example of application for a representative case study. Nevertheless, the procedure deserves to be assessed also against alternative approaches, e.g. by expressing the probability of damage as a function of the ground motion including site amplification, and validated versus case studies where on-site seismic records and more detailed inventories of the damage mechanisms are available.

Acknowledgements. This work was carried out as part of WP16.3 "Soil-Foundation-Structure Interaction" in the framework of the research programme funded by Italian Civil Protection through the ReLUIS Consortium (DPC-ReLuis 2019-2021).

References

1. Independent Evaluation Group (IEG): Development Actions and the Rising Incidence of Disasters. The World Bank, Washington, DC (2007)
2. Giardini, D., et al.: Seismic hazard harmonization in Europe (SHARE): Online Data Resource 2013 (2013). https://doi.org/10.12686/SED-00000001-SHARE
3. Crowley, H., et al.: The European seismic risk model 2020 (ESRM 2020). In: ICONHIC2019 2nd International Conference on Natural Hazards & Infrastructure, Chania, Greece (2019)
4. Augenti, N., Parisi, F.: Learning from construction failures due to the 2009 L'Aquila, Italy, Earthquake. J. Perform. Constr. Facil. **24**, 536–555 (2010). https://doi.org/10.1061/(asce)cf. 1943-5509.0000122
5. Dolce, M., Di Bucci, D.: Comparing recent Italian earthquakes. Bull. Earthq. Eng. **15**, 497–533 (2017). https://doi.org/10.1007/s10518-015-9773-7
6. Vlachakis, G., Vlachaki, E., Lourenço, P.B.: Learning from failure: damage and failure of masonry structures, after the 2017 Lesvos earthquake (Greece). Eng. Fail. Anal., 117 (2020). https://doi.org/10.1016/j.engfailanal.2020.104803
7. Daniell, J.E., Khazai, B., Wenzel, F., Vervaeck, A.: The CATDAT damaging earthquakes database. Nat. Hazards Earth Syst. Sci. (2011). https://doi.org/10.5194/nhess-11-2235-2011
8. Kausel, E.: Early history of soil-structure interaction. Soil Dyn. Earthq. Eng. **30**, 822–832 (2010). https://doi.org/10.1016/j.soildyn.2009.11.001
9. Casolo, S., Diana, V., Uva, G.: Influence of soil deformability on the seismic response of a masonry tower. Bull. Earthq. Eng. **15** (2017). https://doi.org/10.1007/s10518-016-0061-y
10. de Silva, F.: Influence of soil-structure interaction on the site-specific seismic demand to masonry towers. Soil Dyn. Earthq. Eng. **131**, 106023 (2020). https://doi.org/10.1016/j.soildyn.2019.106023
11. Bayraktar, A., Hökelekli, E.: Influences of earthquake input models on nonlinear seismic performances of minaret-foundation-soil interaction systems. Soil Dyn. Earthq. Eng. **139** (2020). https://doi.org/10.1016/j.soildyn.2020.106368
12. de Silva, F., Ceroni, F., Sica, S., Silvestri, F.: Non-linear analysis of the Carmine bell tower under seismic actions accounting for soil–foundation–structure interaction. Bull. Earthq. Eng. **16**, 2775–2808 (2018). https://doi.org/10.1007/s10518-017-0298-0
13. de Silva, F., Pitilakis, D., Ceroni, F., Sica, S., Silvestri, F.: Experimental and numerical dynamic identification of a historic masonry bell tower accounting for different types of interaction. Soil Dyn. Earthq. Eng. (2018). https://doi.org/10.1016/j.soildyn.2018.03.012
14. Karatzetzou, A., Pitilakis, D., Kržan, M., Bosiljkov, V.: Soil–foundation–structure interaction and vulnerability assessment of the Neoclassical School in Rhodes, Greece. Bull. Earthq. Eng. **13** (2015). https://doi.org/10.1007/s10518-014-9637-6

15. Fathi, A., Sadeghi, A., Emami Azadi, M.R., Hoveidae, N.: Assessing the soil-structure interaction effects by direct method on the out-of-plane behavior of masonry structures (case study: Arge-Tabriz). Bull. Earthq. Eng. **18**, 6429–6443 (2020). https://doi.org/10.1007/s10518-020-00933-w

16. Güllü, H., Jaf, H.S.: Full 3D nonlinear time history analysis of dynamic soil–structure interaction for a historical masonry arch bridge. Environ. Earth Sci. **75** (2016). https://doi.org/10.1007/s12665-016-6230-0

17. Brunelli, A., et al.: Numerical simulation of the seismic response and soil–structure interaction for a monitored masonry school building damaged by the 2016 central Italy earthquake. Bull. Earthq. Eng. **19**, 1181–1211 (2021). https://doi.org/10.1007/s10518-020-00980-3

18. Cavalieri, F., Correia, A.A., Crowley, H., Pinho, R.: Seismic fragility analysis of URM buildings founded on piles: influence of dynamic soil – structure interaction models. Bull. Earthq. Eng. **18**, 4127–4156 (2020). https://doi.org/10.1007/s10518-020-00853-9

19. Brunelli, A., de Silva, F., Cattari, S.: Site effects and soil-foundation-structure interaction: derivation of fragility curves and comparison with codes-conforming approaches for a masonry school. Soil Dyn. Earthq. Eng. **154**, 107–125 (2022)

20. Gazetas, G.: Formulas and charts for impedances of surface and embedded foundations. J. Geotech. Eng. (1991). https://doi.org/10.1061/(ASCE)0733-9410(1991)117:9(1363)

21. Amendola, C., de Silva, F., Vratsikidis, A., Pitilakis, D., Anastasiadis, A., Silvestri, F.: Foundation impedance functions from full-scale soil-structure interaction tests. Soil Dyn. Earthq. Eng. **141**, 106523 (2021). https://doi.org/10.1016/j.soildyn.2020.106523

22. Tileylioglu, S., Stewart, J.P., Nigbor, R.L.: Dynamic stiffness and damping of a shallow foundation from forced vibration of a field test structure. J. Geotech. Geoenviron. Eng. **137**, 344–353 (2011). https://doi.org/10.1061/(asce)gt.1943-5606.0000430

23. Pais, A., Kausel, E.: Approximate formulas for dynamic stiffnesses of rigid foundations. Soil Dyn. Earthq. Eng. **7** (1988). https://doi.org/10.1016/S0267-7261(88)80005-8

24. Liou, G.S.: Impedance for rigid square foundation on layered medium. Doboku Gakkai Rombun-Hokokushu/Proc. Jpn. Soc. Civ. Eng. (1993) https://doi.org/10.2208/jscej.1993.471_47

25. Pitilakis, D., Moderessi-Farahmand-Razavi, A., Clouteau, D.: Equivalent-linear dynamic impedance functions of surface foundations. J. Geotech. Geoenviron. Eng. **139** (2013). https://doi.org/10.1061/(asce)gt.1943-5606.0000829

26. Cremer, C., Pecker, A., Davenne, L.: Cyclic macro-element for soil-structure interaction: material and geometrical non-linearities. Int. J. Numer. Anal. Methods Geomech. **25** (2001). https://doi.org/10.1002/nag.175

27. Cremer, C., Pecker, A., Davenne, L.: Modelling of nonlinear dynamic behaviour of a shallow strip foundation with macro-element. J. Earthq. Eng. **6** (2002). https://doi.org/10.1080/13632460209350414

28. Veletsos, A.S., Meek, J.W.: Dynamic behaviour of building-foundation systems. Earthq. Eng. Struct. Dyn. (1974). https://doi.org/10.1002/eqe.4290030203

29. Maravas, A., Mylonakis, G., Karabalis, D.L.: Simplified discrete systems for dynamic analysis of structures on footings and piles. Soil Dyn. Earthq. Eng. (2014). https://doi.org/10.1016/j.soildyn.2014.01.016

30. Piro, A., de Silva, F., Parisi, F., Scotto di Santolo, A., Silvestri, F.: Effects of soil-foundation-structure interaction on fundamental frequency and radiation damping ratio of historical masonry building sub-structures. Bull. Earthq. Eng. **18**, 1187–212 (2020). https://doi.org/10.1007/s10518-019-00748-4

31. Gaudio, D., Rampello, S.: On the assessment of seismic performance of bridge piers on caisson foundations subjected to strong ground motions. Earthq. Eng. Struct. Dyn. **50**, 1429–1450 (2021). https://doi.org/10.1002/eqe.3407

32. D'Ayala, D., Speranza, E.: Definition of collapse mechanisms and seismic vulnerability of historic masonry buildings. Earthq. Spectra **19**, 479–509 (2003). https://doi.org/10.1193/1.1599896
33. Augenti, N., Parisi, F.: Teoria e tecnica delle strutture in muratura. Hoepli, Milan, Italy (2019). (in Italian)
34. CEN. Eurocode 8: Design of structures for earthquake resistance – Part 1: general rules, seismic actions and rules for buildings (2004)
35. Augenti, N., Parisi, F.: Constitutive models for tuff masonry under uniaxial compression. J. Mater. Civ. Eng. **22**, 1102–1111 (2010). https://doi.org/10.1061/(ASCE)MT.1943-5533.0000119
36. Capatti, M.C., et al.: Implications of non-synchronous excitation induced by nonlinear site amplification and of soil-structure interaction on the seismic response of multi-span bridges founded on piles. Bull. Earthq. Eng. **15**, 4963–4995 (2017). https://doi.org/10.1007/s10518-017-0165-z
37. Seed H.B., Idriss I.M.: Soil moduli and damping factors. Report No. UCB/EERC-70/l0, University of California, Berkeley, December 1970. https://ci.nii.ac.jp/naid/10022026877/
38. Vucetic, M., Dobry, R.: Effect of soil plasticity on cyclic response. J. Geotech. Eng. (1991). https://doi.org/10.1061/(ASCE)0733-9410(1991)117:1(89)
39. Piro, A.: Soil - structure interaction effects on seismic response of masonry buildings. PhD thesis in Structural and Geotechnical Engineering and Seismic Risk. University of Naples Federico II (2021)
40. MIT Ministero delle Infrastrutture e dei Trasporti. Circolare 21 gennaio 2019, n. 7 Istruzioni per l'applicazione dell'«Aggiornamento delle "Norme tecniche per le costruzioni"». Gazzetta Ufficiale della Repubblica Italiana (2019). (in Italian)
41. Itasca. FLAC 7.0 – Fast Lagrangian Analysis of Continua – User's Guide, Itasca Consulting Group, 3 Minneapolis (2011)
42. Kuhlemeyer, R., Lysmer, J.: Finite element method accuracy for wave propagation problems. J. Soil Mech. Found. Div. (1973). https://doi.org/10.1061/jsfeaq.0001885
43. Gallipoli, M.R., et al.: Evaluation of soil-building resonance effect in the urban area of the city of Matera (Italy). Eng. Geol. **272**, 105645 (2020). https://doi.org/10.1016/j.enggeo.2020.105645
44. Tragni, N., et al.: Sharing soil and building geophysical data for seismic characterization of cities using Clara Webgis: a case study of Matera (southern Italy). Appl. Sci. **11** (2021). https://doi.org/10.3390/app11094254
45. Jalayer, F., De Risi, R., Manfredi, G.: Bayesian cloud analysis: efficient structural fragility assessment using linear regression. Bull. Earthq. Eng. **13**, 1183–1203 (2015). https://doi.org/10.1007/s10518-014-9692-z
46. Smerzini, C., Galasso, C., Iervolino, I., Paolucci, R.: Ground motion record selection based on broadband spectral compatibility. Earthq. Spectra **30**, 1427–1448 (2014). https://doi.org/10.1193/052312EQS197M
47. Holtz, R.D.: Stress distribution and settlement of shallow foundations. In: Foundation Engineering Handbook, pp. 166–222 (1991). https://doi.org/10.1007/978-1-4757-5271-7_5 (1991)
48. Polshin, D.E., Tokar, R.A.: Maximum allowable non-uniform settlement of structures. In: 4th International Conference on Soil Mechanics and Foundation Engineering, vol. 1, pp. 402–5 (1957)
49. Lagomarsino, S.: Seismic assessment of rocking masonry structures. Bull. Earthq. Eng. **13**, 97–128 (2015). https://doi.org/10.1007/s10518-014-9609-x
50. De Felice, G.: Out-of-plane seismic capacity of masonry depending on wall section morphology. Int. J. Archit. Herit. **5**, 466–482 (2011). https://doi.org/10.1080/15583058.2010.530339

51. Petridis, C., Pitilakis, D.: Fragility curve modifiers for reinforced concrete dual buildings, including nonlinear site effects and soil–structure interaction. Earthq. Spectra (2020). https://doi.org/10.1177/8755293020919430

52. Aisa, E., De Maria, A., De Sortis, A., Nasini, U.: Analisi del danneggiamento di Onna (l'Aquila) durante il sisma del 6 Aprile 2009. Ingegneria Sismica **28**, 63–74 (2011). (in Italian)

53. Bindi, D., et al.: Ground motion prediction equations derived from the Italian strong motion database. Bull. Earthq. Eng. **9** (2011). https://doi.org/10.1007/s10518-011-9313-z

54. Evangelista, L., Landolfi, L., d'Onofrio, A., Silvestri, F.: The influence of the 3D morphology and cavity network on the seismic response of Castelnuovo hill to the 2009 Abruzzo earthquake. Bull. Earthq. Eng. (2016). https://doi.org/10.1007/s10518-016-0011-8

55. Grünthal, G.: European Macroseismic Scale 1998 (EMS-98). Cahiers du Centre Européen de Géodynamique et de Séismologie 15, vol. 15 (1998)

56. Gruppo di Lavoro MS–AQ. Microzonazione sismica per la ricostruzione dell'area aquilana. Regione Abruzzo – Dipartimento della Protezione Civile, L'Aquila, 3 vol. e Cd-rom.5. Volume II, pp. 199–221 (2010). (in Italian)

Coseismic and Post-seismic Slope Instability Along Existing Faults

Ikuo Towhata[✉]

Kanto Gakuin University, Yokohama 236-8501, Japan
towhata.ikuo.ikuo@gmail.com

Abstract. The world in the 21st Century faces increased hazards of gigantic earthquakes and heavy rains as compared with the previous century. Under this circumstance, coseismic landslide is one of the important topics in geotechnical earthquake engineering and is promoted by the increased precipitation both before and after the gigantic earthquakes. This paper sheds light on this topic, paying attention not only to the effects of strong shaking but also to the important role played by faults, which are either active or inactive. The fault action consists of the rock fissures along the fault and possible water ejection from the fault plane. The rock fissures made by the fault dislocation cause the long-term instability of mountains slopes for years or for centuries after strong shaking.

Keywords: Earthquake-induced landslides · Slope instability · Rock rupture · Fault

1 Introduction

The traditional concept on coseismic landslides has been that the shear stress induced by the seismic inertial effects in conjunction with the gravity exceeds the material shear strength of the mountain body and that the hazard extent has to be assessed without detailed subsurface investigation. The second issue is simply due to the financial limitation that cannot take care of the vast vulnerable mountain areas in which the topography and material properties as well as geo-hydraulic conditions are highly variable. Because of this limitation, there is always a possibility of unexpected coseismic landslide disaster during real earthquakes. In the recent times, we are aware that the types of coseismic landslide disaster is not limited to the fall of materials during shaking. The newly recognized problems, although they did occur and were problems in the past, are the landslide dams, the compound effects of seismic shaking and heavy rain and the long-term slope instability and sediment disasters that last for years or for decades after the causative earthquakes. Moreover, the fault rupture is recognized as a threat to the human community nowadays, in contrast to the past when the fault rupture was just a target of scientific/geological studies. This situation is a consequence of the spreading human activities in mountainous and hilly terrains where human did not live. With these changing situations in mind, the author summarized the latest situations of coseismic landslides in [1]. Because the content of this publication is vast, the present paper attempts to pick up what is related with fault and show details.

L. Wang et al. (Eds.): PBD-IV 2022, GGEE 52, pp. 195–213, 2022.
https://doi.org/10.1007/978-3-031-11898-2_10

2   Effects of Rainfall on Coseismic Landslides

Science and technology achieved profound development from 1970s to the end of the 20th Century when people started to demand more safety under natural disasters and the public sectors are more supportive of improved safety. It appears that many important principles and design codes were made in those decades and are still used widely at the present time. We have to know, however, that the situations in those great decades and today may not be identical and that the previous way of thinking may not be valid today. The difference from the past is not limited to the growth of human population and spreading of human habitations into difficult natural conditions. The natural conditions are changing in the meantime as well.

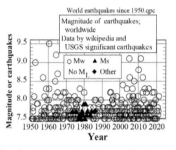 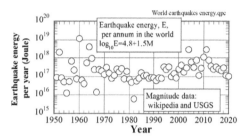

Fig. 1. History of major earthquakes in the world.

Fig. 2. History of annual earthquake energy in the world.

Figure 1 plots the seismic magnitude of major earthquakes in the world that happened since 1950s. For most events, the moment magnitude was plotted in this diagram, while, for some events, the moment magnitude value is not available and other types are used here. It is evident in this figure that gigantic earthquakes with their magnitude greater than 8.5 did not occur during the aforementioned decades (1970s, 80s and 90s) of technology development in contrast to the years in the 21st Century. Figure 2 employs the same dataset and indicates the change of annual earthquake energy release all over the world. The earthquake magnitude was converted to the energy by using the equation by [2], which is shown in this figure. It is evident again that the earth was relatively quiet in those decades of technology development, while getting more active since the beginning of the 21st Century. This increasing trend is stressed by the two gigantic earthquakes in 2004 (in Indian Ocean) and 2011 (Tohoku, Japan) but, even without them, the increasing trend in the first 10 years of the Century is evident. Although it is difficult to say further about this trend, it is reasonable that we may not be able to fully rely on the safety frameworks that were developed during the relatively quiet years of the earth and to expect more safety than before. In other words, the experiences in the 20th Century should be reviewed and, if necessary, revised in the 21st Century.

This situation is worsened by the recent increasing trend of precipitation that is a very important cause of flood and landslide disasters. The recent increase of rainfall is possibly a consequence of the global climate change but I cannot make any definite remark on this issue. Only what I can do is to present data recorded and reported by established institutions. While many institutions have been publishing the annual or extreme daily rainfall records that are increasing, I collected the only precipitation records that induced flood and sedimentary disasters in Japan since 1945 [3]. Because a heavy but short rainfall event does not trigger disaster, Fig. 3 indicates the nation's No. 1 rainfall record of one-day (or recently 24-h) precipitation as well as the total rainfall during one disaster event. It is shown here that the upper bound of the disastrous rainfall has been increasing since 1990s, implying that the risk of significant rainfalls has been aggravating for the past 30 years. By combining the views on earthquakes and rainfalls, it can be said that the risk of the compound disasters as induced by earthquakes followed by heavy rain or, conversely, the antecedent rainfall followed by earthquakes is becoming significant nowadays.

(a) No. 1 24-hour precipitation   (b) No.1 total rainfall in one disaster event

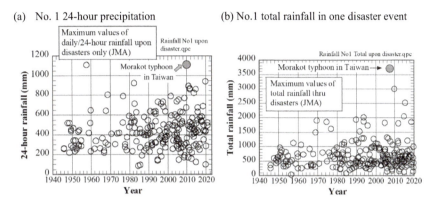

Fig. 3. History of rainfall records in Japan that triggered disasters since 1945.

The effects of the antecedent rainfall are captured by looking at Figs. 4 and 5 that illustrate the less important occurrence of coseismic landslide during the 1995 Kobe earthquake (Mw = 6.9) together with the rainfall history together with the similar information of the 2004 Niigata-ken-Chuetsu earthquake (Mw = 6.6) (for their locations, see Fig. 6) which was preceded by a heavy rainfall and was associated with many coseismic landslides.

(a) Kobe in 1995   (b) Niigata-ken-Chuetsu in 2004

Fig. 4. Coseismic landslides induced by two different earthquakes with similar magnitudes.

An interesting phenomenon was reported by Oike [4] who investigated the correlation between the onsets of small but many earthquakes (tremors) and rainfalls along the

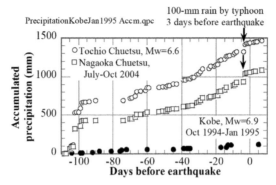

Fig. 5. Antecedent rainfall records prior to the 1995 Kobe and the 2004 Niigata-ken-Chuetsu earthquakes.

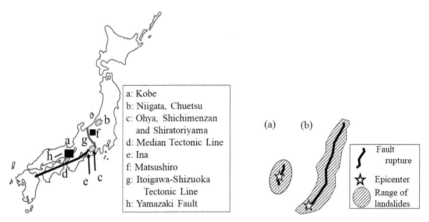

Fig. 6. Location of the sites prone coseismic slope instability.

Fig. 7. Schematical illustration of the effects of fault length on landslides distribution.

Yamazaki Fault in Japan. It was demonstrated that more tremors occurred after heavy rainfalls and high level of ground water. This implies that the pore water pressure in the ground increased after infiltration of rain water, leading to reduced effective stress and shear strength in the earth crust. Accordingly, the shear strength in the underlying faults decreased and the earthquakes were made easier to occur.

The knowledge obtained from the reality shows that water effect is significant in stability of earth crust and most probably the mountains slopes. It deserves attention that local rainfall may trigger minor earthquake along an existing fault.

3 Distribution of Coseismic Landslides Along Existing Faults

3.1 General Issue

Discussion in this section begins with the distribution of coseismic landslides along the causative faults that generate the seismic shaking. In the past, the studies on earthquake-induced landslides were interested in the effects of epicentral distance on the extent of landslides [5–7 and 8]. In contrast, the gigantic earthquakes in the twenty-first century (the 2005 Kashmir earthquake of Mw = 7.6 and the 2008 Wenchuan earthquake of Mw = 7.9) demonstrated that many coseismic landslides were distributed along the fault rupture. This difference is probably because the studies in the 20th Century collected data from earthquakes of smaller magnitude and shorter fault length, while the events in the 21st Century had greater seismic magnitude and longer fault length; see Fig. 7. Although the intensity of shaking varies with the distance from the fault rupture plane, the greater fault length situates the landslides geometrically along the fault rupture. Figure 8 shows the coseismic landslides during the 2005 Kashmir earthquake. Noteworthy is that the 2016 Kaikoura earthquake of Mw = 7.8 triggered a number of coseismic landslides along the fault rupture, while the epicenter at the southernmost part of the rupture was out of the landslides area [9]. Therefore, the "distance" is unlikely the only major factor that controls the likelihood of coseismic landslides.

Fig. 8. Landslides along the fault valley caused by the 2005 Kashmir earthquake (south of Balakot).

3.2 Ohya and Shichimenzan Landslides in Central Japan

The present paper calls attention upon the effects of existing fault, which is not necessarily the causative fault, on the landslides induced by an earthquake. Figures 9 and 10 show the landslides at Ohya and Shichimenzan in central Japan (Fig. 6). Both sites have been unstable for many centuries. Legends say that the Ohya landslide was triggered by the 1707 Hoei earthquake of Mw = 8.7–9.3 with the volume of the failed sediment = 94 million m³ [10]. The other one at Shichimenzan is old as well, having been unstable at the latest since the 13th Century as historical documents in AD 1278 refers to it [11].

Fig. 9. Landslide at Ohya, Shizuoka, Japan.   **Fig. 10.** Landslide at Shichimenzan, Yamanashi, Japan.

What is interesting is that both landslide sites are located close to the Itoigawa-Shizuoka Tectonic Line (Fig. 6) that transversally crosses the Honshu Island of Japan in N-S direction and is associated with many local faults in the parallel direction (Fig. 11 as well as [12]), which implies a complicated history of tectonic stress in this region. Consequently, the rock mass has been undergoing the effects of both the movement of the nearby faults and the regional tectonic stress, and are fractured profoundly. The unstable situation in these landslides sites can be captured in Figs. 9 and 10. The disrupted rock (Fig. 12) and the rock mass folding (Fig. 13) imply the highly unstable condition of the rock mass at Ohya site. This material property is most probably related with the creep deformation of the mountain body at both sites (Figs. 14 and 15). Because of the unstable condition, the slopes of Ohya and Shichimenzan sites have been undergoing continuous instability and landslides over centuries.

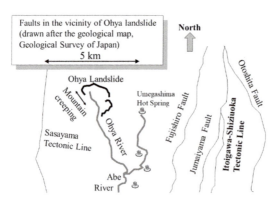

Fig. 11. Location of local faults around the Ohya landslide site (drawn after a geological map, Geological Survey of Japan).

Fig. 12. Fractured rock of Ohya landslide site.   **Fig. 13.** Folded rock mass of Ohya.

Fig. 14. Creep deformation at Ohya.   **Fig. 15.** Small lake formed by creep deformation/depression of Shichimenzan Mountain.

3.3 Shiratori Yama, Japan

To the south of Shichimenzan (Fig. 6), there is another unstable mountain body that is called Shiratoriyama Mountain (Fig. 16). Although low (568 m in altitude), this mountain caused coseismic landslide disasters twice in the recent history; in 1707 upon the Hoei earthquake of Mw = 8.7–9.3 and in 1854 upon the Ansei Tokai earthquake of Mw = 8.4–8.6. During both events, the landslide mass blocked the Fuji River channel at the bottom (Fig. 17) and, most probably, the dam breaching caused further disasters. It is thought that the 1707 landslide left a substantial amount of material at the top which partially fell down in 1854 [13]. Still there is a big mass of earth at the top.

Shiratoriyama mountain appears very fragile, comprising fractured or broken pieces of rock: Fig. 18 illustrates the mountain surface near the headscarp where there are many cobbles that are said to be a product of geological procedure. It is noteworthy in Fig. 16 that this mountain is located near the southern tip of Minobu Fault that has recently been registered as an active fault [14]. There are possibly the effects of fault movement on the quality of the rock mass of the mountain. It should be stressed that rock fracturing effect of a fault can remain for a long time even if the fault has ceased its activity and is not considered active anymore.

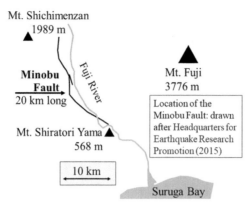

Fig. 16. Location of Shiratoriyama Mountain and Minobu Fault [14].

Fig. 17. Shiratoriyama slope that failed in 1707.

Fig. 18. Cobbles on the flank of Shiratoriyama near the headscarp of the failure in 1707.

3.4 Landslides Along the Longmenshan Fault After the 2008 Wenchuan Earthquake, China

The long-term instability of mountain slopes was a serious problem in Sichuan Province of China after the Wenchuan earthquake [15]. In addition to the coseismic landslides in May, 2008, the heavy rainfall in September triggered additional landslides in the fault-affected area. Figure 19 illustrates one of the examples of the post-seismic disasters in which the debris came from the coseismically disrupted mountain slope behind the town (Fig. 20). Figure 21 shows another landslide caused by the post-seismic heavy rain. This landslide occurred in a slope made of mud stone that had been most probably prone to deterioration by infiltration of rain water (or slaking) and could not remain stable. Obviously, there are two kinds of post-earthquake instability caused by 1) mountain slopes disrupted by strong shaking (generation of cracks and fissures) as shown in Fig. 20, and 2) rainfall erosion of soil deposits on slopes and in valley bottoms. The former case implies the vulnerability of mountain slopes along faults under strong shaking. It took 5–7 years for the mountain slopes to be stabilized and recover vegetation cover (Fig. 22).

Fig. 19. Buildings in Yinchanggou Gully destroyed by debris flow during post-seismic rainfall.

Fig. 20. Disrupted mountain slope as the source of debris flow.

Fig. 21. Landslide to the west of Qushan Zhen induced by post-seismic rain.

Fig. 22. Recovery of vegetation on slopes that was failed in 2008 (picture taken on May 15 2018).

3.5   Landslides Along the Ina part of the Median Tectonic Line, Japan

The Median Tectonic Line (MTL) is a long and old fault that runs in the approximately E-W direction in the western part of Japan ("d" and "e" in Fig. 6). It started its activity in the Cretaceous Period [16] and the dislocation to date has accumulated to 60 km [17] or more. Figure 23 shows one of the MTL outcrops in the Ina region. It is hence reasonable to expect the fault-induced fracture of rocks along it. While very little is known about the activity of MTL (causing earthquakes) during the historic times (in the past 1500 years or so), its Ina section triggered two damaging earthquakes in 8th Century (probably in AD 715) and then in 1718. Figure 24 illustrates one of the sites of the historical coseismic landslides in Tohyama Village of Ina. The more serious problem along MTL is the instability of mountain slopes undergoing heavy rainfalls. Figure 25 shows the site of Ohnishi Yama landslide (1961) and Fig. 26 indicates the ongoing instability at Tobigasu.

In line with the previously-described sites, it is likely that the MTL dislocation in the past affected the mechanical property of the mountain slopes in the Ina region. This idea seems supported to a certain extent by the landslide disasters as shown in

Fig. 23. Exposed fault dislocation at Itayama within the Ina section of MTL.

Fig. 24. Site of historical landslide triggered by Tohyama earthquake of Magnitude = 7 in 1718.

Fig. 25. Ohnishi Yama landslide site Immediately upon MTL.

Fig. 26. Tobigasu slope failure along MTL.

Figs. 24 and 25. In this perspective, Matsushima [18] assembled the historical data on earthquakes and landslides in Ina, which are both coseismic and non-seismic. The Author added a few more data to his study and plotted the results in Fig. 27. Although the severity of earthquakes and landslides in this figure is prone to the subjective judgment of researchers, it appears that floods and landslides were more frequent after the strong seismic shakings in early 18th and middle 19th Centuries.

3.6 Landslide Induced by Water Ejection and Seismic Tremors in Matsushiro, Japan

Matsushiro in Central Japan (Fig. 6) was affected by seismic tremors in middle 1960s (Fig. 28). This event was characterized by substantial water ejection and induced land-slides. JMA [23] presents photographs of significant water ejection from hot springs in the area. Eventually, a huge landslide was triggered by water ejection. The mechanism of water ejection-induced landslide is simple and understandable from the soil-mechanic viewpoints; pressurized water came up from the deep elevation and increased the pore water pressure in the slope, while the total stress was unchanged, thus reducing the effective stress and shear strength in the slopes. Later, water injection tests were conducted in Matsushiro [24] in order to verify that the increased water pressure triggered the fault

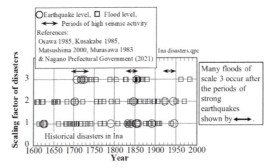

Fig. 27. Possible correlation between severe seismic effects and landslide disasters during their aftermath (data by [18–22]).

rupture at depths by reducing the effective stress. The landslide site remains vacant today due to the possible risk of reactivation of water ejection and landslide (Fig. 29).

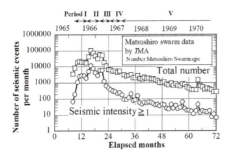

Fig. 28. History of tremors in Matsushiro.

Fig. 29. Present view of the Makiuchi landslide site.

4   Long-Term Instability of Slopes After Big Earthquakes

Section 3.4 addressed the slope instability that started after the earthquake and lasted for many years. Similar problem occurred after the 1999 Chi-chi earthquake in Taiwan (Mw = 7.6) and Pakistan in 2005 (Mw = 7.6). Figure 30 shows the extent of slope instabilities within a selected area along the East-West Crossing Highway in the mountainous area of Taiwan Island. It appears that the size of slopes affected by typhoons and rainfalls after the Chi-chi earthquake increased drastically but then decreased slowly towards the original level. This decreasing process took twenty years. During this process, however, there was a positive correlation between the amount of rainfall and the increment of landslide area (%) (Fig. 31), suggesting that the decreasing landslide size in Fig. 30 was possibly the consequence of less amount of rainfall during the post-earthquake years. To investigate the rainfall effect quantitatively and illustrate the vulnerability, the vulnerability index was defined by

$$Vulnerability\ index = \left[Areal\ increment\ ratio\ (\%)\ in\ Fig.\ 31\right] / \left[Total\ rainfall\ (mm)\right] \tag{1}$$

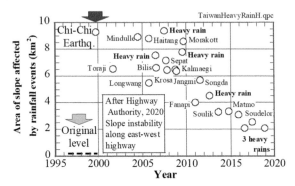

Fig. 30. Summary of slope instability in Central Mountain of Taiwan drawn after data of Highway Office of Taiwan [25].

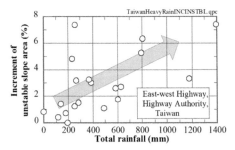

Fig. 31. Relationship between ratio of rainfall-induced increment of landslide area and total rainfall during disasters in Central Taiwan [25].

and plotted in Fig. 32. It is now evident that the vulnerability decreased with the elapsed time after the earthquake, implying that the stabilization of the mountain slopes took twenty years after the Chi-chi earthquake. Noteworthy is that the mountain slopes in Sichuan Province, China, took shorter time for recovery (Fig. 22). This is probably because of the rainfall difference between two regions as well as the geology; Taiwan Island is prone to heavier rainfall and geological weak rocks are subject to more weathering. Moreover, the island is more or less 30 million years old "only" and rock is not so lithified yet.

In clear contrast, the Ohya and Shichimenzan slopes (Figs. 8 and 9) have been unstable for centuries. One possible reason for this is many and frequent strong seismic events caused by the 1707 (Mw = 8.7–9.3) and 1854 (Mw = 8.4–8.6) earthquakes as well as many events in the past. The other reason is the weak geology as indicated by

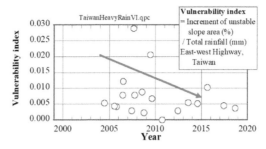

Fig. 32. Temporal change of rainfall vulnerability of slopes in Taiwan [25].

the fractured rock mass (Fig. 12) as a consequence of the accumulated damaging effects of the fault action.

The fault effect mentioned above is otherwise called the process zone and its width (Fig. 33) was originally supposed to be 1% of the fault length on the basis of field inspection [26]. Figure 34 illustrates this 1% correlation together with the results of more literature information collected by the author. This figure shows that "1%" is a good value for the upper bound of the process zone width while there are narrower width at several

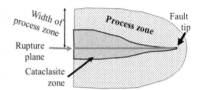

Fig. 33. Definition of rock damage around the fault plane (drawn after [26]).

places. Note that the judgement of the process zone is subject to personal judgment and difference.

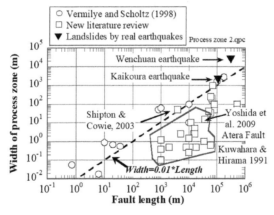

Fig. 34. Empirical relation between the width of process zone and the length of the causative fault [26–29].

5   Slope Instability Along Pull-Apart Mechanism of Strike-Slip Faults

The pull-apart mechanism of a strike-slip fault produces a basin where a unique geotechnical problem occurs. Figure 35 illustrates schematically the mechanism of a pull-apart basin where the lateral extension of the earth crust results in vertical compression and depression. One of the examples of this type of geomorphology is the Izmit Bay of Turkey whose southern coast line was affected by subsidence (Fig. 36) and submarine landslides [30].

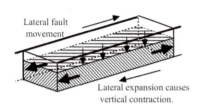

Fig. 35. Basin formation by pull-apart mechanism.

Fig. 36. Subsidence of the coast line of the Izmit Bay (after the 1999 Kocaeli earthquake).

In a more striking manner, many landslides occurred on September 28th, 2018, in the pull-apart basin of Palu, Central Sulawesi of Indonesia, after an earthquake of Mw = 7.5 (Fig. 37). This basin is the product of highly active Palu-Koro Fault with the mean rate of dislocation being 30 mm/year or more [31]. The following section discusses briefly the causative mechanism of the landslides.

Figure 38 shows one of the four big landslides that occurred in the Balaroa area in the southwestern suburb of Palu City. The size of this landslide was 0.85 km in length and 0.58 km in width, making the area 0.39 km^2, while the ground gradient was merely 2% on average. Similar to the other three landslides of this type, the surface soil of about 3–5 m in thickness suffered from liquefaction and the slope instability was induced. The induced soil displacement in the horizontal direction was 500 m or more, which was notably greater than those of other liquefaction-induced displacement in gentle slopes (typically 1–2% gradient) that were merely several meters [32]. The similar extent of displacement occurred in other three big landslides in Palu Basin as well. The notably long distance of displacement implies that the disaster in Palu Basin had something special in its mechanism.

Fig. 37. Sites of liquefaction-affected landslides around Palu City in 2018.

Fig. 38. Landslide at Balaroa.

Fig. 39. Water spring observed at the bottom of Balaroa landslide on November 2nd, 2018.

The second special feature of the landslides was the existence of substantial ground water ejection as evidenced by the water springs recorded five weeks after the disaster in the exposed slip plane (Fig. 39). Springs were found in other landslide sites as well. It should be recalled that the ample ejection of ground water during the Matsushiro swarm triggered landslides as mentioned before. The source of this ample ground water is not necessarily the irrigation channel or leakage from local sewage system because the Balaroa site does not have either such a channel or evidence of leakage (Figs. 40 and 41). Accordingly, there seems to be very special feature in the local geology, geohydrology and soil conditions in the four landslide sites. They are special because occurrence of similar landslide is not known elsewhere in the world.

Fig. 40. Lack of irrigation channel above the headscarp of Balaroa landslide.

Fig. 41. Dry headscarp of Balaroa landslide.

Extensive study has been made of the causative mechanism of these landslides. Among several ideas, the undrained softening of loose sand (Fig. 42) was declined first because the laboratory tests on undisturbed specimens did not exhibit it. The second mechanism to be studied was the formation of water films at the interface between less pervious silty layer and the underlying liquefied sand (Fig. 43). When the excess pore water pressure dissipates after liquefaction, the drained pore water remains below the layer interface and forms a "film" with null shear resistance, thus making large shear deformation easy [33–36]. This mechanism is possible in Sulawesi [37] but a question arises why the same disaster has not been known elsewhere despite that the interbedding of more and less pervious layers is not uncommon in alluvial fans over the world. Another issue is that the water film mechanism requires a continuous and smooth film of water over the entire landslide area, which may not be very likely.

Fig. 42. Schematic illustration of undrained stress paths of sand.

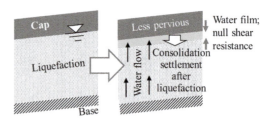

Fig. 43. Mechanism of water film formation during pore pressure dissipation after liquefaction.

The Author pays attention to the two features that are special to the Palu Basin. First, the four big landslides associated with liquefaction occurred in slightly low spaces between bigger alluvial fans (Fig. 44). This suggests that the long-distance mass movement was a process of new fan formation by which the existing fans were produced long time ago. It is speculated that the pressurized water aquifers as suggested by many local springs were destroyed by the earthquake and the out-blow of water uplifted the surface soil crust, enabling its long-distance flow. Because the alluvial fans are several km long, the induced flow displacement was long as well.

The second possible mechanism [38] is related with the underground fault plane [39]. Because of the pull-apart depression, there are many normal faults parallel to the edge of the basin. There are many springs and hot springs in the basin as well, some of which are under artesian pressure. All in all, it is supposed that a substantial amount of water blew out through the normal fault planes at the time of the

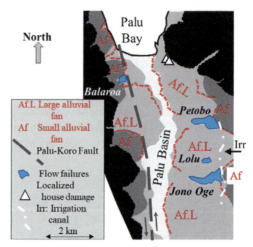

Fig. 44. Location of landslides between existing alluvial fans abbreviated as "Af"

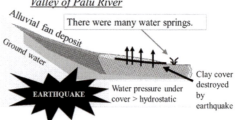

Fig. 45. Mechanism of water out-blow from under-ground pressurized aquifer.

earthquake, making the entire landslide mass slip immediately. In other words, it might be possible that the high pore water pressure in the fault plane induced the earthquake, similar to the Matsushiro swarm (Fig. 28).

To date, no conclusion has been achieved yet and discussion is going on. The difficulty is that there is no other example of such a landslide in clean sand both internationally and in the past in Sulawesi.

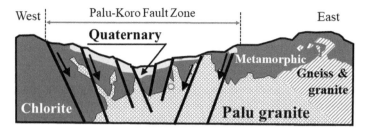

Fig. 46. East-west geological cross section of Palu Basin (drawn after [39]).

6   Conclusion

The magnitude of earthquakes and rainfalls are greater in the twenty-first century than in the late twentieth century when the current practice of disaster-mitigation engineering was developed. It is therefore important to attempt a new and different perspective on natural disasters in the new century. In the traditional approach of soil mechanics to coseismic landslide problems, the effect of ground water was considered just in terms of the formula of "total stress – pore water pressure = effective stress." However, the recent experiences of real coseismic landslides suggest that the ground water is producing the pore pressure effect through many dynamic or mysterious mechanisms. Such a mechanism seems more intensified along faults, whether active or inactive, where rocks are fractured by the continuous fault dislocation. The author attempted to introduce the examples of such a situation to the readers and invite them to further studies.

Acknowledgements. The comprehensive investigation of the landslide disaster in Sulawesi, 2018, was supported by JICA (Japan International Corporation Agency). This aid is deeply appreciated by the author.

References

1. Towhata, I.: Earthquake-induced landslides and related problems. In: Towhata, I., Wang, G., Xu, Q., Massey, C.: Coseismic Landslides - Phenomena, Long-Term Effects and Mitigation. Springer, Singapore (2022)
2. Gutenberg, B.: The energy of earthquakes. Q. J. Geol. Soc. Lond. **112**(1–4), 1–14 (1956)
3. Towhata, I.: Views on recent rainfall-induced slope disasters and floods. In: Hazarika, H. (ed.) CREST 2021 1st International Symposium on Construction Resources for Environmentally Sustainable Technologies, March, Fukuoka, Keynote Lecture (2021)
4. Oike, K.: On the relation between rainfall and the occurrence of earthquakes. Disaster Prev. Res. Inst. Ann. **20**(B-1), 35–45 (1977). (in Japanese)

5. Keefer, D.K.: Landslides caused by earthquakes. Geol. Soc. Am. Bull. **95**(4), 406–421 (1984)

6. Keefer, D.K.: The importance of earthquake-induced landslides to long-term slope erosion and slope-failure hazards in seismically active regions. Geomorphology **10**(1–4), 265–284 (1994)

7. Rodríguez, C.E., Bommer, J.J., Chandler, R.J.: Earthquake-induced landslides: 1980–1997. Soil Dyn. Earthq. Eng. **18**(5), 325–346 (1999)

8. Yasuda, S.: Zoning for slope instability, manual for zonation of seismic geotechnical hazards. In: Technical Committee 4, International Society for Soil Mechanics and Foundation Engineering, pp. 48–71 (1993)

9. Dellow, S., Massey, C., McColl, S., Townsend, D., Villeneuve, M.: Landslides caused by the 14 November 2016 Kaikoura earthquake, South Island, New Zealand. In: Proceedings of 20th NZGS Symposium, Napier (2017)

10. Tsuchiya, S., Imaizumi, F.: Large sediment movement caused by the catastrophic Ohya-Kuzure landslide. J. Disaster Res. **5**(3), 257–263 (2010)

11. Nagai, O., Nakamura, H.: The large-scale landslide of Mt. Shichimen - the study on the history of the landslide and the enlargement process. J. Jpn. Landslide Soc. **37**(2), 20–29 (2000). (in Japanese)

12. Yamashita, N.: Fossa Magna. Tokai University Press, 25–29 (1995). (in Japanese)

13. Inokuchi, T., Yagi, H.: Shiratori-yama landslide caused by the Hoei earthquake in 1707. J. Jpn. Landslide Soc. **50**(3), 45–46 (2013). (in Japanese)

14. Headquarters for Earthquake Research Promotion: Long-term assessment of Minobu fault (2015). (in Japanese)

15. Zhang, S., Zhang, L., Lacasse, S., Nadim, F.: Evolution of mass movements near epicentre of Wenchuan earthquake, the first eight years. Sci. Rep. **6**(1), 1–9 (2016)

16. Taira, A.: Tectonic evolution of the Japanese Island arc system. Annu. Rev. Earth Planet. Sci. **29**(1), 109–134 (2001)

17. Matsushima, N.: The median tectonic line in the Akaishi mountains. In: Sugiyama, R. (ed.) Median Tectonic Line, pp. 9–27. Tokai University Press (1973). (in Japanese)

18. Matsushima, N.: History of debris flow disasters (called "Mansui") in the Ina Valley − in relation to periods of seismic activity. Nat. Hist. Rep. Inadani **1**, 11–15 (2000). (in Japanese)

19. Kusakabe, A.: Earthquake in Ina recorded in historical documents. Ina **33**(2), 65–73 (1985). (in Japanese)

20. Nagano Prefectural Government: List of damaging historical earthquakes. Appendix 1 for disaster mitigation plan (2021). (in Japanese)

21. Osawa, K.: Toyama earthquake in Ina. Ina **23**(2), 74–77 (1985). (in Japanese)

22. Murasawa, T.: Natural Disasters and Poor Harvest in Ina, 2nd edn. Kitahara Technical Office (1983). (in Japanese)

23. JMA (Japan Meteorological Agency). https://www.data.jma.go.jp/svd/eqev/data/matsushiro/mat50/disaster/onsenido/onsenido4.jpg. Accessed 20220120

24. Ohtake, M.: Seismic activity induced by water injection at Matsushiro, Japan. J. Phys. Earth **22**, 163–176 (1974)

25. Highway Office of Taiwan: Field survey for reconstruction of East-West No. 8 Highway. Transportation Authority, Taiwan (2020) (in Chinese).

26. Vermilye, J.M., Scholz, C.H.: The process zone: a microstructural view of fault growth. J. Geophys. Res. **103**(B6), 12223–12237 (1998)

27. Kuwahara, T., Hirama, K.: Study on engineering material evaluation of fault and sheared zone. Report of the Technical Research Institute, vol. 43, pp. 99–106. Ohbayashi-Gumi Ltd. (1991). (in Japanese)

28. Shipton, Z.K., Cowie, P.A.: A conceptual model for the origin of fault damage zone structures in high-porosity sandstone. J. Struct. Geol. **25**(3), 333–344 (2003)

29. Yoshida, E., Oshima, A., Yoshimura, K., Nagamoto, A., Nishimoto, S.: Fracture characterization along the fault - a case study of 'damaged zone' analysis on Atera fault, central Japan. J. Jpn. Soc. Eng. Geol. **50**(1), 16–28 (2009). (in Japanese)
30. Kanatani, M., et al.: Damages on waterfront ground during the 1999 Kocaeli earthquake in Turkey. Soils Found. **43**(5), 29–40 (2003)
31. Bellier, O., et al.: High slip rate for a low seismicity along the Palu-Koro active fault in central Sulawesi (Indonesia). Terra Nova **13**(6), 463–470 (2001)
32. Hamada, M., Yasuda, S., Isoyama, R., Emoto, K.: Generation of permanent ground displacements induced by soil liquefaction. In: Proceedings of the JSCE(376/III-6), pp. 211–220 (1986). (in Japanese)
33. Arulanandan, K., Yogachandran, C., Muraleetharan, K.K., Kutter, B., Chang, G.S.: Seismically induced flow slide on centrifuge. ASCE J. Geotech. Eng. **114**(12), 1442–1449 (1988)
34. Boulanger, R.W., Truman, S.P.: Void redistribution in sand under post-earthquake loading. Can. Geotech. J. **33**(5), 829–834 (1996)
35. Kokusho, T.: Water film in liquefied sand and its effect on lateral spread. ASCE J. Geotech. Geoenviron. Eng. **125**(10), 817–826 (1999)
36. Kokusho, T.: Mechanism of water film generation and lateral flow in liquefied sand layer. Soils Found. **40**(5), 99–111 (2000)
37. Hazarika, H., et al.: Large distance flow-slide at Jono-Oge due to the 2018 Sulawesi earthquake. Soils Found. **61**(1), 239–255 (2021)
38. Yasuda, S.: Possibility of water ejection from underground faults in Palu Basin. Personal communication (2020)
39. Kadarusman, A., Van Leeuwen, T., Sopaheluwakan, J.: Eclogite, peridotite, granulite, and associated high-grade rocks from the Palu region, Central Sulawesi, Indonesia: an example of mantle and crust interaction in a young orogenic belt. In: Proceedings JCM Makassar 2011 the 36th HAGI and 40th IAGI Annual Convention and Exhibition, Makassar, Indonesia (2011)

Static Liquefaction in the Context of Steady State/Critical State and Its Application in the Stability of Tailings Dams

Ramon Verdugo$^{(\boxtimes)}$

CMGI, Santiago, Chile
`rverdugo@cmgi.cl`

Abstract. Mining operations produce large quantities of tailings that must be conducted and stored safely in a technical framework that reconciles economic restrictions and environmental sustainability. Unfortunately, the history of tailings deposits is marked by episodes of catastrophic failures that have caused many victims and have cost enormous material losses. This is confirmed by the recent catastrophic collapses of tailings dams that occurred in Canada (Mount Polley), Australia (Cadia) and Brazil (Samarco and Brumadinio), where the tailings that flowed downstream severely affected the environment and, in the Brazilian case, caused many casualties. Due to this alarming empirical evidence left by the mining industry worldwide, there is international concern about the stability of tailings dams and a demand to build these deposits safely from every point of view. In this scenario, the correct evaluation of the actual shear strength of the deposited tailings is crucial. The critical state soil mechanic (CSSM) and/or the steady state of deformation provide the conceptual framework for evaluating the ultimate strength mobilized by particulate materials. In the context of a performance-based design, these states are discussed, and based on them, procedures and a flow index, I_f, are presented to analyze the physical stability of tailings.

Keywords: Tailings · Liquefaction · Steady state

1 Introduction

Mine tailings are generated from mineral ore processing and correspond to the waste material that remains after the valuable ore has been extracted from the processed rock. The expansion of the world economy and the increase in the world population has produced significant growth in the utilization of mineral resources. Therefore, mineral extraction has increased notably, as shown in Fig. 1 (Baker et al. 2020). Since the beginning of this century, there has been a sustained and strong increase in the extraction of metals and minerals, which ultimately has resulted in a substantial rise in mine residues (tailings), that must be managed and stored safely. Additionally, it is necessary to mention that in most mines, the ratio between the valuable mineral and waste material is decreasing, increasing the generation of tailings. Indeed, the annual global generation of tailings is currently of the order of eight billion tons (https://worldminetailingsfailures.org/). The current world

© The Author(s), under exclusive license to Springer Nature Switzerland AG 2022
L. Wang et al. (Eds.): PBD-IV 2022, GGEE 52, pp. 214–235, 2022.
https://doi.org/10.1007/978-3-031-11898-2_11

demand for minerals automatically implies the need to build tailings deposits with a large storage capacity, which means designing, building, and operating tailings dams of essential dimensions in length and height.

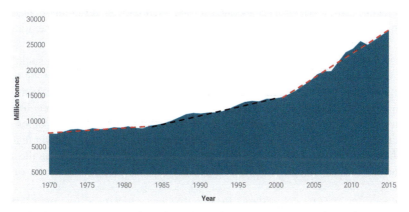

Fig. 1. The global extraction of metal ores from 1970 to 2017 (Baker et al. 2020)

For example, in Chile, there are several mining projects requiring tailings sand dams of more than 100 m in height such as, as Ovejería (120 m in height), Las Tórtolas (170 m), Quillayes (198 m), and El Mauro (237 m). Therefore, it is possible to point out that many tailings dams are among the largest earth structures. The design, construction, and operation of a tailings storage facility (TSF) must guarantee its physical stability at all events. However, the empirical fact is that the number of disasters caused by tailings dam failures is surprisingly high and increases. Figure 2 shows the number of serious and very serious failures occurring per decade, as reported from 1958 to 2017.

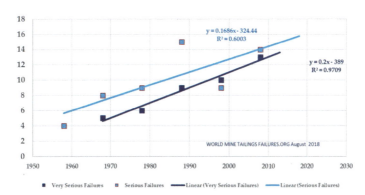

Fig. 2. Serious and very serious TSF failures from 1958 to 2017 (https://worldminetailingsfail ures.org/).

The trend of severe tailings dam failure events is confirmed by the recent catastrophic tailings dam failures in Canada (Mount Polley in 2014), Australia (Cadia in 2018) and

Brazil (Samarco in 2015 and Brumadinio in 2019). These failures have caused the flow of tailings to affect the environment severely. Unfortunately, the failures also resulted in many casualties in the cases of the Brazilian failures. A systematic increase in tailings dam failures is unacceptable to both society and the mining industry. The continued mismanagement of tailings dams has generated strong reactions from different organizations and demands to significantly improve the design, analysis, construction, operation, and closure of tailings deposits.

Several of the tailings dam failures have resulted from the occurrence of static lique-faction of the stored tailings. Therefore, identification of the initial states of the tailings and the undrained shear strength of the saturated tailings is crucial. Consequently, this paper focuses on this geotechnical aspect of tailings, which can be conceptualized and evaluated in the framework provided by the steady state of deformation (Poulos 1971) or the critical state soil mechanics (Schofield and Wroth 1968). Considering that there are still some discrepancies between the equivalence or differences between the steady state and the critical state, it is deemed necessary to start with a brief review of these soil states.

The amount of tailings generated by the different minerals is presented in Fig. 3 (Baker et al. 2020), where it is observed that more than 75% of the tailings are produced from the processes of recovering copper (46%), gold (21%) and iron (9%). A summary of tailings types and geotechnical classification is presented in Fig. 4 (ICOLD 2019). It can be observed that majority of the aforementioned tailings are described as sandy silts of non to low plasticity.

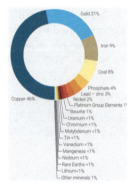

Tailings Type	Symbol	Description (compare)	Example of mineral/ore
Coarse tailings	CT	Silty SAND, non-plastic	Salt, mineral sands, coarse coal rejects, iron ore sands
Hard Rock tailings	HRT	Sandy SILT, non to low plasticity	Copper, massive sulphide, nickel, gold
Altered Rock tailings	ART	Sandy SILT, trace of clay, low plasticity, bentonitic clay content	Porphyry copper with hydrothermal alteration, oxidized rock, bauxite. leaching processes
Fine tailings	FT	SILT, with trace to some clay, low to moderate plasticity	Iron ore fines, bauxite (red mud), fine coal rejects, leaching processes, metamorphosed/weathered polymetallic ores
Ultra Fine tailings	UFT	Silty CLAY, high plasticity, very low density and hydraulic conductivity	Oil sands (fluid fine tailings), phosphate fines; some kimberlite and coal fines

Fig. 3. Amount of tailings (Baker et al. 2020).

Fig. 4. Tailings type and geotechnical classification (ICOLD 2019).

Therefore, the most significant quantities of tailings are from copper, gold, and iron mines, which in general, vary from fine sands to sandy silt. Thus, the observed behavior of sandy soils applies to these tailings.

Experimental results on the Japanese Toyoura sand from laboratory tests performed at the University of Tokyo, Japan, are presented in this paper (Ishihara et al. 1991; Verdugo 1992; Ishihara 1993; Verdugo and Ishihara 1996). The tested Toyoura sand has a specific gravity, $G_s = 2.65$, $D_{50} = 0.17$ mm, and maximum and minimum void ratios of $e_{max} = 0.977$ and $e_{min} = 0.597$, respectively.

2   The Ultimate State of Soils

2.1   Steady State and Critical State

The concept of critical void ratio was first postulated by Casagrande in 1935 at a Boston Civil Engineering Society conference and published a year later in the same society's journal (Casagrande 1936). Using direct shear tests, Casagrande observed that during shearing, dense sands expand, increasing their void ratio, while very loose sands contract and reduce their volume and consequently their void ratio. Based on this observation and his intuition, Casagrande developed the concept of the critical void ratio in which dense and loose sands, when sheared in a drained condition, change the void ratio until a unique constant void ratio is finally reached, that Casagrande called the critical void ratio. In this state, the soil continues to deform under constant strength and constant volume, hence the soil behaves as a frictional fluid. Initially, Casagrande thought that the critical void ratio was a unique constant value for a given soil.

In the same direction, Roscoe and co-workers (Roscoe et al. 1958) studied the ultimate conditions reached by soil samples sheared to large deformation under drained and undrained loading conditions. They confirmed that in the p' - e - q space, both for drained and undrained tests, there is a unique line where the loading paths converge, which they called the critical voids ratio line (CVRL). In the case of a drained test, the critical void ratio state is reached when any further increment of shear distortion does not change the voids ratio. In undrained tests, the effective mean stress, p', is modified to bring the sample into an ultimate state such that the shear stress and the effective stress become constant. Roscoe obtained the CVRL experimentally for clayey soils, sandy soils, glass beads, and steel balls. For Weald Clay, this line is shown in Fig. 5.

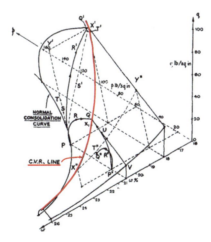

Fig. 5.   Critical voids ratio line for Weald Clay (Roscoe et al. 1958).

Later, Schofield and Wroth (1968) published the book that can be considered the official birth of the concept known as Critical State Soil Mechanics (CSSM). The authors in this book indicate that soil and other granular materials, if continuously distorted until they flow as a frictional fluid, will come into a well-defined critical state determined by the following two equations:

$$q = Mp'$$ (1)

$$e = \Gamma - \lambda \cdot \mathrm{Ln}(p')$$ (2)

where $q = (\sigma_1' - \sigma_3')$ corresponds to the deviator stress, $p' = (\sigma_1' + 2\,\sigma_3')/3$ represents the effective mean stress, e is the void ratio, M is the shear stress ratio q/p at critical state, λ is the slope of the CVRL in the e vs. $\ln(p')$ plot, and Γ is the void ratio of the CVRL for $p' = 1$ stress unit.

The first equation of the critical states determines the magnitude of the deviator stress q needed to keep the soil flowing continuously. The second equation states that the void ratio of flowing particles will decrease as the logarithm of the effective pressure increases. Schofield and Wroth (1968) emphasized that the total change from any initial state to an ultimate critical state where the soil flowed as a frictional fluid can be predicted with these two equations. It is interesting to note that, strictly speaking, there are no requirements regarding soil behavior. The definition of the critical state is pragmatic since it simply depends on empirical data that can fit the two equations described above.

Later in 1971, following both the concept of critical void ratio initially postulated by Casagrande and the experimental results reported by Castro (1969), which showed how a sufficiently loose sand collapses and deforms continuously under an undrained stress-controlled test, Poulos (1971) proposed the concept of the steady state of deformation. He defined it as the state in which a particulate material of any composition or particle shape deforms continuously under a constant state of effective stress at constant velocity and at a constant void ratio. In his monograph, Poulos explicitly states that the term "steady state" is used for two reasons: 1) it conveys correctly (by analogy to steady state flow of liquids) the concept of flow, and 2) the term "critical state" has been applied by other researchers to a completely different condition of soils. The steady state concept was then used by Castro (1975) and Castro and Poulos (1977) to explain the difference between "true" liquefaction and cyclic mobility. Later, Poulos (1981) stated that the steady state of deformation is achieved only after all particle orientation has reached a statistically steady-state condition and after all particle breakage, if any, is complete, so that the shear stress needed to continue deformation and the velocity of deformation remain constant. In addition, it is indicated that during the steady state of deformation, the original structure of the specimen is completely destroyed and reworked into a new "flow structure."

According to these definitions, the steady state of deformation has the following conditions that must be satisfied by the soil under continuous deformation:

- Constant effective stress
- Constant shear stress
- Constant velocity
- Constant void ratio
- Particle orientation has reached a statistically steady-state condition

- All particle breakage, if any, is complete
- Original structure is reworked into a new flow structure

These requirements are of such practical implication that the critical state can be clearly differentiated from the steady state of deformation depending on the soil type. Considering that the critical state is a state that does not pay attention to the rate of deformation or the orientation of particles, its actual application to clayey materials is somewhat limited, as shown below.

2.2   Behavior of Clayey Soils in Relation to Their Ultimate State

A comprehensive set of tests are reported for the Japanese Kawasaki clay (Gs = 2.69, LL = 55.3, PI = 29.4, FC = 83.9%), which has been tested at different rates of deformations (Nakase and Kamei 1986). Figure 6 presents the results of a series of undrained triaxial tests started from a K_0-consolidation and loaded in compression and extension. Three different rates of deformations were applied during loading: $7 \cdot 10^{-1}$, $7 \cdot 10^{-2}$ and $7 \cdot 10^{-3}$%/min.

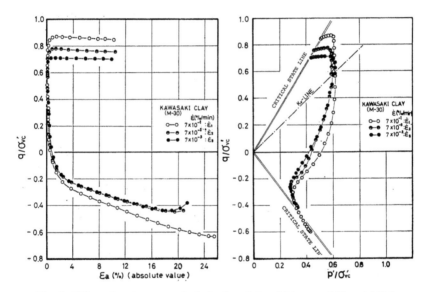

Fig. 6. Effect of strain rate on the behavior of clay (Nakase and Kamei 1986).

These results clearly show that the rate of deformation has an important effect on both the stress-strain curves and the undrained shear strength; the higher the rate of deformation the higher the undrained strength. For clayey soils, different researchers have reported the dependence of the undrained strength on the rate of strain (Bjerrum 1969; Sheahan et al. 1996; Díaz-Rodríguez et al. 2009; Chow et al. 2012), among many others). However, the angle of friction mobilized at the ultimate state or the critical state is not influenced by the rate of deformation. This means that the critical state line in the

q-p' plane is unaffected by the rate of deformation, but the critical state line in the e-log p' plane is definitely influenced by the rate of deformation.

On the other hand, with respect to the particle orientation, this phenomenon has been reported in clayey material with an important amount of platy particles of clays. The re-orientation of the platy particles in the direction of the applied shear stress has made possible to explain the reduction of the resistance to a "residual strength" that has been measured at significantly large deformations under a considerable low rate of deformation (Bishop et al. 1971; Lupini et al. 1981; Skempton 1985). As example, in Fig. 7 the stress-displacement curve obtained from a ring shear test is presented, where after about 30 days of shearing, a residual strength associated with a friction angle, ϕ_r = 8.6°, is mobilized (Skempton 1985).

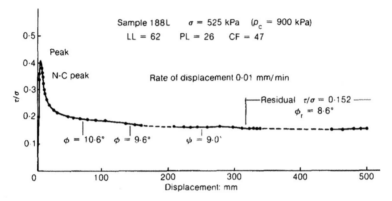

Fig. 7. Residual strength on clayey soil (Skempton 1985).

In this matter, Schofield and Wroth (1968) explicitly show that the residual strength developed by clayey soils with platy particles differs from the strength established at the critical state, as indicated in Fig. 8, where for consistency the critical state should be at point D, but the actual ultimate state is at point E.

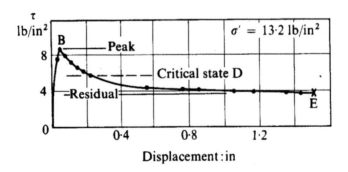

Fig. 8. Critical state (D) and residual strength (E) (Schofield and Wroth 1968).

Therefore, in the particular case of clayey materials and those that behave as clay-like soils, the requirements imposed by the steady state are definitely not satisfied, unless different steady states are accepted depending on the strain rate and the level of deformation. In this context, it is important to point out that the critical state, by not imposing requirements for the ultimate state, somehow allows a free choice of what will be used as critical state, adjusting the variables to the particular problem under analysis.

2.3   Behavior of Sandy Soils in Relation to Their Ultimate State

In the case of sandy soils, the rate of deformation is significantly less important in the response at large deformations, as can be inferred from the experimental results reported by Been et al. (1991) as reproduced in Fig. 9. Contractive samples under stress-controlled tests collapse, developing a high rate of deformation during the collapse, which is several orders of magnitude higher than the common rate of deformation imposed in strain-controlled tests. Nevertheless, it is apparent in Fig. 9 that the same ultimate state is achieved in contractive samples under load-controlled and strain-controlled tests.

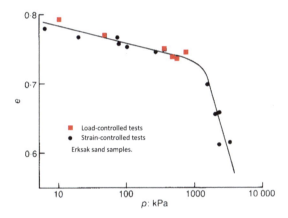

Fig. 9. Critical state line Erksak sand (Been et al. 1991).

The phenomenon of re-orientation of particles, which can significantly reduce the shear strength at extremely large deformation, needs an important presence of platy particles that do not exist in sandy soils.

When the ultimate state is reached in sandy soils, the particle breakage, if any, is definitely complete. Otherwise, the pore pressure would increase under undrained loads, or equivalently, the volumetric strains would show contraction under drained loads, which would not be an ultimate state. Regarding Poulos's concept, it is difficult to verify that the original structure is reworked into a new flow structure in the ultimate state. However, experimental results reported by Verdugo and Ishihara (1996) may provide some insight that suggests at least the development of a unique fabric during the ultimate state. Figure 10 shows the results of stress-strain curves and effective stress-paths of a series of undrained triaxial tests carried out at effective confining pressures of 0.1, 1, 2, and 3 MPa on samples with the same void ratio of $e = 0.735$ after isotropic consolidation. Once the ultimate condition was reached, each sample was unloaded to zero shear stress. Besides the fact that these results confirm that, under undrained load, the void ratio controls the ultimate state, it is essential to realize that during unloading, the effective stress paths of the four tested samples converge into one, independently of the confining pressure of each sample. The development of a unique effective stress path during unloading suggests that the induced fabric of the four samples was the same when the unloading started from the ultimate state.

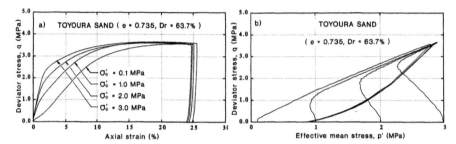

Fig. 10. Unloading from the steady state (Verdugo and Ishihara 1996).

Other potential evidence about the creation of a particular structure during the steady state is presented in Fig. 11 (Verdugo 1992), where the results of two samples prepared at the same density by different procedures such as wet tamping and water sedimentation, show different stress-strain curves but identical ultimate shear strength. Furthermore, the effective stress paths during unloading are practically the same.

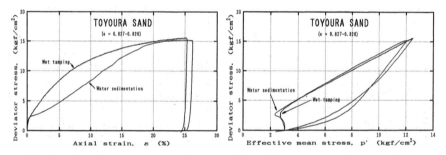

Fig. 11. Undrained tests on samples prepared by different methods (Verdugo 1992).

These experimental results suggest that the different original fabrics were modified during loading so that upon reaching the ultimate state, a new and similar fabric was developed in these samples, which is manifested through the same shear strength and the same effective stress-path during unloading.

2.4 Critical and Steady States Are Different

– Following what was presented above, there are similarities, but also differences, between the critical state and the steady state, which can be described as follows:

– Both states are associated with the ultimate condition reached at large deformations by a soil subjected to shear stress.
– The critical state has no special requirements to establish how it has been reached. Simply, in the case of drained loading, it is reached when any further increment of shear distortion does not change the void ratio. In the case of undrained loading, it is reached when the ultimate state is such that both the shear stress and the effective stress become constant under continuous shear distortion.
– The steady state establishes conditions that are definitely not specified by the critical state. These conditions are constant velocity, particle orientation, particle breakage ended, and modification of the original structure into a new flow structure.

Considering that the mechanical response of clayey materials can be seriously affected by both the rate and the amount of deformation, they should develop different ultimate states according to the characteristics of the applied loads. Accordingly, the critical state is more applicable because this state can be established for the particular problem under analysis. On the other hand, the steady state is appropriate in sandy soils because its requirements are satisfied by granular soils. Obviously, these soils also satisfy the less rigorous framework imposed by the critical state.

In summary, the steady state of deformation corresponds to the ultimate state that any particulate material reaches when it is sheared at large deformations. All the aspects observed in the behavior of soils are considered. On the other hand, in a more practical application, the critical state of soils corresponds to the ultimate state reached at large strains, when there is no change of void ratio in drained condition, and the effective stresses become constant in undrained load condition.

In the case of several tailings, its behavior can be considered sand-like materials. Accordingly, the ultimate state of these tailings can be referred to as steady state or critical state. However, because the requirements of the steady state are more rigorous, this concept is used in this article as a synonym for ultimate state.

3   Characteristics of the Contractive States

3.1   Condition of Minimum Shear Strength in Sandy Soils

As shown in Fig. 12, sandy soils in a contractive state may develop under undrained loading a stress-strain curve with a clear peak followed by the development of a minimum strength at a medium level of strain. Thereafter, an increase in the strength toward the end of the test is observed.

This condition of minimum shear resistance has been called quasi-steady state (QSS) because it is a transient state in which the steady state requirements are mostly satisfied (Alarcon-Guzman et al. 1988; Ishihara 1993). In the experimental results presented in Fig. 13, the minimum strength can be as low as 16% of the strength mobilized at the critical state. Therefore, from a practical point of view, it is important to pay attention to this transient state due to the significant difference between the minimum strength and the strength mobilized at large deformation where the steady state is achieved.

The state of minimum resistance is associated with a rather significant level of deformation, which could be severe for some structures. However, when analyzing a flow

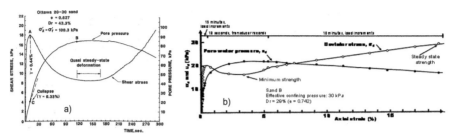

Fig. 12. Stress-strain curves with development of a minimum strength (a) Alarcon-Guzman et al. 1988, b) Castro 1969).

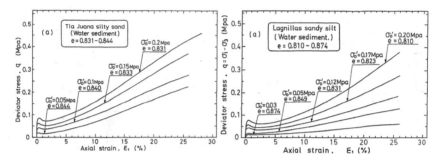

Fig. 13. Minimum strength and steady state strength (Ishihara 1993)

failure, exceptionally large deformations are necessarily involved, which are definitely not manifested during the condition of minimum strength. Therefore, this condition does not represent a true flow failure. Nevertheless, it is important to understand how these limited deformations and excess pore pressures may interact negatively in the soil mass.

Unlike what is observed in the steady state, the development of the minimum strength is affected by the initial fabric of the soil and is relatively affected by the level of confining pressure. This state of minimum strength is obtained in load-controlled tests (Castro 1969) as well as in strain-controlled tests (Alarcon-Guzman et al. 1988; Ishihara 1993) and in triaxial tests with enlarged and lubricated ends that allow a more uniform deformation of the samples (Verdugo 1992). It is not clear what is the cause of this transient state. Still, it is important to be aware of its real existence. Depending on the project, it can have an important effect on the global stability of a soil mass as illustrated below.

It is important to point out that there is a special condition in which the static shear can be greater than the minimum strength (i.e., the quasi-steady state strength), but smaller than the steady state strength, as is illustrated in Fig. 14. A load that triggers the undrained soil response can generate soil deformations and excess pore pressure that reduces the mean effective stress from the initial state at point A to the point B. Therefore, in those initial states where the soil can develop a quasi-steady state condition and the associated minimum resistance is less than the acting static shear, a significant increase of both deformations and pore pressure can be generated. The interaction of deformation and dissipation of pore pressure with the rest of the soil mass can trigger a failure mechanism, which depends on the initial state of static shear, void ratio, and mean effective stress of the soil mass.

The phenomenon of minimum strength, or QSS, is not captured by the state parameter, since the latter is based on the ultimate condition or steady state. In fact, the state parameter in these cases is negative, meaning the soil response is dilatant and stable.

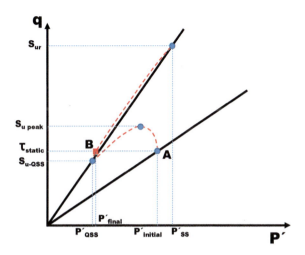

Fig. 14. Static shear stress greater than QSS strength ($S_{u\text{-QSS}}$) and smaller than SS strength (S_{ur})

3.2   States with Loss of Strength

For Toyoura sand, the steady state line established as the ultimate state reached at large strains, in the order of 25%, is presented in Fig. 15a. In comparison, the quasi steady state is presented in Fig. 15b (Ishihara 1993). In both cases, the initial states of each test are indicated by white squares, making it clear that the QSSL can only be obtained from a contractive initial condition.

Additionally, in Fig. 16 are presented the SSL, the QSSL and the so-called IDL (Initial Dividing Line). The IDL is the line above which the initial states that develop loss of strength are located.

It is important to note that the IDL is below the SSL in a significant range of effective mean pressure. Therefore, it is not entirely true that only those initial states located above the SSL are contractive and with evidence of instability.

3.3   Peak Undrained Strength

The undrained response of sandy soils in a contractive state, with a drop of strength, shows a maximum value of shear stress, commonly called undrained peak strength or yield strength. The sketch of Fig. 17 shows how a medium amplitude cyclic load could

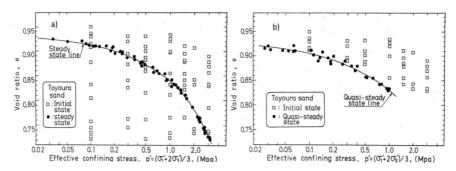

Fig. 15. a) Steady state line, b) Quasi-steady state line. Toyoura sand (Ishihara 1993).

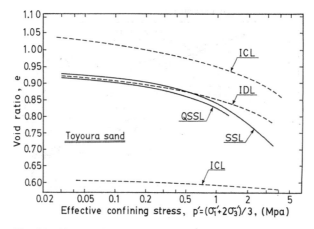

Fig. 16. Characteristic lines in the e-p' plane (Ishihara 1993).

trigger the ultimate resistance, without the need of mobilizing the peak strength (Poulos 1988). Experimental evidence confirming that a cyclic load can bypass the peak strength is presented in Fig. 18.

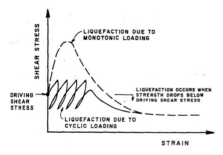

Fig. 17. Cyclic and monotonic loadings (Poulos 1988).

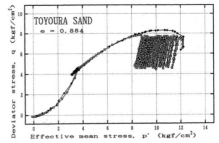

Fig. 18. Cyclic and monotonic loadings. Tests on Toyoura sand (Verdugo 2009).

From the above, it can be concluded that the peak strength is not a reliable parameter in cyclic stability analyses. However, it is useful to establish the collapse surface as it is explained below.

3.4 Flow Failure or Static Liquefaction

Even when a soil mass is in an initial contractive state such that it may suffer a loss of its strength, an additional condition is still necessary for a flow failure (or static liquefaction) to occur. This initial condition is that the driving shear stresses (static shear stress) must be greater than the residual undrained strength. Therefore, the static shear stress also plays an essential role in the occurrence of a flow failure, as shown in Fig. 19.

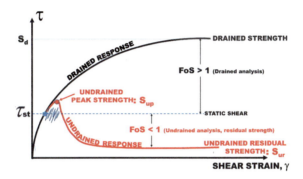

Fig. 19. Concept of static liquefaction.

The application of an initial load under a drained condition is considered until a static shear stress τ_{st} is reached. If the loading continues under a drained condition, theoretically, the soil element will move along the black curve until it mobilizes the drained resistance S_d. However, if the undrained response is triggered, for instance due to a fast disturbance, the strength of the soil would drop from S_d to S_{ur}. In Fig. 19, the static shear stress is greater than the mobilized residual undrained strength, S_{ur}, which means that a flow failure would occur. It is of great importance to mention again that the peak undrained strength can be bypassed by the soil response.

3.5 The Collapse Boundary

The collapse surface concept establishes in the q-p$'$ plane the existence of a boundary that effective stress paths cannot cross without the collapse of the soil, reaching the steady state condition. Initially, the collapse surface was defined in the q-p' plane based on the monotonic effective stress path of a contractive sample with loss of strength. The collapse surface was identified as the straight line between the peak resistance and the ultimate state as is shown in Fig. 20a (Sladen et al. 1985). Figures 20b and 18 presents experimental results that confirm the existence of the collapse boundary (Verdugo 2009).

It is observed in Fig. 20b that the monotonic effective stress path acts as a boundary of the cyclic load. These tests were performed under a strain-controlled regime, which allows the measurements during the post-peak soil response.

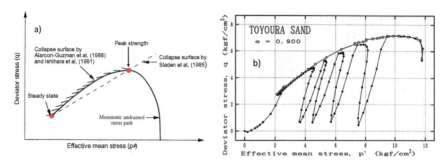

Fig. 20. a) Collapse surface (from Sladen et al. 1985), b) Test results on Toyoura sand.

For Toyoura sand, the monotonic effective stress paths obtained for the loosest states that were achieved are presented in Fig. 21. It is seen that a straight line can be drawn through the peak strength points. The line can be considered as a lower boundary above which the soil would collapse. Also shown in Fig. 21 are the monotonic effective stress paths of those initial states that present just a marginal loss of strength. These stress paths also allow to establish a line that can be seen as an upper boundary of the states that exhibiting a strength loss. The two boundaries shown in Fig. 21 represent the zone in the e-p' plane associated with the collapse of this soil. Between these lines, there are many are other lines with slopes that are functions of the initial void ratio.

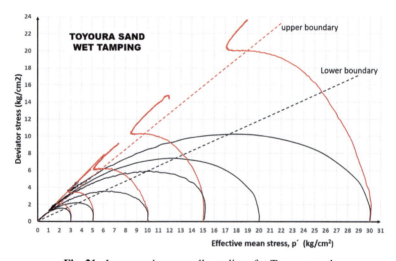

Fig. 21. Lower and upper collapse lines for Toyoura sand.

Experimental evidence suggests that the monotonic effective stress path is the boundary in the e-p' plane that controls the onset of soil collapse. However, from any practical point of view, the maximum shear strength locus established in the e-p' plane can be considered as a good approximation to define the limit that triggers soil collapse, which is a function of the void ratio.

Due to the field the conditions are load-controlled, two practical stress paths can lead the soil collapse (Fig. 22):

1) Rise of the water table or the seepage flow that decreases the effective mean stress while keeping the static shear constant (Sasitharan et al. 1993).
2) Initially drained load, which combination of q and p' passes the collapse line. In this case, there is a sudden change of the drained load into an undrained response of the soil.

The in-situ condition of load-controlled impedes the application of drained stress beyond the collapse line. The soil responds with a sudden increase of both strains and pore pressure that triggers the undrained response and the collapse of the soil.

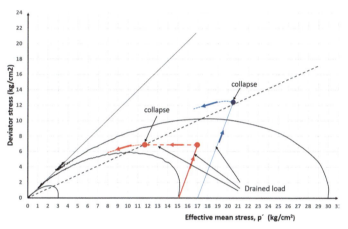

Fig. 22. Examples of initial drained stress paths that trigger the collapse of soil in a contractive state. (in red: constant shear with decreasing mean effective stress; in blue: high ratio q/p')

4 Characterization of the Initial States of a Soil

4.1 State Parameter Ψ

Using the steady state line as the reference state, Been and Jefferies (1985) introduced the so-called state parameter, ψ, which considers both the degree of packing and the pressure effect. The state parameter is defined in the e-p' plane as the difference between the current void ratio of the soil and the void ratio at the steady state, at the same mean effective stress. This means that initial states that have the same distance to the steady state line in the e-p' plane would have a similar response. Indeed, this observation is similar to the conclusion reported by Roscoe and Poorooshasb (1963): any two samples of a given soil, when subjected to geometrically similar stress path, have the same strains provided that the difference between the initial void ratio and the void ratio at the critical state at the same normal stress is the same for each sample. In this statement, Roscoe and Poorooshasb have referred to the critical state instead of the steady state, but for the present purpose these states can be considered identical.

The positive values of the state parameter are associated with initial states that are located above the steady state line in the e-p' plane, which means that the initial state is contractive. The negative values of the state parameter correspond to states below the steady state line, which represent initial dilatant states. Although the state parameter is widely used, it has a severe limitation of incorporating the potential instability induced by the static shear, as explained in detail below.

4.2 State Index Is

The state index I_s proposed in the e-p' plane by Ishihara (1993) is considered a more rigorous parameter to characterize the response of sandy materials and low-plasticity silts as tailings. I_s uses two reference lines: the quasi steady state line (QSSL) and the isotropic consolidation line ICL of the loosest state. Along the ICL, the state index takes the value $I_s = 0$, and along the QSSL the state index takes the value $I_s = 1$. Then, for a given effective mean pressure p', the state index is defined as the linear interpolation between the current void ratio e, the void ratio e_o at the ICL, and the void ratio e_s at QSSL. Figure 23 gives the definition of I_s including the zone of zero strength.

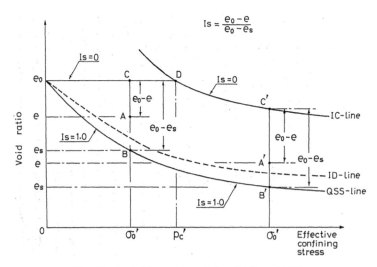

Fig. 23. Definition of state index (Ishihara 1993).

It is important to highlight that using the QSSL as a reference line allows considering the minimum resistance condition that can generate significant increases in pore pressure and deformation. This condition depends on the soil fabric, which cannot be captured when the ultimate state is used as a reference line.

The state index I_s can be a good predictive capacity of soil behavior according to the initial state of density and confining pressure. However, the fact that it does not consider the shear stress means that it also suffers from the same deficiency as the state parameter.

5   The Importance of Static Shear in the State of a Soil

The state parameter ψ and the state index I_s are based on isotropic triaxial consolidated tests. For this initial isotropic stress condition, the distance, in terms of void ratio to the SSL or QSSL, is a reasonable approximation to the expected soil response. However, when initial static shear stress is included, or in other words, when the stress paths to reach the initial permanent condition are different, these parameters can give a misleading interpretation of the soil response. Note that if a given combination of e and p′ is reached, but from totally different stress paths, the response to a new given stress path could be different depending on the previous stress paths.

To illustrate the effect of the initial static shear stress on the undrained responses, two different samples tested in the triaxial equipment are considered for simplicity, but the concept is valid for real situations in situ. Figure 24 illustrates how the static shear stress can drastically modify the soil behavior from a stable undrained contractive-dilative response to an unstable fully contractive response that can experience flow failure.

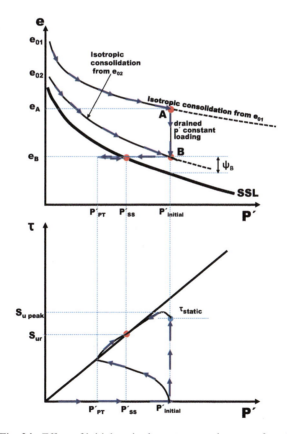

Fig. 24. Effect of initial static shear stress on the state of a soil

First, a loose sample prepared with a void ratio, e_{01}, is considered, which is isotropically consolidated up to a pressure $p'_{initial}$, that generates a void ratio, e_A. Then, this hypothetical sample is loaded under a drained p' constant condition up to a shear stress identified as τ_{static}. This drained load decreases the void ratio from e_A to the value e_B. A second hypothetical sample prepared with a lower void ratio, e_{o2}, is considered. This void ratio is such that after the sample is isotropically consolidated to $p'_{initial}$, the void ratio achieved is e_B. These two samples have the peculiarity of having the same void ratio and the same effective mean stress, which implies that they have the same residual undrained strength, S_{ur}, and the same state parameter, ψ_B. However, one of the samples has high static shear stress that is greater than the undrained residual strength, which means it can develop a flow failure. Whereas the other sample has no static shear stress, and additionally, its undrained response does not present any instability.

The main conclusion of this example is that the state parameter does not capture the existence of any static shear. In other words, any given distance to the SSL in the e-p' plane can be associated with a fairly significant number of levels of static shear stress. In some cases, the static shear stress can be less than the undrained residual strength, but in others, it may be the other way around, and therefore, the state of the soil can be stable or ready for a flow failure.

6 Flow State Index

It is recognized that a flow failure requires that the existing static shear stress be greater than the undrained residual strength. On the contrary, if the static shear does not exceed the undrained residual strength, a flow failure cannot be triggered, unless there is migration of water from one sector to another, which is a more complex phenomenon beyond the present analysis. Therefore, an index that directly considers this situation would be useful. Accordingly, the Flow State Index I_f, defined below, is proposed:

$$I_f = \frac{S_{u-min}}{\tau_{static}} \tag{3}$$

As it has been shown, there are cases where the difference between the minimum strength and the one mobilized at the ultimate state can be drastically different; it is estimated that in problems of stability of soil masses against flow failure, it is appropriate (conservative) to use the undrained minimum strength, S_{u-min}, which can be evaluated by means of in-situ testing in the case of existing tailings dams, for example, using vane shear tests with a high speed of rotation to ensure the undrained response of the tested soils or by CPT-u measurements (Roberson 2010), or just the sleeve resistance (Verdugo et al. 2014).

The driving static shear stresses, τ_{static}, can be estimated through numerical analysis. In general, it can be indicated that the assessment of the static stress field in earth structures is not sensitive to the constitutive model used, which is not valid for the strain field. In tailings dams, the basic resistance and stiffness properties of the different materials involved in the stability can be reasonably estimated. Also, the static stress field in the different parts of a tailings dam can be numerically evaluated with an acceptable accuracy using relatively simple stress-strain constitutive models. In addition, it is necessary to measure and/or calculate the seepages to estimate the pore pressure distribution, and accordingly, the effective stresses. Therefore, using a rather simple numerical analysis it is possible to estimate both the effective mean stresses and shear stresses existing in the different zones of a tailings dam.

Accordingly, the proposed flow index I_f is evaluated considering field data and numerical analysis. Obviously, values of I_f less than one mean that the permanent static shear is greater than the minimum undrained shear strength, and therefore, the state of the soil is such that a flow failure can be triggered. Conversely, values of I_f greater than one represent soil states that are not associated with a potential failure due to static liquefaction.

The question that naturally arises is what would be the extension of the zone where the values of I_f are less than one, in such a way that a flow failure of the soil mass can be triggered. The answer to this question involves more sophisticated numerical analyses. Still, the fundamental point is that the proposed parameter I_f makes possible to identify the existence of soil states that can collapse and be aware of their existence, if they exist.

7 Concluding Remarks

Mining operations worldwide produce vast amounts of tailings that need to be stored safely, reconciling economic constraints and environmental sustainability. However, the recent history of tailings deposits is marked by episodes of catastrophic failures that have caused many victims and have cost enormous material losses. One of the leading causes of these failures is the static liquefaction of the stored tailings.

The concepts of steady state and critical state, which allow the analysis of static liquefaction, have been reviewed and discussed, resulting in the steady state concept being more rigorous to study static liquefaction. It is possible to think that after so many years it is not acceptable to still be stuck with the potential differences between SS and CS. It is true that at the end it does not matter whether CS and SS are different or the same; in practice it might be irrelevant. However, behind this discussion, which could be seen simply as a matter of semantics, it is the real behavior of the soil, which in many aspects is different for clays and sands and that in practice can seriously affect its modeling and analysis. It was shown that the behavior of clays is strongly affected by the strain-rate, also clays can undergo reorientation of the platy particles that results in a significant reduction of the drained strength. These phenomena are not relevant in sandy soils, but the behavior of the sands can present a state of minimum resistance, followed by significant dilation and a strong recovery to the ultimate strength that could be useless in some practical problems. Therefore, what really matters is to specify which conditions are considered in soil characterization and modeling, and which are left out.

The occurrence of static liquefaction requires the soil to be in a contractive state with a loss of strength. This condition is currently addressed using parameters that provide information on the state of the soil, in terms of its void ratio and confining pressure, and its location concerning the steady state line or quasi-steady state line. However, this approach has severe limitations because it does not consider the driving static shear stress, or the stress paths to reach the initial states.

A very simple but straightforward index has been proposed that gives a clear indication of the state of a soil with respect to its potential for flow failure.

Acknowledgements. The author wishes to acknowledge the review generously made by my colleague David Solans and by Professor Marte Gutierrez, who corrected and improved the original work. The support provided by CMGI during the preparation of this work is also appreciated.

References

Alarcon-Guzman, A., Leonards, G.A., Chameau, J.: Undrained monotonic and cyclic strength of sands. J. Geotech. Eng. **114**(10), 1089–1109 (1988)

Baker, E., Davies, M., Fourie, A., Mudd, G., Thygesen, K.: Mine tailings facilities: overview and industry trends. In: Towards Zero Harm. A Compendium of Papers Prepared for the Global Tailings Review, Chap. II (2020)

Been, K., Jefferies, M.G.: A state parameter for sands. Geotechnique **35**(2), 99–112 (1985)

Been, K., Jefferies, M.G., Hachey, J.: A critical state of sands. Geotechnique **41**(3), 365–381 (1991)

Bishop, A.W., Green, G.E., Garga, V.K., Andresen, A., Brown, J.D.: A new ring shear apparatus and its application to the measurement of residual strength. Geotechnique **21**(4), 273–328 (1971)

Bjerrum, L.: Effect of rate of strain on undrained shear strength of soft clays. In: Second Contribution to Panel Discussion, Main Session 5, 7th International Conferenceon Soil Mechanics and Foundation Engineering, Mexico (1969)

Casagrande, A.: Characteristics of cohesionless soils affecting the stability of slopes and earth fills. J. Boston Soc. Civ. Eng. **23**, 13–32 (1936)

Castro, G.: Liquefaction of Sands. Harvard Soil Mechanics Series, vol. 81. Harvard University, Cambridge (1969)

Castro, G.: Liquefaction and cyclic mobility of saturated sands. J. Geotech. Eng. **101**(GT6), 551–569 (1975)

Castro, G., Poulos, S.: Factors affecting liquefaction and cyclic mobility. J. Geotech. Eng. **103**(GT6), 501–516 (1977)

Chow, S., Alonso-Marroquin, F., Airey, D.: Viscous rate effects in shear strength of clay. In: Proceedings of the 11th Australia New Zealand Conference on Geomechanics (2012)

Díaz-Rodríguez, J., Martínez-Vasquez, J., Santamarina, J.: Strain-rate effects in Mexico City soil. J. Geotech. Geoenviron. Eng. **135**(2), 300–305 (2009)

ICOLD: Tailings Dam Design, Technology Update. ICOLD Committee on Tailings Dams (2019)

Ishihara, K.: 33th Rankine Lecture: liquefaction and flow failure during earthquakes. Geotéchnique **43**(3), 351–415 (1993)

Ishihara, K., Verdugo, R., Acacio, A.: Characterization of cyclic behavior of sand and post-seismic stability analysis. In: Proceedings of the IX Asian Regional Conference on Soil Mechanics and Foundation Engineering, Bangkok, Thailand (1991)

Lupini, J.F., Skinner, A.E., Vaughan, P.R.: The drained residual strength of cohesive soils. Geotechnique **31**(2), 181–213 (1981)

Nakase, A., Kamei, T.: Influence of strain rate on undrained shear characteristics of K_0-consolidated cohesive soils. Soils Found. **26**, 85–95 (1986)

Poulos, S.J.: The Stress-Strain Curves of Soils. Geotechnical Engineers, Inc., Winchester (1971)

Poulos, S.J.: The steady state of deformation. J. Geotech. Eng. Div. **107**(GT5), 553–562 (1981)

Poulos, S.J.: Liquefaction and related phenomena. In: Jansen, R.B. (ed.) Advance Dam Engineering for Design, Construction, and Rehabilitation, pp. 256–320. Van Nostrand Reinhold, New York (1988)

Robertson, P.: Evaluation of flow liquefaction and liquefied strength using the cone penetration test. J. Geotech. Geoenviron. Eng. **136**(6), 842–853 (2010)

Roscoe, K., Poorooshasb, H.: A fundamental principle-of similarity in model test for earth pressure problems. In: Proceedings of the 2nd Asian Conference Soil Mechanics, Japan, vol. 1, pp. 134–140 (1963)

Roscoe, K.H., Schofield, A.N., Wroth, C.P.: On yielding of soils. Geotéchnique **8**(1), 22–53 (1958)

Sasitharan, S., Robertson, P., Sego, D., Morgenstern, N.: Collapse behavior of sand. Can. Geotech. J. **30**(4), 569–577 (1993)

Skempton, A.: Residual strength of clays in landslides, folder strata and laboratory. Geotechnique **35**(1), 3–18 (1985)

Sladen, J., D'Hollander, D., Krahn, J.: The liquefaction of sands, a collapse surface approach. Can. Geotech. J. **22**, 564–578 (1985)

Schofield, A., Wroth, C.: Critical State Soil Mechanics. McGraw-Hill, London (1968)

Sheahan, T., Ladd, C., Germaine, J.: Rate-dependent undrained shear behavior of saturated clay. J. Geotech. Eng. **122**(2), 99–108 (1996)

Verdugo, R.: Characterization of sandy soil behavior under large deformations. Ph.D. thesis, University of Tokyo, Japan (1992)

Verdugo, R.: Seismic performance based-design of large earth and tailings dams. In: Proceedings of International Conference on Performance-Based Design in Earthquake Geotechnical Engineering – from Case History to Practice, Tsukuba, Japan (2009)

Verdugo R., Echevarría J., Peters G., Caro G.: Feasibility evaluation of converting a conventional tailings disposal into a thickened tailings deposit. In: Jewell, R.J., Fourie, A.B., Wells, S., van Zyl, D. (eds.) Proceedings of Paste 2014, Canada (2014)

Verdugo, R., Ishihara, K.: The steady state of sandy soils. Soils Found. **36**(2), 81–91 (1996)

Seismic Behaviour of Retaining Structures: From Fundamentals to Performance-Based Design

Giulia M. B. Viggiani[1]([⊠]) and Riccardo Conti[2]

[1] University of Cambridge, Cambridge, UK
gv278@cam.ac.uk
[2] Università Niccolò Cusano, Rome, Italy
riccardo.conti@unicusano.it

Abstract. This lecture summarises research carried out by the Authors on the seismic behaviour of displacing or yielding retaining structures, *i.e.*, structures that can undergo permanent displacements during strong earthquakes without failing. For these systems, energy dissipation on shaking, leading to reduced inertia forces, can be achieved by allowing the activation of ductile plastic mechanisms. These must be correctly identified to guarantee the desired strength hierarchy, and depend on the specific retaining structure under examination. It will be shown that the critical acceleration, or the smallest value of acceleration corresponding to the activation of the critical plastic mechanism, is a key ingredient for performance based design of yielding retaining structures. In fact, the critical acceleration controls both the maximum internal forces on the structural elements and the magnitude and trend of post-seismic permanent settlements and rotations, required for quantitative serviceability and post-earthquake operability assessment of infrastructure. Based on a clear understanding of the physical mechanisms governing the dynamic behaviour of these systems, pseudostatic limit equilibrium solutions and simplified dynamic methods can be developed for their seismic design. Theoretical results are validated against data from reduced scale centrifuge models and the results of pseudo-static and fully dynamic numerical analyses. Finally, all the results presented in the paper, including experimental, numerical and theoretical findings, are used to provide suggestions for the performance-based design of retaining structures.

Keywords: Retaining structures · Seismic performance based design · Displacements

1 Introduction

In contrast to other types of retaining structures, such as basement walls or bridge abutments, yielding walls are characterized by the absence of kinematic constraints preventing the occurrence of displacements in the retained soil, where active limit conditions can then develop, both under static and dynamic conditions. This class of retaining structures includes gravity and semi-gravity (cantilevered) walls, as well as embedded cantilevered, single-propped and anchored walls.

Field observations during recent earthquakes (Verdugo et al. 2012; Wagner and Sitar 2016) have shown that the overall performance of these structures is generally satisfactory and that the majority of damages affected either waterfront structures, due to the onset of liquefaction phenomena within the saturated backfill, or structures on slopes. However, even in these cases, it was recognised that, rather than catastrophic failures, many quay walls suffered from excessive displacements and deformations that compromised the serviceability of port facilities (Koseki et al. 2012).

Following the seminal works by Okabe (1924) and Mononobe and Matsuo (1929), the seismic design of earth-retaining structures is conventionally carried out using a force-based approach, in which dynamic actions are represented as pseudostatic forces proportional to an equivalent acceleration and the performance of the system is quantified conventionally in terms of a static safety factor against an assumed collapse mechanism (PIANC 2000). In this framework, several studies have tackled the problem of computing pseudostatic earth pressures on retaining structures with theoretical (Lancellotta 2007; Mylonakis et al. 2007), experimental (Atik and Sitar 2010) or numerical methods (Evangelista et al. 2010). For yielding walls, the activation of plastic mechanisms within the soil-structure system makes the dynamic interaction problem a strength-driven rather than a deformability-driven problem. As a result, elastic solutions for the computation of the dynamic soil thrust (*e.g.*, Brandenberg et al. 2015) are of little applicability in this context.

The displacement-based approach, developed within the Performance-Based Design (PBD) methodology, provides a more rational approach to the seismic design of yielding structures. Following this approach, it is admitted that the soil-wall system can undergo permanent deformations under the design earthquake, provided that its overall behaviour is ductile and the residual displacements do not exceed admissible thresholds, prescribed based on the limit state under consideration. Broadly speaking, the rationale behind this approach is that in highly seismic areas it would be too expensive (if not impossible) to design a wall capable of withstanding the design earthquake without moving (Pender 2019). Even when designing with the simpler force-based approach, the PBD methodology is implemented in modern codes (NTC 2018; EC8 2004) by linking the equivalent acceleration to be used in a pseudostatic calculation to the maximum displacement that the structure can sustain, with respect to different levels of the design earthquake.

In this context, the horizontal permanent displacement of the wall under a given design earthquake is usually taken as a performance indicator of the whole system (PIANC 2000). Therefore, following the pioneering works by Newmark (1965) and Richards and Elms (1979), many works have been devoted to the computation of wall displacements (Ling 2001; Huang et al. 2009; Conti et al. 2013; Cattoni et al. 2019; Callisto 2019), most of them based on Newmark's rigid-block analysis.

Within this framework, the critical acceleration, a_c, corresponding to the full mobilisation of the strength of the system, turns out to be a key ingredient for both the geotechnical and structural design of the wall. In fact, a_c controls both the amount of permanent displacements of the wall at the end of the earthquake and the maximum internal forces that the structure may ever experience during an earthquake. In fact, the internal forces in the wall remain approximately constant for accelerations larger than the critical acceleration, whereas no permanent displacements occur for accelerations lower than the critical value, at least for the case of sliding gravity and semi-gravity walls (Conti et al. 2013).

This lecture summarizes recent research on the dynamic behaviour of yielding retaining structures. Based on a clear understanding of the physical mechanisms governing the seismic response of these systems, pseudostatic limit equilibrium solutions and simplified dynamic methods were developed for their seismic design. Theoretical solutions were validated against data from reduced scale centrifuge models and the results of pseudostatic and fully dynamic numerical analyses. Finally, all the observations presented are used to provide suggestions for the performance based design of retaining structures.

The numerical analyses discussed in this work, for gravity, semi-gravity and embedded retaining structures were carried out in plane-strain conditions using the finite difference code FLAC v5 (Itasca 2005). The soil was always modelled as an elastic-perfectly plastic material with a Mohr-Coulomb failure criterion and a non-associative flow rule, with zero dilatancy, combined with a hysteretic model making use of the Masing (1926) rules to describe the unloading-reloading behaviour under cyclic loading. Structural elements were modelled using elastic solid (gravity walls), beam (semi-gravity and embedded walls) and cable (tie-rods) elements. A pair of retaining walls facing each other were modelled in all dynamic analyses, in order to simplify the definition of lateral constraints. A horizontal acceleration time history was applied to the bottom nodes of the grid, while standard periodic constraints were applied to the lateral boundaries of the mesh. Both simple Ricker wavelets and real acceleration time histories were used as input, the latter covering a significant range of amplitudes and frequency contents.

2   Gravity and Semi-gravity Cantilever Walls

In addition to sliding, field observations after the Kobe earthquake in 1995 revealed significant tilting of gravity and semi-gravity walls, clearly pointing at the activation of bearing capacity failure under dynamic loading (Anderson et al. 2009). Similar conclusions were drawn by Huang et al. (2009) and Conti et al. (2015), based on experimental data of shaking table and dynamic centrifuge tests on concrete gravity walls, respectively, and by Smith and Cubrinovski (2011) and Conti and Caputo (2019), based on the results of pseudostatic and dynamic numerical analyses.

In the light of Newmark's approach, translational (Ling 2001; Conti et al. 2013) rotational (Zeng and Steedman 2000) and bearing capacity (Huang 2005) failure mechanisms have been considered in the literature to compute permanent displacements. However, only very recently attention was given to the relative importance of the three failure mechanisms in computing the critical acceleration of the wall (Kloukinas et al. 2015; Conti et al. 2015; Pender 2019; Conti and Caputo 2019). Centrifuge dynamic tests and numerical studies have shown that Newmark's rigid-block analysis provides good results when applied to gravity retaining structures, either sliding or rotating on their base (Zeng and Steedman 2000; Conti et al. 2013). On the other hand, predictions turn out to be less satisfactory in the case of combined sliding and rotating plastic mechanism, induced by a bearing capacity failure of the foundation soil (Conti and Caputo 2019).

Regarding the seismic behaviour of cantilevered walls, two further issues are the applicability of the Mononobe-Okabe (MO) theory in computing the dynamic soil thrust and the possible phase shift between the maximum value of the soil thrust and the inertia forces into the wall-soil system. Both issues affect the computation of maximum structural internal forces and result in quite controversial recommendations for the structural design of these structures under seismic conditions (AASHTO 2012).

2.1 Plastic Mechanisms and Critical Acceleration

For the sake of simplicity, theoretical computation of the critical acceleration and of the earthquake-induced permanent displacements are typically carried out neglecting both the vertical input acceleration, $a_v(t)$, and any possible change in the system geometry during the applied earthquake. Indeed, assuming a pure horizontal input acceleration ($k_v = 0$) and a fixed geometry of the soil-wall system, the yield acceleration of the wall, related to the activation of a given plastic mechanism, $a_y = k_y g$, is a property of the sole system and does not depend on the applied input acceleration.

Regarding the first assumption, $a_v(t)$ is generally out of phase with and has a different frequency content than the horizontal acceleration, $a_h(t)$, with the corresponding peak values never occurring simultaneously. Therefore, positive and negative contributions from $a_v(t)$, on average, have little effect on the dynamic response of retaining structures and can be reasonably overlooked without significant loss of accuracy (Garini et al. 2011). In relation to the second assumption, changes in the system geometry during shaking have no relevant effects on the computation of the final permanent displacement, at least for ratios of the critical acceleration of the wall and the maximum horizontal input acceleration, $a_c/a_{max} \geq 0.2$, i.e., for well-designed walls (Stamatopoulos et al. 2006). Conti et al. (2013) came to a similar conclusion, showing that the failing soil wedge remains approximately constant during sliding, and the same as at the onset of sliding.

To provide a consistent theoretical framework for the computation of the critical acceleration of gravity and cantilever walls, Conti and Caputo (2019) and Pender (2019), showed that these structures accumulate permanent displacements either by sliding on their base or by bearing capacity failure of the foundation soil. As a result, a_c can be computed as:

$$a_c = \min\left(a_{y,SLID}, a_{y,QLIM}\right) \tag{1}$$

where $a_{y,SLID}$ and $a_{y,QLIM}$ are the pseudostatic yield accelerations corresponding to which the strength of the system is fully mobilised under a pure sliding mechanism (Richards and Elms 1979) and a bearing capacity failure (Conti 2018), respectively.

The type of plastic mechanism effectively mobilised within the soil-wall system depends ultimately on the strength properties of the backfill and the foundation soil. In the case of walls resting on cohesionless soils, experimental data (Conti et al. 2015; Kloukinas et al. 2015; Koseki et al. 2003), numerical results (Smith and Cubrinovski 2011; Conti and Caputo 2019) and theoretical studies (Pender 2019) have shown that the bearing capacity failure is likely to be the 'natural' failure mechanism. More in general, Fig. 1 shows the results of a parametric study carried out by Viggiani and Conti (2016), highlighting the role played by the cohesion of the foundation soil on the critical failure mechanism of gravity walls. Also, numerical and theoretical results for cantilever walls on fine-grained soils indicate that the sliding mechanism usually controls the dynamic behaviour of the wall in undrained conditions (Conti and Caputo 2019).

A powerful tool for investigating the behaviour of the soil-wall system under critical (limit equilibrium) conditions is provided by numerical pseudostatic analyses, which permit to investigate both the shape of the plastic mechanism induced by a uniform horizontal acceleration field and the value of the critical pseudostatic coefficient, $k_c = a_c/g$, corresponding to which the mechanism is activated.

Conti and Caputo (2019) conducted an extensive numerical parametric study on semi-gravity cantilever walls, highlighting the behaviour of such systems under both pseudostatic and dynamic (earthquake) excitations. The Authors considered three different geometries for the wall and two different soil deposits, including a cohesionless and a cohesive soil layer immediately beneath the foundation, with drained (D) and undrained (UD) behaviour respectively.

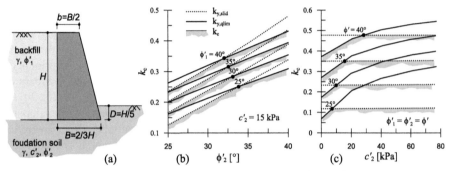

Fig. 1. Gravity walls: (a) wall layout; (b) dependence of k_c on ϕ'_1 and ϕ'_2, for $c'_2 = 15$ kPa; (c) dependence of kc on c'_2, for $\phi'_1 = \phi'_2 = \phi'$

Figure 2 shows one of the layouts considered in their work (W2), designed to have a proper static safety factor under drained (D) and undrained (UD) conditions. For this layout, Fig. 3 shows the contours of shear strains computed in pseudostatic critical conditions, together with the critical failure mechanism predicted by theoretical methods. In drained conditions (Fig. 3a) the plastic mechanism involves the development of shear deformations within the supporting soil, related to the bearing failure of the foundation, eventually leading to both sliding and rotation of the wall. On the other hand, in undrained conditions (Fig. 3b), the critical mechanism corresponds essentially to pure sliding of the wall along its base. Numerical and theoretical results show some discrepancies in terms of the shape of the slip surfaces, but the predicted values of k_c are in very good agreement, both in drained and undrained conditions.

The results indicate that the critical acceleration controls the dynamic behaviour of semi-gravity walls under real earthquake inputs. Figure 4 shows the maximum acceleration computed at mid height of the vertical stem ($a_{max.MID}$) against the maximum free-field acceleration ($a_{max.ff}$), for the drained (a) and the undrained (b) analyses, together with the theoretical values of the critical acceleration. Maximum rightwards ($a_{max.MID}$, $a_{max.ff}$) and leftwards ($|a_{min.MID}|$, $|a_{min.ff}|$) accelerations are considered for the right and the left wall respectively. As expected, once the critical threshold is attained (i.e., as soon as a plastic mechanism develops in the soil-wall system), the absolute acceleration of the system remains approximately constant, starting to deviate from the free-field excitation. Theoretical predictions of a_c are in good agreement with the numerical dynamic results.

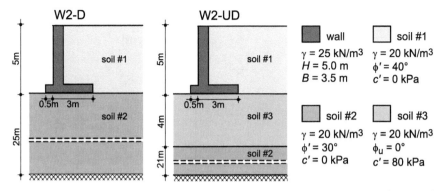

Fig. 2. Semi-gravity cantilever walls: layout analysed in pseudostatic and dynamic numerical analyses (modified after Conti and Caputo 2019)

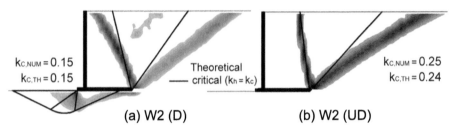

Fig. 3. Pseudostatic numerical analyses of semi-gravity cantilever walls: contours of shear strains at the onset of critical conditions and comparison with theoretical predictions: layout W2 in drained (a) and undrained (b) conditions (modified after Conti and Caputo 2019)

2.2 Structural Internal Forces

Figure 5(a) shows the forces acting on the soil-wall system under the horizontal ($a_h = k_h g$) and vertical ($a_v = k_v g$) pseudostatic accelerations, even if the latter are usually neglected. Both the dynamic active soil thrust acting on the vertical plane AV, S_{AE}, its inclination on the horizontal, δ_S, and the inclination of the two failure surfaces, ω_α and ω_β, were derived by Kloukinas and Mylonakis (2011) by rigorous plasticity solutions. When dealing with the internal stability of the vertical stem (structural design), the soil thrust effectively acting on its back (S_E) must be taken into account, resulting from the dynamic interaction between the soil volume above the heel and the wall (Fig. 5(b)). Table 1 reports two possible approximate solutions for $S_{E,h}$. In the first case (S1), it is assumed that no shear stresses develop at the contact between the heel and the soil above it ($T_E = 0$), and hence the soil thrust S_{AE} and the inertia forces $k_h W_s$ are entirely transferred to the vertical stem. The second solution (S2) assumes that the soil volume above the heel is in active limit state conditions and that the presence of the horizontal stem does not alter the resulting soil thrust. In this condition, S_E can be computed using the Mononobe-Okabe (MO) theory.

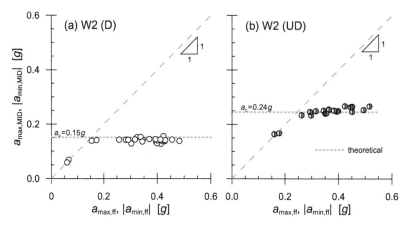

Fig. 4. Dynamic numerical analyses of cantilever walls: Maximum wall horizontal acceleration as a function of the maximum free-field acceleration, for layouts: (a) W2 (D) and (b) W2 (UD) (modified after Conti and Caputo 2019)

Figure 6 shows, for layouts W2-D and W2-UD, the maximum normalised bending moment computed under the applied earthquakes, $M_{max}/\gamma H^3$, as a function of the maximum free-field acceleration (Conti and Caputo 2019). The approximate solutions $S1(k_h)$, $S1(k_c)$ and $S2(k_h)$ are also plotted for comparison. The maximum bending moment can increase even for $a_{max,ff} > a_c$, even though the absolute acceleration of the system remains constant (see e.g., Fig. 4), due to an internal redistribution of stresses leading to a reduction of T_E. Nonetheless, the solution $S1(k_c)$, corresponding to $T_E = 0$, always defines the upper bound for M_{max}. Moreover, as long as this limiting condition is not achieved, the solution $S2(k_h = a_{max,ff}/g)$ provides a reasonable estimate of the maximum internal forces in the stem, with a maximum relative scatter of about 20% with respect to the numerical values.

Based on numerical results, Conti and Caputo (2019) drew the following conclusions regarding the forces acting on the wall and the accelerations in the soil-wall system:

• During a dynamic event, the maximum values of $S_{E,h}$ and of the structural bending moment are always in phase and occur when the inertia forces into the system are directed away from the backfill.
• Possible phase shifts can occur between free field and wall accelerations, when the wall undergoes permanent displacements. Moreover, the actual average acceleration of the soil-wall system can differ significantly from the free-field one, its physical upper bound corresponding to the critical value a_c.
• Up to the critical condition ($k_h \leq k_c$, where $k_h = a_{max,ff}/g$), MO pseudostatic solution provides a good estimate of the active soil thrust acting on the vertical stem.

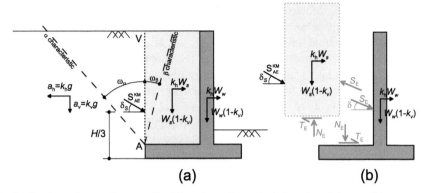

Fig. 5. Semi-gravity cantilever walls: (a) system of pseudostatic external forces; (b) assessment of the internal stability (modified after Conti and Caputo 2019)

Table 1. Semi-gravity cantilever walls: approximate theoretical solutions for the horizontal force acting on the vertical stem

Solution	$S_{E,h}$	M_{max}
$S1(k_h)$	$S_{AE,h}^{KM}(k_h) + k_h W_s$	$S_{AE,h}^{KM}(k_h) \cdot H/3 + k_h(W_{w,stem} + W_s) \cdot H/2$
$S2(k_h)$	$S_{AE,h}^{MO}(k_h)$	$S_{AE,h}^{MO}(k_h) \cdot H/3 + k_h W_{w,stem} \cdot H/2$

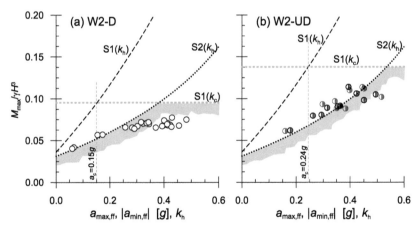

Fig. 6. Dynamic numerical analyses of semi-gravity cantilever walls: Maximum bending moment on the vertical stem as a function of the maximum free-field acceleration, for layouts: (a) W2 (D) and (b) W2 (UD) (modified after Conti and Caputo 2019)

- As soon as a plastic mechanism develops within the soil-wall system ($k_h > k_c$), the average absolute acceleration of the system remains constant. Therefore, also the maximum inertia force that the system can ever experience during an earthquake is bounded by its critical value. Moreover, the pseudostatic solution S1(k_c) (corresponding to T_E = 0) defines the upper bound for $S_{Eh,max}$ and M_{max}.

- The maximum internal forces in the wall ($S_{Eh,max}$, M_{max}) can increase even for $k_h > k_c$, even though the absolute acceleration of the system remains constant, due to an internal redistribution of stresses leading to a reduction of T_E. This behaviour cannot be predicted within a perfect plasticity framework, as it depends on the amounts of shear deformations at the contact between the heel and the soil above it. However, from a practical point of view, the solution S2(k_h) provides a reasonable estimate of the structural internal forces until the limiting condition S1(k_c) is achieved.

2.3 Displacements

The plastic mechanism activated within the soil-wall system governs both the value of the critical acceleration and the pattern of permanent displacements that the wall may undergo during an applied earthquake. Any theoretical model for the computation of permanent wall displacements must take into account properly the kinematics effectively activated within the soil-wall system once the critical acceleration is exceeded.

Conti et al. (2013) proposed a Newmark's like theoretical model for the computation of permanent displacements of gravity and semi-gravity walls sliding on a rigid base. Exploiting the analogy between the wall-soil wedge system and an elementary system comprising two-rigid blocks sliding on an inclined plane, the Authors showed that the dynamic equilibrium of the system can be written as:

$$-\ddot{u}_r(t) = \eta \cdot [k_h(t) - k_c]g \tag{2}$$

where $\ddot{u}_r$ is the horizontal relative acceleration of the wall, $k_h(t) = a_h(t)/g$ is the base horizontal acceleration and $k_c = a_c/g$ is the critical acceleration corresponding to a sliding mechanism, which can be computed as:

$$k_c = \frac{W_w tan\phi_b - S_{AE}[cos(\delta + \beta) - sin(\delta + \beta)tan\phi_b]}{W_w} \tag{3}$$

Equation (3) corresponds with the implicit equation proposed by Richards and Elms (1979). Coefficient η in Eq. (2) depends solely on the mechanical and geometrical properties of the system, and can be computed as:

$$\eta = \frac{cos\phi_b cos(\delta + \beta + \phi - \alpha) + (W/W_w)cos(\phi - \alpha)cos(\delta + \beta + \phi_b)}{cos\phi_b cos(\delta + \beta + \phi - \alpha) + (W/W_w) \cdot \frac{cos\phi cos\beta cos(\delta + \beta + \phi_b)}{cos(\alpha - \beta)}} \tag{4}$$

where the quantities α, β, δ and W_w are defined in Fig. 7 for gravity (a) and semi-gravity (b) walls, respectively.

The ability of the proposed model to describe the dynamic behaviour of gravity walls was tested against the results of two plane-strain finite difference analyses of a pair of gravity walls retaining a 4 m thick ideal layer of dry sand overlying a stiffer soft rock deposit (Conti et al. 2013). The acceleration time history is a simple wavelet with a maximum acceleration of 0.3 g in the first analysis, while it corresponds to a real rock

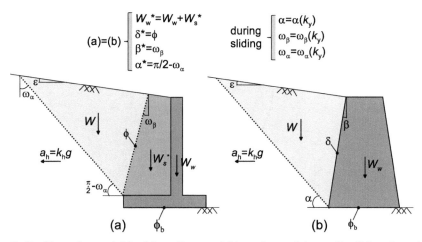

Fig. 7. Double wedge model for (a) cantilever and (b) gravity retaining walls, sliding along their base

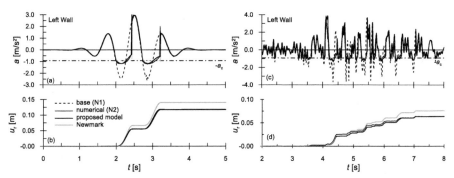

Fig. 8. Gravity walls: Comparison of numerical and model predictions: (a) absolute horizontal acceleration and (b) relative displacements of the left wall. Wavelet (left) and Tolmezzo (right)

outcrop earthquake recording in the second case (Tolmezzo earthquake: $a_{max} = 0.35$ g, mean period $T_m = 0.4$ s, duration $T_{5\text{-}95} = 4.2$ s).

According to the model described above, it is: $k_c = 0.09$ and $\eta = 0.83$. Figure 8 shows the results obtained from the numerical analyses in terms of computed absolute acceleration and relative displacements of the left wall. These are compared with the predictions obtained with the model described above, using the free field acceleration as the input base motion. The results show that the absolute acceleration of the wall coincides with the base acceleration until the seismic coefficient is smaller than the critical value. Once the base acceleration exceeds k_c, the wall slides on its base under an absolute acceleration that varies with time, both in the numerical and theoretical models, while in the standard Newmark's approach ($\eta = 1$) the absolute acceleration of the wall during sliding is taken to be constant and equal to its critical value. The agreement between numerical and theoretical predictions is extremely good, both in magnitude and trend; on the other hand, a prediction obtained by direct application of Newmark's method over predicts the numerical displacement by about 17%.

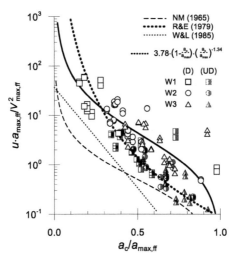

Fig. 9. Semi-gravity walls. Results of numerical dynamic analyses (Conti ad Caputo 2019): normalised final horizontal displacement of the top of the wall as a function of $a_c/a_{max,ff}$

Newmark's like methods based on the sliding block assumption do not provide reliable predictions when the plastic mechanism comprises both translation and rotation of the wall, as in the case of a bearing failure of the foundation soil. This fact is evident by inspection of Fig. 9, collecting all the numerical results obtained by Conti and Caputo (2019) for cantilever wall layouts subjected to real earthquakes. Specifically, Fig. 9 shows the final normalised displacement of the top of the wall, $u \cdot a_{max,ff}/V^2_{max,ff}$, where $V_{max,ff}$ is the maximum velocity computed in free-field conditions, as a function of the ratio $a_c/a_{max,ff}$ between the critical acceleration and the maximum free-field acceleration. Figure 9 also reports some of the interpolating functions proposed in the literature, all derived from the application of the Newmark's sliding block procedure (Newmark 1965; Richards and Elms 1979; Whitman and Liao 1985), together with the best fit of the numerical results. Full and open symbols refer to the undrained (UD) and open symbols (D) foundation soil, where the activated plastic mechanism is pure sliding and bearing failure, respectively. As expected, the computed displacements reduce with increasing $a_c/a_{max,ff}$. However, while the numerical results for the undrained (UD) foundation soil are in satisfactory agreement with the equation proposed by Richards and Elms (1979) for sliding walls, those corresponding to the drained (D) profile are always above the empirical relationships. This observation suggests that the available equations, all derived from the analysis of a rigid block sliding over a horizontal plane, can lead to non-conservative results if applied to retaining walls for which the permanent displacement stems from a combination of both sliding and rotation.

2.4 Suggestions for Design

As far as the geotechnical design is concerned, the concept of an admissible wall displacement has been widely accepted within the PBD philosophy. However, the possibility of admitting wall tilting, related to a temporary attainment of the bearing resistance, is still a controversial issue. Indeed, many provisions and codes of practice still recommend to design the wall ensuring an adequate safety margin with respect to a bearing failure of the foundation and assuming the sliding mechanism as the critical one. The rationale behind it is twofold: (1) excessive wall tilting could induce brittle collapse of the wall by overturning; (2) no reliable procedures are available to accommodate a mixed sliding-rotational failure mode within the well-established Newmark's approach (Pender et al. 2019).

With this respect, a rational seismic design of gravity and semi-gravity walls should contemplate the possible activation of both mechanisms, instead of excluding a priori the expected rotation. In fact, the temporary mobilization of the soil shear strength beneath the foundation would not lead to a fragile failure of the system, provided that tilting of the wall is not excessive. This is a fundamental difference with respect to a pure overturning mechanism, which is indeed an intrinsically brittle mechanism. On the other hand, further research is required to develop reliable (and simple) theoretical models, capable of handling combined tilting and sliding failure modes as, in this case, a direct application of the Newmark's sliding block procedure can lead to significant under-prediction of the final displacement (Conti and Caputo 2019).

Moving to the structural design of cantilever retaining walls, a simple three-step procedure can be defined to take into account, though approximately, the possible contribution of the horizontal stem to the overall dynamic equilibrium, as observed in numerical dynamic analyses (Conti and Caputo 2019):

1. compute the critical acceleration of the wall;
2. use $S1(k_c)$ to compute the maximum internal forces that the wall could ever experience during an earthquake;
3. for a given design earthquake, corresponding to which $k_h = a_{max,ff}/g$, use the $\min[S2(k_h), S1(k_c)]$ to compute the internal forces in the wall.

3 Embedded Cantilevered and Anchored Walls

A number of factors make the seismic behaviour of yielding embedded retaining structures different from what observed in the case of gravity and semi-gravity walls, also affecting their seismic design:

- Both numerical analyses (Conti et al. 2014; Caputo et al. 2019a, b) and experimental results (Conti et al. 2012; Fusco et al. 2019; Fusco 2022), indicate that the stability of embedded retaining walls is guaranteed by the soil passive strength below dredge level. This always requires some permanent displacement in order to be fully mobilised, thus making the rigid perfectly plastic assumption not conservative in most cases.

- The definition of the critical plastic mechanism is not trivial, given that no simple equivalent "sliding block" can be identified, particularly when different failure mechanisms can be recognised for the soil-wall system, each one characterised by its own value of the yielding acceleration (Caputo et al. 2021).
- The collapse mechanism is often characterised by diffuse plastic deformations, and the associated displacement field cannot be reduced to a simple translational or rotational rigid body motion (Oliynyk et al. 2022).
- The soil volume interacting with the wall can be composed of layers of different mechanical properties and phase-shift of accelerations can occur along the height of the wall during the earthquake. In the majority of cases, however, both aspects are of less relevance for cantilevered and single-restrained embedded walls, with typical heights in the range of 10–20 m.
- Computation of internal forces is always a fundamental task for the structural design of these structures, requiring simple and reliable methods for the evaluation of pseudo-static contact earth-pressure distributions. These methods provide also the theoretical basis for the assessment of the critical acceleration for the soil-wall system.

The following sections will present the limit equilibrium methods proposed by Conti and Viggiani (2013) and Caputo et al. (2021) for the analysis of embedded cantilevered and anchored walls, inspired by the behaviour observed in numerical pseudostatic and dynamic simulations, overcoming limitations of standard approaches available in the literature. The attention will be placed on Steel Sheet Pile walls with shallow passive Anchors (ASSP) in the form of continuous sheet piles, rather than grouted anchors and propped walls, even if they show essentially the same dynamic behaviour of ASSP walls. The reader can refer to Conti et al. (2012) and Callisto and Dal Brocco (2015) for a thorough discussion on these structures.

Final considerations on the permanent displacements accumulated by these structures during earthquakes, and on available theoretical methods for their estimation, will be also provided.

3.1 Embedded Cantilevered Walls

Numerical dynamic analyses carried out by Conti et al. (2014) and Conti (2017) led to the following main conclusions:

- the passive resistance of the soil in front of the wall is mobilised progressively during the earthquake, starting from dredge level downwards: the stronger the applied acceleration, the greater the depth down to which the passive resistance is fully mobilised;
- in the time instants when the internal forces in the wall attain their maximum, the accelerations in the soil below dredge level are only a small fraction of the maximum value computed on the retained side, and always lower than about 0.1 g;
- permanent displacements of the wall correspond to an approximately rigid rotation around a pivot point located at a depth of between $0.8 \times d$ and $0.9 \times d$, where d is the embedment depth.

Figure 10 shows the horizontal earth pressure distribution assumed in the limit equilibrium method proposed by Conti and Viggiani (2013). For a given $k_h < k_c$ (continuous line), the soil on the retained side is in active limit state down to d_0, and in passive limit state below the rotation point. On the excavated side, the passive resistance of the soil is fully mobilised down to a depth $\overline{d}$; the soil is in active limit state below the pivot point, and the horizontal contact stress decrease linearly with depth between $\overline{d}$ and d_0. Moreover, it is assumed that the passive earth pressure coefficient takes its static value, K_P.

In critical conditions ($k_h = k_c$) the passive resistance of the soil below the excavation is mobilised completely down to the pivot point, i.e. $\overline{d} = d_0$, with d_0 about $0.9 \times d$ (dashed line in Fig. 10).

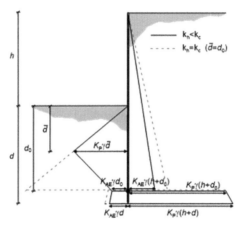

Fig. 10. Embedded cantilevered wall: theoretical pseudostatic earth pressure distribution (from Conti and Viggiani 2013)

A comparison between numerical results and limit equilibrium calculations is shown in Fig. 11, where the normalised maximum bending moments on the walls are plotted against the maximum accelerations computed behind the walls during the earthquakes (Conti et al. 2014). The proposed method is in good agreement with the numerical data and always provides conservative values of the maximum (critical) bending moment. The results show that for increasing strength of the soil-wall system, that is for increasing values of the soil and soil-wall friction angle and for increasing embedded depth, both the critical acceleration and the maximum bending moment increase. In other words, a stronger soil-wall system will experience smaller displacements during the earthquake, but this is paid for by increasing internal forces in the wall.

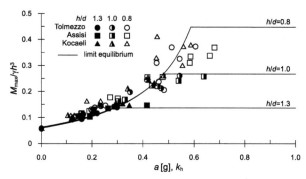

Fig. 11. Embedded cantilevered wall: maximum normalised bending moments computed from dynamic numerical simulations and limit equilibrium ($\phi' = 35°$, $\delta = 20°$, $\gamma = 20$ kN/m³, $h = 4$ m) (from Conti and Viggiani 2013)

3.2 Anchored Steel Sheet Pile (ASSP) Walls

In the case of ASSP walls, the potential plastic mechanisms to examine involve a full mobilization of the shear strength in the soil, as their seismic design does not rely (typically) upon the ductility of the structural members. Specifically, three failure mechanisms can be identified: (1) attainment of the anchor capacity (anchor failure, $a_{y,AF}$); (2) complete mobilisation of the soil passive strength below dredge level (toe failure, $a_{y,TF}$); and (3) activation of a global mechanism, involving the wall, the anchor plate and the soil volume interacting with them (global failure, $a_{y,GF}$). Once the accelerations corresponding to each mechanism are computed, then the critical acceleration is given by:

$$a_c = \min(a_{y,AF}, a_{y,TF}, a_{y,GF}) \tag{5}$$

The possible occurrence of different plastic mechanisms was confirmed by numerical analyses (Caputo et al. 2021) and experimental results from dynamic centrifuge tests on reduced scale models of ASSP walls embedded in dry sand (Fusco et al. 2019). As an example, Figs. 12(a) and (b) report the contours of vertical displacements measured at the end two earthquakes applied in Tests AF03 and AF04, whose layouts, according to limit equilibrium calculations, should correspond to anchor failure and global failure, respectively. Indeed, in Test AF03, the soil horizontal displacements occurred mostly in front of the anchor and in a wedge behind the main wall, showing the typical features of a local anchor failure. In contrast, during Test AF04, the soil included between the anchor, the main wall, and a curved surface extending between the two, experienced a nearly uniform horizontal displacement, confirming the predicted global failure.

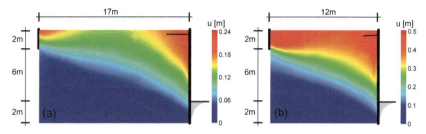

Fig. 12. ASSP walls: contours of horizontal displacements measured in two dynamic centrifuge tests (Fusco et al. 2019): (a) Test AF03, at the end of earthquake EQ3; (b) Test AF04, at the end of earthquake EQ2

Figure 13 shows the theoretical earth pressure distributions assumed in the pseudostatic limit equilibrium method developed by Caputo et al. (2021) to assess (a) the maximum internal forces in the structural members and (b) the anchor capacity, T_{lim}. The former is consistent with a quasi-rigid rotation of the wall around a pivot point in the embedded portion. Specifically, for a given k_h, the soil is in active (retained side) and passive (below dredge level) limit condition down to a depth D^*. At larger depths, contact pressures are no longer related to a limit state condition and a simplified linear distribution is assumed.

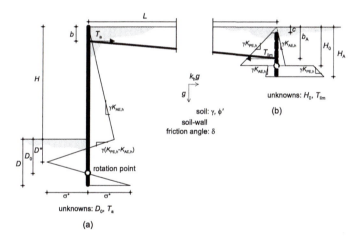

Fig. 13. ASSP walls: typical layout and theoretical pseudostatic earth pressure distribution (from Caputo et al. 2021)

The predicted capabilities of the new limit equilibrium method were assessed against the results of pseudostatic and dynamic numerical analyses. Figure 14 shows the normalized maximum bending moment, $M_{\text{max}}/\gamma H^3$, and axial force, $T_a/\gamma H^2$, as a function of k_h, for one of the layouts analysed by the Authors ($H = 10$ m; $b_A/H_A = 1/3$; $L/H = 2.1$; $D/H = 0.4$; $\phi' = 35°$; $\gamma = 20$ kN/m³). For $k_h < k_c$, LE predictions slightly overestimate the bending moment and underestimate the anchor force, while the method predicts correctly the critical values. Once again, maximum internal forces are basically independent of the earthquake intensity and turn out to be a property of the system.

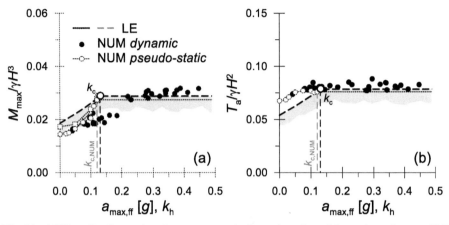

Fig. 14. ASSP walls: Comparison between numerical pseudostatic and dynamic analyses and LE predictions in terms of (a) $M_{max}/\gamma H^3$ and (b) $T_a/\gamma H^2$ (modified after Caputo et al. 2021)

3.3 Displacements of Embedded Retaining Structures

Figure 15 shows the numerical values of the final permanent displacement computed at the top of embedded cantilevered (a) and ASSP (b) walls subjected to real earthquakes, as a function of $a_c/a_{max,ff}$ (Conti et al. 2014; Caputo et al. 2019a, b). Contrary to what observed for gravity and semi-gravity walls, significant displacements occur for $a_c/a_{max,ff} > 1$, that is before the strength of the system is fully mobilized. This is consistent with the experimental observations by Conti et al. (2012), indicating that embedded walls may accumulate rigid permanent displacements concurrently with an increase of the internal forces in the structural members, that is before attaining the critical acceleration. This behaviour is due to a stress redistribution and a progressive mobilisation of the soil strength on the passive side of the wall produced by the earthquake (Conti et al. 2014; Conti 2017).

The results shown in Fig. 15 have two relevant implications for design. Firstly, allowable displacements less than about 0.1–0.2 m would result in a ratio $a_c/a_{max,ff} > 1$: that is, the wall should be designed to have a critical acceleration larger than the maximum acceleration expected at the site (*i.e.*, using an equivalent acceleration that is larger than the maximum free-field acceleration). This is completely different from the performance-based design of gravity retaining walls, as they will experience permanent displacements only if $a_c/a_{max,ff} < 1$. Secondly, application of Newmark's like procedures, based on the rigid perfectly-plastic soil assumption, would yield displacements that are much smaller than observed, as the analysis would overlook the displacements experienced by the wall before the acceleration reaches its critical value.

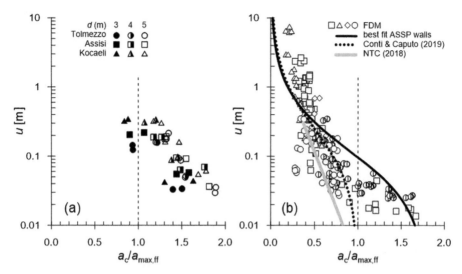

Fig. 15. Results of numerical dynamic analyses for (a) Embedded Cantilevered and (b) ASSP walls: final horizontal displacement of the top of the wall as a function of the ratio $a_c/a_{max,ff}$

4   Conclusions

Research carried out by the authors on the seismic behaviour of displacing or yielding retaining structures shows that the key ingredient for their performance-based design is the critical acceleration, controlling both the maximum internal forces that a retaining structure may experience during the earthquake and its permanent displacements at the end of the earthquake.

The critical acceleration may be defined as the smallest value of acceleration required to activate a (ductile) plastic mechanism in the soil-retaining structure system. Potential plastic mechanisms must be correctly identified and depend on the layout of the retaining structure under examination.

Gravity and semi-gravity cantilevered walls accumulate permanent displacements either by sliding on their base or by bearing capacity failure. Ultimately, the type of plastic mechanism effectively mobilised within the soil-wall system depends on the strength of the backfill and of the foundation soil. The available evidence, including results of physical and numerical modelling and theoretical analyses, indicates that bearing capacity failure eventually leading to both sliding and rotation of the wall is likely to be the natural failure mechanism for walls resting on cohesionless foundation soils, whereas a mechanism of pure sliding controls the dynamic behaviour of the wall in undrained conditions. In either case, once the critical acceleration is attained, *i.e.*, as soon as a plastic mechanism develops in the soil-wall system, the absolute acceleration of the system remains approximately constant, even for increasing free-field acceleration.

The structural design of semi-gravity cantilevered walls must account for the soil thrust effectively acting on the stem of the wall, resulting from the dynamic interaction between the soil volume above the heel and the wall. Up to the critical condition, Mononobe-Okabe pseudo-static solution provides a good estimate of the active soil thrust acting on the vertical stem. As soon as a plastic mechanism develops within the soil-wall system, the average absolute acceleration of the system remains constant at the critical value. However, the maximum shear force and bending moment in the stem of the wall can still increase due to an internal redistribution of stresses leading to a reduction of the shear stresses at the contact between the heel and the soil above it. A pseudo-static solution for the soil thrust acting on the vertical stem obtained neglecting any shear stress at the contact between the heel and the soil above it and using the critical seismic coefficient defines an upper bound for both the maximum shear force and the maximum bending moment in the stem of the wall.

A modified version of Newmark's method in which the wall-soil wedge system is assimilated to an elementary system comprising two-rigid blocks sliding on an inclined plane can be used to compute the permanent displacements of gravity and semi-gravity walls sliding on a rigid base. Further work is required to develop simple and reliable methods to predict the permanent displacements for walls where the critical mechanism corresponds to the attainment of bearing capacity of the foundation, as direct application of the Newmark's sliding block procedure can lead to significant under-prediction of the final displacement when the plastic mechanism comprises both translation and rotation of the wall.

Several factors make the seismic design of yielding embedded retaining structures more complex than that of gravity and semi-gravity walls. Full mobilisation of soil passive strength below dredge level, required to guarantee the stability of the wall, requires finite permanent displacements of the wall making the assumption of rigid perfectly plastic behaviour not conservative. Moreover, the definition of the critical plastic mechanism is not trivial given that no simple equivalent sliding block can be identified and the collapse mechanism is often characterises by diffuse plastic deformation. In particular, for anchored steel sheet pile walls, it is possible to identify at least three failure mechanisms involving full mobilization of the shear strength in the soil. Two of these are local failure mechanisms, corresponding to the attainment of anchor capacity or to the full mobilisation of the strength of the soil below dredge level, respectively. The third one corresponds to the activation of a global mechanism, involving the wall, the anchor plate and the soil volume interacting with them.

For embedded walls, the calculation of maximum shear forces and bending moments is always required for their structural design, preferably by simple limit equilibrium methods for the evaluation of both the pseudo-static contact earth-pressure distributions and the critical acceleration. Based on a clear understanding of the physical mechanisms governing the dynamic behaviour of these systems, pseudostatic limit equilibrium solutions have been illustrated for the analysis of both embedded cantilevered and anchored walls overcoming limitations of standard approaches available in the literature. Both for cantilevered walls and for anchored walls experiencing local failures the maximum internal forces are basically independent of the earthquake intensity and turn out to be a property of the system, with a tendency for the maximum internal forces to increase with the strength, and hence the critical acceleration, of the system. For anchored walls experiencing global failure, the bending moment in the wall and the anchor forces may increase above the values computed for the critical acceleration.

The important issue of the calculation of permanent displacements of embedded walls has been dealt with only marginally in the paper. However, it is essential to recognise that, contrary to what happens for gravity and semi-gravity walls, significant displacements occur before the strength of the system is fully mobilised. Direct application of Newmark's like procedures, based on the rigid perfectly-plastic soil assumption, would yield displacements that are much smaller than observed, as the analysis would overlook the displacements experienced by the wall before the acceleration reaches its critical value.

References

Al Atik, L., Sitar, N.: Seismic earth pressures on cantilever retaining structures. J. Geotech. Geoenviron. Eng. **136**(10), 1324–1333 (2010)

American Association of State Highway and Transportation Officials (AASHTO): LRFD Bridge Design Specifications, 6th edn. AASHTO, Washington, D.C. (2012)

Anderson, D.G., Martin, G.R., Lam, I., Wang, J.N.: Seismic analysis and design of retaining walls, buried structures, slopes, and embankments. NCHRP Report 611, Transportation Research Board, Washington, D.C. (2009)

Brandenberg, S.J., Mylonakis, G., Stewart, J.P.: Kinematic framework for evaluating seismic earth pressures on retaining walls. J. Geotech. Geoenviron. Eng. **141**(7), 04015031 (2015)

Callisto, L.: On the seismic design of displacing earth retaining systems. In: Proceedings of the 7th International Conference on Earthquake Geotechnical Engineering, Rome, Italy (2019)

Callisto, L., Del Brocco, I.: Intrinsic seismic protection of cantilevered and anchored retaining structures. In: Proceedings of the SECED Conference on Earthquake Risk and Engineering Towards a Resilient World, Cambridge, UK (2015)

Caputo, G., Conti, R., Viggiani, G., Prüm, C.: Improved method for the seismic design of anchored steel sheet pile walls. J. Geotech. Geoenv. Eng. **147**(2) (2021). https://doi.org/10.1061/(ASCE)GT.1943-5606.0002429

Caputo, G., Conti, R., Viggiani, G.M.B., Prüm, C.: Theoretical framework for the seismic design of anchored steel sheet pile walls. In: Proceedings of the 7th International Conference on Earthquake Geotechnical Engineering, Rome, Italy (2019a)

Caputo, G., Conti, R., Viggiani, G.M.B., Prüm, C.: Improved method for the seismic design of anchored steel sheet pile walls. J. Geotech. Geoenviron. Eng. **147**(2) (2019b)

Cattoni, E., Salciarini, D., Tamagnini, C.: A Generalized Newmark Method for the assessment of permanent displacements of flexible retaining structures under seismic loading conditions. Soil Dyn. Earthq. Eng. **117**, 221–223 (2019)

Conti, R.: Numerical modelling of centrifuge dynamic tests on embedded cantilevered retaining walls. Ital. Geotech. J. **51**(2), 31–46 (2017)

Conti, R., Madabhushi, S.P.G., Viggiani, G.M.B.: On the behaviour of flexible retaining walls under seismic actions. Géotechnique **62**(12), 1081–1094 (2012)

Conti, R., Viggiani, G.M.B.: A new limit equilibrium method for the pseudostatic design of embedded cantilevered retaining walls. Soil Dyn. Earthq. Eng. **50**, 143–150 (2013)

Conti, R., Viggiani, G.M.B., Cavallo, S.: A two-rigid block model for sliding gravity retaining walls. Soil Dyn. Earthq. Eng. **55**, 33–43 (2013)

Conti, R., Viggiani, G.M.B., Burali d'Arezzo, F.: Some remarks on the seismic behaviour of embedded cantilevered retaining walls. Géotechnique **64**(1), 40–50 (2014)

Conti, R., Madabhushi, G.S.P., Mastronardi, V., Viggiani, G.M.B.: Centrifuge dynamic tests on gravity retaining walls: an insight into bearing vs sliding failure mechanisms. In: Proceedings of the 6th International Conference on Earthquake Geotechnical Engineering, Christchurch, New Zealand (2015)

Conti, R.: Simplified formulas for the seismic bearing capacity of shallow strip foundations. Soil Dyn. Earthq. Eng. **104**, 64–74 (2018)

Conti, R., Caputo, G.: A numerical and theoretical study on the seismic behaviour of yielding can-tilever walls. Géotechnique **69**(5), 377–390 (2019)

European Committee for Standardization. EN 1998-5:2004: E. Eurocode 8: Design of structures for earthquake resistance-Part 5: Foundations, retaining structures and geotechnical aspects, Brussels (2004)

Evangelista, A., Scotto di Santolo, A., Simonelli, A.L.: Evaluation of pseudostatic active earth pressure coefficient of cantilever retaining walls. Soil Dyn. Earthq. Eng. **30**(11), 1119–1128 (2010)

Fusco, A., Viggiani, G.M.B., Madabhushi, G.S.P., Caputo, G., Conti, R., Prüm, C.: Physical modelling of anchored steel sheet pile walls under seismic actions. In: Proceedings of the 7th International Conference on Earthquake Geotechnical Engineering, Rome, Italy (2019)

Fusco, A.: Seismic behaviour of anchored steel sheet pile walls in sand. Ph.D. thesis, University of Cambridge, UK (2022)

Garini, E., Gazetas, G., Anastasopoulos, I.: Asymmetric 'Newmark' sliding caused by motions containing severe 'directivity' and 'fling' pulses. Géotechnique **61**(9), 733–756 (2011)

Huang, C.C.: Seismic displacement of soil retaining walls situated on slope. J. Geotech. Geoenviron. Eng. **131**(9), 1108–1117 (2005)

Huang, C.C., Wu, S.H., Wu, H.J.: Seismic displacement criterion for soil retaining walls based on soil strength mobilization. J. Geotech. Geoenviron. Eng. **135**(1), 74–83 (2009)

Itasca: FLAC Fast Lagrangian Analysis of Continua v. 5.0. User's Manual (2005)

Kloukinas, P., Mylonakis, G.: Rankine solution for seismic earth pressures on L-shaped retaining walls. In: 5ICEGE, Santiago, Chile (2011)

Kloukinas, P., Scotto di Santolo, A., Penna, A., et al.: Investigation of seismic response of cantilever retaining walls: limit analysis vs shaking table testing. Soil Dyn. Earthq. Eng. **77**, 432–445 (2015)

Koseki, J., Tatsuoka, F., Watanabe, K., et al.: Model tests of seismic stability of several types of soil retaining walls. In: Ling, H.I., Leshchinsky, D., Tatsuoka, F. (eds.) Reinforced Soil Engineering: Advances in Research and Practice, pp. 317–358. ACM, New York (2003)

Koseki, J., Koda, M., Matsuo, S., Takasaki, H., Fujiwara, T.: Damage to railway earth structures and foundations caused by the 2011 off the Pacific Coast of Tohoku Earthquake. Soils Found. **52**(5), 872–889 (2012)

Lancellotta, R.: Lower-bound approach for seismic passive earth resistance. Géotechnique **57**(3), 319–321 (2007)

Ling, H.I.: Recent applications of sliding block theory to geotechnical design. Soil Dyn. Earthq. Eng. **21**(3), 189–197 (2001)

Mononobe, N., Matsuo, H.: On the determination of earth pressure during earthquake. In: Proceedings of the 2nd World Engineering Conference, vol. 9, pp. 177–185 (1929)

Mylonakis, G., Kloukinas, P., Papantonopoulos, C.: An alternative to the Mononobe-Okabe equations for seismic earth pressures. Soil Dyn. Earthq. Eng. **27**(10), 957–969 (2007)

Newmark, N.M.: Effects of earthquakes on dams and embankments. Géotechnique **15**(2), 139–160 (1965)

NTC: Aggiornamento delle Norme Tecniche per le Costruzioni. Ministero delle Infrastrutture e dei Trasporti, Rome, Italy (2018)

Oliynyk, K., Conti, R., Viggiani, G.M.B., Tamagnini, C.: A Generalized Newmark Method with displacement hardening for the prediction of seismically–induced permanent deformations of diaphragm walls. Submitted to Géotechnique (2022)

Okabe, S.: General theory of earth pressure and seismic stability of retaining wall and dam. J. Jpn. Soc. Civ. Eng. **12**(1) (1924)

Pender, M.J.: Foundation design for gravity retaining walls under earthquake. Proc. Inst. Civ. Eng. Geotech. Eng. **172**(1), 42–54 (2019)

Pender, M.J., Conti, R., Caputo, G., Viggiani, G.M.B.: Discussion: foundation design for gravity retaining walls under earthquake. Proc. Inst. Civ. Eng. Geotech. Eng. (2019). https://doi.org/10.1680/jgeen.18.00125

PIANC (Permanent International Association of Navigation Congresses): Seismic Design Guidelines for Port Structures. Balkema, Rotterdam (2000)

Richards, R., Elms, D.G.: Seismic behaviour of gravity retaining walls. J. Geotech. Eng. Div. **105**(4), 449–464 (1979)

Smith, C.C., Cubrinovski, M.: Pseudo-static limit analysis by discontinuity layout optimisation: application of seismic analysis of retaining walls. Soil Dyn. Earthq. Eng. **31**(10), 1311–1323 (2011)

Stamatopoulos, C.A., Velgaki, E.G., Modaressi, A., Lopez-Caballero, F.: Seismic displacement of gravity walls by a two-body model. Bull. Earthq. Eng. **4**, 295–318 (2006)

Verdugo, R., et al.: Seismic performance of earth structures during the February 2010 Maule, Chile, earthquake: dams, levees, tailings dams, and retaining walls. Earthq. Spectra **28**(S1), S75–S96 (2012)

Viggiani, G.M.B., Conti, R.: On the behaviour of gravity retaining structures under seismic actions. In: Proceedings of the 1st International Conference on Natural Hazards and Infrastructure, Chania, Greece (2016)

Wagner, N., Sitar, N.: On seismic response of stiff and flexible retaining structures. Soil Dyn. Earthq. Eng. **91**, 284–293 (2016)

Whitman, R.V., Liao, S.: Seismic design of retaining walls. Miscellaneous Paper GL-85-1, U.S. Army Engineer Waterways Experiment Station, Vicksburg, Mississippi (1985)

Zeng, X., Steedman, R.S.: Rotating block method for seismic displacement of gravity walls. J. Geotech. Geoenviron. Eng. **126**(8), 709–717 (2000)

Study and Practice on Performance-Based Seismic Design of Loess Engineering in China

Lanmin Wang[1]([⊠]), Jinchang Chen[2], Ping Wang[1], Zhijian Wu[3], Ailan Che[2], and Kun Xia[1]

[1] Lanzhou Institute of Seismology, China Earthquake Administration, Lanzhou 730000, China
wanglm@gsdzj.gov.cn
[2] School of Naval Architecture, Ocean and Civil Engineering, Shanghai Jiao Tong University, Shanghai 200040, China
[3] College of Transportation Science and Engineering, Nanjing Tech University, Nanjing 211816, Jiangsu, China

Abstract. Loess is a kind of special soil with porous structure and weak cohesion, which widely deposits in China with an area of 640,000 km^2. Especially, it is continuously distributed in the Loess Plateau of China with an area of 440,000 km^2 and a thickness ranging from tens meters to more than 500 m, where is a region with the biggest thickness and the most complicated topography of loess deposit in the globe. On the other hand, the Loess Plateau is a strong earthquake-prone region, where 120 earthquakes with Ms $\geq$ 6.0 and 7 events of Ms $\geq$ 8.0 occurred in history. These earthquakes killed more than 1.4 million people in the region. The investigation shown that so large casualties should be attributed to a large scale of landslides, liquefaction, subsidence and amplification of ground motion. In this paper, the characteristics of engineering geology and seismic hazards with different probabilities of exceedance in the Loess Plateau of China is introduced. Based on the field investigation and exploration, observation data, in-situ tests, laboratory tests, and numerical analysis, the method of evaluating amplification of ground motion on topography and deposit of loess sites is proposed. The method of risk assessments of seismic landslides is developed. Seismic design for engineering loess slopes with single-step and multi-steps is provided. The methods of evaluating liquefaction and seismic subsidence of loess ground and its treatment measurements are respectively presented. Moreover, the seismic design methods of pile foundation in loess ground are proposed for considering the negative friction on piles due to seismic subsidence and horizontal pushing force on piles under liquefaction. The above-mentioned methods and techniques have been adopted by a national code and a provincial code, which have been proved to be practical, efficient and rational through a large number of major engineering projects and buildings' construction.

Keywords: Loess sites · Seismic design · Seismic landslides · Liquefaction · Subsidence · Amplification

L. Wang et al. (Eds.): PBD-IV 2022, GGEE 52, pp. 258–287, 2022.
https://doi.org/10.1007/978-3-031-11898-2_13

1 Introduction

Earthquakes occur frequently in the loess region of China. By 2021, a total of 120 earthquakes with Ms6 or higher were recorded in the region, including 7 events with Ms8 and above. All of them have caused serious loess earthquake disasters such as seismic landslide, seismic subsidence and liquefaction, with more than one million casualties. In the former Soviet Union, the 1989 Tajikistan M 5.5 earthquake caused extensive liquefaction of saturated loess layer, resulting in large-scale mud flow, burying villages and killing 220 people. In the loess region of the Middle East of the United States, the New Madrid M 7.8 earthquake from 1887 to 1888 caused large-area liquefaction, land subsidence, and the Mississippi River poured into the subsided region to form a great lake. Table 1 presents some typical earthquake disaster examples in the loess area of China, which can be triggered under the effect of medium and strong earthquakes, resulting in huge casualties and property losses.

Table 1. Typical earthquake disasters in loess area

Time	Epicenter	Magnitude (M_S)	Maximum intensity	Main seismic geotechnical disasters	Casualties
1303	Hongtong, Shanxi Province	8	XI	Liquefaction, landslide and ground motion amplification	200,000
1556	Huaxian, Shaanxi Province	8	XI	Landslide, seismic subsidence, liquefaction and ground motion amplification	830,000
1654	Tianshui, Gansu Province	8	XI	Landslide, collapse and ground motion amplification	31,000
1659	Linfen, Shanxi Province	8	XI	Liquefaction, landslide and ground motion amplification	52,600
1718	Tongwei, Gansu Province	7.5	X	Landslide and seismic subsidence	40,000

(*continued*)

Table 1. (*continued*)

Time	Epicenter	Magnitude (M_S)	Maximum intensity	Main seismic geotechnical disasters	Casualties
1920	Haiyuan, Ningxia Province	8.5	XII	Landslide, liquefaction, seismic subsidence and ground motion amplification	270,000
1927	Gulang, Gansu Province	8	XI	Landslide, collapse and ground motion amplification	40,000
1995	Yongdeng, Gansu Province	5.8	VIII	Landslide, seismic subsidence and ground motion amplification	12
2013	Minxian, Gansu Province	6.6	VIII	Landslide, liquefied mud flow and ground motion amplification	95

The important progress of disaster research in loess area is that after the 1980s, the main research work was carried out in China, the United States, the former Soviet Union and Japan. From 1982 to 1998, a preliminary study on the liquefaction of loess in the central United States was conducted, and an in-depth study on the liquefaction standard, physical property index and particle size distribution of loess in this area was performed [1–3]. In 1990, Professor Ishihara of Japan conducted a detailed investigation and report on the liquefaction of saturated aeolian loess through the field investigation of large-scale mud flow caused by the M 5.5 earthquake in Tajikistan in 1989 [4]. In China, the research on the dynamic characteristics of loess mainly focuses on two respects. The first respect is the research on dynamic characteristics of loess, involving in elastic modulus, damping ratio, constitutive model, shear wave velocity and earth tremor according to the needs of site seismic zoning; The second respect is the research on seismic geotechnical disasters of loess, such as seismic landslide, seismic subsidence and liquefaction of loess. On the first respect, a lot of research on the stress-strain relationship, strength, elastic modulus and damping ratio of loess under constant amplitude sinusoidal cyclic loading on the dynamic triaxial apparatus have been done [5–9]; The dynamic constitutive model, elastic modulus, damping ratio and dynamic strength of loess under random seismic loading were widely studied [10–12], and the dynamic parameters of loess under different seismic loads were compared; The change law of loess strength under dynamic loadings were studied by means of laboratory simulated dynamic compaction tests [13]; The characteristics of shear modulus and damping ratio of loess in Northwest China were tested [14]. All these studies provide a lot of important data for the seismic microzonation of large and middle-sized cities in the loess region. On the second respects, field investigation and research on large-scale loess seismic landslide caused by the Haiyuan 8.5 earthquake in 1920 were conducted [15]. Aerial photos of large-area loess landslides caused by Tongwei earthquake in 1718 and Tianshui earthquake in 1654 were interpreted [16]. The seismic engineering geological problems of loess site was studied and the corresponding evaluation methods were put forward [17]. Seismic subsidence characteristics of undisturbed loess during humidification were studied [18]. The random search method to determine the most dangerous slip surface of loess slope was developed [19]. The prediction methods of seismic landslide, liquefaction and seismic subsidence of loess under random seismic loading were put forward [20–23]. A fuzzy comprehensive evaluation on the distribution law of landslides in Gulang and Haiyuan areas was made [24–26]. The intensity attenuation law and the development and distribution characteristics of seismic landslide and collapse in loess areas of China were studied [27]. The characteristics of high-speed sliding of loess landslide induced by earthquake were also studied [28]. The large-scale sliding of loess deposit caused by liquefaction of saturated loess layer during the Haiyuan 8.5 earthquake in 1920 was studied [29, 30]. Experimental research on liquefaction potential of dynamic compacted loess ground and dynamic liquefaction of saturated loess were conducted [31, 32]. The liquefaction of sand mixtures with different loess contents were studied [33]. It is found that the loess slows down or even prevents the dissipation of pore water pressure, which is caused by the large deformation and long-term high pore water pressure of fine-grained sand during liquefaction.

Although the above research works have achieved a lot of valuable research results in loess dynamics and seismic disasters in loess area, these results rarely involve seismic design of loess slopes and loess ground. However, the large-scale urbanization construction, infrastructure construction and construction of major engineering projects in the Loess plateau urgently need study the seismic design methods of loess engineering sites against seismic landslides, liquefaction, subsidence and amplification effect. In this paper, the authors present the methods of evaluating amplification of ground motion on topography and deposit of loess sites, assessing risk of seismic landslides, seismic design for engineering loess slopes with single-step and multi-steps, evaluating liquefaction and seismic subsidence of loess ground and its treatment measurements. seismic design of pile foundation in loess ground considering the negative friction on piles due to seismic subsidence and horizontal pushing force on piles under liquefaction. The achievements have been widely popularized and applied in the major engineering projects including high-speed railways, dams, airports in the loess region. The methods have been adopted by the National Standard for Seismic Design of Underground Structures (GB/T51336-2018) and the Seismic Design Code for Buildings in Gansu Province (GB62/T25-3055-2011).

2 Characteristics of Engineering Geology and Seismic Hazards in the Loess Plateau

2.1 Characteristics of Engineering Geology

Chinese loess area mainly distributed in the upstream and midstream of the Yellow River, involving in 7 provinces of Qinghai, Gansu, Ningxia, Shaanxi, Shanxi, Henan, Hebei, where is the Loess Plateau (Fig. 1). It has scattered distribution in other provinces. The thickness of loess rangs from tens meters to more than 500 m with the most complicated topography of loess deposit in the globe.

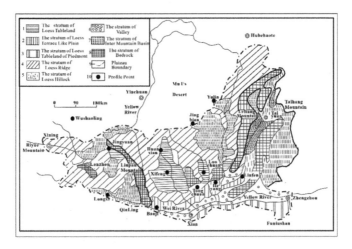

Fig. 1. Distribution of Loess Plateau and loess landforms (Teng 1990)

The typical loess landforms in the Loess Plateau are mainly loess tableland, loess ridge, and loess round hill (Fig. 2(a)(b)(c)). The microstructures of loess are respectively trellised porous structure, intergranular porous structure and coagulated porous structure from the western region, middle region, to eastern region of the Loess Plateau (Fig. 2(d)). The former two structures have weak cohesion and the latter one has a little strong cohesion. The partiles of loess are dominated by silt ranging from 60–70%. The other part is composed of sand and clay, which are 11%-3% and 18–29% respectively from the western region to eastern region of the plateau (Table 2). Such a special soil with porous microstructure, weak cohesion and silt dominated particles have a high vulnerability of seismic landslides, subsidence and liquefaction in the different landforms under the effect of strong earthquakes.

(a) Loess table-land   (b) Loess ridge

(c)Loess round hill   (c) Microstructures of loess in the Loess Plateau

Fig. 2. The typical landforms and microstructures of loess in the Loess Plateau

Table 2. Particles composition of loess in the difference regions

Sampling location	Sand	Silt	Clay
Western region	10.94	70.92	18.14
Central region	8.7	65.45	25.75
Eastern region	3.8	67.1	29.10

2.2   Seismic Hazards in the Loess Plateau

Factor superposition method is a common method for medium-scale disaster zoning, which has the advantages of clear significance and strong operability. There are many influencing factors of loess seismic disasters, but the influence degree of each factor on disasters is not the same. Therefore, determining scientifically the weight of each influencing factor will make the zoning results more reasonable. In the compilation of this zoning map for the Loess Plateau, we adopted the method of factor weighted superposition. The steps of this method are: ① Identification of factors affecting subsidence, liquefaction and landslide hazards; ② Figure digitization of influencing factors; ③ Classification assignment and weight determination of influence factors; ④ GIS weighted overlay analysis; ⑤ Classification of divisions; ⑥ Mapping. The workflow diagram is shown in Fig. 3.

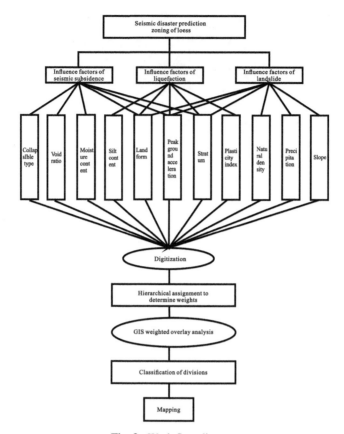

Fig. 3. Work flow diagram

Seismic Landslide in Loess Plateau

When the exceedance probability in 50 years is 2% (Fig. 4(a)), serious landslides are mainly distributed near Tianshui, Haiyuan and Baiya. Moderate landslides are mainly concentrated in southern Ningxia and the eastern part of the Baiyin area, Tianshui area, the Weihe basin and around Taiyuan to Linfen. Mild landslides are distributed in the loess ridge and loess hill areas of the Loess Plateau. Non-landslide areas are mainly in the loess tableland.

When the exceedance probability in 50 years is 10% (Fig. 4(b)), there is no serious landslide area. Moderate landslide areas are mainly distributed near Haiyuan and Tianshui. Mild landslides are mainly distributed in southern Ningxia, Baiyin, Tianshui, eastern and southwestern of Ordos block and Henan Province. The non-landslide area is mainly distributed in the Ordos block.

When the exceeding probability in 50 years is 63.5% (Fig. 3(c)), there is no serious landslide area. The moderate landslide area has sporadic distribution in Tianshui area. Mild landslide areas concentrate in Tianshui and southern Ningxia.

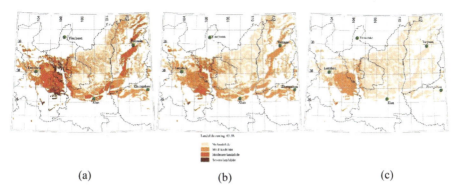

(a)   (b)   (c)

Fig. 4. Zoning maps of seismic landslides with different exceeding probabilities in 50 years ((a) 2%; (b) 10%; (c) 63.5%)

Seismic Liquefaction in Loess Plateau

Since the liquefaction zoning is based on the assumption of loess saturated by underground water, rainfall water, and irrigation water, the liquefaction zoning results only represent the liquefaction potential, not the zoning results under the natural state.

When the exceeding probability in 50 years is 2% (Fig. 5(a)), there are mainly two regions with severe liquefaction potential. One is Tianshui to Huining area, the other is southern Ningxia and Jingyuan border area. It is also distributed near Gulang and Tianzhu in Gansu and Taiyuan in Shanxi. The areas with medium liquefaction potential are mainly distributed in the south of Ningxia, Tianshui area, Baiyin area, eastern part of Dingxi area, western part of Pingliang area. The areas with slight liquefaction potential are mainly located in the peripheral areas of the areas with medium liquefaction potential, and the area is large. The areas without liquefaction potential are mainly located in the northern part of Shanxi Province.

When the exceeding probability in 50 years is 10% (Fig. 5(b)), the area with serious liquefaction potential is distributed in Haiyuan area. The area with medium liquefaction potential is distributed in the south of Ningxia, east of Lanzhou, west of Liupan Mountain, and the belt area between Taiyuan and Linfen in Shanxi. The areas with slight liquefaction potential are mainly distributed in the eastern, western, southern margin of Ordos block and Henan province. Non-liquefaction potential areas are mainly distributed in Ordos block.

When the exceeding probability in 50 years is 63.5% (Fig. 5(c)), there is no severe liquefaction potential area and medium liquefaction potential area. The areas with slight liquefaction potential are mainly distributed in the south of Ningxia, the east of Lanzhou, the west of Liupan Mountain, and the northwest of Shanxi. There is also a small amount of distribution near Fuping in Shaanxi and Luoyang in Henan. Most loess areas have no liquefaction potential.

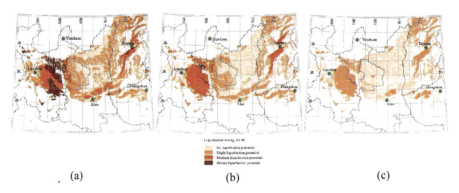

(a)   (b)   (c)

Fig. 5. Zoning maps of liquefaction potential with different exceeding probabilities in 50 years ((a) 2%; (b) 10%; (c) 63.5%)

Seismic Subsidence in Loess Plateau

When the exceeding probability in 50 years is 2% (Fig. 6(a)), the areas with the most serious loess seismic subsidence disasters are near Tianshui City, Haiyuan and Guyuan at the junction of Ningxia and Gansu Province, which reach the level of Grade IV seismic subsidence with 60 cm and above. The Grade III seismic subsidence area with 30–60 cm is distributed on the south of Ningxia and the southeast of Gansu. The Grade II seismic subsidence area with 7-30 cm is widely distributed in the western edge, southern edge and eastern edge of the Ordos block, showing a zonal distribution. The Grade I seismic subsidence with less than 7 cm and non-seismic subsidence area is mainly distributed in the southern Ordos block.

When the exceedance probability in 50 years is 10% (Fig. 6(b)), the distribution of Grade III and Grade II seismic subsidence areas is roughly equivalent to that of Grade IV and Grade III seismic subsidence areas when the exceedance probability is 2% in 50 years. Grade I seismic subsidence area is widely distributed in the east and west sides of Grade II seismic subsidence area. Non-seismic subsidence area is mainly distributed in the northern of Shaanxi.

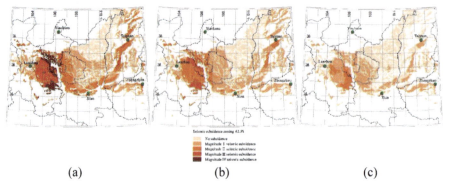

(a)   (b)   (c)

Fig. 6. Zoning maps of seismic subsidence with different exceeding probabilities in 50 years ((a) 2%; (b) 10%; (c) 63.5%)

When the exceedance probability in 50 years is 63.5% (Fig. 6(c)), there is no seismic subsidence hazards in the most areas.

Since ground motion, stratum and landform are the common influencing factors in seismic landslides, liquefaction and subsidence zoning and the dynamic vulnerabilities of loess in the different region of the plateau, the weight of ground motion appears mostly high. Therefore, the zoning of the three kinds of disasters have the similar overall trend. However, three kinds of disaster zoning have their own unique impact factors, so it will reflect differences in local distribution.

3   Seismic Design Methods of Ground Motion Amplification

3.1   Field Investigation

The Wenchuan Ms8.0 earthquake in 2008 not only had a great influence on structures near the fault area, but also made a significant impact on the loess tableland far from the epicenter due to predominate amplification of ground motion. In Gansu Province, Dazhai Village in Pingliang City is located at the top of the 100-m-thick loess plateau platform where is 500 km away from the epicenter. The houses in the village are mainly wooden structures, masonry structures and adobe structures. 70% of the houses collapsed or seriously damaged during the earthquake and the seismic intensity reached their fortification intensity of VII degree, which is 1° higher than that of the area below the tableland where the houses were in good condition (Fig. 7).

Fig. 7. The seismic intensity is VII-degree at Dazhai village on the loess tableland

3.2   On-Site Observation

After the Wenchuan main shock, the investigation team monitored the ground motion on the slope of a loess tableland edge in Wenxian County, Gansu Province ("Wenxian tableland slope" for short) (Fig. 8). The gradient of Wenxian tableland slope is 35° ~ 45° and the relevant hight is about 50 m. Three observation stations of strong ground motion were set up in different locations of the slope with basically the same geological conditions: station-1 at the bottom of slope (N32.94°, E104.70°, altitude is 927 m), station-2 at the middle of slope (N32.95°, E104.67°, altitude is 960 m) and station-3 at the top of slope (N32.94°, E104.67°, altitude is 969 m). The recorded PGA of the aftershocks and related information are shown in Table 3. The observation station at the bottom of the slope was selected as the reference station to analyze the amplification effect of slope on the loess tableland. Totally 9 aftershocks were recorded at the observatory stations. The time histories of ground motion shown that the amplification coefficient of horizontal peak ground acceleration ranges from 1.1–2.7 with the manitudes of Ms4.1-5.7.

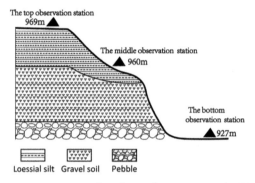

Fig. 8. The temporary strong motion observation array near the Wenxian County

In addition, the analysis on the recorded time histories of peak ground motion of the Wenchuan Ms8.0 mainshock at 4 observatory stations with a distance of 65 km and 120 km away from the epicenter shown that the amplification coefficient of horizontal peak ground acceleration due to loess deposit is 1.3 and 4.1 respectively. The recorded data of the Mixian-Zhangxian Ms6.6 earthquake at the same 4 stations shown that the amplification coefficient due to loess deposit is 2.1 and 5.0 respectively.

3.3   The Seismic Design Method of PGA Ampification

In order to evaluate the amplification effect of loess deposit thickness and slope hight on ground motion, a series of large-scale shaking-table tests and numerical simulations were carried out. Furthermore, the seismic design method of PGA amplification is proposed based on the field investigation, observation data, shaking-table tests and numerical simulation comprehensively. The design PGA and characteristic period at engineering sites with different loess thickness may be respectively determined by the local fortification PGA plus an amplification coefficient shown as Table 3 and Table 4. The design PGA at engineering slope sites with relevant hight of 50 m and above may be determined by the local fortification PGA plus an amplification coefficient for different slope angles shown as Table 5. The characteristic period at engineering slope sites with relevant hight of 50 m and above may be determined by Table 4, in which the relevant hight of slope is equivalent to thickness of loess.

Table 3. The amplification coefficient of PGA with different thicknesses of loess sites.

Thickness/m	20	40	60	80	100	120	160	200	≥240
≤0.05 g	1.1	1.2	1.3	1.3	1.3	1.4	1.4	1.5	1.6
0.10 g	1.1	1.1	1.2	1.2	1.2	1.3	1.3	1.4	1.4
0.15 g	1.05	1.05	1.1	1.1	1.1	1.2	1.2	1.2	1.2
0.20 g	1	1.05	1.05	1.05	1.05	1.1	1.1	1.1	1.1
0.30 g	1	1	1	1	1	1	1.05	1.05	1.05
≥0.40 g	1	1	1	1	1	1	1	1	1

Table 4. Characteristic periods (s) adjusted with different thicknesses of loess site.

Thickness/m	20	40	60	80	100	120	160	200	≥240
≤0.05 g	0.54	0.63	0.69	0.7	0.7	0.7	0.7	0.7	0.75
0.10 g	0.55	0.64	0.69	0.7	0.7	0.7	0.7	0.7	0.75
0.15 g	0.57	0.65	0.69	0.7	0.7	0.7	0.7	0.7	0.75
0.20 g	0.59	0.67	0.69	0.7	0.7	0.7	0.7	0.75	0.75
0.30 g	0.61	0.68	0.7	0.7	0.7	0.7	0.7	0.75	0.75
≥0.40 g	0.63	0.69	0.7	0.7	0.7	0.7	0.7	0.75	0.75

Table 5. Amplifying factor with loess slope effects

Slope (50 m high)	20° ~ 30°	30° ~ 45°	45° ~ 60°	60° ~ 70°
PGA amplifying factor	1.1–1.3	1.3–1.5	1.5–1.7	1.7–2.0

4 Seismic Design of Engineering Slopes of Loess

Based on the investigation of a large number of loess seismic landslides, the prediction and calculation of the maximum sliding distance and disaster range of loess seismic landslides, and considering the amplification effect of slope ground motion, the seismic safety design method of loess slope is developed.

When the high-rise building is located at the slope crest, the distance between the building and the edge of the slope crest should not be less than 20 m; The distance between the edge of the foundation slab and the edge of the slope bottom shall not be less than 1.25 times of the slope height below the foundation slab (H_1). See Fig. 9(a) In the figure, d is the buried depth of the foundation.

When the multi-storey building is located at the crest of the slope, the distance between the building and the edge of the slope crest should not be less than 10 m; The distance between the base edge and the slope bottom edge should not be less than the slope height below the base (H_1). See Fig. 9(b).

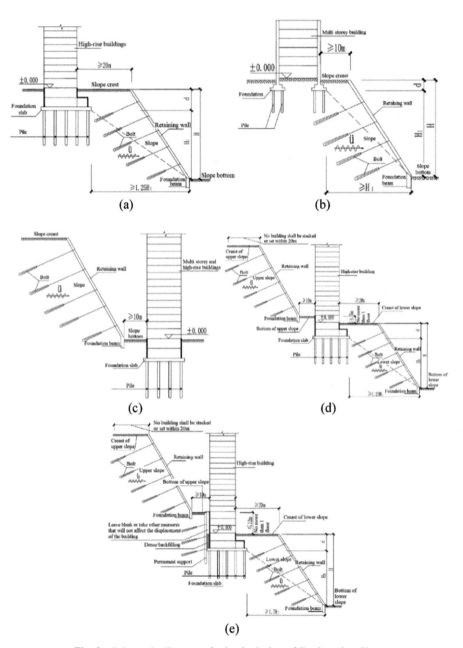

Fig. 9. Schematic diagram of seismic design of Engineering Slope

When multi-storey and high-rise buildings are located at the slope bottom, the distance between the buildings and the edge of the slope bottom should not be less than 10 m, as shown in Fig. 9(c).

When the high-rise building is located on the platform in the middle of the slope and the ground height difference on both sides of the building is no more than 4.5 m and no more than 1 floor, the distance from the building to the bottom edge of the upper slope should not be less than 10 m; The distance from the edge of the lower slope crest shall not be less than 20 m; The distance between the edge of the foundation slab and the bottom edge of the lower slope should not be less than 1.25 times of the slope height below the foundation slab (H_1); No building shall be stacked or set within 20 m from the top edge of the upper slope; The surface soil layer of the lower slope shall be compacted. See Fig. 9(d).

When the high-rise building is located on the platform in the middle of the slope, and the ground height difference on both sides of the building is no more than 12 m and no more than 3 floors, in addition to meeting the above requirements, the following requirements shall also be met: The distance between the edge of the foundation slab and the bottom edge of the lower slope should not be less than 1.5 times of the slope height below the foundation slab (H_1); Permanent support facilities shall be set at the bottom of the upper slope near the outside of the building, and the design service life of the support facilities shall not be less than the service life of the building, and must be separated from the building; The deformation joint below the embedded end shall be backfilled tightly, and the space above the embedded end shall be left blank or other measures that will not affect the displacement of the building shall be taken. See Fig. 9(e).

The above research results have been fully adopted by the code 'Specification for seismic design of buildings' in Gansu Province (GB62/T25-3055-2011) (2021 revised verion), in which the first author is one of chief editors.

5   Evaluation and Prevention of Liquefaction of Loess Ground

5.1   Preliminary Discrimination Method

Based on the evaluation of factors affecting loess liquefaction including soil saturation, soil depth, sedimentary age, particle composition, physical property, and microstructure type, we proposed the preliminary discrimination method of liquefaction potential of loess ground with engineer practicality. When saturated loess (including loess-like soil) meets one of the following conditions, it can be preliminarily judged as non-liquefaction circumstance or that the impact of liquefaction may not be considered.

(1) When the geological age is the early Middle Pleistocene loess (Q_2^1) and before, it can be considered that no liquefaction would occur at seismic intensity of VII (0.15 g) VIII (0.20 g, 0.30 g) and IX (0.4 g)degree.
(2) Loess with saturation less than 60%.
(3) When the content percentage of clay particles (particle size < 0.005 mm) of loess is not less than 12%, 15%, and 18%, the impact of liquefaction may be ignored at seismic intensity of VII, VIII, and IX degree respectively.

(4) For buildings with shallow buried natural foundation, when the thickness of over-lying non-liquefied soil layer and the depth of groundwater level meet one of the following conditions, the impact of liquefaction may be not considered:

$$d_u > d_0 + d_b - 2$$

$$d_w > d_0 + d_b - 3$$

$$d_w + d_u > 1.5d_0 + 2d_b - 4.5$$

where, d_u (m) is thickness of overlying non-liquefied soil layer (excluding silt and mudding soil layer); d_w (m) is depth of groundwater level, taken as the annual average maximum water level within the design reference period or the recent annual maximum water level; d_b (m) is buried depth of foundation, taken as 2 m if less than 2 m; d_0 (m) is characteristic depth of loess liquefaction.

5.2 Detailed Discrimination Method

The research on detailed discrimination method of saturated loess ground liquefaction involves liquefaction shear stress ratio, liquefaction index, microstructure characteristic parameter, and SPT (standard penetration test) counts. Herein detailed discrimination standard is discussed based on the SPT counts.

The results of frequentness statistics of SPT counts in saturated loess sites are shown in Fig. 10(a). It can be seen that about 50% of the sites have SPT counts lower than 6, while 85% of the sites have SPT counts lower than 9, and only 16% of the saturated loess sites have SPT counts higher than 9.

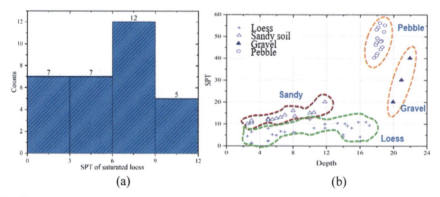

(a)   (b)

Fig. 10. (a) SPT counts frequentness histogram of saturated loess sites; (b) SPT counts of different saturated sites

The SPT counts of saturated loess and other saturated soils are further compared, and the SPT counts of different saturated sites is shown in Fig. 10(b). It can be noticed that the SPT counts increases from loess to sandy soil, gravel, and pebble. The SPT counts of

Table 6. SPT counts of different saturated soil

Soil	Minimum SPT counts	Maximum SPT counts	Average SPT counts
Loess	1.8	12	6.3
Sandy soil	10	21	13.1
Gravel	20	40	30
Pebble	39	55	47.4

loess is close to that of sandy soil, but there are still obvious differences between them. The SPT counts value of different saturated soils is listed in Table 6.

The average SPT counts of saturated loess is 6.3, while that of saturated sandy soil is 13.1, which is approximately twice the difference. However, it will become easy for engineers to make the detailed discrimination of loess liquefaction if we still use the discrimination formula of sand liquefaction from the national code with the reference value of SPT counts of saturated loess.

In National Code for Seismic Design of Buildings, the critical value of SPT counts for sand liquefaction discrimination within 20 m depth is calculated as formula (1):

$$N_{cr} = N_0\beta[\ln(0.6d_s + 1.5) - 0.1d_w]\sqrt{3/\rho_c} \tag{1}$$

where, N_{cr} is critical value of SPT counts for liquefaction discrimination; N_0 is reference value of SPT counts for liquefaction discrimination; d_s (m) is SPT depth of saturated soil; d_w (m) is depth of groundwater level; ρ_c is percentage of clay content, taken as 3 if less than 3 or sandy soil; β is adjustment coefficient, taken as 0.80 for seismic design first group, 0.95 for seismic design second group, and 1.05 for seismic design third group.

The comparative results of field test and laboratory test show that formula (1) is applicable to the discrimination of liquefaction potential of natural saturated loess, but the reference value of SPT counts needs correction. Table 7 provides the reference value of SPT counts for loess liquefaction discrimination.

Table 7. Reference value of SPT counts for loess liquefaction discrimination

Seismic design acceleration (g)		0.10	0.15	0.20	0.30	0.40
Reference value of SPT	National code	7	10	12	16	19
	Provincial specification	7	8	9	11	13

The comparison between measured SPT counts and calculated critical value of SPT counts for liquefaction discrimination is shown in Fig. 11.

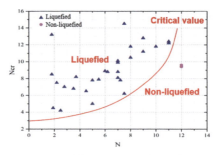

Fig. 11. Comparison between measured SPT counts and calculated critical value of SPT counts for liquefaction discrimination

For the saturated loess ground defined by Formula (1), the liquefaction index of each borehole can be calculated using Formula (2) to comprehensively divide the liquefaction grade of ground.

$$I_{lE} = \sum_{i=1}^{n} [1 - \frac{N_i}{N_{cri}}] d_i W_i \tag{2}$$

where, N_i is SPT counts of soil layer; di is thickness of soil layer; N_{cri} is critical value of SPT counts for liquefaction discrimination; Wi is depth weight, calculated using Formula (3):

$$W_i = \begin{cases} 10 \, (d_s < 5\,\text{m}) \\ \frac{10}{3} \cdot d_s - \frac{20}{3} (5 < d_s \leq 20) \end{cases} \tag{3}$$

According to the result of Formula (2), the liquefaction characteristics of loess site can be determined by the comprehensive classification standard of liquefaction potential of natural loess given in Table 8.

Table 8. Classification standard of liquefaction potential of natural loess

Liquefaction grade	Slight	Medium	Serious
Liquefaction index (I_{1E})	$0 < I_{1E} \leq 6$	$6 < I_{1E} \leq 18$	$I_{1E} > 18$

5.3 Technology and Standard of Anti-liquefaction Treatment of Loess Ground

Dynamic Compaction Method (DCM)

Dynamic compaction ground of 330 kV Dongjiao substation in Lanzhou is selected as the research subject. The liquefaction resistance performance of loess ground before and after dynamic compaction is studied. The site is in Liugouping, 500 m east of Taoshuping, Donggang Village, Lanzhou.

The curve of strain and pore pressure during loess liquefaction after DCM are shown in Fig. 12. The loess after dynamic compaction still has large dynamic strain. Although pore pressure ratio is only about 0.5, when pore pressure increases to the maximum, the strain growth accelerates suddenly. Therefore, the loess ground after DCM still has a certain liquefaction potential.

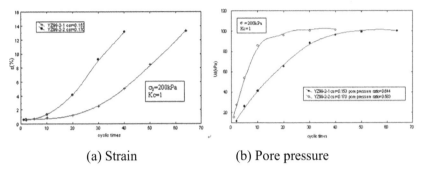

(a) Strain   (b) Pore pressure

Fig. 12. Curve of strain and pore pressure during loess liquefaction after DCM

The dry density and liquefaction stress ratio of ground soil are improved after DCM. Due to the close relation between loess liquefaction and the characteristics of particle composition and structural cementation in loess, although DCM cannot eliminate the liquefaction potential of loess, it is of positive significance to improve the liquefaction resistance of loess ground. Other anti-liquefaction measures should be considered in combination with DCM to ensure the safety of buildings in engineering activities.

Compaction Pile Method (CPM)
Research site is the primary school of Lanzhou Refinery and lime soil CPM is used for ground treatment. After lime soil CPM treatment, the dry density of ground soil has been improved, generally more than 1.53 g/cm^3, with an average of 1.58 g/cm^3. To study the effect of lime soil CPM on reducing the liquefaction potential of loess ground, dynamic triaxial liquefaction tests are conducted on soil samples taken before and after treatment, and test results are shown in Fig. 13.

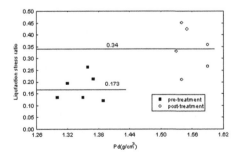

Fig. 13. Variation of liquefaction stress ratio before and after lime soil CPM treatment

Results show that lime soil CPM is effective in dealing with the collapsibility of loess, but there is limitation in reducing the liquefaction potential of loess ground. After lime soil CPM treatment, the liquefaction stress ratio of loess ground is improved, and the average liquefaction stress ratio of loess sample after treatment is twice that before treatment. Therefore, it can be considered that lime soil CPM can increase the liquefaction stress ratio of loess by about 0.1–0.2.

Above calculation indicates that the liquefaction triggering seismic intensity of loess ground after treatment can be increased by more than one degree compared with that before treatment. However, under the seismic intensity of VIII degree and above, the loess ground treated with lime soil CPM still has the possibility of liquefaction. Lime soil CPM treatment can not eliminate liquefaction potential of loess ground, but it can improve the liquefaction resistance of loess ground.

In conclusion, the liquefaction potential of saturated loess ground can be reduced or eliminated using DCM or lime soil CPM treatment under condition of seismic intensity less than VIII degree. Anti-liquefaction performance of ground soil depends on the compactness of ground soil after treatment. The above treatment methods can be approximately selected according to the importance of construction project. Under the condition of seismic intensity of IX, compaction treatment can enhance the liquefaction resistance of loess ground, but it has a certain limitation on eliminating liquefaction potential of saturated loess ground. For important buildings or projects, some other treatment (like chemical grouting) should be adopted in combination with compaction treatment to further improve the seismic safety of saturated loess ground.

5.4 Performance-Based Seismic Design of Anti-liquefaction Treatment Standard and Technology of Loess Ground

According to the important grade of buildings and the seismic liquefaction degree of loess ground, we proposed the performance-based seismic design of anti-liquefaction treatment technology and standard of loess ground shown in Table 9.

I buildings is the major engineering projects involving national public security and probably inducing serious secondary disasters by earthquakes with caterstrophical consequence, which is special fortification. II buildings mainly refer to buildings that use functions cannot be interrupted or need to be restored as soon as possible, and will cause significant social impact and major losses to the national economy, which is important fortification; III buildings are that have general impact after earthquake damage and other buildings that do not belong to I, II, and IV, which is regular fortification; IV buildings are that its damage or collapse will not affect people's life and slight social impact and economic loss, which is proper fortification.

The treatment techniques against loess liquefaction include soil replacement; composite foundation using cement, flyash or gravel pile; pile foundation crossing the potential liquefaction layer, and superstructure strengthing measures can be adopted to eliminating liquefaction potencial and relative deformation. In addition, water proof measures are important for preventing liquefaction potential of loess ground.

Table 9. Treatment standards and measures of liquefaction of loess ground

Seismic liquefaction degree		Slight	Medium	Serious
Treatment standards of loess ground	II buildings	Partial elimination of liquefaction settlement	All or partial elimination of liquefaction settlement	Elimination of all liquefaction settlement
	III buildings	No measures	Foundation and superstructure need treatment, or higher requirements	All or partial elimination of liquefaction settlement and foundation and superstructure need treatment

* I, II, III and IV buildings are the seismic fortification category of buildings based on the importance of buildings.

5.5 Seismic Design of Loess Pile Foundation Considering Liquefaction

The land subsidence caused by saturated loess liquefaction lags behind the earthquake action. Therefore, considering the liquefaction of saturated loess, the seismic design of loess pile foundation should be divided into two stages: before liquefaction and after liquefaction.

Before liquefaction, the horizontal bearing capacity and horizontal displacement of piles are calculated under seismic action. The calculation diagram is shown in Fig. 14(a). According to the experimental study of pile on saturated loess ground in Gansu Province, the proportional coefficient of horizontal resistance coefficient of saturated loess is m = $3 \sim 6$ MN/m^4, and the calculation formula of horizontal bearing capacity is:

$$H_{Eik} \leq 1.2 R_{ha} \tag{4}$$

In the formula, H_{Eik} is the horizontal seismic force acting on the top of single pile under the standard combination of loads effect; R_{Eha} is the characteristic value of horizontal bearing capacity of single pile.

The control value of horizontal displacement is:

$$X_{oa} = 15 \, \text{mm} \tag{5}$$

In the formula, X_{oa} is the allowable horizontal displacement of pile top.

After liquefaction, the horizontal seismic action is nearly zero, the liquefied soil around the pile is flowing, and the constraint effect on the pile is very small. The pile is close to the horizontal displacement column embedded in the top hinge bottom. Therefore, the stability of the pile under pressure under vertical force should be verified, and the calculation diagram is shown in Fig. 14(b).

$$N \leq \varphi \psi_a f_c A_{Pa} \tag{6}$$

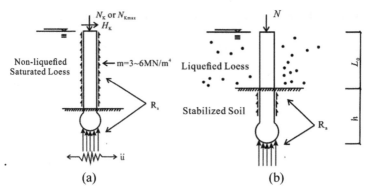

Fig. 14. Calculation diagram of loess pile foundation considering liquefaction ((a) Before earthquake; (b) After earthquake)

where, N is the axial pressure design value of pile top under the basic combination of load effect; φ is the stability coefficient, which can be determined according to the calculated length L of the pile and the design diameter of the pile. The calculated length of the pile can be determined according to the constraints on the top of the pile, L_0 is the free length of the surface of the stable soil layer exposed by the pile, h is the length of the pile in the stable soil layer, φ is the stability coefficient, ψ_a is the pile forming process coefficient; f_c is the design value of axial compressive strength of concrete; A_{Pa} is the cross-sectional area of the pile.

6   Evaluation and Prevention of Seismic Subsidence of Loess Ground

6.1   Preliminary Discrimination Method

Based on laboratory test research, field test and on-site investigation of earthquake damage, the seismic subsidence of loess ground can be judged by loess stratum, natural water content, void ratio, dry density, soil depth, compaction coefficient, plasticity index, shear wave velocity, site predominant period and so on. The discriminant value is shown in Table 10. Generally speaking, Holocene loess (Q_4) and late Pleistocene loess (Q_3) have seismic subsidence, while the middle Pleistocene loess (Q_2) and early Pleistocene loess (Q_1) do not. The loess, which has elevated pore microstructure, natural water content higher than shrinkage limit, void ratio more than 0.8, dry density lower than 16.3kN/m^3 (or compaction coefficient less than 0.95), shear wave velocity less than 300 m/s and standard penetration number less than 18 times, may have seismic subsidence. The water content has an obvious influence on the seismic subsidence of loess. When the water content exceeds the shrinkage limit of loess (the shrinkage limit of Lanzhou loess is generally between 5.8%), the seismic subsidence of loess increases significantly with the increase of water content. The void ratio of loess is generally more than 1.0. Under the condition of optimal water content, the seismic collapsibility of compacted loess decreases with the increase of dry density, and when the dry density is higher than 16.3 kN/m^3, it is very small and may not be considered (Fig. 15). The adoption of the compaction coefficient is to provide convenience for the engineering foundation treatment, which in principle belongs to the compaction index and is related to the dry density. Compared with the seismic code, it can be seen that when the loess site is below medium-soft, it will generally have seismic subsidence, but when the site is above medium-hard, only part of the

Table 10. Initial judgement index of seismic subsidence of loess

Initial judgement index	Layer	Microstructure type	W (%)	e	γ_d (kN/m^3)	Compaction coefficient
Discriminant value of seismic subsidence	Q_3, Q_4	Elevated pore microstructure	$> W_S$	>0.8	<16.3	<0.95
Initial judgement index	V_S (m/s)	Predominant period	PGA (gal)	Soil depth	Number of standard penetration hits	Plasticity index
Discriminant value of seismic subsidence	≤ 200	≥ 0.15	≥ 100	≤ 20	<18	<15

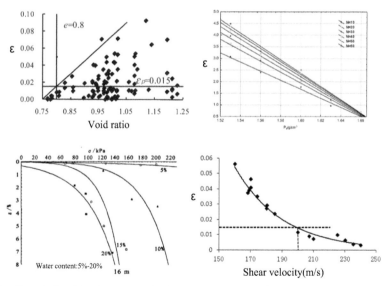

Fig. 15. The relationships between seismic subsidence coefficient and void retio, dry density, water content, and shear velocity.

site will have seismic subsidence. Through the earth tremor test of a large number of loess sites, the predominant periods of different sites are obtained. Generally speaking, the softer the loess site is, the longer the predominant period is. When the predominant period of the loess site is less than 0.15 s, it reflects that the soil of the site is very hard and generally does not cause seismic subsidence. PGA index is the minimum ground horizontal peak acceleration of seismic subsidence in loess site. Laboratory experiments and calculation analysis show that its value is 100 gal, and the corresponding seismic intensity is equal to 7°. The on-site investigation of historical earthquakes confirmed that the seismic subsidence of loess occurred in the area with intensity above 7°. The lower limit depth of earthquake subsidence soil layer reflects that when the loess soil layer is buried deep, its structure is relatively stable and earthquake subsidence is not easy to occur.

The above indexes are usually on the safe side. However, for the projects of different importance, it is still necessary to adopt reasonable principles when using the above initial judgment indexes to distinguish the site seismic subsidence. It is suggested that for III and IV buildings with no more than 10 storeys, if the above discriminant indexes reach the standard of non-seismic subsidence soil, the problem of seismic subsidence may not be considered. However, for important projects, high-rise buildings and other sites with big additional stress, it is necessary to reach at least two relatively independent indexes at the same time in order to judge that the loess site does not have seismic subsidence. Otherwise, a special research and evaluation should be carried out.

6.2 Detailed Discrimination Method

The detailed discrimination of seismic subsidence of loess ground can be carried out by dynamic triaxial tests. The seismic subsidence coefficient is defined as the residual strain, ε_p (N), under a certain cyclic times (N) shown as formula (7):

$$\varepsilon_p = \frac{h_0 - h_1(N)}{h_0} \tag{7}$$

where, $\varepsilon_p(N)$ is seismic subsidence coefficient (%); h_0 denotes the hight of a specimen before applying dynamic loading (mm); $h_1(N)$ denotes the hight of a specimen after applying dynamic loading (mm). Based on dynamic triaxial tests, ε_p (N) can be determined by the recorded time histories of dynamic axial stress and strain shown as Fig. 16(a). Through a series of dynamic triaxial tests with a certain confining stress for 4–5 specimens to be subjected to different amplitudes of dynamic stress, the curves of dynamic stress, σd, versus residual strain, ε_p (seismic subsidence coefficient), can be figured out (Fig. 16(b)).

The quantity of cumulative seismic subsidence of loess ground can be calculated by the formula (8)

$$Sd = \Sigma \Delta h_i * \varepsilon_{pi} \tag{8}$$

where, Sd is the quantity of cumulative seismic subsidence of loess ground under a certain fortification PGA; Δh_i is the thickness of loess layer i; and ε_{pi} denotes the seismic subsidence coefficient of loess layer i. Seismic subsidence of loess ground are evaluated

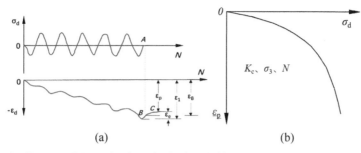

(a)   (b)

Fig. 16. The diagram of determination of seismic subsidence coefficient ((a) The recorded time histories of dynamic axial stress and strain; (b) the curves of dynamic stress versus seismic subsidence coefficient).

with 3 grade, light, medium, serious, according to the value of Sd under a certain fortification PGA shown as Table 11. Thus, loess ground can be evaluated according to its accumulative quantity of seismic subsidence under a certain fortification PGA: 0 < Sd ≤ 20 cm, slight seismic subsidence; 20 cm < Sd < 80 cm, medium seismic subsidence; Sd > 80 cm, serious seismic subsidence.

Table 11. The detailed discrimination of seismic subsidence of loess ground

Seismic subsidence grade	Slight	Medium	Serious
Sd (cm)	0 < Sd ≤ 20	20 < Sd ≤ 80	Sd > 80

6.3 Technology and Standard of Anti-subsidence Treatment of Loess Ground

Dynamic Compaction Method (DCM)
The dynamic triaxial tests of loess seismic subsidence at the depths of 4 m in 3000 kN•m tamping area and 2 m, 4 m and 6 m in 6000 kN•m tamping area were carried out respectively, and compared with the seismic subsidence test of undisturbed loess before treatment. The result is shown in Fig. 17. It shows that the seismic subsidence curve of loess in the effective depth range of dynamic compaction changes from non-linear to linear after dynamic compaction. No matter 3000 kN•m or 6000 kN•m energy level dynamic compaction, the seismic subsidence coefficient is less than 0.01 (1%) under the action of dynamic stress of 220 kPa. And with the increase of dynamic stress, the increase of seismic subsidence coefficient decreases. After dynamic compaction, the seismic subsidence coefficient of loess does not increase with the increase of cyclic times. The residual strain mainly occurs in the first 10 cyclic times, and then increases more slowly with the increase of cyclic times, and the seismic subsidence curve is very close.

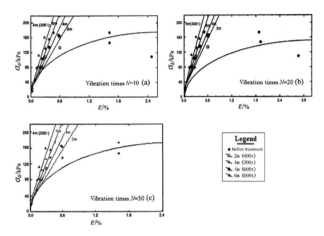

Fig. 17. Seismic subsidence curve of loess site before and after dynamic compaction

To sum up, the method of dynamic compaction to deal with the collapsibility of loess ground can completely eliminate the seismic subsidence under the optimal water content in the range of effective depth. The dry density of the treatment should be increased with water content. The dry density standard should be 1.63 g/cm^3 at the optimal water content (about 16%). This value can be reduced at a lower water content, but it has to be determined according to the water content.

Compaction Pile Method (CPM)
The research site is the first Primary School of Lanzhou Refinery and the No. 15 Depot of Lanzhou Grain Depot Reconstruction and extension Project directly under the State Grain Reserve Bureau. Lime-soil compaction piles are used to eliminate the collapsibility of loess in the range of 9 m and 7 m respectively. The filling material in the hole is lime-soil at 2:8. The length of lime-soil pile is 9 m and 7 m respectively, and the diameter of the pile is 400 mm, and the distance between piles is 1 m. The technology of pile sinking is adopted. The backfill height of each layer of lime soil is 100 cm, which is compacted by layers. After treatment, the collapsibility of loess is eliminated in the depth of 9 m and 7 m respectively.

It can be seen from Fig. 18 that due to the improvement of the treatment standard of lime-soil compaction pile, the compactness of ground loess has been greatly increased, and the seismic subsidence curve has changed from a nonlinear curve to an approximate straight line. The seismic subsidence of loess within 7 m depth has been eliminated. This shows that under certain conditions, to improve the standard of collapsibility treatment of loess ground by lime-soil compaction pile, while collapsibility is eliminated, seismic subsidence can also be eliminated. However, by comparing the results of the first section, it can be found that the seismic subsidence coefficient under 200 kPa is still larger than that under static state. The former is about 0.012 and the latter is 0.004. Therefore, the treatment of seismic subsidence of loess ground should consider the intensity of ground motion that may be suffered in the future.

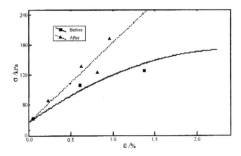

Fig. 18. Comparison of seismic subsidence of loess ground before and after lime-soil compaction pile treatment

6.4 Performance-Based Seismic Design of Anti-seismic Subsidence Treatment Technology and Standard of Loess Ground

According to the important grade of buildings and the seismic subsidence degree of loess ground, we proposed the performance-based seismic design of seismic subsidence treatment technology and standard of loess ground. The contents are shown in Table 12.

Table 12. Treatment measures of seismic subsidence of loess ground

Seismic subsidence grade		Slight	Medium	Serious
Treatment methods		No measures	The collapsibility and seismic subsidence of loess ground should be considered together in the design, and comprehensive measures combining ground treatment, waterproof measures and structural measures should be taken	
Treatment standard	II buildings	No measures	The whole or local cushion, dynamic compaction and other composite foundation shall be used to eliminate all or part of the settlement of the ground, or the pile foundation shall be used to transfer the building load to the deeper non-collapsible and non-seismic subsidence soil layer	
	III buildings	No measures	No measures	Ditto

6.5   Seismic Calculation of Loess Pile Foundation Considering Seismic Subsidence

We carried out a large-scale field tests and observation for negative friction resistance of large-diameter concrete cast-in-situ piles (pile diameter 0.8 m, length 20 m) in Q3 loess ground with seismic subsidence induced by blasting simulation in the southern margin of Lijiawanping, Lintao County, Gansu Province [34]. The measured results show that:

(1) The maximum seismic subsidence of loess site is 33 mm, which is far less than that of soil settlement when loess site is immersed and collapsible. However, the average negative friction stress is 54 kPa, and the corresponding total negative friction force is 1654 kN, which is much larger than the observed value of negative friction under loess collapsibility. Therefore, the negative friction caused by seismic subsidence of large-diameter concrete cast-in-place pile foundation in unsaturated loess area cannot be ignored

(2) The seismic subsidence of loess below 20 m in Q3 loess layer is negligible.

The large-diameter concrete pile foundation on Q3 loess ground should consider the two situations that is before and after the earthquake to carry out the seismic calculation of the pile foundation.

Before earthquake, the calculation diagram is shown in Fig. 19(a). The design criterions are presented in formula (9) and (10).

$$N_k \leq 1.2R_a \tag{9}$$

$$N_{k\,max} \leq 1.2R_a \tag{10}$$

In the formulae, N_k is the average vertical force of single pile under the axial vertical force of standard combination of load effect; N_{kmax} is the maximum vertical force of pile top under eccentric vertical force of standard combination of load effect; R_a is the characteristic value of vertical bearing capacity of single pile.

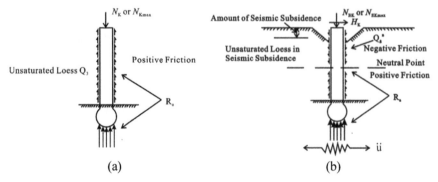

Fig. 19. Calculation diagram of loess pile foundation considering seismic subsidence ((a) Before earthquake; (b) After earthquake)

After earthquake, the calculation diagram is shown in Fig. 19(b). The design criterions are presented in formula (11) and (12).

$$N_{EK} + Q_g^n \leq 1.25R_a \tag{11}$$

$$N_{EK\,\text{max}} + Q_g^n \leq 1.5R_a \tag{12}$$

In the form, N_{Ek} is the average vertical force of single pile under the standard combination of seismic effect and load effect; N_{EKmax} is the maximum vertical force of single pile under standard combination of seismic effect and load effect; Q_g^n is the pull-down load of single pile caused by negative friction.

Drawdown loads of negative friction pile foundation based on experimental results:

$$Q_g^n = u \cdot q_{sk}^n \cdot L_s \tag{13}$$

In the formula, u is the perimeter of the pile (m); q_{sk}^n is the average negative friction characteristic value (kPa) of single pile under seismic subsidence; L_s is the thickness of loess layer (m).

7 Conclusions

Loess is a kind of silt dominated special soil with porous microstructure and weak cementation, which makes it have high seismic vulnerability and water sensitivity. Loess Plateau is a strong earthquake-prone region, where large casualties were caused by earthquakes. Large casualties attributed to a large scale of landslides, liquefaction, seismic subsidence and amplification of ground motion at loess sites.

The distribution of the three kinds of geotechnical disasters has the similarity of overall trend. The most serious loess earthquake disasters involve in Ningxia, Gansu, Shaanxi and Shanxi provinces. The region with lower risk of loess earthquake disasters is Ordos block.

The amplification effect of deposit thickness and topography of loess sites in the Loess Plateau on seismic ground motion is remarkable, which may amplify PGA by 1.1–2.2 times of the value at rock base and predominately enlonge characteristic period. In seismic design of engineering loess sites, it should be into account.

The risk assessments of seismic landslides and seismic design methods for engineering loess slopes with single-step and multi-steps are provided. The distance between the building and the edge of the slope crest and the distance between the edge of the foundation slab and the edge of the slope bottom should be determined by the type of building and the height of slope.

The preliminary discrimination and detailed discrimination methods of evaluating liquefaction and seismic subsidence of loess sites and its treatment measurements are respectively presented. Seismic treatment standards of loess ground should be adopted based on the importance of buildings and grade of liquefaction and subsidence of loess ground.

The pile foundation design considering seismic subsidence in unsaturated loess ground and liquefaction in saturated loess ground should distinguish before and after the earthquake. The vertical bearing capacity of pile foundation should consider the negative friction caused by seismic subsidence after the earthquake. The stability of the pile without surrounding soil constraints shall be checked in liquefied loess ground after the earthquake.

Acknowledgment. This research work is financially supported by the Key Project of National Natural Science Foundation of China (No. U1939209).

References

1. Prakash, S., Puri, V.K.: Liquefaction of loessial soils. In: Third International Conference on Seismic Microzonation, Seattle, USA, pp. 1101–1107 (1982)
2. Prakash, S., Sandoval, J.A.: Liquefaction of low plasticity silts. J. Soil Dyn. Earthq. Eng. **71**(7), 373–397 (1992)
3. Prakash, S., Guo, T.: Liquefaction of silts and silt-clay mixtures. In: Geotechnical Earthquake Engineering and soil Dynamics III, Seattle, Washington, pp. 337–346. ASCE (1998)
4. Ishihara, K., Okuss, S., Oyagi, N., Ischuk, A.: Liquefaction-induced flow slide in the collapsive loess deposit in Soviet Tajik. Soils Found. **30**(4), 73–89 (1990)
5. Xie, D.Y.: Soil Dynamics. Xi'an Jiaotong University Press, Xi'an (1988)
6. Xie, D.Y., Zhang, Y.H.: The influence of coupling change of static stress on the development law of pore water pressure under dynamic load. In: Proceedings of the 4th National Academic Conference on Soil Dynamics, pp. 13–20. Zhejiang University Press, Hangzhou (1994)
7. Wu, Z.H., Fang, Y.: Seismic subsidence characteristics of undisturbed loess during humidification. In: Proceedings of the third National Soil Dynamics Conference. Tongji University Press, Shanghai (1990)
8. Duan, R.W.: Experimental study on dynamic characteristics of Lanzhou loess. J. Northwest Seismol. **1**(2), 43–49 (1979)
9. Duan, R.W., Zhang, Z.Z., Li, L., Wang, J.: Further study on dynamic characteristics of loess. J. Northwest Seismol. **12**(3), 72–78 (1990)
10. Wang, J., Wang, L.M., Li, L.: Effects of different seismic loads on dynamic modulus and damping ratio of loess. J. Nat. Disasters **1**(4), 75–79 (1992)
11. Wang, L.M., Zhang, Z.Z., Wang, J., Li, L.: Test method for dynamic strength of loess under random seismic load. J. Northwest Seismol. **13**(3), 50–55 (1991)
12. Wang, L.M., Zhang, Z.Z., Wang, J., Li, L.: Experimental study on dynamic constitutive relationship of loess under random seismic load. J. Northwest Seismol. **14**(3), 61–68 (1992)
13. Hu, R.L., Li, Z.F., Wang, S.J., Zhang, L.Z., Li, X.Q.: Study on strength characteristics and structural change mechanism of loess under dynamic load. J. Geotech. Eng. **22**(2), 174–181 (2000)
14. Shi, Y.C., Zheng, M.J.: Characteristics of dynamic shear modulus and damping ratio of loess in Northwest China. J. Northwest Seismol. **19**, 144–150 (1997)
15. Zhang, Z.Z., Sun, C.S., Duan, R.W., Wang, L.M.: Loess Earthquake Disaster Prediction. Seismological Publishing House, Beijing (1999)
16. Liu, B.C., Zhou, J.X., Li, Q.M.: Interpretation of aerial photos of Tongwei earthquake in 1718 and Tianshui earthquake in 1654. Earthq. Sci. Res. **1**, 1–9 (1984)
17. Chen, B.W., Zhang, J.S.: Problems of earthquake engineering geology of loess sites and their estimating methods. In: Aspects of Loess Research, pp. 403–409. China Ocean Press (1987)

18. He, G., Zhu, H.B.: Study on seismic subsidence of loess. J. Geotech. Eng. **17**(6), 99–103 (1995)
19. Huang, Y.H.: Application of random search method for the most dangerous slip surface of soil slope in earthquake landslide disaster prediction. In: Proceedings of the Fourth National Academic Conference on Soil Dynamics. Zhejiang University Press, Hangzhou (1994)
20. Wang, L.M., Wang, J., Li, L., Shi, X.F., Wu, J.H.: Prediction method of liquefaction of urban loess site. J. Northwest Seismol. **19**, 7–12 (1997)
21. Wang, L.M., Yuan, Z.X., Shi, Y.C., Sun, C.S.: Study on indexes and methods of seismic disaster zoning in loess. J. Nat. Disasters **8**(3), 87–92 (1999)
22. Wang, L.M., Yuan, Z.X., Wang, J., Sun, C.S.: Influence of dry density on seismic subsidence of compacted loess. Earthq. Eng. Eng. Vib. **20**(1), 75–80 (2000)
23. Wang, L.M., Liu, H.M., Li, L., Sun, C.S.: Experimental study on liquefaction mechanism and characteristics of saturated loess. J. Geotech. Eng. **22**(1), 89–94 (2000)
24. Zou, J.C., Shao, S.M.: Fuzzy comprehensive evaluation of landslide distribution law in Gulang Haiyuan area. J. Seismol. **1**, 1–6 (1994)
25. Zou, J.C., Shao, S.M.: Gulang earthquake landslide and its relationship with fault zone. J. Northwest Seismol. **16**(3), 60–64 (1994)
26. Zou, J.C., Shao, S.M.: Discussion on Haiyuan earthquake landslide and its distribution characteristics. Inland Earthq. **10**(1), 1–6 (1996)
27. Sun, C.S., Cai, H.W.: Development and distribution characteristics of landslide and collapse geological disasters during historical earthquakes in China. J. Nat. Disasters **6**(1), 25–30 (1995)
28. Qin, H.Y.: Research and discussion on high speed loess landslide induced by earthquake. In: Soil Dynamics and Geotechnical Earthquake Engineering, pp. 236–246. China Construction Industry Press, Beijing (2002)
29. Bai, M.X., Zhang, S.M.: Liquefaction movement of loess stratum during high intensity earthquake. Eng. Invest. **6**, 1–5 (1990)
30. Wang, J.D., Zhang, Z.Y.: Study on Mechanism of earthquake induced high speed loess landslide. J. Geotech. Eng. **21**(6), 670–674 (1999)
31. Liu, H.L., She, Y.X., Wang, L.M.: Experimental study on liquefaction of dynamic compaction loess ground. In: The 6th National Conference on Soil Dynamics, pp. 218–224. China Construction Industry Press, Beijing (2002)
32. She, Y.X., Liu, H.L., Wang, L.M.: Experimental study on pore water pressure growth model of loess. In: The 6th National Conference on Soil Dynamics, pp. 225–230. China Construction Industry Press, Beijing (2002)
33. Wang, G.H.: Study on liquefaction of the sandy soil mixed with loess. In: Proceedings of Fourth International Conference on Recent Advances in Geotechnical Earthquake Engineering and Soil Dynamics, University of Missouri-Rolla, San Diego, USA, Paper No. 4.37 (2001)
34. Wang, L.M., Sun, J.J., Huang, X.F., Xu, S.H., Shi, Y.C.: A field testing study on negative skin friction along piles induced by seismic subsidence of loess. Soil Dyn. Earthq. Eng. **31**, 45–58 (2011)

Large-Scale Seismic Seafloor Stability Evaluation in the South China Sea Incorporating Soil Degradation Effects

Yuxi Wang⬥, Rui Wang(✉)⬥, and Jian-Min Zhang

Tsinghua University, Beijing 100084, China
wangrui_05@tsinghua.edu.cn

Abstract. Submarine landslides are severe threats to the safety of offshore facilities. Earthquake is considered as a main trigger of submarine landslides. This study characterizes the large-scale seismic seafloor stability in the South China Sea with consideration for the earthquake induced degradation of soil. The digital elevation models, Peak Ground Acceleration (PGA) maps, and 3-D continuous soil models are used for the large-scale analysis. A seismic soil degradation model is developed to simulate the degradation of soil subjected to earthquake loading. The infinite slope model with bidirectional seismic load is adopted to evaluate seafloor stability. The results indicate that the degradation effects of soil can significantly reduce seafloor stability, especially in areas of high PGA. The findings emphasize the importance of soil degradation characteristics when performing large-scale seismic seafloor stability evaluation.

Keywords: South China Sea · Large-scale submarine landslides · Earthquake · Soil degradation

1 Introduction

The stability of the seafloor is important to the safety of the ocean engineering facilities (Roesner et al. 2019; Locat and Mienert 2012; Locat and Lee 2005). Evidence suggests that earthquake is a main contributor to large-scale submarine landslides (Hance 2003; Brink et al. 2009; Haflidason et al. 2004). Hence, it is significant to evaluate large-scale seismic seafloor stability.

The South China Sea is an area with uneven seismicity (Chen et al. 2018; Li et al. 2018). Guo et al. (2019) evaluated the large-scale seafloor stability of the northern continental slope of the South China Sea assuming uniform seismicity. Liu and Wang (2014) first took the nonuniform seismicity in the South China Sea into consideration and analyzed the stability of the Zhujiang River Mouth Basin. Wang et al. (2021a, b) analyzed the large-scale seismic seafloor stability in the South China Sea with characterization of the real ground motion level derived from sufficient historical earthquake data via the Chinese Probability Seismic Hazard Analysis (CPSHA) method. The influence of horizontal and vertical bidirectional shaking has yet to be assessed.

However, besides the real ground motion level, the characterization of soil properties under dynamic loads is also important for large-scale seismic seafloor stability evaluation. Under dynamic loads, a series of complex phenomena may occur on the soil, such as the degradation of the soil stiffness and strength (Zhang and Wang 2012). Nian et al. (2019) suggest that the dynamic behavior of marine soil can be more complicated than its on land counterpart. Currently, few large-scale analyses have taken soil degradation effects into account.

This study evaluates the large-scale seismic seafloor stability in the South China Sea incorporating soil degradation effects. The models for topography, seismicity, and soil properties are first adopted from Wang et al. (2021a, b). An infinite slope model with bidirectional seismic load is constructed to evaluate submarine slope stability. The seismic soil degradation effects are characterized with a simple strength degradation model. Finally, the large-scale seismic seafloor stability evaluation in the South China Sea is performed and the difference caused by the consideration of soil degradation effects is analyzed.

2 Methodology

2.1 Models for Topography, Seismicity, and Soil Properties

The models for topography, seismicity, and soil properties are adopted from Wang et al. (2021a, b).

Wang et al. (2021a, b) constructed the topography model of the South China Sea using the geographic information system (GIS). The topography model of the South China Sea research area is illustrated in Fig. 1 and denoted with the dark red rectangle. The resolution of the model is $15'' \times 15''$ (approximately 400 m $\times$ 400 m). The research area ranges from 113° E to 120° E and 19° N to 22.5° N. The water depth of the research area ranges from 0 to 5000 m. Most of the seafloor in the research area is gentle while the gradient of some local area can reach 30°.

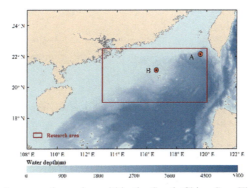

Fig. 1. The topography map of a region within the South China Sea. The deep dark rectangle denotes the research area. Site A and B are used to analyze the seafloor stability and will be introduced in Sect. 3 (Modified after Wang et al. 2021a, b).

Wang et al. (2021a, b) calculated the Peak Ground Acceleration (PGA) maps for the South China Sea under different exceedance probabilities (EPs) via the Chinese Probability Seismic Hazard Analysis (CPSHA) method. Three EPs were adopted including 63%, 10%, and 2% in 50 years. The PGA maps are exhibited in Fig. 2. From the maps, the PGA level of the South China Sea is distinctly uneven, being relatively weak in the western and central parts while intense in the eastern part. The highly nonuniform PGA level can cause a significant difference in the seismic seafloor stability in the research area.

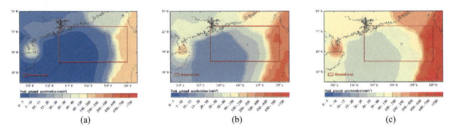

Fig. 2. PGA maps for the South China Sea region under different exceedance probabilities of seismicity, (a) 63% in 50 years, (b) 10% in 50 years, (c) 2% in 50 years. (Modified after Wang et al. 2021a, b).

Wang et al. (2021a, b) constructed the 3D continuous models for the soil properties through a proposed best-fit distribution-based 3D Kriging method. The models of undrained shear strength s_u and unit weight γ are illustrated in Fig. 3. s_u is under 12 kPa in the research area, indicating that the soil is soft. γ ranges from 12 kN/m^3 to 17 kN/m^3.

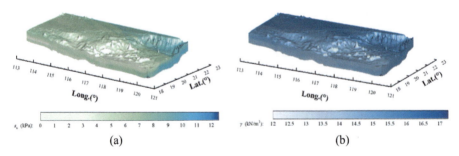

Fig. 3. 3-D models for soil properties of the South China Sea, (a) shear strength s_u, (b) unit weight γ. (Modified after Wang et al. 2021a, b).

2.2 Infinite Seafloor Slope Model

The infinite slope model is adopted to evaluate the seismic seafloor stability, as many researchers have demonstrated the appropriateness of the model for the analysis of large-scale submarine slope stability (Prior and Suhayda 1979; Ikari et al. 2011; Baeten et al.

2014). By using the safety factor *FS*, which is defined as the ratio of the resisting force f_τ to the sliding force f_s along the sliding surface ($FS = f_\tau/f_s$), the stability of the submarine slope can be quantified. Large *FS* indicates the submarine slope is safe while $FS < 1$ implies the submarine slope is unstable.

In the analysis of seismic seafloor stability, the seismic effect is usually treated as "quasi-static" loads (Nian et al. 2019; Wang et al. 2021a, b; Guo et al. 2019) as illustrated in Fig. 4. However, Nian et al. (2019) pointed out that the vertical seismic effect is important which had been overlooked in most studies (Ingles et al. 2006; Chen et al. 2015). In addition, the American Petroleum Institute (API) suggested that the vertical seismic load should be taken as half of the horizontal seismic load on the study of fixed offshore platforms (RP2A-WSD API 2000). In this study, bidirectional seismic loads are considered. The forces on the submarine slope are illustrated in Fig. 4.

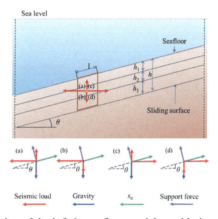

Fig. 4. Illustration of the infinite seafloor model considering seismic action.

There are four situations for seismic loads on the slope ((a), (b), (c), and (d)), of which the most dangerous situation is situation (b). The safety factor *FS* of situation (b) and be calculated through:

$$FS = \min_{1 \leq j \leq n} (FS_j) = \min_{1 \leq j \leq n} \left(\frac{s_{uj}}{\sin\theta \cdot \cos\theta \cdot \sum_{i=1}^{j} \gamma_i' h_i + \cos^2\theta \cdot k_h \sum_{i=1}^{j} \gamma_i h_i + \lambda \cdot \sin\theta \cdot \cos\theta \cdot k_h \sum_{i=1}^{j} \gamma_i h_i} \right)$$

(1)

where θ is the gradient of the slope; k_h is the horizontal seismic coefficient ($k_h = PGA/g$); *j* is the seafloor soil layer index; *n* is the total number of layers; γ' is the buoyant unit weight. λ is the ratio of the vertical seismic load to the horizontal seismic load and is set as 0.5.

2.3 Improved Infinite Seafloor Slope Model Incorporating Soil Degradation Effects

Seismic load is a dynamic load that has complex effects on the soil, the degradation of the soil stiffness and strength can occur during loading (Wang et al. 2014, 2021a, b; Xue et al. 2021). These soil degradation effects are usually quantified through sophisticated constitutive models for soil. However, for large-scale seafloor stability analysis, it is often impossible to acquire enough samples for the parameters of the sophisticated models to be calibrated. In this study, the soil degradation model for undrained shear strength s_u under seismic loads by Nian et al. (2019) is modified to quantify the soil degradation effects.

The soil degradation model is illustrated in Fig. 5. With the increase of k_h, the complex effects of seismic load on the soil are quantified through a simple strength degradation factor ζ. ζ is a piecewise linear function of k_h. For simplicity, $\zeta = 0.4$ is adopted when $k_h > 0.4$.

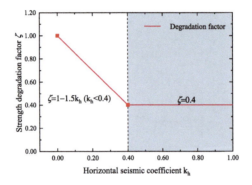

Fig. 5. Illustration of the soil degradation model. (Modified after Nian et al. 2019)

Adopting the soil degradation model, the infinite slope model is improved with ζ:

$$FS = \min_{1 \leq j \leq n} (FS_j) = \min_{1 \leq j \leq n} \left(\frac{\zeta \cdot s_{uj}}{\sin \theta \cdot \cos \theta \cdot \sum_{i=1}^{j} \gamma_i' h_i + \cos^2 \theta \cdot k_h \sum_{i=1}^{j} \gamma_i h_i + \lambda \cdot \sin \theta \cdot \cos \theta \cdot k_h \sum_{i=1}^{j} \gamma_i h_i} \right)$$

(2)

3 Results

The large-scale seismic seafloor stability FS maps considering soil degradation effects can be calculated based on the models in Sect. 2.1 via Eq. (2), and the results are illustrated in Fig. 6.

From Fig. 6, it can be found that with the increase of seismicity hazards (i.e., the decrease of exceedance probability), FS of the South China Sea research area decreases

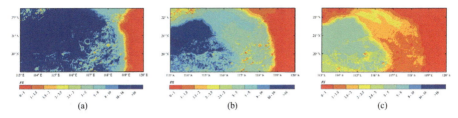

Fig. 6. Seafloor stability safety factor *FS* maps for the South China Sea region under different exceedance probabilities of seismicity, (a) 63% in 50 years, (b) 10% in 50 years, (c) 2% in 50 years.

dramatically. Hence, it is important to evaluate the seismic seafloor stability considering the earthquake probability. For each map, *FS* is uneven, indicating the spatial nonuniformity of seafloor stability. The stability of the central (between 116.5° E and 119° E) and western (west of 116.5° E) parts are much higher than the eastern part (east of 119° E). Moreover, *FS* of the eastern part (east of 119° E) is almost below 1 under all three seismic hazards, implying that the eastern part is unstable. While for the western part (west of 116.5° E), the seafloor is stable under all three seismic hazards. For the central part, *FS* shows a significant difference under all three seismic hazards. These large-scale *FS* maps can assist the engineering design in the South China Sea.

Figure 6 is calculated based on the soil degradation model. To quantify the soil degradation effects, the *FS* maps of the research area without consideration of soil degradation effects by Wang et al. (2021a, b) are used as a comparison. Here only *FS* of two sites A and B (illustrated in Fig. 1) are used to illustrate the difference. Information for the two sites is listed in Table 1. *FS* of the two sites with and without the consideration of soil degradation effects are plotted in Fig. 7. From Fig. 7, *FS* of site B has a little discrepancy with and without the consideration of soil degradation effects while the discrepancy of site A is significant under all three seismic hazards. The discrepancy increases with the increase of seismicity hazards. Under seismic hazard of 2% exceedance probability (50 years), the soil degradation effects can even cause *FS* to drop by 60% compared to the *FS* without consideration of soil degradation effects. From the PGA in Table 1, it can be found that the PGA of site A is far bigger than that of site B, which is the main contributor for the difference. The soil degradation effects are significant, especially for the area with high seismicity. Hence, it is important to evaluate the seismic seafloor stability considering the soil degradation effects.

Table 1. Information of site A and site B.

	Long. (°)	Lat. (°)	Depth (m)	Gradient (°)	s_u (kPa)	γ (kN/m^3)	PGA (cm/s^2)		
							EP 63%	EP 10%	EP 2%
A	119.58	22.13	2005	0.49	9.51	15.15	168.78	389.98	652.12
B	116.63	21.09	305	0.76	7.24	1.30	5.31	28.94	79.16

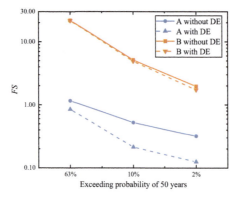

Fig. 7. Seafloor stability safety factor *FS* for site A and site B under different exceedance probabilities of seismicity without and with consideration of soil degradation effects (DE).

4 Conclusions

This study evaluates the seismic seafloor stability within the South China Sea considering bidirectional shaking and soil degradation effects. The main findings are listed as follow:

(1) The large-scale seafloor stability in the South China Sea research area is highly uneven, and the research area can be divided into the eastern, central, and western parts based on their stability. In all three seismic hazards (63%, 10%, and 2% exceedance probability in 50 years), the western part is rather stable while the eastern part is unstable, and the central part has different safety levels.

(2) Soil degradation can dramatically reduce the safety level of the seafloor, especially for the area with high seismicity. It is important to consider the soil degradation effects for the evaluation of seismic seafloor stability.

The soil degradation model adopted in this study is simple, which may not comprehensively reflect the soil degradation effects. With the accumulation of soil properties data and the development of seafloor stability evaluation models, the soil degradation effects can be quantified more precisely.

Acknowledgements. The authors would like to thank the National Natural Science Foundation of China (No. 52022046 and No. 52038005) and the State Key Laboratory of Hydroscience and Hydraulic Engineering (No. 2021-KY-04) for funding this study.

References

Baeten, N.J., et al.: Origin of shallow submarine mass movements and their glide planes—sedimentological and geotechnical analyses from the continental slope off northern Norway. J. Geophys. Res. Earth Surf. **119**(11), 2335–2360 (2014)

Brink, U.S.T., Lee, H.J., Geist, E.L., Twichell, D.: Assessment of tsunami hazard to the U.S. east coast using relationships between submarine landslides and earthquakes. Mar. Geol. **264**(1–2), 65–73 (2009)

Chen, B., Wang, D., Li, H., Sun, Z., Shi, Y.: Characteristics of earthquake ground motion on the seafloor. J. Earthq. Eng. **19**(6), 874–904 (2015)

Chen, L., Feng, Y., Okajima, J., Komiya, A., Maruyama, S.: Production behavior and numerical analysis for 2017 methane hydrate extraction test of Shenhu, South China Sea. J. Nat. Gas Sci. Eng. **53**, 55–66 (2018)

Guo, X.S., Zheng, D.F., Nian, T.K., Lv, L.T.: Large-scale seafloor stability evaluation of the northern continental slope of South China Sea. Mar. Georesour. Geotechnol. **4**, 1–14 (2019)

Haflidason, H., et al.: The Storegga Slide: architecture, geometry and slide development. Mar. Geol. **213**(1–4), 201–234 (2004)

Hance, J.J.: Development of a database and assessment of seafloor slope stability based on published literature. Master's thesis, University of Texas at Austin (2003)

Ikari, M.J., Strasser, M., Saffer, D.M., Kopf, A.J.: Submarine landslide potential near the megasplay fault at the Nankai subduction zone. Earth Planet. Sci. Lett. **312**(3–4), 453–462 (2011)

Ingles, J., Darrozes, J., Soula, J.C.: Effects of the vertical component of ground shaking on earthquake-induced landslide displacements using generalized newmark analysis. Eng. Geol. **86**(2–3), 134–147 (2006)

Li, J.F., et al.: The first offshore natural gas hydrate production test in South China Sea. China Geol. **1**(1), 5–16 (2018)

Liu, K., Wang, J.: A continental slope stability evaluation in the Zhujiang River Mouth Basin in the South China Sea. Acta Oceanologica Sinica **33**(11), 155–160 (2014). https://doi.org/10.1007/s13131-014-0565-8

Locat, J., Lee, H.J.: Subaqueous debris flows. In: Jakob, M., Hungr, O. (eds.) Debris-flow Hazards and Related Phenomena. Springer, Heidelberg (2005). https://doi.org/10.1007/3-540-271 29-5_9

Locat, J., Mienert, J.: Submarine Mass Movements and Their Consequences: 1st International Symposium. Springer, Dordrecht (2012)

Nian, T.K., Guo, X.S., Zheng, D.F., Xiu, Z.X., Jiang, Z.B.: Susceptibility assessment of regional submarine landslides triggered by seismic actions. Appl. Ocean Res. **93**, 101964 (2019)

Prior, D.B., Suhayda, J.N.: Application of infinite slope analysis to subaqueous sediment instability, Mississippi delta. Eng. Geol. **14**(1), 1–10 (1979)

Roesner, A., et al.: Impact of seismicity on nice slope stability—Ligurian Basin, SE France: a geotechnical revisit. Landslides **16**, 23–35 (2019). https://doi.org/10.1007/s10346-018-1060-7

RP2A-WSD API: Recommended Practice for Planning, Designing and Constructing Fixed Offshore Platforms–Working Stress Design. American Petroleum Institute, Houston (2000)

Wang, R., Cao, W., Xue, L., Zhang, J.M.: An anisotropic plasticity model incorporating fabric evolution for monotonic and cyclic behavior of sand. Acta Geotechnica **16**, 43–65 (2021a). https://doi.org/10.1007/s11440-020-00984-y

Wang, R., Zhang, J.M., Wang, G.: A unified plasticity model for large post-liquefaction shear deformation of sand. Comput. Geotech. **59**, 54–66 (2014)

Wang, Y., Wang, R., Zhang, J.M.: Large-scale seismic seafloor stability analysis in the South China Sea. Ocean Eng. **135**, 109334 (2021b)

Xue, L., Yu, J.K., Pan, J.H., Wang, R., Zhang, J.M.: Three-dimensional anisotropic plasticity model for sand subjected to principal stress value change and axes rotation. Int. J. Numer. Anal. Meth. Geomech. **45**(3), 353–381 (2021)

Zhang, J.M., Wang, G.: Large post-liquefaction deformation of sand, part I: physical mechanism, constitutive description and numerical algorithm. Acta Geotechnica **7**(2), 69–113 (2012). https://doi.org/10.1007/s11440-011-0150-7

Seismic Inspection of Existing Structures Based on the Amount of Their Deformation Due to Liquefaction

Susumu Yasuda$^{(\boxtimes)}$

Tokyo Denki University, Hatoyama, Saitama, Japan
yasuda@g.dendai.ac.jp

Abstract. If structures such as houses, or structures with very long lengths, such as river dikes, railway embankments, and sewers, have not been constructed with liquefaction in mind, it is necessary to extract repair their parts that could be easily damaged in the event of a future earthquake. In inspecting these structures, it is necessary to quantitatively estimate the points where countermeasures are required based on the amount of settlement or the amount of uplift of the structures. For that purpose, it is necessary to properly consider the process of settlement or uplift of the structures due to liquefaction based on past cases of damage and the results of model tests, and develop a method to easily estimate the amount of settlement or uplift. In Japan, there are many cases of structural damage due to liquefaction, and many model tests using shaking tables have been conducted. Simple estimation methods have been developed based on these and are being used for inspections. Furthermore, measures are being taken based on the inspection results.

Keywords: Liquefaction · Seismic inspection · River dike · Low-rise house · Sewer manhole

1 Introduction

During the last half century, methods for predicting the occurrence of liquefaction and countermeasures against damage due to liquefaction have been vigorously developed. These methods and countermeasures are now being considered when constructing large, new structures. However, they have not been applied to many previously constructed structures, and they should be to prevent damage stemming from future earthquakes. In these cases, the damage likely to be caused by liquefaction must be estimated. For important structures, such as bridges, mid-rise buildings, and tanks, detailed examinations, such as seismic response analysis, can be conducted to estimate the degree of damage. However, it is impossible to examine in detail all structures, such as houses in big cities, or very long structure, such as river dikes, railway embankments, and sewers. Therefore, simple methods are needed to estimate the damage to these structures and to extract repair parts of long structures that are likely to suffer damage.

As will be described later, the degree of actual damage to a house can be quantitatively evaluated by its penetration settlement and inclination angle. The damage to river dikes and railway embankments can be evaluated by the amount of their subsidence. Damage to sewers can be evaluated by the amount of manhole uplift. A simple method for estimating these values can be derived by setting parameters related to damage and creating empirical formulae based on actual damage and model experiment results. This method can be easily created as long as there is data on past damage. However, the processes of the settlement and uplift of structures due to liquefaction are complicated, as described later, so it is important to derive simple methods that properly consider these processes.

These processes are clarified by examining the damage caused by past earthquakes in detail and by conducting model tests. In Japan, earthquakes that cause liquefaction have occurred once every one to two years, and many model tests using shaking tables have been conducted. Based on the damage caused by these earthquakes and on model test results, simple methods for estimating the amount of settlement or uplift have been developed and are being used for inspections. Furthermore, measures to prevent damage are being taken based on the inspection results. The current status of such efforts in Japan is introduced below.

2 Embankments

2.1 Earthquake Damage Cases Where the Amount of Settlement Due to Liquefaction Was Measured

There are various types of embankments, such as earth dams, tailings dams, river dikes, road embankments, railway embankments, and filled housing lots, in Japan. These are damaged by sliding and/or subsidence due to liquefaction of their foundation ground and of the embankments. If it is estimated that a large slip collapse is likely to occur due to liquefaction, countermeasures must be taken, even for existing structures. On the other hand, if the amount of estimated subsidence is small enough that the structure could maintain its function, countermeasure are not necessary. For example, in the case of a river dike, it is a serious problem that the river water overflows into adjacent areas when the dike subsides, as shown in Fig. 1 (1). However, even if subsidence occurs, countermeasures are not necessary if the top of a river dike is higher than any probable river water level. In the case of railway and road embankments, as long as subsidence does not prevent trains or cars from crossing the embankment, countermeasures are not needed. Figure 1 (2) diagrams a railway embankment that is dysfunctional because it has subsided so much that a train cannot cross a bridge girder. Therefore, for long embankments, it is necessary to easily estimate the distribution of subsidence due to liquefaction during future earthquakes and to determine the priority of the locations where countermeasures will be taken.

About 70% of Japan's land area is mountainous, and the country is surrounded by the sea. Furthermore, the annual rainfall is about 1,700 mm, which is about twice as high as the world average, so there are more than 30,000 rivers. Furthermore, due to substantial seismic activity, some river dikes have been damaged every time an earthquake occurs. These damaged river dikes were usually restored just by filling the dikes with soil again, and earthquake countermeasures were hardly taken until the 1995 Hyogoken-nambu (Kobe) Earthquake. The 1995 Kobe Earthquake caused the Yodo River dike to subside significantly, as shown in Photo 1, and the large city of Osaka was about to be flooded by river water. Therefore, since then, many river dikes have been subjected to seismic inspection and seismic retrofitting. Nonetheless, many river dikes were damaged by liquefaction during the 2011 Great East Japan (Tohoku) Earthquake. In Japan, river dikes are managed separately by the national and local governments, but the rivers managed by the national government alone damaged 939 dikes in the Kanto region and 1,195 in the Tohoku region, for a total of about 2,000 dikes. Figure 2 shows the rivers in the Kanto region where dikes were damaged. Of these dikes, 55 were heavily damaged, of which, 51 were damaged by liquefaction. As shown in Fig. 3, there were three patterns of dike liquefaction: liquefaction of the foundation ground, liquefaction of the dike, and liquefaction of both the foundation ground and the dike, and these patterns occurred in about 70%, 10%, and 20%, respectively, of the damaged dikes.

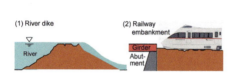

Fig. 1. Dysfunctional embankments.

Photo 1. Damaged dike.

Fig. 2. Damaged rivers by the Tohoku Earthquake.

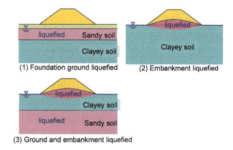

Fig. 3. Three patterns of liquefaction.

Fewer road and railway embankments have been damaged by earthquake liquefaction than the number of damaged river embankments. Nevertheless, road and railway embankments were damaged by earthquakes such as the 1983 Nihonkai-Chubu Earthquake and the Tohoku Earthquake.

2.2 A Large Shaking Table Test to Investigate the Time History of Excess Pore Water Pressure in an Embankment Due to Liquefaction

At the Public Works Research Institute of the Ministry of Construction in Japan, a shaking table test was conducted on the effects of liquefaction on dikes and semi-buried roads using a large soil container with a width of 8 m (Taniguchi et al. [1]). Figure 4 shows the cross section of the dike. The model ground was 8 m wide, and the model dike was 50 cm high. Sand with a mean particle size of $D_{50} = 0.28$ mm and a uniformity coefficient of $U_C = 2.91$ was used for the sand of the ground and dike. The ground and dike were constructed with relative densities of $Dr = 59\%$ and 53%, respectively. Seventeen accelerometers and 40 pore water pressure transducers were installed in the ground. Settlement of the dike was also measured. The model ground was shaken at a sine wave with a frequency of 2 Hz and at an acceleration of 231 cm/s^2. Figure 5 shows the time histories of excess pore water pressure measured by pore water pressure transducers in the ground. The vertical broken line in each time history indicates the time when the shaking is over. If the inside of the ground is roughly divided into four zones: the deepest A zone, the intermediate depth B zone under the dike, the shallowest C zone under the dike, and the D zone outside the dike, the time history of excess pore water pressure was different in each zone. In Zone B, the pore water pressure increased with the start of shaking and decreased with the end of shaking. In Zone A, the pore water pressure gradually increased with the start of shaking and rapidly decreased with the end of the shaking. In the C zone, the pore water pressure increased at the start of shaking as in the A zone. However, after a large drop in the middle of shaking, it soon returned to its original state, then, it decreased with the end of shaking. In the D zone, the pore water pressure increased considerably during the shaking, and after the shaking was over, the pore water pressure increased for a few more seconds and then began to decrease.

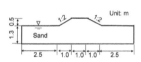

Fig. 4. Dike model [1].

Fig. 5. Time histories of excess pore water pressure.

Figure 6 shows the distribution of excess pore water pressure 6 s after the start of shaking. High excess pore water pressure was generated in the B zone and the C zone. Figure 7 shows the time history of subsidence at the crest of the dike. It gradually subsided from the start of shaking, and subsided to about 15 cm. Judging from this, together with Figs. 5 and 6, it is considered that the pore water pressure increased and decreased and the dike subsided according to the following process from the start of shaking.

(1) Pore water pressure increased in the B, C, and D zones with the start of shaking.
(2) As the pore water pressure increased, the dike began to slip and subsidence began to occur. In the C zone positive dilatancy of the soil occurred due to the large deformation of the soil and the pore water pressure decreased once. After that, the pore water pressure increased due to the propagation of excess pore water pressure from Zone B.
(3) Since the effective confining pressure in the D zone was smaller than that in the B and C zones, the excess pore water pressure due to shaking rose less in the D zone than in the B and C zones. However, since the excess pore water pressure from the B and C zones propagated, the pore water pressure in the D zone increased even after the shaking was ended.
(4) Zone A was the deepest part of the ground, so it was slightly difficult to liquefy, and pore water pressure was generated later than in the other zones. When the shaking

was ended, the excess pore water pressure dissipated toward the upper part, so that the pressure decreased rapidly.

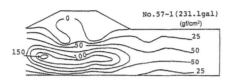

Fig. 6. Distribution of Excess pore water pressure.

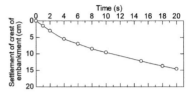

Fig. 7. Time history of subsidence the dike.

2.3   Development of a Simple Method for Estimating the Amount of Subsidence of River Dikes

The amount of subsidence of river dikes due to liquefaction is thought to be affected by several factors, such as the height of the dike, the type and density of the dike soil and the surrounding ground, the groundwater level, and the seismic motion. However, if the height of the dike and the soil type of the ground are relatively constant in a certain area, it is considered that the thickness of the liquefied layer and the intensity of liquefaction greatly affect the amount of subsidence. Then the relationship between the liquefaction potential index P_L and the amount of subsidence was investigated for the dikes of three large rivers flowing near Nagoya City, Japan that were damaged during the 1944 Tohnankai Earthquake, as shown in Fig. 8 (Nakamura et al. [2]). Although the plotted points are a little scattered, the amount of subsidence increased as P_L increased, so it could also be used to estimate subsidence in future earthquakes. The liquefaction potential index P_L is calculated by the following formula based on the safety factor for liquefaction F_L, and is generally used to roughly estimate the magnitude of damage to a structure due to liquefaction (Iwasaki et al. [3])

$$P_L = \int_0^{20} (1 - F_L)(10 - 0.5z)dz$$

$$(1 - F_L) = 0 \quad \text{(for } F_L > 1) \tag{1}$$

On the other hand, in analyzing the stability of the dike slope, the safety factor against sliding F_S can be expressed by the following equation, considering the horizontal seismic force k_h and the excess pore water pressure Δu.

$$F_{sd} = \frac{\sum \{c \cdot l + \{(W - u_0 \cdot b - \Delta u \cdot b) \times \cos\alpha - k_h \cdot W \cdot \sin\alpha\} \cdot \tan\phi\}}{\sum (W \cdot \sin\alpha + k_k \cdot W \cdot y/r)} \tag{2}$$

If F_S is small, the amount of subsidence at the crest of the dike may be large. As mentioned above, the time history of excess pore water pressure in the ground is complicated, and it is necessary to obtain the distribution of excess pore water pressure in the cross section at a certain point in time and consider it in the calculation. Since this is time-consuming, it is common practice to simplify and estimate the distribution of excess pore water pressure from the distribution of F_L at the peak of cyclic shear stress, ignoring the time history of excess pore water pressure. After the 1995 Kobe Earthquake, using this simplified method, the safety factor F_S for slips was calculated for 27 river dikes damaged during six past earthquakes and compared with the settlement rate of the dikes (settlement/height of dike), as shown in Fig. 9 (Ministry of Construction [4], Matsuo [5]). However, in Eq. (2), both the safety factor F_{sd} (Δu) when only the excess pore water pressure is considered and the safety factor F_{sd} (k_h) when only the horizontal seismic intensity is considered are calculated, and the smaller one is shown in this figure. Although the plotted points are scattered, the settlement rate increased as the safety factor decreased. Therefore, based on the upper limit shown in the figure, damages to river dikes due to level 1 seismic motion (medium-scale seismic motion) have been inspected, and countermeasures have been applied.

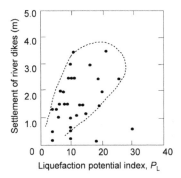 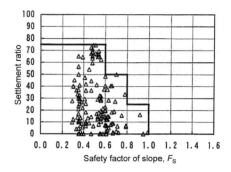

Fig. 8. Relationship between P_L and S [2].   **Fig. 9.** Relationship between F_S and S/H [4].

After that, it became necessary to investigate the seismic resistance of the river dikes against Level 2 seismic motion (large-scale seismic motion), design countermeasures, if necessary, and implement the countermeasures (Ishihara and Sasaki [6]). In order to estimate the amount of subsidence of the dike here, instead of the indirect method of estimating from F_S, methods of directly analyzing the deformation of the dike have been adopted. The optimal analysis method is two-dimensional seismic response analysis that can consider liquefaction. However, since this requires detailed soil tests and costly analysis, the applicability of a simpler static analysis method was examined. Then, ALID was taken up as a static analysis method, and it was used to design countermeasures that were included in a guidance in 2016 (Public Works Research Institute [7]). ALID is a residual deformation analysis method using the static finite element method. For details, see the references (Yasuda et al. [8]). Figure 10 shows the results of the reproduction analysis of the Yodo River dike in Photo 1. Figure 11 shows the results of the 27 river dikes damaged by the past six earthquakes, analyzed by ALID and compared with the actual subsidence (Wakinaka et al. [9]). The amount of subsidence can be reproduced well even though ALID is a static analysis method. As shown in the schematic diagram in Fig. 12, the guidance covers the compaction method, the solidification method, the method using steel sheet piles, and the drain method that lowers the groundwater level inside the dike.

These methods were designed by analysis using ALID. Currently, measures are being taken according to this guide. When using the compaction or solidification methods, improving the ground under the vicinity of the toe of an dike, as shown in Fig. 12, will increase F_S, though F_S does not increase if only the area directly under the crest of the dike is improved. However, according to the analysis using ALID, the amount of subsidence will be dramatically reduced if only the area directly under the dike is improved. Therefore, in addition to improving the ground under the toes of dikes, construction work is being carried out to improve the ground under the crests of dikes.

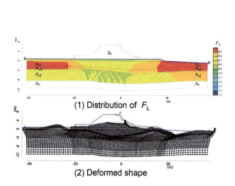

(1) Distribution of F_L

(2) Deformed shape

Fig. 10. Analyzed results of Yodo River dike.

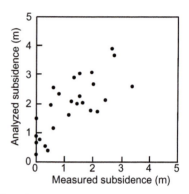

Fig. 11. Measured and analyzed subsidence [9].

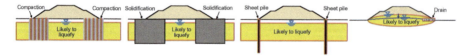

Fig. 12. Countermeasures against the subsidence of river dikes introduced in the guidance.

2.4 Development of a Simple Method for Estimating the Amount of Subsidence of Railway Embankments

Railway embankments are compacted well while being constructed, so excessive pore water pressure is unlikely to occur in these embankments during an earthquake. Therefore, it is considered that the shear rigidity of these embankments does not decrease during an earthquake, and that a decrease in the shear rigidity of the ground and the weight of the embankment cause subsidence. Then, a method for estimating the amount of subsidence was developed based on the results of several previous shaking table tests (Sawada et al. [10]). Factors that affect the subsidence of a railway embankment include the characteristics of the embankment, the characteristics of the ground beneath and around the embankment, and the seismic motion, but experimental results suggest that the characteristics of the embankment have little effect. On the other hand, since it was clarified that there is a relationship between the thickness of the liquefaction layer and the layer's excess pore water pressure and amount of subsidence, the liquefaction potential index P_L was used as a representative index. If P_L is used, the influence of seismic motion can also be taken into consideration. Figure 13 (1) shows the relationship between P_L and the settlement ratio when the relative density is less than 60% and when it is 60% or more. The smaller the relative density, the larger the settlement. In Fig. 13 (2), the data on soil with a relative density of less than 60% are further divided by the number of shakings. The amount of subsidence increased as the number of shakings increased. These figures have been introduced into railway construction standards and are used to estimate the amount of embankment subsidence due to liquefaction (Railway Technical Research Institute [11]). Naganawa et al. [12] studied railway, roads and river embankments that were damaged due to liquefaction during eight past earthquakes and investigated the relationship between the P_L and the settlement ratio of these embankments. Their findings supported the relationships shown in Fig. 13.

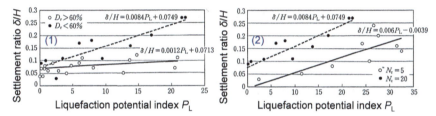

Fig. 13. Relationship between P_L and δ/H introduced into railway design standards [10, 11].

3 Low-Rise Housing

3.1 Earthquake Damage Cases in Which the Amount of Settlement and Inclination Angle of Buildings Due to Liquefaction Were Measured

When the ground liquefies, it softens, so low-rise houses and mid-rise buildings settle into the ground under their own weight, and in some cases tilt, as shown in Fig. 14. The ground also subsides due to volume shrinkage as the excess pore water pressure disappears after liquefaction occurs. Therefore, it is necessary to distinguish between i) the absolute settlement of a building, ii) the penetration settlement of the building into the ground, and iii) the subsidence of the ground. If the thickness of the layer liquefied in a certain area is about the same in plane, the ground subsides uniformly in that area, so the building does not incline. Therefore, the amount of land subsidence does not damage the building much, but the amount of penetration settlement greatly affects damage.

Liquefaction has caused much damage to mid-rise buildings all over the world. The settlement and tilt angles of many buildings were measured in Niigata, Japan in 1964, in Dagupan, the Philippines in 1990, and in Adapazarı, Turkey in 1999 (Yasuda, et al. [13]). Since the settlement was measured locally as the difference from the surrounding ground surface, the amount of settlement corresponds to the amount of penetration settlement. In the 1964 Niigata earthquake, 340 reinforced concrete buildings were damaged by liquefaction. Figure 15 shows the frequency distribution of the average settlement of buildings by taking out only the buildings that are spread foundations and whose settlement was measured. The maximum settlement was about 250 cm, and many buildings settled more than 100 cm. Figure 16 shows the relationship between the settlement and the inclination. Though the data are scattered, settlement increased as the inclination angle increased. Figure 16 plots a similar relationship for Dagupan and Adapazarı, but in the case of Adapazarı, settlement was smaller than in the other two cities. The cause of this is not clear, but the liquefied soil in Adapazarı contained many fine particles, and it may have been difficult to settle.

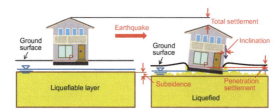

Fig. 14. Diagram of liquefaction-induced settlement of a house and the ground.

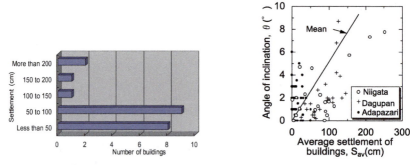

Fig. 15. Settlement of buildings in Niigata City [13].   **Fig. 16.** Inclination of buildings [13].

The liquefaction of ground beneath low-rise buildings with one or two floors has caused substantial damage all over the world. For example, in February 2011, such houses in Christchurch, New Zealand suffered a lot of damage, and in March 2011, the Tohoku Earthquake damaged about 27,000 houses. Low-rise houses suffered less subsidence and inclination due to liquefaction than mid-rise buildings and did not appear to be greatly damaged, so damage to low-rise houses has not received much attention in Japan. However, during the 2000 Tottori-ken Seibu Earthquake, it became clear that even a slight tilt would cause health problems, such as dizziness and nausea. Therefore, two months after the Tohoku Earthquake, the Cabinet Office provided new criteria, shown in Table 1, for judging the degree of damage from the amount of settlement and the angle of inclination. Consequently, the settlement and tilt angles of many houses in cities where liquefaction occurred were measured. Figures 17 (1) and 17 (2) show the relationship between the amount of penetration settlement and the inclination angle measured in four cities (Yasuda [14]).

Table 1. New criteria for judging the degree of damage to a house.

Grade of damage		Totally collapsed	Large-scale half collapsed	Half collapsed	Partially damaged
Evaluation method	Inclination	> 50/1000	16.7/1000 to 50/1000	10/1000 to 16.7/1000	<10/1000
	Settlement	1m above the floor	Up to floor	25cm to the top of footing	

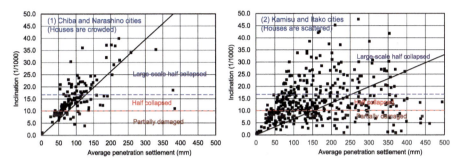

Fig. 17. Relationship between the penetration settlement and the inclination angle [14].

3.2 Shaking Table Tests to Investigate the Processes of Settlement and Tilting Due to Liquefaction

In order to investigate the process of settlement and inclination of low-rise houses due to liquefaction, many shaking table tests and analyses of disaster cases have been conducted. The timing of the start of settlement of buildings due to liquefaction during an earthquake was studied based on the testimony of eyewitnesses and on the results of a large shaking table experiment (Yasuda et al. [15]). According to Mr. Yuminamohi, who was at the airport terminal at the time of the 1964 Niigata Earthquake, about one and a half minutes after he felt the shaking, he heard a voice saying "The building is sinking!" He hurriedly started to take motion picture, and it was said that water began to spout from the side of the terminal building at Niigata Airport. In the shaking table test using a large soil container with a depth of 5 m, the ground liquefied after shaking for 30 s, and a concrete block of 1.86 tons in weight began to sink about 1 min after the shaking ended. Therefore, it appears that the penetration settlement of a light, low-rise house does not start during the main shaking motion, but occurs slowly after liquefaction.

Regarding the process of subsidence due to liquefaction, an experiment was conducted by placing a large laminar soil container with the cross section shown in Fig. 18 (1) on a large shaking table after the Tohoku Earthquake (Kaneko and Yasuda [16]). Figure 18 (2) and 18 (3) shows the time histories of excess pore water pressure for Case 2 of the experiment. Pore water pressure between the two house models and at GL-1.5m under the house model reached the maximum value in about 5 s, indicating liquefaction at these sites. On the other hand, at GL-0.3 m the house model did not generate excess pore water pressure until 5 s, and this pressure slowly increased to the maximum value after 32 s. It is considered that the increase after 5 s was due to the propagation of excess pore water pressure from the lower layer. Figure 19 shows the time history of penetration settlement with photos at two points of time for Case 2. Based on case histories from past earthquakes and on model tests, the authors concluded that a structure such as a building or a house does not sink into a hole that was produced by spewing water, but penetrates the ground due to a decrease in the shear modulus of the surface layer following the outside lateral flow of the ground under the house and the heaving of ground surrounding the house, as schematically shown in Fig. 20. The ground surface settled slowly because the liquefied layer under and around the house densified gradually due the spewing of the pore water.

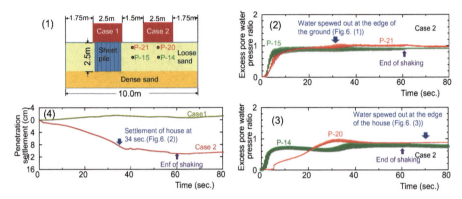

Fig. 18. Time histories of pore water pressure ratio and penetration settlement [16].

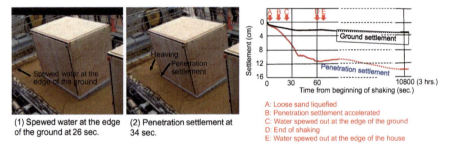

Fig. 19. Process of liquefaction, settlement and water spouting.

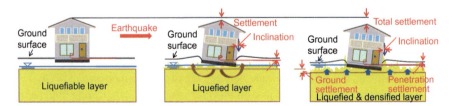

Fig. 20. Diagram of the mechanism of liquefaction-induced settlement of a house and the ground.

Regarding the process of tilting of houses in a densely populated urban area, there was a tendency to tilt inward when two houses were close to each other, as shown in

Photo 2, taken after the 2000 Tottoriken-seibu Earthquake (Yasuda and Ishikawa [17]). However, when two houses were farther apart, they tended to tilt away from each other, as shown in Fig. 21. Figure 22 schematically illustrates these tendencies. These tendencies were also obtained by analysis. Figure 23 shows an example of analysis by ALID. The direction of inclination and the angle of inclination differed depending on the distance between the two houses. In addition, the analysis show that the groundwater level has a large effect on the inclination angle.

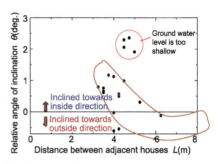

Photo 2. Adjacent houses tilted towards inside.

Fig. 21. Relationship between L and θ [17].

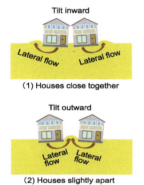

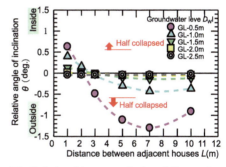

Fig. 22. Mechanism of tilting.

Fig. 23. Relationship among L, θ, and D_W analyzed by ALID [17].

3.3 Simple Method to Estimate the Penetration Settlement and Inclination Angle of Low-Rise Houses

In Japan, although many low-rise houses were severely damaged by the Tohoku Earthquake, measures to prevent liquefaction damage are not being applied for new houses, for three main reasons: i) such measures are not required by law for low-rise houses, ii) the possibility of housing damage due to liquefaction is not fully recognized by residents and house makers, and iii) the probable damage to houses cannot be estimated concretely from the liquefaction hazard maps currently being created. Therefore, as stated in Yasuda [18], the Ministry of Land, Infrastructure, Transport and Tourism issued a "Liquefaction

Hazard Map Guide for Risk Communication" in order to promote the adoption of countermeasures. This guide suggests the following procedure to easily estimate the probable amount of penetration settlement and the angle of inclination of a house:

(1) The local government creates a liquefaction hazard map specialized for low-rise houses.
(2) Residents and house makers look at the hazard map, and if they are in a gray zone, conduct a simple soil investigation, such as screw weight sounding, to estimate the depth distribution of F_L.
(3) The house maker calculates the amount of penetration settlement of the house based on Stainbrenner's method (Fig. 24), which calculates the settlement due to the weight of the house based on the elasticity theory. The shear rigidity of the liquefied layer is set according to F_L using Fig. 25 (1) used in ALID. However, while ALID assumes that the stress-strain relationship after liquefaction is bilinear, as shown in Fig. 25 (2), the elasticity theory can only consider a straight-line relationship with a slope of G_1. Therefore, it was set that the decrease in shear rigidity due to liquefaction should be up to 1/300, so the amount of settlement does not become excessive.
(4) Estimate the inclination angle from the amount of settlement based on Fig. 17, depending on the density of the houses.
(5) Residents and house makers will judge the degree of likely damage based on Table 1 and take measures, if necessary.

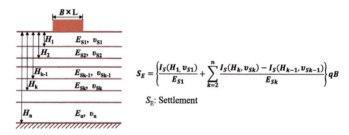

Fig. 24. Stainbrenner's method to calculate the amount of penetration settlement.

4   Sewer Manholes and Pipes

4.1   Earthquake Damage Cases in Which the Amount of Uplift of Sewer Manholes and Pipes Buried Pipelines Due to Liquefaction Was Measured

Buried pipelines include water, sewer, gas, and telecommunications pipelines. Many sewer manholes and pipes are buried deeper than the groundwater level, and because the apparent unit volume weight of these pipes is lighter than the apparent unit volume weight of other buried pipes, they tend to float due to liquefaction. Liquefaction can also cause buried pipes to bend and to disconnect at their joints, but the following will focus on damage caused by the uplift of sewer manholes and pipes.

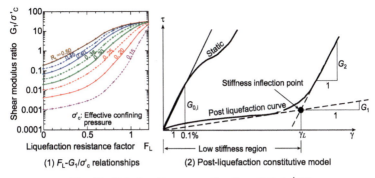

Fig. 25. Relationship among F_L, R_L and G_1/σ'_c [8].

In Japan, the 1964 Niigata Earthquake and the 1983 Nihonkai-Chubu Earthquake caused uplift damage to buried water and gasoline tanks on liquefied ground. On the other hand, in peat ground that was not liquefied by the 1993 Kushiro-oki Earthquake, sewer manholes and sewer pipes were uplifted, as shown in Photo 3, which surprised Japanese researchers. When the damaged site was excavated and investigated, as shown in Photo 4, it became clear that the sand backfilled after excavating and installing manholes and pipes, as shown in Fig. 26, had liquefied. In the 2003 Tokachi-oki Earthquake, many manholes uplifted on peat ground. Of these, in Onbetsu Town, as shown in Fig. 27, the extent of manhole raising varied greatly, depending on the location, and while manholes along the B-B' survey line surfaced more than 60 cm, those along the A-A' survey line did not surface at all. When the cause was investigated, the former was clayey ground and the latter was sandy ground. The 2014 Niigata-ken-chuetsu Earthquake caused the uplift of about 1,400 manholes. Based on these cases, the authors consider the influence of the surrounding ground on the amount of rising, as shown in Fig. 28 (Yasuda and Kiku [19]). However, while the 2011 Tohoku Earthquake liquefied soil over a wide area in cities along Tokyo Bay, the manholes in this area did not rise much. Instead, as shown in Fig. 29 (1), prefabricated manholes were displaced horizontally, pipe joints were disconnected, and liquefied sand entered the pipes, causing severe damage. The author thinks that this is because the surrounding ground was also widely liquefied and the duration of the earthquake motion was long, so the liquefied ground continued to swing for 1 to 2 min (Yasuda et al. [20]). However, in places where the surrounding ground with cohesive soil was not liquefied, as shown in Fig. 29 (2), liquefaction of the backfill soil caused uplifting damage.

As shown in Photo 5, the Niigata-ken-chuetsu Earthquake caused damage, such as a car hitting a floating manhole. Therefore, the impact of a rise of sewer manholes on the lives of residents is not only that sewage water cannot flow after the earthquake, but also that the manholes become an obstacle to traffic. They can be a particularly serious obstacle when residents have to evacuate in an emergency due to a fire or tsunami caused by an earthquake, or when emergency vehicles have to pass through. For example, when a hearing was held with the staff of the fire department, they answered that they can't pass if it rises from the road surface by about 10 cm to 20 cm as shown in Fig. 30.

Photo 3. Uplifted sewer manhole in Kushiro Town.

Photo 4. Investigation work.

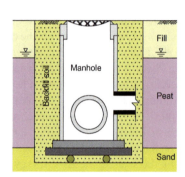

Fig. 26. Diagram of the cross section.

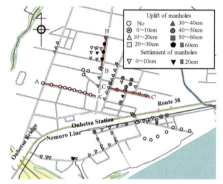

Fig. 27. Uplift of manholes in Onbetsu Town [19].

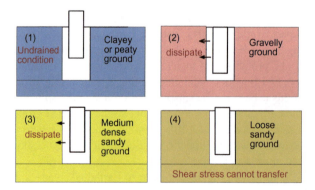

Fig. 28. Proposed relationship between uplift of a manhole and soil condition of ground [19].

4.2 Shaking Table Tests to Investigate the Process of Uplift of Manholes and Buried Pipes Due to Liquefaction

After the Kushiro-oki Earthquake, the author and his colleagues conducted tests in which a small soil container, described schematically in Fig. 31, was placed on a shaking table and vibrated, in order to investigate the process of uplift of manholes and buried pipes

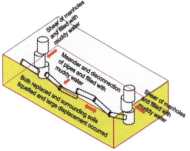 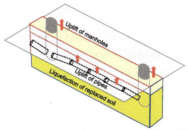

(1) Damage due to the swing of liquefied ground   (2) Damage due to the liquefaction of backfill soil

Fig. 29. Patterns of damage to manholes and pipes caused by the 2011 Tohoku Earthquake [20].

 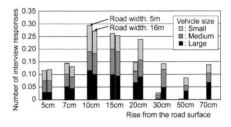

Photo 5. Car that collided with a manhole.   **Fig. 30.** Levitation that a fire engine can pass
through.

[Yasuda and Kiku [19]]. Toyoura sand was used as the soil for the test, and pipes and manhole models were installed in the ground. There were two types of model ground: i) homogeneous ground with a relative density of about 30%, and ii) ground with only the backfill soil having a relative density of about 30% and the surrounding ground having a relative density of 90%, as shown in Fig. 31. In the latter case, the backfill soil was wrapped in a vinyl sheet so that the excess pore water pressure generated in the backfill soil would not propagate to the surrounding non-liquefied ground. Then, tests were conducted under many conditions by changing the width of the backfill soil, the groundwater level, the pipe diameter, and the thickness from the bottom of the manhole to the bottom of the backfill soil.

Since the front of this soil container was made of glass, "udon (Japanese soft noodle) chips" were installed at constant intervals between the glass and the ground. Then, the movement of soil particles was estimated from the displacement of the "udon chips". The soil container was shaken with a large force of 3 Hz and 500 cm/s^2 so that liquefaction would occur in a few seconds. Figure 32 shows the time histories of the amount of rise when a pipe with a diameter of 11.4 cm was used and the width of the backfill soil was 25 cm, 30 cm, and 70 cm. Looking at the test results when the width was 70 cm, the pipe first rises a little with shaking (Stage 1), then the speed of uplifting slows down (Stage 2), and then it rises to the ground surface at once (Stage 3). Figure 33 shows the displacement amount of the udon chips up to the time points of Stage 2 and Stage 3. From these, it was considered that the uplift occurred according to the following process:

(1) First, the ground around the pipe was liquefied, and buoyancy was generated in the pipe, which floated about 5 cm at a rapid speed of about 4 cm/s. The uplift force of the pipe pushed the ground above the pipe upward, raising the ground surface a little. (Stage 1)
(2) When the soil on the pipe was pushed upward, it was sheared and deformed to generate a negative water pressure, and the shear rigidity of the liquefied soil was slightly restored. As a result, the uplift was temporarily stopped and the speed slowed down. (Stage 2)
(3) After that, excess pore water pressure from the surrounding soil propagated to the upper part of the pipe, and the soil above the pipe also returned to a liquefied state. For this reason, the soil on the upper part and sides of the pipe wrapped around the lower part of the pipe and floated it up at a speed of about 2 cm/s. (Stage 3)

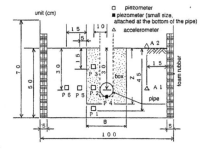

Fig. 31. Soil container used for tests [21].

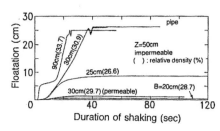

Fig. 32. Time histories of the amount of rise of a pipe.

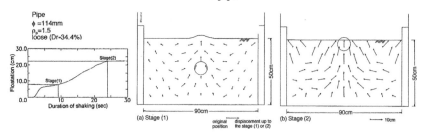

Fig. 33. Movement of "udon chips" up to the time points of Stage 2 and Stage 3.

When the width of the backfill soil was 30 cm, the speed of rising from the beginning was as slow as about 1 cm/s, and the pipe floated monotonously to the ground surface only in Stage 3 without Stages 1 and 2. Also, when the width became 25 cm, the pipe only floated about 8 cm and did not rise to the ground surface. Furthermore, in the test in which the vinyl sheet between the backfill and the surrounding dense ground was removed ("permeable data" in the figure), even if the width was 30 cm, the pipe hardly floated. It is considered that this was because the excess pore water pressure generated in the backfill soil was dissipated to the surrounding ground, as shown in Fig. 28 (3).

Figure 34 shows the time histories of the rise when the width of the backfill soil was 20 cm, 25 cm, and 30 cm using a model manhole. In each case, the manhole gradually floated at an almost constant speed of 0.5 cm/s to 2 cm/s. The narrower the backfill soil, the slower the rise speed. The reason why there was no Stage 1 and 2 and only Stage 3 was because there was no soil on the manhole. The movement of the "udon chips" is shown in Fig. 35. The manhole floated as the surrounding soil wrapped under it. In addition, the size of pipes and manholes, soil density, backfill width, density of surrounding ground, degree of liquefaction, duration, etc. had an effect on the amount and speed of uplift of pipes and manholes.

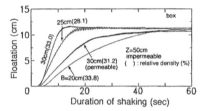

Fig. 34. Time histories of the rise for a manhole.

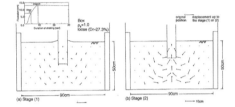

Fig. 35. Movement of "udon chips."

4.3 Simple Methods for Estimating the Amount of Uplift for Sewer Manholes

The safety factor against uplift due to liquefaction F_S can be calculated by the balance between the weight of the structure, the frictional force with the ground, and the floating force. Figure 36 shows how to calculate F_S for sewer manholes and utility tunnels in Japan. This safety factor determines whether or not these structures float. However, the smaller this value, the smaller the amount of ascent may be. Koseki et al. [21] calculated the safety factor for manholes that uplifted due to the Kushiro-oki Earthquake and compared it with the observed uplift displacement. Figure 37 shows the result when the calculation is performed assuming that the backfill soil was liquefied. As can be seen in the figure, no clear relationship was found between the two.

In the case of sewer manholes, as described above, various factors affect the amount of uplift, and the connection of pipes to the manhole also reduces the amount of levitation. However, the maximum possible amount of uplift without these constraints can be calculated from the balance between buoyancy and gravity. For manholes that were raised in Toyokoro Town during the 2003 Tokachi-oki Earthquake, the maximum amount of uplift was calculated from the manhole installation depth, groundwater level, unit volume weight of the ground, and the manhole's own weight and compared with their actual uplift displacements, as shown in Fig. 38. As can be seen in the figure, the amount of uplift that occurred was less than 1/4 of the calculated maximum amount of uplift.

Using this relationship of 1/4, the amount of uplift of a manhole in the Kitasenju district of Tokyo during a future earthquake was estimated. This area is one of the most densely populated residential areas in Tokyo, and there is a risk of fire in the event of an

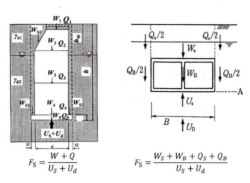

$$F_S = \frac{W+Q}{U_S + U_d} \qquad\qquad F_S = \frac{W_S + W_B + Q_S + Q_B}{U_S + U_d}$$

Fig. 36. Safety factor against uplift F_S in specifications for sewar manholes and utility tunnels.

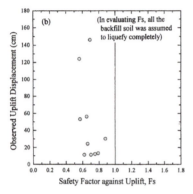

Fig. 37. Relationships between F_S and uplift (Koseki et al. [21]).

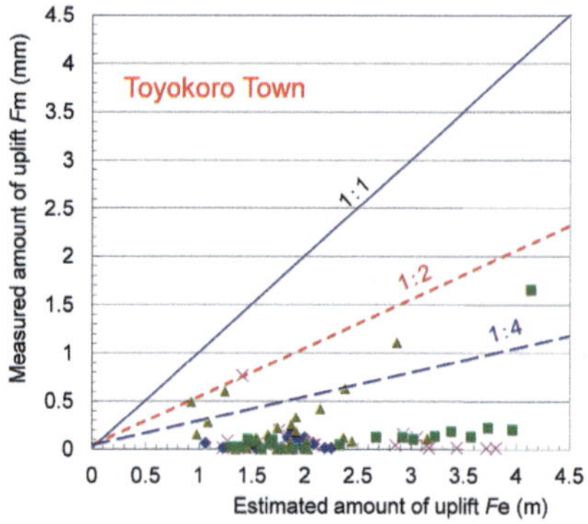

Fig. 38. Fe vs. Fm in Toyokoro.

earthquake, and if a manhole rises, there is a risk of hindering the evacuation of residents and the passage of fire engines. Figure 39 shows the estimation results. The estimated amount of uplift differs depending on the area due to the influence of the groundwater level and the burial depth, but there are some areas where large uplift of about 60 cm is likely. In this district, the author and his colleagues created 250 m mesh soil profile models and created a liquefaction hazard map using the liquefaction potential index P_L (Yasuda and Ishikawa [22]). On the map, the areas where the manholes shown in Fig. 39 were calculated to rise by 60 to 80 cm are shown again by blue circles in Fig. 40. Comparing the two, the places where manholes were projected to rise by 60 to 80 cm have $P_L < 5$, and conversely, little rise was projected for areas of $P_L > 15$. In other words, the results were not related to the possibility of liquefaction of the ground. This is because Fig. 39 assumes floating due to liquefaction of the backfill soil. Therefore, it should be noted that the liquefaction of backfill soil may cause manhole floating damage even in zones that are judged not to be liquefied in the hazard map. In Japan, including the Kitasenju area, measures are being taken to prevent manhole floating, such as placing weights on the manholes and dissipating excess pore water pressure near them, for manholes buried in major roads.

Fig. 39. Estimated uplift height of manholes.

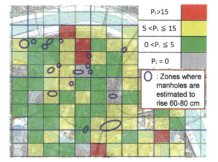

Fig. 40. Comparison with hazard map.

5   Concluding Remarks

The current situation in which inspections are being conducted in Japan for areas that require countermeasures against liquefaction, targeting river dikes, railway embankments, low-rise houses, and sewer manholes is introduced. The Ministry of Land, Infrastructure, Transport and Tourism is inspecting river dikes and taking measures. As for low-rise housing, an inspection guide has just been created, and since residents and house makers are not aware of the damage caused by liquefaction, it is necessary to promote the adoption of countermeasures. Measures against uplift of manholes are being taken for existing sewer manholes, but inspection methods still need to be developed.

References

1. Taniguchi, E., Morishita, T., Koga, Y., Yasuda, S., Umeuchi, K.: Shaking table tests on aseismicity of semi-buried structure near embankments. In: Proceedings of the 20th Japan National Conference on SMFE, pp. 687–690 (1985). (in Japanese)
2. Nakamura, Y., Murakami, Y.: A study on the seismic stability of dikes of three big rivers in Nobi Plain. In: Proceedings of the 34th Technical Conference of Ministry of Construction, pp. 96–104 (1980). (in Japanese)
3. Iwasaki, T., Tokida, K., Tatsuoka, F., Watanabe, S., Yasuda, S., Sato H.: Microzonation for soil liquefaction potential using simplified methods. In: Proceedings of the 3rd International Conference on Microzonation, vol. 3, pp. 1319–1330 (1982)
4. River Improvement and Management Division, River Bureau, Ministry of Construction: Manual for Seismic Inspection of Levee (1995). (in Japanese)
5. Matsuo, O.: Seismic design of river embankments. Tsuchi-to-kiso **47**(6), 9–12 (1999). (in Japanese)
6. Ishihara, M., Sasaki, T.: Approaches to improve design method of countermeasure for liquefaction of levee. Civ. Eng. J. Public Works Res. Cent. **60**(4), 20–23 (2018). (in Japanese)
7. Public Works Research Institute: Guidance of countermeasure for liquefaction of levee, Technical Note of PWRI, No. 4332 (2016). (in Japanese)
8. Yasuda, S., Yoshida, N., Adachi, K., Kiku, H., Ishikawa, K.: Simplified evaluation method of liquefaction-induced residual displacement. J. Jpn. Assoc. Earthq. Eng. **17**(6), 1–20 (2017)
9. Wakinaka, K., Ishihara, M., Sasaki, T.: Recurrence numerical analysis of liquefaction damage of river bank in consideration of effects such as age of ground. In: Proceedings of the 49th Japan National Conference on SMFE, pp. 1643–1644 (2014). (in Japanese)

10. Sawada, R., Tanamura, S., Nishimura, A., Koseki, J: Simplified method to subsidence of embankment on liquefiable ground. In: Proceedings of the 34th Japan National Conference on SMFE, pp. 2091–2092 (1999). (in Japanese)
11. Railway Technical Research Institute: The Design Standards for Railway Structures and Commentary (Seismic Design). Supervised by Ministry of Land, Infrastructure and Transportation, Maruzen, (2012). (in Japanese)
12. Naganawa, T. Achiwa, H., Machida, F., Morimoto, I., Tamamoto, H.: Examination on grasping the degree of liquefaction damage of soil structures (Part 2). In: Proceedings of the JSCE 2002 Annual Meeting, vol. 3, pp. 1033–1034 (2002). (in Japanese)
13. Yasuda, S., Irisawa, T., Kazami, K.: Liquefaction-induced settlements of buildings and damages in coastal areas during the Kocaeli and other earthquakes. In: Proceeding of the Satellite Conference on Lessons Learned from Recent Strong Earthquakes, 15th ICSMGE, pp. 33–42 (2001)
14. Yasuda S. New liquefaction countermeasures for wooden houses. In: Soil Liquefaction During Recent Large-Scale Earthquakes, pp. 167–179. CRC Press, Taylor & Francis Group, A Balkema Book (2014)
15. Yasuda, S., Ishikawa, K.: Study on the mechanism of the liquefaction-induced settlement of structures by case histories and model tests. In: Proceedings of 16th Asian Regional Conference on SMGE, Paper No. ATC3-008 (2019)
16. Kaneko, M., Yasuda, S.: Experimental research on a reduction method for liquefaction damage to house using thin sheet piles. In: Proceedings of the 2nd ECEES, pp. 147–154 (2014)
17. Yasuda, S., Ishikawa, K.: Appropriate measures to prevent liquefaction-induced inclination of existing houses. Soil Dyn. Earthq. Eng. **115**, 652–662 (2018)
18. Yasuda, S.: History of liquefaction hazard map development and a new method for creating hazard maps for low-rise houses, PBD-IV (2022, submitted)
19. Yasuda, S., Kiku, H.: Uplift of sewage manholes and pipes during the 2004 Niigataken-chuetsu earthquake. Soils Found. **46**(6), 885–894 (2006)
20. Yasuda, S., Suetomi, I., Ishikawa, K.: Effect of long duration of the main shock and a big aftershock on liquefaction-induced damage during the 2011 Great East Japan Earthquake. In: Ansal, A., Sakr, M. (eds.) Perspective on Earthquake Geotechnical Engineering, Geotechnical, Geological and Earthquake Engineering, vol. 37, pp. 343–364. Springer, Cham (2015). https://doi.org/10.1007/978-3-319-10786-8_13
21. Koseki, J., Matsuo, O., Ninomiya, Y.: Uplift of sewer manholes during the 1993 Kushiro-oki Earthquake. Soils Found. **37**(1), 109–122 (1987)
22. Yasuda, S., Ishikawa, K.: Liquefaction-induced damage to wooden houses in Hiroshima and Tokyo during future earthquakes. In: Proceedings of 16th ECEE, Paper No. 11014 (2018)

Invited Theme Lectures

Site Characterization for Site Response Analysis in Performance Based Approach

Atilla Ansal[1($\boxtimes$)] and Gökçe Tönük[2]

[1] Ozyegin University, Istanbul, Turkey
atilla.ansal@ozyegin.edu.tr
[2] MEF University, Istanbul, Turkey

Abstract. The local seismic hazard analysis would yield probabilistic uniform hazard acceleration response spectrum on the engineering bedrock outcrop. Thus, site-specific response analyses need to produce a probabilistic uniform hazard acceleration response spectrum on the ground surface. A possible performance based approach for this purpose requires a probabilistic estimation of soil stratification and engineering properties of encountered soil layers in the soil profile. The major uncertainties in site-specific response analysis arise from the variabilities of (a) local seismic hazard assessment, (b) selection and scaling of the hazard compatible input earthquake time histories, (c) soil stratification and engineering properties of encountered soil and rock layers, and (d) method of site response analysis. Even though the uncertainties related to first two items have primary importance on the outcome of the site-specific response analyses, the discussion in this article focuses on the observed variability and level of uncertainty in site conditions, related to soil stratification, thickness and type of encountered soil layers and their engineering properties, depth of ground water table and bedrock and properties of the engineering bedrock. Thus, one option may be conducting site response analyses for large number of soil profiles produced by Monte Carlo simulations for the investigated site to assess probabilistic performance based design acceleration spectra and acceleration time histories calculated on the ground surface based on 1D, 2D, or 3D site response analysis with respect to different performance levels.

Keywords: Soil stratification · Site response analysis · Uniform hazard acceleration spectrum · Performance based design

1 Introduction

Site characterization for geotechnical earthquake engineering applications involve many variables and related uncertainties [1, 7]. These uncertainties may include soil and rock properties, as well as measurement errors and statistical uncertainty in a very large domain. Therefore, it may be suitable to narrow the approach and focus on the site specific response analysis to evaluate the uncertainties to achieve a probabilistic methodology in defining a performance based approach [13, 17]. One of the controlling issues in the performance based design involves the definition of performance objectives. In the

L. Wang et al. (Eds.): PBD-IV 2022, GGEE 52, pp. 319–326, 2022.
https://doi.org/10.1007/978-3-031-11898-2_16

case of site response analysis, the performance objectives may be considered as the probabilistic definition of the uniform acceleration hazard spectrum and acceleration time histories calculated on the ground surface.

The general purpose of site response analysis is to calculate the uniform acceleration hazard spectrum and acceleration time histories on the ground surface based on the probabilistic uniform acceleration hazard spectrum on the rock outcrop calculated considering the local seismicity and source models for the region. For the calculated uniform hazard acceleration response spectrum on the ground surface, the major uncertainties are due to (a) local seismic hazard assessment, (b) selection and scaling of the hazard compatible input earthquake time histories, (c) soil stratification and corresponding engineering properties of encountered soil and rock layers, and (d) method of site response analysis.

The uncertainties related in the selection and scaling of the hazard compatible input earthquake time histories was reviewed previously [1]. The discussion in this manuscript will start with the uncertainties related to variabilities in site conditions and engineering properties of soil layers. For a probabilistic methodology, the additional challenge involves probabilistically acceptable definition of soil profiles. One option is to conduct site response analyses for large number of soil profiles generated using Monte Carlo Simulations to assess effects of variability with respect to different performance levels [3].

2 Probabilistic Evaluation of Site Response Analysis

Important factors controlling site response analyses are soil stratification with respect to soil types, layer thickness, shear wave velocity for each soil layer and the depth of engineering bedrock. Monte Carlo simulation scheme has been adopted to study the effect of variability of assigned shear wave velocities and layer thicknesses for all soil layers encountered in 209 soil borings in the Zeytinburnu microzonation project [4]. The effect of variability is studied by generating Monte Carlo simulations (MCS) for soil profiles assuming that initially measured shear wave velocities are mean values with the range of possible variation is ±20% of the mean and observed layer thickness are mean values with the range of possible variation is ±10% of the mean, thus 100 soil profiles were generated for each 209 soil profiles.

Total of 459,800 site response analysis for 100 Monte Carlo simulations for 209 soil profiles and for 22 acceleration records were conducted. Based on the observed discrete distribution approach, the 90% percentile value for each period level for acceleration spectra is calculated corresponding to 10% exceedance in 50 years or for 475 year return period and 98% percentile spectrum for corresponding to 2% exceedance in 50 years or for 2475 years return period as shown for two soil profiles in Fig. 1.

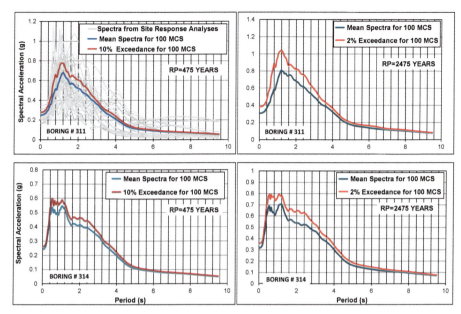

Fig. 1. 10% and 2% exceedance uniform hazard spectra on the ground surface based on Monte Carlo Simulations

3   Acceleration Time Histories for Performance Based Design

It appears possible to calculate uniform hazard acceleration spectra on the ground surface that can be adopted in the performance based design of engineering structures. However, in the case of dynamic response analysis using acceleration time histories for the super structures, the question is still remains open for the general performance based design. Thus, the one possibility may involve use of acceleration time histories calculated by site specific response analysis [8, 10, 16].

The state-of-practice in earthquake engineering design has moved toward the use of dynamic non-linear time history analysis. The critical issue in performance based analysis is the selection of acceleration of time histories for nonlinear time-history analysis of engineering structures. One option is to use recorded accelerations in real earthquakes scaled to match design code spectrum or uniform hazard spectra. However, when selecting time history records from the data banks, the effects of site conditions are totally neglected. One possible approach at this stage may be to adopt the calculated acceleration time histories on the ground surface for the nonlinear time-history analysis of engineering structures. In this case the reliability of the calculated acceleration time histories on the ground surface become an important issue.

In a parametric study conducted previously, acceleration spectra and acceleration time histories calculated based on the modified version of Shake91 [5] are compared with acceleration response spectra and acceleration time histories recorded by the Ataköy SM station in Istanbul during 1999 $M_w = 7.4$ Kocaeli Earthquake as shown in Fig. 2 and Fig. 3. The higher frequency motions are more dominant during the whole duration in comparison to the recorded time histories. In the light of the similarities between recorded and calculated acceleration response spectra (Fig. 2), and the similarity between recorded and calculated acceleration time histories, it is acceptable of the calculated acceleration time histories as an input to structural analysis.

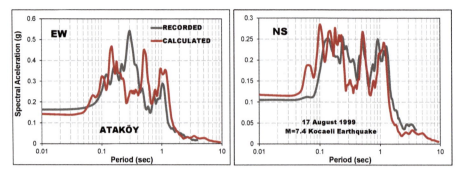

Fig. 2. Calculated and recorded acceleration spectra at Ataköy SM station for the 17/8/1999 Mw = 7.4 Kocaeli Earthquake

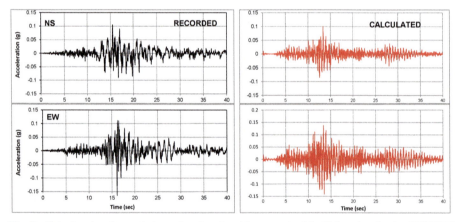

Fig. 3. Calculated and recorded acceleration time history at Ataköy SM station for the 17/8/1999 Mw = 7.4 Kocaeli Earthquake.

4 2D Site Effects

When considering site conditions, it is very clear that everything is 3D, the variation of soil stratification, and properties of soil layers, as well as 3D effects of the earthquake induced seismic source and seismic wave action. Therefore, one performance criteria for geotechnical earthquake engineering may need to be considered in comparison to 1D and 2D site characterization and response analysis [9, 12, 14].

The horizontal variations of the layer thickness and soil properties in 1D site response analysis cannot be modelled accurately. 2D site response analyses may be used to account horizontal variability to have a better estimate of the 2D earthquake characteristics on the ground surface. There have been significant number of studies reported in the literature that used different methodologies to evaluate 2D site analysis and site amplification. Within this framework, Kaklamanos et al. [11] evaluated the critical parameters affecting site response results. They observed that 1D site response in general may yield lower spectral accelerations excluding the short period range, however, within the range of maximum shear strain $\gamma_{max} \approx 0{:}1\%$ to 0.4%, the accuracy of equivalent-linear site response analysis may be realistic.

In a previous study conducted by the authors [2, 17] using QUAKE software [15], it was observed that topographic irregularity of the ground surface and the thickness of the soil layers, the difference between 1D and 2D may be significant. As shown in Fig. 4 and Fig. 5, the variation of soil stratification along the south-north profile in the European side of Istanbul indicate significant variations in the thickness of the soil layers, thus 1D and 2D results are not similar.

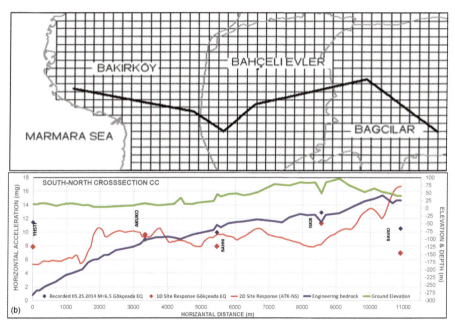

Fig. 4. 2D Soil profile (a) in map view (b) as soil cross-section for calculated PGA by 1D and 2D site response analysis at IRRN stations on the ground surface.

As shown in Fig. 5, The thickness of the soil profile from south to north increases continuously introducing a 2D effect that may not be modelled by 1D site response analysis that is also reflected in Fig. 6(b).

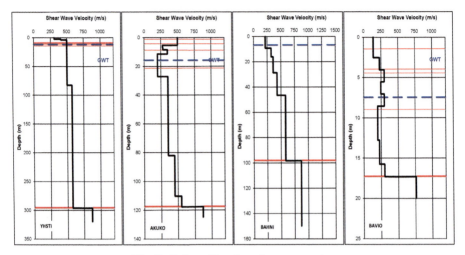

Fig. 5. Soil profiles along the cross section

For some SM stations, PGAs as well acceleration response spectra calculated by 1D and 2D as in Fig. 6(a) the match between all seems almost perfect however, as in Fig. 6(b) 2D results compared to 1D appears slightly better in modeling the recorded results.

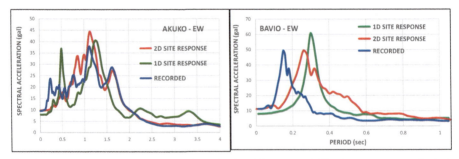

Fig. 6. Acceleration response spectra for calculated by 1D and 2D site response analysis in comparison to the recorded acceleration response spectra

In a study conducted by Bonilla et al. [6] for a basin structure, it was observed that amplification from 2D analysis is higher than 1D that was observed in similar studies. However, in the case of 2D analysis conducted by the authors [2] where there no basins, such effects were not observed, justifying 1D to be sufficient to model or estimate possible earthquake characteristics on the ground surface.

For sites with complex and irregular stratigraphy, two-dimensional (2D) and three-dimensional (3D) ground response is preferred over 1D wave propagation for more realistic evaluation of ground response under seismic load.

Unidimensional analyses are a standard procedure in practical SRA, however these analyses do not involve the effects of the topography and existence of surface and underground structures, so two or three-dimensional models provide an adequate alternative to perform the same analysis with the advantage to include more complex conditions as well the possibility to include the three components of the seismic input motion.

5 Conclusions

Site characterization for a geotechnical earthquake engineering application involves many variables and uncertainties. One of the controlling issues in the performance based design involves the definition of performance objectives. In the case of site response analysis, the performance objectives may be considered as the probabilistic definition of the uniform acceleration hazard spectrum and acceleration time histories calculated on the ground surface.

The variation of peak ground and spectral accelerations recorded on the ground surface by some of the Istanbul Rapid Response Network (IRRN) stations are assessed with respect to soil stratification, ground surface and bedrock topography. 1D and 2D site response analyses were conducted to model the observed PGAs recorded during the $M_L = 6.5$ Gökçeada 24/5/2014 earthquake for the soil profile along south-north direction to see the effect of two dimensional modelling with respect to 1D and 2D site response analysis and recorded acceleration records. It is observed in this preliminary investigation to specify the performance base objectives for site response analysis may require at least 2D site response analysis to be on more realistic approach for performance based approach.

Acknowledgement. The authors would like to express their appreciation to Prof. Yasin Fahjan of Istanbul Technical University for his support during the preparation of this manuscript.

References

1. Ansal, A., Tönük, G., Kurtuluş, A.: Implications of site specific response analysis. In: Pitilakis, K. (eds.) Recent Advances in Earthquake Engineering in Europe, GGEE, vol. 46, pp. 51–68. Springer, Cham (2018). https://doi.org/10.1007/978-3-319-75741-4_2
2. Ansal, A., Fercan, Ö., Kurtuluş, A., Tönük, G.: 2D site response analysis of the istanbul rapid response network. In: Theme Lecture, PBD III, Vancouver, Canada (2017)
3. Ansal, A., Tönük, G., Kurtuluş, A.: A probabilistic procedure for site specific design earthquake. In: 6th International Conference on Earthquake Geotechnical Engineering, Theme Lecture, Christchurch, New Zealand (2015)
4. Ansal, A., Kurtuluş, A., Tönük, G.: Seismic microzonation and earthquake damage scenarios for urban areas. Soil Dyn. Earthq. Eng. **V30**, 1319–1328 (2010)
5. Ansal, A., Kurtuluş, A., Tönük, G.: Earthquake damage scenario software for urban areas. In: Papadrakakis, M., Charmpis, D.C., Lagaros (eds.) Computational Structural Dynamics and Earthquake Engineering. Structures and Infrastructures Series, vol. 2, pp. 377–391 (2009)

6. Bonilla, F.B., Liu, P.C., Nielsen, S.: 1D and 2D linear and nonlinear site response in the Grenoble area. In: Third International Symposium on the Effects of Surface Geology on Seismic Motion, Grenoble, France. Paper Number: 082/S02 (2006)

7. Cavallaro, A., Grosso, S., Maugeri, M.: The role of site uncertainty of soil parameters on performance-based design in geotechnical earthquake engineering. In: International Conference on Performance-Based Design in Earthquake Geotechnical Engineering - from Case History to Practice, Tokyo, Japan (2009)

8. Fahjan, Y., Kara, I.F., Mert, A.: Selection and scaling time history records for performance-based design. In Performance-Based Seismic Design of Concrete Structures and Infrastructures, pp. 1–35. IGI Global (2017)

9. Flores Lopez, F.A., Ayes Zamudio, J.C., Vargas Moreno, C.O., Vázquez Vázquez, A.: Site Response Analysis (SRA): a practical comparison among different dimensional approaches (2015)

10. Hu, Y., Lam, N., Khatiwada, P., Menegon, S.J., Looi, D.T.W.: Site-specific response spectra: guidelines for engineering practice. CivilEng **2**, 712–735 (2021)

11. Kaklamanos, J., Bradley, B.A., Thompson, E.M., Baise, L.G.: Critical parameters affecting bias and variability in site-response, analyses using KiK-net downhole array data. Bull. Seismol. Soc. Am. **103**(3), 1733–1749 (2013)

12. Nautiyal, P., Raj, D., Bharathi, M., Dubey, R.: Ground response analysis: comparison of 1D, 2D and 3D approach. In: Patel, S., Solanki, C.H., Reddy, K.R., Shukla, S.K. (eds.) Proceedings of the Indian Geotechnical Conference 2019. LNCE, vol. 138, pp. 607–619. Springer, Singapore (2021). https://doi.org/10.1007/978-981-33-6564-3_51

13. Pappin, J.W., Lubkowski, Z.A., King, R.A.: The significance of site response effects on performance based design. In: 12th World Conference on Earthquake Engineering, Auckland, New Zealand (2000)

14. Pehlivan, M., Rathke, E.M., Gilbert, R.B.: Influence of 1D and 2D spatial variability on site response analysis. In: 15th World Conference on Earthquake Engineering, Lisbon, Portugal, Paper no. 3419 (2012)

15. QUAKE/W: Finite Element Dynamic Earthquake Analysis, Software, Geo-Slope Office (2016)

16. Turkish Earthquake Code, Turkish Republic, Interior Ministry, Disaster and Emergency Management Presidency (2018). (in Turkish)

17. Tönük, G., Ansal, A., Kurtuluş, A., Çetiner, B.: Site specific response analysis for performance based design earthquake characteristics. Bull. Earthq. Eng. **12**(3), 1091–1105 (2014)

Seismic Landslide Susceptibility Assessment Based on Seismic Ground Motion and Earthquake Disaster Analysis

Ailan Che[(✉)], Hanxu Zhou, Jinchang Chen, Yuchen Wu, and Ziyao Xu

School of Naval Architecture, Ocean and Civil Engineering, Shanghai Jiao Tong University, 800 Dongchuan Road, Shanghai 200240, China
alche@sjtu.edu.cn

Abstract. Seismic motion is one of the significant factors triggering slope insta-bility. Landslides induced by intense earthquakes pose a great threat on the public security and traffic safety. About 720 landslides were caused by the 2014 Ms6.5 Ludian earthquake. The seismic ground motion records of Ludian earthquake and the influencing factors of seismic landslides are analyzed, and the susceptibility of landslides was evaluated in combination with machine learning models (BP neural network and SVM). The characteristics of Ludian earthquake motion records in time domain, frequency domain and time-frequency domain reveal that, the max-imum value of PGA reached 949.2 cm/s^2 (EW), and the duration of seismic wave within the nearest station to epicenter was 5.0 s (EW), 4.9 s (NS) and 4.3 s (UD), respectively. The result of Hilbert Huang Transform suggests that large instan-taneous and cumulative energy concentrated in low frequency region (0–5 Hz), which can be considered to be related to large area landslides. Twelve factors are selected as the influencing factors of seismic landslides under Ludian earthquake, and the spatial correlation analysis indicates that the relation between factors and seismic landslide is characterized by strong nonlinear properties, of which ground motion parameters show a strong positive correlation with landslide distribution. The receiver operating characteristic (ROC) curve is adopted to compare the per-formance of two models. The results show that the AUC values of two curves are 90.1% and 89.5%, respectively, showing that BP neural network has higher accuracy of seismic landslide susceptibility results, and illustrating dependency on spatial distribution of seismic ground motion parameters.

Keywords: Ludian earthquake · Seismic landslides · Hilbert Huang Transform · BP neural network · Support vector machine · Landslide susceptibility

1 Introduction

Earthquake is a main factor inducing landslides. The statistical data show that earth-quakes with magnitude greater than 4.0 could induce landslides [1], and the seismic disasters would be severe under earthquakes with magnitude above 6.0. For example, the Mw 6.7 Northridge earthquake in the United States in 1994 induced more than 11000

© The Author(s), under exclusive license to Springer Nature Switzerland AG 2022
L. Wang et al. (Eds.): PBD-IV 2022, GGEE 52, pp. 327–341, 2022.
https://doi.org/10.1007/978-3-031-11898-2_17

landslides in an area of nearly 10000 km^2 [2]. The Mw 6.9 Iwate Miyagi inland earthquake in Japan in 2008 produced 4161 landslides, including 3779 shallow mudslides [3, 4]. Approximately 56000 landslides and collapses were triggered by Ms 8.0 Wenchuan earthquake in China in 2008, which directly led to nearly 30000 deaths, accounting for about 1/3 of the total number of earthquake deaths [5–9]. The M_w 6.6 Hokkaido earthquake in 2018 produced about 3000 landslides and 41 people died in the earthquake, of which 36 were caused by large-scale landslide [10]. Therefore, it is of great significance to realize landslides susceptibility assessment, which could provide reference for emergency rescue work and post-earthquake reconstruction.

With the development of seismic monitoring instruments, data transmission and processing technology, it has gradually become a reality to quickly calculate the seismic parameters and evaluate the seismic intensity according to the seismic ground motions. After the Kanto earthquake in 1923, in order to solve the problem of seismic resistance of buildings, it was proposed to develop strong motion observation instruments. In 1932, the United States first deployed a strong seismograph, and the strong motion records of the Long Beach earthquake in California were obtained in the same year. In 1940, the first record with PGA (peak ground acceleration) > 0.3 g (El Centro) was obtained [11]. Strong motion observation in China started in 1962, Xin-feng River dam construction. Since 2000, strong motion observation in China has developed rapidly [12]. Based on the strong motion data, the source location, source depth and magnitude can be determined, and seismic intensity could be evaluated rapidly. The rapid evaluation of seismic intensity can be carried out through the empirical statistical model, based on the attenuation law or on the quantitative relationship between seismic intensity and ground motion parameters [13].

The seismic disasters are closely related to the characteristics of seismic ground motion. Many cases show that moderate magnitude earthquake can cause large scale disasters and the disasters induced by the earthquakes with close magnitude present completely different characteristics, which are closely related to the characteristics of seismic ground motions and local site effect. For example, a Mw 5.6 earthquake occurred at Ionian offshore, which caused the high-level seismic hazard and huge disasters in Sicily (Italy) [14]. Two large earthquakes (Ms 6.6 Jinggu Earthquake and the Ms 6.5 Ludian Earthquake) occurred in Yunnan province, China in 2014. The magnitude of the two earthquakes is close and local site conditions are similar, but the disasters have a big difference. Large landslides and buildings with 3 and more floors are destroyed in Ludian earthquake, however, few landslides or damaged structures were induced by Jinggu earthquake [15]. Seismic landslides depend on the frequency components and peak ground acceleration of input ground motions [16–21]. In theory, high frequency seismic ground motion with shorter wave-length than dimensions of slopes unlikely trigger landslides because high frequency seismic wave drive different parts of slope in the opposite direction simultaneously, so, the whole slope would remain stable. On the other hand, low frequency seismic wave with longer wavelength than dimensions of slopes can lead slope move in the same direction, which more likely induce landslides [22, 23]. Similar with landslides, damage characteristics of structures also closely related to the input seismic ground motions. The pulse-like ground motions can increase the inelastic displacement demand of SDOF (single degrees of freedom) systems, especially when the ratio of normalized shaking period T to velocity pulse period Tp is less than one [24, 25]. So, it is need to consider the characteristics of ground motion in seismic landslide susceptibility evaluation.

Seismic landslide susceptibility evaluation initially focused on the mechanism of landslides, then statistical models and machine learning methods are implemented with the development of computer science. According to whether a mechanical model with definite physical meaning is established during the assessment, landslide susceptibility evaluation methods can be divided into two categories: deterministic method and non-deterministic method. Deterministic method includes slope limit equilibrium analysis, numerical simulation, Newmark permanent displacement method, etc. The Newmark model is the most widely used deterministic method, which calculates the displacement of the slope to decide whether the slope is stable [26]. However, it requires detailed geological data and spatial parameters as support. So, when the model is applied in region, assumptions and simplifications are made which makes the result less accurate [27]. And many methods have been developed to solve this problem such as building an integrated Spectral Element Method (SEM)-Newmark model [28], combing non-deterministic model with Newmark method [29, 30] and so on. Among non-deterministic methods, mathematical and statistical models are used to explore the characteristics of data and make decisions on landslide susceptibility. Support Vector Machine model (SVM) is a binary model, which has good applicability for imprecise and small sample size problems [31, 32]. Artificial neural network (ANN) model is also a widely used model in landslide susceptibility [33]. Except for the basic neural network model, many improvements have been made to obtain higher evaluation accuracy [34, 35]. Recently, with the developing of study, many scholars begin to combine different methods for better application [36, 37]. Also, various models are implemented in the same area to compare the accuracy of evaluation result for a better understanding of model performance [38–40].

This work investigated the amplitude, duration and spectrum (in time and time-frequency domains) of typical seismic ground motion in Ludian earthquake at first. Then, the characteristics of seismic disasters and the conditioning factors of disaster were analyzed. Considering the characteristics mentioned above, seismic landslide susceptibility evaluation based on machine learning method (BP neural network and SVM) were performed at last. This work has practical significance for major infrastructure construction and disaster prevention and reduction planning.

2 Seismic Ground Motion and Disaster Characteristics in Ludian Earthquake

Earthquake frequently occur in the Yunnan Province, China. Since the detailed earthquake records, 23.6% of the earthquakes with M ≥ 7.0 occurred in Yunnan Province, which accounts for only 4.1% of the land area of China. From 1949 to 2020, there were 8 strong earthquakes with M ≥ 7.0 in Yunnan Province. Recent strong earthquakes in Yunnan Province include Ludian Ms 6.5 earthquake in 2014, Jinggu Ms 6.6 earthquake in 2014, Yingjiang Ms 6.1 earthquake in 2014 and Yangbi Ms 6.4 earthquake in 2021.

Among them, Ludian earthquake caused the largest disasters and casualties. The epicenter intensity of Ludian earthquake up to IX and the focal depth is approximately 12 km, which killed 617 people, injured 2400 and destroyed 86000 buildings. In this section, we mainly introduced the typical strong motion in amplitude, duration and spectrum at first. Then, disaster characteristics and the possible influencing factors of it were analyzed.

2.1 Seismic Ground Motion Characteristics

The distribution of intensity map is shown in Fig. 1. Within the intensity map, 7 strong motion stations were triggered, including 1 in the IX region, 1 in VII region and 5 in VI region. The basic information of each station was shown in Table 1. Two of them (HYC and QJT) at rock site. 53LLT with 8.3 km epicentral distance and 53HZH with epicentral distance 65.9 km are the nearest and farthest stations from the epicenter respectively.

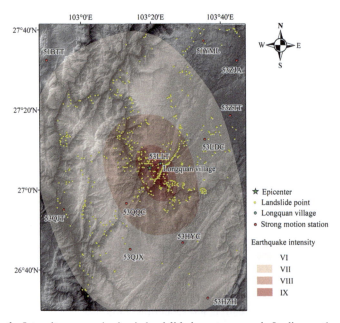

Fig. 1. Intensity map and seismic landslide inventory map in Ludian earthquake

The specific wave forms of 53LLT in three directions are shown in Fig. 2(a). The whole length of time is approximately 130 s. The PGA (peak ground motion) in three directions is 949.2 cm/s² (EW), 705.8 cm/s² (NS) and 504.4 cm/s² (UD), respectively. PGA shows a decreasing trend with the increase of epicentral distance, but it is not linear decreasing especially in 53QJX, where PGA has a great increase, as shown in Fig. 2(b). So, the PGA decays relatively slowly.

The duration of seismic ground motion can reflect the cumulative effect of earthquake on the stability of slope or structures. The duration includes absolute duration and relative duration. We used Kawashinat bracketed duration to reflect the effect, which is a kind

Table 1. Strong motion stations in Ludian earthquake (within the intensity map)

Event	Stations	Site conditions	Epicentral distance (km)	Intensity scale	PGA (cm/s²)		
					EW	NS	UD
Ludian earthquake	53LLT	Soil	8.3	IX	949.2	705.8	504.4
	53QQC	Soil	18.7	VII	146.0	140.3	52.8
	53LDC	Soil	32.5	VI	45.9	44.8	25.6
	53QJX	Soil	38.1	VI	135.2	133.4	65.0
	53HYC	Rock	39.6	VI	88.3	87.7	47.5
	53QJT	Rock	46.4	VI	24.8	20.6	−13.2
	53HZH	Soil	65.9	VI	50.2	69.2	16.0

of relative duration. It takes the duration between the first and last two times reaching or exceeding a fraction (1/5–1/2) of the PGA as the duration. It is taken 0.3 in this work. The duration of each seismic ground motion is listed in Table 2. 53LLT has the shortest duration compared with other motions, which are 5.0 (EW), 4.9 (NS) and 4.3 (UD). With the increase of epicentral distance, the duration shows an increasing trend. It reflects that the 53LLT motion has a large and instantaneous inertia force to the slope or structures.

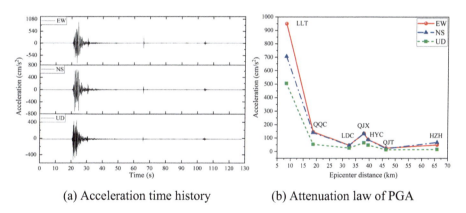

(a) Acceleration time history   (b) Attenuation law of PGA

Fig. 2. Acceleration time history of 53LLT and attenuation law of PGA

In order to clarify the frequency characteristics of the motions, FFT (Fast Fourier Transform) was done and the Fourier spectrum was obtained. Fourier spectrums of 53LLT in three directions are shown in Fig. 3. It shows that the energy (amplitude) is concentrated within 35 Hz. Three predominate frequency in EW and NS directions are same, which are within 0–5 Hz, 5–10 Hz and 10–15 Hz respectively. The first predominate frequency (0–5 Hz) is clear than other predominate frequency. The distribution of predominate frequency in UD direction is different from EW and NS directions.

Table 2. Duration of acceleration time history (within the intensity map)

Events	Station	Duration (s)		
		EW	NS	UD
Ludian earthquake	53LLT	5.0	4.9	4.3
	53QQC	11.4	12.1	14.4
	53LDC	11.5	15.3	15.2
	53QJX	13.3	10.1	13.4
	53HYC	12.1	12.1	9.0
	53QJT	16.6	20.3	21.3
	53HZH	6.8	4.2	16.7

Three predominate frequency are within 0–5 Hz, 5–15 Hz and 15–20 Hz and the second predominate frequency is the most outstanding.

The time-frequency characteristics of motions are investigated by Hilbert spectrum in the study. In order to acquire the spectrum, the original signal was decomposed into series of intrinsic mode function (IMF) by Variational Mode Decomposition (VMD) at first, because the original signal does not meet the conditions of Hilbert transform. Each IMF is well behaved in Hilbert transform. The Hilbert spectrum is acquired by composing the spectrum of each IMF. Hilbert spectrums of 53LLT in three directions are shown in Fig. 4. Hilbert spectrum reflects the distribution characteristics of instantaneous energy. Instantaneous energy mainly distributed at 20–30 s. The shape of the spectrum presents multi peaks. The energy in EW and NS directions mainly distributed within three frequency range which are 0–5 Hz, 5–10 Hz and 10–15 Hz. Peak instantaneous energy is in 0–5 Hz. Compared with EW and NS directions, the frequency range of main instantaneous energy in UD directions has a slight increase and the peak energy is lived in the second and third predominate frequency.

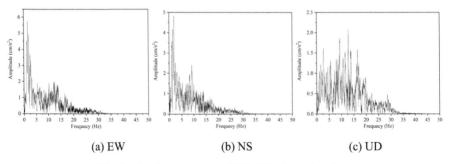

(a) EW   (b) NS   (c) UD

Fig. 3. Fourier spectrums of 53LLT in three directions

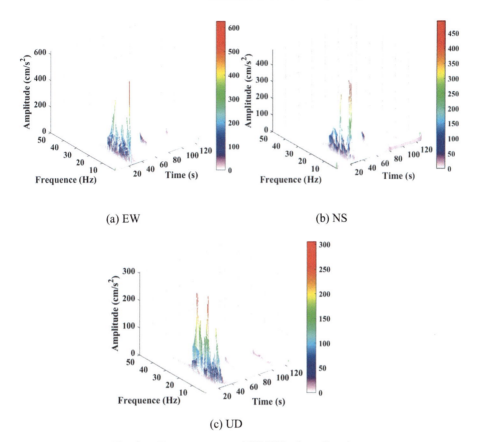

(a) EW

(b) NS

(c) UD

Fig. 4. Hilbert spectrums of 53LLT in three directions

Marginal spectrum is the integral of Hilbert spectrum over time. It reflects the distribution characteristics of cumulative energy at different frequencies. Marginal spectrums of 53LLT in three directions are shown in Fig. 5. The shape of marginal spectrums is close to the Fourier spectrum, however, the predominate frequency shown in marginal spectrum is clearer than Fourier spectrum. Three predominate frequencies in EW direction

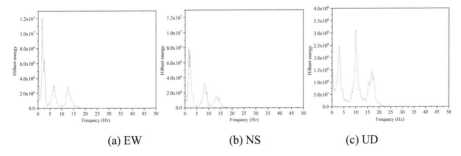

(a) EW

(b) NS

(c) UD

Fig. 5. Marginal spectrums of 53LLT in three directions

are 1.625 Hz, 6.625 Hz and 12.625 Hz. Three predominate frequencies in NS direction are same to that in EW direction. In UD direction, predominate frequencies are 3.125 Hz, 10.125 Hz and 16.625 Hz, and the peak energy occurs at the second frequency.

Above results show that the ground motion of 53LLT has large PGA, short duration, large instantaneous and cumulative energy concentrated in low frequency region (0–5 Hz), and decay of PGA is relatively slow, which caused the huge disasters including large landslides and seriously destroyed building in local area.

2.2 Factor Analysis of Seismic Landslides

Seismic landslides have the characteristics of wide influence area, complex formation mechanism and various influencing factors. In order to evaluate the susceptibility of seismic landslides, it is quite essential to analyze the condition factors related to the landslide first.

The landslide inventory of Ludian earthquake was mapped based on the data recorded during field investigation. A total number of 714 seismic landslides are compiled, as show in Fig. 1. In presented study, 12 conditioning factors of Ludian earthquake are considered to explore the relationship between factors and the susceptibility of seismic landslides. Among these factors, the internal factors contain elevation, slope, slope aspect, surface curvature, profile curvature, distance to fault, distance to river, NDVI, and lithology. The external factors consist of PGA, PGV and epicenter distance.

To evaluate the landslide sensitivity of each causative factor, frequency ratio method is implemented. The frequency ratio can be calculated by the following formula:

$$FR = \frac{N_i/A_i}{N/A}$$

where: for each factor k, FR is the frequency ratio of class i of factor k, Ni is the number of landslides within class i, and Ai is the area of class i, N is the total number of landslides, and A is the total area of the study area. It can be seen that the larger the ration of frequency is, the higher the landslide sensitivity of the factor is. This method excludes the influence of the class area itself on the assessment of relationship between landslide and each factor.

The frequency ratios of each class of 12 factors are calculated and the results are shown in Fig. 6. For the internal factors, Fig. 6(a) reveals that landslides are concentrated in the area with an elevation of about 1400–1800 m, there is no single positive correlation between landslide sensitivity and elevation. And Fig. 6(b) reflects probability of landslide increases with the increasement of slope degree. The class of 65–70° has the highest landslide sensitivity. As can be seen from Fig. 6(f, g), the landslide sensitivity of fault distance and river distance decreases monotonically with the increase of distance on the whole. The landslide sensitivity of river distance within 3 km and fault distance within 4 km are the highest. Figure 6(h) shows that the sensitivity of NDVI to landslide is similar in all sections, and the sensitivity is most prominent when the value ranges from 0.09 to 0.12. Lithology is also in strong relation with landslides, and landslides are often concentrated in specific strata. It can be seen from the Fig. 6(i) that the Ordovician and Devonian strata are significantly more sensitive to landslides than other strata. As for seismic parameters, the rule is more obvious. Co-seismic landslides are strongly correlated with PGA and PGV. Figure 6(j, k, l) indicate that in general, the landslide sensitivity shows a good relationship with the increase of PGA and PGV and the decrease of epicenter distance.

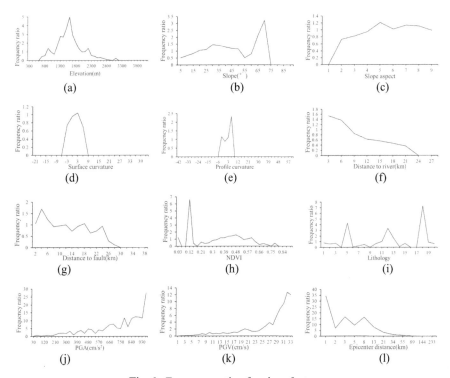

Fig. 6. Frequency ratio of various factors

3 Seismic Landslide Susceptibility Analysis

3.1 Methodology

Landslide susceptibility analysis is able to evaluate the spatial probability of seismic landslide. The factor analysis reveals that the relation between factors and landslide susceptibility is characterized by high dimensions and nonlinear properties. In this study, two non-deterministic methods including BP neural network and support vector machine are used to analyze the susceptibility of seismic landslides, and the accuracy of the assessment results of the models is compared.

3.1.1 BP Neural Network

BP neural network is a feedforward multilayer network model, which can establish any nonlinear mapping from input to output. Data is transmitted forward from input layer to output layer, and the error between the calculated data and the real data is fed back from the output layer to the input layer. The structure of a simple BP neural network can be divided into input layer, hidden layer and output layer, as shown in Fig. 7. The mapping relation between input and output is established with the activation function and threshold in hidden layer. A classical three-layer network including one hidden layer is adopted. Except for the output layer, each layer could have different numbers of neurons. The neuron number of input layer corresponds to the number of influencing factors, and the neuron number of hidden layer would be optimized based on the error of testing set. The network building process is composed of three steps, including forward calculation, error back propagation and weight update.

3.1.2 SVM

SVM (support vector machine) is a commonly used machine learning algorithm following the principle of structural rick minimization. It has been widely applied in decision making and data forecasting of various fields. The two fundamental components of SVM are the kernel function and the optimal classification hyperplane (Fig. 8). The kernel function could transform the input data into high dimensional space, to separate the data linearly. The commonly used kernel functions include linear kernel function, polynomial kernel function, radial basis function and sigmoid kernel function. Radial basis function is selected as the kernel function in the present research. And the hyperplane helps to discrete the input samples in the high dimensional space and maximize the classification margin between hyperplane and vector samples.

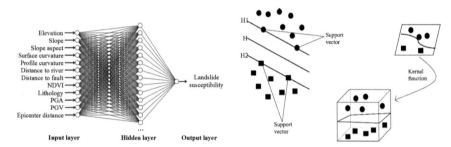

Fig. 7. BP neural network structure diagram   **Fig. 8.** Basic principles of SVM

3.2 Seismic Landslide Susceptibility Result

70% of all samples constitute the test set and the rest 30% are categorized into verification set for network training. The label value of landslide data and non-landslide data are set as 1 and −1, respectively. And 0 is determined as the label value threshold to distinguish the category of landslide and non-landslide. The label value of susceptibility results greater than 0 is classified as landslide.

The comparison between actual landslide inventory and susceptibility results from different methods is shown in Fig. 9. It can be found that two machine learning algorithms produce relatively good susceptibility map with smooth transition of landslide among hilly area. Compared to the actual landslide points from inventory map, the results of BP neural network show fewer landslides, which are concentrating in the area near the epicenter. However, in the actual situation, there are some landslides occurred far from the epicenter. The results of BP neural network show dependency on spatial distribution of seismic ground motion parameters like PGA, PGV, and epicenter distance. The results of SVM show more discrete evaluation results of seismic landslide susceptibility. Some regions far from epicenter also have landslide area, which shows consistence with the actual distribution of seismic landslides.

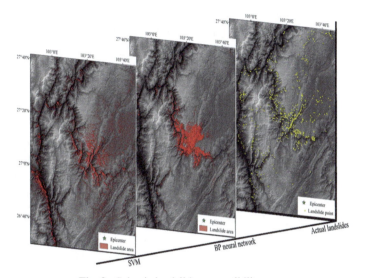

Fig. 9. Seismic landslide susceptibility map

To quantitatively assess the performance of two machine learning algorithms, the Receiver Operating Characteristic (ROC) curve is adopted. The closer the curve is to the upper left corner of the coordinate system, the higher the proportion of landslide judged as positive results, indicating that the evaluation result of the model is more accurate. The area under the curve is defined as AUC (area under curve) value, which can be used to quantitatively compare the accuracy. The ROC curves of BP neural network and SVM are shown in the Fig. 10. The AUC values of two curves are 90.1% and 89.5%, respectively, showing that BP neural network has higher accuracy of seismic landslide susceptibility results. It can be observed that although the distribution trend of SVM result is similar to the actual landslide points, a large number of non-landslide points are evaluated as landslide points, greatly increasing the number of landslide points and reducing the accuracy. In addition, rainfall could be an important factor related to the seismic landslide susceptibility under Ludian earthquake. Ignoring the presence of rainfall factor might decrease the accuracy of results.

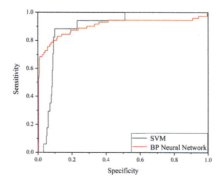

Fig. 10. ROC curves of SVM and BP neural network

4   Discussion

4.1   Comparison with Yangbi Earthquake in Typical Ground Motion

On May 21, 2021, a Ms 6.4 earthquake occurred in Yangbi County, Yunnan Province. The magnitude with 0.1 smaller than Ludian earthquake, but the disaster shows big difference. Few landslides occurred in Yangbi earthquake.

53YBX station with epicentral distance 8.6 km shows that the PGA in three directions is 379.8 cm/s^2, 720.3 cm/s^2 and 448.4 cm/s^2 respectively, as shown in Fig. 11 and the corresponding duration (Kawashinat bracketed duration with ratio 0.3) is 4.9 s, 2.1 s and 5.1 s. Marginal spectrum of the station is shown in Fig. 12. Compared with 53LLT in Ludian earthquake, cumulative energy is evenly distributed in high (10–15 Hz) and low frequency (approximately 2 Hz) and the energy of high frequency in NS direction is larger than low frequency. The results are corresponding to the theory that high frequency seismic ground motion with shorter wavelength than dimensions of slopes unlikely trigger landslides and low frequency seismic wave with longer wave-length than dimensions of slopes can lead slope move in the same direction, which more likely induce landslides [37, 38].

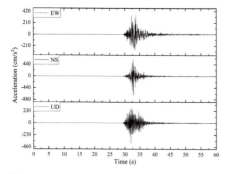
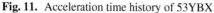
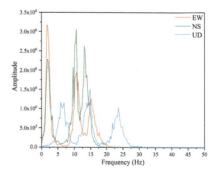

Fig. 11. Acceleration time history of 53YBX   **Fig. 12.** Marginal spectrum of 53YBX

4.2 Landslide Susceptibility Evaluation Considering Characteristics of Ground Motion

This work indicates that the size and the number of landslides is closely related to the characteristics of ground motion especially in frequency domain. However, landslide susceptibility evaluation methods at present only take PGA or PGV into consideration to reflect the effect of earthquake. Obviously, the evaluation result is inconsistent with the actual situation. So, it is need to develop a susceptibility evaluation method that can reflect the frequency characteristics of seismic ground motion, which can improve the accuracy of evaluation result and have meaningful guiding significance for disaster prevention and reduction planning.

5 Conclusion

1. Ground motion in the epicenter of Ludian earthquake shows special characteristics. In time domain, PGA in EW direction up to 949.2 cm/s^2, and the corresponding duration is 5.0 s which show short duration characteristic. In time-frequency domain, large instantaneous and cumulative energy based on Hilbert spectrum and marginal spectrum respectively concentrate in low frequency region (0–5 Hz), which easily trigger large landslides.
2. Approximately 720 landslides or rockfalls distribute within 60 km of the epicenter of Ludian earthquake, with sliding distance ranges from 20 to 850 m. The distribution of landslides influenced by elevation, slope, slope aspect, curvature, river, NDVI, PGA and PGV and Lithology. Among them, landslide sensitivity shows a good relationship with the increase of PGA and PGV and the decrease of epicenter distance, and the sensitivity reaches the largest in the class with the largest PGA and PGV and the class closest to the epicenter.
3. Seismic landslide susceptibility method based on machine learning (BP neural network and SVM) can well determine the distribution of landslides. The actual distribution of landslides is closer to the result of SVM. However, the BP neural network has higher accuracy than SVM. This is because that a large number of non-landslide points are determined as landslide points by SVM.

Acknowledgments. This work is financially supported by the National Key R&D Program of China (2018YFC1504504).

References

1. Huang, R., Fan, X.: The landslide story. Nat. Geosci. **6**(5), 325–326 (2013)
2. Calais, E., Minster, J.B.: GPS detection of ionospheric perturbations following the January 17, 1994, Northridge Earthquake. Geophys. Res. Lett. **22**(9), 1045–1048 (2013)
3. Aydan, Ö.: Some considerations on a large landslide at the left bank of the Aratozawa dam caused by the 2008 Iwate-Miyagi intraplate earthquake. Rock Mech. Rock Eng. **49**(6), 2525–2539 (2016)

4. Moratto, L., Vuan, A., Saraò, A.: A hybrid approach for broadband simulations of strong ground motion: the case of the 2008 Iwate-Miyagi Nairiku earthquake. Bull. Seismol. Soc. Am. **105**(5), 2823–2829 (2015)

5. Lai, J., et al.: Characteristics of seismic disasters and aseismic measures of tunnels in Wenchuan earthquake. Environ. Earth Sci. **76**(2), 1–19 (2017). https://doi.org/10.1007/s12 665-017-6405-3

6. Guo, X., Cui, P., Li, Y., Zou, Q., Kong, Y.: The formation and development of debris flows in large watersheds after the 2008 Wenchuan Earthquake. Landslides **13**(1), 25–37 (2014). https://doi.org/10.1007/s10346-014-0541-6

7. Zhang, Y., Zhang, J., Chen, G., Zheng, L., Li, Y.: Effects of vertical seismic force on initiation of the Daguangbao landslide induced by the 2008 Wenchuan earthquake. Soil Dyn. Earthq. Eng. **73**, 91–102 (2015)

8. Chen, Q., Cheng, H., Yang, Y., Liu, G.X., Liu, L.Y.: Quantification of mass wasting volume associated with the giant landslide Daguangbao induced by the 2008 Wenchuan earthquake from persistent scatterer In-SAR. Remote Sens. Environ. **152**, 125–135 (2014)

9. Changwei, Y., Xinmin, L., Jianjing, Z., Zhiwei, C., Cong, S., Hongbo, G.: Analysis on mechanism of landslides under ground shaking: a typical landslide in the Wenchuan earthquake. Environ. Earth Sci. **72**(9), 3457–3466 (2014). https://doi.org/10.1007/s12665-014-3251-4

10. Yamagishi, H., Yamazaki, F.: Landslides by the 2018 Hokkaido Iburi-Tobu earthquake on September 6th. Landslides **15**, 2521–2524 (2018)

11. Trifunac, M.D.: 75th anniversary of strong motion observation – a historical review. Soil Dyn. Earthq. Eng **29**(4), 591–606 (2009)

12. Lu, D.W., Yang, Z.B.: Strong motion observation from recent destructive earthquakes in China mainland. AMM **724**, 358–361 (2015)

13. Wang, D.C., Ni, S.D., Li, J.: Research status of rapid assessment on seismic intensity. Prog. Geophys. **28**(4), 1772–1784 (2013). (in Chinese)

14. Panzera, F., et al.: Correlation between earthquake damage and seismic site effects: the study case of Lentini and Carlentini, Italy. Eng. Geol. **240**, 149–162 (2018)

15. Jia, H., Chen, F., Pan, D.: Comparison of two large earthquakes in China: the Ms 6.6 Yunnan Jinggu Earthquake and the Ms 6.5 Yunnan Ludian Earthquake in 2014. Int. J. Disaster Risk Reduct. **16**, 99–107 (2016)

16. Bozzano, F., Lenti, L., Martino, S., Paciello, A., Scarascia Mugnozza, G.: Self-excitation process due to local seismic amplification responsible for the 31st October 2002 reactivation of the Salcito landslide (Italy). J. Geophys. Res. **113**, B10312 (2008)

17. Bozzano, F., Lenti, L., Martino, S., Montagna, A., Paciello, A.: Earthquake triggering of landslides in highly jointed rock masses: reconstruction of the 1783 Scilla rock avalanche (Italy). Geomorphology **129**, 294–308 (2011)

18. Bozzano, F., Lenti, L., Martino, S., Paciello, A., Scarascia Mugnozza, G.: Evidences of landslide earthquake triggering due to self-excitation process. Int. J. Earth Sci. **100**, 861–879 (2011)

19. Del Gaudio, V., Wasowski, J.: Advances and problems in understanding the seismic response of potentially unstable slopes. Eng. Geol. **122**, 73–83 (2010)

20. Lenti, L., Martino, S.: The interaction of seismic waves with step-like slopes and its influence on landslide movements. Eng. Geol. **126**, 19–36 (2011)

21. Martino, S., Scarascia Mugnozza, G.: The role of the seismic trigger in the Calitri landslide (Italy): historical reconstruction and dynamic analysis. Soil Dyn. Earthq. Eng. **25**, 933–950 (2005)

22. Kramer, S.L., Smith, M.W., et al.: Modified newmark model for seismic displacements of compliant slopes. J. Geotech. Geoenviron. Eng. **123**(7), 635–644 (1997)

23. Jibson, R.W., Tanyaş, H.: The influence of frequency and duration of seismic ground motion on the size of triggered landslides—A regional view. Eng. Geol. **273**, 105671 (2020)

24. Baker, J.W., Cornell, C.A.: Vector-valued intensity measures for pulse-like near-fault ground motions. Eng. Struct. **30**(4), 1048–1057 (2008)
25. Iervolino, I., Chioccarelli, E., Baltzopoulos, G.: Inelastic displacement ratio of nearsource pulse-like ground motions. Earthq. Eng. Struct. Dyn. **41**(15), 2351–2357 (2012)
26. Shinoda, M., Miyata, Y.: Regional landslide susceptibility following the Mid NIIGATA prefecture earthquake in 2004 with NEWMARK'S sliding block analysis. Landslides **14**(6), 1887–1899 (2017). https://doi.org/10.1007/s10346-017-0833-8
27. Chen, X., Shan, X., Wang, M., et al.: Distribution pattern of coseismic landslides triggered by the 2017 Jiuzhaigou Ms 7.0 earthquake of China: control of seismic landslide susceptibility. ISPRS Int. J. Geo-Inf. **9**(4), 198 (2020)
28. Huang, D., Wang, G., Du, C., et al.: An integrated SEM-Newmark model for physics-based regional coseismic landslide assessment. Soil Dyn. Earthq. Eng. **132**, 106066 (2020)
29. Han, L., Ma, Q., Zhang, F., et al.: Risk assessment of an earthquake-collapse-landslide disaster chain by Bayesian network and Newmark models. Int. J. Environ. Res. Public Health **16**(18), 3330 (2019)
30. Wang, Y., Song, C., Lin, Q., et al.: Occurrence probability assessment of earthquake-triggered landslides with Newmark displacement values and logistic regression: the Wenchuan earthquake, China. Geomorphology **258**, 108–119 (2016)
31. Huang, Y., Zhao, L.: Review on landslide susceptibility mapping using support vector machines. CATENA **165**, 520–529 (2018)
32. Hong, H., Pradhan, B., Bui, D.T., et al.: Comparison of four kernel functions used in support vector machines for landslide susceptibility mapping: a case study at Suichuan area (China). Geomat. Nat. Hazards Risk **8**(2), 544–569 (2017)
33. Shahri, A.A., Spross, J., Johansson, F., et al.: Landslide susceptibility hazard map in southwest Sweden using artificial neural network. CATENA **183**, 104225 (2019)
34. Xi, W., Li, G., Moayedi, H., et al.: A particle-based optimization of artificial neural network for earthquake-induced landslide assessment in Ludian county, China. Geomat. Nat. Hazards Risk **10**(1), 1750–1771 (2019)
35. Moayedi, H., Mehrabi, M., Kalantar, B., et al.: Novel hybrids of adaptive neuro-fuzzy inference system (ANFIS) with several metaheuristic algorithms for spatial susceptibility assessment of seismic-induced landslide. Geomat. Nat. Hazards Risk **10**(1), 1879–1911 (2019)
36. Hong, H., et al.: A novel hybrid integration model using support vector machines and random subspace for weather-triggered landslide susceptibility assessment in the Wuning area (China). Environ. Earth Sci. **76**(19), 1–19 (2017). https://doi.org/10.1007/s12665-017-6981-2
37. Abbaszadeh Shahri, A., Maghsoudi Moud, F.: Landslide susceptibility mapping using hybridized block modular intelligence model. Bull. Eng. Geol. Env. **80**(1), 267–284 (2020). https://doi.org/10.1007/s10064-020-01922-8
38. Dou, J., Yunus, A.P., Merghadi, A., Wang, X.-K., Yamagishi, H.: A comparative study of deep learning and conventional neural network for evaluating landslide susceptibility using landslide initiation zones. In: Guzzetti, F., Mihalić Arbanas, S., Reichenbach, P., Sassa, K., Bobrowsky, P.T., Takara, K. (eds.) Understanding and Reducing Landslide Disaster Risk, pp. 215–223. Springer, Cham (2021). https://doi.org/10.1007/978-3-030-60227-7_23
39. Polykretis, C., Chalkias, C.: Comparison and evaluation of landslide susceptibility maps obtained from weight of evidence, logistic regression, and artificial neural network models. Nat. Hazards **93**(1), 249–274 (2018). https://doi.org/10.1007/s11069-018-3299-7
40. Sadighi, M., Motamedvaziri, B., Ahmadi, H., Moeini, A.: Assessing landslide susceptibility using machine learning models: a comparison between ANN, ANFIS, and ANFIS-ICA. Environ. Earth Sci. **79**(24), 1–14 (2020). https://doi.org/10.1007/s12665-020-09294-8

SEM-Newmark Sliding Mass Analysis for Regional Coseismic Landslide Hazard Evaluation: A Case Study of the 2016 Kumamoto Earthquake

Zhengwei Chen and Gang Wang[✉]

Hong Kong University of Science and Technology, Kowloon, Hong Kong
gwang@ust.hk

Abstract. Coseismic landslides have been observed to cause severe damage during many historic earthquakes. Numerical simulation of fault rupture process, wave propagation and triggering of landslides considering realistic topography and geological conditions helps identify the key factors in the landslide triggering and evaluate the potential slope instability in seismically active areas.

In this study, a physics-based regional coseismic landslide evaluation framework is constructed by integrating the flexible sliding analysis into spectral element model (SEM). The framework combines advantages of SEM model for its capability of simulating complex 3D large-scale wave propagation and a flexible mass sliding analysis for capturing local soil response in coseismic slope stability evaluation. Besides, equivalent linear method is implemented to incorporate the effect of soil nonlinearity.

The developed model is adopted to simulate the landslides in the Aso Volcano area triggered by the 2016 Kumamoto earthquake. Realistic digital elevation model, site conditions and fault rupture model of simulated region (51 km × 43 km) are used in the simulation. The resulting sliding response is compared with the inventory of the triggered landslides to validate the proposed model.

Keywords: Coseismic landslide · Spectral element method · Flexible sliding analysis · Equivalent linear model · 2016 Kumamoto earthquake

1 Introduction

Earthquake-induced landslides have caused catastrophic damage to human society. For example, the 1994 Northridge earthquake caused more than 57 deaths and over 9000 injuries, and more than 82,000 residential and commercial units were damaged. It is important to estimate the location and severity of triggered landslides given earthquake and geologic conditions. In the seismic hazard evaluation procedure, prediction of the permanent sliding displacement is commonly conducted, and it is crucial for the design of infrastructure and protection of human lives. There are many factors affecting seismic response of natural terrains, such as topography, site condition, soil properties, and

© The Author(s), under exclusive license to Springer Nature Switzerland AG 2022
L. Wang et al. (Eds.): PBD-IV 2022, GGEE 52, pp. 342–352, 2022.
https://doi.org/10.1007/978-3-031-11898-2_18

characteristics of earthquake motions. These factors can significantly affect regional wave propagation and local ground-motion amplification, which need to be considered in the prediction of coseismic landslides.

A typically used method for the sliding displacement calculation is Newmark sliding block method [1]. It models landslide as a rigid block sliding on an inclined slope; A yield acceleration is determined to represent the resistance of the block against sliding. The sliding displacement is calculated by double integrating the difference between ground-motion acceleration and the yield acceleration in the time domain, until the sliding velocity becomes zero. Many studies have extended this method to flexible sliding masses by accounting for the response of soil layers [2–5], and many regressed predictive equations have been developed [6–9]. However, these studies are based on 1D local site response analysis without considering regional wave propagation and 3D topographic effects.

In this study, a physics-based regional coseismic landslide evaluation framework is constructed by integrating the rigid/flexible sliding analysis into wave propagation simulation. The wave propagation simulation is conducted using spectral element method (SEM) that is capable of simulating complex large-scale 3D wave propagation. Its local site response is used in the sliding analysis to calculate potential sliding displacement. Besides, an equivalent linear method is implemented to incorporate the effects of soil nonlinearity. The developed model is applied to simulate the coseismic landslides in the Aso Volcano area triggered by the 2016 Kumamoto earthquake.

2   Development of Regional Coseismic Landslide Evaluation Framework

Figure 1 shows the overall framework of the proposed model. We use the realistic digital elevation data to construct a 3D SEM model for a regional scale wave propagation simulation. During the simulation, shear stress acting on the sliding interface is used to drive the sliding mass at each time step.

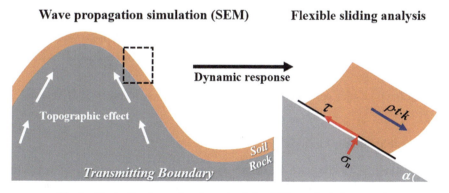

Fig. 1. Illustration for regional coseismic landslide analysis framework

2.1 Wave Propagation Simulation

For the regional scale wave propagation simulation, the Spectral Element Method (SEM) is adopted to model the complex topography and site conditions. SEM is efficient and accurate for wave propagation simulation [10]. Using high-order interpolation functions, the SEM is capable of capturing a wavelength within one mesh. To improve the computational efficiency, we used a graded mesh in soil and rock, such that the numerical model is accurate to simulate a wave frequency up to 5 Hz.

2.2 Implementation of Equivalent Linear Model

Accurate modeling of nonlinear soil behavior is important for the seismic hazard assessment, where variation of soil stiffness and energy dissipation due to soil nonlinearity should be considered [11]. The equivalent-linear method (ELM) is commonly used to approximate the soil nonlinearity in site response analysis [12]. Figure 2a shows the implementation of ELM in SEM by iteratively adjusting soil properties to reach a strain-compatible modulus (G) and damping ratio (λ) in each soil element. Firstly, the small-strain soil modulus and damping are chosen, and a new strain field is derived after the wave propagation analysis; Then, G and λ are updated according to the new strain field, and this procedure is repeated until the relative error of the maximum shear strain (γ_{max}) is less than 2%; finally, the strain-compatible soil properties are achieved and the dynamic response can be applied in the sliding analysis procedure. Note that the effective strain is adopted for the updating of soil modulus and damping, which is typically 65% of the maximum strain.

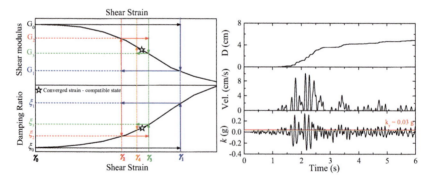

Fig. 2. (a) Iteration procedure of ELM (b) Newmark analysis algorithm in SEM model

2.3 Flexible Sliding Analysis

The Newmark-type sliding analysis method is commonly used in the evaluation of natural slopes or earthquake-induced landslides, because it provides a simple index of the seismic slope performance. The calculation requires two parameters: acceleration-time history (k) representing the driving force, and a yield acceleration (k_y) indicating resistance on the sliding interface. In this study, k is determined by the shear stress on sliding interface divided by the mass of the sliding column, which also represents the averaged acceleration within the soil mass (Fig. 1). As illustrated in Fig. 2b, sliding displacement (D) is accumulated by double integration of $(k - k_y)$ until the sliding velocity becomes zero.

The sliding calculation requires a pre-defined sliding direction, and it is assumed along the dip direction of the slope. The aspect and slope angles can be calculated using the digital elevation data of the simulated region. Then, shear stress τ along the sliding direction (also called tangential direction) can be obtained by projection of the stress tensor. Accordingly, the seismic coefficient k, can be derived as follow:

$$k = \frac{\tau}{\rho t} \tag{1}$$

where ρ is the density of the soil; t is the thickness of the sliding layer; α is the slope angle.

Conventional Newmark sliding block method represents the shear resistance by the yield acceleration k_y, which is a function of static factor of safety (FS) and slope angle (α) as follow [7]:

$$k_y = (FS - 1) \cdot \text{g} \cdot \sin\alpha \tag{2}$$

Using a limit-equilibrium model of an infinite slope, FS is derived as follow:

$$FS = \frac{c'}{\rho g t \sin\alpha} + \frac{\tan\varphi'}{\tan\alpha} - \frac{m\rho_w \tan\varphi'}{\rho \tan\alpha} \tag{3}$$

where c' and φ' are the effective cohesion and friction angle; ρ and ρ_w are soil and water densities; t is soil thickness in the slope-normal direction; m is the proportion of soil that is submerged, which in this study is set to zero. By introducing the Eq. (3) to Eq. (2), the k_y can be derived as:

$$k_y = \frac{c'}{\rho t} + \left(\tan\varphi' \cos\alpha - \frac{m\rho_w \tan\varphi' \cos\alpha}{\rho} - \sin\alpha\right) \cdot \text{g} \tag{4}$$

3   Coseismic Landslides in Aso Volcano Area Triggered by the 2016 Kumamoto Earthquake

On April 14, 2016, an M_w 7.0 earthquake occurred in Kumamoto and Oita Prefectures, central Kyushu, Japan. More than 100 people were killed, 2000 were injured and 38,000 houses were destroyed due to surface rupture, strong ground motion and landslides [13]. This earthquake caused more than 3000 landslides with a sliding area of around 7 km^2 [14]. Most of landslides triggered by this event were shallow, disrupted failures with a few flow-type slides and large rock/soil avalanches. The most impressive characteristic of the landslides is that they were highly concentrated around the Aso volcano area.

In this study, the SEM model is constructed to reproduce the full process of this earthquake, including the earthquake rupture process, wave propagation and induced coseismic landslides in a study area of 51 km × 43 km around the Aso volcano, as illustrated in Fig. 3. The highest elevation is more than 1.5 km above the sea level atop Aso caldera.

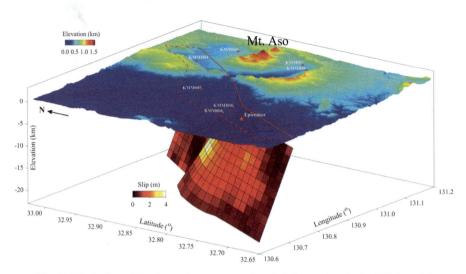

Fig. 3. Illustration of the Aso volcano area and underlying rupture in the SEM model.

3.1 Source Rupture Process of the 2016 Kumamoto Earthquake

Through multiple-time-window linear waveform inversion, the rupture process of the M_w 7.0 main event is derived [13]. As shown in Fig. 4a, the fault can be modelled as a curved fault plane divided into 28 subareas along the strike direction and 12 subareas along the dip direction, each with a size of approximately 2 km × 2 km. Figure 4b shows total slip in each subarea. Large fault rupture mainly occurred in three regions, including the Kumamoto region, the region north of Mt. Aso, and the Oita region. The surface rupture has extended approximately 30 km and almost reached the caldera of Mt. Aso.

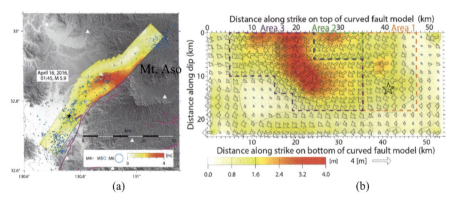

Fig. 4. (a) Map projection of the total slip distribution for the main even (epicenter is the star); (b) Planar projection of the total slip distribution (the vectors denote the direction and the slip amount on the hanging wall side) [13].

The rupture process is represented by a number of point sources in SEM, each source has its own time history referring to the given slip velocity time functions [13]. Moment tensor of each subarea is calculated with its final slip, slip, dip and rake directions.

3.2 Velocity Structure and Strength Parameters

The velocity structure adopted in this study refers to the 3D subsurface structure model of this region [15] and soil condition data explored at K-net and KiK-net stations provided by National Research Institute for Earth Science and Disaster Resilience (NIED) of Japan (Fig. 5).

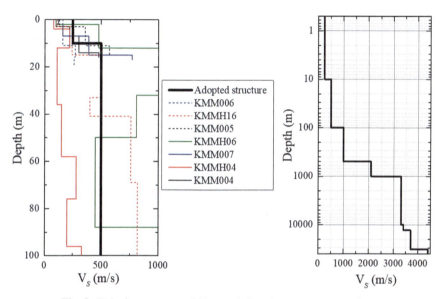

Fig. 5. Velocity structure within (a) shallow layer and (b) deep layer

Information on the soil strength parameters within the Aso volcano is very limited, as not many laboratory or in-situ tests have been conducted. The strength constants obtained by a simple test using a soil strength probe gives a friction angle of approximately 30° and cohesion of 3.5 kPa for the pumice around this area, while these two parameters are 2.0 kPa and 15° for volcanic ash [16]. The vane cone shear test was performed by [17] on a site in the Sanno-Tanigawa area in the southwest part of Mt. Eboshidake and obtained a cohesion between 4 and 10 kPa and a friction angle within 20 to 40°. They also showed that for pumice in the Takanodai area, the cohesion is in the range of 5 to 15 kPa and friction angle of 25 to 45°. A back calculation for strength parameters around this area using Newmark method was conducted to fit the realistic triggered landslides in [18], and a detailed summary for strength parameters for different geologic classifications is given. In this study, we set strength constants by referring to available experimental data, and for those not available we adopted the inferred values by [18]. Besides, sliding depths of 3 m around the central cones and 10 m around the caldera edge are identified as potential sliding depth, where shear stresses are recorded to calculate the seismic coefficient k.

3.3   Wave Propagation and Ground Motion Distribution

Figure 6 illustrates the ground velocity in the East–West direction at different times. In this simulation, the fault rupture was triggered at 0 s at the hypocenter located around 13 km in depth. It took less than 6 s for the wave to reach the ground, and then radiated outwards (see Fig. 6a). The rupture propagated to the northeast direction and the strong-motion directivity effect can be observed, as demonstrated by strong velocity as great as 1.5 m/s in northeast area (see Fig. 6b). The whole rupture process lasted around 16 s, while the entire simulation lasted 30 s until the seismic wave field propagated out of the computational domain.

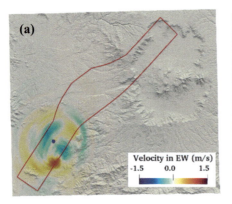 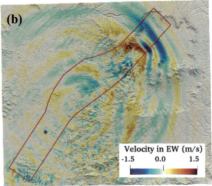

Fig. 6. Ground velocity in the East–West direction at (a) $t = 6$ s and (b) $t = 14$ s

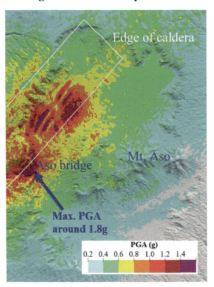 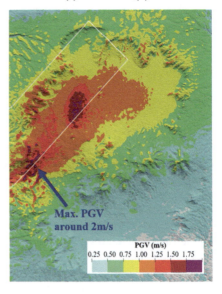

Fig. 7. PGA and PGV maps around Aso Volcano

The PGA and PGV shakemaps around Aso volcano are shown in Fig. 7. It can be seen that large PGA and PGV are concentrated on the projected area of the fault plane, i.e., the hanging wall side of the fault plane. The strong motions in this region are mainly contributed by the directivity effect and larger rupture slip (as shown in Fig. 4). Obvious topographic amplification was also observed around the edge of the Aso caldera, and we simulated a significantly large PGA of 1.8 g in the Aso bridge area due to this local topographic effect [19, 20]. Similarly, we simulated a PGV as large as 2 m/s in this location. The central cones of Aso caldera experienced relatively moderate ground motions, with PGA values between 0.2 g and 0.6 g and PGV values in the range of 0.25 m/s to 0.75 m/s. As the rupture distance increases, the peak ground motion decreases gradually. The distribution of these ground motions is closely related to the triggered landslide pattern, as will be discussed in the following part.

3.4 Predicted Landslide and Its Comparison with Inventory

The SEM-Newmark sliding analysis method evaluates the difference between the seismic coefficient k (calculated from Eq. 1) and sliding resistance k_y (calculated from Eq. 4), and it leads to sliding episodes and accumulation of permanent sliding displacements in the downslope direction. In this study, the field of sliding resistance k_y was pre-calculated using the estimated soil strengths. Then, during the SEM simulation of wave propagation, the driving force (shear stress along the sliding interface) at each time step was recorded to calculate the relative sliding displacement at each location. Sliding was considered to occur when the sliding displacement was greater than 15 cm. Finally, a sliding map was derived based on the SEM-Newmark prediction, as shown in Fig. 8a.

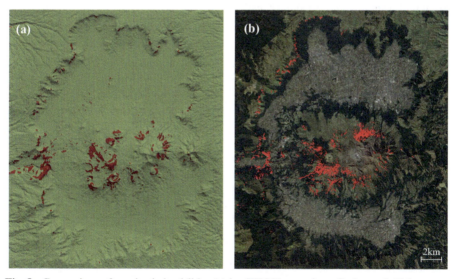

Fig. 8. Comparison of coseismic landslides (a) by SEM-Newmark prediction and (b) inventory data (compiled from [21] and [22])

The landslide inventory data is used to verify our prediction based on the landslide survey conducted by the Forest Agency of the Ministry of Agriculture, Forestry, and Fisheries, Japan [21]. They carried out high-density aerial laser measurements to identify the occurrence of landslides and cracks in the slope after the 2016 Kumamoto earthquake. As reported, the inventory map was precise after analyzing the detailed topographic map and orthorectified aerial photographs. Sliding data in some areas inside the Aso caldera are missing, and we refer to another investigation conducted by NIED [22]. These two inventories complemented each other and provided the final landslide inventory map adopted in this study (cf. Fig. 8b).

Overall, our predicted landslide distribution generally matches the inventory data. It can be found that landslides were mainly concentrated on central cones and the western edge of the caldera. A great amount of pumice with low cohesion and friction angle were located around the central cones, leading to significant amount of landslides in that area due to weak soil strength. Note that the ground motions within this region are quite strong, with PGA around 0.5 g and PGV around 75 cm/s. The intensive ground motions along the western edge of the caldera also lead to great amount of landslides. One of the most notable destroyed the Aso bridge [14]. In addition, the northwestern edge of the caldera also suffered severe damage.

Figure 9 illustrates the distribution of slope angles and the percentage of landslides by statistical analysis using grid cells of 5 m × 5 m. Note that the inventory data includes both the sliding sources and deposits, while the Newmark method only evaluates the potential sliding source based on a threshold value. In this study, we identify the sliding sources as the upper half of the sliding mass in the inventory based on common practice. Figure 9 shows that around 38% of the landslide source occurred on slope with angles of 10–20°, and 32% occurred on slope angles of 20–30°. The corresponding percentages of the predicted landslides by Newmark analysis are also reported for comparison using different threshold values of sliding displacements.

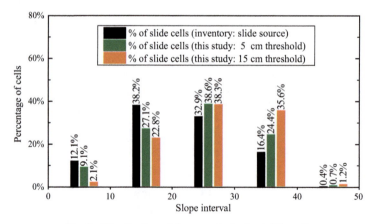

Fig. 9. Distribution of slope angle for landslide cells

Finally, we evaluate the accuracy of the SEM-Newmark prediction by one-to-one comparison of the predicted and measured landslides using the 5 m × 5 m grid cells. If a threshold sliding displacement of 15 cm is adopted, the SEM-Newmark model can capture 37.4% of the sliding sources from the inventory, and the model over-predicted the landslide area by a factor of 4.6. If a more conservative threshold of 5 cm is adopted, the SEM-Newmark model can capture up to 50.5% of the sliding sources, however, the over-prediction ratio increases to 7.7.

4 Summary and Discussion

In this study, a computational framework is proposed by integrating 3D physics-based wave propagation simulation with the Newmark-type sliding mass analysis for a regional evaluation of coseismic landslides. The model is examined by simulating wave propagation and coseismic landslides in Kumamoto Prefecture during the 2016 Kumamoto earthquake. This study is first of its kind to evaluate regional coseismic landslide (51 km × 43 km) considering realistic rupture process, wave propagation, topographic-site effect, and local site conditions. The rupture directivity and ground motion amplification effects are clearly observed, as demonstrated in the ground motion maps. The sliding analysis satisfactorily identifies the potential sliding areas, with prediction accuracy up to 50% (using 5 cm as threshold).

The accurate prediction of coseismic landslides is attributed to reasonable strength values adopted for various geologic classifications in this region, which helps identify areas susceptible to seismic hazards. Besides, accurate simulation of the wave propagation is also very important, because they can be significantly amplified due to local topography and altered by the pattern of fault rupture. Despite the good performance of the proposed framework, it still remains as a challenge to preciously predict all sliding areas especially for such a large region. This is because of the existence of many uncertainties and variabilities. For example, velocity structure of the domain, crucial to the wave propagation, is treat as uniformly distributed based on limited geologic data. The rupture process obtained from the waveform inversion also included many sources of uncertainties. Other factors like soil strength, material nonlinearity, sliding depth and plant root reinforcement also affect the prediction. These effects will be investigated in a future study.

Acknowledgements. The study is supported by Hong Kong Research Grants Council (Grant No. 16214322).

References

1. Newmark, N.M.: Effects of earthquakes on dams and embankments. Geotechnique **15**, 139–159 (1965)
2. Seed, H.B., Martin, G.R.: The seismic coefficient in earth dam design. J. Soil Mech. Found. Div. **92**(3), 25–58 (1966)
3. Bray, J.D., Rathje, E.M.: Earthquake-induced displacements of solid-waste landfills. J. Geotech. Geoenviron. Eng. **1243**, 242–253 (1988)
4. Wang, G.: Efficiency of scalar and vector intensity measures for seismic slope displacements. Front. Struct. Civ. Eng. **6**(1), 44–52 (2012)
5. Rathje, E.M., Antonakos, G.: A unified model for predicting earthquake-induced sliding displacements of rigid and flexible slopes. Eng. Geol. **122**(1–2), 51–60 (2011)

6. Saygili, G., Rathje, E.M.: Empirical predictive models for earthquake-induced sliding displacements of slopes. J. Geotech. Geoenviron. Eng. **134**(6), 790–803 (2008)
7. Jibson, R.W.: Regression models for estimating coseismic landslide displacement. Eng. Geol. **91**(2–4), 209–218 (2007)
8. Bray, J.D., Travasarou, T.: Simplified procedure for estimating earthquake-induced deviatoric slope displacements. J. Geotech. Geoenviron. Eng. **1334**, 381–392 (2007)
9. Watson-Lamprey, J., Abrahamson, N.: Selection of ground motion time series and limits on scaling. Soil Dyn. Earthq. Eng. **265**, 477–482 (2006)
10. Komatitsch, D., Vilotte, J.P.: The spectral element method: an efficient tool to simulate the seismic response of 2D and 3D geological structures. Bull. Seismol. Soc. Am. **88**(2), 368–392 (1998)
11. Kaklamanos, J., Bradley, B.A.: Critical parameters affecting bias and variability in site-response analyses using KiK-net downhole array data. Bull. Seismol. Soc. Am. **103**(3), 1733–1749 (2013)
12. Zalachoris, G., Rathje, E.M.: Evaluation of one-dimensional site response techniques using borehole arrays. J. Geotech. Geoenviron. Eng. **141**(12), 04015053 (2015)
13. Kubo, H., Suzuki, W., Aoi, S., Sekiguchi, H.: Source rupture processes of the 2016 Kumamoto, Japan, earthquakes estimated from strong-motion waveforms. Earth Planets Space **68**(1), 1–13 (2016)
14. Xu, C., Ma, S., Tan, Z., Xie, C., Toda, S., Huang, X.: Landslides triggered by the 2016 Mj 7.3 Kumamoto, Japan, earthquake. Landslides **15**(3), 551–564 (2018)
15. Fujiwara, H., et al.: A study on subsurface structure model for deep sedimentary layers of Japan for strong-motion evaluation. Technical Note 337, National Research Institute for Earth Science and Disaster Prevention, Tsukuba, Japan (2009)
16. Kanai, T., Asai, K., Sasaki, Y., Norimizu, S.: Evaluation of slope stability based on the soil strength probe. In: Proceeding of Japan Society of Engineering Geology, pp. 169–170 (2016)
17. Fukunaga, E., Shimizu, O.: Soil layer structure and soil strength in the vicinity of the slip surface in the collapsed area of the Aso central volcano induced by the 2016 Kumamoto earthquake. In: Proceeding of Japan Society of Erosion Control Engineering, pp. 676–678 (2017)
18. Shinoda, M., Miyata, Y., Kurokawa, U., Kondo, K.: Regional landslide susceptibility following the 2016 Kumamoto earthquake using back-calculated geomaterial strength parameters. Landslides **16**(8), 1497–1516 (2019)
19. Huang, D., Wang, G., Du, C., Jin, F., Feng, K., Chen, Z.: An integrated SEM-Newmark model for physics-based regional coseismic landslide assessment. Soil Dyn. Earthq. Eng. **132**, 106066 (2020)
20. Chen, Z., Huang, D., Wang, G., Jin, F.: Topographic amplification on hilly terrain under oblique incident waves. In: Tournier, J.P., Bennett, T., Bibeau, J. (eds.) Sustainable and Safe Dams Around the World, ICOLD Proceedings, vol. 2, pp. 2778–2786. CRC Press, Leiden (2019)
21. Forest Agency of the Ministry of Agriculture, Forestry, and Fisheries: Report of aviation laser measurement work in the forest area (2016)
22. NIED. https://www.bosai.go.jp/mizu/dosha.html. Accessed 30 Oct 2021

Understanding Excess Pore Water Dissipation in Soil Liquefaction Mitigation

Siau Chen Chian[1]([⊠]) and Saizhao Du[2]

[1] National University of Singapore, Singapore 117576, Singapore
sc.chian@nus.edu.sg
[2] Beijing Jiaotong University, Beijing 100044, China

Abstract. Soil liquefaction has conventionally been studied in cyclic laboratory tests as a pure undrained condition, assuming that excess pore water pressure is unable to dissipate during rapid shearing. As such, dissipation of excess pore water pressure generated during shearing is often not modelled in classic cyclic simple shear and triaxial tests. In contrast, dissipation of excess pore pressure does occur in the field following major earthquakes in the form of severe ground settlement and sand boils where fine granular soils were found. In this theme lecture, two aspects of soil liquefaction are discussed. First, a study on the interaction of excess pore water pressure generation and dissipation on liquefiable clean sand in the cyclic triaxial test setup is conducted. Results show that pore water dissipation at merely a fraction of the permeability of the soil can have a significant effect to the soil's susceptibility to liquefaction. This calls for future cyclic laboratory tests to adopt similar near-perfect undrained condition to better reflect a more realistic representation of soil liquefaction in the field. Second, a critical cyclic shear stress ratio is obtained, where excess pore pressure dissipation dominates excess pore pressure generation, beyond which the soil is no longer susceptible to liquefaction. Complementing this ratio is the identification of a critical void ratio in which the excess pore pressures do not build up despite considerable shearing amplitude, implying the soil is no longer susceptible to liquefaction at that degree of densification. Such information would serve as a guidance to mitigate soil liquefaction in soil densification operations.

Keywords: Soil liquefaction · Pore pressure · Generation · Dissipation · Cyclic shear stress ratio · Void ratio

1 Introduction

Damage caused by soil liquefaction following strong earthquakes persist to date as evident in recent earthquake events such as the 2016 Muisne [1], 2011 Tohoku [2], 2011 Christchurch [3], 2010 Maule [4] and 2009 Padang [5]. Earthquake induced soil liquefaction can lead to damaged relating to excessive ground settlement, lateral spreading, foundation failures and uplift of underground infrastructure. The phenomenon of soil liquefaction is caused by the buildup of excess pore water pressure when loose saturated granular soil is subjected to rapid shearing in which the pore water is unable to

L. Wang et al. (Eds.): PBD-IV 2022, GGEE 52, pp. 353–362, 2022.
https://doi.org/10.1007/978-3-031-11898-2_19

escape from the collapsing voids within the soil particles. As such, permeability during the shearing is a crucial parameter to consider in the development of excess pore water pressure.

At present, single element laboratory studies on soil liquefaction using cyclic simple shear and triaxial shear testing are often limited to undrained condition. This implies that dissipation of pore water is ignored. In contrast, observations from geotechnical centrifuge modelling [6] and video footages of the ground following the 2011 Tohoku Earthquake [7] do indicate that pore pressure generation and dissipation do take place concurrently, often in the form of sand boils. The importance of pore water drainage is also inferred by most liquefaction induced failures taking place during the dissipation phase [8]. In this paper, small amount of pore water drainage in cyclic triaxial testing to mimic some dissipation of excess pore water pressure during shearing is introduced to offer a closer presentation of the field condition.

2   Cyclic Triaxial Setup with Drainage

2.1   Soil Properties

A series of cyclic triaxial testing were carried out to investigate the influence of small amount of drainage (i.e. a near perfect undrained condition) on the impact of soil lique-faction. The granular soil used in these tests was fine grained uniform silica sand with particle size distribution presented in Fig. 1. The physical geotechnical properties of the sand are presented in Table 1.

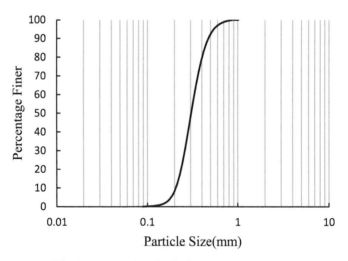

Fig. 1. Particle size distribution of silica sand used

Table 1. Physical properties of silica sand

Properties	Values
Φ_{crit}	34°
G_S	2.65
e_{max}	0.86
e_{min}	0.48
D_{10}	0.22 mm
D_{50}	0.26 mm
D_{90}	0.48 mm

2.2   Triaxial Setup

The cyclic triaxial testing was carried out on a GDS enterprise level dynamic triaxial testing system (ELDYN), based on an axially-stiff load frame with a beam mounted electro-mechanical actuator. Two sets of pressure controllers, each connecting to the ends of the cylindrical soil samples, were adopted to record the amount of water flowing into or out of the sample. This permits measurement of permeability as well as computation of void ratio changes during testing. Figure 2 shows the setup of the modified triaxial system.

Cylindrical soil samples of 76 mm and 38 mm in height and diameter were prepared using air pluviation in a rubber membrane with porous stones and filter paper on each end. In order to facilitate pluviation of sand into the membrane, vacuum was applied on the outside of the membrane to keep the membrane upright. Higher densities of soil samples were achieved by raising the fall height while pouring sand into the membrane. After completion of sand pouring, the dry sand sample was saturated with de-aired water using the back pressure inlet of the triaxial setup. Thereafter, the sample was consolidated isotropically to the target final effective isotropic confining pressure. This marks the end of sample preparation and the sample ready to be sheared.

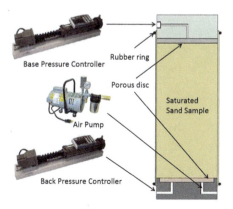

Fig. 2. Setup of modified triaxial setup

Triaxial permeability tests were carried out using the same sand and similar density and effective confining pressure. At static condition, the coefficient of permeability for a fully saturated sand sample was about $5 \times 5 \times 10^{-4}$ m/s (or flow rate of about 7400 mm^3/s for a 38 mm diameter sample) under a pressure difference of 10 kPa. In order to study the impact of pore water dissipation, 3 different flowrates (5, 10 and 20 mm^3/s) far lower than the static permeability of the sand to ensure a near-perfect undrained condition.

2.3 Test Programme

The cyclic triaxial tests involved shearing the sand undrained till liquefaction to maintain the height and diameter of sample, before transiting to stress-controlled shearing with allowance of little drainage outflow to take place as illustrated in Fig. 3. Soil samples were consolidated under an initial effective isotropic confining stress of about 75 kPa and sheared at a frequency of 0.2 Hz. In Table 2, the tests were to investigate the dissipation profile resembling consolidation on self-weight and dependent on permeability of the liquefied soil (i.e. Test D-0-5) as well as study the impact of pore water expulsion from the ground (i.e. small amount of drainage) during earthquake shaking which is prevalent in the field. The permeability of the sand obtained from triaxial permeability test was 5×10^{-4} m/s or flow rate of 7400 mm^3/s for a 38 mm diameter sample. Hence, discharge rates in Table 2 are comparatively low.

Table 2. Details of cyclic triaxial tests conducted

Amplitude (kN)	σ (kPa)	Q (mm^3/s)	Dissipation stage (start with $r_u \geq 0.95$)			
			Test ID	e_0	$D_{r,0}$ (%)	R
0	0	5	D-0-5	0.714	38.3	0.00
0.01	8.8	5	D-0.01-5	0.698	42.6	0.06
		10	D-0.01-10	0.694	43.7	0.06
		20	D-0.01-20	0.688	45.3	0.06
0.05	44.1	5	D-0.05-5	0.692	44.2	0.30
		10	D-0.05-10	0.696	43.2	0.30
		20	D-0.05-20	0.700	42.1	0.30
0.1	88.2	5	D-0.1-5	0.710	39.5	0.59
		10	D-0.1-10	0.702	41.6	0.59
		20	D-0.1-20	0.697	42.9	0.59
0.15	132.3	5	D-0.15-5	0.714	38.4	0.88
		10	D-0.15-10	0.708	40.1	0.88
		20	D-0.15-20	0.705	40.8	0.88

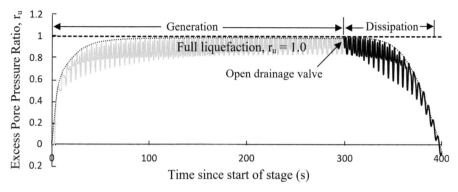

Fig. 3. Cyclic triaxial test with generation and dissipation

3 Results and Discussion

3.1 Effect of Discharge on Excess Pore Pressure Dissipation

Based on Table 2, for thCe test involving no shearing during dissipation (Test D-0-5), a parabola behaviour in excess pore pressure dissipation was observed as expected in Fig. 4a. In the case of continual shearing during dissipation, full liquefaction was maintained for few loading cycles before excess pore pressure ratio commenced its decline as demonstrated in Fig. 4b–e. This implies that prior to the decline while partial drainage was permitted, the rate of excess pore pressure generation was larger than dissipation during those few loading cycles. However, over time, the allowance of drainage would lead to the reduction in volume of the soil sample by means of collapsing of large voids. Eventually, the soil sample would possess a greater resistance to liquefaction due to densification which is demonstrated with the increasing rate of decline in excess pore pressure ratio observed. It can also be observed that a larger cyclic stress amplitude would prolong the full liquefaction regime since larger stress amplitudes are expected to generate greater excess pore pressure generation. Such greater excess pore pressure generation is also supported by the gentler slope of the excess pore pressure decline for larger stress amplitude tests.

Figure 5 shows the effect of discharge flowrate on the dissipation of excess pore water pressure. It can be observed that the discharge flowrate has a similar but opposite effect as compared to the cyclic stress amplitude. A larger discharge flowrate lead to a more rapid rate and commencement of excess pore pressure reduction. This is despite the flowrate being excessively low as compared to the static permeability of the soil, which infers that the assumption of undrained condition of earthquake induced soil liquefaction may lead to overdesign of infrastructure due to prolong effects of soil liquefaction in the simulation.

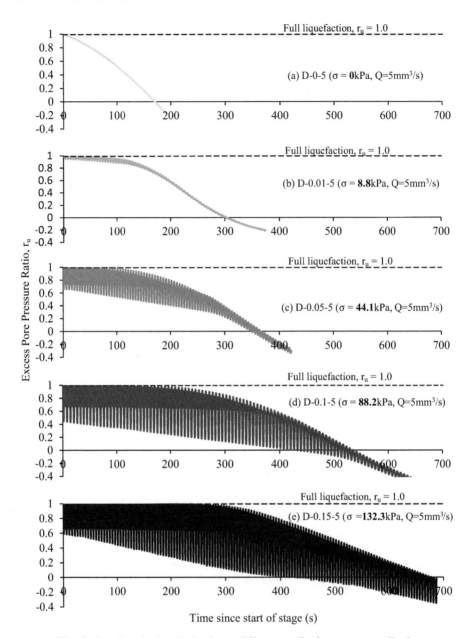

Fig. 4. Post liquefaction dissipation at different cyclic shear stress amplitude σ

In order to make comparison between tests, the juncture of transition from an excess pore water pressure generation to excess pore water pressure dissipation dominant behaviour is identified as the point in time where maximum pore pressure change per unit cycle (i.e. the point of steepest slope of $\frac{\Delta u}{t}$) is observed from a near linear trend (see Fig. 5).

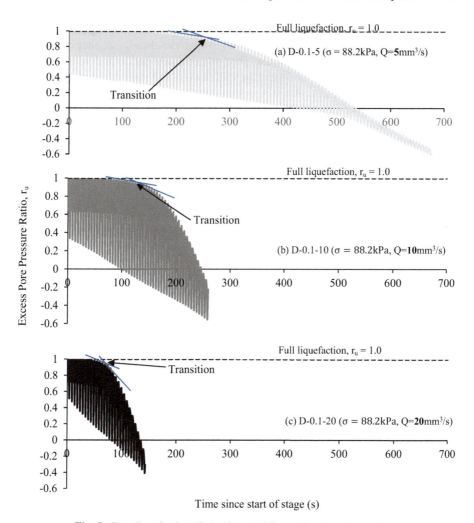

Fig. 5. Post liquefaction dissipation at different discharge flowrate Q

Figure 6 shows the relationship between shear stress amplitude, flowrate and volumetric strain. These data points refer to the volumetric strains where excess pore pressure profiles began to fall under coupled shear loading and draining (i.e. transition point). It is interesting to note that these points show a sharp rise in shear stress amplitude at the juncture of about 0.02% volumetric strain. This persists even at different flowrate, thereby demonstrating similar limiting packing of the sand grains where liquefaction susceptibility is reduced dramatically.

Since a limiting condition exists with respect to the packing of sand grain, a detailed analysis with the use of void ratio is conducted to ascertain the significance of the limit volumetric strain observed. Their excess pore pressure dissipation and void ratios are presented in Fig. 7. The figure shows that the void ratios which marks the commencement of decline of excess pore pressure (i.e. at transition point) were similar in magnitude at 0.688 for these tests, despite differing initial void ratios and stress amplitudes. This indicate the critical void ratio to be about 0.688 where excess pore pressure generation can no longer surpass the dissipation to support the high excess pore pressure in the sand.

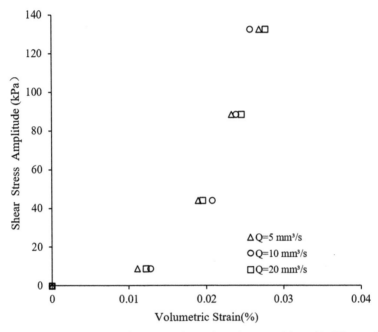

Fig. 6. Cyclic shear stress amplitude versus volumetric strain at transition with different discharge flowrate

The minimum degree of densification is influenced by both arrangement of sand grain and cyclic shear stress ratio. As shown in Fig. 5, higher flowrate requires less time to dissipate the entire excess pore pressure and vice versa. The outcome results in consistent amount of water flowed out of sample, i.e. consistent reduction in void ratio e at similar volumetric strain, suggesting that minimum degree of densification exists under similar cyclic shear stress amplitude and confining pressure condition. Determining such critical void ratio may well provide guides on the minimum degree of densification of the sand as part of conventional soil liquefaction mitigation measures.

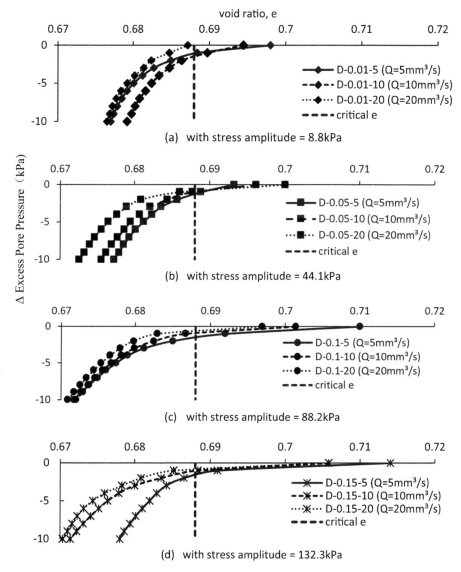

Fig. 7. Critical void ratio on pore pressure dissipation with different stress amplitude

4 Conclusion

Field observations have demonstrated that a liquefied ground does not necessarily behave undrained during an earthquake. Near perfect undrained triaxial tests were conducted and showed that stronger shaking in the form of higher cyclic shear amplitude prolongs full liquefaction, while higher permeability soil with high rate of pore water discharge discourages sustained liquefaction. These are in line with conventional undrained laboratory tests. However, in the case of these near perfect undrained tests, the progressive discharge of pore water would lead to the collapse of large voids which respond alike to soil densification. Liquefaction susceptibility is hence reduced as a result. The existence of a critical void ratio was also identified where further shearing does not lead to soil liquefaction. Such identification of critical void ratio would suggest the desired degree of densification in soil liquefaction mitigation measures.

References

1. Franco, G., et al.: The April 16 2016 Mw7.8 Muisne earthquake in Ecuador – preliminary observations from the EEFIT reconnaissance mission of May 24 – June 7. In: 16th World Conference on Earthquake Engineering, Paper No. 4982 (2017)
2. Chian, S.C., Tokimatsu, K., Madabhushi, S.P.G.: Soil liquefaction-induced uplift of underground structures: physical and numerical modeling. J. Geotech. Geoenviron. Eng. **140**(10), 04014057 (2014)
3. Earthquake Engineering Research Institute (EERI): The M 6.3 Christchurch, New Zealand. Earthquake of EERI February 22, 2011. EERI Special Earthquake Report (2011)
4. Geoengineering Extreme Events Reconnaissance (GEER): Geo-engineering reconnaissance of the 2010 Maule, Chile earthquake. Report of the NSF Sponsored GEER Association Team (2010)
5. Wilkinson, S.M., Alarcon, J.E., Whittle, J., Chian, S.C.: Observations of damage to building from Mw 7.6 Padang earthquake of 30 September 2009. Nat. Hazards **63**, 521–547 (2012)
6. Chian, S.C., Madabhushi, S.P.G.: Influence of fluid viscosity on the response of buried structures under earthquake induced liquefaction. In: 7th International Conference on Physical Modelling in Geotechnics, Zurich, pp. 111–115 (2010)
7. Goldberg, S.: See the ground actually open up and move. YouTube website. https://www.youtube.com/watch?v=TzlodnjPAuc. Accessed 30 Dec 2021
8. Kramer, S.L.: Geotechnical Earthquake Engineering. Prentice Hall, New York (1996)

Buried Pipeline Subjected to Ground Deformation and Seismic Landslide: A State-of-the-Art Review

Deepankar Choudhury$^{(\boxtimes)}$ ⬥ and Chaidul H. Chaudhuri ⬥

Department of Civil Engineering, Indian Institute of Technology Bombay, Powai, Mumbai 400076, India
dc@civil.iitb.ac.in, chaudhuri.ch@iitb.ac.in

Abstract. Pipelines are commonly used for transporting different materials namely water, gas, sewage, and oil from one place to another. The different past earthquakes (1923 Kanto earthquake, 1971 San Fernando earthquake, 1994 Northridge earthquake, 2010 Chile earthquake) induced hazards such as landslide, fault movement, liquefaction, etc., resulting in the damage of buried pipelines. These hazards induced ground deformations are known as permanent ground deformation (PGD), and the deformation resulting from wave propagation is called transient ground deformation (TGD). Further, soil can move along or normal to the pipe axis, and accordingly, it can be further categorized as axial and transverse ground deformation respectively. Apart from seismic excitation, ground deformation and vibration can also be generated from other sources like pipe bursting, underground explosion etc. Failure of pipelines due to ground deformation can cause the source of firing, contamination to the environment, explosion, economic loss etc. Therefore, it is vital to design the buried pipeline incorporating the effect of possible ground movement on buried pipelines. Thus, the focus of the present review study is to understand the various possible patterns of ground deformation, estimation of additional forces on pipeline due to ground deformation, and their influence on the response of buried pipeline, which can be implemented in practice to carry out performance-based design of buried pipelines subjected to earthquake loadings.

Keywords: Buried pipe · Ground deformation · Seismic landslide

1 Introduction

Buried pipelines may be subjected to ground movement or vibration either resulting from earthquake induced hazards like fault movements [1–4], soil spreading [5–11], liquefaction [12–17], or pipe bursting induced ground movement [18–23], surrounding explosion [24–28], etc. Underground pipe failure may disrupt the whole pipe network and even be the source of a disaster based on the substance carried by the pipeline. Figure 1 shows some past failures of pipeline. Hence, it is pivotal to understand the response of buried pipeline under such kind of ground deformation or vibration. Pipe-soil system can

© The Author(s), under exclusive license to Springer Nature Switzerland AG 2022
L. Wang et al. (Eds.): PBD-IV 2022, GGEE 52, pp. 363–375, 2022.
https://doi.org/10.1007/978-3-031-11898-2_20

be modelled using various available techniques such as beam on elastic foundation, shell model, plane-strain model, and Hybrid model [29]. Each model has its own merits and demerits. For instance, beam on spring foundation concept will not be able to capture the buckling and fracture phenomenon of the pipe. Still, the method has advantages in terms of less time-consuming, input parameters, and complexity. Hence, before performing a detailed continuum-based numerical or rigorous experimental study, beam on elastic foundation model can be used in the initial design stage. Further, Psyrras and Sextos [30] mentioned four types of mode of failure (such as shell-mode buckling, beam-mode buckling, tensile failure, and cross-section ovalization) of buried steel pipes under seismic loads. Shell-mode buckling is generally found for large diameter pipe buried at greater depth—such type of buckling induced from pure bending or compressive load. Beam-mode buckling is observed for small diameter pipes buried at shallow depth. Such bending occurs under compression. Pipes under tensile force lead to a tensile mode of failure. When pipe is subjected.

(a)

(b)

(c)

Fig. 1. (a) Buckling of steel pipe during 1971 Sanfernando earthquake (after Chenna et al. [31]) (b) Failure of pipe during 1999 Kocaeli Earthquake (after Chenna et al. [31]) and (c) Water leakage from pipe near IIT Bombay main building (3 June, 2019).

to bending stress, and it leads to the change of pipe diameter from circular to oval shape, it is called cross-section ovalization. The present study highlights past studies performed in the area of pipe subjected to ground deformation, mainly focused on some of the past analytical and semi-analytical works to estimate the response of pipe subjected to ground deformation.

2 Pipe Under Horizontal Transverse Ground Deformation

The horizontal transverse ground deformation patterns can be divided in two ways such as abrupt horizontal transverse ground deformation and spatially distributed horizontal transverse ground deformation [32, 33]. A typical representation of both types of ground deformation patterns are shown in Fig. 2.

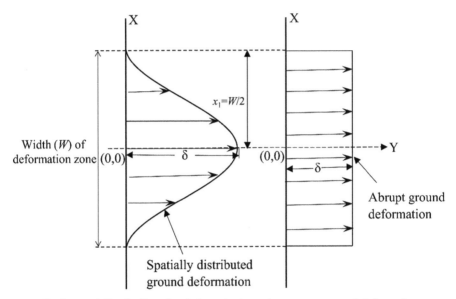

Fig. 2. Spatially distributed and abrupt horizontal transverse ground deformation.

O'Rourke and Lane [12] used an improved beta function to represent lateral ground deformation pattern obtained from liquefaction. The effect of this lateral ground movement pattern on underground pipeline was obtained by UNIPIP coding [34]. Miyajima and Kitaura [35] investigated the influence of transverse permanent ground deformation on buried pipeline through a theoretical approach based on the theory of beam on spring foundation without considering the axial tension of the pipe. Later on, Chaudhuri and Choudhury [36] presented a semi-analytical approach adopting the pipe axial tension for estimating the response of buried pipe under lateral ground deformation. A spatially distributed cosine function was used to represent the pattern of lateral ground deformation. The soil and pipe were idealized as single parameter Winkler spring foundation and conventional Euler Bernoulli's beam respectively. The study was further extended by Chaudhuri and Choudhury [37] considering pipe as Timoshenko beam to take care of the effect of transverse shear deformation and soil as Winkler springs along with the shear interaction between individual springs. The study also used the cosine function to represent the ground deformation pattern (w_g) as shown in Eq. (1) (modified after O'Rourke [38])

$$w_g = \frac{\delta}{2}\left(1 + \cos\frac{\pi x}{x_1}\right) \tag{1}$$

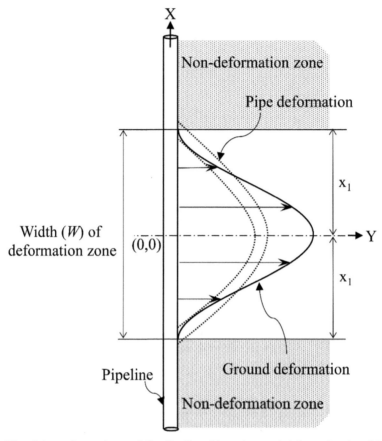

Fig. 3. Pipe deformation under spatially distributed lateral ground deformation (modified after Chaudhuri and Choudhury [37]).

Chaudhuri and Choudhury [37] assumed that the ground deformation pattern is symmetric (refer Figs. 2 and 3). Hence, only half of the model is considered to obtain the governing differential equations and their corresponding solutions. The pictorial representation of the deformation pattern of pipe and soil under spatially distributed lateral ground deformation is shown in Fig. 3. Axial resistance within the ground deformation zone was neglected because the resistance against pipe deformation will mainly be provided by transverse soil springs in the deformation zone. However, in the non-deformation area, both the axial and transverse resistance were taken into account. ALA guidelines [39] were used to model the Winkler spring stiffness and axial soil resistance. The plasticity part of the bi-linear soil-springs were not considered in the study. ALA guidelines provide the following expressions to model the lateral soil springs.

$$P_u = N_{ch}cD + N_{qh}\overline{\gamma}HD \qquad (2)$$

$$N_{ch} = a + bx + \frac{c}{(x+1)^2} + \frac{d}{(x+1)^3} \leq 9 \qquad (3)$$

$$N_{qh} = a + b(x) + c(x^2) + d(x^3) + e(x^4) \qquad (4)$$

where, P_u is the peak soil force in the lateral direction per unit length of the pipe. N_{ch} and N_{qh} are the lateral bearing capacity factors for clay and sand respectively. c is the cohesion of soil, D is the exterior dia of the pipe, H is the burial depth up to pipe center, and $\overline{\gamma}$ is the effective unit weight of soil. The value of x depends on the ratio of H/D. a, b, c, d, and e depend on the soil friction angle mentioned by ALA guidelines. Displacement at P_u is $0.04 (H + D/2) \leq 0.10D$ to $0.15D$.

Further, soil force in the axial direction can be evaluated using the following expressions [39]

$$T_u = \pi Dac + \pi DH\overline{\gamma}\frac{1 + K_0}{2}\tan \delta \qquad (5)$$

$$\alpha = 0.608 - 0.123c - \frac{0.274}{c^2 + 1} + \frac{0.695}{c^3 + 1} \qquad (6)$$

where, K_0 is the earth pressure coefficient at rest condition, α is the adhesion factor, δ is the interface friction angle for pipe and soil. Displacement at T_u is 10 mm, 8 mm, 5 mm, and 3 mm for soft clay, stiff clay, loose sand, and dense sand respectively.

The governing differential equations of pipe deflection (w) and pipe rotation (φ) considering pipe as Timoshenko beam subjected to ground deformation as shown in Fig. 3 can be represented as follows [37]:

$$D'\frac{d^4w}{dx^4} - \frac{D'}{C}\frac{d^2q(x)}{dx^2} - T\frac{d^2w}{dx^2} + q(x) = 0 \qquad (7)$$

$$\varphi = \frac{D'}{C}\frac{d^3w}{dx^3} + \frac{dw}{dx} - \frac{D'}{C^2}\frac{dq(x)}{dx} - \frac{T}{C}\frac{dw}{dx} \qquad (8)$$

where, D' is the bending stiffness of pipe, C is the shear stiffness of pipe, T is the pipe tension, and $q(x)$ is the soil pressure acting on the pipe. From the end boundary conditions and continuity conditions between the deformation and normal zone of Fig. 3, the complete solution of prior mentioned differential equations is obtained. From the detailed parametric study, it was noticed that after a specific range of peak ground deformation, the maximum pipe deformation started reducing with respect to peak ground deformation. Further, beyond the ground deformation zone pipe deformation was also

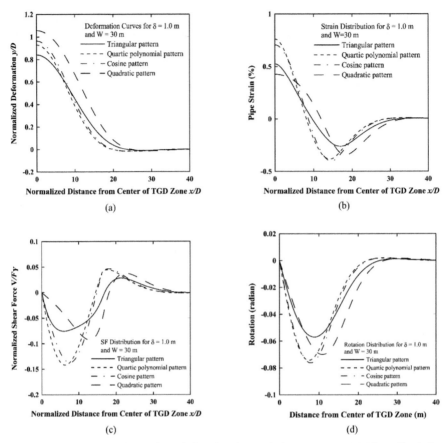

Fig. 4. (a) Normalized deformation of pipe (b) Pipe strain in percentage (c) Normalized shear force, and (d) Pipe rotation in radian, distribution for various shapes of ground deformation (after Chaudhuri and Choudhury [37]).

observed up to a certain extent. Pipe stability against such lateral ground movement can be increased by reducing pipe diameter, increasing pipe wall thickness, or providing loose soil around the pipe. Further, along with the cosine pattern of ground deformation, additional three types of pattern namely triangular, quartic polynomial, and quadratic were taken for the semi-analytical study. It was recognized that pipe responses change substantially with varying the expected shape of ground deformation, as depicted in Fig. 4.

3   Pipeline Under Seismic Landslide

Landslide is under the category of permanent ground deformation (PGD) [30]. Zheng et al. [40] mentioned two accidents of buried pipeline in the landslide area in Zhejiang province of China (Yuyao city and Ningbo city) and failure segment of buried pipeline is detected by electromagnetic induction. Field investigation shows deflection is non-uniform and mainly in the horizontal direction. The non-uniform distribution of horizontal deflection is close to the quartic polynomial curve. Zheng et al. [40] and Luo et al. [41] carried out FEM-based numerical analysis to investigate the influence of seismic landslide on buried pipeline. In the numerical analysis, a quartic polyno-mial distribution of lateral displacement was applied on the soil to simulate the seismic landslide. The piping system is deformed under the given ground-induced actions. Ma et al. [42] and Zheng et al. [40] performed the numerical analysis considering soil as linear-elastic material whereas, Luo et al. [41] adopted Drucker-Prager model for soil to simulate seismic landslide problem. Later on, Chaudhuri and Choudhury [43] proposed a theoretical solution to obtain the impact of seismic landslide on buried pipeline. In the-oretical solution, pipe was assumed as conventional Euler Bernoulli's beam and soil was represented by Winkler foundation. Following quartic polynomial function was used to simulate the pattern of ground deformation induced by seismic landslide (modified after Luo et al. [41]):

$$w_g = \frac{2\delta}{x_1^2}(x - x_1)^2 - \frac{\delta}{x_1^4}(x - x_1)^4 \quad (0 \leq x \leq x_1) \tag{9}$$

Chaudhuri and Choudhury [43] also performed numerical analysis using FEM based program Abaqus 3D. In the numerical analysis, block pattern ground deformation was simulated and obtained the impact of the ground deformation on the buried pipeline. A typical 3D model of a soil-pipe system subjected to block pattern ground deformation and pipe response is shown in Fig. 5(a)–(c).

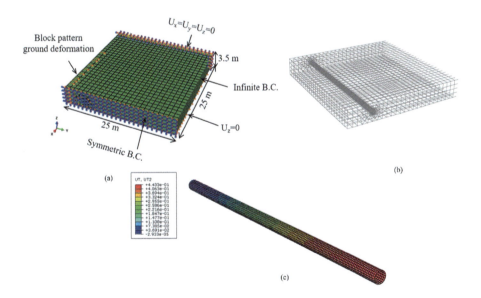

Fig. 5. Finite element model of soil-pipe system with (a) Boundary conditions (b) Mesh discretization, and (c) Deformed buried pipeline

4 Pipe Under Static Pipe Bursting Underneath

Nowadays, the static pipe bursting technique gained popularity over conventional cut and cover method for laying the new pipe or replacing the existing one due to the involvement of minimum excavation and disturbance on the ground surface during pipe bursting operation. During this operation, an outward force will act surrounding the expander (which is used for pipe bursting operation) and hence ground heave will generate. Pipe bursting induced ground movement may have severe effects on the surrounding existing structures, if any. Cholewa et al. [21] performed a laboratoty experiment, where a polyethylene pipe was used to replace an existing unreinforced concrete pipe using static pipe bursting technique. Another PVC (Polyvinyl chloride) pipe was laid in cross-sectional direction above the concrete pipe. During pipe bursting process, the response of surrounding PVC pipe in terms of pipe strain and displacement were recorded. Later on, Shi et al. [23] performed prior mentioned study through a simplified numerical analysis using FEM based Abaqus program. In the numerical analysis, pipe and soil were modelled by hollow beam and PSI (pipe-soil interaction) elements respectively. From literature, it is realized that pipe bursting induced upward ground heave can be well simulated by a Gaussian function [20, 21]. Chaudhuri and Choudhury [44] investigated the behaviour of an adjacent buried pipe exposed to pipe bursting underneath through an analytical study. The mathematical formulations were proposed considering soil as Pasternak foundation and adjacent pipe as both Bernoulli's beam and Timoshenko beam. Figure 6 replicates the simplified assumptions in the analytical model. The Gaussian error function was used to simulate the ground heave pattern obtained from static pipe bursting operation. The following error function was used in the study:

$$S(x) = S_{max} \exp\left[-\frac{x^2}{2(\frac{i}{\sin\theta})^2}\right] \tag{10}$$

where, i and θ are the distance from the pipe center to the inflection point of the error function and the intersection angle between the pipes respectively.

The additional load $P(x)$ acting on the neighboring pipe owing to pipe bursting underneath can be calculated as:

$$P(x) = K_{eq}S(x) - G\frac{d^2}{dx^2}S(x) \tag{11}$$

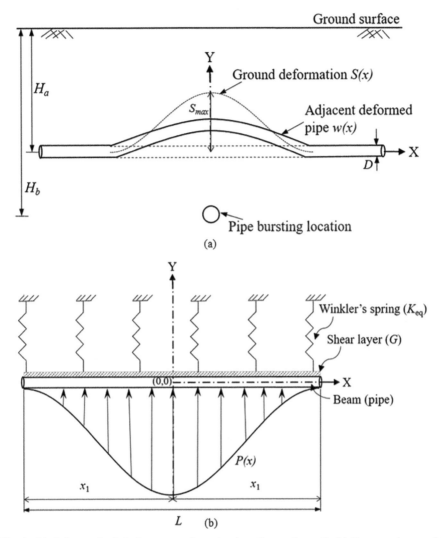

Fig. 6. (a) Adjacent buried pipe exposed to pipe bursting underneath (b) Beam-spring model (modified after Chaudhuri and Choudhury [44]).

Chaudhuri and Choudhury [44] found that the closed-form solution built on Timoshenko beam formulation provides more appropriate results compared to Euler Bernoulli's beam formulation from a comparative study as shown in Fig. 7. Further, it was noticed that for flexible pipe, pipe will move along the ground movement and for rigid pipe, pipe movement is significantly less but moment is high. For the particular case adopted by Shi et al. [23] and Chaudhuri and Choudhury [44], Maximum pipe curvature was recorded at 30° and 35° intersection angles respectively.

Hence, it is essential to estimate the influence of intersection angles between the pipes on adjacent pipe's response.

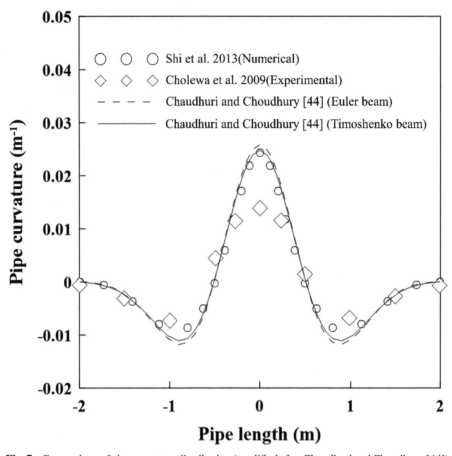

Fig. 7. Comparison of pipe curvature distribution (modified after Chaudhuri and Choudhury [44]).

5 Conclusions

The current study highlights past studies performed on burier pipelines subjected to ground deformation induced from various sources. However, rigorous experimental works or three-dimensional continuum based numerical analysis provides more accurate results and insights of the problem than simplified mathematical analysis based on the beam-springs model. But from the economic point of view, mathematical solution has advantages as the method involves fewer input parameters, less complexity and less time consuming and can provide reasonable results with moderate accuracy. Hence, for the initial stage of the design mathematical solution can be used to get quick and reliable results. In the beam-springs model, it is better to model the pipe as Timoshenko beam over Euler Bernoulli's beam and soil as two-parameter model instead of single parameter Winkler's model. Timoshenko beam can capture the shear behaviour of the pipeline and two-parameter foundation provides shear interaction between the individual Winkler springs. To estimate the response of pipe subjected to ground deformation, it is important to identify the peak ground deformation, width of deformation, and deformation pattern. Pipe stability against ground movement can be increased by providing loose backfill, or rising the pipe wall thickness. For the case of the adjacent buried pipe subjected to static pipe bursting underneath, the critical intersection angle is not always 90°. It can vary with the input parameters. The present review study shows that the analytical works can further be extended by incorporating the plasticity and non-linearity of pipe and soil material.

References

1. Karamanos, S.A., Sarvanis, G.C., Keil, B.D., Card, R.J.: Analysis and design of buried steel water pipelines in seismic areas. J. Pipeline Syst. Eng. Pract. **8**(4), 04017018 (2017)
2. Trifonov, O.V., Cherniy, V.P.: A semi-analytical approach to a nonlinear stress–strain analysis of buried steel pipelines crossing active faults. Soil Dyn. Earthq. Eng. **30**(11), 1298–1308 (2010)
3. Joshi, S., Prashant, A., Deb, A., Jain, S.K.: Analysis of buried pipelines subjected to reverse fault motion. Soil Dyn. Earthq. Eng. **31**(7), 930–940 (2011)
4. O'Rourke, T.D., Jung, J.K., Argyrou, C.: Underground pipeline response to earthquake-induced ground deformation. Soil Dyn. Earthq. Eng. **91**, 272–283 (2016)
5. O'Rourke, M.J., Nordberg, C.: Longitudinal permanent ground deformation effects on buried continuous pipelines. Technical report NCEER-92-0014 (1992)
6. O'Rourke, M.J., Liu, X., Flores-Berrones, R.: Steel pipe wrinkling due to longitudinal permanent ground deformation. J. Transp. Eng. **121**(5), 443–451 (1995)
7. O'Rourke, T.D., O'Rourke, M.J.: Pipeline response to permanent ground deformation: a benchmark case. In: Proceedings of the 4th U.S. Conference on Lifeline Earthquake Engineering, TCLEE, San Francisco, California, pp. 288–295. ASCE (1995)
8. Rajani, B.B., Robertson, P.K., Morgenstern, N.R.: Simplified design methods for pipelines subject to transverse and longitudinal soil movements. Can. Geotech. J. **32**(2), 309–323 (1995)
9. Liu, X., O'Rourke, M.J.: Behaviour of continuous pipeline subject to transverse PGD. Earthq. Eng. Struct. Dyn. **26**(10), 989–1003 (1997)
10. Lim, Y.M., Kim, M.K., Kim, T.W., Jang, J.W.: The behavior analysis of buried pipeline considering longitudinal permanent ground deformation. In: Pipelines 2001: Advances in Pipeline Engineering and Construction (2004)
11. Wham, B.P., Davis, C.A.: Buried continuous and segmented pipelines subjected to longitudinal permanent ground deformation. J. Pipeline Syst. Eng. Pract. **10**(4), 04019036 (2019)
12. O'Rourke, T.D., Lane, P.A.: Liquefaction hazards and their effects on buried pipelines. Technical report NCEER-89-0007 (1989)
13. Wang, L.R.L., Shim, J.S., Ishibashi, I., Wang, Y.: Dynamic responses of buried pipelines during a liquefaction process. Soil Dyn. Earthq. Eng. **9**(1), 44–50 (1990)
14. O'Rourke, T.D., Stewart, H.E., Gowdy, T.E., Pease, J.W.: Lifeline and geotechnical aspects of the 1989 Loma Prieta earthquake. In: Proceedings of the Second International Conference on Recent Advances in Geotechnical Earthquake Engineering and Soil Dynamics, St. Louis, Missouri, 11–15 March 1991, Paper No. LP04 (1991)
15. Ling, H.I., Mohri, Y., Kawabata, T., Liu, H., Burke, C., Sun, L.: Centrifugal modeling of seismic behavior of large-diameter pipe in liquefiable soil. J. Geotech. Geoenviron. Eng. **129**(12), 1092–1101 (2003)
16. Sumer, B.M., Truelsen, C., Fredsøe, J.: Liquefaction around pipelines under waves. J. Waterw. Port Coast. Ocean Eng. **132**(4), 266–275 (2006)

17. Roy, K., Hawlader, B., Kenny, S., Moore, I.: Upward pipe-soil interaction for shallowly buried pipelines in dense sand. J. Geotech. Geoenviron. Eng. **144**(11), 04018078 (2018)
18. Rogers, C.D.F., Chapman, D.N.: An experimental study of pipe bursting in sand. Proc. Inst. Civ. Eng. Geotech. Eng. **113**(1), 38–50 (1995)
19. Saber, A., Sterling, R., Nakhawa, S.A.: Simulation for ground movements due to pipe bursting. J. Infrastruct. Syst. **9**(4), 140–144 (2003)
20. Lapos, B., Brachman, R.W., Moore, I.D.: Laboratory measurements of pulling force and ground movement during a pipe bursting test. NASTT, No-Dig, 22–24 March 2004
21. Cholewa, J.A., Brachman, R.W.I., Moore, I.D.: Response of a polyvinyl chloride water pipe when transverse to an underlying pipe replaced by pipe bursting. Can. Geotech. J. **46**(11), 1258–1266 (2009)
22. Rahman, K., Moore, I., Brachman, R.: Numerical Analysis of the Response of Adjacent Pipelines During Static Pipe Bursting. North American Society for Trenchless Technology, Washington, DC (2011)
23. Shi, J., Wang, Y., Ng, C.W.: Buried pipeline responses to ground displacements induced by adjacent static pipe bursting. Can. Geotech. J. **50**(5), 481–492 (2013)
24. De, A., Zimmie, T.F., Vamos, K.E.: Centrifuge experiments to study surface blast effects on underground pipelines. In Pipelines 2005: Optimizing Pipeline Design, Operations, and Maintenance in Today's Economy, pp. 362–370 (2005)
25. Nourzadeh, D., Takada, S., Bargi, K.: Response of buried pipelines to underground blast loading. In: The 5th Civil Engineering Conference in the Asian Region and Australasian Structural Engineering Conference, p. 233. Engineers Australia (2010)
26. Abedi, A.S., Hataf, N., Ghahramani, A.: Analytical solution of the dynamic response of buried pipelines under blast wave. Int. J. Rock Mech. Min. Sci. **88**, 301–306 (2016)
27. Jiang, N., Gao, T., Zhou, C., Luo, X.: Effect of excavation blasting vibration on adjacent buried gas pipeline in a metro tunnel. Tunn. Undergr. Space Technol. **81**, 590–601 (2018)
28. Zhang, J., Zhang, H., Zhang, L., Liang, Z.: Buckling response analysis of buried steel pipe under multiple explosive loadings. J. Pipeline Syst. Eng. Pract. **11**(2), 04020010 (2020)
29. Datta, T.K.: Seismic response of buried pipelines: a state-of-the-art review. Nucl. Eng. Des. **192**(2–3), 271–284 (1999)
30. Psyrras, N.K., Sextos, A.G.: Safety of buried steel natural gas pipelines under earthquake-induced ground shaking: a review. Soil Dyn. Earthq. Eng. **106**, 254–277 (2018)
31. Chenna, R., Terala, S., Singh, A.P., Mohan, K., Rastogi, B.K., Ramancharla, P.K.: Vulnerability assessment of buried pipelines: a case study. Front. Geotech. Eng. **3**(1), 24–33 (2014)
32. O'Rourke, M.J., Liu, X.: Response of buried pipelines subject to earthquake effects. Multidisciplinary Center for Earthquake Engineering Research, New York (1999)
33. Yiğit, A., Lav, M.A., Gedikli, A.: Vulnerability of natural gas pipelines under earthquake effects. J. Pipeline Syst. Eng. Pract. **9**(1), 04017036 (2017)
34. O'Rourke, T.D., Tawfik, M.S.: Analysis of pipelines under large ground deformations. Cornell University, Ithaca (1986)
35. Miyajima, M., Kitaura, M.: Effects of liquefaction-induced ground movement on pipeline. In: Proceedings of the 2nd US-Japan Workshop on Liquefaction, Large Ground Deformation and Their Effects on Lifelines, pp. 386–400. National Center for Earthquake Engineering Research, Taipei (1989)
36. Chaudhuri, C.H., Choudhury, D.: Effect of earthquake induced transverse permanent ground deformation on buried continuous pipeline using winkler approach. In: Geo-Congress 2020: Geotechnical Earthquake Engineering and Special Topics, pp. 274–283. American Society of Civil Engineers, Reston (2020)
37. Chaudhuri, C.H., Choudhury, D.: Semianalytical solution for buried pipeline subjected to horizontal transverse ground deformation. J. Pipeline Syst. Eng. Pract. **12**(4), 04021038 (2021)

38. O'Rourke, M.J.: Approximate analysis procedures for permanent ground deformation effects on buried pipelines. In: Proceedings of the 2nd US-Japan Workshop on Liquefaction, Large Ground Deformation and Their Effects on Lifeline Facilities. Multidisciplinary Center for Earthquake Engineering Research, New York (1989)
39. ALA (American Lifelines Alliance).: Guidelines for the design of buried steel pipe. ALA, Washington, DC (2001)
40. Zheng, J.Y., Zhang, B.J., Liu, P.F., Wu, L.L.: Failure analysis and safety evaluation of buried pipeline due to deflection of landslide process. Eng. Fail. Anal. **25**, 156–168 (2012)
41. Luo, X., Ma, J., Zheng, J., Shi, J.: Finite element analysis of buried polyethylene pipe subjected to seismic landslide. J. Press. Vessel Technol. **136**(3), 031801 (2014)
42. Ma, J., Shi, J., Zheng, J.: Safety investigation of buried Polyethylene pipe subject to seismic landslide. In ASME 2012 Pressure Vessels and Piping Conference, pp. 233–242. American Society of Mechanical Engineers Digital Collection (2012)
43. Chaudhuri, C.H., Choudhury, D.: Buried pipeline subjected to seismic landslide: a simplified analytical solution. Soil Dyn. Earthq. Eng. **134**, 106155 (2020)
44. Chaudhuri, C.H., Choudhury, D.: Buried pipeline subjected to static pipe bursting underneath: a closed-form analytical solution. Géotechnique, 1–10 (2021). https://doi.org/10.1680/jgeot.20.p.167

Performance-Based Assessment and Design of Structures on Liquefiable Soils: From Triggering to Consequence and Mitigation

Shideh Dashti[1]([⊠]), Zachary Bullock[2], and Yu-Wei Hwang[3]

[1] University of Colorado Boulder, Boulder, CO, USA
Shideh.Dashti@colorado.edu
[2] University of British Columbia, Vancouver, BC, Canada
[3] University of Texas at Austin, Austin, TX, USA

Abstract. Effective liquefaction mitigation requires an improved fundamental understanding of triggering in terms of excess pore pressures in realistically stratified deposits that experience cross-layer interactions as well as performance-based consequence procedures that account for 3D soil-structure interaction (SSI), all mechanisms of deformation, total uncertainty, and the impact of mitigation. In this paper, we first present a series of centrifuge experiments to evaluate site response and pore pressure generation in layered liquefiable deposits, soil-structure interaction (SSI), and the impact of ground densification as a mitigation strategy on SSI and building performance. Second, experimental results are used to validate 1D and 3D, fully-coupled, nonlinear, dynamic finite element analyses of layered sites and soil-foundation-structure systems with and without mitigation. Third, numerical parametric studies (exceeding 167,000 1D and 63,000 3D simulations) are used to identify the functional forms for predicting liquefaction triggering in the free-field based on the capacity cumulative absolute velocity (CAV_C) required to achieve a threshold excess pore pressure ratio (r_u), settlement of unmitigated structures, and the relative impact of ground densification on foundation's permanent settlement. And finally, a limited case history database is used validate the triggering and consequence models, accounting for field complexities not captured numerically or experimentally. This integrative approach yields a set of procedures that are the first to consider variations in soil layering and geometry, layer-to-layer cross interactions, foundation and structure properties (in 3D), contribution of all mechanisms of deformation below unmitigated structures, geometry and properties of densification, ground motion's cumulative characteristics, total inherent model uncertainties, and the explicit conditionality of structural settlement on free-field triggering—which are necessary to realize the benefits of performance-based engineering in liquefaction assessment.

Keywords: Liquefaction triggering · Mitigation · Soil-structure interaction

1 Introduction and Background

Probabilistic predictive models for the triggering (e.g., Seed et al. 2003; Boulanger and Idriss 2012; Maurer et al. 2017) and consequences (e.g., Bray and Macedo 2017;

L. Wang et al. (Eds.): PBD-IV 2022, GGEE 52, pp. 376–396, 2022.
https://doi.org/10.1007/978-3-031-11898-2_21

Bullock et al. 2019a,b) of soil liquefaction are becoming increasingly common in the design of structures founded on such deposits. However, the application of triggering and consequence procedures remains primarily disconnected, because their methodological approaches and implementation remain separate. The most recent probabilistic consequence models (e.g., Bullock et al. 2019a,b), which are capable of predicting the settlement and tilt of isolated shallow-founded structures on layered soil systems with evolutionary ground motion intensity measures (IMs), do not explicitly depend on evaluation of triggering (e.g., in terms of FS_{liq}). However, through inclusion of case histories in their formulation, these models remain implicitly conditioned on the occurrence of liquefaction or its surface manifestation. This implicit conditioning has the potential to lead to overestimation of the risk of liquefaction consequences.

The existing models for predicting the occurrence of liquefaction have their own set of limitations. They are often based on empirical observations of surface manifestation of liquefaction (e.g., Cetin and Seed 2004; Idriss and Boulanger 2008) or mechanistic prediction of $r_u = 1$; where $r_u = \Delta u/\sigma'_{vo}$. Meanwhile, it is known that partial or full liquefaction may occur without necessarily generating surface manifestations in the form of sand boils or ejecta. These procedures typically focus on individual soil layers and do not consider layer cross interactions (Beyzaei et al. 2018), which can influence the manifestation and propagation of damage. The procedures typically rely on the cyclic stress ratio (CSR) at the surface to define the demand by interpolating PGA at non-liquefied free-field sites, which is not necessarily an optimum choice of IM and ignores the degree of softening and pore pressure generation in soil. Lastly, the mechanistic procedures do not acknowledge that significant softening and deformations can still occur at r_u values less than 1.0.

After evaluating the likelihood of triggering and consequence, an engineer often needs to consider different forms of ground improvement and mitigation. None of the existing procedures for ground improvement consider the presence and/or interaction of a structure with soil, the active mechanisms of deformation near a structure, or the underlying uncertainties. In designing ground densification for instance, engineers commonly rely on empirical triggering or volumetric settlement estimations in the free-field, or at best semi-empirical probabilistic consequence models that do not consider densification dimensions in relation to and in interaction with the overlying structure. Additionally, since mitigation is often designed based on free-field triggering calculations (i.e., FS_{liq} in various soil layers), the evaluation of post-mitigation triggering and consequence may also be implicitly conditioned on the occurrence of liquefaction in free-field conditions. This conditioning exists because the size of the mitigation is determined using results from free-field triggering analyses, not because the performance of the mitigation is necessarily tied to triggering in the free-field.

In this study, we first propose probabilistic models for the cumulative absolute velocity required to trigger liquefaction (CAVc), based on data from a comprehensive 1D numerical site response parametric analysis (detailed by Bullock 2020 and Bullock et al. 2022), here referred to as the first parametric study (PS-I). The numerical models in PS-I were calibrated and validated with element level and centrifuge free-field experiments. The models consider the influence of layer, profile, ground motion properties, as well as system-level effects caused by low-permeability strata within the soil profile. These models describe the uncertainty in evaluating the probability of liquefaction according to a threshold r_u value (e.g., 1.0 in the traditional definition of liquefaction or 0.9 in that of Olson et al. 2020). The results are compared with observations of liquefaction or no liquefaction in the 2010 Darfield, 2011 Christchurch, and 2016 Valentine's Day earthquakes in Canterbury (Geyin et al. 2021), together with other existing predictive models of triggering.

We subsequently extend the triggering model to predict liquefaction-induced permanent settlement (including volumetric and deviatoric deformations) beneath an unmitigated, mat-founded structure based on the centrifuge-calibrated numerical analyses described by Karimi et al. (2018) and Bullock et al. (2019a): PS-II. The initial functional form developed based on the numerical database was then compared with and adjusted based on a case history database consisting of 50 cases from six earthquakes. The adjustment helped account for sedimentation and ejecta effects that were not effectively captured numerically or experimentally in unmitigated models (see Bullock et al. 2019a for additional details). This model incorporates the influence of the soil profile, the presence and 3D properties of the structure, and cumulative characteristics of the ground motion. The model is improved from Bullock et al. (2019a) through removal of the implicit conditioning on the occurrence of liquefaction by applying the Bullock et al. (2022) triggering model for CAVc.

In the end, we extend the numerical database in PS-III to include ground densification around structures (770, 3D, fully-coupled, FEA through quasi-Monte Carlo sampling of key input parameters), which was also calibrated and validated with related centrifuge models. This database is subsequently used to propose a probabilistic predictive procedure for foundation's permanent settlement on liquefiable soils that are improved with ground densification. The models consider realistic, nonlinear, 3D structures on mat foundations or basement, SSI, interlayering and layer cross interactions, ground densification properties and geometry, and ground motion characteristics. Similar to the settlement models for unmitigated cases, we use nonlinear regression with lasso-type regularization to estimate model coefficients. The reliability of the predicted model is subsequently evaluated by comparing the trends with a limited number of centrifuge and field case histories. With the small number of well-documented case histories available on mitigated buildings, no further model adjustment was possible in this case. However, this was judged acceptable at this time, due to the relatively small error observed in case history predictions.

The proposed set of models are the first of their kind to comprehensively consider the triggering, consequences, and mitigation of soil liquefaction near structures and the underlying uncertainties in a unified and explicit manner, aiming to guide a more reliable and performance-based treatment of liquefiable soils near structures.

2   Case History Collection

The case histories related to each of the three analysis phases (PS-I through PS-III) are summarized in Table 1. In PS-III, the case history database involving ground densification around shallow-founded structures was too small for a meaningful statistical evaluation and adjustment. Hence, the limited case histories (see Table 1) that included mitigation and some information about the soil and structure were only used to evaluate the performance of the predictions from the proposed settlement model on densified ground. The representative case history was collected during the 1964 Niigata, Japan Earthquake. In Niigata City, liquefaction-induced damage on mitigated ground was documented (Watanabe 1966), including one oil tank and one shallow-founded building. The mitigated site near Hakusan district (Kawakami and Asada 1966) included a 14 m-thick, fine sand layer with a D_r of about 90% (in average), followed by a 10 m-thick, sandy soil with a $D_r \approx 60\%$ ($N_{1,60}$ ranging from 10 to 20). Subsequently, both layers were overlaid by 6 m of a silty sand layer deposit with a $D_r \approx 30\%$ ($N_{1,60}$ ranging from 3 to 6). The liquefiable layer was identified (Ishihara and Koga 1977) from a depth of about 3 m to 13 m below the ground surface, based on the cyclic triaxial tests on undisturbed specimens. For additional details on the case history listed in Table 1, please refer to Hwang et al. (2022b).

Table 1 Source and quantity of case histories related to each parametric study (PS)

PS	Source	Earthquake	No. of cases
I[a]	Geyin et al. 2020, 2021	Darfield, 2010	997
		Christchurch, 2011	3846
		Valentine's Day, 2016	155
II[b]	Yoshimi and Tokimatsu (1977)	Niigata, 1964	15
	Acacio et al. (2001)	Luzon, 1990	17
	Bray and Sancio (2009)	Kocaeli, 1999	3
	Unutmaz and Cetin (2010)	Kocaeli and Düzce, 1999	27
	Bertalot et al. (2013)	Chile, 2010	21
	Bray et al. (2014)	Christchurch, 2011	4
III[c]	Yoshida (2000)	Kobe, 1995	1
	Watanabe (1966)	Niigata, 1964	2

[a] Additional details inBullock et al. (2022)
[b] Additional details in Bullock et al. (2019a).
[c] Additional details in Hwang et al. (2022a).

3 Centrifuge Modeling

Five centrifuge tests were conducted under a centrifugal acceleration of 70 g at the University of Colorado Boulder's 5.5 m-radius, 400 g-ton centrifuge facility (Olarte et al. 2017, 2018a; and Paramasivam et al. 2018), to investigate seismic site response in a layered, liquefiable, free-field soil profile (with no structures present) and the influence of ground densification on shallow-founded, potentially inelastic structures on liquefiable soils. The tests including structures examined two types of moment-resisting frame structures (a 3-story Structure A on a 1 m-thick mat foundation and a heavier, taller, and more flexible, 9-story Structure B on a 1-story basement foundation) and one layered, liquefiable soil profile with and without ground densification, as shown in Fig. 1. Structures A_{NM} or B_{NM} in these tests represented Structures A or B on an unmitigated soil profile, while A_{DS} or B_{DS} represented the same structures and soil profile, now including ground densification.

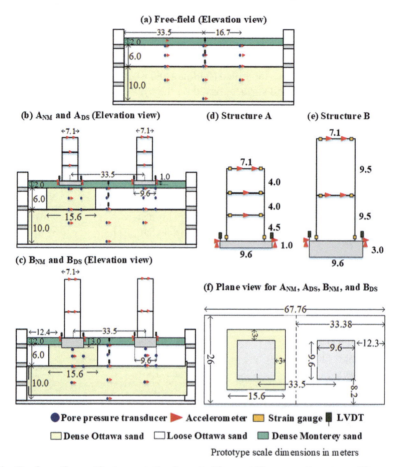

Fig. 1. Configuration and instrumentation layout of the centrifuge experiments used for numerical model validation. (All units are in prototype scale meters at 70 g of centrifugal acceleration.)

In all centrifuge tests, a 10 m-thick (in prototype scale) dense Ottawa sand F65 layer was dry-pluviated to attain a relative density (D_r) of approximately 90% ($e_{min} = 0.53$ and $e_{max} = 0.81$), prior to testing. Subsequently, a 6 m-thick loose Ottawa sand layer was dry pluviated to reach a D_r of 40%. Above this layer, a 2 m-thick Monterey sand layer was dry pluviated to attain a $D_r \approx 90\%$ ($e_{min} = 0.54$ and $e_{max} = 0.84$) as a dense, draining crust. The groundwater table in all centrifuge tests was located at the ground surface. The mitigated (densified) zone was attained by dry pluviating Ottawa sand with a $D_r \approx 90\%$ around Structures A_{DS} or B_{DS}. The entire thickness of looser Ottawa sand was treated (i.e., 6 m), with the densification width beyond the foundation edges selected as half of the densification depth (i.e., 3 m), based on JGS (1998).

All model specimens were spun to a nominal centrifugal acceleration of 70 g. A series of four one-dimensional (1D) horizontal earthquake motions was applied to each model after saturation (with a methylcellulose solution 70 times more viscous than water). For numerical validation in the free-field or below the unmitigated or mitigated structure, we only used the first major motion, prior to which the soil properties and geometry were known with greater accuracy. All units presented in this paper are in the prototype scale. The subsequent numerical simulations of these tests were also performed in prototype scale units. Detailed discussions of all centrifuge tests were provided by Olarte et al. (2017, 2018a) and Paramasivam et al. (2018), which are not repeated here for brevity.

4   Numerical Modeling Details

Three-dimensional (3D), fully coupled, FE simulations were performed in the object-oriented, parallel computation FE platform OpenSees (Mazzoni et al. 2006). The pressure-dependent, multi-yield surface, version 2, soil constitutive model (Elgamal et al. 2002 and Yang et al. 2008), PDMY02, was used to simulate the nonlinear response of saturated, granular soils and consequences of liquefaction-induced ground deformations (Karimi et al. 2016a,b; Ramirez 2018; and Hwang et al. 2021).

The PDMY02 model parameters used for Ottawa sand at a handful of relative densities (30%, 40%, 50%, 60%, 70%, 80%, and 90%) were adopted from Hwang et al. (2021). The calibration process was based on: i) a series of fully drained and undrained, monotonic and cyclic triaxial tests (Badanagki 2019); ii) a boundary value problem involving free-field site response as recorded in centrifuge with the same soil types but no structure or mitigation involved (Fig. 1a, Ramirez et al. 2018); and iii) empirical observations of liquefaction triggering in terms of number of cycles to liquefaction (NCEER 1997). The best-fitting PDMY02 soil constitutive model parameters were consequently identified by minimizing the sum of root mean squared error (RMSE) among all numerical results and experimental or field observations equally. Figures 2a-c compare the experimental data (e.g., cyclic triaxial test, free-field site response, and CSR to trigger liquefaction in 15 cycles) and numerically computed response (by adopting the best-fitting parameters) to demonstrate that the calibrated PDMY02 parameters could roughly capture the dynamic behavior of Ottawa sand from an element level to a boundary value problem and to field observations of liquefaction. Additional data and comparisons (e.g., monotonic triaxial tests) are provided by Hwang et al. (2021; 2022a,b).

After calibration, the numerical models of the entire soil-foundation-structure system with and without densification were validated with the corresponding centrifuge experiments (e.g., Figs. 3b-3c). 3D, 20–8 nodes, serendipity, brick elements with the u-p formulation (Zienkiewicz et al. 1990) were used to model the soil domain. The fluid bulk modulus was assumed to be 2×10^6 kPa at atmospheric pressure. The element size was calculated at each depth based on the soil's empirical small-strain shear wave velocity profile (Seed and Idriss 1970; Bardet et al. 1993; and Menq 2003) and the minimum wavelength of interest (detailed by Ramirez et al. 2018 and Hwang et al. 2021).

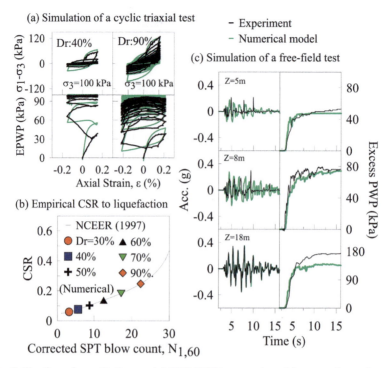

Fig. 2. Calibration of constitutive model (PDMY02) parameters: (a) comparison of numerical simulations with a representative strain-controlled cyclic triaxial experiments on Ottawa Sand (isotropic consolidation stress = 100 kPa; axial strain amplitude = 0.2%); (b) comparison of numerical simulations with empirical observations in terms of CSR to trigger liquefaction in 15 cycles for different relative densities ($N_{1,60} = 35D_r^2$; Skeptom 1986); (c) comparison of numerical simulations with a free-field centrifuge experiment using the same soil profile (see Fig. 1a).

The foundations were modeled with 20–8 node brick u-p elements and the linear-elastic material. Note that the fluid mass density was set to zero in the foundation to avoid generation of excess pore pressures. The foundation-soil interface was modelled with an equal degrees-of-freedom (DOF) connection to the soil (Karimi and Dashti 2016a,b and Hwang et al. 2021). To allow for relative settlement of foundation with respect to the surrounding soil, the nodes around the foundation's lateral perimeter were tied to soil only in the horizontal direction (e.g., x- and y-directions).

The beams and columns in Structures A and B were modeled with elastic beam-column elements and a linear-elastic material, while the reduced section "fuses", where structural nonlinearity localized, were modeled with a nonlinear fiber section and the uniaxial steel material. A damping ratio of 0.8% and 1.2% was assigned to Structures A and B, respectively. These properties were calibrated with component tests as well as fixed-base hammer impact tests. Details of the design and calibration of structure parameters were reported by Olarte (2017) and Ramirez (2019).

In models involving the soil-foundation-structure system, half of the model container (in the direction perpendicular to shaking) was simulated by taking advantage of

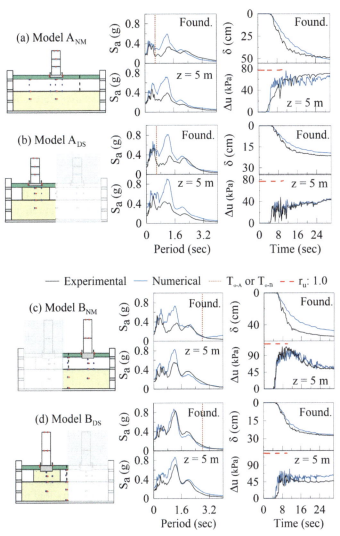

Fig. 3. Comparison of experimental and numerical results in terms of 5%-damped acceleration response spectra and settlement time histories of foundations as well as excess pore pressure–time histories and acceleration response spectra under the center of isolated structures in the middle of the critical layer (z = 5 m) during the Kobe-L motion recorded in the centrifuge.

symmetry (Karimi and Dashti 2016a, b; Ramirez 2019; and Hwang et al. 2021). The lateral boundary nodes located at the same elevation were tied to move together in both horizontal directions, roughly representing the conditions in a flexible-shear-beam container. The bottom boundary of the soil domain was fixed in all translational directions (i.e., x-, y-, and z-directions) when modeling the centrifuge tests, to simulate the rigid container base. The length and width of the soil domain were, however, increased compared to the container, in order to reduce the adverse effects of reflecting waves from the boundaries and displacement constraints near the center of the domain (where the structure was placed). The minimum numerical domain length (parallel to shaking) was, accordingly, determined as 6B (where B is the foundation width) and the domain width (perpendicular to shaking) as 3B through an initial sensitivity study. Note that this same domain size (6B in length and 3B in width) was adopted for the subsequent numerical parametric studies PS-I through III. The excess pore pressures were only allowed to dissipate through the domain surface. The acceleration time history recorded at the base of the container in centrifuge during the first major motion was applied directly to the base nodes of the numerical models after reaching static equilibrium.

Comparisons of experimental and numerical results in terms of 5%-damped acceleration response spectra and settlement time history of foundations as well as excess pore pressure–time history and acceleration response spectra in the middle of the critical layer ($z = 5$ m) are provided in Fig. 3. These results show that overall, numerical simulations of unmitigated and mitigated isolated structures like A and B roughly captured spectral accelerations near the structure's fundamental period, peak excess pore pressures, and foundation's average permanent settlement observed experimentally (difference of less than about 9, 24, and 6% in each response, respectively). However, the numerical models overestimated the dilation cycles (as sharp drops in excess pore pressures or spikes in acceleration) within the looser (unmitigated) Ottawa sand layer, particularly below the lighter Structure A. This was due to an over-estimation of shear strain excursions because of soil softening, encouraging cycles of re-stiffening and acceleration spike at lower periods. Accordingly, accelerations were slightly over-estimated on the foundation and roof of Structure A near its fundamental period ($T_{o-A} \approx 0.56$ s) and the motion's mean period ($T_m \approx 0.9$ s). The differences were reduced substantially for mitigated (densified) cases with smaller shear strains and pore pressures.

In general, the treated soil below Structures A_{DS} and B_{DS} experienced smaller excess pore water pressures and settlements compared to their unmitigated counterparts, both experimentally and numerically. This is because increasing soil relative density (Dr) increases its stiffness and strength, reducing its contractive tendencies and hence, net generation of excess pore water pressures. A reduction in the degree of soil strength loss reduced the contribution of volumetric and shear deformations below the structure. In general, 3D, fully-coupled, effective-stress, finite element models with a well-calibrated constitutive model and numerical setup (e.g., with use of higher order elements and sufficient boundary size) could capture the primary experimental patterns. Considering all sources of error and uncertainty in both experimental and numerical models, we judged the comparisons to be reasonable, particularly for mitigated or densified conditions, and moved on with the design of three parametric studies.

5   Numerical Parametric Studies

The schematic view of typical 3D models used in the numerical parametric studies PS-I through III are shown in Fig. 4. The numerical framework, selection of elements, and loading conditions were consistent with the results and conclusions presented in the previous section. However, in the parametric studies, our goal was also to account for variations in structural type and properties, soil layering, rock's elastic properties, and ground motion characteristics.

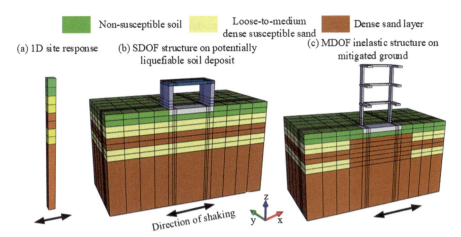

Fig. 4. Schematic drawings showing the numerical simulation of: (a) a 1D single soil column for nonlinear site response analyses in PS-I; (b) a SDOF, isolated, elastic structure on a layered liquefiable deposit in PS-II; and (c) an MDOF, isolated, potentially inelastic structure on a layered liquefiable deposit improved with ground densification in PS-III.

For developing the liquefaction triggering models in the free-field, the simulations in PS-I focused on free-field conditions (using single-column, 1D, nonlinear site response analyses). The PS-II for the consequence model without mitigation involved 3D, linear-elastic, single-degree-of-freedom (SDOF) structures on mat foundations. A linear and elastic structural response was judged reasonable given the large degree of nonlinearity and damping in the underlying unmitigated soil. In contrast, the PS-III for ground improvement focused on inelastic (damageable), multi-degree-of-freedom (MDOF) structures representative of the building stock in seismically active regions of the U.S. founded on mat foundations and layered, liquefiable, soil profiles treated with ground densification. The considered soil profiles included liquefiable, loose- to medium-dense, saturated, clean sand layers as well as dense sand interlayers and in some models, a thin silt capping layer. Ground densification was only applied to portions of the most critical, loose to medium dense granular layer(s) in the parametric study.

To ensure all input parameters (IPs) characterizing the soil-foundation-structure system and when applicable, the mitigation geometry, were adequately and uniformly represented in the parameter space, a quasi-Monte Carlo sampling procedure was adopted to select the initial model suites in each PS (see Table 2). Each IP was initially drawn from either a uniform or a triangular distribution. These ranges and distributions were informed by a combination of: (1) observed values of these parameters in the Bullock et al. (2019a) case history database of buildings on liquefiable soils; and (2) ranges considered and shown as influential in the numerical database previously generated by Karimi et al. (2018). Subsequently, the corresponding parameter combinations were determined by taking the inverse cumulative distribution functions from a quasi-random set of numbers between 0 and 1 for each IP. In total, in PS-III a suite of 70 basic model configurations was designed by quasi-Monte Carlo sampling to account for variations in soil-foundation-structure systems and mitigation design. PS-II included a suite of 420 model configurations designed to facilitate sensitivity analysis for each independently-varied parameter. PS-I included 125 model configurations, determined using quasi-Monte Carlo sampling, which were each analyzed using three soil constitutive models to facilitate quantification of epistemic uncertainty. For additional details on the parameter distributions for soil layer, structural, rock, densification, and ground motion properties in each of the three PS's, please refer to Bullock et al. (2019a), Bullock et al. (2022), and Hwang et al. (2022b).

Table 2. The main input parameters (IPs) in the numerical parametric study (PS) I, II, and III.

Parametric study ID	The main IPs in the numerical parametric study
PS-I	· Number, thickness, and density of the loose to medium-dense sand layers · Total deposit thickness above bedrock · Bedrock shear wave velocity · Presence of low-permeability strata throughout the profile · Ground motion intensity measures
PS-II	· The same main IPs as PS-I · Foundation bearing pressure, footprint size, and embedment depth · Height and inertial mass of elastic structure
PS-III	· The same main IPs as PS-I and PS-II · Nonlinear structural properties including wood-frame, reinforced concrete moment frame, and steel moment frame buildings · Ground densification design in terms of depth and width beyond the foundation edge

6   Statistical Modeling and Initial Functional Forms

The Bullock et al. (2019a) settlement model consists of a base model developed using a database of 63,000 numerical analyses of soil-foundation-structure systems (PS-II) and adjusted according to a database of 50 empirical case histories. The base model captures the relationships among soil, foundation, and structure-specific parameters and settlement, as well as the ground motion intensity and its interaction with other parameters. The functional form of the base model is given by:

$$ln(\delta) = f_{soil} + f_{fnd} + f_{str} \tag{1}$$

where $ln(\delta)$ is the natural logarithm of the median predicted foundation settlement (in units of mm) and f_{soil}, f_{fnd}, and f_{str} are functions that reflect how a shallow-founded structure's settlement is influenced by properties of the soil, foundation, and structure, respectively.

Equations 2–6 describe f_{soil}, where $H(\cdot)$ is the Heaviside step function of the argument, $H_{S,i}$ is the thickness of the i-th liquefaction-susceptible layer (in units of m), ε is an infinitesimal positive quantity, CAV is the outcropping rock cumulative absolute velocity (in units of cm/s), $q_{c1N,i}$ is the normalized CPT cone tip resistance of the i-th liquefaction-susceptible layer, $D_{S,i}$ is the depth from the foundation to the center of the i-th liquefaction-susceptible layer (in units of m), and F_{LPC} is a flag that is equal to 1 if there is a low-permeability cap at the top of the soil profile and 0 otherwise.

$$f_{soil} = [\Sigma_i H(H_{S,i} - 1 + \varepsilon) f_{S,i} f_{H,i}] + [c_0 + c_1 ln(CAV)] F_{LPC} + s_0 ln(CAV) \quad (2)$$

$$f_{S,i} = a_0 \text{ for } q_{c1N,i} < 112.4 \quad (3)$$

$$f_{S,i} = a_0 + a_1(q_{c1N,i} - 112.4) \text{ for } 112.4 \le q_{c1N,i} < 140.2 \quad (4)$$

$$f_{S,i} = a_0 + 27.8 a_1 \text{ for } 140.2 \le q_{c1N,i} \quad (5)$$

$$f_{H,i} = b_0 H_{S,i} exp[b_1 (max(D_{S,i}, 2)^2 - 4)] \quad (6)$$

Equations 7–9 describe f_{fnd}, where q is the foundation bearing pressure (in units of kPa), $D_{S,1}$ is the depth from the foundation to the center of the top-most liquefaction-susceptible layer in the soil profile (in units of m), B is the foundation width (in units of m), L is the foundation length (in units of m), and D_f is the foundation embedment depth (in units of m). The foundation width (B) is defined as its shorter dimension (i.e., $L/B \ge 1$).

$$f_{fnd} = f_q + f_{B,L} \quad (7)$$

$$f_q = \{d_0 + d_1 ln[min(CAV, 1000)]\} ln(q) \times exp\{d_2 min[0, B - max(D_{S,1}, 2)] \quad (8)$$

$$f_{B,L} = \{e_0 + e_1 ln[max(CAV, 1500)]\} ln(B)^2 + e_2 L/B + e_3 D_f \quad (9)$$

Equation 10 describes f_{str}, where h_{eff} is the structure's effective height (in units of m) and M_{st} is the structure's inertial mass (in units of kg).

$$f_{str} = \{f_0 + f_1 ln[min(CAV, 1000)]\} h_{eff}^2 + f_2 min[M_{st}/10^6, 1] \quad (10)$$

Equation 11 describes the adjustment to the base model's prediction of settlement, where $ln(\delta_{adj})$ is the adjusted prediction of settlement (in units of mm) and $H_{S,1}$ is the thickness of the top-most liquefaction-susceptible layer (in units of m). In the case history database, $H_{S,1}$ is usually also the thickest liquefaction-susceptible layer, and we may therefore consider $H_{S,1}$ to be the thickness of the "critical" layer, which can be identified as the layer with the highest value of $f_{S,i} f_{H,i}$ as described above. This approach treats the critical layer as the one contributing the most to settlement.

$$ln(\delta_{adj}) = ln(\delta) + k_0 + k_1 min(H_{S,1}, 12)^2 + k_2 min(q, q_c) + k_3 max(q - q_c, 0) \ge ln(\delta) \quad (11)$$

Table 3 Model coefficients for Bullock et al. (2019a)

Parameters	Value	Parameters	Value
a_0	1.000	e_3	−0.2148
a_1	−0.0360	f_0	−0.0137
b_0	0.3026	f_1	0.0021
b_1	−0.0205	f_2	0.1703
c_0	1.3558	s_0	0.4973
c_1	−0.1340	k_0	−1.5440
d_0	−1.3446	k_1	0.0250
d_1	0.2303	k_2	0.0295
d_2	0.4189	k_3	−0.0218
e_0	−0.8727	q_c	61
e_1	0.1137	σ_{adj}	0.6746
e_2	−0.0947		

Bullock et al. (2019a) also provided the logarithmic standard deviation of a lognormal distribution around δ_{adj}, denoted as σ_{adj}. Table 3 provides the model coefficients for Bullock et al. (2019a):

The lognormal distribution defined by δ_{adj} and σ_{adj} can be used to formulate the probability of exceeding settlement thresholds of interest (i.e., those related to certain repair actions). Here, we denote a threshold value of settlement as δ_{thresh}. However, note that these probabilities are implicitly conditioned on the occurrence of liquefaction due to the correction based on case history observations that included observations of liquefaction surface manifestation. Equation 12 shows the calculations needed to obtain a conditional probability of exceedance based on the Bullock et al. (2019a) model, where *Liq* is a flag that indicates the triggering of liquefaction and $\Phi[\cdot]$ is the standard normal cumulative distribution function evaluated at the argument.

$$P(\delta_{adj} > \delta_{thresh}|Liq) = 1 - \Phi[(ln(\delta_{thresh}) - ln(\delta_{adj}))/\sigma_{adj}] \qquad (12)$$

As discussed above, this conditioning arises because (i) the case history adjustment is based only on cases where liquefaction damage was observed; and (ii) the numerical database was filtered to include only observations with $\delta \geq 10$ mm. A probabilistic liquefaction triggering model is needed to remove this conditioning.

Bullock et al. (2022) developed liquefaction triggering models in the free-field based on *CAV* using the results of PS-I. These models give estimates of the capacity cumulative absolute velocity (CAV_c) in units of cm/s. CAV_c is defined as the input outcropping rock *CAV* needed to generate a certain r_u value in a given soil element. Equation 13 describes CAV_c:

$$ln(CAV_c) = a_0 + a_1\sqrt{z} + a_2[f_{lps}/(z_{lps} + z)] + a_3(q_{c1N}/100) \qquad (13)$$

where $ln(CAV_c)$ is the natural log of the median predicted capacity CAV, z is the depth from the ground surface to the soil element (in units of m), f_{lps} is a flag that indicates whether there is a low-permeability stratum anywhere above the soil element in the profile, and z_{lps} is the depth from the soil element to the nearest low-permeability stratum above (in units of m). Table 4 provides the model coefficients for Bullock et al. (2022) for CAV_c corresponding to an r_u threshold of 0.9 [i.e., *liquefaction* according to Olson et al. (2020)].

Table 4 Model coefficients for estimating CAV_c in Bullock et al. (2022)

Parameters	Value	Parameters	Value
a_0	4.640	a_3	0.273
a_1	0.167	σ_c	0.675
a_2	−0.877	ρ_{dc}	0.249

CAV_c can be combined with the outcropping rock CAV to obtain a factor of safety against liquefaction triggering (FS_{liq}):

$$FS_{liq} = CAV_c/CAV \tag{14}$$

$$ln(FS_{liq}) = ln(CAV_c) - ln(CAV) \tag{15}$$

The Bullock et al. (2022) free-field liquefaction triggering model can also be used to obtain probabilities of liquefaction that are consistent with the functional form and approach used in the consequence model proposed by Bullock et al. (2019a). Equation 15 can be rewritten in terms of variables from Bullock et al. (2019a) to obtain the median FS_{liq} in the center of the critical layer as follows:

$$ln(FS_{liq,1}) = a_0 + a_1\sqrt{D_{S,1} + D_f} + a_2[f_{lps}/(z_{lps} + D_{S,1} + D_f)] + a_3(q_{c1N,1}/100) - ln(CAV) \tag{16}$$

where $ln(FS_{liq,1})$ is the natural logarithm of the median predicted factor of safety in the center of the critical layer and $q_{c1n,1}$ is the normalized cone tip resistance in that layer.

Bullock et al. (2022) provides the correlation coefficient between the uncertainties around CAV_c and CAV, denoted as ρ_{dc}. The logarithmic standard deviation of the factor of safety, σ_f, is given by Eq. 17, where σ_c is the logarithmic standard deviation around the median CAV_c and σ_d is the logarithmic standard deviation around the median outcropping rock CAV.

$$\sigma_f = \sqrt{\sigma_c^2 + \sigma_d^2 + 2\rho_{dc}\sigma_c\sigma_d} \tag{17}$$

We can obtain the probability of liquefaction triggering (i.e., the probability of FS_{liq} less than 1.0) using the lognormal distribution:

$$P(Liq) = \Phi[-ln(FS_{liq,1})/\sigma_f] \tag{18}$$

This equation can be combined with Eq. 12 to obtain the marginal probability of settlement exceeding a threshold value (i.e., the probability without conditioning on *Liq*):

$$P(\delta_{adj} > \delta_{thresh}) = P(\delta_{adj} > \delta_{thresh}|Liq)P(Liq) \tag{19}$$

Following the development of triggering and consequence models described above, a total of 770 numerical simulations using three-dimensional (3D), fully-coupled, non-linear finite element analyses of the soil-foundation-structure systems enabled a comprehensive evaluation of the seismic performance of shallow-founded structures on treated soils. The detailed properties of all numerical analyses are provided in Hwang et al. (2022a). The nonlinear regression with lasso-type (Tibshirani 1996) regularization was used to develop the predictive model for permanent settlement of structures on densified soils.

Equation 20 shows the general functional form for estimating permanent settlement of structures on liquefiable deposit improved by ground densification:

$$lnln(\delta_{DS}) = \alpha_0 + f_{FN-ST} + f_{SL-DS} + f_{UD} + f_{GM} \tag{20}$$

where $lnln(\delta_{DS})$ is the natural logarithm of the median predicted foundation settlement (in units of mm) on densified soils, α_0 is a constant intercept. $f_{FN-ST}, f_{SL-DS}, f_{UD}$, and f_{GM} are functions that reflect how structure's settlement is influenced by the foundation (FN)-structure (ST) system, layered soil profile (SL) and densification properties (DS), the properties of any remaining undensified loose- to medium-dense soil layer (UD) below the densified zone within foundation's influence zone, and an optimum ground motion intensity measure (GM). The functional forms for $f_{FN-ST}, f_{SL-DS}, f_{UD}$, and f_{GM} are described by Eqs. 21–25.

$$f_{FN-ST} = \alpha_1 D_f + ln\alpha_2 ln(B) + ln\alpha_3 ln(q) + ln\alpha_4 ln\left(\frac{H}{B}\right) + \alpha_5 lnln\left(\frac{L}{B}\right) \tag{21}$$

$$f_{SL-DS} = ln\alpha_6 ln\left(H_{dep}\right) + +\alpha_8 ln(\frac{D_{DS}}{H_L}) + \alpha_9 D_{r,DS} \tag{22}$$

$$f_{UD} = \alpha_{10} f_{D_r(H_{UD})} + \alpha_{11} H_{UD} \tag{23}$$

$$f_{D_r(H_{UD})} = \{0, H_{UD} = 0$$

$$\{D_r ofundensifiedlayer, H_{UD} > 0 \tag{24}$$

$$f_{GM} = \{\alpha_{12} lnln(CAV), CAV < 450cm/s$$

$$\{\alpha_{12} ln(450) + \alpha_{13} lnln\left(\frac{CAV}{450}\right), CAV \geq 450cm/s \tag{25}$$

where D_f is the foundation embedment depth from the ground surface (in m), B is the foundation width (in m), q is the bearing pressure of the foundation-structure system (in kPa), H/B is the unitless actual structure height to width ratio, L/B is the unitless mat foundation length (L) to width (B) aspect ratio, H_{dep} is the total deposit depth above bedrock (in m), H_L is the total cumulative thickness of critical, loose- to medium-dense granular soil layers (i.e., the summation of layer thickness with $D_r \leq 70\%$ within foundation's influence zone (defined as 1.5B in this model) in m, W_{DS} is the densification width beyond the edges of the foundation (in m), D_{DS} is the densification depth from ground surface (in m), $D_{r,DS}$ (in percentage) is the target relative density of the treated soil, where $f_{D_r(H_{UD})}$ is a flag that is equal to the average relative density (%) of the undensified layer if the densification depth does not reach the full depth of the deepest critical layer within the foundation's influence zone; or is zero otherwise. H_{UD} is the thickness of the remaining undensified, loose- to medium-dense granular soil layer within the foundation's influence zone (in m). CAV is calculated on outcropping bedrock (in cm/s), while the threshold value of CAV was determined as 450 cm/s through nonlinear regression. All regression coefficients, $\alpha_0 - \alpha_{13}$, are listed in Table 5. The predictive model (Eq. 21) can be characterized with a lognormal distribution by passing the Lilliefors (1967) test at the 5% significance level. This indicates that the uncertainty in foundation residual settlement on densified sites could be characterized with a lognormal distribution and a standard deviation (σ_{DS}) of 0.54.

Table 5. Model coefficients for predicting foundation settlement on mitigated ground

Parameters	Value	Parameters	Value
α_0	-4.49	α_8	-0.1556
α_1	-0.1998	α_9	-0.001
α_2	-0.3333	α_{10}	-0.0047
α_3	0.8604	α_{11}	0.0861
α_4	-0.1169	α_{12}	0.4766
α_5	0.2632	α_{13}	1.386
α_6	0.6577	σ_{DS}	0.54
α_7	-0.0711		

7   Predictive Model Validation with Case Histories

The Bullock et al. (2019a) settlement models were validated and adjusted with case history data. The case history adjustment described above used a database of 50 case histories of structures on mat foundations that were affected by liquefaction-induced settlement. The adjusted model was then used to predict the settlement of 37 additional shallow-founded structures with other foundation types that were treated as equivalent mat foundations (i.e., mat foundations with dimensions equal to the building footprint). The model's predictions for those cases, which were excluded from the development of the adjustment, were unbiased and their residuals pass a two sample Kolmogorov–Smirnov test in comparison to the lognormal distribution obtained while regressing the adjustment.

The Bullock et al. (2022) models were developed using only data from the PS-I numerical study, with no case history adjustments. As discussed above, the soil models used in this study were calibrated and validated using element-level laboratory tests, centrifuge tests, and field observations. The models for CAV_c themselves were used to develop an LPI-style manifestation model that was shown to perform similarly to existing models for predicting observations of free-field surface liquefaction manifestations.

Figure 5 shows the exceedance probability curve of foundation settlement for the oil tank and building on densified ground in Niigata city, Japan, during the 1964 Niigata earthquake. These results compared well with the range of observed settlements in the field with an average difference of about 6.3%. The model validation with the very limited case history observations may suggest that our probabilistic model roughly captures the settlement patterns expected in realistic soil profiles. In PS-III, the contribution of sedimentation and ejecta (mechanisms not well captured numerically) were expected to be less significant than unmitigated conditions. For example, Hausler (2002) showed 18 case histories related to ground densification with various qualities/extents of densification, with little to no evidence of ground failure or major damage during previous earthquakes (but no details were reported). However, we acknowledge that further adjustment (similar to the approach in PS-II) should be performed when more well-documented case histories of mitigated structures become available in the future. More detailed model limitations were discussed in Hwang et al. (2022b), which are not repeated here for brevity.

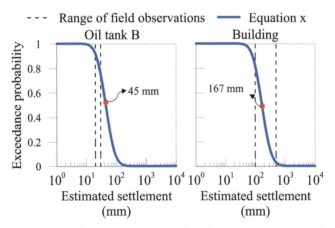

Fig. 5. Exceedance probability curves of structural settlement based on the Niigata City case history example.

8 Concluding Remarks

In this paper, we introduce a comprehensive, unified, and performance-based framework for predicting triggering of soil liquefaction in interlayered deposits, its consequences in terms of seismic settlement near shallow-founded structures, and its mitigation using ground densification. Effective liquefaction mitigation requires an improved fundamental understanding of triggering as a function of excess pore pressures in realistically stratified deposits that experience cross-layer interactions. It also requires improved probabilistic consequence procedures that account for 3D soil-structure interaction (SSI), all mechanisms of deformation, total uncertainty, and the impact of mitigation.

First, a series of centrifuge experiments were performed to evaluate site response and pore pressure generation in layered liquefiable deposits, the dominant mechanisms of deformation in the free-field and near shallow-founded structures, SSI, and the impact of ground densification as a mitigation strategy on SSI. Second, experimental results were used to validate 1D and 3D, fully-coupled, nonlinear, dynamic finite element analyses of layered sites and soil-foundation-structure systems with and without ground densification in OpenSees. Third, three sets of numerical parametric studies (PS-I through PS-III) were used to identify the functional forms for predicting: (i) liquefaction triggering in the free-field based on the capacity cumulative absolute velocity (CAV_c) corresponding to a threshold excess pore pressure ratio (r_u); (ii) settlement of unmitigated shallow-founded structures on potentially liquefiable soils; and (iii) the relative beneficial impact of ground densification with varying dimensions on foundation's permanent average settlement. In the end, a limited case history database helped validate the triggering and consequence models, accounting for field complexities that are not captured numerically or experimentally. The proposed set of models are the first of their kind to comprehensively consider the triggering, consequences, and mitigation of soil liquefaction near structures and the underlying uncertainties in a unified and explicit manner, aiming to guide a more reliable and performance-based treatment of liquefiable soils near structures.

Acknowledgements. The authors acknowledge support from the US National Science Foundation (NSF) under Grant no. 1454431. Any opinions, findings, and conclusions or recommendations expressed in this material are those of the author(s) and do not necessarily reflect the views of the NSF. This work utilized the Summit supercomputer, which is supported by the National Science Foundation (awards ACI-1532235 and ACI-1532236), the University of Colorado Boulder, and Colorado State University.

References

Acacio, A.A., Kobayashi, Y., Towhata, I., Bautista, R.T., Ishihara, K.: Subsidence of building foundation resting upon liquefied subsoil: case studies and assessment. Soils Found. **41**(6), 111–128 (2001)

Bardet, J.P., Huang, Q., Chi, S.W.: Numerical prediction for model no 1. In: Proc., Int. Conference on the Verification of Numerical Procedures for the Analysis of Soil Liquefaction Problems, edited by Arulanandan, K., Scott, R.F., pp. 67–86. A.A. Balkema, Rotterdam, Netherlands (1993)

Bertalot, D., Brennan, A.J., Villalobos, F.A.: Influence of bearing pressure on liquefaction-induced settlement of shallow foundations. Géotechnique **63**(5), 391–399 (2013)

Bray, J.D., Sancio, R.B.: Assessment of the liquefaction susceptibility of fine-grained soils. J. Geotechnical geoenvironmental Eng. **132**(9), 1165–1177 (2006)

Bray, J., Cubrinovski, M., Zupan, J., Taylor, M.: Liquefaction effects on buildings in the central business district of Christchurch. Earthq. Spectra **30**(1), 85–109 (2014)

Boulanger, R.W., Idriss, I.M.: Probabilistic standard penetration test–based liquefaction–triggering procedure. J. Geotech. Geoenviron. Eng. **138**(10), 1185–1195 (2012). https://doi.org/10.1061/(ASCE)GT.1943-5606.0000700

Beyzaei, C.Z., Bray, J.D., Cubrinovski, M., Riemer, M., Stringer, M.: Laboratory-based characterization of shallow silty soils in southwest Christchurch. Soil Dyn. Earthq. Eng. **110**, 93–109 (2018)

Bray, J.D., Macedo, J.: 6th Ishihara lecture: Simplified procedure for estimating liquefaction-induced building settlement. Soil Dyn. Earthquake Eng. **102**(Nov), 215–231 (2017) https://doi.org/10.1016/j.soildyn.2017.08.026.

Bullock, Z., Dashti, S., Karimi, Z., Liel, A., Porter, K., Franke, K.: Probabilistic models for residual and peak transient tilt of mat-founded structures on liquefiable soils. J. Geotech. Geoenviron. Eng. **145**(2), 04018108 (2019a). https://doi.org/10.1061/(ASCE)GT.1943-5606.0002002

Bullock, Z., Karimi, Z., Dashti, S., Porter, K., Liel, A.B., Franke, K.W.: A physics-informed semi-empirical probabilistic model for the settlement of shallow-founded structures on liquefiable ground. Géotechnique **69**(5), 406–419 (2019b). https://doi.org/10.1680/jgeot.17.P.174

Bullock, Z.: A framework for performance-based evaluation of liquefaction effects on buildings. Ph.D. dissertation, Dept. of Civil, Environmental, and Architectural Engineering, Univ. of Colorado at Boulder (2020)

Bullock, Z., Dashti, S., Liel, A.B., Porter, K.A., Maurer, B.W.: Probabilistic liquefaction triggering and manifestation models based on cumulative absolute velocity. J. Geotechnical Geoenvironmental Eng. **148**(3), 04021196 (2022)

Badanagki, M.: Influence of dense granular columns on the seismic performance of level and gently sloped liquefiable sites. Ph.D. dissertation, Dept. of Civil, Environmental and Architectural Engineering, Univ. of Colorado Boulder (2019)

Cetin, K.O., Seed, R.B.: Nonlinear shear mass participation factor (rd) for cyclic shear stress ratio evaluation. Soil Dyn. Earthq. Eng. **24**(2), 103–113 (2004)

Elgamal, A., Yang, Z., Parra, E.: Computational modeling of cyclic mobility and post-liquefaction site response. Soil Dyn. Earthquake Eng. **22**(4), 259–271 (2002). https://doi.org/10.1016/S0267-7261(02)00022-2

Geyin, M., Maurer, B., Bradley, B.A., Green, R., van Ballegooy, S.: CPT-based liquefaction case histories resulting from the 2010–2016 Canterbury, New Zealand, earthquakes: A curated digital dataset (version 2). DesignSafe-CI (2020) https://doi.org/10.17603/ds2-tygh-ht91

Geyin, M., Maurer, B.W., Bradley, B.A., Green, R.A., van Ballegooy, S.: CPT-based liquefaction case histories compiled from three earthquakes in Canterbury, New Zealand. Earthquake Spectra 8755293021996367 (2021) https://doi.org/10.1177/8755293021996367

Hwang, Y.W., Ramirez, J., Dashti, S., Kirkwood, P., Liel, A., Camata, G., Petracca, M.: Seismic interaction of adjacent structures on liquefiable soils: insight from centrifuge and numerical modeling. J. Geotechnical Geoenvironmental Eng. **147**(8), 04021063 (2021a)

Hwang, Y.W., Dashti, S., Kirkwood, P.: Impact of ground densification on the response of urban liquefiable sites and structures. J. Geotechnical Geoenvironmental Eng. **148**(1), 04021175 (2022a)

Hwang, Y.W., Bullock, Z., Dashti, S., Liel, A.: A probabilistic predictive model for the settlement of shallow-founded structure on liquefiable soils improved with ground densification. ASCE J. Geotech. Geoenviron. Eng. (accepted and in press) (2022b)

Idriss, I.M., Boulanger, R.W.: Soil Liquefaction During Earthquakes. Earthquake Engineering Research Institute, Berkeley, CA (2008)

JGS, Japanese Geotechnical Society. Remedial measures against soil liquefaction. A.A. Balkema, Rotterdam (1998)

Karimi, Z., Dashti, S.: Numerical and centrifuge modeling of seismic soil-foundation-structure interaction on liquefiable ground. J. Geotech. Geoenviron. Eng. **142**(1), 04015061 (2016). https://doi.org/10.1061/(ASCE)GT.1943-5606.0001346

Karimi, Z., Dashti, S.: Seismic performance of shallow founded structures on liquefiable ground: validation of numerical simulations using centrifuge experiments. J. Geotech. Geoenviron. Eng. **142**(6), 04016011 (2016). https://doi.org/10.1061/(ASCE)GT.1943-5606.0001479

Karimi, Z., Dashti, S., Bullock, Z., Porter, K., Liel, A.: Key predictors of structure settlement on liquefiable ground: a numerical parametric study. Soil Dyn. Earthq. Eng. **113**, 286–308 (2018)

Maurer, B.W., van Ballegooy, S., Bradley, B.A.: Fragility functions for performance-based ground failure due to soil liquefaction. In: Proceedings, PBD-III: Performance Based Design in Geotechnical Engineering III. Vancouver, BC: International Society for Soil Mechanics and Geotechnical Engineering (2017)

Menq, F.Y.: Dynamic properties of sandy and gravelly soils. Ph.D. dissertation, Dept. of Civil, Architectural and Environmental Engineering, Univ. of Texas at Austin (2003)

Mazzoni, S., McKenna, F., Scott, M., Fenves, G.: Open system for earthquake engineering simulation user command-language. Network for Earthquake Engineering Simulations, Berkeley, CA (2006)

NCEER: Proceedings of the NCEER worskhop on evaluation of liquefaction resistance of soils. Technical Report NCEER-97–0022. National Center for Earthquake Engineering Research (1997)

Olarte, J., Paramasivam, B., Dashti, S., Liel, A., Zannin, J.: Centrifuge modeling of mitigation-soil-foundation-structure interaction on liquefiable ground. Soil Dyn. Earthquake Eng. **97**, 304–323 (2017). https://doi.org/10.1016/j.soildyn.2017.03.014

Olarte, J., Dashti, S., Liel, A.: Can ground densification improve seismic performance of the soil-foundation-structure system on liquefiable soils. Earth. Eng. Struct. Dyn. 1–19 (2018a) https://doi.org/10.1002/eqe.3012.

Paramasivam, B., Dashti, S., Liel, A.: Influence of prefabricated vertical drains on the seismic performance of structures founded on liquefiable soils. J. Geotech. Geoenvirom. Eng. **144**(10), 04018070 (2018) https://ascelibrary.org/doi/abs/https://doi.org/10.1061/(ASCE)GT.1943–5606.0001950

Ramirez, J., Barrero, A.R., Chen, L., Ghofrani, A., Dashti, S., Taiebat, M., Arduino, P.: Site response in a layered liquefiable deposit: evaluation of different numerical tools and methodologies with centrifuge experimental results. J. Geotech. Geoenvirom. Eng. **144**(10), 04018070 (2018). https://doi.org/10.1061/(ASCE)GT.1943-5606.0001947

Ramirez, J.: Numerical modeling of the influence of different liquefaction remediation strategies on the performance of potentially inelastic structures. Ph.D. dissertation, Dept. of Civil, Environmental and Architectural Engineering, Univ. of Colorado Boulder (2019)

Seed, H.B., Idriss, I.M.: Soil moduli and damping factors for dynamic response analyses. Technical Rep. No. EERRC-70–10, Univ. of California, Berkeley, CA (1970)

Seed, R.B., et al.: Recent advances in soil liquefaction engineering: a unified and consistent framework. In: Proceedings of the 26th Annual ASCE Los Angeles Geotechnical Spring Seminar: Long Beach, CA (2003)

Olson, S.M., Mei, X., Hashash, Y.M.: Nonlinear site response analysis with pore-water pressure generation for liquefaction triggering evaluation. J. Geotech. Geoenviron. Eng. **146**(2), 04019128 (2020) https://doi.org/10.1061/(ASCE)GT.1943-5606.0002191.

Unutmaz, B., Cetin, K.: Seismic performance of mat foundations on potentially liquefiable soils after 1999 Turkey earthquakes. Technical Report METU/GTENG 10/09–02 (2010)

Watanabe, T.: Damage to oil refinery plants and a building on compacted ground by the Niigata earthquake and their restoration. Soils Found. **6**(2), 86–99 (1966)

Yang, Z., Lu, J., Elgamal, A.: OpenSees Soil Models and Solid Fluid Fully Coupled Elements: User's Manual. Dept. of Structural Engineering, Univ. of California, San Diego (2008)

Yoshida, N.: Liquefaction of improved ground in Port Island and its effect on vertical array record. In: Proc. of 12th WCEE, Vol. 1509 (2000)

Yoshimi, Y., Tokimatsu, K.: Settlement of buildings on saturated sand during earthquakes. Soils Found. **17**(1), 23–38 (1977)

Zienkiewicz, O.C., Chan, A.H., Pastor, M., Paul, D.K., Shiomi, T.: Static and dynamic behaviour of soils: a rational approach to quantitative solutions. I. Fully saturated problems. Proceedings of the Royal Society of London. A. Mathematical and Physical Sci. **429**(1877), 285–309 (1990)

Calibration and Prediction of Seismic Behaviour of CFRD Dam for Performance Based Design

Barnali Ghosh[1](✉), Vipul Kumar[1], and Sergio Solera[2]

[1] Mott MacDonald, London, UK
barnali.ghosh@mottmac.com
[2] Mott MacDonald, Cambridge, UK

Abstract. Dams have an important role in ensuring climate change resilience with limited available resources while balancing the needs for water and energy security for future. In this paper the behaviour of a tall CFRD dam located in a highly seismic area will be discussed. During the first impoundment, this concrete-faced rockfill dam (more than 150 m high) showed cracking of the parapet wall and damage to the joints between the parapet wall and the upstream face slabs. The broad objectives of this paper are to evaluate the available monitoring data and create a three-dimensional Finite Element (FE) model for dynamic behaviour. An extensive review of the deformation material parameters combined with the monitoring data during construction and impoundment was used to calibrate a stress-deformation 3D FE model. Material properties from both geophysical and geotechnical site investigations were used to develop the inputs for the non-linear time history analyses to estimate the permanent earthquake-induced deformations and stability within the dam body. Performance based criteria were used to confirm that the seismic displacements (horizontal and vertical) within the body of the dam are acceptable.

Keywords: CFRD dam · Seismic · Deformation

1 Introduction

It is recognized that dams located in highly seismic regions need to be designed using a performance-based matrix for different earthquake levels. However, in reality, design engineers make several assumptions during design. In general, tall Concrete Faced Rockfill Dams (CFRDs) have performed well under strong earthquakes although, some damage has been observed as noted in recent case histories such as the performance of Zipingpu dam during the Wenchuan strong earthquake (2008). This paper will highlight that it should be valuable to set a predefined way to state the assessment strategy and to justify the level of detail using a risk register.

In this paper the static and seismic behaviour of a tall CFRD located in a highly seismic area will be discussed. The reservoir behind the CFRD dam began filling in June 2015. During the first impoundment, the dam settled more than expected resulting in cracks of the parapet wall and damaged joints in between the parapet wall and the

© The Author(s), under exclusive license to Springer Nature Switzerland AG 2022
L. Wang et al. (Eds.): PBD-IV 2022, GGEE 52, pp. 397–411, 2022.
https://doi.org/10.1007/978-3-031-11898-2_22

concrete upstream face slabs. Due to the important deformations observed at the dam's crest, the reservoir was not reaching its full potential, operating below capacity. The client needed assistance to face this issue in order to raise the reservoir up to Full Supply Level (FSL) and have full operational conditions, producing more energy for the country. As the site is located in a seismic area, understanding the behaviour of the dam under a seismic event or a dynamic load was also required. It is noted that the authors were not the designers of this dam but were providing specialist technical advice and due diligence to ensure safety performance of the dam.

The broad objectives of this paper are to evaluate the available monitoring data and the dam's behaviour during construction and under reservoir loading, to create a three-dimensional Finite Element (FE) model of this CFRD for static and dynamic analyses. An extensive review of the deformation material parameters combined with the monitoring data during construction and impoundment is used to calibrate a stress-deformation 3D FE model will be presented. Material properties from both geophysical and geotechnical site investigations were used to develop the inputs for the non-linear time history analyses to estimate the permanent earthquake-induced deformations and stability within the dam body. Performance based criteria were used to confirm that the seismic displacements (horizontal and vertical) within the body of the dam are acceptable. The numerical model was used to understand the behaviour of the CFRD and estimate the additional settlements and displacements for the different reservoir operation levels under static and seismic influence.

1.1 Seismic Performance of CFRD Dams

It is recognized that the dam is in highly seismic region and is exposed to the occurrence of earthquakes affecting its performance. To define the seismic loads for the analysis of dams, several agencies (The Bureau of Reclamation (U.S. Department of the Interior), U.S. Army Corps of Engineers for Earthen dams and "Guide for Stability Analysis of Tailing Dams" (MINEM-PERU)), recommend the use of Maximum Credible Earthquake (MCE) or Safety Evaluation Earthquake (SEE). The MCE/SEE is the maximum level of ground motion for which the dam should be designed or analyzed.

The Operating Basis Earthquake (OBE) represents the ground motion for which minor damage is acceptable. The dam, structures should remain functional, and damage should be easily repairable. In many cases it is advisable to use a minimum return period of 145 years. It is noted that the dam stability would be typically examined for the OBE case by pseudo-static methods of analysis, with a "no damage" criteria and the MCE/SEE case by dynamic analysis methods, with the acceptance of some damage but no risk to the integrity of the dam itself.

In general, CFRD dams have performed well in earthquakes. However, some damage has been observed in previous earthquake. The Zipingpu dam is 156 m high, and 663.7 m long is one of the largest CFRDs in China and was designed for a peak ground acceleration of 0.26 g. This dam was shaken by the Wenchuan earthquake (Mw = 8.0) and is located at 17 km from the epicentre. However, during the earthquake, the reservoir was low with a volume of 300 Mn^3. The dam crest recorded accelerations greater than 2 g suggesting the behaviour of the concrete elements were different from the behaviour of the rock fill material. After the earthquake the maximum settlement at the dam crest was 735 mm and the horizontal deflection in downstream direction was 180 mm as reported by Chen et al. [1]

Due to this situation, it is difficult to estimate the dam behaviour when the reservoir is full and thus this aspect needs to be carefully considered. However, there are certain features with these dams, which should not be overlooked as discussed by Wieland and Brenner [2] and Weiland [3]:

1. Crack development in the concrete face slab: Due to deformations in the rockfill the concrete face slab will experience cracking. By properly compacting rockfill deformations can be minimised as well as crack development. Zones in the upstream part of the embankment in particular, including fine and coarse-grained transition zones under the slab, should be constructed of material of low compressibility.
2. Deformations in CFRDs: The long-term settlement of well compacted rockfill is in the range of about 0.1 to 0.2% of the dam height. Strong ground shaking can produce settlements in the order of 0.5 to 1 m.
3. Provision of generous freeboard which accounts for seismic deformations and impulse waves in the reservoir. The minimum freeboard allowance for dams more than 100 m height is 2.5 m (MOC 2015)
4. Presence of wide crest which improves safety of crest region of dam and increases resistance against overtopping from impulse waves.
5. Proper material selection in dam body with proper zoning (rockfill shall allow free draining of water leaking through the concrete face).
6. The dam body is made up of well compacted rockfill

1.2 Seismic Methodology

The seismic response of an earth or rockfill dam depends on many factors, such as the quality of rockfill compaction, the dam geometry, the narrowness of the canyon, the irregularity of the abutment, the flexibility of the canyon rock, the ground motion intensity, its frequency characteristics, spatial variability as reported by Dakoulas [4]. Table 1 shows the parameters and the level of analysis required for predicting the seismic behaviour of the rockfill dam based on available data. This should be the first attempt to pave a way for the analysis framework. This will ensure that analysis is following the different design phases including Feasibility, Tender design and Detailed design phases.

The ICOLD [5] guidelines do not explicitly define the seismic level which the non-critical appurtenant structures and components should be designed to withstand with accumulation of damage but without failure. The seismic design level should be selected based on.

- Life cycle of the structure/facility
- Consequences to public and environmental safety
- Potential lifeline and property losses

As per on a project level the considering all the hazards and the performance level of the assets, it was decided to have the following performance level.

- Normal – not related to safety; designed as ordinary structures to withstand design basis (DBE) ground motion intensities with PoE = 10% in 100 yrs (return period 950 yrs)
- Critical – critical for the safety of the dam and the downstream; designed to withstand the Safety Evaluation Earthquake (SEE) with PoE = 1% in 100 yrs (return period ≈10,000 yrs)

Table 1. Requirement and scope for seismic analysis of rockfill dams

Parametres	Simplified	Detailed	Complete analysis
Seismic Data	Design spectrum, Earthquake magnitude, maximum acceleration	Acceleration Time History	
Material data	Static resistance to shear, Dynamic rigidity variation and damping, Poisson ratio	Additionally, cyclic resistance to shear, excess generation of pore pressure	Additionally, residual resistance, nonlinear constitutive model
Analysis scope	Dynamic response, seismic stability, seismic permanent deformation	Additionally, post-seismic stability, post seismic permanent deformation, other post seismic damage	

2   Calibration with Monitoring Data

Rockfill material during dam construction and operation undergoes complex primary loading, unloading/reloading scenarios and possible change in the aggregate morphology due to environmental factors such as wetting of rockfills due to precipitation. As such, the calibration process focusses on the key timelines during the dam construction and operation to estimate the predict the deformations and stress within the dam body from numerical analyses. This section summarises the results from the calibration process undertaken for the monitoring data at the dam. The modelling was done using the Finite element software PLAXIS [6]. Specifically, the calibration has followed the following steps.

- Calibrate the Hardening Soil (HS) model parameters for the rockfill materials in the CFRD dam against the published data for other rockfill and CFRDs.
- Model the construction stages in PLAXIS 3D at specific dam sections using the construction lifts information available.
- Calibrate the 3D models against the End of Construction monitoring data (vertical displacements) at specific sections.

2.1 Selection of constitutive model

The current dam a high rockfill dam with a maximum height of 150 m at mid-section. In the case of high rockfill dams, the compaction of granular material during the construction stage of the dam causes density variation of rockfills. The broad stress range experienced by the rockfill materials in the construction of a high rockfill dam highlights the necessity of a constitutive model which can reasonably capture:

- Stress-dependency of stiffness.
- Non-linear and stress-strain behaviour in the framework of plasticity
- Shear induced dilatancy in the rockfill
- Stress states of rockfill materials including loading, unloading and reloading associated with complicated construction sequence and variation of water level in the reservoir of high dams

Duncan-Chang model (1970) [7] has been used in the past to model rockfill behaviour due to its ease of implementation and obtaining model parameters. However, this model is insufficient to capture the behaviour of rockfills under deviatoric loading as discussed by Szostak-Chrzanowski et al. [8]. This necessitates the requirement of a more robust nonlinear model which can capture the behaviour of dense rockfills under high confining stress levels.

The Hardening Soil (HS) model has been used to predict the deformational behaviour of rockfill dams and comparison of analysis predictions with monitoring data for 133 m high Kurtun Dam discussed by Özkuzukiran, S et al. [9] and a 182 m high rockfill dam by Pramthawee P et al. [10]. This model has been used for analysis here. The HS model can be easily adapted for performing dynamic analysis in its small-strain stiffness formulation. The hyperbolic law for small-strain and larger strain problems proposed by Santos, et al. [11] is used:

$$\frac{G_s}{G_o} = \frac{1}{1 + 0.385 \left| \frac{\gamma}{\gamma_{0.7}} \right|}$$

where,

- G_o is the initial or very small-strain shear modulus
- G_s is the secant shear modulus
- $\gamma_{0.7}$ is the shear strain level at which the secant shear modulus G_s is reduced to about 70% of G_o

The HS model in its small-strain formulation can generate hysteric damping under cyclic loading conditions. The amount of hysteresis depends on the magnitude of the corresponding strain amplitude. The hysteretic behaviour in unloading/reloading cycles is governed by Masing's (1962) criteria. A Consolidated Drained Triaxial test was simulated within PLAXIS 2D/3D at different confining stress levels. In the following example, the HS model is calibrated against CD-Triaxial results for Zone 3C as reported by Pramthawee P et al. [10] as shown in Fig. 1. It can be seen that in general the variation of deviatoric stress against axial strain is reasonably predicted by the PLAXIS HS model for low to medium stress levels.

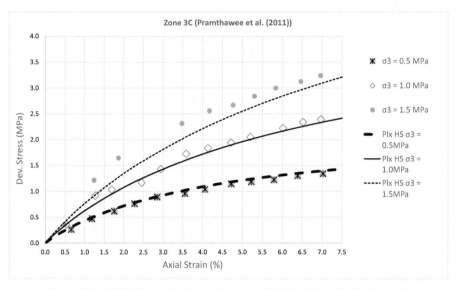

Fig. 1. PLAXIS HS model calibrated against CD-Triaxial results for Zone 3C

2.2 3D model Calibration

Horizontal displacement at the Dam Crest and Top of Parapet walls were recorded during the period Jun 2016 to Feb. 2019. The horizontal movement at dam crest is recorded from a total of 20 instruments. A 3D Finite Element model was developed in PLAXIS for the dam body and the valley as shown in Fig. 2. The dam model geometry is internally divided to simulate the construction sequence. The construction of the dam involved placement of the quarried rockfill material in vertical lifts such that the upstream face of the dam body was always higher or at the same elevation as the downstream face.

In the direction parallel to the dam crest at crest elevation 718 msnm, the model boundary is placed at 200 m and 235 m from the Right Abutment and Left Abutment, respectively. In the direction normal to the dam crest, the model boundaries are placed at about 140 m away from the downstream and upstream toe levels. These dimensions were taken approximately as 0.93–1.3 times the max. dam height of 150 m to ignore the boundary effects. A nominal thickness of 60 m of bedrock was considered below the dam body. The valley slopes off the Right and Left Abutment have been assumed to be parallel to the slope of upstream and downstream faces of the dam. This assumption was made to have a good quality mesh and smooth distribution of elements around the abutments and the valley. The 3D FE model has about 159,000 s-order tetrahedral elements.

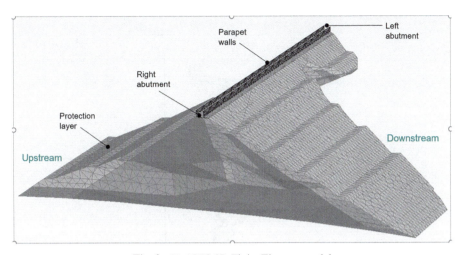

Fig. 2. PLAXIS 3D Finite Element model

The reservoir loading was modelled as hydrostatic load applied as a triangular pressure acting on the upstream face of the dam. The weight density of water for calculation of hydrostatic pressure was taken as 10 kN/m^3. In this model upstream face slab, starter slabs, plinth and corresponding joints are not modelled as the purpose of this modelling was to understand the geotechnical behaviour of the rockfill material. The different rockfill zones and bedrock foundation were modelled with HS and linear elastic model, respectively.

Most of the rockfill material was assigned friction angle of 42 ° and stiffness based on the laboratory testing data available. Further assumptions have been made in the 3D model to understand the behaviour of the dam in the static conditions. This included the following.

- Upstream face slab, starter slabs, plinth and corresponding joints are not modelled as the intention of the modelling was to understand the geotechnical settlement characteristics of the dam.
- The effect of siltation in increasing the unit weight of water held in the reservoir has not been considered.

The analysis was run multiple times by adjusting the rockfill parameters in HS model until a good match was obtained with the settlement monitoring data at RMV 3 within the dam body due to the construction of the dam as seen in Fig. 3. The settlements shown here are cumulative in nature during the construction of the dam. The monitoring data suggests that a max. cumulative settlement of about 0.750 m occurred around Elev. 680 msnm. This is about 0.5% of the dam height at this section. Simulations were also performed for other monitoring data.

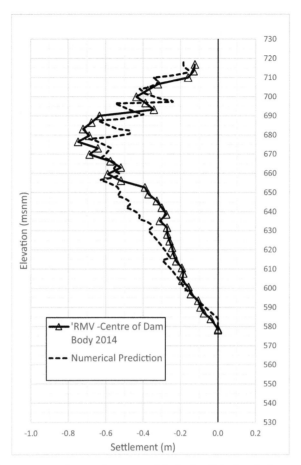

Fig. 3. Settlement comparison at the center of the body of the dam before reservoir filling

3 Selection of Input Time Histories

A site-specific seismic hazard study has been performed for the dam and the Peak Ground acceleration (PGA) is estimated to be 0.4 g for return period of 1000 years. The dynamic analysis of the dam would require suitable earthquake ground motions. In general, the selected acceleration time histories should match the bedrock response spectrum. The most common way to perform this is to scale existing acceleration time histories so that their spectra are close to the design spectrum, which is called 'target spectrum. In order to minimize the amount of artificial manipulation of the frequency content of the record using spectral matching software, individual components of the earthquake (2 × horizontal, 1 × vertical), has been scaled independently to best match the target spectra. The scale factor was determined by considering several points on the target spectrum. The PEER NGA database was used for the ground motion selection. This database is available on the internet (http://peer.berkeley.edu/nga/) and provides meta-data which are well-suited for automated selection procedures. It contains mostly earthquake records from North American events, as well as a large number from the Chi-Chi event in Taiwan and its aftershocks.

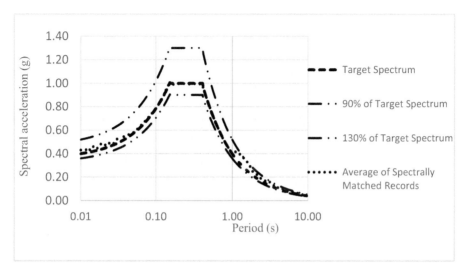

Fig. 4. Spectral matching of selected records used for analysis

Spectra matching of time histories were performed by following a systematic app-roach using RSPmatch 2005b programme. The resulting ground motion histories shall have characteristics that reasonably represent the input motion expected for the location. Hancock et al. [12] presents a set of criteria for selecting seed motions for RSPMatch05. The most important consideration is the matching of the initial spectral shape and other ground motion characteristics are used to filter the initial database. For spectral matching codal requirements ASCE 4–16 Sect. 2.6.2 and ASCE/SEI 43–05 Sect. 2.4 specify that.

- The computed 5% damped response spectrum of the accelerogram or of the average of all accelerograms (if a suite of motions is used) shall not fall more than 10% below the target spectrum at any one frequency.
- The mean of the computed 5% damped response spectra (if a suite of motions is used) shall not exceed the target spectrum at any frequency by more than 30%)

Figure 4 presents the average of the spectral matching for the selected records and minimum and maximum codal requirements. The selected records match this requirement.

- Whittier Narrows
- Kern County, "Santa Barbara Courthouse, Magnitude 7.35, 1952"
- Coalinga

4 Dynamic Parameters

4.1 Geophysical Survey

Geophysical surveys were carried out to determine the elastic parameters and deformation moduli of the rockfill (within the small-strain range) using Seismic Refraction, Seismic Tomography and MASW methods. In general, the P-wave data acquisition and data processing was generally reliable. However, the S-wave data acquisition is not as effective as for the P-waves as a dedicated shear-wave source was not used. 3D effects and depth of investigation limited the reliability and application of MASW results.

For bulk of the dam, V_p = 1077–1499 m/s and V_s = 605–671 m/s; for shallow materials, V_p = 603–917 m/s and V_s = 215–292 m/s based on Seismic Tomograms. The MASW results were more in line with previously published values for rockfill dams.

4.2 Small-Strain Stiffness (Gmax) for Rockfill

G_{max} (small-strain shear modulus) was obtained from the V_s inferred from the data obtained along Line 2 and Line 3 using the following formula,

$$V_s = \sqrt[2]{\frac{G_{max}}{\rho}} \tag{1}$$

where, ρ is the mass density of the soil.

Dynamic shear modulus at very small strain, G_{max}, is associated with the confining pressure and relative density of gravels. Based on numerous tests on a spectrum of disturbed and undisturbed soils, the shear modulus decreases, with increasing strain amplitude. The following empirical formula proposed by Hardin-Drnevich [13] is adopted here:

$$G_{max} = GoP_aF(e)(\frac{\sigma c'}{Pa})^m \tag{2}$$

where, Go is a modulus coefficient. Go is taken equal to 360; F(e) denotes an influencing function of void ratio. F(e) is calculated for angular rockfills with an assumed void ratio of 0.4; σc' is the mean effective principal stress and m is a modulus exponent whose value was taken as 0.65. It is assumed that the Ko is equal to 0.5 and there is no water inside the dam body when calculating the effective stress.

Gmax from Hardin-Drnevich's equation is used to calculate a best-estimate fit for the range of Gmax calculate from geophysics data. Figure 5 shows the comparison of best-fit Gmax from Hardin-Drnevich's equation as compared to the geophysics data. It is assumed that Best-fit Gmax derived using Hardin-Drnevich equation for geophysics data obtained for the downstream face of the dam is applicable for the entire dam body.

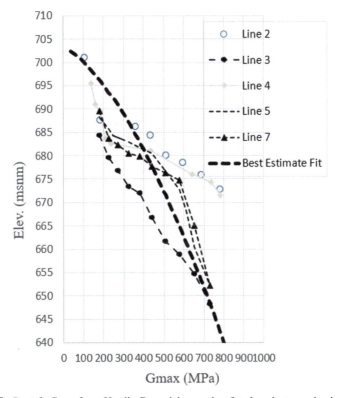

Fig. 5. Best-fit G_{max} from Hardin-Drnevich equation fitted against geophysics data

4.3 Dynamic Properties of Rockfill

The behaviour of rockfill under seismic loading is characterised by degradation of shear modulus with increasing shear strains ($G/Gmax - \gamma$) and generation of associated hysteretic damping ($\xi - \gamma$). Since the applied input ground acceleration is in horizontal direction (vertically propagating, horizontally polarised shear waves) only, the degradation of shear modulus and constrained modulus is considered to be identical. Due to the lack of site-specific cyclic triaxial, resonant column or torsional shear tests, reference is made to $G/Gmax - \gamma$ and $\xi - \gamma$ curves obtained from statistical analysis of 122 rockfill specimens Yu Feng et al. [14] was used in the analysis.

5   3D FE Dynamic Model

The 3D FE model for dynamic analysis of the dam is based on the geometry for the static analysis. The model extents of the FE model for dynamic analysis were extended compared to the static analysis to reduce the effects of wave reflection from the boundary; the geometry of the dam body is kept consistent with that for the static analysis. The 3D FE model for dynamic analysis has about 150,000 elements. Model boundaries for seismic analysis need special treatment as compared to the static analysis.

Seismic loading is applied at the base of the model as acceleration time-histories. The primary case for dynamic analysis is with reservoir at Elev. 720 msnm. The reservoir loading would cause a bias in seismic displacement of the dam towards the downstream. Therefore, to have a conservative estimate of seismic displacement within the dam body, the orientation of applied acceleration time-histories at the base of the model is from upstream to downstream direction and in the direction normal to dam crest. The PLAXIS 3D dynamic analysis does not explicitly model the fluid–structure interaction between the reservoir and the dam body. Instead, hydrodynamic loading from the reservoir is modelled as a time-invariant pressure load applied directly to the upstream face of the dam.

5.1 Constitutive Model and Calibration

The degradation of shear modulus (G) and generation of associated hysteretic damping (ξ) with cyclic shear strains (γ) during dynamic analysis is modelled with HS_{small} constitutive model within PLAXIS. HS_{small} is similar to the HS-type of soil model used during static analysis of the dam. Application of HS_{small} for rockfill under dynamic loading requires calibration against the $G/Gmax - \gamma$ and $\xi - \gamma$ curves.

5.2 Reservoir Load

The effect of the reservoir in dynamic analysis is considered as a cumulative effect of hydrostatic and hydrodynamic pressure acting on the upstream face of the dam. The hydrostatic pressure is calculated for the reservoir at Elev. 720 msnm applied as a pressure load. The hydrodynamic pressure is calculated for the reservoir at Elev. 720 msnm using an approach proposed by Zangar [15]. The total pressure is applied on the upstream face of the dam for 3D dynamic analysis.

5.3 Horizontal Acceleration at Crest

Horizontal acceleration time histories for the selected locations at Left Abutment, Central section and Right Abutment were obtained for all 3 input ground motions. The acceleration time histories at the selected nodes have either a peak towards downstream (positive) or towards upstream (negative) direction during the entire duration of shaking. Figure 6 shows the maximum of (absolute magnitude) of peak horizontal component (either positive or negative) of acceleration at the selected locations. The average of maximum horizontal acceleration at Left Abutment section is about 1.25 g, at Central section is about 0.98 g and at Right Abutment section is about 0.95 g. Acceleration at Left Abutment location is greater or equal to acceleration at the locations of Central section and Right Abutment. The higher acceleration at the Left Abutment could be attributed to:

- Left Abutment is separated from the main dam body and is 'nestled' between valley features which form a narrow canyon (very stiff material with low damping ratio) around the geometry of Left Abutment. This could potentially have the effect of concentrating the seismic waves within the Left Abutment geometry.

- Less rockfill thickness at Left Abutment means that the seismic waves reaching crest would undergo lower attenuation (due to damping within rockfill) as compared to seismic waves reaching crest at the central section of the dam.

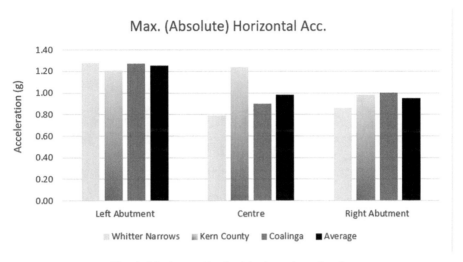

Fig. 6. Maximum (absolute) horizontal acceleration

The average maximum horizontal acceleration for the entire length of dam crest can be taken as an average of these three values, equal to 1.06 g. Rockfill deformation during impounding and earthquake loading depends on intact particle strength, rock-fill gradation, mineralogy, foundation conditions as well as construction technique and loading. The results obtained is compared with the published database to ensure that we are predicting the correct deformation and amplification. Figure 7 shows the amplified acceleration factor at crest (AFC). The amplified acceleration at crest for the dam is obtained as:

$$AFC = \frac{PCA}{PGA} = \frac{1.06}{0.4} = 2.65$$

This is slightly higher than that what is predicted by the equation presented by Yu et al. [16] but this is likely due to the effect of side valleys captured on 3D models.

5.4 Freeboard

The performance of a dam during and at the end of seismic loading is the allowance of available freeboard. The free board can be calculated as 3.4 m where the ratio of available freeboard to dam height is about 2.3% which is within the performance matrix developed for the project.

B. Ghosh et al.

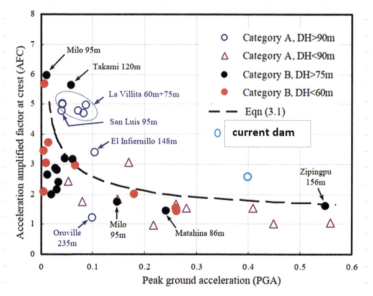

Fig. 7. Acceleration amplified at Crest from numerical analysis (Yu, et al. 2012)

6 Conclusions

Due diligence carried out for the design of a CFRD dam led to the development of 2D and 3D FE numerical models simulating detailed construction sequence and reservoir loading and unloading during the lifetime of the dam. Safety Assessment for existing dams in Static and Seismic conditions were performed using advanced constitutive soil/rockfill models performing non-linear analysis in time-domain. The analyses are performed using PLAXIS 3D finite element software. HS$_{small}$ model is used to model the degradation of stiffness with increasing strain and associated hysteretic damping. Best-estimate small-strain shear modulus profile (G_o) were developed from the geophysics data used to represent the full 3D dam. The effect of the reservoir in dynamic analysis is considered as a cumulative effect of hydrostatic and hydrodynamic pressure acting on the upstream face of the dam. Three earthquake time histories were used in the analysis. These records were spectrally matched to a target spectrum using wavelet techniques. Horizontal and vertical displacements are provided in the critical sections to assess the overall performance of the dam in seismic conditions. Amplified acceleration at Crest is calculated and compared with existing case histories.

Performance rather than safety is the main concern, as it is widely accepted that the effect of seismic loading on CFRDs is plastic deformation and settlement but not a slope failure in its classical sense. No slope failure was observed from the analysis completed for this work. The long-term settlement of well compacted rockfill is in the range of about 0.1 to 0.2% of the dam height. In this case, the maximum of average residual settlement for the 3 input ground motions is about 0.7 m which is about 0.45% of the dam height. The minimum freeboard allowance for dams more than 100 m height is 2.5 m (MOC 2015). Usually, the freeboard allowance for earthquake induced settlement should not exceed

1% of the dam's height. A potential risk framework should be developed for managing the overall performance of the dam in seismic conditions with adequate monitoring arrangements.

References

1. Chen, Y., et al.: Seismic Analysis of the Zipingpu Concrete-Faced Rockfill Dam Response to the 2008 Wenchuan, China, Earthquake. Journal of Performance of Constructed Facilities, **29**(5), 04014129 (2015)
2. Wieland, M., Brenner, R.P., Bozovic, A.: Potentially active faults in the foundations of large dams, Part I: Vulnerability of dams to seismic movements in dam foundation, Special Session S13. In: Proceedings 14th World Conference on Earthquake Engineering, Beijing, China (2008)
3. Wieland, M.: First International Symposium on Rockfill Dams, held in Chengdu, China from (2009)
4. Dakoulas: Nonlinear seismic response of tall concretefaced rockfill dams in narrow canyons. Soil Dynamics Earthquake Eng. **34**, 11–24 (2012)
5. ICOLD Bullletin 148: Selecting Seismic Design Parameters for Large Dams Guidelines. Bulletin 72, 2010 Revision, s.l.: International Committee on Large Dams (2010)
6. PLAXIS BV (2017), PLAXIS Material Models Manual
7. Duncan, J.M., Chang, C.Y.: Nonlinear Analysis of Stress and Strain in Soils. Journal of Soil Mechanics and Foundation Division, American Society of Civil Engineers (ASCE). 96(5), 1629–1653 (1970)
8. Szostak-Chrzanowski, A., Deng, N., Massiera, M.: Monitoring and Deformation Aspects of Large Concrete Face Rockfill Dams. In: 13th International Federationof Surveyors (FIG) Symposium on Deformation Measurement and Analysis, Lisbon, Portugal, May 12–5, pp. 1–10 (2008)
9. Özkuzukiran, S., Özkan, M.Y., Özyazicioğlu, M., Yildiz, G.S.: Settlement behavior of a concrete faced rockfill dam. J. Geotechnical Geological Eng. 24, 1665–1678 (2006)
10. Pramthawee, P., Jongpradist, P., Kongkitku, W.: Evaluation of hardening soil model on numerical simulation of behaviors of high rockfill dams. Songklanakarin J. Sci. Technol. **33**(3), 325–334 (2011)
11. Santos, J.A., Gomes Correia, A.: Reference threshold shear strain of soil. Its application to obtain an unique strain-dependent shear modulus curve for soil. In: XV International Conference on Soil Mechanics and Geotechnical Engineering, Istanbul, Turkey (2001)
12. Hancock, J., Watson-Lamprey, J., Abrahamson, N.A., Bommer, J.J., Markatis, A., McCoy, E., Mendis, R.: An improved method of matching response spectra of recorded earthquake ground motion using wavelets. J. Earthquake Eng. **10**(SPEC. ISS. 1), 67–89 (2006)
13. Hardin, B.O., Drnevich, V.P.: Shear modulus and damping in soil: design equations and curves. J. Soil Mech. Found. Div. **98**(7), 667–692 (1972)
14. Yu Feng, J., Shi Chun, C.: Application of Rockfill Dynamical Characteristic Statistic Curve in Mid-small Scale Concrete Face Dam Dynamic Analysis, pp. 15574–15582. Portugal, Sociedade Portuguesa de Engenharia Sismica (SPES), Lisbon (2012)
15. Zangar, C.: Hydrodynamic pressures on dams due to horizontal earthquakes, Cambridge, Massachusetts: Proc. Soc. Exp. Stress Anal. (1953)
16. Yu, L., Kong, X., Xu, B.: Seismic Response Characteristics of Earth and Rockfill Dams. Portugal, Sociedade Portuguesa de Engenharia Sismica (SPES), Lisbon (2012)

Uncertainties in Performance Based Design Methodologies for Seismic Microzonation of Ground Motion and Site Effects: State of Development and Applications for Italy

Salvatore Grasso[✉] and Maria Stella Vanessa Sammito

Department of Civil Engineering and Architecture, University of Catania, Catania, Italy
sgrasso@dica.unict.it

Abstract. Performance-Based Design (PBD) is a more rational and general design approach, particularly for seismic regions, in which the design criteria are expressed in terms of achieving stated performance objectives. In this approach it is relevant the performance required to structures and to geotechnical works subjected to stated levels of seismic hazard, as well as the geotechnical constitutive model used to predict the performance. The parameters of the constitutive models are related in turn to soil properties. Earthquake hazard zonation in urban areas is the first and most important step towards a seismic risk analysis in densely populated Regions. The Seismic Microzonation is nowadays a world-wide accepted tool for the mitigation of seismic risk. It is a complex process involving different disciplines ranging from Geology and Applied Seismology to Geotechnical and Structural Engineering. The aim achieved in seismic hazard microzonation studies throughout the last 20 years performed at the presented typical case histories in Italy was to quantify the spatial variability of the site response on some typical historical scenario earthquakes that would be expected in the area. In order to quantify the expected ground motion, the manner in which the seismic signal is propagating through the subsurface was defined.

Propagation was particularly affected by the local geology and by the geotechnical dynamic ground conditions. Large amplification of the seismic signals generally occurs in areas where layers of low seismic shear wave velocity overlie material with high seismic wave velocity, i.e. where soft sediments cover bedrock or more stiff soils. Therefore, essential key issue here is to obtain a good understanding of the local subsurface conditions. The study builds on the recent experience of seismic microzonation studies in Sicily (Italy), after the effects of the 2018 seismic sequence. Examples of ground response analysis are presented by using some 1-D and 2-D codes, including methodologies taking into account soil uncertainties for site characterisation.

Keywords: Seismic microzonation studies · Spectrum compatibility · 1D and 2D numerical analyses

© The Author(s), under exclusive license to Springer Nature Switzerland AG 2022
L. Wang et al. (Eds.): PBD-IV 2022, GGEE 52, pp. 412–427, 2022.
https://doi.org/10.1007/978-3-031-11898-2_23

1 Introduction

This paper describes the Performance Based Design (PBD) methodology for the Seismic Microzonation (SM) of ground motion and site effects. One of the most important aspects is the definition of the input motions at the seismic outcropping bedrock based on the reference response spectra provided by the Italian seismic code [1] from probabilistic seismic hazard analysis. The suites of spectrum-compatible accelerograms can be used for 1D or 2D ground response analyses.

Another important aspect is the knowledge of the geological, geotechnical and geophysical characteristics of the areas under consideration in order to define the subsoil models [2]. On the basis of this methodology, it is possible to obtain a detailed delineation of the spatial variability in seismic responses, which can be used for the seismic microzonation mapping [2, 3].

Although probabilistic hazard analyses could be helpful in selecting sets of accelerograms, a deterministic evaluation is preferred in areas where large and rare earthquakes are observed [4–7]. The use of advanced methods capable of generating synthetic seismograms can give a valuable insight into the evaluation of a seismic ground motion scenario [4].

After the intense Etna seismic activity that began on 23 December 2018 and characterized by about seventy events with a magnitude $M > 2.5$, seismic microzonation studies were performed in Sicily (Italy) with the aim of supporting the structural design of buildings and the urban planning.

The studies have been carried out in several municipalities located in the region of Sicily (Riposto, Acireale, Piedimonte Etneo, Zafferana Etnea, Milo, Termini Imerese, Cefalù, Finale di Pollina, Trabia, Campofelice di Roccella, Alì Terme, Salaparuta, Balestrate and Caronia).

2 Historical Earthquakes in Sicily

Sicily is one of the Italian regions with high seismic risk. It was shaken in the past by large destructive events (1169, 1542, 1693, 1818, 1848, and 1990) [8, 9]. A repetition of events with similar characteristics would provide the additional risk of a damaging tsunami, as well as, liquefaction phenomena around the coastal areas of Sicily [10, 11].

The Val di Noto earthquake of January 11, 1693, struck a vast territory of southeastern Sicily and caused the partial, and in many cases total, destruction of 57 cities and 60,000 casualties [12, 13]. The Etna earthquake that took place on February 20, 1818, was a moderate earthquake, but its effects were noticed over a vast area [14–16]. The largest damage was observed in Augusta, where almost two-thirds of the buildings collapsed [9, 17]. On December 1908, a devastating earthquake occurred along the Strait of Messina between eastern tip of Sicily and the western tip of Calabria in the south of Italy. The Southern Calabria-Messina earthquake (Intensity MSC XI, Mw 7.24) was the strongest seismic event of the 20th century in Italy with the most ruinous in term of casualties (at least 80,000) [18].

More recently, the 26 December, 2018, earthquake (Mw 4.9) occurred at shallow depth (<1 km) on the eastern flank of Mount Etna. It was the most energetic event that hit the volcanic area during the last 70 years [19]. Figure 1 shows the intensity maps for the 1693, 1818, 1908 and 2018 earthquakes.

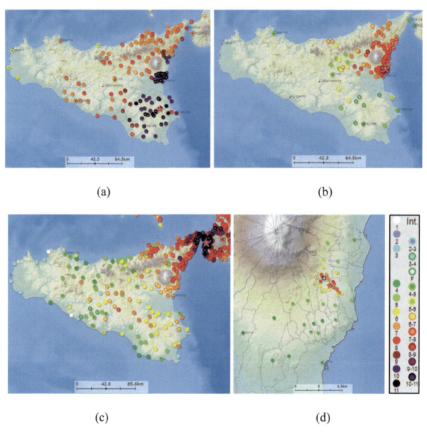

(a)   (b)

(c)   (d)

Fig. 1. Intensity maps: a) earthquake of 11 January, 1693; b) earthquake of 20 February, 1818; c) earthquake of 28 December, 1908; d) earthquake of 26 December, 2018 [20, 21]

3   Geology and Site Characterization Programme

Sicily, located on the Pelagian promontory of the African plate, is formed by the Iblean foreland, the Gela foredeep, the thick Sicilian orogen and the thick-skinned Calabrian-Peloritani wedge. The topography and geomorphology of Sicily is the result of constructive and destructive forces following the collision between the African and European plates [22]. The geological map of Sicily is reported in Fig. 2.

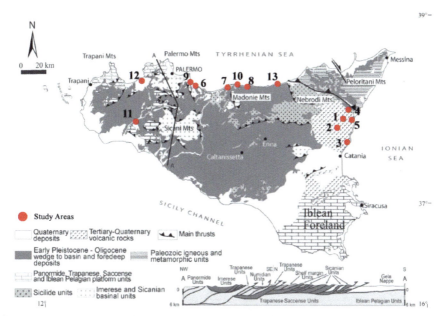

Fig. 2. Geological map of Sicily (data compiled from various Authors [23, 24] and simplified from [22]; modified) and location of the study areas.

The geotechnical and geophysical characterization of the study areas, whose locations are shown in the Fig. 2, has required an accurate field investigation that reached the depth of 30 m. Moreover, laboratory tests have been also performed on undisturbed samples. Laboratory and in situ investigations include the following tests: Boreholes, Down Hole Tests (DH), Multichannel Array Surface Wave Tests (MASW), Seismic Tomographies, Horizontal to Vertical Spectral Ratio Tests (HVSR) and Direct Shear Tests (DST).

Shear wave velocity, V_s, plays a fundamental role in Seismic Microzonation Studies. Therefore, in order to summarize V_s profiles against depth obtained by geophysical tests, the equivalent shear wave velocity ($V_{s,eq}$) has been calculated for each test sites according to the Italian seismic code [1]. $V_{s,eq}$ values and the corresponding soil types [1] are reported in Table 1.

Table 1. Equivalent shear wave velocities and soil types obtained from geophysical tests according to the NTC 2018[1].

	Test sites	$V_{s,eq}$ [m/s]	Soil Types
1	Milo	384	B
2	Zafferana Etnea	449	B
3	Acireale	343	C
4	Piedimonte Etneo	504	B
5	Riposto	458	B
6	Termini Imerese	466	B
7	Campofelice di Roccella	389	B
8	Finale di Pollina	397	B
9	Trabia	486	B
10	Cefalù	365	B
11	Salaparuta	291	C
12	Balestrate	261	C
13	Caronia	388	B

4   Selection of Sets of Input Ground Motions

The Italian seismic code [1] adopts the performance-based seismic design for the calculation of seismic actions on structures from probabilistic seismic hazard analysis (PSHA). The structural performance is verified against ground motions (GM) that have predetermined exceedance return periods at the site of interest. The corresponding GM is represented by the uniform hazard spectrum (UHS) [25].

The Italian seismic code [1] is based on the work of the National Institute of Geophysics and Volcanology (INGV) [26]. The INGV (http://esse1-gis.mi.ingv.it) evaluated the probabilistic seismic hazard for each node of a regular grid that covers the Italian territory. This resulted in hazard curves that are lumped in nine probabilities of exceedance in 50 years (2%, 5%, 10%, 22%, 30%; 39%, 50%, 63% and 81%) [26, 27]. The interactive seismic hazard map of Sicily (Fig. 3) has been obtained from http://esse1-gis.mi.ingv.it considering the 10% probability of exceedance in 50 years (return period of 475 years) that is the case of design for life-safety structural performance (SLV) [1].

The disaggregation data have been also derived (Fig. 4) considering the latitudine and longitudine of each test site. Disaggregation, expressed in terms of magnitude (M), source to site distance (R) and standard deviations (ε), identifies the values of some earthquake characteristics providing the largest contributions to the hazard in terms of exceeding a specified spectral ordinate threshold [26].

The software REXEL v. 3.5 (http://www.reluis.it) [27] allows to search for suites of waveforms, from the European Strong-motion Database, compatible to the reference spectrum in a defined range of periods according to the Italian seismic code [1]. A combination of seven accelerograms has selected for each test site in a way that their average is in an interval between 10% (lower threshold) and 30% (upper threshold) of the reference spectrum. The range of periods is equal to 0.1–1.1 s that is representative of the vibration periods of the structures target. The Magnitude-Distance Criterion has been used considering the disaggregation data derived from http://esse1-gis.mi.ingv.it website. The accelerograms have not been scaled.

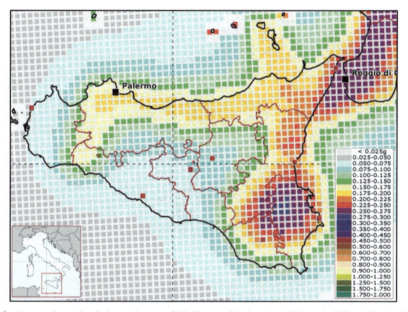

Fig. 3. Interactive seismic hazard map of Sicily considering the 10% probability of exceedance in 50 years (from http://esse1-gis.mi.ingv.it; modified)

Table 2 and Fig. 5 report the suites of waveforms compatible to the reference spectra (Soil Class A) [1] for Campofelice di Roccella (lat. 38.005; lon. 13.905), as an example, considering the 10% probability of exceedance in 50 years, the minimum event magnitude $M_{min} = 4.0$, the maximum event magnitude $M_{max} = 6.5$; the minimum epicentral distance $R_{min}[km] = 0$ and maximum epicentral distance $R_{max}[km] = 20$.

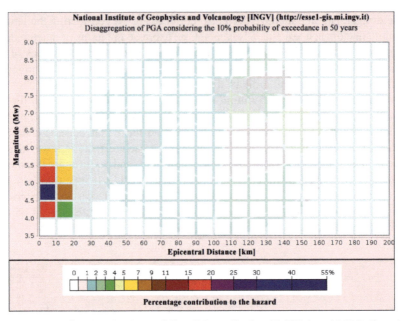

Fig. 4. Disaggregation data for Campofelice di Roccella (lat. 38.005; lon. 13.905) (from http://esse1-gis.mi.ingv.it; modified)

Table 2. Combination of waveforms obtained from REXEL v. 3.5. For Campofelice di Roccella (lat. 38.005; lon. 13.905)

Waveform ID	Earthquake ID	Station ID	Earthquake Name	Date	Mw	R[km]
982	72	ST309	Friuli (aftershock)	16/09/1977	5.4	9
4675	1635	ST2487	South Iceland	17/06/2000	6.5	13
242	115	ST225	Valnerina	19/09/1979	5.8	5
5079	1464	ST2552	Mt. Hengill Area	04/06/1998	5.4	6
6115	2029	ST1320	Kozani	13/05/1995	6.5	17
2025	710	ST1357	Kremidia (aftershock)	25/10/1984	5	16
4674	1635	ST2486	South Iceland	17/06/2000	6.5	5
Mean					5.9	10.1

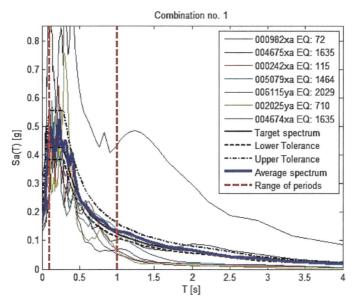

Fig. 5. Combination of waveforms obtained from REXEL v. 3.5. For Campofelice di Roccella (lat. 38.005; lon. 13.905)

A deterministic evaluation has been preferred for a slope located in Caronia test site, also because it can better account the complex heterogeneity of the area. Seven different inputs have been used, consisting of six synthetic seismograms related to the Messina and Reggio Calabria 1908 earthquake (Bottari et al. [29]; Tortorici et al. [30]; Amoruso et al. [31]; DISS Aspromonte Est; DISS Gioia Tauro; DISS Messina Strait) and one related to the 1693 Catania scenario earthquake [32, 33]. The seismograms have been scaled to the PGA of 0.176 g provided by the NTC2018 [1].

5   Definition of the Subsoil Models

The definition of the subsoil model requires the knowledge of the depth of bedrock and the characterization of the soil layers in terms of geometry, geophysical and geotechnical properties. Moreover, laws of shear modulus and damping ratio against strain have to be considered in order to take into account the soil non linearity. For the estimation of the local site responses, 1D and 2D numerical analyses have been carried out.

5.1 1D Numerical Analyses

Local site response analyses have been performed using 1-D linear equivalent code STRATA [34] assuming the geometric and geological models as 1-D physical models. The equivalent linear analysis consists in the execution of a sequence of complete linear analysis with subsequent update of the parameters of stiffness and damping until the satisfaction of a predetermined convergence criterion [35].

The V_S profiles used for soil response analyses have been derived from geophysical tests reported in Paragraph 3. The value of the other parameters have been taken from the geotechnical characterization obtained through in situ and laboratory tests. In order to consider the degradation of the shear modulus and the increase of the damping ratio with the shear strain levels, Eqs. (1) and (2) suggested by Yokota et al. [36], calibrated to the soil under consideration, have been used.

$$\frac{G(\gamma)}{G_o} = \frac{1}{1 + \alpha\gamma(\%)^\beta} \tag{1}$$

where: $G(\gamma)$ = strain dependent shear modulus; γ = shear strain; α, β = soil constant.

$$D(\gamma)(\%) = \eta \cdot \exp\left[-\lambda \cdot \frac{G(\gamma)}{G_o}\right] \tag{2}$$

where: $D(\gamma)$ = strain dependent damping; γ = shear strain; η, λ = soil constant. Table 1 shows the obtained soil constants. The G-γ and D-γ curves are reported in Table 3.

Table 3. G- γ and D- γ curves used for site response analyses

Curves	α	β	η	λ
1	7.5	0.897	90	4.5
2	6.9	1	23	2.21
3	16	1.2	33	2.4
4	linear	linear	linear	linear
5	9	0.815	80	4
6	20	0.87	19	2.3
7	22	1.05	10	1.05

5.2 2D Numerical Analyses

To evaluate topographic and stratigraphic effects, 2-D equivalent-linear site response analyses have been also performed by the QUAKE/W [37] code based on a finite element formulation with a direct integration scheme in the time domain.

Local seismic amplification has been studied for a slope located in Caronia area in order to determine the topographical contribution on the ground acceleration.

Figure 6 shows the cross-section used for the finite element analyses. Three-surface points have been monitored: (1) the first behind the crest of the slope, (2) the second at the crest of the slope, (3) the third at a point downstream sufficiently far from the foot of the slope.

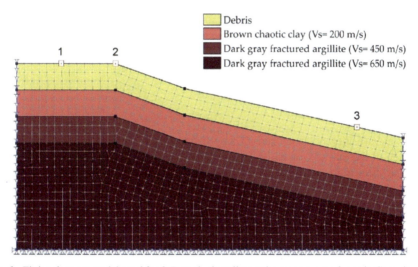

Fig. 6. Finite element model used for 2-D equivalent-linear site response analyses in Caronia area

6 Main Results

In the following, the main results obtained by numerical analyses have been summarized for all of the test sites considering the mean of the values calculated using the sets of seven input ground motions. Moreover, two case studies have been chosen (Campofelice di Rocella and Caronia) to outline specific results.

The soil dynamic response has been investigated in terms of accelerations. The values of the mean surface maximum accelerations and mean soil amplification factors, R, for all of the test sites, are reported in Table 4. The stratigraphic amplification values, S_s, provided by Italian technical code [1], are also presented for each test sites.

Table 4. Mean surface maximum accelerations and the mean soil amplification factors.

	Test sites	$PGA_{m,i}$	$PGA_{m,o}$	$R = PGA_{m,o}/PGA_{m,i}$	Ss
1	Milo	0.282 g	0.410 g	1.453	1.164
2	Zafferana Etnea	0.282 g	0.337 g	1.195	1.164
3	Acireale	0.282 g	0.439 g	1.554	1.345
4	Piedimonte Etneo	0.282 g	0.480 g	1.701	1.164
5	Riposto	0.282 g	0.476 g	1.688	1.164
6	Termini Imerese	0.173 g	0.262 g	1.513	1.200
7	Campofelice di Roccella	0.184 g	0.269 g	1.460	1.200
8	Finale di Pollina	0.184 g	0.345 g	1.877	1.200
9	Trabia	0.173 g	0.270 g	1.561	1.200
10	Cefalù	0.188 g	0.322 g	1.710	1.200
11	Salaparuta	0.178 g	0.310 g	1.743	1.463
12	Balestrate	0.148 g	0.274 g	1.849	1.500

In Fig. 7, results are presented for Campofelice di Roccella in terms of mean response spectrum at the surface, obtained by setting a structural damping of 5%. The elastic response spectrum provided by the Italian seismic code [1] is also shown. Moreover, according to the guidelines for Seismic Microzonation studies [38], FA factors have been determined for each test sites from the input and output spectra (Table 5) according to the following procedure:

1. Calculation of the period corresponding to the maximum spectral acceleration of the input spectrum ($S_{e,max,i}$): TA_i;
2. Calculation of the period corresponding to the maximum spectral acceleration of the output spectrum ($S_{e,max,o}$): TA_o;
3. Calculation of the mean value of the input spectrum in $0.5TA_i$ and $1.5TA_i$: $SA_{m,i}$:

$$SA_{m,i} = \frac{1}{TA_i} \int_{0.5TA_i}^{1.5TA_i} SA_i(T)dt \qquad (3)$$

4. Calculation of the mean value of the output spectrum in $0.5TA_o$ and $1.5TA_o$: $SA_{m,o}$:

$$SA_{m,o} = \frac{1}{TA_o} \int_{0.5TA_0}^{1.5TA_0} SA_o(T)dt \qquad (4)$$

where $SA_i(T)$ is the input spectrum and $SA_o(T)$ is the output spectrum;
5. Calculation of $FA = SA_{m,o}/ SA_{m,i}$

Figure 8 shows the amplification functions, A(f), for Campofelice di Roccella evaluated as the ratio between the Fourier spectra at the surface level and the Fourier spectra of the input motions applied at the base of the model. The main resulting frequencies f(I) and f(II) obtained by numerical analyses for each test site are reported in Table 6.

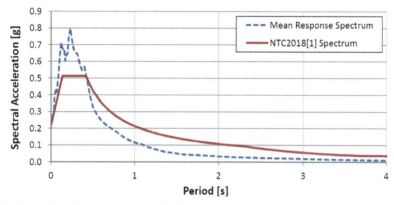

Fig. 7. Comparison between the mean elastic response spectrum obtained by numerical analyses and the same provided by NTC 2018 [1]

Table 5. Values of FA factors

Test sites	$Se_{max,i}[g]$	$TA_i[s]$	$Se_{max,o}[g]$	$TA_o[s]$	$SA_{m,i}$	$SA_{m,o}$	FA
1	0.73	0.20	0.99	0.40	0.62	0.87	1.41
2	0.73	0.20	0.91	0.43	0.62	0.72	1.16
3	0.73	0.20	1.07	0.43	0.62	0.87	1.41
4	0.73	0.20	1.52	0.20	0.62	1.09	1.75
5	0.73	0.20	1.21	0.20	0.62	1.04	1.69
6	0.52	0.22	0.81	0.22	0.45	0.69	1.52
7	0.51	0.22	0.79	0.23	0.46	0.68	1.46
8	0.52	0.24	1.34	0.22	0.46	0.98	2.13
9	0.52	0.22	0.72	0.12	0.44	0.64	1.44
10	0.47	0.19	0.83	0.06	0.39	0.53	1.36
11	0.50	0.12	0.86	0.15	0.40	0.72	1.79
12	0.36	0.15	0.76	0.20	0.33	0.63	1.88

Topographic seismic effects have been evaluated for the slope located in Caronia area. Three measures of amplifications have been computed [39]: the topographic amplification, A_t; the site amplification, A_a; the apparent amplification, A_a. The following

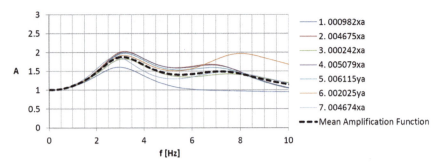

Fig. 8. Amplification functions obtained by numerical modelling for Campofelice di Roccella

Table 6. Main resulting frequencies f(I) and f(II) obtained by numerical analyses

	Test sites	f(I)[Hz]	f(II)[Hz]
1	Milo	2.59	8.34
2	Zafferana Etnea	2.89	12.65
3	Acireale	1.99	6.19
4	Piedimonte Etneo	6.30	15.35
5	Riposto	4.02	11.06
6	Termini Imerese	5.26	11.57
7	Campofelice di Roccella	3.10	7.08
8	Finale di Pollina	4.81	15.81
9	Trabia	6.10	13.62
10	Cefalù	2.99	6.77
11	Salaparuta	2.29	5.66
12	Balestrate	2.28	5.58

equations have been used:

$$A_t = \frac{a_{max} - a_{ffc}}{a_{ffc}} \tag{5}$$

$$A_s = \frac{a_{ffc} - a_{fft}}{a_{fft}} \tag{6}$$

$$A_a = \frac{a_{max} - a_{fft}}{a_{fft}} \tag{7}$$

where a_{fft} is the maximum free field acceleration in front of the toe; a_{ffc} is the maximum free field acceleration behind the crest and a_{max} is the maximum crest acceleration (Fig. 9). The results of numerical analyses are shown in the Table 7.

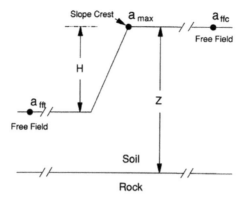

Fig. 9. Ashford et al. [39] model for topographical amplification evaluation

Table 7. Measures of amplifications obtain by numerical analyses

Input	A_t	A_s	A_a
Bottari et al. [29]	0.10	4.29	4.81
Tortorici et al. [30]	0.29	2.76	3.83
Amoruso et al. [31]	0.65	0.76	1.91
DISS Messina Strait	0.22	1.88	2.52
DISS Gioia Tauro	0.01	0.87	0.89
DISS Aspromonte Est	0.09	2.16	2.43
1693 Catania Earthquake	0.16	1.24	1.59

7   Concluding Remarks

In this paper, the Performance Based Design (PBD) methodology for the Seismic Micro-zonation (SM) of ground motion and site effects has been described. The study has been carried out in three main phases: the definition of sets of seven input ground motions for each test site, the definition of the subsoil models and the analysis of the results. The procedure adopted for selecting the suits of waveforms for outcropping and rock conditions has been explained in detail. For the definition of subsoil model in-hole geo-physical tests, surface tests and laboratory tests have been carried in order to characterize the soil layers in terms of geometry, geophysical and geotechnical properties. Ground response analyses have been presented by using 1-D and 2-D codes. The main results obtained by numerical analyses have been summarized considering the mean of the values calculated using sets of seven input ground motions. The dynamic response has been investigated in terms of accelerations, response spectra, and amplification functions. Soil amplification factors obtained from site response analyses are high and greater than amplification values provided by Italian technical code [1]. The outcome is presented in terms of preliminary results for the definition of the seismic microzoning map that represents an important tool for the seismic improvement of structures, indispensable for the assessment of seismic hazard.

References

1. NTC D.M. New Technical Standards for Buildings. (2018). https://www.gazzettaufficiale.it/eli/gu/2018/02/20/42/so/8/sg/pdf
2. Pergalani, F., et al.: Seismic microzoning map: approaches, results and applications after the 2016–2017 Central Italy seismic sequence. Bull. Earthq. Eng. **18**(12), 5595–5629 (2019)
3. Pagliaroli, A., Pergalani, F., Ciancimino, A., et al.: Site response analyses for complex geological and morphological conditions: relevant case-histories from 3rd level seismic microzonation in Central Italy. Bulletin Earthquake Eng. **18**, 5741–5777
4. Castelli, F., Cavallaro, A., Grasso, S., Lentini, V.: Seismic microzoning from synthetic ground motion earthquake scenarios parameters: the case study of the city of Catania (Italy). Soil Dyn. Earthq. Eng. **88**(2016), 307–327 (2016)
5. Grasso, S., Maugeri, M.: The seismic microzonation of the City of Catania (Italy) for the maximum expected scenario earthquake of january 11, 1693. Soil Dyn. Earthq. Eng. **29**(6), 953–962 (2009). https://doi.org/10.1016/j.soildyn.2008.11.006
6. Grasso, S., Maugeri, M.: The seismic microzonation of the city of Catania (Italy) for the Etna Scenario Earthquake (M=6.2) of February 20, 1818. Earthquake Spectra **28**(2), 573–594 (2002). https://doi.org/10.1193/1.4000013
7. Grasso, S., Maugeri M.: Seismic microzonation studies for the city of Ragusa (Italy). Soil Dynamics Earthquake Eng. **56**(2014), 86–97 (2014). ISSN: 0267-7261.
8. Bonaccorso, R., Grasso, S., Lo Giudice, E., Maugeri, M.: Cavities and hypogeal structures of the historical part of the City of Catania. Adv. Earthq. Eng. **14**, 197–223 (2005)
9. Castelli, F., Grasso, S., Lentini, V., Sammito, M.S.V.: Effects of soil-foundation-interaction on the seismic response of a cooling tower by 3D-FEM analysis. Geosciences **11**, 200 (2021) https://doi.org/10.3390/geosciences11050200
10. Grasso, S., Massimino, M.R., Sammito, M.S.V.: New stress reduction factor for evaluating soil liquefaction in the Coastal Area of Catania (Italy). Geosciences **11**, 12 (2021). https://doi.org/10.3390/geosciences11010012
11. Maugeri, M., Grasso, S.: Liquefaction potential evaluation at Catania Harbour (Italy). Wit Trans Built Environ. **1**, 69–81 (2013). https://doi.org/10.2495/ERES130061
12. Castelli, F., Cavallaro, A., Grasso, S.: SDMT soil testing for the local site response analysis. In Proceedings of the 1st IMEKO TC4 International Workshop on Metrology for Geotechnics, Benevento, Italy, 17–18 March 2016; pp. 143–148 (2016)
13. Castelli, F., Grasso, S., Lentini, V., Massimino, M.R.: In situ measurements for evaluating liquefaction potential under cyclic loading. In: Proceedings of the 1st IMEKO TC4 International Workshop on Metrology for Geotechnics, Benevento, Italy, 17–18 March 2016; pp. 79–84 (2016)
14. Boschi, E., Guidoboni, E., Mariotti, D.: Seismic effects of the strongest historical earthquakes in the Syracuse area. Ann. Geofis. **38**, 223–253 (1995)
15. Barbano, M.S., Rigano, R.: Earthquake sources and seismic hazard in Southeastern Sicily. Ann. Geofis. **44**, 723–738 (2001)
16. Imposa, S., Lombardo, G.: The Etna earthquake of February 20, 1818. In: Atlas of Isoseismal Maps of Italian Earthquakes; Postpischl, D.: (Ed); Quaderni de La Ricerca Scientifica: Bologna, Italy, pp. 80–81 (1985)
17. Barbano, M.S., Rigano, R., Cosentino, M., Lombardo, G.: Seismic history and hazard in some localities of south-eastern Sicily. Boll. Geof. Teor. Appl. **42**, 107–120 (2001)

18. Castelli, F., Cavallaro, A., Ferraro, A., Grasso, S., Lentini, V., Massimino, M.R.: Dynamic characterisation of a test site in Messina (Italy). Ann. Geophys. **61**(2), 222 (2018)
19. Emergeo Working Group: Photographic collection of the coseismic geological effects originated by the 26th December 2018 Etna (Sicily) earthquake. Misc. INGV **48**, 176 (2019)
20. Rovida, A., Locati, M., Camassi, R., Lolli, B., Gasperini, P., Antonucci, A.: Catalogo Parametrico dei Terremoti Italiani (CPTI15), versione 3.0. Istituto Nazionale di Geofisica e Vulcanologia (INGV) (2021). https://doi.org/10.13127/CPTI/CPTI15.3
21. Rovida, A., Locati, M., Camassi, R., Lolli, B., Gasperini, P.: The Italian earthquake catalogue CPTI15. Bull. Earthq. Eng. **18**(7), 2953–2984 (2020). https://doi.org/10.1007/s10518-020-00818-y
22. Di Maggio, C., Madonia, G., Vattano, M., Agnesi, V., Monteleone, S.: Geomorphological evolution of western Sicily. Italy. Geologica Carpathica **68**(1), 80–93 (2017). https://doi.org/10.1515/geoca2017-000
23. Catalano, R., Merlini, S., Sulli, A.: The structure of western Sicily, Central Mediterranean. Pet. Geosci. **8**, 7–18 (2002)
24. Catalano, R., et al.: Sicily's fold-thrust belt and slab roll-back: The SI.RI.PRO. seismic crustal transect. J. Geol. Soc., London **170**(3), 451–464 (2013)
25. Cito, P., Iervolino, I.: Peak-over-threshold: quantifying ground motion beyond design. Earthq Eng Struct Dyn. **49**(5), 458–478 (2020)
26. Iervolino, I., Chioccarelli, E., Convertito, V.: Engineering design earthquakes from multimodal hazard disaggregation. Soil Dynamics Earthquake Eng. **31**(9), 1212–1231 (2011)
27. Stucchi, M., Meletti, C., Montaldo, V., Crowley, H., Calvi, G.M., Boschi, E.: Seismic Hazard Assessment (2003–2009) for the Italian Building Code. Bull. Seismol. Soc. Am. **101**(4), 1885–1911 (2011). https://doi.org/10.1785/0120100130
28. Iervolino, I., Galasso, C., Cosenza, E.: REXEL: computer aided record selection for code-based seismic structural analysis. Bull. Earthquake Eng. **8**, 339–362 (2010). https://doi.org/10.1007/s10518-009-9146-1
29. Bottari, A., et al.: The 1908 Messina strait earthquake in the regional geostructural framework. J. Geodyn. **5**, 275–302 (1986)
30. Tortorici, L., Monaco, C., Tansi, C., Cocina, O.: Recent and active tectonics in the Calabrian Arc (Southern Italy). Tectonophysics **243**, 37–55 (1995)
31. Amoruso, A., Crescentini, L., Scarpa, R.: Source parameters of the 1908 Messina Strait, Italy, earthquake from geodetic and seismic data. J. Geophys. Res. **107**, ESE 4–1–ESE 4–11 (2002)
32. Valensise, G., Pantosti, D.: Database of potential sources for earthquakes larger than M 5.5 in Italy. Ann. Geophys. **44**, 1–964 (2001)
33. Castelli, F., Cavallaro, A., Grasso, S.; Ferraro, A. In situ and laboratory tests for site response analysis in the ancient city of Noto (Italy). In: Proceedings of the 1st IMEKO TC4 International Workshop on Metrology for Geotechnics, Benevento, Italy, 17–18 March 2016; pp. 85–90 (2016)
34. Kottke, A.R., Rathje, E.M.: Technical Manual for STRATA. PEER Report 2008/10; Univ. of California: Berkeley, CA, USA (2008)
35. Ferraro, A., Grasso, S., Massimino, M.R.: Site effects evaluation in Catania (Italy) by means of 1-D numerical analysis. Ann. Geophys. **61**(2), SE224 (2018)
36. Yokota, K., Imai, T., Konno, M.: Dynamic deformation characteristics of soils determined by laboratory tests. OYO Tec. Rep. **3**, 13–37 (1981)
37. Krahn, J.: Dynamic Modeling with QUAKE/W: An Engineering Methodology. GEO-SLOPE International Ltd., Calgary, AB, Canada (2004)
38. Dipartimento della Protezione Civile e Conferenza delle Regioni e delle Province Autonome: Indirizzi e criteri per la microzonazione sismica https://www.protezionecivile.gov.it
39. Ashford, S.A., Sitar, N., Asce, M.: Simplified method for evaluating seismic stability of steep slopes. J. Geotech. Geo-Environ. Eng. **128**, 119–128 (2002)

Mechanisms of Earthquake-Induced Landslides: Insights in Field and Laboratory Investigations

Ivan Gratchev[✉] [iD]

Griffith University, Southport, QLD 4222, Australia
i.gratchev@griffith.edu.au

Abstract. In this lecture, the author will share his experience in investigating several earthquake-induced landslides that occurred in Japan, Indonesia, and China, and discuss the mechanisms of slope failures and contributing factors. This work will present data from reconnaissance field surveys conducted at landslide sites and analysis of geological conditions related to slope instability. It will also discuss the results from a series of laboratory tests performed on soil specimens subjected to various loading modes and issues that may occur during investigations. This lecture will be of interest to researchers in the field of soil dynamics as well as engineers and decision-makers who are interested in the causes and mechanisms of earthquake-triggered landslides.

Keywords: Earthquake-triggered landslides · Landslide mechanism · Investigation

1   Introduction

Earthquake-triggered landslides are one of the most dangerous natural disasters that claim many casualties and cause significant damage to infrastructure. Accounts of destruction caused by the 2004 Mid Niigata Prefecture Earthquake, Japan [1], 2009 Sumatra Earthquake, Indonesia [2], and 2008 Wenchuan earthquake, China [3] have been documented in the literature. The understanding of this natural phenomenon shall assist engineers and decision-makers in building more resilient infrastructure for safer communities.

This lecture will present the major factors that cause landslides and discuss different methods of landslide investigations. A few earthquake-triggered landslides whose mechanisms the author has investigated through field and laboratory work are used as examples. The author will also share his knowledge and experience, and discuss some issues that can arise during similar studies. Finally, using the data from field and lab investigations, common mechanisms of earthquake-triggered landslides will be discussed in relation to site geology, geomorphology and soil properties. For more detailed studies, the interested reader is referred to the wealth of the literature on landslides that is easily accessible via national or university libraries. It is noted that some observations and insights presented by the author in this paper may not be applicable to all landslide sites, as the mechanism of this natural phenomenon is rather complex. Yet, it is believed that

© The Author(s), under exclusive license to Springer Nature Switzerland AG 2022
L. Wang et al. (Eds.): PBD-IV 2022, GGEE 52, pp. 428–436, 2022.
https://doi.org/10.1007/978-3-031-11898-2_24

this lecture will be of interest to students, researchers in the field of soil dynamics as well as to engineers and decision-makers who are interested in the causes and behavior of earthquake-triggered landslides.

2 Reconnaissance Surveys and Laboratory Work

The complex mechanism of earthquake-induced landslides involves several factors, which can be identified and studied using different methods of site investigation as well as a series of laboratory tests. This section will discuss the major aspects of such investigations.

2.1 Reconnaissance Surveys

Reconnaissance surveys are essential in collecting vital information on landslide characteristics and soil mass conditions which can shed light on the mechanism of landslides. The field work may include a series of boreholes (which is desirable but not always feasible) and field tests such as SPT and CPT. However, due to time and budget constraints and technical issues such as access to the landslide site, reconnaissance surveys are rather limited to detailed examinations of landslide area and collection of soil/rock samples for laboratory testing. In addition, portable field tests such as dynamic cone penetrometer [2] or portable cone penetrometer tests [4] are often performed to obtain estimations of landslide mass strength. Unfortunately, most field tests are commonly performed on disturbed soil masses after the landslide event (or a site next to the landslide), and for this reason, they don't provide an accurate estimation of soil mass conditions prior to earthquake. Thus, engineering judgement must be exercised when using the obtained results.

Fig. 1. Landslide sites and debris material that buried a settlement during the 2009 Sumatra Earthquake, Indonesia.

A challenging task during a reconnaissance survey is to obtain reliable data on groundwater conditions before the earthquake. In such cases, the groundwater characteristics and degree of soil mass saturation can be roughly estimated based on the historical precipitation data from the nearest Bureau of Meteorology station. Witness accounts can also provide important information which will help to better understand the landslide mechanism. For example, witness accounts of significant amount of water coming out of the landslide site at a rural area in Sumatra (Fig. 1) helped the investigation team [2] to justify the assumption that the landslide mass was saturated (or close to saturation) prior to the 2008 Sumatra earthquake.

It is noted that water not only weakens the soil strength, but it also deteriorates rock properties, a process of chemical weathering that can also lead to slope instability. For example, highly acidic groundwater was believed to be responsible for the formation of heavily weathered volcanic rock at the Aratozawa landslide site, making the dominant rock (pumice) extremely loose and weakened [4]. For this reason, it is recommended to obtain data on surface and ground water pH when possible.

2.2 Soil Examination and Laboratory Testing

Forensic investigations in the mechanism of earthquake-related landslides also include a careful examination of soil (or rock) material from the landslide mass and soil testing in the laboratory by means of conventional triaxial, shear box, and/or ring-shear apparatuses. The experimental programs shall include a series of tests with both cyclic and monotonic stress applications to soil samples so that the soil strength before, during and after an earthquake event can be obtained. There are several issues discussed in the following sections which need to be considered and addressed to obtain reliable laboratory data.

3 Factors Related to Earthquake-Triggered Landslides

The major factors contributing to the occurrence of landslides can be divided into two categories: 'triggering' factors such as earthquake magnitude and its epicenter; and 'causing' factors which are related to geomorphological and geological settings of the area, and soil/rock properties of the landslide mass.

3.1 Earthquake Characteristics

Earthquake magnitude and the epicenter location have a strong effect on the occurrence of landslides. It is logical to assume that the closer the studied area to the epicenter, the greater the damage caused by the earthquake can be expected. Keefer (1984) [5] conducted a first systematic investigation on relations between landslide distribution and seismic parameters such as magnitudes and distance from epicenter and reported that the landslide concentration tends to increase towards the epicenter. Although this relationship seems to work for most past earthquakes, recent studies also suggest that there are other factors that can affect the landslide distribution. For example, Gorum et al. (2011) [3] did not obtain any obvious relation between the distance from the epicenter and concentration of landslides triggered by the 2008 Sichuan earthquake, China. The investigators reported that the landslide distribution was primarily controlled by the geological fault rupture when most of events occurred within a 10 km range from the fault. This would lead to the next important factor, site geology, which is described in the following section.

3.2 Geological and Geomorphological Factors

It is already well-known that sound knowledge and understanding of the local geology can assist with landslide hazard assessment. Site geology and geological structures will be a key factor in determining the type of landslides. For example, a geological structure known as caldera was associated with earthquake-induced landslides triggered by the 2008 Iwate-Miyagi Nairiku earthquake, Japan [4] and 2009 Sumatra Earthquake, Indonesia [2]. The caldera-like structures are commonly made of combination of relatively weak and weathered sedimentary debris deposits overlain by volcanic rocks. Such geological setup tends to 1) amplify the ground motion of the rock during earthquake, and 2) provide a longer period of vibrations, which are two factors that can contribute to the relatively larger number of earthquake-triggered landslides within this caldera structure [6].

It is common that geological and geomorphological conditions are related. For example, steeper terrains/slopes are typically made of relatively harder geological material such as rock (note that rock can be of different levels of weathering) (Fig. 2), while relatively gentle slopes (about 30–35°) can be mostly made of coarse-grained material [7].

Fig. 2. Omigawa landslide triggered by the 2007 Chuetsu Oki earthquake, Japan. The landslide mass was formed in weathered sandstone on a steep slope of 55–60°.

The gradient of valleys adjoining to steep slopes as well as the presence of water can be crucial to the formation of debris flow. Jiufengchun landslide (Fig. 3), which was triggered by the 2008 Sichuan earthquake, buried more than 60 local people in Jiufengchun village. According to eyewitness accounts and the site survey conducted by [8], it was initiated as rockslide in weathered, relatively dry granite; however, as the landslide mass plunged in a valley that contained water springs, it immediately turned in a rapid debris flow that buried a village within 1 min [8]. The width of the landslide at the toe part was estimated as 300 m with about 1,500 m distance from the landslide source area to the toe.

Fig. 3. Jiufengchun landslide triggered by the 2008 Sichuan earthquake, China. It started as rock fall in weathered granite (see exposed slopes in the background) and turned in catastrophic debris flow.

The thickness of the landslide mass can vary depending on geology and material properties. Experience shows that most of earthquake-triggered landslides are relatively shallow, with the average thickness of 2–3 m. Thicker landslide masses are typically attributed to coarse-grained deposits, which can be a product of sandstone/siltstone weathering as observed during the 2004 Niigata earthquake [1, 9, 10]. It is noted that such large landslides may be part of ancient landslide mass that was re-activated by a strong earthquake shaking [1].

3.3 Soil and Rock Properties

The data on soil mass properties shall include information on soil type, the level of saturation, and in-situ soil characteristics such as void ratio and strength. Numerous studies have indicated that the resistance of soil to earthquake depends on soil capacity to generate excess pore water pressures during cyclic loading. Experience indicates that a soil structure that forms significant amount of pore space and the presence of water in these pores may contribute to this. As water plays a key role in the occurrence of land-slides, the generation of excess pore water pressures during earthquake can significantly decrease the shear strength of soil to a value that is no longer sufficient to withstand the shear stresses acting on soil mass. For coarse-grained soil mass (where sand fraction is dominant), the term 'liquefaction' is commonly used as a reason for strength reduction and consequent slope instability. It should be mentioned that although there is lots of evidence showing that relatively loose saturated sand can liquefy in the lab (effective stress drops to 0), that may not be the case in the field. However, significant decrease in the effective stress can still lead to substantial reduction in shear strength of soil and slope instability.

It has been reported by [9] and [11] that the presence of fines may increase soil resistance to liquefaction, and soil with a relatively large plasticity index [12] have much smaller potential for generation of excess pore water pressures, and thus it may not liquefy. In addition, the fine-grained matrix in such soils may provide some additional strength against liquefaction. As a result, the cyclic shear strength of plastic fine-grained soils may not significantly decrease during an earthquake. Yet, there are still some landslides (as shown in Fig. 2) reported in the literature that occurred in plastic fine-grained soils. Gratchev and Towhata (2008) [13] noted that the factor leading to earthquake-triggered landslides in plastic fine-grained soils can include: 1) large initial shear stresses originated from the high slope inclination and 2) permanent displacements of large magnitudes developed in soil mass during the earthquake loading.

4 Mechanism of Earthquake-Triggered Landslides

The mechanism of earthquake-triggered landslides can be explained to a certain level using the principles of soil mechanics. It is noted that this explanation involves a few assumptions and simplifications which would make it possible to apply the laboratory data from a small-scaled test to a large-sized landslide site.

4.1 Assumptions and Simplifications used to Study the Mechanism

Some of the main assumptions, which are related to data collection, soil mass inhomogeneity, and laboratory specimen testing techniques, are discussed below:

- Landslide sites can be quite large, and it is challenging to collect a good number of samples which will be sufficient to accurately reproduce the soil conditions prior to earthquake. Even within an area of a few meters, the soil conditions may vary significantly with the depth and width. Also, it is rather difficult to collect undisturbed samples that would have an intact soil structure that existed before the earthquake event. For this reason, it is not uncommon that disturbed samples are used during a forensic investigation because disturbed samples can still provide some idea about the soil conditions. Nevertheless, caution is needed when such results are analyzed as the strength properties of undisturbed and disturbed samples of the same soil material can be significantly different as shown by [4].
- It can be difficult to accurately represent earthquake loading in laboratory tests due to technical issues. To deal with this, a procedure developed by Seed et al. (1975) [14] may be used to estimate the number of load cycles based on the earthquake record.
- While performing lab tests, it is required by the relevant testing standard to saturate soil specimens to the highest level of saturation possible, which is commonly defined as $B > 0.95$, where B is the saturation parameter. However, it is still debatable that the landslide mass would be saturated, which is another common assumption that requires scientific judgment. Unfortunately, there is still limited amount of research in this field that can clearly show relationships between the degree of saturation and cyclic behavior of various types of soil.

– It should be mentioned that the pore water chemistry can affect the shear strength of fine-grained soil as shown by [15]. Although the pore water chemistry is generally ignored during forensic investigations, it can have some effect on both cyclic and monotonic strength, especially for those types of soil that contain expansive clay minerals such as smectite or montmorillonite.

4.2 Mechanism of Earthquake-Triggered Landslides through Soil Mechanics Principles

To better understand the mechanism of earthquake-triggered landslides, the data from ring-shear box tests (adapted from [10]) will be analyzed. Figure 4 presents a time series data from ring-shear box tests on coarse-grained material. The initial normal and shear stresses applied to the specimen before and after the cyclic loading represent the static field stress conditions prior to and after earthquake. The cyclic shear stress is then applied to the specimen (cyclic loading), reproducing the changes in shear stresses caused by the earthquake.

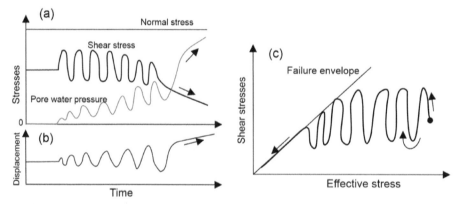

Fig. 4. Time series data of normal stress, shear stress, pore water pressure (a) and displacements (b) before, during and after earthquake loading. (c) Stress path of the test.

As the cyclic loading stage continues, the excess pore water pressure generates with every loading cycle and the specimen develops shear displacements (Fig. 4b). In Fig. 4c, a decrease in the effective stress due to the excess pore water pressure can be seen through the stress path moving to the left towards the failure envelope. If the excess pore water pressure generated by the cyclic loads is high enough to make the stress path touch the failure envelope line, then failure occurs. As a result, the shear strength of soil decreases, and at some point, it becomes insufficient to withstand the shear stresses acting on the soil mass after the earthquake. It can be seen in Fig. 4a when the shear stress plot goes down after the end of cyclic loading. This drop in soil strength is followed by increased values of shear displacement (Fig. 4b). Such rapid increases in shear displacement would represent the landslide movement in the field.

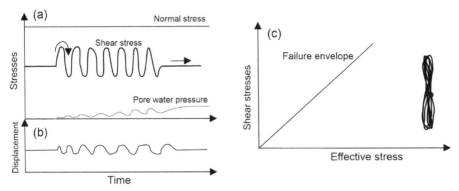

Fig. 5. Time series data of normal stress, shear stress, pore water pressure (a) and displacements (b) before, during and after earthquake loading. (c) Stress path of the test. Compared to Fig. 4, no failure has occurred in this test.

Figure 5 presents similar test results with the exception that the excess pore water pressure generated during the cyclic loading phase of the experiment was not high enough to decrease the effective stress acting on the soil specimen (Fig. 5c). Although the shear displacements were generated during the cyclic loading (Fig. 5b), they ceased after this phase of experiment was over. No failure occurred. Such soil mass behavior during an earthquake event would be expected from stable slopes.

5   Concluding remarks

This lecture summarizes the author's experience in regards to earthquake-induced land-slides and discusses issues that may arise during site and laboratory investigations. The main features of earthquake-related landslides are discussed in relation to site geology and soil mass conditions. The following main observations can be drawn:

– The complex mechanism of earthquake-induced landslides involves a combination of triggering and causing factors. The triggering factors are related to the earthquake characteristics while the causing factors are related to geology, geomorphology of the site, and the in-situ properties of landslide mass. The combination of these factors will determine the size of landslides, slope inclination, and type of movement.
– Earthquake-triggered landslides may occur in different soil types (coarse-grained or fine-grained soils) and weathered rock mass. The capacity of soil mass to generate excess pore water pressures is essential in the process of landslide initiation and move-ment. The soil types that generate sufficient amount of excess pore water pressures can be associated with earthquake-triggered landslides more often than the soil mass with smaller potential to do so.

References

1. Chigira, M., Yagi, H.: Geological and geomorphological characteristics of landslides triggered by the 2004 Mid Niigta prefecture earthquake in Japan. Eng. Geol. **82**(4), 202–221 (2006)

2. Gratchev, I., Irsyam, M., Towhata, I., Muin, B., Nawir, H.: Geotechnical aspects of the Sumatra earthquake of September 30, 2009. Indonesia. Soils and foundations **51**(2), 333–341 (2011)
3. Gorum, T., Fan, X., van Westen, C.J., Huang, R.Q., Xu, Q., Tang, C., Wang, G.: Distribution pattern of earthquake-induced landslides triggered by the 12 May 2008 Wenchuan earthquake. Geomorphology **133**(3–4), 152–167 (2011)
4. Gratchev, I., Towhata, I.: Geotechnical characteristics of volcanic soil from seismically induced Aratozawa landslide. Japan Landslides **7**(4), 503–510 (2010)
5. Keefer, D.K.: Landslides caused by earthquakes. Geol. Soc. Am. Bull. **95**, 406–421 (1984)
6. Kamiyama, M.: Evaluation of the damage caused by the 2008 Iwate Miyagi Nairiku earthquake. In: Special presentation at the 44th Japanese Geotechnical Society Conference. Yokohama, Japan (2009)
7. Gratchev, I.B., Towhata, I.: Analysis of the mechanisms of slope failures triggered by the 2007 Chuetsu Oki earthquake. Geotech. Geol. Eng. **29**(5), 695 (2011)
8. Wang, F., Cheng, Q., Highland, L., Miyajima, M., Wang, H., Yan, C.: Preliminary investigation of some large landslides triggered by the 2008 Wenchuan earthquake, Sichuan Province. China. Landslides **6**(1), 47–54 (2009)
9. Gratchev, I.B., Sassa, K., Osipov, V.I., Sokolov, V.N.: The liquefaction of clayey soils under cyclic loading. Eng. Geol. **86**(1), 70–84 (2006)
10. Sassa, K., Fukuoka, H., Wang, F., Wang, G.: Dynamic properties of earthquake-induced large-scale rapid landslides within past landslide masses. Landslides **2**(2), 125–134 (2005)
11. Bray, J.D., Sancio, R.B.: Assessment of the liquefaction susceptibility of fine-grained soils. J. Geotechnical Geoenvironmental Eng. **132**(9), 1165–1177 (2006)
12. Gratchev, I.B., Sassa, K., Fukuoka, H.: How reliable is the plasticity index for estimating the liquefaction potential of clayey sands? J. Geotechnical Geoenvironmental Eng. **132**(1), 124–127 (2006)
13. Gratchev, I., Towhata, I.: Analysis of a slope failure triggered by the 2007 Chuetsu Oki Earthquake. In: Proceedings of first world landslide Forum, pp. 227–230. Tokyo (2008)
14. Seed, H.B., Idriss, I.M., Makdisi, F., Banerjee, N.: Representation of irregular stress time histories by equivalent uniform stress series in liquefaction analyses. In: EERC 75–29, Earthquake Engineering Research Center, University of California, Berkeley (1975)
15. Gratchev, I.B., Sassa, K.: Cyclic shear strength of soil with different pore fluids. J. Geotechnical Geoenvironmental Eng. **139**(10), 1817–1821 (2013)

Regionalization of Liquefaction Triggering Models

Russell A. Green[✉]

Virginia Tech, Blacksburg, VA 24061, USA
rugreen@vt.edu

Abstract. The stress-based simplified liquefaction triggering procedure is the most widely used approach to assess liquefaction potential worldwide. However, empirical aspects of the procedure were primarily developed for tectonic earthquakes in active shallow-crustal tectonic regimes. Accordingly, the suitability of the simplified procedure for evaluating liquefaction triggering in other tectonic regimes and for induced earthquakes is questionable. Specifically, the suitability of the depth-stress reduction factor (r_d) and magnitude scaling factor (MSF) relationships inherent to existing simplified models is uncertain for use in evaluating liquefaction triggering in stable continental regimes, subduction zone regimes, or for liquefaction triggering due to induced seismicity. This is because both r_d, which accounts for the non-rigid soil profile response, and MSF, which accounts for shaking duration, are affected by the characteristics of the ground motions, which can differ among tectonic regimes, and soil profiles, which can vary regionally. Presented in this paper is a summary of ongoing efforts to regionalize liquefaction triggering models for evaluating liquefaction hazard. Central to this regionalization is the consistent development of tectonic-regime-specific r_d and MSF relationships. The consistency in the approaches used to develop these relationships allows them to be interchanged within the same overall liquefaction triggering evaluation framework.

Keywords: Liquefaction triggering · Simplified procedure · Earthquakes

1 Introduction

The stress-based "simplified" liquefaction evaluation procedure is the most widely used approach to evaluate liquefaction triggering potential worldwide. Analysis of fifty well-documented liquefaction case histories from the 2010-2011 Canterbury, New Zealand, earthquake sequence showed that the three commonly used Cone Penetration Test (CPT)-based simplified liquefaction triggering models (i.e., [1–3]), performed similarly, with the Idriss and Boulanger [3] model performing slightly better than the others. The same conclusion was obtained from the analysis of several thousand case studies from the Canterbury earthquake sequence, wherein the models were used in conjunction with surficial liquefaction manifestation severity indices (e.g., Liquefaction Potential Index: LPI [4]) to evaluate the severity of liquefaction [5, 6].

© The Author(s), under exclusive license to Springer Nature Switzerland AG 2022
L. Wang et al. (Eds.): PBD-IV 2022, GGEE 52, pp. 437–451, 2022.
https://doi.org/10.1007/978-3-031-11898-2_25

Despite the conclusions from the comparative studies using the New Zealand data, the suitability of the existing variants of the simplified models for use in evaluating liquefaction triggering in stable continental or subduction zone tectonic regimes is questionable. Additionally, the suitability of existing models for evaluating liquefaction potential due to induced earthquakes is also questionable. This is because the simplified procedure is semi-empirical, with the empirical aspects of it derived from data from tectonic earthquakes in active shallow-crustal tectonic regimes (e.g., California, Turkey, and portions Japan and New Zealand). As a result, existing variants of the simplified procedure may not be suitable for use in evaluating liquefaction triggering when the geologic profiles/soil deposits or ground motion characteristics differ significantly from those used to develop some of the empirical aspects of the models. Specifically, the depth-stress reduction factor (r_d) and magnitude scaling factor (MSF) relationships inherent to existing variants of the simplified liquefaction evaluation models are significantly influenced by the characteristics of the geologic profiles and ground motions.

In this paper, ongoing efforts by the author and his collaborators to develop region-specific r_d and MSF relationship are summarized. Relationships are being developed for tectonic earthquakes in stable continental regimes (e.g., the central-eastern US: CEUS) and subduction zone events (e.g., the Pacific Northwest, PNW of the US), and induced earthquakes in the Groningen region of the Netherlands due to natural gas extraction and in Midwest of the US (MWUS: Oklahoma, Texas, and Kansas) due to deep waste water disposal. The significance of the differences in these regional relationships is shown by the ratios of the factors of safety against liquefaction triggering computed using the various preliminary region-specific relationships and those from active shallow-crustal tectonic regimes.

2   Background

2.1   Overview of the Simplified Model

In the simplified procedure, the seismic demand is quantified in terms of Cyclic Stress Ratio (CSR), which is the cyclic shear stress (τ_c) imposed on the soil at a given depth in the profile normalized by the initial vertical effective stress (σ'_{vo}) at that same depth. The word "simplified" in the procedure's title originated from the proposed use of a form of Newton's Second Law to compute τ_c at a given depth in the profile, in lieu of performing numerical site response analyses [7, 8]. The resulting "simplified" expression for CSR is given as:

$$CSR = \frac{\tau_c}{\sigma'_{vo}} = 0.65(\frac{a_{max}}{g})(\frac{\sigma_v}{\sigma'_{vo}})r_d \tag{1}$$

where: a_{max} = maximum or peak horizontal acceleration at the ground surface; g = acceleration due to gravity; σ_v and σ'_{vo} = total and initial effective vertical stresses, respectively, at the depth of interest; and r_d = depth-stress reduction factor that accounts for the non-rigid response of the soil profile.

Additional factors are applied to Eq. 1, the needs for which were largely determined from results of laboratory studies, to account for the effects of the shaking duration

(MSF: Magnitude Scaling Factor, where the reference motion duration is for a moment magnitude, M_w, 7.5 earthquake in an active shallow-crustal tectonic regime), initial effective overburden stress (K_σ, where the reference initial effective overburden stress is 1 atm), and initial static shear stress (K_α, where the reference initial static shear stress is zero, e.g., level ground conditions). The resulting expression for the normalized CSR (i.e., CSR*: CSR normalized for motion duration for a $M_w 7.5$ active shallow-crustal event, 1 atm initial effective overburden stress, and level ground conditions) is given as:

$$CSR^* = \frac{CSR}{MSF \cdot K_\sigma \cdot K_\alpha} = 0.65(\frac{a_{max}}{g})(\frac{\sigma_v}{\sigma'_{vo}})r_d \frac{1}{MSF \cdot K_\sigma \cdot K_\alpha} \qquad (2)$$

Case histories compiled from post-earthquake investigations were categorized as either "liquefaction" or "no liquefaction" based on whether evidence of liquefaction was or was not observed. The normalized seismic demand (or normalized Cyclic Stress Ratio: CSR*) for each of the case histories is plotted as a function of the corresponding normalized in-situ test metric, e.g., Standard Penetration Test (SPT): $N_{1,60cs}$; Cone Penetration Test (CPT): q_{c1Ncs}; or small strain shear-wave velocity (V_S): V_{S1}. In this plot, the "liquefaction" and "no liquefaction" cases tend to lie in two different regions of the graph. The "boundary" separating these two sets of case histories is referred to as the Cyclic Resistance Ratio ($CRR_{M7.5}$) and represents the capacity of the soil to resist liquefaction during the reference event and initial stress conditions (i.e., a $M_w 7.5$ active shallow-crustal event, 1 atm initial effective overburden stress, and level ground conditions). This boundary can be expressed as a function of the normalized in-situ test metrics.

The factor of safety against liquefaction (FS_{liq}) is defined as the capacity of the soil to resist liquefaction for the reference conditions divided by the normalized seismic demand:

$$FS_{liq} = \frac{CRR_{M7.5}}{CSR^*} \qquad (3)$$

The ability of the soil to resist liquefaction during earthquake shaking can be considered an inherent property of the soil, independent of earthquake shaking characteristics. However, because $CRR_{M7.5}$ was developed from case histories from active shallow-crustal tectonic earthquakes, the seismic demand imposed on the soil needs to be "corrected" so that it is consistent with the reference conditions. In the simplified liquefaction evaluation procedure, the potential differences in ground motion characteristics manifest in the r_d and MSF relationships used to compute CSR* (Eq. 2). Both r_d and MSF are discussed in detail in the following.

2.2 Depth-Stress Reduction Factor

As stated above, r_d is an empirical factor that accounts for the non-rigid response of the soil profile. For illustrative purposes, the r_d relationship developed by Idriss [9] and used in both the Idriss and Boulanger [3] and the Boulanger and Idriss [10] simplified liquefaction evaluation models is shown in Fig. 1. The Idriss [9] r_d relationship is a function of earthquake magnitude and depth, with r_d being closer to one for all depths for larger magnitude events (note that $r_d = 1$ for all depths corresponds to the rigid response of the profile). This is because larger magnitude events have longer characteristic periods (e.g., [11]) and hence, longer wavelengths. As a result, even a soft profile will tend to respond as a rigid body if the characteristic wavelength of the ground motions is significantly longer than the height of the profile. Accordingly, the correlation between earthquake magnitude and the frequency content of the earthquake motions significantly influences the r_d relationship. Additionally, the dynamic response characteristics of the geologic profile, to include the impedance contrast between bedrock and the overlying soil, also influences r_d. This raises questions regarding the universality of existing r_d relationships that were developed using motions recorded during moderate to major tectonic events ($5 < M_w < 8$) in active shallow-crustal regimes and geologic profiles characteristic of California. Specifically, it is uncertain regarding whether or not such relationships are suitable for evaluating liquefaction triggering in other tectonic regimes, for induced earthquakes, or geologic profiles that significantly differ from those used to develop the existing r_d relationships.

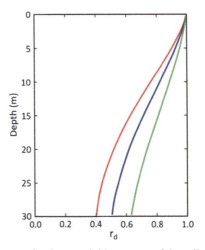

Fig. 1. r_d factor used to account for the non-rigid response of the soil column. The red, blue, and green lines were computed using the Idriss [9] r_d relationship for $M_w5.5$, $M_w6.5$, and $M_w7.5$ events, respectively.

2.3 Magnitude Scaling Factor: MSF

As stated above, MSF account for the influence of the strong motion duration on liquefaction triggering. For historical reasons, the influence of ground motion duration on liquefaction triggering is expressed relative to of that of a $M_w7.5$ shallow-crustal event in an active tectonic regime. Towards this end, MSF have traditionally been computed as the ratio of the number of equivalent cycles for the reference event ($n_{eq\,M7.5}$) to that of the event of interest ($n_{eq\,M}$), raised to the power b [i.e., $MSF = (n_{eq\,M7.5}/n_{eq\,M})^b$]. Commonly, the Seed et al. [12] variant of the Palmgren–Miner (P-M) fatigue theory is used to compute $n_{eq\,M}$ and $n_{eq\,M7.5}$ (collectively referred to as n_{eq} henceforth) from earthquake motions recorded at the surface of soil profiles. Additionally, the value of b is commonly obtained from laboratory test data. b is the negative slope of a plot of $\log(CSR)$ vs. $\log(N_{liq})$, as shown in Fig. 2; N_{liq} is the number of cycles required to trigger liquefaction in a soil specimen subjected to sinusoidal loading having an amplitude of CSR, typically determined using cyclic triaxial or cyclic simple shear tests.

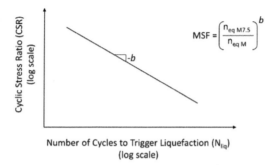

Fig. 2. For liquefaction evaluations, the Seed et al. [12] variant of the P-M fatigue theory has most commonly been used to compute the equivalent number of cycles (n_{eq}). Per this approach, the negative of the slope of a CSR vs. N_{liq} curve (or b-value) developed from laboratory tests is used to relate the "damage" induced in a soil sample from a pulse having one amplitude to that having a different amplitude. The b-value is also used to relate n_{eq} and MSF

There are several shortcomings inherent to the Seed et al. [12] variant of the Palmgren–Miner (P-M) fatigue theory used to compute n_{eq} and how the b-value used to compute MSF is determined. These include:

- Because the Seed et al. [12] approach for computing n_{eq} is based on motions recorded at the surface of the soil profile, both the magnitude and uncertainty in n_{eq}, and hence MSF, are assumed to be constant with depth. However, Green and Terri [13] have shown that n_{eq} can vary with depth in a given profile, and Lasley et al. [14] showed that while the median values of n_{eq} computed for a large number of soil profiles and ground motions remains relatively constant with depth, the uncertainty in their values varies with depth.
- Pulses in the acceleration time history having an amplitude less than $0.3 \cdot a_{max}$ are assumed not to contribute to the triggering of liquefaction, and thus are not considered in the computation of n_{eq}. Using a relative amplitude criterion to exclude pulses is contrary to the known nonlinear response of soil which is governed by the absolute amplitude of the imposed load, among other factors. The use of a relative amplitude exclusion criterion with tectonic earthquake motions may inherently bias the resulting MSF, limiting its validity for use with motions having different characteristics (e.g., motions from induced earthquakes).
- Each of the two horizontal components of ground motion is treated separately, inherently assuming that both components have similar characteristics. However, analysis of recorded motions has shown this is not always the case, particularly for motions

in the near fault region of tectonic events (e.g., [15–17]). Also, the horizontal components of the induced motions recorded in the Groningen region of the Netherlands have been shown to exhibit very strong polarization [18].

- b-values used to compute MSF are commonly derived from multiple laboratory studies performed on various soils, and it is uncertain whether all these studies used a consistent definition of liquefaction in interpreting the test data. As a result, the b-values entail a considerable amount of uncertainty [19]. Additionally, previously used b-values are not necessarily in accord with those inherent to the shear modulus and damping degradation curves used in the equivalent linear site response analyses to develop the r_d and n_{eq} relationships (elaborated on subsequently).
- Recent studies have shown that the amplitude (e.g., a_{max}) and duration (e.g., n_{eq}) of earthquake ground motions are negatively correlated (e.g.,[14, 20, 21]). Few, if any, of the MSF relationships developed to date have considered this.

Some of the above listed shortcomings likely will be more significant to the liquefaction hazard assessment than others, but it is difficult to state a-priori which ones these are. Furthermore, even for tectonic earthquakes in active shallow-crustal regimes, the validation of MSF is hindered by the limited magnitude range of case histories in the field liquefaction databases, with the majority of the cases being for events having magnitudes ranging from $M_w 6.25$ to $M_w 7.75$ [22]. Specific to liquefaction hazard assessment for induced earthquakes, MSF for small magnitude events are very important, particularly given that published MSF values vary by a factor of 3 for $M_w 5.5$ [23], with this factor increasing if the proposed MSF relations are extrapolated to lower magnitudes.

3   Development of Regional r_d and MSF Relationships

3.1   Regional-Specific Relationships

Region-specific r_d and MSF relationships are being developed using approaches outlined below. The approaches are consistently being implemented across tectonic regimes and for tectonic vs. induced events. As a result, any bias inherent to the relationships should be consistent across all relationships, which is essential to allow use of these relationships in conjunction with the same $CRR_{M7.5}$ curve.

3.2   Regional r_d Relationships

Region-specific r_d relationships are being developed using an approach similar to that used by Cetin [24]. Equivalent linear site response analyses are being performed using soil profiles and ground motions representative of: (1) active shallow-crustal events (e.g., western US: WUS); (2) shallow-crustal events in the stable continental setting (e.g., CEUS); (3) mega-thrust subduction zone events (e.g., PNW); (4) induced events resulting from deep waste water disposal in the MWUS; and (5) induced events resulting from natural gas production in the Groningen region of the Netherlands. In all cases ground motions representative of the respective source mechanisms and soil profiles representative of the respective regions are being compiled. Several functional forms for the region-specific r_d relationships are being examined in regressing the results from the site response analyses for each region, with the final form of regressed r_d relationship for each region being selected by balancing simplicity and low standard deviation of the regressed equation.

3.3 Regional MSF Relationships

The development of region-specific MSF relationships that overcome all the shortcomings listed in Sect. 2.3 is not as straightforward as developing the new r_d relationships. The reason for this is that there are many more issues with how n_{eq} and MSF relationships have been developed than there are for how r_d relationships have been developed. As a result, new approaches for computing n_{eq} and MSF need to be developed, as opposed to implementing an existing approach using a more comprehensive dataset and a more rigorous regression analysis.

Well-established fatigue theories have been proposed for computing n_{eq} for materials having varying phenomenological behavior; reviews of different approaches for computing n_{eq} are provided in Green and Terri [13], Hancock and Bommer [25], and Green and Lee [26], among others. Developed specifically for use in evaluating liquefaction triggering, the approach proposed by Green and Terri [13] is selected herein for developing the region-specific n_{eq} relationships. This approach is an alternative implementation of the P-M fatigue theory that better accounts for the nonlinear behavior of the soil than the Seed et al. [12] variant. In this approach, dissipated energy is explicitly used as the damage metric. n_{eq} is determined by equating the energy dissipated in a soil element subjected to an earthquake motion to the energy dissipated in the same soil element subjected to a sinusoidal motion of a given amplitude and a duration of n_{eq}. Dissipated energy was selected as the damage metric because it has been shown to correlate with excess pore pressure generation in saturated cohesionless soil samples subjected to undrained cyclic loading (e.g., [27, 28]). Furthermore, from a microscopic perspective, the energy is thought to be predominantly dissipated by the friction between sand grains as they move relative to each other as the soil skeleton breaks down, which is requisite for liquefaction triggering.

Conceptually, the Green and Terri [13] approach for computing n_{eq} is shown in Fig. 3. Stress and strain time histories at various depths in the soil profile are obtained from a site response analysis. By integrating the variation of shear stress over shear strain, the cumulative dissipated energy per unit volume of soil can be computed (i.e., the cumulative area bounded by the shear stress-shear strain hysteresis loops). n_{eq} is then determined by dividing the cumulative dissipated energy for the entire earthquake motion by the energy dissipated in one equivalent cycle. For historical reasons, the shear stress amplitude of the equivalent cycle (τ_{avg}) is taken as $0.65 \cdot \tau_{max}$ (where τ_{max} is the maximum induced cyclic shear stress, τ_c, at a given depth), and the dissipated energy associated with the equivalent cycle is determined from the constitutive model used in the site response analysis.

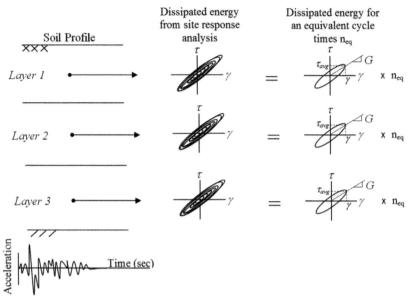

Fig. 3. Illustration of the proposed procedure to compute n_{eq} [13]. In this procedure, the dissipated in a layer of soil, as computed from a site response analysis, is equated to the energy dissipated in an equivalent cycle of loading multiplied by n_{eq}

The same soil profiles and ground motions being compiled for developing the region-specific r_d relationships are being used to develop the region-specific n_{eq} relationships, thus ensuring consistency between the relationships. As with the region-specific r_d relationships, several functional forms for the n_{eq} relationships are being examined in regressing the results from the site response analyses for each region, with the final form of regressed n_{eq} relationship for each region being selected by balancing simplicity and low standard deviation of the equation.

As mentioned previously, $MSF = (n_{eq\ M7.5}/n_{eq\ M})^b$ where the value of b is commonly obtained from laboratory test data (i.e., b is the negative slope of a plot of $\log(CSR)$ vs. $\log(N_{liq})$, as shown in Fig. 2). However, it is also possible to compute b-values from contours of constant dissipated energy computed using modulus reduction and damping (MRD) curves [29], as illustrated in Fig. 4. Assuming that the relationship between CSR and N_{liq} is a contour of constant dissipated energy, the b-value representing this relationship can be computed by estimating CSR for a range of N_{liq} values from the dissipated energy per unit volume of soil for a $M_w 7.5$ active shallow-crustal event ($\Delta W_{M7.5}$). Towards this end, the same approach used in equivalent-linear site response analysis can be used to compute $\Delta W_{M7.5}$: using a visco-elastic constitutive model in conjunction with MRD curves (e.g., Ishibashi and Zhang [30]). This approach results in the following equation:

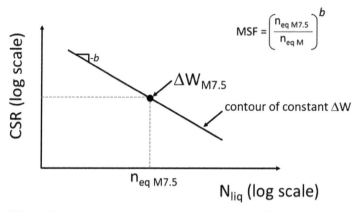

Fig. 4. A CSR vs. N_{liq} curve can be computed from shear modulus and damping degradation curves assuming the curve is a contour of constant dissipated energy. $\Delta W_{M7.5}$ can be computed using Eq. 4 and the remaining portions of the curve can be computed for different amplitudes of loading by simply computing the number of cycles for the assumed loading amplitude required for the dissipated energy to equal $\Delta W_{M7.5}$.

$$\Delta W_{M7.5} = \frac{2\pi D_\gamma \left(CRR_{M7.5} \cdot K_\gamma \cdot \sigma'_{vo} \right)^2}{G_{max} \left(\frac{G}{G_{max}} \right)_\gamma} n_{eqM7.5} \qquad (4)$$

where K_γ accounts for the overburden per Green et al. [31] and is analogous to the overburden correction factor, K_σ; G_{max} is the small-strain shear modulus; and D_γ and $(G/G_{max})_\gamma$ are the damping and shear modulus ratios, respectively, associated with a given value of shear strain, γ.

Because the relationship between CSR vs. N_{liq} is assumed to be a contour of constant dissipated energy, the remaining portions of the curve can be computed for different amplitudes of loading (i.e., CSR) by simply computing the number of cycles for the assumed loading amplitude required for the dissipated energy to equal $\Delta W_{M7.5}$. In this approach, the ΔW for one cycle of loading (i.e., ΔW_1) having amplitude CSR is computed as:

$$\Delta W_1 = \frac{2\pi D_\gamma \left(CSR \cdot \sigma'_{vo} \right)^2}{G_{max} \left(\frac{G}{G_{max}} \right)_\gamma} \qquad (5)$$

and the N_{liq} corresponding to this CSR amplitude is $\Delta W_{M7.5}/\Delta W_1$.

Although an estimate of G_{max} is necessary to compute $\Delta W_{M7.5}$ or ΔW_1 individually, because it appears in both the numerator and the denominator when computing $N_{liq} = \Delta W_{M7.5}/\Delta W_1$, it cancels out. Accordingly, the value of N_{liq} (and b-value) computed from MRD curves in this manner is not contingent on G_{max}.

Values of D_γ and $(G/G_{max})_\gamma$ can be determined using any published, applicable MRD curves. The Ishibashi and Zhang [30] curves, hereafter denoted IZ, are being used in this study. These MRD curves are dependent on the initial mean effective stress, σ'_{mo}, and soil type or plasticity index, PI. For liquefiable soils, PI is being assumed to equal zero and σ'_{mo} is being computed as a function of the at-rest lateral earth pressure coefficient, K_o, which is assumed to equal 0.5.

Using this procedure, b-values are being regressed for a range of q_{c1Ncs} and σ'_{vo} values. Values of q_{c1Ncs} were correlated to D_r using the following expression [3, 32] [33]:

$$D_r = 0.478(q_{c1Ncs})^{0.264} - 1.063 \tag{6}$$

In general, b-values remain relatively constant with increasing D_r and σ'_{vo}. An average b-value from the IZ MRD curves for $\sigma'_{vo} = 100$ kPa is 0.28. Note that although the b-values showed some sensitivity to changes in D_r and σ'_{vo}, the ranges of b-values (0.25–0.31) have only a mild impact on MSF. It is noted that using the same MRD curves to compute b as are also being used in the site response analyses to develop the r_d and n_{eq} relationships, ensuring consistency among all the relationships.

4   Ratios of FS_{liq}

To assess the significance of the region-specific r_d and MSF relationships on the liquefaction triggering predictions, the ratios of the FS_{liq} computed using the region-specific relationships and relationships for the reference tectonic regime (i.e., active shallow-crustal regimes) are examined. Because it is assumed that the ability of the soil to resist liquefaction during earthquake shaking is an inherent property of the soil, the ratio of FS_{liq} for the CEUS and WUS, for example, becomes:

$$FS_{liq\,ratio} = \cfrac{\cfrac{CRR_{M7.5}}{0.65\frac{a_{max}}{g}\frac{\sigma_v}{\sigma_{vo}}r_{d\,CEUS}\frac{1}{MSF_{CEUS}\cdot K_\gamma \cdot K_\alpha}}}{\cfrac{CRR_{M7.5}}{0.65\frac{a_{max}}{g}\frac{\sigma_v}{\sigma_{vo}}r_{d\,WUS}\frac{1}{MSF_{WUS}\cdot K_\gamma \bullet K_\alpha}}} = \frac{r_{d\,WUS}}{r_{d\,CEUS}} \bullet \frac{MSF_{CEUS}}{MSF_{WUS}} \tag{7}$$

Plots of $FS_{liq\,ratio}$ computed using preliminary region-specific r_d and MSF relationships are shown in Fig. 5. Specifically, $FS_{liq\,ratio}$ are shown for tectonic events in stable continental regimes (e.g., CEUS: Fig. 5a), tectonic subduction zone events (e.g., PNW: Fig. 5b), induced earthquakes in the MWUS (i.e., Oklahoma, Texas, and Kansas: OTK) (Fig. 5c), and induced earthquakes in the Groningen region of the Netherlands (Fig. 5d).

Values of $FS_{liq\,ratio}$ that are less than one imply that the use of the simplified models developed for tectonic events in active shallow-crustal regimes will over-predict the liquefaction hazard. In contrast, values of $FS_{liq\,ratio}$ that are greater than one imply that the use of the simplified models developed for tectonic events in active shallow-crustal

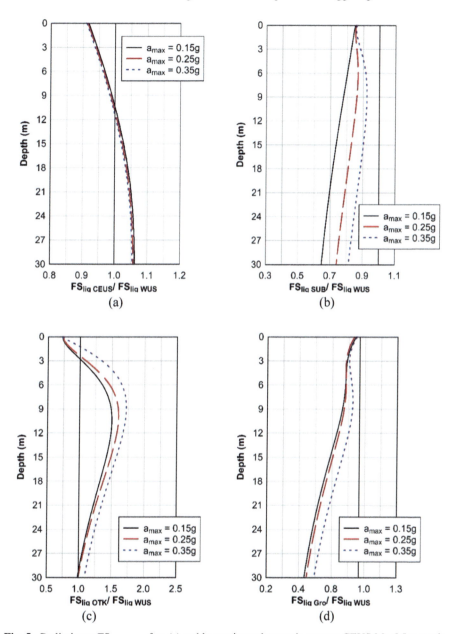

Fig. 5. Preliminary $FS_{liq\ ratios}$ for: (a) stable continental tectonic events: CEUS $M_w6.5$ tectonic event; (b) subduction zone tectonic events: PNW $M_w8.5$ tectonic event; (c) MWUS (i.e., Oklahoma, Texas, and Kansas: OTK): $M_w5.25$ induced event; and (d) Groningen region of the Netherlands: $M_w5.25$ induced event. In all cases, soil deposits were assumed comprising loose uniform sand deposits with the ground water table at a depth of ~ 2 m

regimes will under-predict the liquefaction hazard. Accordingly, based on the plots in Fig. 5, use of liquefaction triggering models developed for active shallow-crustal tectonic regimes will result in a higher computed liquefaction hazard for subduction zone events in the PNW and for induced earthquake in the Groningen region of the Netherlands than the actual liquefaction hazard. Additionally, use of liquefaction triggering models developed for active shallow-crustal tectonic regimes will result in a higher computed liquefaction hazard for tectonic events in the CEUS and for induced earthquake in the MWUS at shallower depths and lower liquefaction hazard at deeper depths than the actual liquefaction hazard. In some of the cases, the liquefaction is significantly under- or over-predicted. These trends provide the impetus to develop and use regionally applicable liquefaction triggering models in assessing liquefaction hazards.

While the plots of the computed $FS_{liq\ ratio}$ shown in Fig. 5 show trends in over- or under-prediction of the liquefaction hazard in various tectonic regimes or tectonic vs. induced earthquakes using models developed for active shallow-crustal tectonic events, region-specific r_d and MSF relationships can be used to directly evaluate the liquefaction hazard in a given region. Specifically, the region-specific r_d and MSF relationships can be used to compute CSR*, which in turn can be used in conjunction with the $CRR_{M7.5}$ curve determined from case histories from active shallow-crustal tectonic events, to compute FS_{liq} in a given region. This is because $CRR_{M7.5}$ is considered to be an inherent property of the soil and because consistent approaches are being used to develop all the region-specific relationships. As a result, any bias inherent to these relationships will be consistent among all relationships, which is essential to allow use of the region-specific relationships in conjunction with the same $CRR_{M7.5}$ curve. The exceptions to the assumption that $CRR_{M7.5}$ is an inherent property of the soil is if the mineralogy or age of the soil of interest differs from that represented in the case history databased used to develop the $CRR_{M7.5}$ curve. In these cases, corrections need to be applied to the $CRR_{M7.5}$ curve, but for the purpose of this study, these exceptional conditions are not considered.

5 Summary and Conclusions

The stress-based "simplified" liquefaction evaluation procedure is the most widely used approach to evaluate liquefaction triggering potential worldwide. However, the suitability of the existing variants of the simplified models for use in evaluating liquefaction triggering in tectonic regimes other than active shallow-crustal regimes is questionable. Additionally, the suitability of existing models for evaluating liquefaction potential due to induced earthquakes is also questionable. This is because the simplified procedure is semi-empirical, with the empirical aspects of it derived from data from tectonic earthquakes in active shallow-crustal tectonic regimes (e.g., California, Turkey, and portions Japan and New Zealand). As a result, existing variants of the simplified procedure may not be suitable for use in evaluating liquefaction triggering when the geologic profiles/soil deposits or ground motion characteristics differ significantly from those used to develop some of the empirical aspects of the models.

In this paper, ongoing efforts by the author and his collaborators to develop region-specific r_d and MSF relationship were summarized. Relationships are being developed for tectonic earthquakes in the central-eastern US (CEUS) and subduction zone events in the Pacific Northwest (PNW) of the US, and induced earthquakes in the Groningen region of the Netherlands due to natural gas extraction and in Midwest of the US (MWUS: Oklahoma, Texas, and Kansas) due to deep waste water disposal.

The significance of the region-specific relationships is shown by the ratios of the factors of safety against liquefaction triggering computed using the various preliminary region-specific relationships and those from active shallow-crustal tectonic regimes. Based on preliminary relationships developed by the author and his collaborators, use of liquefaction triggering models developed for active shallow-crustal tectonic regimes will result in a higher computed liquefaction hazard for subduction zone events in the PNW and for induced earthquake in the Groningen region of the Netherlands than the actual liquefaction hazard. Additionally, use of liquefaction triggering models developed for active shallow-crustal tectonic regimes will result in a higher computed liquefaction hazard for tectonic events in the CEUS and for induced earthquake in the MWUS at shallower depths and lower liquefaction hazard at deeper depths than the actual liquefaction hazard. In some of the cases, the liquefaction is significantly under- or over-predicted. These trends provide the impetus to use develop and use regionally applicable liquefaction triggering models in assessing liquefaction hazards.

Acknowledgements. The author has collaborated with many researchers on various aspects of this project, to include: Adrian Rodriguez-Marek, Julian Bommer, Peter Stafford, Jan van Elk, Brett Maurer, Kristin Ulmer, James K. Mitchell, Tyler Quick, Balakumar Anbazhagan, Sam Lasley, Ellen Rathje, and Balakumar Anbazhagan, among others. The input from these collaborators is gratefully acknowledged. This research was funded by National Science Foundation (NSF) grants CMMI-1825189 and CMMI-1937984, U.S. Geological Survey (USGS) awards G18AP00094 and G19AP00093, and NAM. This support is gratefully acknowledged. However, any opinions, findings, and conclusions or recommendations expressed in this paper are those of the author and do not necessarily reflect the views of NSF, USGS, NAM or the listed collaborators.

References

1. Robertson, P.K., Wride, C.E.: Evaluating cyclic liquefaction potential using cone penetration test. Can. Geotech. J. **35**(3), 442–459 (1998)
2. Moss, R.E.S., Seed, R.B., Kayen, R.E., Stewart, J.P., Der Kiureghian, A., Cetin, K.O.: CPT-based probabilistic and deterministic assessment of in situ seismic soil liquefaction potential. J. Geotechnical Geoenvironmental Eng. **132**(8), 1032–1051 (2006)
3. Idriss, I.M., Boulanger, R.W.: Soil liquefaction during earthquakes. Monograph MNO-12, Earthquake Engineering Research Institute, Oakland, CA, p. 261 (2008)
4. Iwasaki, T., Tatsuoka, F., Tokida, K., Yasuda, S.: A practical method for assessing soil liquefaction potential based on case studies at various sites in Japan. In: 2nd International Conference on Microzonation, pp. 885–896. Nov 26-Dec 1, San Francisco, CA, USA (1978)
5. Green, R.A., Maurer, B.W., Cubrinovski, M., Bradley, B.A.: Assessment of the relative predictive capabilities of CPT-based liquefaction evaluation procedures: lessons learned from the 2010–2011 canterbury earthquake sequence. In: Proceedings of the 6th International Conference on Earthquake Geotechnical Engineering (6ICEGE), Christchurch, New Zealand, 2–4 November (2015)

6. Maurer, B.W., Green, R.A., Cubrinovski, M., Bradley, B.: Assessment of CPT-based methods for liquefaction evaluation in a liquefaction potential index framework. Géotechnique **65**(5), 328–336 (2015)
7. Whitman, R.V.: Resistance of soil to liquefaction and settlement. Soils Found. **11**(4), 59–68 (1971)
8. Seed, H.B., Idriss, I.M.: Simplified procedure for evaluating soil liquefaction potential. J. Soil Mechanics Found Division **97**(SM9), 1249–1273 (1971)
9. Idriss, I.M.: An update to the Seed-Idriss simplified procedure for evaluating liquefaction potential. In: Proceedings of the TRB workshop on new approaches to liquefaction, Publication No. FHWA-RD-99- 165, Federal Highway Administration, Washington, DC (1999)
10. Boulanger, R.W., Idriss, I.M.: CPT and SPT based liquefaction triggering procedures. Rep. No. UCD/CGM-14/01. Davis, CA: Univ. of California at Davis (2014)
11. Green, R.A., Lee, J., Cameron, W., Arenas, A.: Evaluation of various definitions of characteristic period of earthquake ground motions for site response analyses. In: Proceedings of the 5th International Conference on Earthquake Geotechnical Engineering, Santiago, Chile, 10–13 January 2011 (2011)
12. Seed, H.B., Idriss, I.M., Makdisi, F., Banerjee, N.: Representation of irregular stress time histories by equivalent uniform stress series in liquefaction analysis. Report Number EERC 75–29, Earthquake Engineering Research Center, College of Engineering, University of California at Berkeley, Berkeley, CA (1975)
13. Green, R.A., Terri, G.A.: Number of equivalent cycles concept for liquefaction evaluations: revisited. J. Geotechnical Geoenvironmental Eng. **131**(4), 477–488 (2005)
14. Lasley, S., Green, R.A., Rodriguez-Marek, A.: Number of equivalent stress cycles for liquefaction evaluations in active tectonic and stable continental regimes. J. Geotechnical and Geoenvironmental Eng. **143**(4), 04016116 (2017)
15. Green, R.A., Lee, J., White, T.M., Baker, J.W.: The significance of near-fault effects on liquefaction. In: Proceedings of 14th world conference on earthquake engineering, Paper no. S26–019 (2008)
16. Carter, W.L., Green, R.A., Bradley, B.A., Cubrinovski, M.: The influence of near-fault motions on liquefaction triggering during the canterbury earthquake sequence. In: Orense, R.P., Towhata, I., Chouw, N. (eds.) Soil Liquefaction during Recent Large-Scale Earthquakes, Taylor & Francis Group, pp. 57–68. England, London (2014)
17. Carter, W.L., Green, R.A., Bradley, B.A., Wotherspoon, L.M., Cubrinovski, M.: Spatial variation of magnitude scaling factors during the 2010 Darfield and 2011 Christchurch, New Zealand. Earthquakes. Soil Dynamics Earthquake Eng. **91**, 175–186 (2016)
18. Bommer, J.J., et al.: Framework for a ground-motion model for induced seismic hazard and risk analysis in the Groningen gas field, the Netherlands. Earthq. Spectra **33**(2), 481–498 (2017)
19. Ulmer, K.J., et al.: A critique of b-values used for computing magnitude scaling factors. In: Proceedings of geotechnical earthquake engineering and soil dynamics V (GEESD V), Austin, TX, 10–13 June) (2018)
20. Bradley, B.A.: Correlation of significant duration with amplitude and cumulative intensity measures and its use in ground motion selection. J. Earthquake Eng. **15**, 809–832 (2011)
21. Bommer, J.J., et al.: Developing an application-specific ground-motion model for induced seismicity. Bull. Seismol. Soc. Am. **106**(1), 158–173 (2016)
22. NRC: State of the art and practice in the assessment of earthquake induced soil liquefaction and consequences. Committee on earthquake induced soil liquefaction assessment, National Research Council, The National Academies Press, Washington, DC (2016)

23. Youd, T.L., et al.: Liquefaction resistance of soils: summary report from the 1996 NCEER and 1998 NCEER/NSF workshops on evaluation of liquefaction resistance of soils. J. Geotechnical and Geoenvironmental Eng. **127**(4), 297–313 (2001)
24. Cetin, K.O.: Reliability-based assessment of seismic soil liquefaction initiation hazard. Ph.D. Thesis, University of California at Berkeley, Berkeley, CA (2000)
25. Hancock, J., Bommer, J.J.: The effective number of cycles of earthquake ground motion. Earthquake Eng. Struct. Dynam. **34**, 637–664 (2005)
26. Green, R.A., Lee, J.: Computation of number of equivalent strain cycles: a theoretical framework. In: Lade, P.V., Nakai, T. (eds.) Geomechanics II: Testing, Modeling, and Simulation, ASCE Geotechnical Special Publication 156, pp. 471–487 (2006)
27. Green, R.A., Mitchell, J.K., Polito, C.P.: An energy-based excess pore pressure generation model for cohesionless soils. In: Smith, D.W., Carter, J.P.: (eds) Proceedings of the John Booker memorial symposium—developments in theoretical geomechanics. A. A. Balkema, Rotterdam, pp. 383–390 (2000)
28. Polito, C.P., Green, R.A., Lee, J.: Pore pressure generation models for sands and silty soils subjected to cyclic loading. J. Geotechnical Geoenvironmental Eng. **134**(10), 1490–1500 (2008)
29. Green, R.A., et al.: Addressing limitations in existing 'simplified' liquefaction triggering evaluation procedures: application to induced seismicity in the Groningen gas field. Bull. Earthq. Eng. **17**(8), 4539–4557 (2019)
30. Ishibashi, I., Zhang, X.: Unified dynamic shear moduli and damping ratios of sand and clay. Soils Found. **33**(1), 182–191 (1993)
31. Green, R.A., Bradshaw, A., Baxter, C.D.P.: Accounting for Intrinsic Soil Properties and State Variables on Liquefaction Triggering in Simplified Procedures. J. Geotech. Geoenvironmental Eng. **148**(7), 04022056 (2022)
32. Salgado, R., Boulanger, R.W., Mitchell, J.K.: Lateral stress effects on CPT liquefaction resistance correlations. J. Geotechnical and Geoenvironmental Eng. **123**(8), 726–735 (1997)
33. Salgado, R., Mitchell, J.K., Jamiolkowski, M.: Cavity expansion and penetration resistance in sands. J. Geotechnical Geoenvironmental Eng. **123**(4), 344–354 (1997)

Different Aspects of the Effects of Liquefaction-Induced Lateral Spreading on Piles, Physical Modelling

S. Mohsen Haeri$^{(\boxtimes)}$

Civil Engineering Department, Sharif University of Technology, Tehran, Iran
smhaeri@sharif.edu

Abstract. Devastating pile failures due to liquefaction induced lateral spreading during or after major earthquakes in gently sloping ground or level grounds with free end, especially the pile foundations located under structures in or near ports and harbors, have been observed and studied for decades. Many physical and numerical modellings have been implemented or developed to study and understand the insight into different aspects of this phenomenon. In this regard, 1-g shake table and N-g dynamic centrifuge tests using both rigid box and laminar shear box have been implemented to physically model the problem and measure the parameters that may affect the impact of lateral spreading on deep foundations. A number of countermeasures have also been examined for tackling this problem. In this paper and theme lecture, the author tries to describe shortly the physical modelling researches and studies that have been conducted by him and his coworkers on this subject in more than a decade, and discuss the various parameters that are involved in physical modelling for studying the behavior of pile foundations subjected to liquefaction induced lateral spreading. A number of limitations involved in such physical modellings are also mentioned and some solutions to the involved challenges are discussed as well.

Keywords: Physical modelling · 1-g Shake table test · Lateral spreading · Liquefaction · Pile foundations · Countermeasure

1 Introduction

During past earthquakes, many important structures supported on pile foundations have been subjected to severe damages due to liquefaction induced lateral spreading. Lateral displacements associated with lateral spreading can be up to several meters which consequently can impose substantial kinematic forces to pile foundations, causing extensive damages. Several observations have been reported during past earthquakes in this regard, among which the 1964 Niigata, Japan, the 1983 Nihonkai-Chubu, Japan, the 1989 Loma Prieta, USA, the 1995 Kobe, Japan, the 2001 Bhuj, India and the 2010 Port-au-Prince, Haiti earthquakes are among the most destructive ones [1–7]. Damaged

---

S. Mohsen Haeri—Theme Lecture to be presented in PBD 2022

piles in these earthquakes revealed that effects of lateral spreading on deep foundations has not been fully understood and therefore several studies have been undertaken by researchers especially in the last two decades. Design codes also were prepared for design of pile foundations against lateral spreading. A simple and efficient method in this respect was suggested by JRA (2002) [8] for pile design against lateral spreading, based on the observations and lessons learned from Kobe 1995 Earthquake. Although the method proposed by JRA (2002) [8] is a very straightforward and useful tool for design of piles against lateral spreading, however, it does not give an insight into many aspects of this problem, such as the effect of the position of piles in a pile group. Physical and numerical modelling, however, are two powerful tools for studying the soil-pile systems. In this respect, many studies implementing physical modelling have been performed to investigate the response of piles to lateral spreading [9–20], however, a comprehensive method for design of piles against lateral spreading is yet in need of more experiments and numerical studies. Mitigation of liquefaction-induced lateral spreading hazard on present piles in liquefiable sloping grounds has been a concern for practitioners and researchers. In this regard using stone columns and micro piles are among those methods. Seed and Booker [21] were the first researchers who investigated the effectiveness of stone columns for mitigation of liquefaction. However, Kavand et al. [22, 23] were the first researchers who conducted a series of shake table experiments to study the effectiveness of stone columns and micropiles to countermeasure the effects of lateral spreading on 3 × 3 flexible and stiff pile groups in a gently sloping ground. They found that stone columns can effectively reduce the kinematic bending moments due to lateral spreading in flexible piles up to a particular level of earthquake magnitude. However, they found that stone columns are ineffective in reducing the effect of lateral spreading on piles in other conditions. Also their test results using picropiles as a mitigation measure showed that the micropiles were not able to reduce effectively the bending moments in the piles while it was able to reduce lateral soil pressures exerted by non-liquefiable crust on the upslope piles of the group. Haeri et el. [24] also conducted two shake table experiments to investigate the efficacy of stone column technique as a mitigation method against liquefaction-induced lateral spreading on two 2 × 2 pile groups in a laminar shear box. Their results illustrate that stone column technique significantly decreased lateral displacemnts in the free field and bending moments in the piles, while increased the acclearions in the pile caps and rate of dissipation of excess pore water pressure in the liquefiable layer. All other researchers, e.g. [25–28] used numerical modelling to study the effectiveness of stone columns as remediation measure against the effects of lateral spreading on piles.

In the present paper, the effects of liquefaction-induced lateral spreading on single piles and pile groups which studied by physical modelling in various conditions are discussed. Also efficacy of some methods to mitigate the impact of lateral spreading on piles are evaluated by physical modelling.

2   Characteristics of Physical Modellings of Piles Subjected to Lateral Spreading

Overall, 16 large scale physical models have been constructed and tested in a long run of studies by the author and his coworkers at Sharif University of Technology (SUT) to investigate various aspects of the response of the piles to lateral spreading. For this purpose, shake table facility of Earthquake Engineering Research Center of SUT has been employed. SUT shaking table is a 4m × 4m, 3DOFS facility, capable of taking vertical loads up to 300 kN with actuators capacities of 1 × 500 kN in longitudinal direction and 2 × 200 kN in transversal direction, and a maximum base acceleration of 2g and a maximum frequency of 50Hz. The first part of this research started in 2009 by design and construction of a rigid box with a transparent side with which 10 large scale tests were carried out and completed in 2013. The results of some of these tests have been reported and published so far [9, 11, 14, 16, 17, 22, 23] and results of a number of tests have not been published yet. The dimensions of the rigid container (box) is 3.5m long, 1.0m wide and 1.5m high. Two Plexiglas windows were provided in one side of the container to be able to view and measure the soil lateral movements by PIV method. As the box was long enough in the direction of lateral spreading, and lateral spreading is rather kinematic in nature, it was assumed that the rigid boundary condition in this study could have minimum effects on the results of the study with respect the effect of lateral spreading on piles. The second stage of the tests which included 6 various experiments started in 2015 by design and built of a large laminar shear box and the tests were completed in 2021. The laminar shear box used in this research has outer dimensions of 4.2 m length, 2.4 m width and 2.0 m height and inner dimensions of 3.06 m length, 1.72 m width and 1.8 m height. Figure 1 is an illustration of these two boxes. Zero height water sedimentation of Firuzkuh no.161 sand, a common soil in experimental studies in Iran, which is a clean uniformly graded sand with mean size of 0.24 mm, was used for modelling the liquefiable layer in both sets of studies. The main physical properties of Firuzkuh sand is outlined in Table 1.

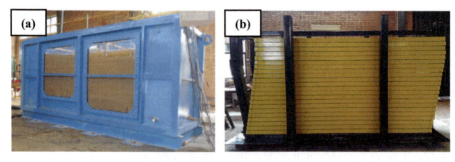

Fig. 1. (a): rigid box, (b): laminar shear box, designed and constructed for the reported studies

Table 1. Properties of Firuzkuh silica sand no. 161

Specific gravity	Mean grain size (D_{50}) (mm)	D_{10} (mm)	Maximum void ratio (e_{max})	Minimum void ratio (e_{min})	Coefficient of uniformity (C_u)
2.670	0.24	0.18	0.884	0.567	1.49

During this fairly long research, the effects of lateral spreading on a group of single piles and various pile groups with different configurations (e.g., 2×2, 3×3, 3×5, etc. with or without superstructure) were studied. The similitude law proposed by Iai [29] and Iai et al. [30] were used to calculate mechanical and geometrical properties of the piles in the models. In most of the reported experiments, the geometrical (prototype/model) scale was selected to be $\lambda = 8$ and only the case studies are different. Model piles were designed based on JRA 2002 [8] to withstand the exerted lateral spreading forces. Aluminum pipes (T6061 alloy) and high density polyethylene (HDPE) pipes were used to similitude steel or concrete piles, respectively. In all models, the piles were sufficiently instrumented with pair strain gauges to detect pure bending moments at various sections. The soil at far field were also instrumented to measure various parameters during and after shaking. Various parameters including acceleration and pore water pressure in different spatial positions of the free fields and near the piles, acceleration at the pile caps, surficial displacements, and bending moments in the piles were measured and evaluated. Also colored sands were formed in a grid pattern at surface of all models as well as in vertical columns at side of the rigid box models behind the Plexiglas windows. In order to monitor the lateral displacement of the soil during lateral spreading, digital high speed cameras and camcorders were implemented both at top and side of the models. In rigid box models the side cameras and camcorders could capture the lateral soil movements though the transparent windows provided especially for this purpose and in laminar shear box they could only record the box shape variation during the shakings. Input shaking consisted of 30 sinusoidal cycles with frequency of 3 Hz with two ramps of three cycles at the start and the end of shakings (Fig. 2). Acceleration amplitude of the input loading was 0.2g and 0.3g for the study of the effects of lateral spreading on piles. Also to study the effect of post liquefaction and post lateral spreading on piles several higher input motions were applied to each physical model. However, this part is not going to be covered in this paper.

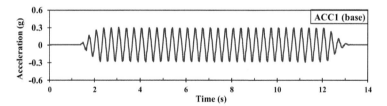

Fig. 2. Input motion applied to the physical models

3   Effects of Lateral Spreading on a Group of Single Piles

The first conducted experiment was on the investigation of the behavior of a group of piles without pile cap in a gently sloping liquefiable ground without a non-liquefiable crust. General test results including time-histories of accelerations, pore water pressures, displacements and bending moments have been presented by Haeri et al. [9]. It was found that the free field soil started to move laterally, right after being liquefied and kept moving towards the downslope until the end of the shaking, while the piles moved with the soil at the early stages of shaking, approaching a maximum displacement at the pile head and then bounced back gradually keeping a residual displacement at the end (Fig. 3) The results indicated that lateral soil pressures exerted from laterally spreading ground vary in individual piles of a group depending on the pile position within the group (Fig. 4). In piles arranged in longitudinal direction, the downslope pile experienced less soil pressure comparing to the upslope one because of the shadow effect. In the transverse pile set, the middle pile received less pressure than the side piles as a result of neighboring effects. The behavior of piles in a group (without pile cap) located in a sloping ground far from a free face can be different from those located behind a quay wall or close to a free face. The reason could be the fact that in sloping ground, the lateral spreading starts from the upslope whereas in free face or quay wall cases the lateral spreading starts from the downslope. As a result, in sloping ground, the front pile is directly pushed by the laterally spreading liquefied soil while the shadow pile is protected by the front pile

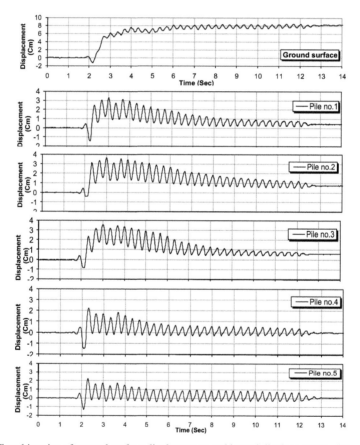

Fig. 3. Time histories of ground surface displacement and lateral displacements of the pile heads [9].

against the direct impact of the liquefied soil hence experiencing less pressure comparing to the front pile. Therefore, the results revealed the importance of the effects of the pile position in the pile group.

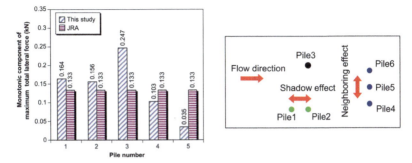

Fig. 4. Comparison between back-calculated monotonic component of maximum total lateral forces from shake table test on a group of single piles and JRA code [9].

In that research, the force-based and displacement-based methods were implemented to evaluate the bending moment profile of the single pile (i.e., Pile 3). Response of single piles to lateral spreading can be obtained by force based method in which the lateral pressures exerted on the piles are modeled as imposed limiting pressures similar to the procedure that JRA code [8] stipulates. This code specifies the induced lateral pressures as 30% of the total overburden pressure of liquefiable layer as shown in Fig. 5. Another procedure for the analysis of piles under lateral spreading is the displacement based approach in which a Beam on Nonlinear Winkler Foundation (BNWF) model is utilized. In the latter approach, as depicted in Fig. 5, the free field lateral soil displacement (Δ_{soil}) was applied at free ends of the p–y springs of the laterally spreading soil. In order to obtain the p–y curve for a liquefied or laterally spreading soil, a reduction factor which is known as p-multiplier is commonly applied to the corresponding p–y curve of the non-liquefied soil. In order to compare the capability of the two methods in predicting the behavior of the single pile in that experiment, bending moment diagram along the pile 3 was calculated based on each method and the results are plotted in Fig. 5. As shown in this figure, displacement based method using p-multiplier value of 0.064 proposed by Brandenburg [31] and p-multiplier value of 0.02 read from the lower bound curve suggested by AIJ [32] underestimate the bending moments in the pile. However, differences between measured and predicted bending moments by AIJ [32] method is much more evident in this regard. Moreover, using a best match p-multiplier value of about 0.115 which is equal to the upper bound value of Brandenburg [31] seems to be more capable of predicting the induced bending moment in the pile. Finally, as it can be observed in Fig. 5, force based approach based on JRA [8], predicted the maximum bending moment in the single pile to be about 45% of the monotonic component of that, recorded in the experiment.

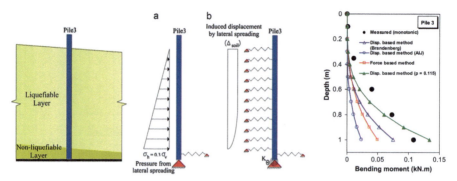

Fig. 5. Numerical modelling of a single pile under lateral spreading: (a) force based approach, (b) displacement based approach, and Comparison between measured and computed bending moments along the model pile [9]

4   Effects of Lateral Spreading on 2 × 2 Pile Groups

Different aspects of the behavior of 2 × 2 pile groups under liquefaction-induced lateral spreading in a medium dense liquefiable layer, located between a crust and bottom non-liquefiable layers, were investigated using the shake table and rigid box at the first stage. The employed physical model consisted of two separate 2 × 2 pile groups. A lumped mass of 12 kg was attached to the cap of one of the pile groups in order to study the effects of superstructure on the response of the pile groups during lateral spreading. The results of this experiment showed that total lateral forces on the piles are influenced by the shadow effect as well as the superstructure mass attached to the pile cap. The presence of superstructure was found to intensify the negative bending moments in the piles. It was also found that the kinematic lateral forces exerted on the piles in the lower half of the liquefied layer are significantly larger than those recommended by JRA code [8] (Fig. 6). This observation can be attributed to the lateral displacement pattern of the soil which were obtained by analyzing the photos taken from the transparent side of the model (Fig. 6). As seen in this figure, the maximum permanent soil displacement was about 20 cm at the end of shaking which occurred near the middle of the liquefiable layer. The maximum lateral soil displacement in non-liquefiable crust layer, which was considerably smaller than the maximum displacement observed in the liquefiable layer, however, occurred at the ground surface. Based on the calculated contribution coefficients of lateral forces, in liquefiable layer, the upslope row of the piles carried larger lateral forces than the downslope one while in non-liquefiable crust layer, the downslope row experienced greater forces. However, contribution coefficient of total lateral force in upslope row was overall greater than that obtained for downslope row.

In next stage of the research, the effects of the existence of non-liquefiable crust layer on the responses of two sets of 2 × 2 pile groups to lateral spreading were studied using shake table and the large laminar shear box. The model without the crust layer consisted of a thick liquefiable layer with a relative density of 15% being underlain by a dense non-liquefiable layer. Time histories of the lateral displacement at the ground surface of the two models with and without crust layer are compared in Fig. 7. As seen

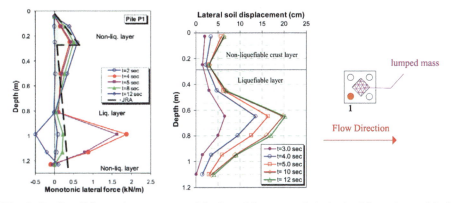

Fig. 6. Profile of kinematic component of the lateral forces on pile1 obtained from the model of 2 × 2 pile group, and profile of lateral soil displacement in free field at upslope side of the model extracted from snapshots during the shaking [16].

in this Figure, existence of the non-liquefiable crust layer increased the maximum soil lateral displacement by 55%. This was in good agreement with the results of bending moments of the piles which showed that the bending moments of 2 × 2 piles increased significantly in the case of existence of the crust layer. Profiles of bending moments on piles in the model with and without the crust layer and superstructure are exhibited in Fig. 8 and Fig. 9, respectively. According to these figures, existence of the crust layer and also existence of the lumped mass increase the maximum bending moment in the piles due to lateral spreading. Between these two parameters, the effect the existence of the crust layer on the responses of the piles is more remarkable.

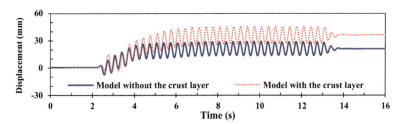

Fig. 7. Time histories of the free field soil displacement of the models with and without crust.

5   Effects of Lateral Spreading on 3 × 3 Pile Groups

The responses of a 3 × 3 pile group to liquefaction-induced lateral spreading was investigated using 1-g shake table test and the rigid box in the first stage of the study. The model ground consisted of a three layers soil profile including a base non-liquefiable layer, a middle loose liquefiable layer and an upper non-liquefiable crust layer. The results showed that lateral forces exerted by lateral spreading were not the same in the

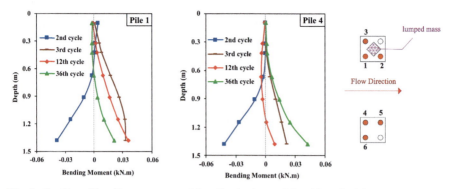

Fig. 8. Profiles of bending moments of the piles in the models with and without superstructure, without the crust layer tested in laminar box.

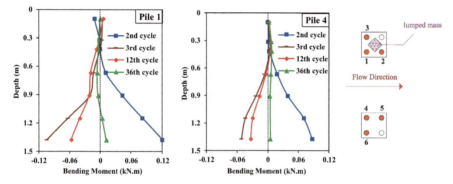

Fig. 9. Profiles of bending moments of the piles in the models with and without superstructure, with the crust layer tested in laminar box.

individual piles of the group, both in transverse and longitudinal directions, depending on the pile position within the group. Monotonic components of maximum total lateral for on the piles were calculated by integrating the lateral soil pressures along the piles. These total lateral forces were separately evaluated for the liquefied layer and the non-liquefiable crust. The calculated forces are displayed in Fig. 10. According to this figure, the amount of total lateral force in pile P2 (located in middle row) is less than the piles located in upslope and downslope rows, i.e. piles P1 and P3. Total lateral force on pile P1 is about 1.24 times that exerted on pile P2. This occurs due to the shadow effect. Since the upslope pile is directly pushed by the laterally spreading soil and acts as a barrier for pile downslope pile, P2. Total lateral force exerted on pile P3 is the largest among all the other piles. Total lateral force on pile P3 is about 1.43 and 1.76 times those of piles Pl and P2, respectively. This can be described by the separation of soil from the downslope side of pile P3 during lateral spreading resulting in lack of lateral support. Comparing total lateral forces in pile Pl (the middle pile in upslope row) and P4 (the side pile in upslope row) shows that the side pile receives larger force than the middle pile by a factor of about 1.27. This phenomenon is called neighboring effect. It was also found that the magnitude of lateral pressure due to lateral spreading on the 3 × 3 pile group of this study, in average, is close to the values recommended by the JRA design code (Fig. 11).

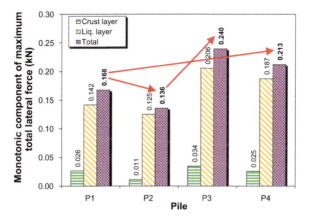

Fig. 10. Comparison of maximum total lateral forces on different piles of the 3 × 3 group in rigid box [11].

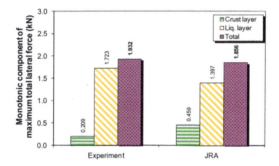

Fig. 11. Comparison between monotonic components of maximum total lateral forces in pile group of this experiment and JRA 2002 recommended values [11].

In the second stage of the experiments, the laminar shear box was employed in physical modelling a 3 × 3 pile group in a gently sloped ground. All the properties of these two models designed and constructed the same except the containers. Bar charts of the maximum kinematic lateral forces on the piles due to lateral spreading are presented in Fig. 12. The maximum lateral force on pile 2 was shown differently in Fig. 12, because the lateral force applied on pile 2 was interpolated from other piles as strain gauges attached to this pile were not enough to be conclusive. As seen in Fig. 12, the kinematic lateral forces of side piles (i.e., piles 1, 2 and 3) are more than corresponding piles in the middle row, i.e. piles 4, 5 and 6, respectively. This issue can be attributed to neighboring effects. Also the downslope piles experienced more kinematic forces due to lateral spreading compared to those in the upslope row of the group. For example, the maximum kinematic lateral force on pile 1 was about 81% higher than that for pile 3. This can be attributed to a more lateral movement of the soil at downstream of the model.

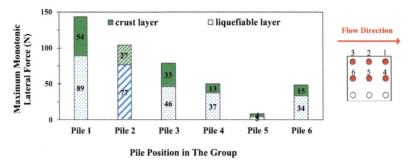

Fig. 12. Bar chart of maximum monotonic lateral forces on different piles of a 3 × 3 pile group tested in laminar shear box.

Maximum monotonic lateral force on the pile group in this experiment was also compared with the values proposed by JRA design code [8] as seen in Fig. 13. According to this figure, the maximum monotonic lateral force obtained in this experiment using laminar shear box, unlike the test results for rigid box, is significantly lower than that estimated by JRA code. Considering the results obtained by two containers it can be assumed that the forces suggested by JRA can be assumed as an upper bound for kinematic force induced by lateral spreading on 3 × 3 piles of a pile group.

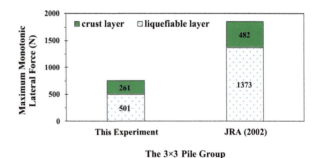

Fig. 13. Bar chart of the maximum kinematic lateral force due to lateral spreading on the pile the 3 × 3 pile group tested in laminar shear box.

6 Effects of Boundary Conditions on Physical Modelling Results

The effects of boundary conditions of the physical models on the results of the experiments using shake table were investigated by comparing the results of two identical experiments with different containers, the rigid box and the laminar shear box as discussed above. The differences between the obtained results of these two experiments can only be attributed to the effect of boundary conditions. It was found that the responses of the model in the rigid box is higher due possibly to wave reflections from the rigid walls and also bouncing back of the liquefied soil from rigid boundaries. For example, time histories of the ground lateral displacement at free filed for these two models are exhibited in Fig. 14. According to Fig. 14 the maximum and final (residual) soil displacement at free field in the rigid box, are about 20% and 40% are more than those of laminar shear box, respectively. This is in good agreement with the results of the bending moments generated in the piles. Time histories of the bending moment of the upstream pile for the models using rigid and laminar shear boxes are compared in Fig. 15. The maximum bending moment of this pile in the model with rigid box was about 2.7 times of that in the other model.

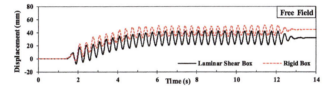

Fig. 14. Time histories of surface displacements in the models using rigid and laminar shear boxes [10].

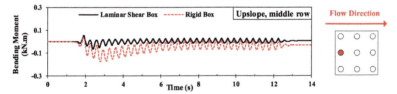

Fig. 15. Time histories of the bending moments of the upstream pile in the models using rigid and laminar shear boxes.

7 Case Studies of Lateral Spreading Effects on Pile Groups

Two physical modelling were carried out to model real cases to study the effect of lateral spreading on two different pile groups, one using rigid box and the other implementing laminar shear box as are introduced and discussed briefly in the following.

7.1 Dolphin-Type Berth

The effects of liquefaction-induced lateral spreading on 3 piles of a dolphin-type berth were investigated using 1-g large scale shake table and the rigid box in stage 1 of these studies. Due to the size of the berth and the box the scaling factor was chosen to be

20. The soil profile included a 1.2 m thick liquefiable layer (loose sand with relative density of 15%) overlying a bottom non-liquefiable dense sand layer having maximum thickness of 25 cm and relative density of about 80%. All the soil layers were inclined by 7% towards downslope. The results indicated that large bending moments were induced in the piles due to lateral spreading. Also the downslope piles of the group received greater bending moments than the upslope one. In upper half of the liquefiable layer, where the soil fully liquefied ($r_u = 1.0$), substantial active lateral pressures were exerted on the piles while in the lower half, in which the soil did not liquefy ($r_u = 0.5$), a passive pressure zone was developed. In Fig. 16, a comparison is made between the maximum monotonic component of lateral soil pressure on model piles and the lateral soil pressure calculated based on JRA code [8]. As shown in Fig. 16, the magnitude and pattern of the lateral pressures on Pile 3 are rather consistent with the prediction of JRA code up to a depth of about 0.6 m corresponding to the upper half of the pile and liquefiable layer. The inconsistency observed in lower depths can be attributed to the fact that the prediction of JRA code was based on the assumption that the lateral soil movement occurred in whole depths of the liquefiable layer. However, smaller lateral soil movement in lower depths created a passive pressure zone in these depths during the experiment. Another finding is that the pattern of lateral soil pressure distribution along the upslope pile (i.e. pile P1) is different from what observed for the downslope one (i.e. pile P3).

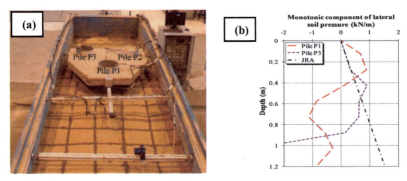

Fig. 16. a): Top view of the physical model of dolphin- type berth on SUT shaking table, b): profiles of monotonic component of maximum lateral soil pressure on model piles along with the corresponding values recommended by JRA design code [14, 17].

7.2 Deep Foundation of a Bridge

The behavior of the deep foundations of Nahang-e-Roogah bridge, a bridge near Bandar-e-Anzali city in North Iran, along Caspian see was studied. For this purpose, a shake table test using laminar shear box with scaling factor of 20 was conducted to study the dynamic response of a 3 × 5 pile group subjected to liquefaction induced lateral spreading in the second stage of the experiments. The results showed that the location of the piles in the group was important. According to Fig. 17, the upslope position of the pile group experienced greater bending moments than downstream piles. This can be attributed to the effects of flow direction in lateral spreading which is known as the "shadow effects". The results also showed that the maximum acceleration of the pile cap amplified being twice the input acceleration amplitude. This is a notable result for design of a superstructure in a liquefiable soil-pile system.

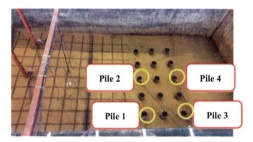 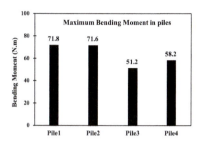

Fig. 17. Top view of physical model of 3 × 5 pile group of Nahang-e-Roogah Bridge, and bar graph of maximum amounts of bending moment of the piles [15].

8 Assessment of Mitigation Methods Against Lateral Spreading

In this research, effectiveness of some mitigation methods (e.g., stone column, micropile) against lateral spreading were investigated using 1-g shake table tests. Main aspects of these models are briefly presented and explained in the following.

8.1 Assessment of Stone Column Method

The effectiveness of stone columns as remedial measures against lateral spreading in a 3 × 3 flexible pile group was investigated by conducting shake table tests using rigid box in first round of the conducted research. In order to evaluate the performance of employed stone columns, comparisons were made between profiles of bending moments in individual piles of the groups obtained for mitigated and unmitigated cases. The stone columns were installed at upslope and downslope sides of the pile group after preparation of one of the unmitigated models (Fig. 18). Construction of the stone columns caused densification of the loose liquefiable layer and associated vertical settlements in an area of about 1.5 m in length extending over the upslope and downslope sides of the stone columns. All settlements were carefully measured after construction of the stone columns and associated volume decrease was subsequently determined to calculate the new relative density of the sand layer which was estimated to change from 15% to about 65% in the affected zone, indicating a reduction in liquefaction potential and subsequent lateral spreading in that area.

Bending moments in piles increased significantly due to lateral spreading in unmitigated model. However, after a peak, due to the loss of shear strength in liquefied soil and rigidity of the piles, the pile groups bounced back towards the upslope shaking continued. Accordingly the bending moments in the piles decreased up to the end of shaking. However, in the model with stone columns, bending moments at different elevations of the

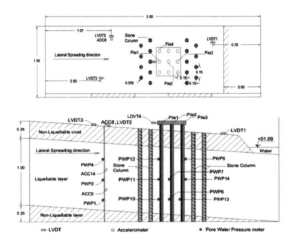

Fig. 18. Plan and cross section views of the physical model along with the stone columns [22].

pile continuously increased during the shaking, due to higher density of the soil and its resistance against elastic rebound of the piles. Therefore the residual bending moments in the piles at the end of the shaking were significant, while the maximum bending moments in the piles slightly showed increase. In fact, during the shaking and lateral pressure exerted by lateral spreading in upstream part of the model to the mitigated zone including pile groups, each cycle of the shaking caused a minor slide in the mitigated ground towards downslope, imposing incremental lateral pressures on the piles similar to the behavior of soils during cyclic mobility compared to the liquefaction and associated lateral spreading around the piles in unmitigated case. As a result, the pile group moved along with the soil towards the downslope so its deflection increased progressively by each loading cycle, leaving a residual lateral displacement as well. However, if the earthquake and associated number of cycles are limited, the stone column can be effective. For instance, during an earthquake with a magnitude of $Mw = 6.6$ (7 cycles of shaking), stone columns could reduce the maximum bending moment of a critical pile in the group, by about 80%. As the magnitude of earthquake and associated number of cycles increased, the effectiveness of the prepared stone columns with such a configuration and method, decreased. Hence, the results showed that the constructed stone columns could effectively reduce the bending moments in piles due to lateral spreading up to a particular level of earthquake magnitude. A series of shake table experiments were also conducted on two set of 2×2 pile groups (with and without superatructure) in the laminar shear box, in the second stage of the study, to investigate the efficacy of stone column technique as a mitigation method against liquefaction-induced lateral spreading and compared the behavior of pile groups in treated and untreated liquefiable layers. In one of the models, 29 stone columns with a diameter of 60 cm at prototype scale and triangular configurations were constructed. The stone columns technique can combine the beneficial effects of densification, reinforcement, and increased drainage of excess pore water pressure. In the study, the stone columns were constructed in a way to minimize densification of the surrounding soil, for observing mainly the drainage and some reinforcement effects of stone columns on liquefaction induced lateral spreading and its effects on the studied pile groups. The results illustrated that stone column technique prepared in this method significantly decreassed lateral displacemnts in the free field and bending moments in the piles, while increased the accelearions in the pile caps and rate of dissipation of excess pore water pressure in the liquefiable layer. Bar charts of the maximum bending moments in the piles of treated and untreated models are presented in Fig. 19. As seen in this figure, stone columns reduced the maximum bending moments of the piles between 51% to 77% in this case.

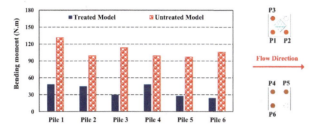

Fig. 19. Bar charts of the maximum bending moments of the piles in treated and untreated models with stone columns [24].

8.2 Assessment of Micropile Method

Effectiveness of a micro-pile system as a countermeasure for the effect of lateral spreading on piles was evaluated using 1g shake table and the rigid box model tested on a 3 × 3 flexible pile group. For this purpose, two physical models including a benchmark model without any mitigation measure and a model remediated with micro-piles were constructed and tested in the first stage of the research. It should be noted that polypropylene pipes were used as model micropiles which were inserted into the liquefiable and the bottom non-liquefiable layers with a spacing of 7.5 cm (center to center) after construction of the physical model. In general, the results of the shake table tests on these models showed that the micropile system deployed in that research did not considerably reduce the kinematic lateral force as well as the induced bending moments due to lateral spreading on the pile group. Besides, the accelerations recorded on the pile cap were adversely amplified compared to the case where no mitigation measure is adopted. The employed micropiles were able to reduce the soil pressure exerted by the upper non-liquefiable layer on the upslope piles of the group. However, they did not effectively decrease the lateral pressures exerted by liquefiable layer on the piles. The soil displacement in the upper parts of the model remediated with micropiles was generally smaller than that in the model with no mitigation measure. This explains that the employed micropiles restricted the lateral displacement of the upper non-liquefiable layer and partially that of the upper depths of the liquefiable layer. However, this restriction was not noticeable in deeper depths of the liquefiable layer. Employing the micropiles in a tighter pattern, adopting stiffer ones or fixing them in the underlying non-liquefiable layer may be considered to more effectively restrict the lateral movement of the liquefied soil thus to reduce the lateral soil pressure on piles subjected to lateral spreading. However, effectiveness of the mentioned solutions needs further investigations in future.

9   Possible Sources of Errors in Physical Modelling

Errors involved in physical modelling can be divided into three main categories: system errors, human errors and environmental errors. The system errors are associated with the errors involved in measuring sensors, data acquisition system (DAS) and the main shaking system. The main shaking system is a high precision 280 bar Instron-Schenck system with a high definition control system which is completely isolated from any virus or intrusion. In the tested physical models it has been delicately tried to minimize errors involved in the measuring sensors by choosing high accuracy sensors and calibrated or checked them against the reference sensors. Also it was tried to minimize the noise associated with the data in the DAS. Some of the errors which may be encountered is due to the installing method of the sensors in which both the human and environmental errors may be involved. For instance very small rotation or dislocation in accelerometers during installation or during shaking and liquefaction or lateral spreading may affect the recorded accelerations. The solution for minimizing this error might be installing the accelrometers on a base with a relatively large area. In addition, strain gauges are very sensitive to temperature and water. To solve the water problem, waterproof sensors should be used or water resistant epoxies should be implemented. Temprature change may affect on the measured data by strain gauges which is an environmental error and erronous result may be deducted. The effect of temperature on strain gauges can be diminished automatically by using half way Watson Bridges. This technique have been used in installationo of strain gauges on all piles in this study. The non-uniformity of the loose sandy layer is another source of error for which a special hopper was designed, constructed and used for water sedimentation of the loose sandy layers in construction of the physical models.

10   Conclusion

Liquefaction-induced lateral spreading can cause severe damages to deep foundations. Two stages of physicsl modelling including several large scale 1-g shake table tests, using both rigid and laminar shear boxes, which have been deployed for investigating various aspects of the effects of lateral spreading on single and pile groups with different configurations of piles, soil profiles and superstructures have been reported and discussed briefly in this paper. Also some conducted experiments to assess the efficacy of different mitigation methods against the effect of lateral spreading on a number of pile groups have been introduced and explained shortly. The results indicated that dynamic soil-pile interactions in liquefied and laterlly spreading grounds are complex problems which can be affected by numerous parameters. Hence, althogh some general results might be deducted from the observations and test results, however, each problem should be solved according to its geometrical and mechanical characteristics of the soil layers and the piles. Therefore, further experiments and numerical studied are still required.

Acknowledgment. The partial financial supported by Construction and Development of Transportation Infrastructures Company and the partial financial support by Transportation Research Institute, Ministry of Roads & Urban Development of Iran for conducting the studies reported in this paper are acknowledged. Also the partial financial support granted by Research Deputy of the Sharif University of Technology is acknowledged. The experiments were conducted at Shake Table Facilities of Civil Engineering Department, Sharif University of Technology. The contribution of all faculty, graduate students and technicians in performing the experiments is acknowledged as well.

References

1. Hamada, M., Isoyama, R., Wakamatsu, K.: Liquefaction Induced Ground Displacement and Its Related Damage to Lifeline Facilities, pp. 81–97. Soils and Foundation, January (Special Issue) (1996)
2. Hamada, M., Yasuda, S., Isoyama, R., Emoto, K.: Study on Liquefaction-Induced Permanent Ground Displacements. Association for the Development of Earthquake Prediction, Japan (1986)
3. Tokimatsu, K., Mizuno, H., Kakurai, M.: Building Damage Associated with Geotechnical Problems. Soils and Foundations, January (Special Issue), 219–234 (1996)
4. Tokimatsu, K., Asaka, Y.: Effects of Liquefaction-Induced Ground Displacements on Pile Performance in The 1995 Hyogoken–Nambu Earthquake. Special Issue of Soils and Foundations, pp. 163–177 (1988)
5. Bardet, J.P., Kapuskar, M.: Liquefaction sand boils in San Francisco during 1989 Loma Prieta earthquake. J. Geotechnical Geoenvironmental Eng. **119**(3), 543–562 (1993)
6. Bhattacharya, S., Sarkar, R., Huang, Y.: Seismic design of piles in liquefiable soils. In: Huang, Y., Wu, F., Shi, Z., Ye, B.: (eds) New Frontiers in Engineering Geology and the Environment. Springer Geology, Springer, Berlin, Heidelberg (2013). https://doi.org/10.1007/978-3-642-31671-5_3
7. Tamura, K.: Seismic design of highway bridge foundations with the effects of liquefaction since the 1995 Kobe earthquake. Soils and Foundations **54**(4), 874–882 (2014)
8. Japan Road Association (JRA): Seismic Design Specifications for Highway Bridges. English version, Prepared by Public Works Research Institute (PWRI) and Ministry of Land, Infrastructure and Transport, Tokyo, Japan (2002)
9. Haeri, S.M., Kavand, A., Rahmani, I., Torabi, H.: Response of a group of piles to liquefaction-induced lateral spreading by large scale shake table testing. Soil Dyn. Earthq. Eng. **38**, 25–45 (2012)
10. Haeri, S.M., et al.: Effects of liquefaction-induced lateral spreading on a 3×3 pile group using 1g shake Table and laminar shear box. In: Proceedings of the 7th international conference on Earthquake Geotechnical Engineering, Rome, Italy (2019a)
11. Haeri, S.M., Kavand, A., Asefzadeh, A., Rahmani, I.: Large Scale 1-g shake table model test on the response of a stiff pile group to liquefaction-induced lateral spreading. In: Proceedings of the 18th International Conference on Soil Mechanics and Geotechnical Engineering, Paris, France (2013)
12. Haeri, S.M., Rajabigol, M., Salaripour, S., Sayaf, S., Pakzad, A., Kavand, A.: Study of dynamic response of a 3×3 pile group to liquefaction-induced lateral spreading using shake table. In: Proceedings of 4th National Conference on Geotechnical Engineering, Tehran, Iran (in Persian, 2019b)

13. Haeri, S.M., et al.: Response of 2×2 pile groups to soil liquefaction in inclined base layer: 1g shake table tests. In: Proceedings of 8th International Conference on Semiology & Earthquake Geotechnical Engineering, Tehran, Iran (2019c)

14. Haeri, S.M., Kavand, A., Raisianzadeh, J., Padash, H., Rahmani, I., Bakhshi, A.: Observations from a large scale shake table test on a model of existing pile-supported marine structure subjected to liquefaction induced lateral spreading. In: Proceedings of the 2nd European Confrence on Earthquake Engineering and Seismology, Istanbul, Turkey (2014)

15. Haeri, S.M., Rajabigol, M., Moradi, M., Zangeneh, M.: A case study of dynamic response of a 3×5 Pile group to liquefaction induced lateral spreading: 1-g shake table test. In: Proceedings of the 12th International Congress on Civil Engineering, Mashhad, Iran (2021)

16. Kavand, A., Haeri, S.M., Asefzadeh, A., Rahmani, I., Ghalandarzadeh, A., Bakhshi, A.: Study of the behavior of pile groups during lateral spreading in medium dense sands by large scale shake table test. Int. J. Civil Eng. **12**(3), 374–439 (2014)

17. Kavand, A., Haeri, S.M., Raisianzadeh, J., Sadeghi Meibodi, A., Afzal Soltani, S.: Seismic Behavior of a dolphin-type berth subjected to liquefaction induced lateral spreading: 1g large scale shake table testing and numerical simulations. Soil Dynamics Earthquake Eng. **140**, 106450 (2021)

18. Su, L., Tang, L., Ling, X., Liu, C., Zhang, X.: Pile response to liquefaction-induced lateral spreading: a shake-table investigation. Soil Dyn. Earthq. Eng. **82**, 196–204 (2016)

19. Ebeido, A., Elgamal, A., Tokimatsu, K., Abe, A.: Pile and pile-group response to liquefaction-induced lateral spreading in four large-scale shake-table experiments. J. Geotechnical Geoenviromental Eng. **145**(10), 04019080 (2019)

20. He., L., Elgamal., A., Hamada., M., Meneses, J.: Shadowing and group effects for piles during earthquake-induced lateral spreading. In: Proceedings of the 14th World Conference on Earthquake Engineering, Beijing, China (2008)

21. Seed, H.B., Booker, J.R.: Stabilization of potentially liquefiable sand deposits using gravel drains. J. Geotechnical Geoenvironmental Eng. **103**(7), 757–768 (1977)

22. Kavand, A., Haeri, S.M., Raisianzadeh, J., Padash, H., Ghalandarzadeh, A.: Performance evaluation of stone columns as mitigation measure against lateral spreading in pile groups using shake table tests. International Conference on Ground Improvement and Ground Control (ICGI 2012), Paris, France (2012)

23. Kavand, A., Haeri, S.M., Raisianzadeh, J., Afzalsoltani. S.: Effectiveness of a vertical micropile system for mitigation of liquefaction-induced lateral spreading effects on Pile foundations: 1g Large Scale Shake Table Tests. Scientia Iranica (2021)

24. Haeri, S.M., Rajabigol, M., Zangeneh, M., Moradi, M.: Assessment of stone column technique as a mitigation method against liquefaction-induced lateral spreading effects on 2×2 pile groups. In: Proceedings of the 4th International. Conference on Performance-based Design in Earthquake. Geotechnical Engineering (PBD-IV) in Beijing, China (2022)

25. Elgamal, A., Lu, J., Forcellini, D.: Mitigation of liquefaction-induced lateral deformation in a sloping stratum: three-dimensional numerical simulation. J. Geotechnical Geoenvironmental Eng. **135**(11), 1672–1682 (2009)

26. Foellini, D., Tarantino, A.M.: Assessment of stone columns as a mitigation technique of liquefaction-induced effects during italian earthquakes (May 2012), Hindawi Publishing Corporation, The Scientific World Journal (2014)

27. Tang, E., Orense, R.P.: Improvement Mechanisms of Stone Columns as A Mitigation Measure Against Liquefaction-Induced Lateral Spreading. New Zealand Society for Earthquake Engineering, Aukland (2014)

28. Liu, J., Kamatchi, P., Elgamal, A.: Using stone columns to mitigate lateral deformation in uniform and stratified liquefiable soil strata. Int. J. Geomechanics **19**(5), 04019026 (2019)

29. Iai, S.: Similitude for shaking table tests on soil–structure–fluid model in 1g gravitational field. Soils Found. **29**(1), 105–118 (1989)

30. Iai, S., Tobita, T., Nakahara, T.: Generalized scaling relations for dynamic centrifuge tests. Géotechnique **55**(5), 355–362 (2005)
31. Brandenburg, S.J.: Behavior of Pile Foundations in Liquefied and Laterally Spreading Ground. PhD thesis, University of California at Davis, CA (2005)
32. Architectural Institute of Japan (AIJ): Recommendations for Design of Building Foundations (in Japanese, 2001)

Hybrid Type Reinforcement of Highway Embankment Against Earthquake Induced Damage

Hemanta Hazarika[1]([✉]), Chengjiong Qin[1], Yoshifumi Kochi[2], Hideo Furuichi[3], Nanase Ogawa[4], and Masanori Murai[5]

[1] Kyushu University, Fukuoka 819-0395, Japan
Hazarika@civil.kyushu-u.ac.jp
[2] K's Lab Inc., Yamaguchi 753-0212, Japan
[3] Adachi Architectural Design Office Co., Ltd., Tokyo 141-0031, Japan
[4] Giken Ltd., Tokyo 135-0063, Japan
[5] Shimizu Corporation, Tokyo 104-8370, Japan

Abstract. In this research, a hybrid type pile system is proposed as a counter-measure of embankment failures during earthquakes. In the proposed technique, inclined piles are added in addition to the rows of vertical steel pipe piles. Inclined piles are expected to reduce ground deformation and settlement due to earthquakes, especially when the embankment widths are large. The focus of the research is to evaluate the stability of embankments by the external force of the earthquake and to elucidate the mechanism of the reinforcement effect by the vertical steel pipe piles and inclined piles. Model tests using shaking table and their numerical simulations were performed to evaluate the performance of the hybrid type reinforcement measures. Model tests were conducted by taking into the consideration the pile rigidity and embedment depth. The numerical simulations were also performed in which the effect of the inclination of the inclined piles was also taken into the consideration. Both the test results and numerical simulations have confirmed that in embankment with hybrid reinforcement measures, the increase in excess pore water pressure during an earthquake could be suppressed, and the settlement of the embankment could be significantly reduced.

Keywords: Shaking table tests · Dynamic effective stress analysis · Embankment settlement · Highway embankment · Hybrid retrofitting technique · Liquefaction

1   Introduction

In the recent earthquakes in Japan, many geotechnical structures, such as river dykes, levees, dams, and road embankments, suffered extensive damage and failures [1, 2]. Highway embankments are strategic infrastructures that play a critical role in connecting and transporting vital rescue components and disaster relief supplies after natural disasters such as earthquakes, tsunami, typhoons, and rainstorms occur. Many highway

L. Wang et al. (Eds.): PBD-IV 2022, GGEE 52, pp. 472–499, 2022.
https://doi.org/10.1007/978-3-031-11898-2_27

embankments in Japan are aging as they were constructed more than six decades ago, and therefore, the Ministry of Land, Infrastructure, Transport, and Tourism (MLIT, Japan) is very much concerned about the vulnerabilities of those during natural disasters. Figure 1 shows an example of liquefaction-induced ground deformation of a highway embankment during the 2016 Kumamoto Earthquake [3]. Few case studies on the seismic failure of the structures and the influence of the liquefiable foundation soil during the recent earthquakes are described in [4–6]. It is of utmost importance to mitigate seismically induced damage to road embankments, and therefore, it is essential to develop sustainable and economic ground improvement techniques that are in line with the current policy of the Japanese government.

Piles are often used to mitigate earthquake-induced damage to embankments. When piles are subjected to intense dynamic loads during earthquakes, soil-structure interaction plays an essential role in the response of piles [7]. Recent observations after major earthquakes have shown that substantial damage and destruction can still be encountered in pile foundations. The behavior of piles in liquefiable soil layers is much more complex than that of the non-liquefiable soil layer due to the decreasing stiffness and shear strength of the surrounding soil due to the increase in pore water pressure during earthquakes. [8, 9]. Many researchers have conducted shaking table tests on the dynamic behavior of pile foundations in liquefiable soils [10–14]. Also, it could be confirmed that restraining the pile heads from rotating increases the maximum shear strength and the piles act as a monolithic structure leading to uniform deformation of piles [15]. Previous studies on beam-pile and anchor-pile supported embankments using shaking table tests have shown that the anchor-pile systems can better support the embankment slopes than that of beam-piles [16]. The purpose of this research is to develop an appropriate deformation countermeasures to protect the existing highway embankments and remediate the associate ground settlements during earthquakes.

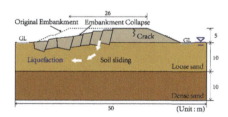

Fig. 1. Failure of road embankment during the 2016 Kumamoto earthquake [3]

A hybrid type pile system is proposed as a countermeasure of embankment failures during earthquakes, as shown in Fig. 2. In this proposed technique, inclined piles are added in addition to the rows of vertical steel pipe piles, as shown in the figure. Inclined piles are expected to reduce ground deformation and settlement, especially when the embankment widths are large. The focus of the research was to evaluate the stability of embankments by the external force of the earthquake and to elucidate the mechanism of the reinforcement effect by the steel pipe piles and inclined piles. Model tests using shaking table and their numerical simulations were performed to evaluate the performance of the reinforcement measures quantitatively. An embankment model with reinforcement measures was created, and a series of shaking table tests were conducted by taking into the consideration the pile rigidity and embedment depth. The numerical simulations were performed using dynamic effective stress analysis.

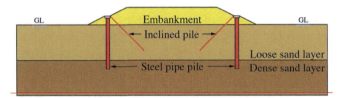

Fig. 2. Proposed countermeasure [17]

2   Shaking Table Tests

In this study, a series of shaking table tests were conducted to confirm the effectiveness of the proposed seismic retrofitting of road embankment. The test series also investigated the effect of repeated load in the form of foreshock and mainshock, such as those observed during the 2016 Kumamoto earthquake. The performances of the unreinforced embankment, existing reinforcement measures, and proposed measures were evaluated.

2.1   Test Conditions

In shaking table tests, a box-shaped soil box with height 800 mm, width 1800 mm, and depth 400 mm was used. The box was made of a transparent acrylic plate, and, therefore, the behavior of the ground during earthquake loadings could be observed. Toyoura sand was used as soil for both embankment and foundation soil. The relative density of the embankment was set to 60%, and the foundation soil was prepared by dividing it into a loose upper layer (relative density 60%) and a lower layer (relative density 90%). The scale ratio was set to 50, and the parameters for the ground, steel pipe pile, and inclined pile were determined as shown in Table 1 using the scaling law [18]. Aluminum pipes were used as vertical piles using the similarity ratio in Table 1. The inclined piles were

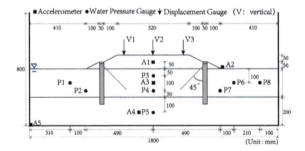

Fig. 3. Instrumentations for the shaking table tests

Table 1. Similitude for 1 g shaking table tests

Parameter	Prototype/model	Scaling factor
Length	N	50
Density	1	1
Stress	N	50
Pore water pressure	N	50
Acceleration	1	1
Displacement	$N^{1.5}$	353.55
Permeability	$N^{0.75}$	18.80
Axial stiffness of piles	$N^{2.5}$	17,677
Bending stiffness of piles	$N^{3.5}$	883,883
Friction	1	1

made of steel. The locations of each measuring device (accelerometer, water pressure gauge, displacement gauge) are shown in the Fig. 3.

Layouts of the four different cases, which were used in the tests, are shown in Fig. 4. Case 1 is an embankment without reinforcement, and Case 2 is a case of existing countermeasure work using only vertical steel pipe piles. Case 3 is a case where geogrid was used for Case 2. Case 4 is a hybrid-type reinforced embankment case using both vertical pile and inclined pile. The vertical pile and the inclined pile were fixed in their heads. The tip of each vertical pile was inserted up to the dense soil layer with a relative density of 90%. In Case 3 and Case 4, to prevent soil movement between the piles, geogrid was used to connect the piles as shown in Figs. 4(c) and 4(d).

The locations of the measuring devices (excess pore water pressure gauges, accelerometers, displacement gauges) are shown in Fig. 4. Dynamic loadings were imparted sequencially in the form of foreshock (=200 Gal), mainshock (=300 Gal), and strong motion (=400 Gal). The frequency of dynamic loadings was determined according to scaling laws and recorded acceleration data at the KMMH16 station during the 2016 Kumamoto earthquake. Tests were conducted using a sine wave with a frequency of 10 Hz and dynamic time of 10 s.

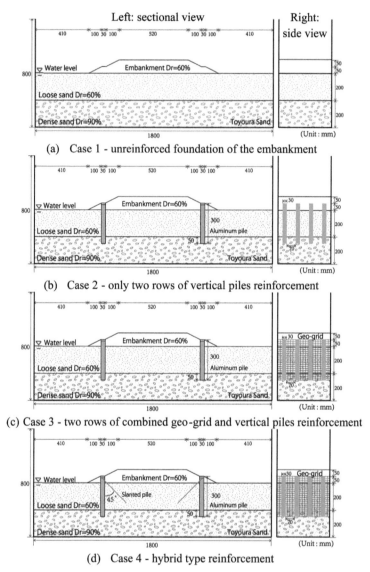

Fig. 4. Layouts of models (left: sectional view and right: side view): (a) Case 1 - unreinforced foundation of the embankment, (b) Case 2 - only two rows of vertical piles reinforcement (c) Case 3 - two rows of combined geo-grid and vertical piles reinforcement and (d) Case 4 - hybrid vertical piles and inclined piles system reinforcement

2.2 Results and Discussions

Excess Pore Water Pressure Ratio ($R_u = \Delta u/\sigma'_v$)

Figures 5 and 6 show the excess pore water pressure ratios at two locations (P3 and P4 in Fig. 4) just below the center of the embankment due to two dynamic loadings (foreshock

= 200 Gal and mainshock = 300 Gal). From these figures, it can be seen that in the case of the unreinforced embankment, the pore water pressure ratio reaches 1.0 in both the foreshock and the mainshock, and the foundation soil is completely liquefied. However, in Case 3 and Case 4, the increase in pore water pressure is minimal in both the foreshock and the mainshock, and therefore, it can be said that these measures could effectively prevent liquefaction. Especially in Case 3, there is a delay in the development of excess pore water pressure. For the proposed measures (Case 4), no increase of the excess pore water pressures could be observed.

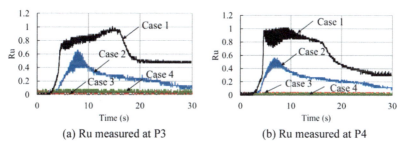

(a) Ru measured at P3   (b) Ru measured at P4

Fig. 5. Time histories of excess pore water pressure ratio during foreshock: (a) Ru measured at P3 and (b) Ru measured at P4

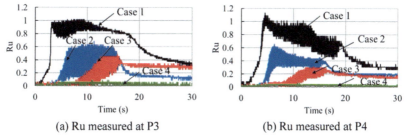

(a) Ru measured at P3   (b) Ru measured at P4

Fig. 6. Time histories of excess pore water pressure ratio during the mainshock: (a) Ru measured at P3 and (b) Ru measured at P4

Figure 7 shows the maximum pore water pressure ratio (Ru) at different depths measured by each water pressure gauge. From Fig. 7, it could be clarified that, in the case of the unreinforced embankment, the value of the pore water pressure ratio exceeded 1.0 in both 200 Gal and 300 Gal in the shallow part, leading to the liquefaction of the foundation soil. On the other hand, in the case of the reinforced embankment (Case 2, Case 3, Case 4), the increase in pore water pressure is suppressed in 200 Gal (Ru < 0.7). In Case 3 and Case 4, although a slight increase of Ru was seen at the mainshock (Ru was 0.5 or less), no liquefaction occured. From these results, it can be said that the developed countermeasure is the most effective in preventing liquefaction of the foundation soil in both the foreshock (200 Gal) and the mainshock (300 Gal).

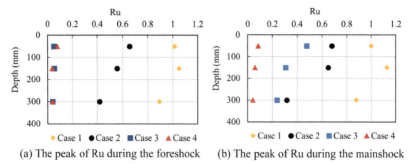

(a) The peak of Ru during the foreshock   (b) The peak of Ru during the mainshock

Fig. 7. The peak of excess pore water pressure ratio during the foreshock and mainshock: P3 (50 mm below ground), P4 (150 mm below ground), and P5 (300 mm below ground)

Settlement of Embankment

Figures 8 and 9 show the time history of the settlement of three locations (V1, V2, and V3) on the embankment during the dynamic loadings (foreshock = 200 Gal, mainshock

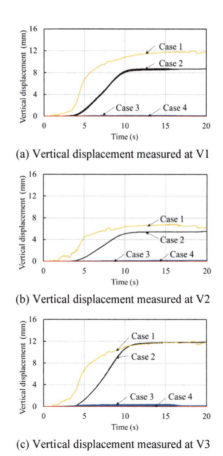

(a) Vertical displacement measured at V1

(b) Vertical displacement measured at V2

(c) Vertical displacement measured at V3

Fig. 8. Time histories of vertical displacement during the foreshock: (a) the left crest – V1, (b) the middle – V2, and (c) the right crest – V3

$= 300$ Gal) measured by three displacement gauges. In the case of the foreshock (Fig. 8), both Case 1 and Case 2 show an increasing trend in the settlement. However, in the case of the mainshock (Fig. 9), all the other cases except for the hybrid type reinforcement measure (Case 4) show a rising trend. In the case of the unreinforced embankment, significant settlement (24 cm on the model scale; 1.2 m on the prototype) occurs during the mainshock. However, the settlements in the case of the reinforced embankments drastically reduced. In particular, in the case of the hybrid type reinforcement, it could be seen that there is almost no settlement in both the foreshock and the mainshock. One particular observation worthy of noting here is that with increasing reinforcing measures (vertical pile, geogrid reinforced vertical pile and hybrid type reinforcement), there is a delay in the occurrence of settlement.

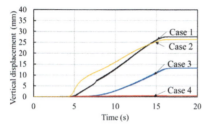

(a) Vertical displacement measured at V1

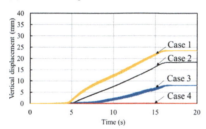

(b) Vertical displacement measured at V2

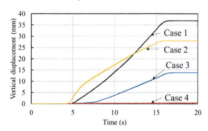

(c) Vertical displacement measured at V3

Fig. 9. Time histories of vertical displacement during the mainshock: (a) the left crest – V1, (b) the middle – V2, and (c) the right crest – V3

Figures 10 and 11 show the seismic resistance of embankments at different dynamic loadings (foreshocks, mainshocks, and strong earthquakes), which were manually measured at the end of each loading. From Fig. 10, it could be confirmed that in the embankment with no reinforcement (Case 1), a large settlement (30 mm) occurs even during foreshock (Fig. 10b). Figure 11a displays that a large deformation in the embankment took place after the mainshock, making further test beyond that impossible. From Fig. 11, it could be seen that embankment without inclined piles (Case 3) shows less settlement (11.6 mm) than the unreinforced embankment (37 mm) during the mainshock. In contrast, embankment with both vertical piles and inclined piles, barely experiences settlement in both the foreshock and the mainshock. In addition, the embankment using hybrid

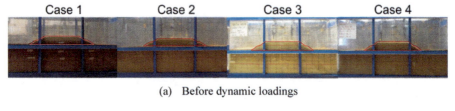

(a)   Before dynamic loadings

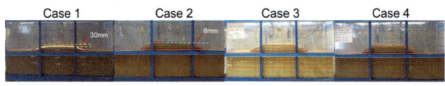

(b)   After the foreshock

Fig. 10. Performance comparison during the foreshock in each case

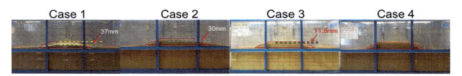

(a)   After the mainshock

(b)   After the strong motion

Fig. 11. Performance comparison during the mainshock and strong motion in each case

type reinforcement shows less settlement (12.6 mm) than the embankment reinforced by only vertical piles (36.5 mm) experiencing motion stronger than main shock (Fig. 11b).

Deformation of Pile Heads

Figure 12 shows the top view of of the test model showing the heads of the piles. Residual deformations of the tips of the piles were measured at the end of each loading for all the cases are plotted in Figs. 13 and 14. From these results, it can be seen that in the case of the hybrid type reinforcement measures (Case 4), the amount of horizontal deformation is the smallest for both the foreshock (200 Gal) and the mainshock (300 Gal).

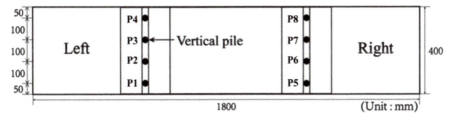

Fig. 12. Top view of the test model

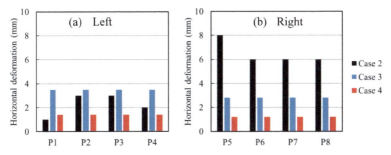

Fig. 13. Residual deformation of pile heads on the left (a) and right (b) during the foreshock

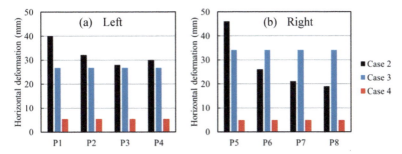

Fig. 14. Residual deformation of pile heads on the left (a) and right (b) during the mainshock

3 Numerical Analysis

The shaking table tests could confirm that the new countermeasure method developed in this research has the effect of suppressing the deformation of the embankment during earthquake. However, in order to elucidate the mechanism, numerical simulations using dynamic effective stress analysis were carried out to reproduce the dynamic behaviors of each case.

3.1 Simulation Models

Figure 15 shows all the analysis models (Case 1, Case 2, Case 3, and Case 4). The embankment and ground conditions are the same as those in the shaking table tests. Case 4 is the condition, in which the inclination angle of the inclined pile was set to 60°,

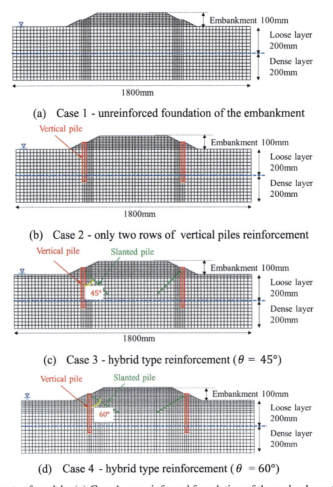

(a) Case 1 - unreinforced foundation of the embankment

(b) Case 2 - only two rows of vertical piles reinforcement

(c) Case 3 - hybrid type reinforcement ($\theta = 45°$)

(d) Case 4 - hybrid type reinforcement ($\theta = 60°$)

Fig. 15. Layouts of models: (a) Case 1 - unreinforced foundation of the embankment, (b) Case 2 - only two rows of vertical piles reinforcement (c) Case 3 - hybrid type reinforcement ($\theta = 45°$) and (d) Case 4 - hybrid type reinforcement ($\theta = 60°$)

to investigate the effect of installation conditions on the reinforcing effect. The lateral and bottom boundaries in the numerical models were fixed to imitate the boundary conditions of the test models. The lateral and bottom boundaries are assumed to be impermeable, whereas the surface of the loose sand layer is a permeable boundary. In order to ensure numerical stability, the time integration increment was selected to be 0.001 s. Figure 16 shows the input acceleration (mainshock), measured on the shaking table. In this analysis, this waveform was used as the input motion, which was imparted to the model's bottom.

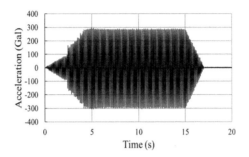

Fig. 16. Input motion

3.2 Materials Parameters

Table 2 shows the parameters used for the dynamic effective stress analysis. Cyclic elasto-plastic model was selected as the constitutive law for the soils in the embankment and foundation. Table 3 shows the material characteristics of the vertical and inclined piles. The inclined piles were modeled as beam elements. For vertical pile, hybrid beam element, consisting of the conventional beam element and elastic solid element, was used to represent the pile volume [19]. The joint elements were introduced along all the interfaces between the soil and piles, which share identical displacements in the horizontal direction.

Table 2. Material parameters of the constitutive model

Material parameters		Loose sand layer $Dr = 60\%$	Dense sand layer $Dr = 90\%$	Embankment
Density	ρ	1.88	2.00	1.88
Coefficient of permeability	k	0.01	0.01	0.01
Initial void ratio	e_0	0.738	0.659	0.738

(*continued*)

Table 2. (*continued*)

Material parameters		Loose sand layer Dr = 60%	Dense sand layer Dr = 90%	Embankment
Compression index	λ	0.02	0.0004	0.02
Swelling index	κ	0.0005	0.00008	0.0005
Failure stress ratio	M_f^*	1.300	1.466	1.300
Phase transformation stress ratio	M_m^*	0.980	0.765	0.980
Initial shear modulus ratio	G_0/σ_m^*	2343	1133	2343
Dilatancy parameters	D_0^*	0.5	0.12	0.5
Hardening parameters	n	5.0	4.0	5.0
	B_0^*	6550	54,000	6550
	B_1^*	65.5	5400	65.5
Reference strain parameter	γ_r^{P*}	0.002	0.03	0.002
	γ_r^{E*}	0.008	0.36	0.008

Table 3. Material parameters of the constitutive model (reinforcing material)

Parameters	Unit	Vertical pile	Inclined pile
Young's modulus	MN/m^2	73,000	206,000
Bending stiffness	$MN \cdot m^2$	3.43×10^{-3}	8.2×10^{-7}

3.3 Simulation Results

Excess Pore Water Pressure Ratio ($R_u = \Delta u/\sigma'_v$)

Figures 17 and 18 show the distribution of excess pore water pressure ratio at the end of foreshock and mainshock. As evident from these figures, in the case of unreinforced embankments and the reinforcement using only vertical piles, the excess pore water pressure ratio reaches 1.0 in both the foreshock and the mainshock, and the foundation soil is completely liquefied. However, in Case 3 and Case 4, the increase in pore water pressure is sufficiently suppressed during the foreshock (200 Gal), where the Ru value is less than 0.3. In the mainshock (300 Gal), Case 4 shows the rise of Ru in some parts of the embankment, however, that could not lead to complete liquefaction. Case 3, with 45 degree inclination of the inclined piles, did not experience any significant rise in the Ru value even during mainshock. Based on these results, it can be said that the developed countermeasure is very effective in preventing liquefaction of the foundation soil during earthquake.

Figure 19 shows the excess pore water pressure ratio observed at location at P4 (refer to Fig. 4) due to the mainshock. The figure reveals that, in Case 1 and Case 2, the excess

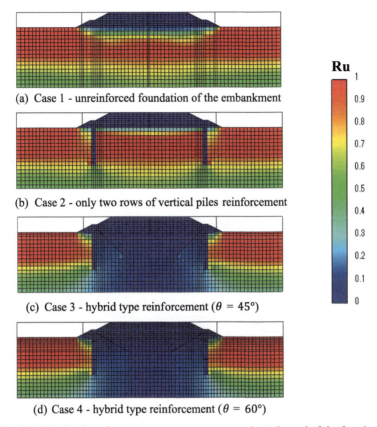

(a) Case 1 - unreinforced foundation of the embankment

(b) Case 2 - only two rows of vertical piles reinforcement

(c) Case 3 - hybrid type reinforcement ($\theta = 45°$)

(d) Case 4 - hybrid type reinforcement ($\theta = 60°$)

Fig. 17. Distribution of excess pore water pressure ratio at the end of the foreshock

pore water pressure ratio rises immediately, and after reaching the peak, it takes sufficient time for the excess pore water pressure to dissipate. However, in Case 4, the increase of Ru was delayed and got dissipiated immediately after the dynamic loading stopped. On the other hand, in Case 3, the increase in excess pore water pressure was relatively small, and it gradually disappeared. Therefore, it can be said that the combination of vertical pile and inclined pile, and the inclination angle play an important role in preventing shear deformation of soils and, therefore, liquefaction. The deformations behaviors of the embankment are discussed in the following sub-section.

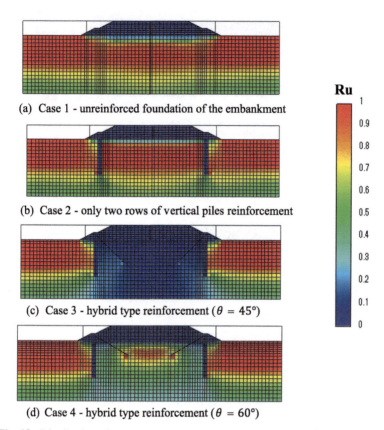

(a) Case 1 - unreinforced foundation of the embankment

(b) Case 2 - only two rows of vertical piles reinforcement

(c) Case 3 - hybrid type reinforcement ($\theta = 45°$)

(d) Case 4 - hybrid type reinforcement ($\theta = 60°$)

Fig. 18. Distribution of excess pore water pressure ratio at the end of the mainshock

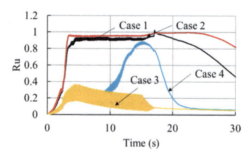

Fig. 19. Time history of excess pore water pressure ratio (P3)

Deformation of Embankment and Foundation Soils

Figures 20 and 21 show the deformed configuration of the four cases at the end of the foreshock and the mainshock. In the case of the unreinforced embankment (Case 1) as well as embankment only with vertical piles (Case 2), large deformation occurs during the mainshock, leading to large embankment settlement. However, in Case 3 and Case 4, deformation of the foundation soil is suppressed, and as a result, embankment settlement could be reduced. In Case 3 (inclination angle = 45°), there is no significant settlement in both the foreshock and the mainshock. In Case 4 (inclination angle = 60°), however, the deformations of the foundation soil and the embankmengt are found to be larger during the mainshock.

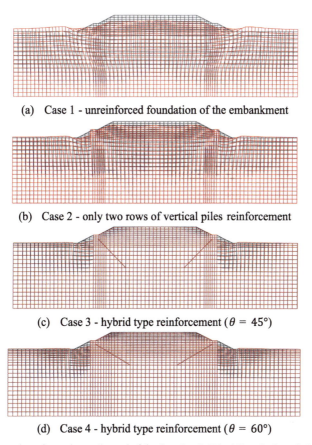

(a)   Case 1 - unreinforced foundation of the embankment

(b)   Case 2 - only two rows of vertical piles reinforcement

(c)   Case 3 - hybrid type reinforcement (θ = 45°)

(d)   Case 4 - hybrid type reinforcement (θ = 60°)

Fig. 20. Deformed configuration at the end of the foreshock (Blackline: Before shaking; Red line: After shaking)

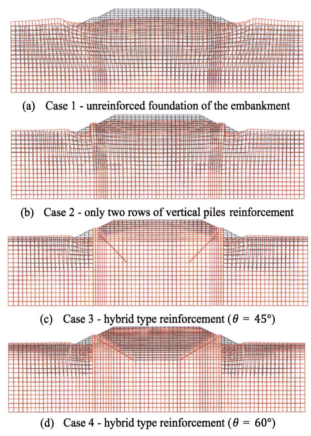

(a)   Case 1 - unreinforced foundation of the embankment

(b)   Case 2 - only two rows of vertical piles reinforcement

(c)   Case 3 - hybrid type reinforcement ($\theta = 45°$)

(d)   Case 4 - hybrid type reinforcement ($\theta = 60°$)

Fig. 21. Deformed configuration at the end of the mainshock (Blackline: Before shaking; Red line: After shaking)

Figure 22 shows the time history of settlements in the central part of the embankment, (similar to V2 in the shaking table model), due to the mainshock for all the four cases. In Case 1 and Case 2, the settlement progresses linearly immediately after the dynamic loading. However, the hybrid type reinforcements (Case 3 and Case 4) show a considerable delay in the occurrence of settlement, and the final settlements were only 1/5 to 1/7 to those of Case 1 and Case 2. These results once again prove that the installation angle of inclined piles has significant contribution towards preventing the settlement of embankment.

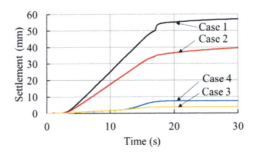

Fig. 22. Time history of the settlement (V2)

4   Comparison of Test and Simulation Results

The results for Case 1 (unreinforced embankment) and Case 3 (hybrid type reinforcement $\theta = 45°$) are selected as typical results for comparisons. The comparisons between analysis (denoted "Ana.") and test (denoted "Tes.") results divulge the predictive capacity of the simulation. The computed results were found to be in good agreement with that observed in the tests, including the settlement and the development of excess pore pressures ratio.

4.1   Excess Pore Water Pressure Ratio (Ru)

The excess pore water pressure ratios (Ru) at the three locations (P3, P4, and P5) directly under the embankment for unreinforced embankment (Case 1), are shown in Fig. 23 and Fig. 24 for the forshock and mainshock respectively. It can be seen that the analysis results and the test results are showing almost the same trend. At P3 and P4, both the analysis and test results show that the loose sand layer liquefied completely. However, no liquefaction is observed in the dense soil layer, as confirmed from the values of Ru (< 0.8) at P5. It is to be noted that the trends of the dissipation phase in the analysis are slightly different from those in the tests. This may be due to some limitations of the constitute models used in the analysis.

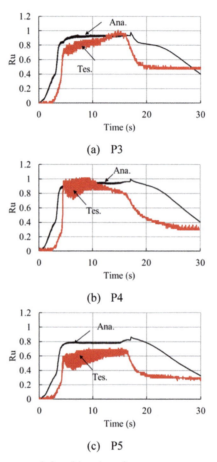

(a) P3

(b) P4

(c) P5

Fig. 23. Computed and measured time histories of excess pore water pressure ratio in Case 1 during the foreshock (a) P3, (b) P4, and (c) P5

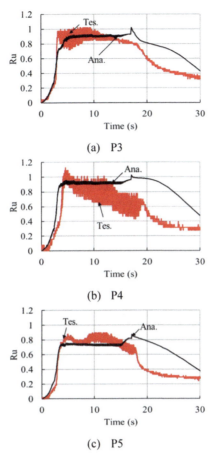

Fig. 24. Computed and measured time histories of excess pore water pressure ratio in Case 1 during the mainshock (a) P3, (b) P4, and (c) P5

Results similar to those in Figs. 23 and 24 are shown in Figs. 25 and 26 for the hybrid type reinforcement (Case 3). These figures reveal that, in comparison with unreinforced foundations, hybrid type reinforcement can limit the development of excess pore water pressure effectively. The Ru values reaches only up to 0.2 to 0.3, and dissipation starts very quickly, confirming the liquefaction prevention capacity of the proposed measure. However, for both the foreshock and mainshock, the analysis results deviate from the test results. This particular aspects needs to be explored further in the future.

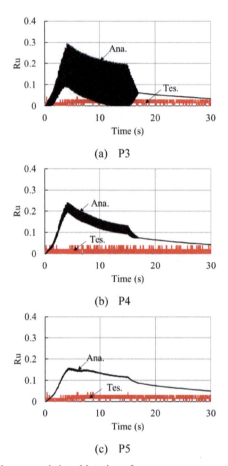

Fig. 25. Computed and measured time histories of excess pore water pressure ratio in Case 3 during the foreshock (a) P3, (b) P4, and (c) P5

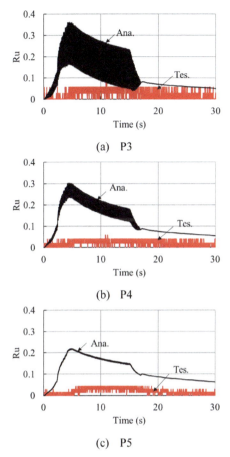

(a)   P3

(b)   P4

(c)   P5

Fig. 26. Computed and measured time histories of excess pore water pressure ratio in Case 3 during the mainshock (a) P3, (b) P4, and (c) P5

4.2   Embankment Settlement

The numerical results are selected at the same locations (V1, V2 and V3) of the shaking table test models (Refer to Fig. 3). Figures 27 and 28 show the comparison between the numerical and test results for the settlement of embankment in the unreinforced case (Case 1) due to the dynamic loading. The time histories of numerical results during both foreshock and mainshock show qualitative agreement with the test results. The computed settlements in the middle of the embankment (V2), however, are larger than the test results.

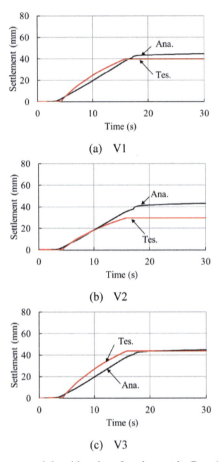

Fig. 27. Computed and measured time histories of settlement in Case 1 during the foreshock (a) V1, (b) V2, and (c) V3

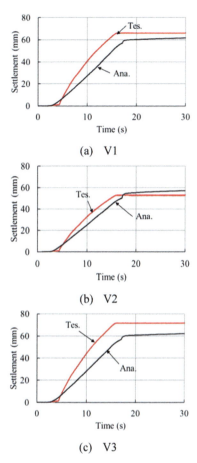

Fig. 28. Computed and measured time histories of settlement in Case 1 during the mainshock (a) V1, (b) V2, and (c) V3

Figures 29 and 30 show the comparison between the numerical and test results of the embankment settlement for hybrid type reinforcement (Case 3). For both the foreshock and the mainshock, the numerical results are much higher than the test results, and this point needs to be re-examined in the future.

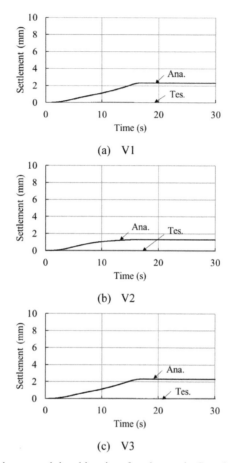

(a) V1

(b) V2

(c) V3

Fig. 29. Computed and measured time histories of settlement in Case 3 during the foreshock (a) V1, (b) V2, and (c) V3

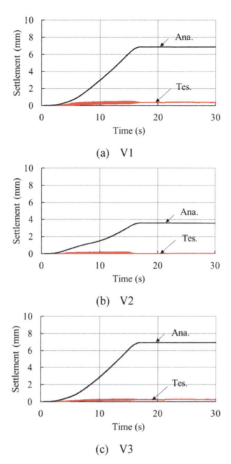

Fig. 30. Computed and measured time histories of settlement in Case 3 during the mainshock (a) V1, (b) V2, and (c) V3

5   Conclusions

In this study, we proposed a hybrid type seismic retrofitting work for road embankment using vertical pile and inclined pile and examined its effectiveness by using shaking table tests and numerical analysis. The main conclusions are described below.

1. The main causes of embankment settlement during an earthquake are large deformation of the foundation soil and resulting liquefaction. Large settlement and horizontal displacement occur even in the foreshock in the case of embankment where no countermeasures against earthquake is adopted.
2. In the embankment with hybrid type reinforcement measures, the increase in excess pore water pressure during an earthquake could be suppressed due to limited shear deformation of the foundation soils, and the settlement of the embankment could be significantly reduced.

3. In hybrid type reinforcement, it is more effective to install the inclined piles at an angle of 45°. It is, however, necessary to further investigate the effect of the inclination and length of inclined piles.
4. The seismic performance of the proposed countermeasure is superior to the reinforcement measures for embankments using only vertical steel pipe piles.

Acknowledgements. A major part of this research was supported by the research grant from Nippon Expressway Company Limited (NEXCO) and affiliated organizations, Japan. Also, a part of this research was funded by Kyushu Regional Management Service Association, Fukuoka, Japan. The authors gratefully acknowledge those financial supports. The authors also would like to thank Mr. Daisuke Matsumoto of Japan Foundation Eng., Co., Ltd., Mr. Takashi Fujishiro of Geo-Disaster Prevention Institute, Mr. Shinichiro Ishibashi of Nihon Chiken Co., Ltd., Dr. Naoto Watanabe of KFC Ltd. and Mr. Shigeo Yamamoto of Chuo Kaihatsu Corporation for their support in this research. Last but not least, the authors are indebted to Mr. Yuichi Yahiro, technical staff of Department of Civil Engineering, Kyushu University, for his continuous advice and cooperation during the shaking table tests.

References

1. Sasaki, Y., et al.: Reconnaissance report on damage in and around river levees caused by the 2011 off the Pacific coast of Tohoku Earthquake. Soils and Foundation **52**(5), 1016–1032 (2012)
2. Hazarika, H., et al.: Geotechnical damage due to the 2016 Kumamoto earthquake and future challenges. Lowland Technol. Int. **19**(3), 189–204 (2017)
3. West Nippon Expressway Company Limited: Kumamoto Earthquake Response Committee WG2 Report: About earthquake resistance performance of highway embankment (2017)
4. Maharjan, M., Takahashi, A.: Liquefaction-induced deformation of earthen embankments on non-homogeneous soil deposits under sequential ground motions. Soil Dyn. Earthq. Eng. **66**, 113–124 (2014)
5. Ishikawa, H., Saito, K., Nakagawa, K., Uzuoka, R.: Liquefaction analysis of a damaged river levee during the 2011 Tohoku earthquake. In: Computer Methods and Recent Advances in Geomechanics: Proceedings of the 14th International Conference of International Association for Computer Methods and Recent Advances in Geomechanics, pp. 673–677. Taylor & Francis, Kyoto (2014)
6. Oka, F., Tsai, P., Kimoto, S., Kato, R.: Damage patterns of river embankments due to the 2011 off the Pacific Coast of Tohoku Earthquake and a numerical modeling of the deformation of river embankments with a clayey subsoil layer. Soils Found. **52**(5), 890–909 (2012)
7. Mylonakis, G., Gazetas, G.: Seismic soil-structure interaction: beneficial or detrimental. J. Earthquake Eng. **4**(3), 277–301 (2000)
8. Lu, C.W., Oka, F., Zhang, F.: Analysis of soil–pile–structure interaction in a two-layer ground during earthquakes considering liquefaction. Int. J. Numer. Anal. Meth. Geomech. **32**(8), 863–895 (2008)
9. Rahmani, A., Pak, A.: Dynamic behavior of pile foundations under cyclic loading in liquefiable soils. Comput. Geotech. **40**, 114–126 (2012)
10. Abdoun, T., Dobry, R.: Evaluation of pile foundation response to lateral spreading. Soil Dyn. Earthq. Eng. **22**(9), 1051–1058 (2002)

11. Bhattacharya, S., Madabhushi, S.P.G., Bolton, M.D.: An alternative mechanism of pile failure in liquefiable deposits during earthquakes. Geotechnique **54**(3), 203–213 (2004)

12. Yao, S., Kobayashi, K., Yoshida, N., Matsuo, H.: Interactive behavior of soil–pile-superstructure system in transient state to liquefaction by means of large shake table tests. Soil Dyn. Earthq. Eng. **24**(5), 397–409 (2004)

13. Suzuki, H., Tokimatsu, K., Sato, M., Abe, A.: Factor affecting horizontal subgrade reaction of piles during soil liquefaction and lateral spreading. In: Seismic performance and simulation of pile foundations in liquefied and laterally spreading ground, pp. 1–10. ASCE (2006)

14. Dungca, J.R., et al.: Shaking table tests on the lateral response of a pile buried in liquefied sand. Soil Dynamic Earthquake Eng. **26**(2), 287–295 (2006)

15. Hazarika, H., Watanabe, N., Sugahara, H., Suzuki, Y.: Influence of placement and configuration of small diameter steel pipe pile on slope reinforcement. In: Proceedings of the 19th International Conference on Soil Mechanics and Geotechnical Engineering, pp. 2151–2154. ICSMGE, South Korea (2017)

16. Ma, N., Wu, H.G., Ma, H.M., Wu, X.Y., Wang, G.H.: Examining dynamic soil pressure and the effectiveness of different pile structure inside reinforced slopes using shaking table tests. Soil Dynamics Earthquakes Eng. **116**, 293–303 (2019)

17. Qin, C.J., et al.: Evaluation of hybrid pile supported system for protecting road embankment under seismic loading. Advances in Sustainable Construction and Resource Management 733–744 (2021)

18. Iai, S.: Similitude for shaking table tests on soil-structure-fluid model in 1g gravitational field. Soils Found. **29**(1), 105–118 (1989)

19. Zhang, F., Kimura, M., Nakai, T., Hoshikawa, T.: Mechanical behavior of pile foundations subjected to cyclic lateral loading up to the ultimate state. Soils Found. **40**(5), 1–17 (2000)

Multiple Liquefaction of Granular Soils: A New Stacked Ring Torsional Shear Apparatus and Discrete Element Modeling

Duruo Huang[1]([✉]), Zhengxin Yuan[2], Siyuan Yang[1], Pedram Fardad Amini[2], Gang Wang[2], and Feng Jin[1]

[1] Tsinghua University, Beijing 100084, China
huangduruo@tsinghua.edu.cn

[2] Hong Kong University of Science and Technology, Clear Water Bay, Kowloon, Hong Kong

Abstract. In this study, we develop an innovative stacked ring torsional shear apparatus for multiple liquefaction tests of fully saturated sands. Stacked rings are built upon an existing hollow cylinder torsional shear apparatus to impose lateral constraint to the soil sample, such that multiple liquefaction can be tested in a fully saturated state. In this project, we used pressure compensation technique to reduce vertical friction between the sample and the ring direct contact force between the membrane and the rings such that the vertical stress will be much more uniformly distributed along the sample height. The new stacked ring device facilitates investigation of fabric evolution and its effect on reliquefaction resistance of sands.

To further study the fundamental mechanism of the multiple liquefaction phenomenon, 3D clumped discrete model is used to construct realistic particle packings, and simulate fabric evolution during liquefaction-reconsolidation-reliquefaction process. The study reveals that strain history significantly influences the number of inter-particle contact and fabric anisotropy after reconsolidation, hence the subsequent liquefaction resistance greatly varies. The increase in relative density due to reconsolidation has only secondary effects. The state-of-the-art DEM simulation provides micromechanical insights into the fundamental mechanism of the multiple liquefaction phenomenon.

Keywords: Multiple liquefaction · Stacked rings · Laboratory tests · Discrete element model

1 Introduction

Liquefaction is a phenomenon in which saturated soils suffer from substantial loss of strength and stiffness induced due to the shear strains during seismic shaking. In the past several decades, liquefaction has always been a significant subject in geotechnical engineering because of the tremendous destructive capacity to infrastructures causing serious economic losses and deaths [1]. Reliquefaction or multiple liquefaction refers to the repeated occurrence (twice or more) of soil liquefaction at the same location during a sequence of earthquake events. The resistance of sands to a subsequent liquefaction event may tend to decrease, even if the sample possesses a higher relative density after being fully consolidated due to the pore pressure dissipation. Multiple liquefaction events have been recorded in all over the world, for example, 1983 Nihonkai-Chubu earthquake [2], 2011 Tohoku earthquake in Japan [3], and from September 2010 to December 2011 in the city of Christchurch, New Zealand [4]. In recent years, studying multiple liquefaction resistance of soils has received great attention within the geotechnical community.

Triaxial apparatus is widely utilized to investigate the influences of various loading histories on the re-liquefaction resistance of sands [5–9]. The development of column-like structure and connected voids in the strain hardening process could be responsible for the reduced reliquefaction resistance [8]. The induced anisotropic structure is extremely unstable and can be developed without much change in the relative density. Moreover, some undrained cyclic triaxial test results indicate that the development of anisotropy is continuous and orderly repeated during the whole liquefaction process [10]. The developed anisotropy is irreversible and remains in subsequent states. Compared with the triaxial device, the hollow cylinder torsional shear apparatus (HCTSA) can simulate more complicated stress paths and strain histories, in such a way that axial load, torque, internal and external pressures can be applied to a hollow cylindrical sample [11]. Recently, a set of torsional shear tests were conducted to investigate the effects of strain histories on re-liquefaction resistance of sands [12]. It was found that the intensity of residual shear strain significantly affects the reliquefaction resistance. For example, for dense samples with a relative density of 70%, the residual shear strain of 0.4% enhances the reliquefaction resistance, while the residual shear strain of 5.0% leads to a significant decrease in the reliquefaction resistance. Unfortunately, the sample is difficult to maintain geometry after experiencing large deformation in post-liquefaction by using flexible membranes in the tests. After two instances of liquefaction, the sample may collapse and the test cannot continue. To simulate multiple liquefaction events, researchers at the University of Tokyo developed a stacked-ring shear equipment [3]. The stacked rings can effectively restrain the lateral deformation of samples during each reconsolidation process. Hence, the soil sample can maintain its original geometry after multiple episodes of liquefaction and reconsolidation. However, the relatively large friction between the sample and stacked rings renders a nonuniform distribution of effective vertical stress in the soil, causing many uncertainty in the experimental analysis. In addition, only dry soil sample can be tested. The effect of undrained liquefaction is approximately by keep a constant volume of the sample.

In this study, a new stacked ring shear apparatus is developed by combining both advantages of an existing HCTSA and stacked rings. A pressure compensation technique is employed to reduce the friction between the membrane and stacked rings, with an aim to generate a uniform stress state to samples. A multiple liquefaction test was conducted to validate the pressure compensation technique and the capability of the new device to simulate multiple liquefaction events for saturated sands.

Although many of previous studies pointed out the important role of soil fabrics in multiple liquefaction process, they cannot be directly observed in laboratory tests. In recent years, Discrete element method (DEM) has become an efficient and direct way to reveal the underlying micro-mechanism associated with the liquefaction responses

[13–15]. A few studies were also devoted to study the multiple liquefaction process [16, 17]. Yet, only simple spherical/circular particles were used in these studies. As one of the key challenges is to realistically simulate soil fabric during multiple liquefaction-consolidation process, in this study, we developed 3D clumped DEM to approximate the irregular shape of Toyoura sand particles, which is critical to realistically characterize fabric evolution and develop fundamental understanding and micromechanical insights into the multiple-liquefaction phenomenon.

2   The Staked Ring Torsional Apparatus

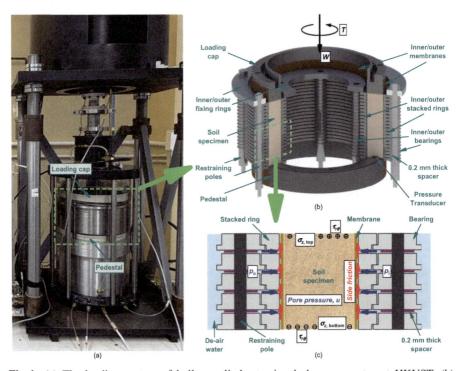

Fig. 1. (a) The loading system of hollow cylinder torsional shear apparatus at HKUST; (b) Schematic diagram of a new stacked-ring device, and (c) stress state of the sand sample placed in the new stacked-ring device.

The stacked ring torsional apparatus mainly includes two parts: an existing HCTSA and a new stacked-ring device. The loading cap and pedestal belong to the loading system of the existing HCTSA at the Hong Kong University of Science and Technology (HKUST), as shown in Fig. 1(a) [12, 18, 19]. Figure 1(b) illustrates the detailed structure of the new stacked-ring device. The new components including inner/outer stacked rings and restraining poles are newly designed according to the HCTSA dimension and fabricated by the Design and Manufacturing Services Facility (DMSF) at HKUST. Each ring of inner/outer stacked rings is made of 5 mm thick stainless steel, and they are supported by specially designed bearings with a flange. Four inner and six outer poles are installed to restrain the bearings and guide the rings move only in the horizontal plane. These bearings only provide a very small friction against rotation of the rings. A stainless-steel spacer is placed between the bearings to keep stacked rings separated vertically by less than 0.2 mm from each other. The soil sample is sealed by inner/outer membranes, with an inner diameter, an outer diameter, and a height of 150, 200, and 100 mm, respectively. The vertical stress at the top $\sigma_{z, top}$ is calculated based on a load cell installed above the main load shaft, while the vertical stress at the bottom $\sigma_{z, bottom}$ is measured through a pressure sensor placed on the bottom platen of the devise.

Figure 1(c) illustrates stresses on the sand specimen in the stacked-ring device. Different from the HCTSA, the side friction generated between the rings and membranes due to the normal contact force, as shown by the red arrows in Fig. 1(c), will result in a large vertical stress reduction and non-uniform distribution within the sample. Similarly, friction in the circumferential direction has an influence on the shear stress distribution. However, the experimental results showed that the circumferential friction is neglectable owing to the bearing system [3]. Therefore, the critical issue is how to effectively reduce the side friction and minimize the reduction on $\sigma_{z, bottom}$. The pressure compensation technique proposed in this study utilizes the application of inner p_i and outer water pressures p_o on the membranes to reduce the direct contact force between the rings and membranes such that the difference between $\sigma_{z, top}$ and $\sigma_{z, bottom}$ shall be smaller and the vertical stress distribution shall be more uniform. Moreover, some lubricants commonly used in the lab, such as oils and greases, shall be applied on the surface of membranes to further eliminate the friction effect.

3 Multiple Liquefaction Tests

3.1 Testing Procedure

Toyoura sand used in all tests is a standard sand with angular to subangular particles. Its mean particle diameter D_{50}, specific gravity G_s, maximum void ratio e_{max}, and minimum void ratio e_{min} are 0.22 mm, 2.65, 0.988, and 0.640, respectively. Toyoura sand was dried at about 105 °C for 24 h and then utilized to prepare the specimen adopting a dry deposition method. The specimen was divided into four layers with equal mass and poured layer by layer into the hollow space between inner and outer stacked rings using a funnel. A rubber hammer was utilized after preparing each layer to tap the inner/outer stacked rings for adjusting the relative density (D_r) to the expectant value. Carbon dioxide and de-air water were circulated through the specimen from the bottom toward the top for 2 h and for 1 h, respectively. A back pressure of 100 kPa was applied to the specimen to obtain full saturation and the B-value of greater than 0.97.

Figure 2 shows the typical shear stress τ and strain γ behaviors of specimens at each stage of multiple-liquefaction tests. Firstly, specimens underwent the initial isotropical consolidation within the effective mean principal stress (EMPS) p' of 100 kPa (from Point A to Point B in Fig. 2(a)). Then, the vertical movement of the load shaft was restricted for simulating simple shear conditions. Cyclic torsional shear tests were carried out at a constant shear strain rate of 0.8%/min under an *undrained* condition and always started in the clockwise (CW) direction. The single-amplitude (SA) shear stress (τ_{SA}) was 20 kPa, and the corresponding cyclic stress ratio (CSR) is 0.2. As shown in Fig. 2(b) and (c), once the double-amplitude (DA) shear strain (γ_{DA}) reached 7.5% (Point C), sand specimens were considered liquefied. After that, the specimens stopped being sheared and started to be reconsolidated isotropically back to the original p' of 100 kPa under a *drained* condition (from Point C&C' to Point D in Fig. 2(a)), and residual shear stress inside the specimens was removed simultaneously (from Point C to Point C' in Fig. 2(b)). The strain between Point C and Point C' is negligible. The subsequent liquefaction stages (the second to sixth stages) were continued through the same procedure as described above.

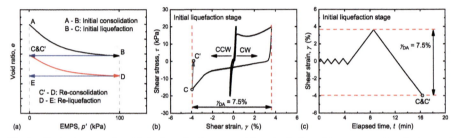

Fig. 2. Typical stage of multiple-liquefaction tests: (a) *e-p'* relation; (b) stress–strain relation; and (c) strain–time relation.

3.2 Testing Results

Figure 3 presents the cyclic liquefaction behaviors at the first four stages of a multiple-liquefaction test with an initial D_r of 46.33%. In the first liquefaction stage, the specimen was in a virgin state, without any pre-shearing history. The sample experienced small-strain deformation until it suddenly reached a double-amplitude strain γ_{DA} of 7.5% due to flow liquefaction. The p' decreased gradually as the cycle number increases due to contractive tendency of the specimen. The relations between the excess pore-water pressure ratio (EPWPR) r_u, , which is defined as the ratio of excess pore-water pressure (EPWP) Δu to the initial p', and γ in the different liquefaction stages are given in Fig. 3(c). Before a Δu of 0.8, the γ increased slowly, and after which the γ accumulated progressively, known as cyclic mobility. After experiencing a flow state and partial hardening state, the first cyclic shearing was ended in the counter-clockwise (CCW) direction.

Due to the cumulative residual strain in previous liquefaction stages, an induced anisotropy inside the specimen affects its liquefaction behaviors in the subsequent liquefaction stages (which will be verified in the DEM study). As shown in Fig. 3(a), in the second liquefaction stage, the specimen was liquefied in only half of the cycle. After the 2nd stage liquefaction, strong anisotropic fabric still remained in the sample, yet the liquefaction resistance began to recover. The butterfly loops can be observed in the stress path of the third and fourth liquefaction stages given in Fig. 3(b). During the butterfly

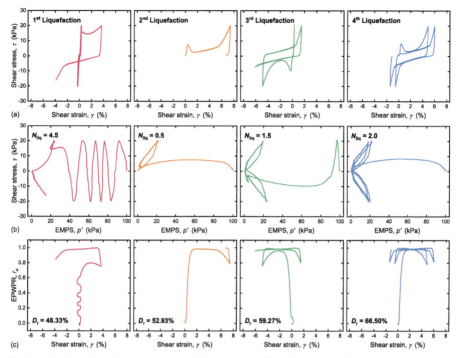

Fig. 3. Multiple-liquefaction behavior of Toyoura sand (the first to fourth liquefaction stages) (a) stress–strain relation; (b) stress path; and (c) EPWPR versus shear strain.

loop, shear strain continues to accumulate upon repeated cyclic stress, termed as "cyclic mobility" after initial liquefaction.

The number of cycles (N_{liq}) to $\gamma_{DA} = 7.5\%$ was used as an index to evaluate the liquefaction resistance of the specimen at each stage. The relation between the N_{liq} and liquefaction stage was summarized in Fig. 4(a). The liquefaction resistance is greatly reduced in the second liquefaction stage, but gradually increases as the liquefaction stage increases. The liquefaction resistance at the sixth stage is even slightly higher than the initial liquefaction resistance.

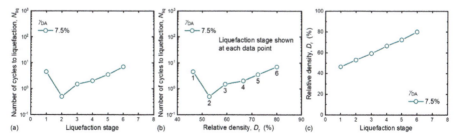

Fig. 4. Variations of (a) number of cycles to liquefaction with liquefaction stage; (b) number of cycles to liquefaction with relative density; and (c) relative density with liquefaction stage.

Figure 4(b) and (c) show the relation between the N_{liq} and the relative density D_r, and the relation between the D_r and liquefaction stage, respectively. As the liquefaction stage progresses, the D_r continuously increases from 46.3% to around 80%. However, the liquefaction resistances at the 2nd to 5th liquefaction stages are smaller than the initial liquefaction resistance, even though their D_r values are significantly higher than the initial value. At the 6th liquefaction stage, D_r is almost twice of the initial D_r, yet, the liquefaction resistance is only slightly increased. Note that increase in N_{liq} is mainly due to increase in cyclic mobility. The resistance to initial liquefaction are very weak in all cases (see Fig. 3).

4   Discrete Element Simulation

4.1   Clumped DEM Packing

In this study, DEM simulations of multiple liquefaction stages are performed using an open-source code, Yade [20]. Figure 5(a) shows a cubic DEM sample generated by 2940 clumped particles, and each clumped particle consists of 14 rigidly connected spheres. The density of each particle is 2650 kg/m^3, and Hertz-Mindlin contact model is used to describe the inter-particle contact relation, with Young's modulus of 70 GPa and Poisson's ratio of 0.15. Figure 5(b) shows some selected examples of clumped particles, which could approximately represent the irregular shape of Toyoura sand. The equivalent diameter of each clumped particle, defined as the diameter of a sphere with the same volume as the corresponding clumped particle, ranges from 0.10 mm to 0.34 m, showing a good agreement with the experimental data [12], as shown in Fig. 5(c).

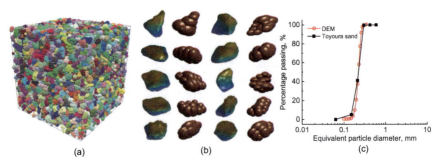

(a)   (b)   (c)

Fig. 5. (a) DEM clumped packing; (b) Examples of Toyoura sand particles models approximated by 14-sphere clumped particles; and (c) Particle size distribution of clumped particles and Toyoura sand particles [12].

4.2 Multiple Liquefaction Simulation

The clumped DEM can qualitatively to quantitatively reproduce the reliquefaction phenomenon. Figure 6 shows stress paths and stress–strain relations of sample with an initial relative density Dr = 46.2% under cyclic stress ratio CSR = 0.15 during the 1st, 2nd, 3rd, 6th and the 10th liquefaction stage. Note that a single-amplitude maximum strain of γ_{max} = 5% was reached before each reconsolidation test. During each cyclic liquefaction test and reconsolidation test, the interparticle friction coefficient is chosen as 0.35.

In the 1st liquefaction stage, it's clear from Fig. 6(a) that the effective mean stress gradually decreases upon cyclic loading in the pre-liquefaction stage. After initial liquefaction, the soil sample is unable to sustain the cyclic loading, thus, large, rapid, and uncontrollable deformation develops, as shown in Fig. 6(b). This phenomenon is referred to as flow liquefaction and often occurs in loose sands. At the 2nd liquefaction stage, although Dr increased to 51.6%, the effective mean stress decreased to zero almost in a half loading cycle (see Fig. 6c), and flow liquefaction occurred after the initial liquefaction state was reached (see Fig. 6d). Thus, liquefaction resistance N_{liq} sigificantly decreased to 0.19. From the 3rd to 10th liquefaction stage, it's clear that initial liquefaction always happens in half a cycle of loading. The relative density increased significantly to

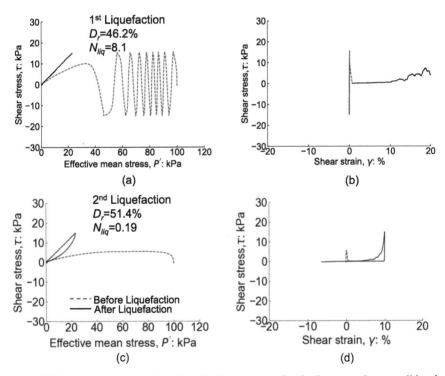

Fig. 6. DEM simulation of multiple-liquefaction response for the loose sand reconsolidated at γ_{max} = 5% at a CSR = 0.15 during the 1st, 2nd, 3rd, 6th and 10th liquefaction stages: (a), (c), (e), (g) and (i): effective stress path; (b), (d), (f), (h) and (j): stress–strain curve.

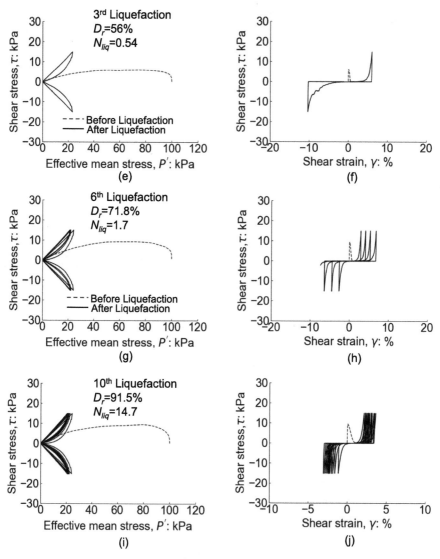

Fig. 6. continued

91.5% after nine reconsolidations, and the soil sample exhibits strong characteristics of cyclic mobility (see Fig. 6h, 6j), similar to the laboratory tests (Fig. 3).

Many DEM tests were conducted under different conditions, and the main results are summarized. Figure 7(a) shows changes in the relative density D_r during each liquefaction stage (after the sample was reconsolidated under various γ_{max}). It is obvious that a larger maximum **shear** strain γ_{max} induces a higher rate of increase in D_r after each reconsolidation test. For example, D_r increased significantly from 46.2% to 91.5% after nine times of reconsolidation under $\gamma_{max} = 5\%$; whereas under a small strain level γ_{max}

= 0.3%, the relative density only slightly increases to 54.4%. The trend of volumetric compression is qualitatively comparable to experimental tests performed on Toyoura sand by Wahyudi et al. (2015), as shown in Fig. 7(a). From this comparison, we concluded that using DEM-Clumps can correctly model the volumetric behavior during multiple liquefaction process, which is one of the key steps to obtain realistic fabrics through DEM simulation.

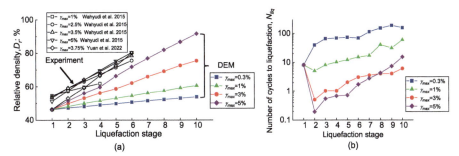

Fig. 7. (a) Relative density D_r; ; and (b) Number of cycles to liquefaction N_{liq} in multiple liquefaction stage under various maximum shear strains γ_{max} at a CSR = 0.15.

Figure 7(b) shows the effects of γ_{max} on liquefaction resistance during multiple liquefaction stage. At a small strain level ($\gamma_{max} = 0.3\%$, 1%), the liquefaction resistance drastically increases at subsequent reliquefaction stages. However, when the soil sample experienced a large shear strain (e.g. $\gamma_{max} = 3\%$, 5%), the liquefaction resistance N_{liq} drops significantly to a very small value at the 2nd stage of liquefaction, and recovers afterwards together with densification of the soil sample in the 2nd to 10th stages of liquefaction/reconsolidation. Referring back to Fig. 6, under $\gamma_{max} = 5\%$, the initial liquefaction always happens in half a cycle of loading, even when Dr increased significantly to 91.5% after nine reconsolidations. The recovery of N_{liq} is mainly due to enhanced cyclic mobility (after initial liquefaction).

4.3 Evolution of Soil Fabrics and Liquefaction Resistance

Using DEM, the coordination number Z can be adopted to quantify the load-bearing structure of the clumped packing, where Z is defined as the averaged number of contacts per particle within the packing. In addition, a scalar quantity a_c can be defined to quantify the fabric anisotropy of the contact vector [13–15], whereas a larger a_c value represents a high anisotropy in load bearing structure. Figure 8 shows the effects of γ_{max} on soil fabrics (Z, a_c) after reconsolidation during multiple liquefaction stages. As shown in Figs. 8(a) and (b), it is clear that when the soil sample experienced a small strain history (e.g. $\gamma_{max} = 0.3\%$), Z value gradually increases in subsequent liquefaction stages, while a_c value slightly reduced or remained largely unchanged. On the other hand, for large maximum shear strains (e.g. $\gamma_{max} = 3\%$ and 5%), Z decreases slightly and a_c increases markedly after the 1st reconsolidation, and then Z gradually increased while a_c was kept at a high value (e.g. when $\gamma_{max} = $ and 5%, a_c value is always greater than 0.35). The above observation demonstrates that when soil sample experienced a small strain history (e.g. $\gamma_{max} = 0.3\%$), the contact number increased with a more isotropic load-bearing structure. However, under large γ_{max}, considerable anisotropic soil fabric was developed and remained in the subsequent liquefaction stage.

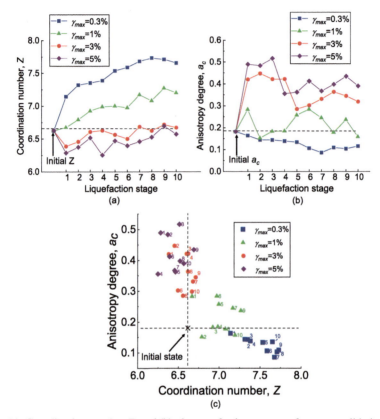

Fig. 8. (a): Coordination number Z; and (b): degree of anisotropy a_c after reconsolidation during multiple liquefaction stages; and (c): Relationship between a_c and Z under different maximum shear strain γ_{max} at a CSR $= 0.15$.

Figure 8(c) summarizes the correlation between coordination number Z and degree of anisotropy a_c of samples at the beginning of each liquefaction stages (right after reconsolidation from the previous liquefied state). Generally, Z and a_c show obvious negative correlation, i.e., an increase in a_c is associated with decrease in Z, and vice versa. These two factors jointly affect the liquefaction resistance of the sample as shown in Fig. 7.

Figures 9(a) and (b) summarize the relations between liquefaction resistance N_{liq} against Z, a_c under CSR $= 0.15$ for samples with different γ_{max}. In general, N_{liq} exhibits a good correlation with Z and a_c after reconsolidation at various maximum shear strains. To be specific, when soil samples experienced a small strain $\gamma_{max} = 0.30\%$, it can be observed from Figs. 9(a) and (b) that liquefaction resistance significantly increases during 2nd -10th liquefaction stage as compared with the 1st liquefaction resistance, which can be attributed higher Z values and lower a_c values as compared with those of the original packing ($Z = 6.62$, $a_c = 0.18$) after reconsolidation. On the other hand, when soil sample experienced a large maximum shear strains (e.g. $\gamma_{max} = 3$ and 5%), liquefaction resistance significantly reduces and this is due to substantial decrease in the number of inter-particles contacts (Z) and formation of a higher degree of fabric anisotropy after reconsolidation.

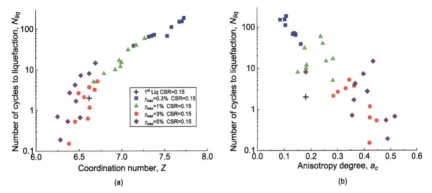

Fig. 9. Liquefaction resistance N_{liq} under CSR $= 0.15$ versus (a): coordination number Z; and (b): degree of fabric anisotropy a_c.

5 Conclusions

In this study, a new stacked-ring torsional shear apparatus was developed to study the multiple liquefaction test of saturated Toyoura sand. 3D clumped DEM simulation were also performed to study the fabric evolution of the soil during the multiple liquefaction process. The main conclusions can be summarized as follows:

1. The developed stacked-ring torsional shear apparatus can be well utilized to study the multiple liquefaction behaviors of saturated sand. More tests need to be conducted to improve the uniform distribution of stress in the sample.
2. From the DEM study, it was observed that strain history due to prior liquefaction and reconsolidation can lead to significant change in soil fabrics, which have strong correlation to the subsequent liquefaction resistance. Under large shear strains, the number of particle contact reduces, and a strongly anisotropic soil structure will be developed and retained in the subsequent liquefaction process, which dramatically reduces the liquefaction resistance. Increase in the relative density after reconsolidation can, to some extent, recover liquefaction resistance mainly through enhancing cyclic mobility in the post-liquefaction stage.

Acknowledgements. The study is supported by research grant No. 52179134 from National Natural Science Foundation of China, and grant No. 16214220 from Hong Kong Research Grants Council.

References

1. Kramer, S.L.: Geotechnical Earthquake Engineering. Pearson Education India (1996)
2. Yasuda, S., Tohno, I.: Sites of reliquefaction caused by the 1983 Nihonkai-Chubu earthquake. Soils Found. **28**(2), 61–72 (1988)
3. Wahyudi, S., Koseki, J., Sato, T., Chiaro, G.: Multiple-liquefaction behavior of sand in cyclic simple stacked-ring shear tests. Int. J. Geomech. **16**(5), C4015001 (2016)
4. Cubrinovski, M., Henderson, D., Bradley, B.: Liquefaction impacts in residential areas in the 2010–2011 Christchurch earthquakes. In: Proc., Int. Symp. on Engineering Lessons Learned from the 2011 Great East Japan Earthquake, pp. 811–824. Japan Association for Earthquake Engineering, Tokyo (2012)
5. Finn, W.D.L., Bransby, P.L., Pickering, D.J.: Effect of strain history on liquefaction of sand. J. Soil Mechanics Found. Div. **96**(6), 1917–1934 (1970)
6. Ishihara, K., Okada, S.: Effects of stress history on cyclic behavior of sand. Soils Found. **18**(4), 31–45 (1978)
7. Suzuki, T., Toki, S.: Effects of preshearing on liquefaction characteristics of saturated sand subjected to cyclic loading. Soils Found. **24**(2), 16–28 (1984)
8. Oda, M., Kawamoto, K., Suzuki, K., Fujimori, H., Sato, M.: Microstructural interpretation on reliquefaction of saturated granular soils under cyclic loading. J. Geotechnical Geoenvironmental Eng. **127**(5), 416–423 (2001)
9. Sitharam, T.G., Vinod, J.S., Ravishankar, B.V.: Post-liquefaction undrained monotonic behaviour of sands: experiments and DEM simulations. Géotechnique **59**(9), 739–749 (2009)
10. Yamada, S., Takamori, T., Sato, K.: Effects on reliquefaction resistance produced by changes in anisotropy during liquefaction. Soils Found. **50**(1), 9–25 (2010)
11. Hight, D.W., Gens, A., Symes, M.J.: The development of a new hollow cylinder apparatus for investigating the effects of principal stress rotation in soils. Géotechnique **33**(4), 355–383 (1983)
12. Fardad Amini, P., Huang, D., Wang, G., Jin, F.: Effects of strain history and induced anisotropy on reliquefaction resistance of toyoura sand. J. Geotechnical Geoenvironmental Eng. **147**(9), 04021094 (2021)
13. Wang, G., Wei, J.T.: Microstructure evolution of granular soils in cyclic mobility and post-liquefaction process. Granular Matter **18**(3), 51 (2016)
14. Wei, J.T., Huang, D.R., Wang, G.: Microscale descriptors for particle-void distribution and jamming transition in pre- and post-liquefaction of granular soils. J. Eng. Mech. **144**(8), 04018067 (2018)
15. Wei, J.T., Huang, D.R., Wang, G.: Fabric evolution of granular soils under multidirectional cyclic loading. Acta Geotech. **15**(9), 2529–2543 (2020)
16. Wang, R., Fu, P., Zhang, J.M., Dafalias, Y.F.: Fabric characteristics and processes influencing the liquefaction and re-liquefaction of sand. Soil Dyn Earthq Eng **125**, 105720 (2019)
17. Bokkisa, S., Wang, G., Huang, D., Jin, F.: Fabric evolution in post-liquefaction and re-liquefaction behavior of granular soils using 3D discrete element modeling. In: Proceedings, 7th Int Conference, on Earthquake Geotechnical Engineering, pp. 1461–1468. Rome, Italy (2019)
18. Yang, Z.: Investigation of fabric anisotropic effects on granular soil behavior. Doctoral dissertation, The Hong Kong University of Science and Technology, Hong Kong (2005)

19. Fardad Amini, P., Huang, D., Wang, G.: Dynamic properties of Toyoura sand in reliquefaction tests. Géotechnique Letters **11**(4), 1–8 (2021)
20. Šmilauer, V., et al.: Yade Documentation. In: Šmilauer V. (ed.) The Yade Project, 1st edn. http://yade-dem.org/doc/ (2010)

Liquefaction-Induced Underground Flow Failures in Gently-Inclined Fills Looser Than Critical

Takaji Kokusho[1]([⊠]), Hazarika Hemanta[2], Tomohiro Ishizawa[3], and Shin-ichiro Ishibashi[4]

[1] Chuo University (Professor Emeritus), Tokyo, Japan
koktak@ad.email.ne.jp
[2] Geo-Disaster Prevention Engineering Research Laboratory, Hemanta Hazarika, Kyushu University, Fukuoka, Japan
[3] National Research Institute for Earth Science and Disaster Resilience, Tsukuba, Japan
[4] Nihon Chiken Company Ltd., Fukuoka, Japan

Abstract. In two similar and unprecedented case histories during recent earthquakes in Hokkaido Japan, liquefied sand strangely flowed underground in gentle man-made fill slopes of a few percent gradient, leaving large surface depression behind. In both of them, a large amount of non-plastic fines was involved in loose fine fill sands. That particular sand with fines content $F_c \approx 35\%$ tested in undrained triaxial tests was found far more contractive with strain-softening and easier to flow than that of the same density deprived of fines. This strongly suggests that high fines content was the major cause of the strange flow failures because it destined the sand flowable on the contractive side of Steady State Line under sustained shear stress. Another series of cyclic simple shear tests on contractive sands with non-plastic fines under initial shear stress indicated that flow failure tends to occur in gentler slopes when the effective stress path comes across a yield line uniquely drawn from the origin on $\tau \sim \sigma_c'$ diagram irrespective of stress paths. Thus, a scenario to realize the unprecedented flow failures has been clarified based on the field observations and test results.

Keywords: Liquefaction · Flow · Gentle slope · Initial shear stress · Contractive · Nonplastic fines

1 Introduction

Since 1964 when geotechnical engineers first recognized the significant effect of earthquake-induced liquefaction on infrastructures during the two earthquakes in Niigata and Alaska, typical failure modes familiar among engineers has been sand boils, soil settlement, lateral spreading, loss of bearing capacity and associated settlement and tilting of foundations relative to adjacent soils. Laboratory tests demonstrated that the failure occurs even in the dilative side of Steady State Line though flow type failure by sustained initial shear stress cannot occur there. The most of earthquake-induced liquefaction cases experienced since then seem to belong to the type on the dilative side.

However, unprecedented but similar failures occurred twice during recent earthquakes in Hokkaido, wherein liquefied sand strangely flowed underground in very gentle man-made fill slopes of about 3% gradient, with no surface fissures and sand boils, leaving large ground depression behind with no surface disturbance.

Here, these two cases are first outlined to highlight their common features including geotechnical properties. Then, undrained triaxial test results on sand sampled from one of the sites are focused to discuss particularly on the role of non-plastic (NP) fines contained in the sand. Furthermore, the flow mechanism is discussed by revisiting torsional shear test results on similar fines-containing sand under initial shear stress again by focusing the role of NP fines in order to clarify the scenario how the unprecedented failures could occur under the given conditions.

2   Two Case Histories of Liquefaction Flow in Gentle Slopes

2.1   Sapporo Case

2018 Hokkaido Iburi-East earthquake ($M = 6.7$) incurred unprecedented liquefaction damage in residential landfill in Satozuka, Kiyota-ward, Sapporo city about 50 km distant from the epicenter (Fig. 1 (a)). The horizontal peak ground acceleration (PGA) near the site was a little lower than 0.2 g with the predominant frequency 2.5 ~ 5 Hz and the duration of major motion around 20 s [1] (Fig. 1 (b)).

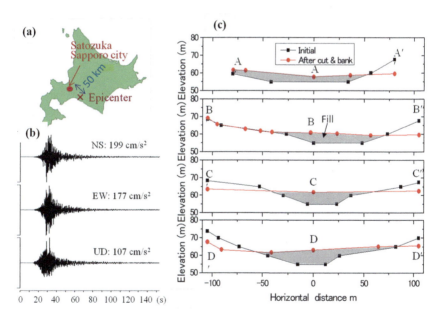

Fig. 1. 2018 Hokkaido Iburi-East earthquake and liquefied site Satozuka in Sapporo city (a), Acceleration records at K-NET Hiroshima 8 km from the site modified from NIED [1] (b), and Cross-section of Satozuka residential landfill (c).

Considerable surface depression occurred locally in a residential landfill, the cross-sections of which are depicted in Fig. 1 (c). Former landscape of the site was undulated hills and lowlands of rice field in between. In early 1980s, the residential land development started by cutting hills (Pleistocene volcanic tuff) and filling lowlands in between with thickness 5 ~ 9 m. In post-earthquake soil investigations (Sapporo City Office 2018 [2]), surprisingly loose soil layers of SPT resistance $N \approx 1$ could be found not only at the depression but also far from it, probably because the landfill was of high fines content ($F_c \approx 35\%$) and very poorly compacted originally.

Figure 2 (a) shows the present map of damaged area including streets and houses, where the depression occurred about 3 m deep maximum, 20 ~ 30 m wide and 200 m long along Point O, A, B, C, D upslope in the shaded filled area. The original surface gradient along Line OD was 2.6% on average, downslope from D to O with the water level GL.-2 ~ -3 m. Huge volume of liquefied sand underneath the depression belt flowed underground laterally to Point O. In Fig. 2 (a), water drainage lines in old rice fields before landfilling are superposed which appear to be coincidental with the depression belt, leading to a suspicion that this might somehow help the depression occur.

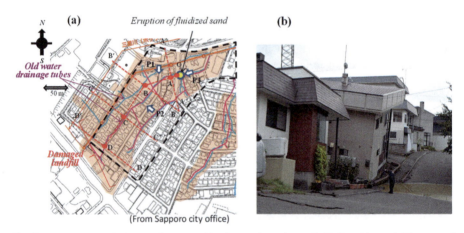

Fig. 2. Town map of damaged area with depression along OABCD (a), and Photograph liquefaction-induced depression in residential land in Satozuka (b).

A number of independent houses along the ground depression belt were significantly influenced as photographed in Fig. 2 (b) and Fig. 3 (a). Note that global building settlement occurred together with the depressed ground while no foundation settlement was observed relative to adjacent ground surface, quite different from normal liquefaction damage experienced so far. Neither fissures nor sand boils occurred in and around the depression. It was because liquefied sands were fluidized and flowed like a liquid exclusively underground all the way, and collectively ejected at a remote fill margin Point O in Fig. 2 (a) into the air and went away as photographed in Fig. 3 (b).

(a)   (b)

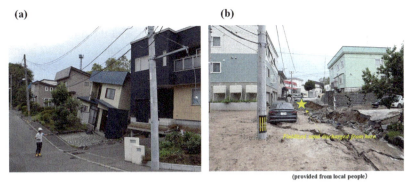

(provided from local people)

Fig. 3. (a) Large depression at P2 in Figure 2 (c), and (b) Sand ejection point where pavements were teared off violently and fluidized sand discharged and flowed downslope.

2.2 Kitami Case

A similar liquefaction case had once occurred in Tanno-cho, Kitami city in Hokkaido, during the 2003 Tokachi-oki earthquake ($M = 8.0$) as reported by Yamashita et al., (2005) [3] and Tsukamoto et al. (2009) [4], though it did not draw attention in general public because it was a rural farmland. The site was about 230 km away from the off-shore hypocenter (Fig. 4 (a)), and the maximum acceleration nearby was only 0.055 g though the major motion lasted long about one minute (NIED 2018) (Fig. 4 (b)). Analogous to the Sapporo case, the farmland, artificially filled with loose volcanic sandy soil and gently inclined (about 3%), liquefied and left a great depression behind. However, the ground surface remained intact with no lateral displacement in furrows and no large fissures and sand boils in the subsided area (Fig. 4 (c)).

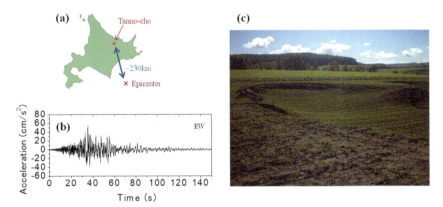

Fig. 4. Similar landfill failure in farmland in Kitami during 2003 Tokachi-oki earthquake: (a) Photo of ground depression, (b) Site and epicenter, and (c) Acceleration record K-NET Kitami (EW) 10 km from the site [1].

As shown in the air-photograph and plan view of Fig. 5 (a), (b), an area of 150 m long and 40 ~ 50 m wide subsided by 3.5 m maximum, and the downslope side was covered by sand ejecta containing NP fines ($F_c = 33\%$) spouted from four points and flowed downstream 1 km along a ditch. This again indicates considerable underground flowability of liquefied sand of maximum distance 150 m to the ejection points along an old shallow valley where the sand fill was 4 ~ 7 m thick and the water table 1 ~ 2 m deep. Unlike the Sapporo case, no drainage had been embedded in the downslope direction (Yamashita 2019 [5]). Soil investigation after the earthquake using Swedish Weight Sounding (SWS) at many points in the figure indicated that SPT N-values converted from SWS was as loose as $N = 1$ in the loosest portions both inside and outside of the depression area (Fig. 5 (c)).

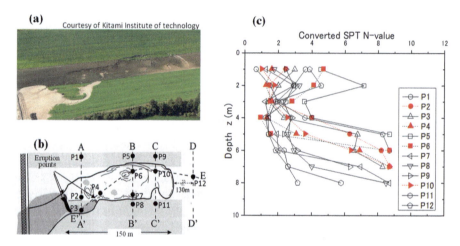

Fig. 5. Air-photograph (a) and plan view with soil investigation points (Tsukamoto et al.2009 [4]) (b), and penetration resistance versus depth (c) in landfill failure in Kitami.

2.3 Similarity of Two Cases

Fluidized sands at the above two cases sedimented in the downstream were sampled to investigate their physical properties. The grain-size curves of the two sands are illustrated in Fig. 6 (a), among which Samples **a, b, c** of the Sapporo sand show almost perfect coincidence; the average grain size was $D_{50} = 0.13$ mm, uniformity coefficient $C_u = 25 \sim 35$, fines content $F_c \approx 35\%$ and non-plastic (NP), despite that they were sampled at different sedimented area, indicating a good homogeneity of the liquefied sand.

Furthermore, the Kitami sand was very similar in physical properties; $D_{50} = 0.2$ mm, uniformity coefficient $C_u = 30$, fines content $F_c = 33\%$ and NP, all of that showed good similarity with the Sapporo sand, despite completely different locations. The specific soil particle densities were both extraordinarily low, $\rho_s = 2.26 \sim 2.28$ t/m^3 in Sapporo and $\rho_s = 2.47$ t/m^3 in Kitami, implying a mixture of large quantity of volcanic pumice (porous and easy to crush).

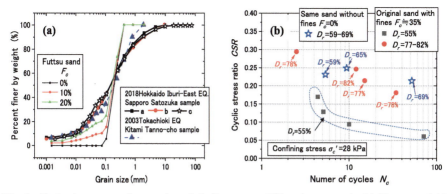

Fig. 6. Grain size curves of erupted sands in Sapporo and Kitami compared with Futtsu sand (a), and *CSR* versus *Nc* plots by undrained cyclic triaxial tests on Sapporo sand (b).

Thus, the two very analogous case histories of unprecedented liquefaction-induced underground flow failure obviously share common features; 1) Both occurred in gently inclined (≈3%) landfills of low density of N-value≈1 with shallow ground water, 2) Both accompanied neither ground surface fissures nor sand boils unlike normal liquefaction manifestations but considerable depressions of 3 m or deeper left behind where liquefied sand traveled long distance downslope and spouted out collectively, and 3) the two sands share very similar physical properties among which NP fines contained more than $F_c = 30\%$ seems to have had the remarkable effects on the flowability. Hence, the effect of NP fines is particularly focused as follows.

3 Undrained Triaxial Tests on Sapporo Sand

In order to know the effect of NP fines on the soil behavior, undrained triaxial tests have been conducted on sampled Sapporo sand. Test specimens of 5 cm in diameter and 10 cm in height were reconstituted by the moist-tamping method. Cyclic loading tests and monotonic loading tests were conducted to compare the results between the original sand and clean sand deprived of fines (smaller than 75 μm) from the original for relative densities $D_r \approx 50$ to 80% after consolidation. The maximum and minimum void ratios to determine D_r were measured by the JGS standardized method (JGS) as $e_{max} = 2.231$, $e_{min} = 1.268$ and $e_{max} = 2.351$, $e_{min} = 1.422$, for the sands with and without fines, respectively.

Figure 6 (b) depicts cyclic stress ratios (*CSR*) versus number of cycles for initial liquefaction (N_L) with solid dots obtained under effective confining stress of $\sigma_c' = 28$ kPa for the original Sapporo sand of relative densities $D_r = 55\%$ and $D_r \approx 80\%$ plotted with close symbols. CRR_{20} (*CSR* for $N_L = 20$) seems to be low (less than 0.1 for $D_r = 55\%$ and nearly 0.2 for $D_r \approx 80\%$) compared to normal sands of the same D_r-values. The open star symbols, the test results for the same sand of $D_r = 59 \sim 69\%$ but deprived of all fines, obviously exhibit higher strength than the original sand of $D_r \approx 80\%$, indicating that the sand was more liquefiable because of the high content of NP fines.

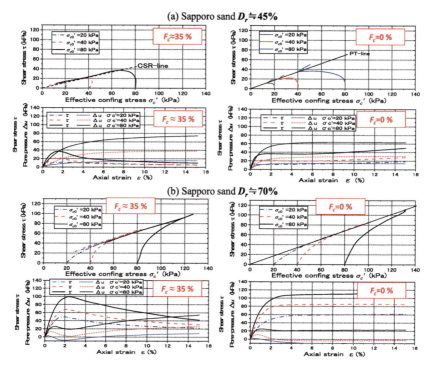

Fig. 7. Effective stress path (top) and stress versus strain curve (bottom) in undrained mon-otonic loading triaxial tests for sands of Fc≈35% and 0%: (a) Dr=45%, (b) Dr =70%.

Figure 7 shows results of undrained monotonic loading triaxial tests in terms of effective stress paths ($\tau - \sigma_c'$) and stress/pore-pressure versus strain curves ($\tau - \varepsilon$ or $\Delta u \, \varepsilon$) for the Sapporo sand of (a) $D_r = 45\%$ and (b) 70%, respectively, in 3-stepwise initial effective confining stresses $\sigma_c' = 20, 40, 80$ kPa. For (a) $D_r = 45\%$, the original sand ($F_c \approx 35\%$) on the left tends to be very contractive in the stress path undergoing monotonic decline of τ after taking peak values (when the stress paths cross a dashed CSR-line mentioned later) for all the σ_c'-values. Correspondingly, the stress -strain ($\tau - \varepsilon$) curves exhibit post-peak strain softening for higher σ_c' in particular.

In contrast, the same sand of $D_r = 45\%$ of $F_c = 0\%$ on the right behaves quite differently with a clear turning point (when the stress paths cross a solid PT line explained later) followed by strain-hardening, and no clear strain softening in the $\tau - \varepsilon$ curves is visible. For (b) $D_r = 70\%$, the stress paths initially behave dilatively with increasing τ to a peak and then take a sharp downturn in the original sand of $F_c \approx 35\%$ on the left in all the confining stresses σ_c' presumably due to collapsibility of soils rich of pumice, and the $\sigma - \varepsilon$ curves have peaks followed by temporary decline. No such post-peak decline both in the stress paths and the $\tau - \varepsilon$ curves occur in the sand of $F_c = 0\%$ on the right.

Thus, the inclusion of NP fines in the Sapporo sand considerably changes the undrained shear behavior from dilative to contractive for $D_r = 45\%$ in particular. Such a remarkable change of dilatancy due to inclusion of fines may characterize not only the

Sapporo sand but presumably the Kitami sand as well having great similarities in many respects.

As for in situ densities of the two sands, the relative density may roughly be estimated as $D_r \approx 20\%$ from SPT $N \approx 1$ for the ground depth of around 3 m (effective vertical overburden $\sigma_v' = 42$ kPa) using the empirical formula; $D_r = 21\{N/[(\sigma_v'/98) + 0.7]\}^{0.5})$ (Meyerhof, 1957 [6]). Though this formula was developed for clean sands without fines, the in situ D_r may well be assumed here to be lower than 45%, indicating that the sands were contractive enough in the light of the above test results.

4 Flow Mechanism Under Initial Shear by Torsional Tests

The effect of NP fines on the undrained cyclic loading failures under initial shear stress was previously discussed by Kokusho (2020) [7] based on a series of monotonic and cyclic torsional simple shear tests. The undrained tests were conducted on reconstituted Futtsu beach sand mixed with NP fines for isotropic effective confining stress of σ_c' = 98 kPa. The grain size curve is compared with the case history sands in Fig. 6 (a), where it is similar to the present case history sands in mean grain size though more poorly-graded.

Effective stress paths $(\tau - \sigma_c')$ and stress $\sim$ strain curves $(\tau - \gamma)$ by undrained monotonic shearing are depicted in the top and bottom of Fig. 8 (a), respectively, for D_r = 24 ~ 26%. The stress path for $F_c = 0\%$ shows dilative response despite the very low density, wherein σ_c' and τ both increase after turning direction at a point PT corresponding to Phase Transformation (Ishihara et al., 1975 [8]). The associated $\tau - \gamma$ curve undergoes strain-hardening after yielding near the PT points. In contrast, the same sand with fines content $F_c = 10\%$ and 20% exhibits contractive strain-softening behavior after taking

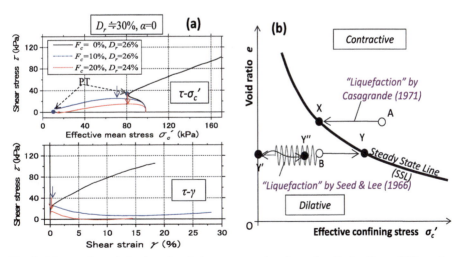

Fig. 8. Monotonic undrained torsional shear test results of varying *Fc* for *Dr* ≈ 30%, *σc'* = 98 kPa (a), and Conceptual chart of dilatancy characteristics on State Diagram with SSL and undrained loading path in contractive and dilative zones (b)

peak stress shown with the arrows in the graph. For $F_c = 20\%$ in particular, the stress path after taking the peak moves toward the origin of zero effective stress, while the $\tau - \gamma$ curve approaches to zero residual strength.

The response of the Futtsu sand by the torsional shear tests in Fig. 8 (a) may be compared to that of the Sapporo sand by the triaxial compression tests depicted in Fig. 7 despite the difference in test methods. The Sapporo sand of $F_c = 0\%$ shows dilative response with no strain-softening behavior for both $D_r \approx 70\%$ and 45%, that is similar to the Futtsu sand of $D_r = 26\%$, $F_c = 0\%$. With fines content changing from $F_c \approx 0$ to 35%, the Sapporo sand of $D_r = 45\%$ becomes contractive with strain-softening, similar to the Futtsu sand of $D_r \approx 30\%$ with changing fines content from $F_c = 0\%$ to 10% or 20%.

As a theoretical background of the above observations, a concept of State Diagram (Casagrande 1971 [9]) is shown in Fig. 8 (b) where the void ratio (e) versus effective confining stress (σ_c') plane is divided by Steady State Line (SSL) into contractive and dilative zones. A soil element on the contractive side of SSL if monotonically sheared in the undrained condition tends to be destabilized with decreasing σ'_c as Point **A** moves horizontally leftward to **X** on the SSL eventually. This corresponds seemingly to Sapporo sand of $D_r = 45\%$, $F_c \approx 35\%$ or Futtsu sand of $D_r = 24 \sim 26\%$, $F_c = 10, 20\%$. If the soil is sheared on the dilative side of SSL, the point **B** moves rightward with increasing σ'_c toward **Y** on SSL where no destabilization but strength increase occurs, corresponding to Sapporo and Futtsu sands of $F_c = 0\%$ despite the almost the same D_r-values.

Obviously, this remarkable change of dilatancy is attributed to the change of F_c from around 35% (Sapporo sand) or 10 ~ 20% (Futtsu sand) to 0%. It was actually shown experimentally by Yang et al. (2006) [10], Papadopoulou and Tika (2008) [11], and Rahman and Baki (2011) [12] that SSL tends to move down-leftward with increasing

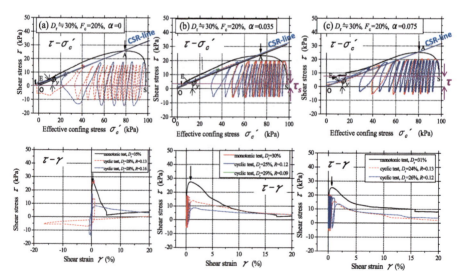

Fig. 9. Torsional shear test results of Futtsu sand in terms of $\tau \sim \sigma c'$ (top) and $\tau \sim \gamma$ (bottom) for $Dr \approx 30\%$, $Fc = 20\%$ for 3 initial shear stress ratios: (a) $\alpha = 0$, (b) $\alpha = 0.035$, (c) $\alpha = 0.075$.

F_c, changing the sand of the same density from dilative to contractive. Thus, it may well be inferred that the increase of NP fines changed the Sapporo sand from dilative to contractive in the same way as the Futtsu sand.

If sand is cyclically sheared, the point **B** even on the dilative side moves to the left due to negative dilatancy during cyclic loading unlike monotonic shearing, and reaches zero-effective stress at **Y′** as in Fig. 8 (b) under the zero-initial shear stress condition in level ground as Seed and Lee (1966) [13] first demonstrated in a cyclic triaxial liquefaction test. However, if sustained shear stress is working there, monotonic shearing starting from zero effective stress retranslates the point rightward to **Y″** reviving the shear resistance. Thus, post-liquefaction flow-type failure by the initial shear stress is difficult to occur in the dilative zone.

In view of the above-mentioned case histories, it is interesting to clarify how the sands presumably on the contractive side of SSL can actually flow in undrained cyclic loading with cyclic stress amplitude (τ_d) under various initial shear stresses (τ_s). In Fig. 9 (a), (b), (c), the torsional shear test results with initial shear stress ratio $\alpha = \tau_s/\sigma_c'$ varying in three steps $\alpha = 0, 0.035, 0.075$, respectively, are addressed [7] in terms of $\tau - \sigma_c'$ (top) and the $\tau - \gamma$ (bottom) for $D_r \approx 30\%$, $F_c = 20\%$. The $\tau - \sigma_c'$ curves for the monotonic test (solid line) indicates the flow-type failure on the contractive side of SSL with decreasing σ_c' all the way from the start (S) to the end (E), where the shear stress τ takes peak at the point A marked with an arrow. The straight line OA drawn from the origin O with the angle ϕ_y represents the CSR-line defined by Vaid & Chern (1985) [14] as the line initiating shear stress decline. The associated $\tau - \gamma$ curves take clear stress peaks corresponding to the point A followed by strain-softening behavior leading to flow failure with monotonically declining shear strength to E.

In the cyclic loading tests with two different τ_d-values, the $\tau - \sigma_c'$ curves (dashed lines) in the top diagrams in Fig. 9 undergo gradual effective stress decrease (or pore-pressure buildup) to certain points marked with * where sudden strain-softening sets off leading to flow failure to the point E. In all the charts, the symbols * are positioned on or nearby the CSR-line AO defined by the corresponding monotonic tests. This indicates

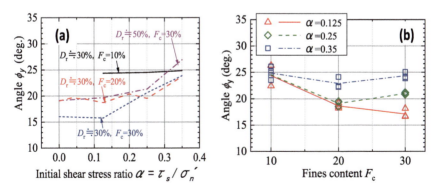

Fig. 10. Angle of flow-triggering CSR-line : (a) plotted versus initial shear stress ratio α, and (b) plotted versus fines content Fc. 0.0 0.1 0.2 0.3 0.4 10 15 20 25 30 35 Dr ≒ 30%,Fc = 20% Dr ≒ 30%,Fc = 10% Dr ≒ 50%,Fc = 30% Dr ≒ 30%,Fc = 30% Initial shear stress ratio α = τs /σn´ Angle

that the CSR-line represents a trigger of flow failure so that whenever the effective stress path comes across, flow failure starts irrespective of loading paths. In the bottom of Fig. 9, it is observed that corresponding strain exerted during cyclic loading is minor in magnitude, and the major strain is attributed to the flow failure by strain-softening. Also note that, except $\alpha = 0$ corresponding to non-flow cyclic failure in the absence of initial shear stress, the cases $\alpha = 0.035$ and 0.075 for very gentle slopes of gradient 3.5% or 7.5% belong to the stress reversal condition ($\tau_s < \tau_d$) wherein the pore pressure tends to build up faster than stress non-reversal condition ($\tau_d < \tau_s$).

In the above discussion, the initiation of flow failure is obviously governed by the CSR-line with the angle ϕ_y with respect to the σ'_c-axis on the $\tau - \sigma'_c$ diagram. In Fig. 10 (a), the angle ϕ_y determined from a set of torsional shear tests on the Futtsu sand of $D_r \approx 30\%$, 50% and $F_c = 10\%$, 20% and 30% are plotted versus the initial shear stress ratio α [7]. The ϕ_y-value tends to be essentially constant against α increasing from zero to a certain limit, thereafter followed individually by an ascending trend.

Figure 10 (b) depicts the variations of ϕ_y against F_c for $\alpha = 0.125, 0.25$ and 0.35 for the same density $D_r \approx 30\%$, wherein the averages of two to three ϕ_y-values for identical F_c-values are connected with straight lines. The ϕ_y-value tends to decrease remarkably with increasing F_c for smaller α in particular, while F_c makes little difference in ϕ_y for the high value of $\alpha = 0.35$. This indicates that flow-type failure tends to be triggered more easily with increasing F_c in gentler slopes (where the stress reversal is more likely to occur) than steeper slopes.

5 Possible Scenario of Underground Liquefaction Flow

The observations above on the flow failure mechanism of the Futtsu sand mixed with NP fines may well be applicable to the in situ sands of Sapporo and Kitami considering the similarity of their physical properties. Thus, a possible scenario of the strange liquefaction-induced failures observed twice in recent decades may be described as follows.

1) In a gently inclined fill, very loose saturated sand containing large amount of NP fines and also crushable volcanic pumice on the contractive side of SSL was cyclically sheared during earthquakes.
2) When the stress path reached the CSR-line, the sand started strain-softening flow behavior under the influence of initial shear stress.
3) Due to unlimited flowability of highly contractive sand leading to ultra-low residual shear resistance, the sand mass started to flow underground due to slope gradient of 3% in downslope direction, and thereby induced suction in its upper boundary compressing the unsaturated non-liquefied surface layer by the atmospheric pressure.
4) Thus, neither ground fissures and sand boils nor relative settlement of building foundations into adjacent soils could occur in/around the depressions unlike normal liquefaction cases.
5) The liquefied sand mass could keep flowing underground, though with low speed because of small driving force in gentle slopes, and ejected downslope from weak points of the fills.

6) The above-mentioned suction seems to cancel pore-pressure buildup, recover effective stress, and thereby interrupt the sand to flow at least near the upper boundary of the liquefied sand. Nevertheless, the major portion of liquefied sand may have been able to flow, though more investigations are certainly needed to substantiate this mechanism more quantitatively.

6 Summary

Two case histories of strange liquefaction-induced flow failures during two recent earthquakes in Hokkaido, Japan are discussed. They are characterized as gently inclined ($\approx 3\%$) landfills of low density with SPT N-value as low as unity in the extreme and shallow ground water. The failures accompanied neither ground fissures nor sand boiling but considerable depression belts of 3 m deep or more, as liquefied sand flowed underground in long distance and erupted downslope collectively. The two sands shared very similar physical properties; almost identical grain size curves, high contents of NP fines $F_c > 30\%$, and extraordinarily low soil particle density.

Undrained triaxial tests conducted on the sand sampled from the site ($F_c > 30\%$) indicated contractive behavior clearly different from that deprived of fines ($F_c = 0\%$). A series of torsional simple shear tests on similar sand demonstrated that flow failure sets off in contractive sand of high F_c when the effective stress path comes across the CSR-line starting from the origin with angle ϕ_y on the $\tau - \sigma_c'$ diagram uniquely determined for both monotonic and cyclic loading. The ϕ_y-value triggering the flow was found to decrease with increasing F_c particularly under small initial shear stress, indicating easier triggering of flow failure of high F_c-sands in gentle slopes in the stress-reversal condition.

Thus, a possible scenario of the two case histories has been constructed by focusing the significant role of NP fines in low density sands in triggering liquefaction-induced underground flow in gentle slopes leaving great depressions behind. In contrast to cyclic failures on the dilative side of SSL often observed so far, these case histories on the contractive side of SSL during earthquakes seem to have been scarce and worth documented and investigated more for mitigating similar future cases.

Acknowledgements. Ex-graduate students of Chuo University, Tokyo, Japan, Takuya Kusaka and Ryotaro Arai, who conducted a series of torsional shear tests and generated the valuable dataset incorporated in this paper are acknowledged for their great contribution. Ex-graduate students of Kyushu University Katsuya Ogo, who carried out the triaxial test of the Sapporo sand is also gratefully appreciated.

References

1. NIED: National Research Institute for Earth Science and Disaster Resilience, Tsukuba, Japan (2021)
2. Sapporo City Office: Publications for Forum to Residents on 2018 Earthquake Damage (2018)

3. Yamashita, S., Ito, Y., Hori T., Suzuki T., Murata, Y.: Geotechnical properties of liquefied volcanic soil ground by 2003 Tokachi-Oki Earthquake. Published with Open Access under the Creative Commons BY-NC Licence by IOS Press (2005)

4. Tsukamoto, Y., Ishihara, K., Kokusho, T, Hara, T., Tsutsumi, Y.: Fluidization and subsidence of gently sloped farming fields reclaimed with volcanic soils during 2003 Tokachi-oki earthquake in Japan. Geotechnical Case History Volume, Balkema, pp. 109–118 (2009)

5. Yamashita, S.: Personal Communication (2019)

6. Meyerhof, G.G.: Discussion. Proc. 4th International Conference on SMFE, Vol.3, 110 (1957)

7. Kokusho, T.: Earthquake-induced flow liquefaction in fines-containing sands under initial shear stress by lab tests and its implication in case histories. Soil Dynamics & Earthquake Engineering, Elsevier, Vol. 130 (2020)

8. Ishihara, K., Tatsuoka, F., Yasuda, S.: Undrained deformation and liquefaction of sand under cyclic stresses. Soils Found. **15**(1), 29–44 (1975)

9. Casagrande, A.: On Liquefaction Phenomena. Geotechnique, London, England, XXI(3), 197-202 (1971)

10. Yang S., Lacasse S., Sandven R.: Determination of the transitional fines content of mixtures of sand and non-plastic fines. Geotech. Test. J., ASTM **29**(2), 102–107 (2006)

11. Papadopoulou, A., Tika, T.: The effect of fines on critical state and liquefaction resistance characteristics of non-plastic silty sands. Soils & Foundations, Japanese Geotechnical Society **48**(5), 713–725 (2008)

12. Rahman, M.M., Lo, S.R., Baki, M.A.L.: Equivalent granular state parameters and undrained behavior of sand-fines mixtures. Acta Geotech. **6**, 183194 (2011)

13. Seed, H.B., Lee, K.L.: Liquefaction of saturated sands during cyclic loading. Journal of SMFD, ASCE **92**(6), 105–134 (1966)

14. Vaid, Y.P., Chern, J.C.: Cyclic and monotonic undrained response of saturated sands, Advances in the art of testing soils under cyclic conditions. Proc., ASCE Convention, Detroit, Mich., pp. 120–147 (1985)

Prediction of Site Amplification of Shallow Bedrock Sites Using Deep Neural Network Model

Duhee Park$^{(\boxtimes)}$ 🆔, Yonggook Lee🆔, Hyundong Roh🆔, and Jieun Kang🆔

Hanyang University, Seoul 04763, Korea
dpark@hanyang.ac.kr

Abstract. Site amplification models are widely used with ground prediction equations to estimate ground motion intensity measures. The time-averaged shear wave velocity of top 30 m (V_{S30}) is the primary site proxy in site amplification models. A large number of models have been developed for a range of site conditions. However, the simplified nature of all models produce large residuals compared with the computed responses. The prediction accuracy of the models can be greatly enhanced through use of machine learning technique. In this study, the outputs of nonlinear one-dimensional site response analyses are used to train the deep neural network (DNN) model. The linear and nonlinear components are separately trained. The comparisons highlight that the DNN model successfully captures the amplification characteristics of the shallow bedrock sites and produces significantly lower residual compared with the available simulation based model.

Keywords: Site amplification · Shallow bedrock · Deep neural network

1 Introduction

In geotechnical earthquake problems, one-dimensional (1D) site response analysis is widely used to estimate the local site amplification effects [1]. The site amplification factors are the important parameters for the design of the geotechnical structures. Most of researches are based on the simulation-based data and utilized the Monte Carlo (MC) simulations for the surface response spectra [2]. In this study, we performed the linear 1D site response analyses and trained all the simulation results using the deep neural network (DNN) model.

2 Site Amplification Prediction Models

2.1 Site Response Analysis

Linear analyses were performed in frequency domain using 1D site response analysis software DEEPSOIL v7 [3]. The number of analyses was 42,840 using 840 shear wave velocity (V_S) profiles and 51 motions (Fig. 1). The simulation results are used to train the proposed DNN model.

© The Author(s), under exclusive license to Springer Nature Switzerland AG 2022
L. Wang et al. (Eds.): PBD-IV 2022, GGEE 52, pp. 527–529, 2022.
https://doi.org/10.1007/978-3-031-11898-2_30

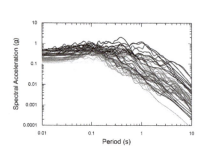

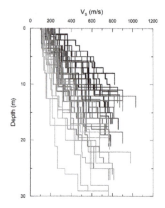

(a) Acceleration response spectra of ground motions

(b) Shear wave velocity profiles

Fig. 1. Input ground motions and shear wave velocity profiles

2.2 Proposed Deep Neural Network Model

Figure 2 shows the architecture of the proposed DNN model. The input features consist of three groups for representing the input ground motions as well as soil properties. The first part of hidden layers for each feature is created to learn and understand the input features. After processing three groups of features, all layers are merged and connected to the fully-connected hidden layers. We labeled the output as the surface spectral acceleration of a 113×1 vector. In all hidden layers, the rectified linear unit, ReLU [4], was applied as an activation function. Before the training, the weights and biases of the DNN model were initialized by Glorot uniform initializer [5] and zero, respectively. The optimization algorithm used in this study is the Adam optimizer [6] and the mean squared error (MSE) was selected as the loss function. The batch size is 128 and the training has been stopped after 1,000 epochs. The whole dataset was split into 80% of dataset as training set and 20% dataset as test set.

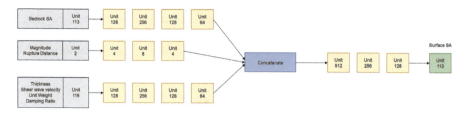

Fig. 2. An architecture of proposed DNN model.

3 Results

The results of the trained network are presented in Table 1 and Fig. 3. The MSE and MAE show that the DNN model does not overfit and the train and test show almost identical values. The DNN model can capture the simulation results of the surface response through the spectral periods. However, the results of conventional method [7] using regression analysis only can predict the median trend of the site amplifications. The DNN model shows the exceptional performance for predicting the linear site amplification.

Table 1. Comparison of MSE and MAE between train and test dataset

Analysis	Dataset	MSE	MAE
Linear	Training	0.0011	0.0226
	Test	0.0012	0.0235

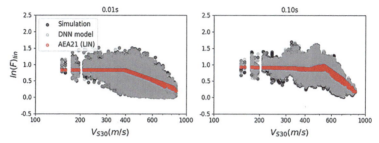

Fig. 3. Comparison of linear amplification components for the spectral periods of 0.01 s (left) and 0.1 s (right).

References

1. Hashash, Y.M., Park, D.: Non-linear one-dimensional seismic ground motion propagation in the Mississippi embayment. Eng. Geol. **62**(1–3), 185–206 (2001)
2. Rota, M., Lai, C., Strobbia, C.: Stochastic 1D site response analysis at a site in central Italy. Soil Dyn. Earthq. Eng. **31**(4), 626–639 (2011)
3. Hashash, Y., Musgrove, M., Harmon, J., et al. DEEPSOIL 7.0, user manual. University of Illinois at Urbana-Champaign (2017)
4. Nair, V., Hinton, G.E.: Rectified Linear Units Improve Restricted Boltzmann Machines. In: presented at: 27th International Conference on Machine Learning; Haifa, Israel (2010)
5. Glorot, X., Bengio, Y.: Understanding the difficulty of training deep feedforward neural networks. In: JMLR Workshop and Conference Proceedings, pp. 249–256 (2010)
6. Kingma, D.P., Adam, B.J.: A method for stochastic optimization. presented at: 3rd International Conference on Learning Representations, ICLR 2015; May 7–9; San Diego, CA, USA (2015)
7. Aaqib, M., Park, D., Adeel, M.B., Hashash, Y.M.A., Ilhan, O.: Simulation-based site amplification model for shallow bedrock sites in Korea. Earthq. Spectra **37**(3), 1900–1930 (2021)

Response of Pumice-Rich Soils to Cyclic Loading

Mark Stringer$^{(\boxtimes)}$ (iD)

University of Canterbury, Christchurch, New Zealand
mark.stringer@canterbury.ac.nz

Abstract. Soils containing pumice are frequently encountered on engineering projects in the North Island of New Zealand. The presence of pumice is known to result in different material behaviours, including the resistance to cyclic loading. In this paper, results of triaxial testing on undisturbed specimens of dense, pumice-rich soils are presented, and examined to identify the apparent effects that differing pumice content has on the observed behaviours. It is shown that significant reductions in the cyclic resistance were observed in these soils compared with expectations for hard-grained materials, but that this effect appears to be fully developed with limited amounts of pumice in the soil. It is further shown that the undrained strength is significantly reduced by increasing amounts of pumice and that typical predictions of post-cyclic reconsolidation strains are unconservative in pumice bearing materials.

Keywords: Pumice · Cyclic loading · Undisturbed testing

1 Introduction

In New Zealand, strong ground shaking associated with earthquakes forms one of the major geotechnical hazards that must be considered throughout the country. The Canterbury Earthquake Sequence (CES) of 2010–2011 served to highlight the disruptive effects of liquefaction to both structures and infrastructure in an urban setting, particularly in recent alluvial deposits and remains the focus of many research efforts. The impacts of the CES, as well as other earthquakes in the past 30 years, have also highlighted the importance of considering the exposure to earthquake-related phenomena across the country.

The North Island of New Zealand is home to a number of active volcanic fields including the Auckland Volcanic Field [1] and the Taupo Volcanic Zone (TVZ) [2]. The TVZ (shown in Fig. 1a) is a North-East trending zone extending from near the centre of the North Island of New Zealand to the Pacific Ocean. The TVZ contains both the Taupo supervolcano, as well as a number of smaller clusters, and is one of the most productive active volcanic areas on the planet. The eruptive products of the TVZ are predominantly rhyolitic [2], producing large volumes of ignimbrites (containing pumice), which are found in both welded and unwelded forms, and cover extensive areas of the central North Island, existing either in their original depositional environment, or having been re-worked and redeposited in river systems. As a result, pumice-rich soils originating

© The Author(s), under exclusive license to Springer Nature Switzerland AG 2022
L. Wang et al. (Eds.): PBD-IV 2022, GGEE 52, pp. 530–544, 2022.
https://doi.org/10.1007/978-3-031-11898-2_31

from the TVZ are frequently encountered in major engineering projects around the central North Island of New Zealand.

Pumice is a frothy foam formed by the expansion of gas in molten magma as it erupts from a volcano. As a result of their foam-type structure, pumice soil grains tend to be extremely light-weight, crushable, angular and have high surface roughness. Examples of a number of pumice grains are shown in the scanning electron microscope (SEM) images in Fig. 1b from which the particular properties of pumice grains can be appreciated. In particular, it has been shown that the crushing strength of pumice grains is typically size-dependent (larger particles are weaker) and approximately 1 order of magnitude lower than typical silica-based materials [3].

The different nature of these materials have led a number of researchers to consider the particular engineering properties of these materials. Studies have shown that pumice based soils tend to develop high friction angles [4–6], lower stiffnesses [7], and higher compressibility compared with hard grained materials. Cyclic strengths of these materials have similarly been investigated, both in studies on undisturbed and reconstituted materials [6–11], which have highlighted the differences in the development of pore pressure and strains in these materials, as well as the differences in engineering properties which can arise depending on the source of the volcanic materials.

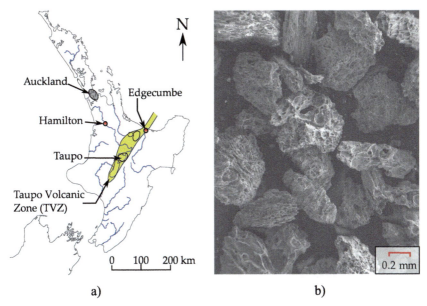

a)   b)

Fig. 1. a) Map of North Island of New Zealand showing the Taupo Volcanic Zone and sampling locations. b) Representative photo of pumice soil grains collected from Hamilton

Characterising pumice-rich soil deposits remains an important and open question. In particular, it has been shown that the penetration resistance is significantly affected by the presence of pumice [12]. Hence, there are significant challenges in understanding not just the behaviours of soils containing volcanically-derived material but also how best to characterise these soil deposits using tools available to practice. As part of ongoing work to address this question, a programme of undisturbed sampling has been undertaken at a number of locations in New Zealand, and initial comparisons of the cyclic resistances to existing simplified procedures are discussed by [7, 11]. In this paper, the behaviours of high-quality specimens obtained as part of this wider study are examined with a focus on the effect of pumice content and stress level on the observed behaviours.

2   Undisturbed Sampling and Laboratory Testing

2.1   Edgecumbe

Undisturbed sampling was undertaken close to Edgecumbe on the Rangitaiki Plains, and approximately 200 m from the present-day Rangitaiki River. At this location, [13] noted that there was only minor to moderate liquefaction during the 1987 Edgecumbe earthquake.

A conventional borelog indicated that the site consisted mainly of sands to a depth of 4 m, and gravelly sands/sandy gravels from 4 to 7 m. Pumice was noted in the log at all depths in both the sand and gravel fractions. Data from both cone penetration testing and shear wave velocity profiling are summarised in Fig. 2 along with the sampling intervals which were targeted at this site. In this paper, the results from the samples in the sandy layers will be considered. Information regarding the cyclic resistance of the deeper layers is described in [7].

Undisturbed samples were obtained using the GP-TR technique. The GP-TR sampler is similar to the Mazier Core Barrel sampler, and introduces a lubricating polymer gel that coats the exterior of the soil sample as it enters the sample liner [14]. Recovery in the layer between depths of 3-4 m was excellent, being close to the theoretical maximum. However it should be noted that the first sample (marked in red in Fig. 2) was very soft immediately after sampling, which was considered indicative of a high degree of disturbance. The samples marked 2 and 3 in this depth interval were stiff after recovery and considered to be of very good quality. The open circles shown in Fig. 2 indicate the laboratory estimates of shear wave velocity at the in-situ stress level. It can be seen that the measurements from samples 2 and 3 generally fall within 85 – 100% of the insitu measurements, while those from sample 1 are between 65 and 85%. Hence it appears that samples 2 and 3 have been recovered in very good condition using the GP-TR technique.

Following sampling, the soil samples were allowed to drain in a vertical orientation prior to uni-axial freezing from top to bottom using dry ice. Samples were kept frozen until testing in the laboratory.

Figure 3 summarises some key results from the characterisation testing performed after triaxial testing. Note that the sub-specimens from each sample are referred to using a letter (i.e. A to G), which is shown in the figure. It can be seen that the particle size distributions (PSD) of the specimens were relatively similar and confirmed that the specimens from samples 2 and 3 were clean sands with a median grain size of 0.2 mm. Pumice content was determined using a sink/swim method [15] and highlighted significant differences in the pumice content of these specimens which was not apparent before triaxial testing had been performed. As shown, specimens located between 2.85 – 3.7 m had much lower pumice contents than those between 2.1 m and 2.85 m. Relative

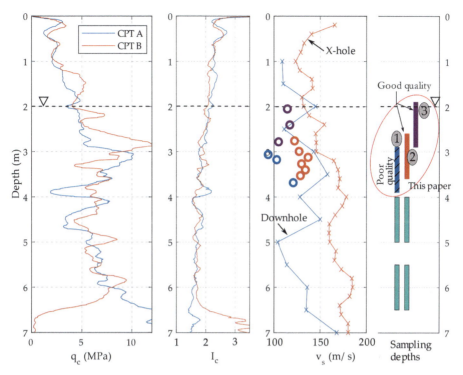

Fig. 2. CPT data, v_s profiles and sampling intervals at the Edgecumbe site.

densities are shown for the specimens after consolidation, with limiting densities based on the Japanese standard method [16]. It is apparent that specimens obtained from samples 2 and 3 were extremely dense. When examining Fig. 2, it is noticeable that the CPT trace shows an increase in the cone tip resistance between approx. 2.7 and 3.7 m in CPT A (shallower in CPT B). It is noticeable that in the post-test characterisation, the relative density of the specimens is quite similar from samples 2 and 3, while there is a marked reduction in pumice content at around 2.85 m. This suggests that the variation in cone penetration resistance may be associated with the change in pumice content, rather than relative density (i.e. [12]).

Frozen specimens of 100 mm length and 500 mm diameter for triaxial testing were prepared using a circular saw, and a core drill (cooled with calcium chloride brine) to reduce the diameter. Small slots were carved in the base and top of the specimens using a hand held drill to accommodate bender elements. Frozen specimens were allowed to defrost in the triaxial apparatus under an isotropic cell pressure of 10 kPa. After defrosting, the cell pressure was raised to 20 kPa. Specimens were saturated by percolating CO_2 and deaired water followed by raising the cell and back pressures (maintaining constant p') until a B-value in excess of 0.97 was achieved.

It was decided to test these specimens at mean effective stresses close to their expected in-situ vertical effective stress, as well as higher stresses to investigate whether larger stress levels affected the cyclic response due to crushing. Assuming a saturated unit

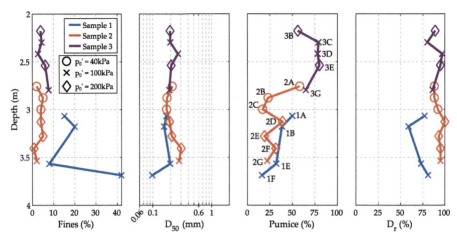

Fig. 3. Particle size, composition and relative density (D_r) of specimens from Edgecumbe site

weight of 14 kN/m^3, the in-situ effective stresses of these specimens are estimated to be in the region of 30 – 35 kPa. Hence specimens were isotropically consolidated to either 40 kPa 100 kPa or 200 kPa. Consolidated specimens were subjected to stress-controlled cyclic loading (CTX), or undrained monotonic loading (CIU). CTX tests were followed by either reconsolidation to the initial effective stress level, or a CIU test phase.

Response to Monotonic Loading The stress paths and stress-strain response measured during the monotonic testing of the Edgecumbe specimens is shown in Fig. 4. The figure shows that the specimens were dilative on shear, as would be expected for sands with high relative densities. In the interpretation of the CIU data, it is important to note that at the end of the experiments, test "3E" had positive excess pore water pressures, while tests "1E" and "2B" had developed significant negative excess pore water pressures so that in these two tests the final pore water pressure was −25 kPa and – 50 kPa. Hence it is possible that the end state of tests "1E" and "2B" are being affected by de-saturation, and that the true end of the stress path would be at higher stresses than shown. Despite this, there is a clear separation in the end strengths of these tests, with tests 1E and 2B displaying similar ultimate strengths which are approximately double those observed in test 3E. It is interesting to note that despite having the highest relative density, test 3E also had the lowest strength, counter to expectations. Referring to Fig. 3, it can be see that specimen 3E had a pumice content around 80% while 1E and 2B had pumice contents of 32% and 23% respectively. Hence it appears that increasing pumice content has the effect of significantly reducing the strength of the specimens in undrained shearing, likely as a direct result of crushing of the soil grains.

In these CIU tests, it is interesting to compare the mobilized friction angles at the maximum deviator stress, which is interpreted here as being representative of the critical state friction angle. Close examination of the data in these plots indicated that the mobilized friction angle in the tests was 39° in tests 1E and 2B, while it was 42° in test 3E. These values are relatively large compared with hard-grained sands, where the

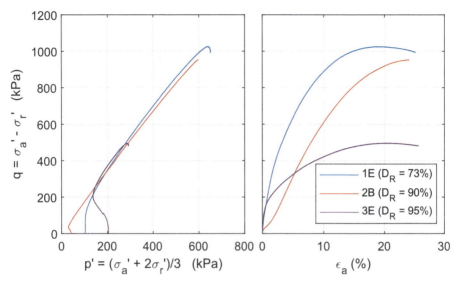

Fig. 4. Response of Edgecumbe specimens to CIU loading.

critical state friction angle might be in the region of 30-35° [17]. This data is however consistent with other studies [4–6, 12, 18], and suggests that increasing pumice content has an impact on the critical state strength of these materials.

Response to Cyclic Loading. Cyclic testing on the specimens was performed at a range of cyclic stress ratios and confining pressures. Given the range of pumice contents in the specimens (found after cyclic testing was concluded), it was expected that there would be significant differences in both the effective stress paths and the stress-strain responses. Figures 5 and 6 show the responses from two tests, which were both conducted with an initial mean effective stress of 100 kPa, similar cyclic stress ratios (CSR)s of approximately 0.375 but with a large difference in the pumice content (22% vs 80%). The responses observed in these two tests are remarkably similar in many regards, with large excess pore pressures being developed relatively early in the loading sequence, while the peak-to-peak axial strains developed gradually over continued cycling. When considering these tests, it is important to note that cyclic failure was defined when the specimens had accumulated 5% double amplitude (DA) strain. Note that while these specimens reached 5% DA strain, and had quickly generated large excess pore water pressures, the behaviour is that of cyclic mobility (consistent with the behaviour of denser sands. i.e.[19]) rather than liquefaction (referring to a collapse of the soil specimen). It is also interesting to note that no localization of strain (i.e. necking) was observed in any of the tests, and strains appeared to develop uniformly throughout the specimens.

When the cyclic responses of the tests of all specimens were compared, it was found that the main difference in the responses came with respect to the inclination of the stress paths once it began the cyclic mobility loops close to the origin. It can be observed that the specimens with high pumice content generally exhibited steeper effective stress paths after phase transformation, indicating more dilative behaviour which may be a result

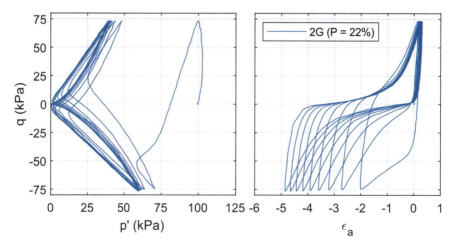

Fig. 5. Cyclic response of Edgecumbe specimen 2G with a low pumice content

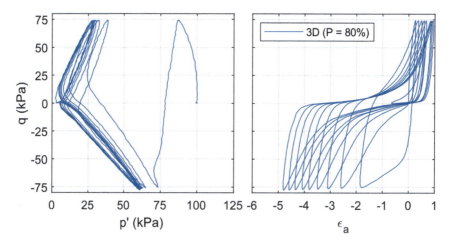

Fig. 6. Cyclic response of Edgecumbe specimen 3D with a high pumice content

of the surface roughness and angularity of these particles creating better interlocking. Across the spread of the tests, it was also noted that there appears to be a trend that the increase in excess pore water pressure ratios (r_u) at 5% DA axial strain (measured at instances when $q \approx 0$ kPa) appeared to decrease slightly with increasing initial effective stress level (Fig. 7). This same effect has been reported in testing by [5]. Surprisingly, no obvious trend appeared for excess pore water pressure ratios at the same instants when compared against the pumice content with this limited data set.

A summary of the number of cycles of uniform load required to cause the development of 5% double amplitude strain is shown in Fig. 8. As shown, there were only two tests undertaken on specimens at 40 kPa, at CSR values of 0.21 and 035. Less than 2% DA axial strain developed after more than 100 loading cycles in both tests. The

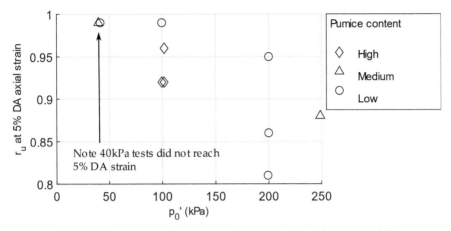

Fig. 7. Excess pore water pressures at 5% DA axial strain and q≈0 kPa

shear wave velocities measured in these specimens were generally consistent with the other specimens from samples 2 and 3 (Fig. 2), and hence did not have clear indications of cementation or other factors which would be significantly different to the other specimens. Hence further testing was not performed at this lowest stress level.

When viewing the results shown in Fig. 8, two sets of data are shown for tests performed with initial mean effective stress, $p_0' = 100$ kPa, corresponding to tests performed on specimens from sample 1 (open markers) and specimens taken from samples 2 and 3 (closed markers). It can be seen that the cyclic resistance of specimens from samples 2 and 3 is significantly higher than those from sample 1, which is due to the disturbance of sample 1 as noted earlier. The significant reduction in the cyclic resistance as a result of disturbance is a known effect (i.e. [20]) and lends confidence to the results from the remaining specimens.

It is clear that the increase in the initial mean effective stress from 40 kPa (close to the insitu vertical stresses) to 100 kPa has significantly reduced the cyclic resistance of the soil specimens. It can be seen that the specimens tested at the highest stress level (i.e. $p_0' = 200$ kPa) had more scatter in the results, but fell largely in the same band as the results from the tests performed at 100 kPa, suggesting that the change in cyclic resistance due to stress level takes place at relatively low stress levels only. The causes for this effect are unknown at this time, but may be due to the inherent crushing strength of the pumice grains being exceeded at the higher stress levels. It might be expected that the increased stress levels would lead to greater crushing taking place, leading to a more contractive response. However, it was previously noted that a reduction in r_u seemed to occur with increasing stress levels, and therefore it is speculated that an additional mechanism causing the reduction in CSR is the localized breakage of the pumice grain cell walls allowing particles to slide past one another more easily.

A final observation to note is that in these results, the pumice content of the specimens (excluding those tested at $p_0' \approx 40$ kPa) does not seem to make a large difference to the observed cyclic resistance, despite specimens having pumice contents varying approximately in the range of 20–80%. It is also noticeable that the cyclic resistance of these specimens (the CSR to cause failure in 15 cycles is approximately 0.31) is extremely low considering the relative density is around 90–100%. [20] reported that frozen samples of Niigata sand with $D_R \approx$ 80–90% was around 0.9. Hence the cyclic resistance ratio (CRR) of the test specimens reported here is very low in comparison with hard-grained material and suggests that the major influences of pumice on the cyclic resistance of the soil takes place at relatively low pumice contents.

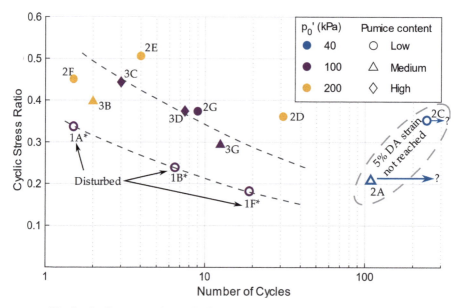

Fig. 8. Cyclic stress ratio required to cause 5% double amplitude axial strain.

2.2 Hamilton

Undisturbed sampling was conducted at a second site within the city of Hamilton (see Fig. 1) approximately 60 m from the Waikato river. CPT and shear wave data collected at this site are shown in Fig. 9 below, as well as the sampling intervals which were targeted. The targeted soils were located in a layer known as the Taupo Pumice Alluvium, which exists close to the river and is up to 30 m thick. This layer was formed as a result of a breakout flooding event after the most recent eruption of the Taupo volcano in approximately AD 250 [21].

The specimens which are discussed here were obtained from between 5 and 6 m below the ground surface, in a layer where cone resistance was approximately constant at 7 MPa. Sampling and testing procedures were identical to those at the Edgecumbe site. A comparison of the shear wave velocities measured in-situ and laboratory estimates are shown in Fig. 9, which suggests that high-quality specimens were obtained.

Post-test characterisation data is shown in Table 1. As shown, these specimens have a relative density of approximately 70%, have extremely high pumice contents, and are reasonably well graded sands with up to 20% fines and 20% gravel contents.

Table 1. Index data for Hamilton specimens

Specimen	Depth (m)	Relative Density (%)	Pumice Content (%)	D_{50} (mm)	Fines Content (%)
HE	5.34	(not meas)	90	0.288	18
HD	5.46	(not meas)	91	0.192	20
HC	5.58	79	83	0.180	15
HB	5.7	67	90	0.147	21
HA	5.82	69	70	(not meas)	(not meas)

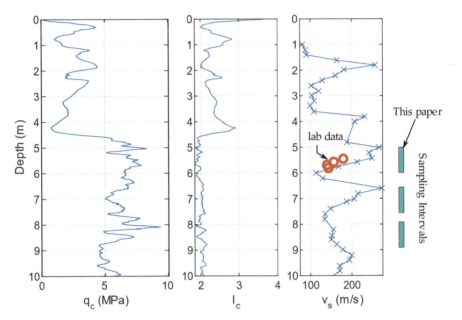

Fig. 9. CPT and v_s data for the Hamilton site

It is estimated that the vertical effective stresses of these specimens were close to 80–85 kPa. Triaxial testing of these specimens was therefore undertaken with mean effective stresses of approximately 90 kPa, and of the five specimens obtained from the sample at this depth, four were tested for cyclic resistance, and one was tested in undrained shear (CIU – specimen "HD"). In the previous section, the details of the monotonic and cyclic responses of Edgecumbe specimens were discussed. The specimens at the Hamilton site showed similar behaviours – the monotonic testing revealed an overall dilative response, as expected given its relative density, and the undrained shear strength of the specimen was comparable to that specimen "3E" in Fig. 4. Under cyclic loading, the specimens from the Hamilton site developed large excess pore water pressures early in the cyclic loading, while the deformation gradually accumulated as a result of cyclic mobility. An overall summary of the cyclic resistances of the specimens from the Hamilton site is shown in Fig. 10, along with the high quality specimens from the Edgecumbe site that were tested at 100 kPa (i.e. comparable to the testing pressures for the Hamilton specimens).

The data in Fig. 10 shows similarity between the cyclic resistances of the specimens from the Hamilton and Edgecumbe sites when tested at similar initial mean effective stress levels, despite the differences in the physical location, gradation and relative density of the specimens. The reasons for the similarity are still being investigated, but given the similarity in the undrained strengths of the materials at the two sites, it is possible that the crushing strength of pumice grains is a controlling factor in the cyclic resistance of these materials.

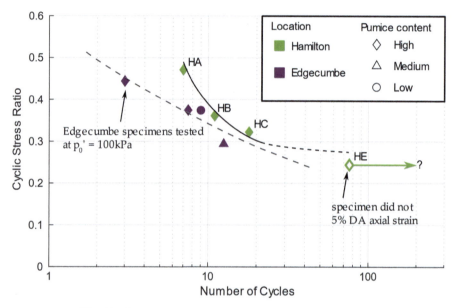

Fig. 10. Cyclic resistance of specimens from the Hamilton site

Post Cyclic Responses of Test Specimens. In the assessment of the potential consequences of liquefaction, it is common to estimate the free field settlement at a given site, often as the integration of expected one dimensional vertical strains due to the reconsolidation of the soil as excess pore water pressures dissipate. It is also the case that the amount of volumetric strain which takes place is of concern as it defines the volume of water that is flowing through the soil, potentially increasing the severity of shear induced deformations in the ground by preventing the recovery of effective stresses. The reconsolidation strains were therefore measured after a number of the cyclic tests, and have been plotted against the maximum cyclic shear strain in Fig. 11, separated into bands of relative density and pumice content. Also shown in the figure are the relationships proposed by [22] for hard-grained soils at the mid-point of each band of relative density. It should be noted that some of the data points shown correspond to tests where multiple loading series were applied (i.e. specimens were subjected to cyclic loading followed by reconsolidation, followed by additional loading etc.). In tests where additional loading sequences were applied, the specimens appeared to get stronger, and displayed none of the problems often encountered with hard-grained soils (i.e. no necking or localized loosening was observed).

The results shown indicate that the existing relationships linking volumetric strains to the maximum shear strain during cyclic loading are unconservative, especially as the relative density of the soil increases. As shown, the majority of the data belonged to the grouping of soils with relative densities higher than 80% (purple markers). Within this grouping, it can also be seen that there is a trend of increasing reconsolidation strains with pumice.

It may be recalled that there is significant variation in the initial mean effective stress of the specimens, and therefore the change in stress which the soil undergoes when it is reconsolidated. However, when the data was closely examined, it was not possible to determine a clear trend in the data which was associated with the different stress levels during testing. Hence it appears that the main factor causing the increase in volumetric strains during reconsolidation is the pumice content of the soil.

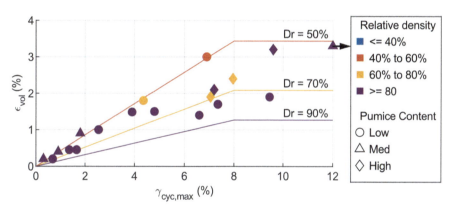

Fig. 11. Volumetric strains during reconsolidation of soil specimens. (solid lines show predicted strains based on [22])

A final aspect of the testing on the undisturbed specimens considered the strength of the soils under undrained conditions. These aspects have been previously discussed in Sect. 2.1, and the stress paths shown in Fig. 4 indicated that the strength of the specimens reduced with increasing pumice content. In addition to the monotonic testing which had previously been discussed, some of the specimens were subjected to undrained shearing after the cyclic loading in order to determine the ultimate strength of the specimens, and to verify whether the accumulated effects of cyclic shearing had any significant impact on the observed strength. The results from this phase of the testing are summarized in Fig. 12, where both the undrained strength (left plot) and the effective stress friction angle at the undrained strength (ϕ'_{cv}) are plotted against the pumice content of the specimen. It can be observed that in both cases, the pumice content is strongly affecting

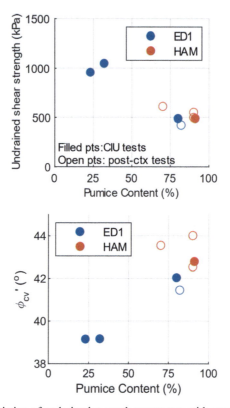

Fig. 12. Variation of undrained strength parameters with pumice content.

these strength parameters; the ultimate strength being reduced strongly as a result of the crushability of the soil grains, while the mobilized friction angle at the point when the undrained strength was attained increases with the pumice content. This second result is likely the effect of the extreme angularity and surface roughness of the pumice particles, which exists for the soil grains in both their original and broken state resulting in the large friction angles with these materials under undrained conditions.

3   Concluding Remarks

In this paper, the results from a series of triaxial tests on high quality undisturbed samples have been presented, and represent the behaviours of dense, alluvial deposits containing a wide range of pumice contents.

It has been shown that there are significant effects arising from the presence of pumice which has affected the undrained strength of the soil, the cyclic resistance and the volumetric reconsolidation strains.

A key result from the testing is that while the undrained strength reduces over a wide range of pumice contents, the cyclic resistance may be affected over a very narrow range of pumice content. In other words, the major changes in cyclic resistance of pumice bearing soils may be observed at relatively low proportions of pumice (i.e. 0% to 30%) in the soil mixture.

Significant reduction in the cyclic resistance of the specimens from Edgecumbe was observed when the initial effective stresses were raised from 40 kPa to 100 kPa and higher, suggesting that careful consideration must be given to stress level when evaluating the behaviour of these soils.

Finally, it should be acknowledged that the results from only two sets of undisturbed samples have been presented here, which have allowed some preliminary observations to be made regarding the effects of pumice content on the behaviour of undisturbed soil deposits. The behaviour of undisturbed samples are invariably affected by many factors which include fabric/structure, ageing and cementation as well as complex stress histories. While these additional aspects have not been investigated as part of this study they may have important influences on the results obtained. Further laboratory testing on pumice-rich soils, both in undisturbed and reconstituted states is therefore recommended to help clarify the effects of pumice content and stress level as well as the specific mechanisms which are responsible for these effects.

Acknowledgements. This project was supported by QuakeCoRE, a New Zealand Tertiary Education Commission-funded Centre. This is QuakeCoRE publication number 0729. The author gratefully acknowledges the collaboration with A. Prof. Rolando Orense as well as the assistance of McMillan Drilling throughout the sampling phase of the project, as well as Dr S. Rees and Dr. M.B. Asadi during the fieldwork and laboratory work.

References

1. Hopkins, J., et al.: Auckland volcanic field magmatism, volcanism, and hazard: a review. NZ J. Geol. Geophys. **64**(2–3), 213–234 (2020)
2. Wilson, C., Houghton, B., McWilliams, M., Lanphere, M., Weaver, S., Briggs, R.: Volcanic and structural evolution of Taupo Volcanic Zone, New Zealand: a review. J. Volcanol. Geoth. Res. **68**(1), 1–28 (1995)
3. Orense, R., Pender, M., Hyodo, M., Nakata, Y.: Micro-mechanical properties of crushable pumice sands. Geotechnique Letters **3**(2), 67–71 (2013)
4. Allely, B., Newland, J.: Some strength properties of weak grained sands. NZ Engineering. **14**, 107–110 (1959)
5. Miura, S., Yagi, K., Asunuma, T.: Deformation strength evaluation by crushable volcanic soils by laboratory and in-situ testing. Soils Found. **43**(4), 47–57 (2003)
6. Shimizu, M.: Geotechnical features of volcanic ash soils in Japan. In: Yanagisawa, E., Moroto, N., Mitachi, T. (eds.) Problematic Soils, pp. 907–927. Balkema, Rotterdam (1999)
7. Orense, R., Asadi, M.S., Asadi, M.B., Pender, M., Stringer, M.: Field and laboratory assessment of liquefaction potential of crushable volcanic soils. In: Silvestri, F., Moraci, N.: Earthquake geotechnical engineering for protection and development of environment and constructions, pp. 442–461. Associazone Geotechnica Italiana, Rome (2019)
8. Asadi, M.S., Asadi, M.B., Orense, R., Pender, M.: Undrained cyclic behaviour of reconstituted natural pumiceous sands. Journal of Geotechnical and Geoenvironmental Engineering. **144**(8), 040180145 (2018)
9. de Cristofaro, M., Olivares, L., Orense, R., Asadi, M.S., Netti, N.: Liquefaction of volcanic soils: undrained behavior under monotonic and cyclic loading. Journal of Geotechnical and Geoenvironmental Engineering **148**(1), 04021176 (2022)

10. Ogo, K., et al.: Fundamental study on liquefaction strength of volcanic ash soil during the 2016 Kumamoto earthquake. In: Silvestri, F, Moraci, N. (eds.) Earthquake geotechnical engineering for protection and development of environment and constructions, pp. 4187–4194. Associazone Geotechnica Italiana, Rome (2019)

11. Orense, R., Asadi, M.B., Stringer, M., Pender, M.: Evaluating liquefaction potential of pumiceous deposits through field testing: case study of the 1987 Edgecumbe Earthquake. Bulletin of the New Zealand Society of Earthquake Engineering **53**(2), 101–110 (2020)

12. Wesley, L., Meyer, V., Pranjoto, S., Pender, M., Larkin, T., Duske, G.: Engineering properties of a pumice sand. In: Vitharana, N., Colman R. (eds.) 8th Australia New Zealand Conference on Geomechanics, vol 2, pp. 901–908. Australian Geomechanics Society, Australia (1999)

13. Pender, M., Robertson, T. (eds). Edgecumbe Earthquake: Reconnaissance Report. Bulletin of the New Zealand Society of Earthquake Engineering **20**(3). 201–249 (1987)

14. Mori, K., Sakai, K.: The GP sampler: a new innovation in core sampling. Australian Geomechanics **51**(4), 131–166 (2016)

15. Stringer, M.: Separation of pumice from soil mixtures. Soils Found. **59**(4), 1073–1084 (2019)

16. Japanese Geotechnical Society (JGS): Test methods for minimum and maximum densities of sands (In Japanese). Soil Testing Standards. 136–138 (2000)

17. Bolton, M.: The strength and dilatancy of sands. Geotechnique **36**(1), 65–78 (1986)

18. Asadi, M.S., Orense, R., Asadi, M.B., Pender, M.: Post-liquefaction behaviour of natural pumice sands. Soil Dyn. Earthq. Eng. **118**, 65–74 (2019)

19. Yoshimi, Y., Tokimatsu, K., Kaneko, O., Makihara, Y.: Undrained cyclic shear strength of dense Niigata sand. Soils Found. **24**(4), 131–145 (1984)

20. Castro, G.: Liquefaction and cyclic mobility of saturated sands. Journal of the Geotechnical Engineering Division. ASCE **101**(GT6), 551–589 (1975)

21. Manville, V., White, J., Houghton, B., Wilson, C.: Paleohydrology and sedimentology of a post-1.8 ka breakout flood from intracaldera Lake Taupo, New Zealand. Geological Society of America Bulletin **111**, 1435–1447 (1999)

22. Ishihara, K., Yoshimine, M.: Evaluation of settlements in sand deposits following liquefaction during earthquakes. Soils Found. **32**(1), 173–188 (1992)

In-Situ Liquefaction Testing of a Medium Dense Sand Deposit and Comparison to Case History- and Laboratory-Based Cyclic Stress and Strain Evaluations

Armin W. Stuedlein[(✉)] [ID] and Amalesh Jana [ID]

Oregon State University, Corvallis, OR 97331, USA
armin.stuedlein@oregonstate.edu

Abstract. Observations of the dynamic loading and liquefaction response of a deep medium dense sand deposit to controlled blasting have allowed quantification of its large-volume dynamic behavior from the linear-elastic to nonlinear-inelastic regimes under *in-situ* conditions unaffected by the influence of sample disturbance or imposed laboratory boundary conditions. The dynamic response of the sand was shown to be governed by the *S-waves* resulting from blast-induced ground motions, the frequencies of which lie within the range of earthquake ground motions. The experimentally derived dataset allowed ready interpretation of the *in-situ* γ-u_e responses under the cyclic strain approach. However, practitioners have more commonly interpreted cyclic behavior using the cyclic stress-based approach; thus this paper also presents the methodology implemented to interpret the equivalent number of stress cycles, N_{eq}, and deduce the cyclic stress ratios, *CSRs*, generated during blast-induced shearing to provide a comprehensive comparison of the cyclic resistance of the *in-situ* and constant-volume, stress- and strain-controlled cyclic direct simple shear (DSS) behavior of reconstituted sand specimens consolidated to the *in-situ* vertical effective stress, relative density, and V_s. The multi-directional cyclic resistance of the *in-situ* deposit was observed to be larger than that derived from the results of the cyclic strain and stress interpretations of the uniaxial DSS test data, indicating the substantial contributions of natural soil fabric and partial drainage to liquefaction resistance during shaking. The cyclic resistance ratios, *CRRs*, computed using case history-based liquefaction triggering procedures based on the SPT, CPT, and V_s are compared to that determined from *in-situ* CRR-N_{eq} relationships considering justified, assumed slopes of the *CRR-N* curve, indicating variable degrees of accuracy relative to the *in-situ CRR*, all of which were smaller than that associated with the *in-situ* cyclic resistance.

Keywords: Liquefaction · In-situ testing · Soil dynamics

L. Wang et al. (Eds.): PBD-IV 2022, GGEE 52, pp. 545–564, 2022.
https://doi.org/10.1007/978-3-031-11898-2_32

1 Introduction

Practitioners rely upon case history- and *in-situ* penetration resistance-based liquefaction triggering procedures owing to the availability of certain subsurface exploration techniques, the results of which can be obtained in the field where evidence of liquefaction has been observed. The basis for commonly used liquefaction triggering procedures rests with the observation that those factors affecting penetration resistance (e.g., relative density, overconsolidation, cementation) also and proportionally affect cyclic resistance (Boulanger and Idriss, 2015). Such procedures provide an approximation of the cyclic resistance ratio, *CRR*, which in reality is complicated by transient, highly irregular multidirectional earthquake loading, inherent soil variability (Bong and Stuedlein, 2018; Stuedlein et al., 2021), redistribution of excess pore pressure (Dobry and Abdoun, 2015; Adamidis and Madabhushi, 2018), and the system response of stratified deposits (Cubrinovski et al., 2019). Sampling soils in an undisturbed state and subsequent laboratory element tests have pointed to the role and importance of natural soil fabric on *CRR* (e.g., Yoshimi et al., 1984). However, sampling soils in an intact, relatively undisturbed state is difficult, particularly for clean and silty sands and gravels, and the true *in-situ* drainage boundary conditions may not be well-simulated in the laboratory (Dobry and Abdoun, 2015). Numerous laboratory tests on reconstituted sand specimens have been conducted to understand how *CRR* varies with such factors as preparation technique, gradation, particle shape, among other variables; however, the major challenges associated with replicating the inherent or natural soil fabric and true stress and drainage boundary conditions in the field remains. Thus, the empirical correlations relating cyclic resistance to *in-situ* penetration resistance (e.g., Youd and Idriss, 2001; Boulanger and Idriss, 2014) and small-strain shear wave velocity, V_s, measurements (e.g., Andrus and Stokoe, 2000; Kayen et al., 2013) continue to serve the profession with the most accessible means for the evaluation liquefaction triggering potential.

Advances in the characterization of the *in-situ* coupled, cyclic shear-induced excess pore pressure and nonlinearity of soil have been made using a mobile shaker truck (Rathje et al., 2001; Cox et al., 2009; Roberts et al., 2016). Mobile shaking of instrumented test panels allows for the direct observation of the soil response to known ground motions and represents an excellent technique for filling the gaps in the understanding of dynamic soil responses. However, the success of the surface loading technique is site-specific and necessarily restricted to shallow depths (typically 4 m or less; van Ballegooy et al., 2015). Another *in-situ* dynamic testing technique, controlled blasting, has been refined to obtain *in-situ* dynamic properties and successfully implemented in the deep medium dense sand deposit (25 m depth; Jana and Stuedlein, 2021a) at the focus of this paper, and a medium-stiff silt deposit (Jana et al., 2021; Jana and Stuedlein, 2021b) at a depth of 10 m. This paper describes the experimental, instrumented Sand Array, the blast liquefaction test programs conducted, the characterization of the observed ground motions, and the framework used to determine the blast-induced shear strains, shear stresses, and the corresponding equivalent number of stress cycles. Thereafter, this paper focuses on characterization of the *in-situ* relationships between shear strain, shear stress, and excess pore pressure generation interpreted within the cyclic stress and cyclic strain frameworks, and compares the *in-situ* responses to the results of cyclic direct simple shear tests conducted on representative reconstituted sand specimens retrieved from the Sand Array.

The *in-situ* liquefaction resistance is shown to exceed that of the laboratory test specimens due to the natural soil fabric and field drainage, despite the application of multidirectional blast-induced ground motions. The paper concludes with a comparison of the *in-situ CRR* to that determined using SPT-, CPT-, and V_s-based liquefaction triggering procedures accompanied by a discussion of the influence of the assumed logarithmic slopes of the *CRR-N* curve implied by certain procedures and selected for the assessment of *in-situ* cyclic resistance. This paper demonstrates the utility of the controlled blasting technique to continue to advance our understanding of the dynamic, *in-situ*, deep liquefaction response of saturated sands.

2   Test Site and Geotechnical Conditions

The test site is situated just south of the Columbia River on the Port of Portland properties in Portland, Oregon (USA) and is underlain by soil deposits that pose potential seismic risk to the facilities owned and operated by the Port. Seismic hazards result in part from the proximity to the Portland Hills fault, located 10 km west, and the Cascadia Subduction Zone, located approximately 150 km west, of the site. Figure 1 presents the experimental layout and subsurface conditions, which consists of dredge sand and silty sand fill in the upper 5 to 6 m, underlain by a $\pm$ 2 m thick layer of native, alluvial, loose, clean sand. The next layer consists of a 5 to 6 m thick alluvial, medium stiff, clayey silt (ML and MH) deposit characterized extensively in terms of its dynamic, *in-situ* and cyclic laboratory responses by Jana and Stuedlein (2021a, 2021b). Extending below the silt layer and to the depth of the explorations lies a deep deposit of alluvial, medium dense sand forming the basis of the current study. The groundwater table depth varied from approximately 3 to 7.3 m due to seasonal fluctuations of the adjacent river and nearby pumping throughout the course of the investigation.

Over the range in depths corresponding to the *in-situ* instrumentation, globally termed the Sand Array and ranging from 23.62 to 26.53 m, the sand layer is charac-terized as medium dense, poorly-graded fine sand (SP) and fine sand with silt (SP-SM), with fines content, *FC*, varying from 3.9% to 12.1% (average *FC* = 6%). The median grain size diameter, D_{50} of the sand ranges from 0.21 to 0.28 mm, with average coeffi-cients of uniformity and curvature of 3.0 and 1.5, respectively. CPT test results indicate that I_c varies from 1.79 to 2.22, with an average I_c = 1.9 within the Sand Array. The stress-normalized equivalent clean sand tip resistance, q_{c1Ncs} (Boulanger and Idriss, 2014) of the sand layer varies from 83 to 108 with an average q_{c1Ncs} = 98. SPT- and CPT-based estimates of relative density indicated a relative density, D_r, that generally ranges from 40 (derived via SPT; Cubrinovski and Ishihara, 1999) to 47% (derived via CPT; Mayne, 2007) over the instrumented depths.

3   Sand Array and Summary of the Dynamic *In-Situ* Test Program

The location of the Sand Array within the saturated, medium dense sand deposit is shown schematically in Fig. 1, whereas the details regarding specific instruments and their geometry is presented in the Fig. 2 inset. Two strings of three triaxial 28 Hz geophone packages (TGPs) accompanied with a six-axis accelerometer gyroscope to capture static tilt and extending from inclinometer casing, were each placed within 200 mm diameter mud-rotary boreholes (B-1 and B-3; Fig. 1) and grouted in place. One borehole was used to install a full-depth inclinometer casing fitted with sondex settlement rings to capture post-shearing volumetric strain (I-1, Fig. 1). Pore pressure transducers (PPTs; Fig. 2 inset) were installed in borehole B-2 (Fig. 1). The calibration of various instruments, installation procedure, borehole deviation survey, and identification of installed TGP locations and their orientations are described in Jana et al. (2021). The Sand Array was designed to form two rectangular elements which facilitated computation of the time-varying shear modulus, shear strain, and excess pore pressure developed within the instrumented soil mass using finite element methodology (Rathje et al., 2001; Cox et al., 2009). Each TGP functioned as a node of the rectangular finite element and allowed the computation of strain using integrated particle velocities, as described below.

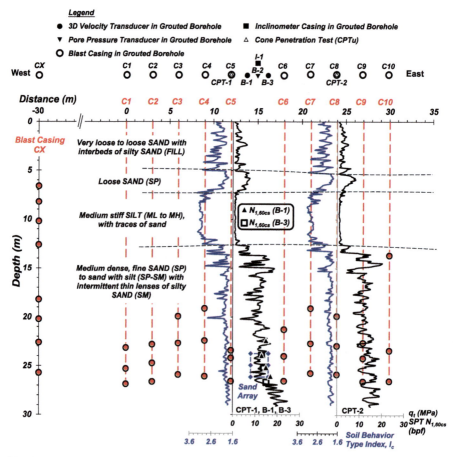

Fig. 1. Experimental layout for the Test (TBP) and Deep Blast Programs (DBP) and subsurface stratigraphy at the test site. Explosive charge locations are shown using red circular markers and the geophones comprising the Sand Array shown using purple diamond markers.

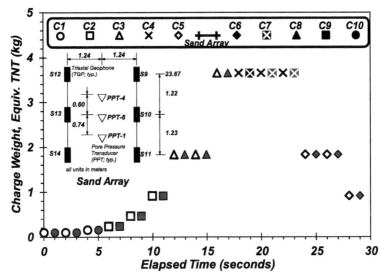

Fig. 2. Thirty-second detonation time history for the Deep Blast Program (DBP) and instruments comprising, and geometry of, the Sand Array centered at a depth of 24.9 m. Refer to Fig. 1 for location in plan and section.

Three *in-situ* dynamic tests using controlled blasting were performed in October 2018: the Test Blast (TBP), Deep Blast (DBP), and Shallow Blast Programs (SBP). The current study mainly focuses on the results from the DBP, the main goal of which was to load the Sand Array dynamically. The interested reader is referred to Jana and Stuedlein (2021a, 2021b) for additional and specific details of each of the blast events. Figure 1 presents a schematic illustrating the as-built position of each 30 charges detonated in the DBP and distributed using three charge decks within blast casings C1 through C10. Figure 2 indicates each detonation location and the sequence and charge weight (ranging from 90 g to 3.65 kg) detonated, illustrating the sequential detonation program alternating from the east to the west of the Sand Array. This alternating pattern was selected to produce reverse dynamic loading of the Sand Array (i.e., alternating the polarity of maximum shear strains for each waveform). Figures 1 and 2 shows that the DBP initiated with small charges ~ 15 m from the center of the Sand Array, which increased in weight as the distance to the array reduced to the maximum charge weight, followed by a reduction in charge weight in proximity to the array at the end of the 30 s detonation program to prevent instrument damage.

4   Characterization and Interpretation of the Blast-Induced Ground Motions

4.1   Ground Motions

Blast-induced ground motions differ somewhat from earthquake-induced ground motions, and depend upon the source-to-site distance and charge weight. Beyond the zone of rapid gas expulsion in proximity to the charge, the ground motions consist of (Jana and Stuedlein, 2021a): (1) a spherical- or cylindrical-shaped compressive shock-wave (i.e., the P-wave) emanating from the charge location, depending on the length of the charge, (2) a longitudinally-propagating, shear or S-wave producing near-field shearing (longitudinal- or x-component dominant) that is generated from the unloading of the expanding shockwave within an anisotropic soil mass, and (3) and a vertically-polarized far-field S-wave (transverse- or z-component dominant) generated at the charge location. The near- and far-field S-wave may be superimposed depending on the ratio of the wavelength and source-to-site distance (Sanchez-Salinero et al., 1986).

Figure 3a presents an example of the vertical, z, and longitudinal, x, particle velocity time histories, V_z and V_x, measured in TGP S11 within the Sand Array (Fig. 2). Velocities increased from 0.033 to 1.002 m/s with reversal in the polarity of the maximum amplitude due to the alternating ray path from the charge locations. The V_z and V_x waveforms for Blast #15 measured in TGP S11 is shown in Fig. 3b, illustrating the P-wave arrival followed by the near- and far-field S-waves. The near-field S-wave exhibits dominant particle motion in the x-direction, rather than the transverse (z) direction owing to its generation at the location of the unloading P-wave (Sanchez-Salinero et al., 1986). The magnitude of displacements, D_x and D_z, generated by the unloading of the P- and near-field S-waves can be smaller or larger that of the far-field S-wave depending on ray path distance and associated attenuation of higher frequencies which serve to reduce the near-field S-wave amplitude. The evolution of frequency, f, content of this blast-induced waveform may be visualized using the normalized Stockwell spectrum (Kramer et al., 2016) shown in Fig. 3c: the predominant frequency of the P-wave is $f_P = 825$ Hz, significantly higher than the near-field S-wave, $f_{S,nf} = 47$ Hz, which is in turn three-fold larger than the far-field S-wave, with $f_{S,ff} = 15$ Hz. Consequential displacements require low frequencies, regardless of the source of the ground motions; hence, the P-wave and its unloading is of little consequence when the charge is located sufficiently far from the point of observation. Furthermore, $f_{S,ff}$ lies within the range of typical earthquake-induced ground motions.

Figure 4a presents the Fourier spectra for the 30 blast-induced, full velocity waveforms of TGP S14z and the corresponding average normalized by their maximum Fourier amplitude. The predominant frequency of each record ranges from approximately 8 to 50 Hz, with higher frequencies occurring earlier in the blast program when the shear modulus of the sand, G, is relatively large (e.g., $G \approx G_{max}$). Note that the Fourier amplitudes for f_P are rather small in comparison to $f_{S,nf}$ and $f_{S,ff}$. The predominant f steadily reduces as the shear stiffness of the sand degrades and excess pore pressure, u_e, is triggered and accumulates (Jana et al., 2021). Figure 4b compares the average normalized Fourier spectra for V_z observed in the TGPs comprising the Sand Array during DBP; the average predominant f is 13.4 Hz, indicative of the S-wave dominance of the blast-induced ground motions. In comparison, the average frequency of the P-waves is 1,185 Hz which travel at an average V_x of 1,559 m/s.

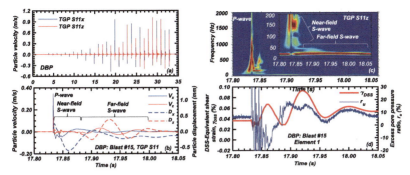

Fig. 3. Ground motions observed within the Sand Array: (a) example 30 s particle velocity time history of TGP S11, and characteristics of DBP Blast # 15 in terms of (b) particle velocity and corresponding displacement, (c) Stockwell spectrum of the of vertical component of motion (TGP S11z), and (d) variation of DSS-equivalent shear strain and excess pore pressure ratio.

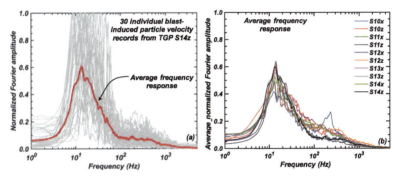

Fig. 4. Frequency content of the blast induced ground motions in the sand array during DBP: (a) normalized Fourier amplitude spectra for the 30 individual blast-induced particle velocity records for TGP S14z and the average response, and (b) average frequency response for the TGPs comprising the Sand Array.

4.2   Computation of Shear Strain

The geometry of the Sand Array allows formation of two isoparametric finite elements, termed Elements 1 and 2, which facilitate the calculation of shear strain, γ, from the integrated velocities. Element 1 is formed by TGPs S10, S11, S13, and S14, whereas Element 2 is formed by TGPs S9, S10, S12, and S13 (Fig. 2 inset). Shear strain is computed using the displacement-based finite element analyses proposed by Rathje et al. (2001) and successfully used in *in-situ* mobile shaking studies (Cox et al., 2009; Roberts et al., 2016). In this formulation, displacements D_x and D_z are used along with appropriate shape functions to deduce the 2D Cauchy strain tensor (i.e., normal strains ε_{xx}, ε_{zz}, and shear strain, γ_{xz}) corresponding to the mid-point of each element and the PPTs (Fig. 2). Although the strains computed using the selected method do not require plane waves, the majority of waveforms generated during the DBP may be assumed to pass as plane waves due to the geometry of the experiment and array (Jana and Stuedlein,

2021a, 2021b). The Cauchy strain tensor is then used to compute the octahedral shear strain, γ_{oct}:

$$\gamma_{oct} = \left(\frac{2}{3}\right)\sqrt{(\varepsilon_{xx})^2 + (-\varepsilon_{zz})^2 + (\varepsilon_{zz} - \varepsilon_{xx})^2 + 6(\frac{\gamma_{xz}}{2})^2} \tag{1}$$

which then allows comparison of the mobilized maximum *in-situ* strain with DSS test data by converting γ_{oct} to the DSS-equivalent, constant-volume shear strain, γ_{DSS}, through the imposition of plane strain boundary conditions on Eq. (1) (Cappa et al., 2017):

$$\gamma_{DSS} = \sqrt{\frac{3}{2}}\gamma_{oct} \tag{2}$$

which is strictly appropriate for 2D plane waves.

Figure 3d presents the computed γ_{DSS} and excess pore pressure ratio, r_u, time histories during Blast #15 in Element 1. The *P*-wave operates with a short wavelength of high frequency, and therefore does not provide an opportunity for movement of porewater during the period of loading and passes in a drained state (Ishihara, 1967). For the experimental conditions in this experiment, the *P*-waves could not produce relative soil movement and corresponding residual excess pore pressure, u_{er} (Martin et al., 1975; Dobry et al., 1982; Jana and Stuedlein, 2021a) within the Sand Array. Immediately following passage of the *P*-wave and coinciding with the unloading-induced near-field *S*-wave, u_e instantaneously returns to the pre-*P*-wave, ambient hydrostatic pressure (which varies over the course of a controlled blasting program as u_{er} accumulates). In contrast, the low frequency *S*-waves produced large displacements and corresponding γ_{DSS} and u_{er}; the excellent correspondence between γ_{DSS} and shear-induced u_{er} is evident in Fig. 3d. During Blast# 15, the maximum γ_{DSS}, $\gamma_{DSS,max}$, was 0.0926%, the maximum shear-induced r_u, $r_{u,max}$, was 17.6% and residual r_u, $r_{u,r}$ following the passage of the full waveform was 9.2%. The development of residual u_e, u_{er}, in the sand is associated with the gross sliding of the soil particles (Martin et al., 1975), which is associated with predominant *S*-wave during the passage of the blast pulse (Jana and Stuedlein, 2021a). Equation (2) allows direct comparison to the strain-controlled cyclic DSS test data prepared from reconstituted specimens retrieved from the Sand Array, described below.

4.3 Computation of Shear Stress and Equivalent Number of Stress Cycles

Figure 5 presents an example waveform (Blast #30) measured at TGP S11z during the DBP. Since the *P*-wave passes in a drained state, and did not produce shear strain or u_{er} owing to its high frequency, the *P*-wave was removed from each particle velocity record using a low pass 70 Hz filter (Fig. 5a). The shear stress, τ, was then calculated from the filtered waveform using the methodology proposed by Joyner and Chen (1975) assuming that the propagating seismic wave can be represented as a plane wave, which was generally the case owing to the relative scales of the body wave front and the array (Jana and Stuedlein, 2021a, 2021b), using:

$$\tau = \rho V V_s \tag{3}$$

where ρ = density, and V = particle velocity (e.g., x- and z- component). The strain dependent V_s was calculated from the crosshole response measured at each of the TGPs during the blast program for both longitudinal and transverse shear, justified by the negligible anisotropy in V_s as documented by Donaldson (2019). The corresponding cyclic (i.e., dynamic) shear stress ratio, CSR, for each component is computed by normalizing the shear stress time history by σ'_{v0}, equal to 256 kPa and 231 kPa in Elements 1 and 2, respectively. Figure 5b presents the resulting CSR time history for Blast #30, indicating a maximum CSR, $CSR_{max} = 0.13$. Individual blast pulses were assessed and reconstituted to form the full 30-s CSR time history for each element from the average resultant CSR vector (i.e., from the longitudinal and transverse particle velocities), as described further below.

An algorithm to determine the equivalent number of stress cycles, N_{eq}, from the blast-induced particle velocities adapted from that developed for earthquake ground motions by Boulanger and Idriss (2004) was scripted within *matlab*. Each positive and negative half cycle, i, of the CSR time history is counted and the absolute maximum CSR_i of each half-cycle is stored. Then, the global maximum CSR_i is stored as CSR_{max}. If the ratio of CSR_i and CSR_{max} is less than 0.1 for any given half-cycle, the script removes the corresponding CSR_i and updates i for which the ratio of CSR_i and the CSR_{max} is more than 0.1 (Boulanger and Idriss, 2004). The user can then input the reference cyclic stress ratio, CSR_{ref}, for which N_{eq} is to be calculated. Note that each CSR_{ref} is associated with a certain number of uniform cycles, N, that corresponds to a cyclic failure criterion (γ_{DSS} = 3%). In the next step, the script requires the exponent b of the power law describing the relationship between the cyclic resistance ratio, CRR, and N, $CRR = a \cdot N^{-b}$, assumed or derived from laboratory tests, as described below. Thereafter, the code computes the equivalent number of stress cycles for the blast pulse measured at a single TGP using (Boulanger and Idriss, 2004):

$$N_{eq} = \frac{1}{2} \sum_{i=1}^{i} \left[\left(\frac{CSR_i}{CSR_{ref}} \right)^{\frac{1}{b}} \right] \qquad (4)$$

Use of $CSR_{ref} = 0.12$ and $b = 0.125$, for example, results in $N_{eq} = 1.81$ for Blast #30 (Fig. 5b); this outcome represents a datapoint on the *in-situ CRR-N* curve. The full equivalent $CRR-N$ curves for a given exponent b and for Elements 1 and 2 were then developed by varying CSR_{ref} to obtain the corresponding average N_{eq} for each CSR_{ref} to obtain sufficient datapairs to construct the equivalent $CRR-N$ curve. This procedure was conducted on the average resultant CSR vector (i.e., from the longitudinal and transverse particle velocities) for each of the four TGPs comprising the element. Further, the process was conducted for $b = 0.125$ and 0.22 to evaluate the role of the logarithmic slope of the assumed $CRR-N$ power law for comparison to the DSS and case history-based cyclic resistances, as described in detail below.

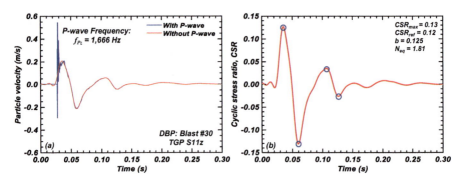

Fig. 5. Conversion of particle velocity to cyclic stress ratio for DBP Blast #30 registered in TGP S11z: (a) full waveform and low pass (70 Hz) filtered z-component particle velocity time histories, and (b) the corresponding CSR time history indicating CSR_{max} and the resulting N_{eq}.

5 *In-Situ* Cyclic Responses of the Medium Dense Sand Deposit and Comparison to Laboratory Behavior and Case History-Based Liquefaction Procedures

5.1 Stress- and Strain-Controlled, Constant-Volume Laboratory Responses

The comparison of similarities and differences between certain dynamic *in-situ* and idealized laboratory element responses is facilitated herein through stress- and strain-controlled, constant-volume, cyclic direct simple shear (DSS) tests performed on reconstituted sand specimens collected from split-spoon samples. The typical height and diameter of sand specimens were 20 and 72 mm, respectively. Dry sand was air-pluviated in the membrane-lined DSS rings and consolidated to the *in-situ* vertical effective stress, $\sigma'_{v0} = \sigma'_{vc} = 240$ kPa to achieve $D_r = 51\%$, similar to that estimated from SPT- and CPT-based measurements and necessary to obtain the same shear wave velocity, $V_s = 218$ m/s, observed using bender elements within the DSS apparatus, as that measured using downhole tests in the Sand Array (Donaldson, 2019; Jana and Stuedlein, 2021a). Following consolidation, stress- and strain-controlled cyclic DSS tests were performed using uniform sinusoids of various constant stress and strain amplitudes at a frequency of 0.1 Hz.

Figures 6a–6c present the results of stress-controlled cyclic DSS tests, indicating the shear stress-shear strain hysteresis in terms of the cyclic stress ratio, $CSR = \tau_{cyc}/\sigma'_{vc}$, development of shear strain, γ_{DSS}, and excess pore pressure ratio, r_u, with the number of loading cycles, N, and the cyclic resistance ratio, CRR with N. All of the specimens exhibited the cyclic failure, with greater CSRs resulting in greater γ_{DSS} and r_u for a given cycle of loading. For example, a specimen with $CSR = 0.185$ experienced a maximum $\gamma_{DSS} = 1.54\%$ to result in a residual excess pore pressure ratio, $r_{u,r}$, of 37.5% in the first cycle, compared to that with $CSR = 0.146$ with maximum $\gamma_{DSS} = 1.01\%$, $r_{u,r} = 30\%$ (Fig. 6b). Herein, $r_{u,r}$ is defined as the ratio of u_e at the end of each loading cycle and σ'_{vc}. These two specimens reached $\gamma_{DSS} = 3\%$ in 2.7 and 11.6 cycles which corresponded to $r_u = 70$ and 90%, and $r_u = 100\%$ at $N = 3.6$ and 12.6, respectively. Figure 6c presents the CRR-N curve developed using all of the stress-controlled cyclic

DSS test specimens, which may be represented using a power law of the form (Idriss and Boulanger, 2008; Xiao et al., 2018:

$$CRR = aN^{-b} \tag{5}$$

where N = number of uniform loading cycles to reach 3% shear strain, and a and b are the fitted coefficient and exponent, respectively, with $a = 0.20$ and $b = 0.125$.

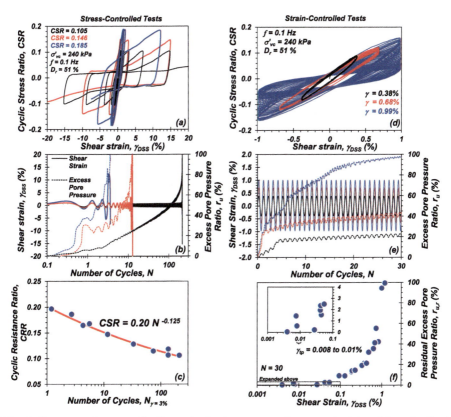

Fig. 6. Constant volume, cyclic direct simple shear test results on sand retrieved from the Sand Array: (a) sample stress-controlled hysteresis, (b) corresponding variation of shear, γ, strain and excess pore pressure ratio, r_u, with the number of cycles, N, (c) variation of cyclic stress ratio, CSR, with N for $\gamma = 3\%$, (d) sample strain-controlled hysteresis, (e) corresponding variation of γ and r_u with N, and (f) variation of γ with r_u.

Figure 6d – f present the strain-controlled cyclic DSS test results conducted using N = 30. Larger imposed γ_{DSS} led to larger degradation in shear stiffness over the course of cyclic testing and corresponding greater r_u (Fig. 6e). The loops clearly exhibit the reduction of the secant shear modulus with N. The degradation is influenced by σ'_{vc} and f, where degradation is larger if σ'_{vc} is smaller and f, is higher (Mortezaie and Vucetic, 2013). Figure 6f presents the variation of γ_{DSS} with $r_{u,r}$ for $N = 30$, indicating the threshold shear strain to trigger u_e, γ_{tp}, equal to about 0.008 to 0.01%, similar to that reported by Dobry and Abdoun (2015). This figure also indicates that $r_{u,r} \approx 100\%$ at $\gamma_{DSS} \approx 1\%$ ($N = 30$).

5.2 *In-Situ* Seismic Response Observed Within the Sand Array

The dataset developed from the Deep Blast Program provides an unprecedented view of the dynamic response of saturated, medium dense sands to blast-induced ground motions. Figure 7 presents examples of the full *CSR* (Fig. 7a and b), DSS-equivalent shear strain (Fig. 7c), and corresponding excess pore pressure (Fig. 7d) time histories observed during the DBP, indicating correspondence between the gradually increasing *CSR*s and the development of γ_{DSS} and r_u. The excess pore pressure time history displays the high frequency *P*- and *S*-wave-induced u_e (termed "dynamic," Fig. 7d) as well as a representation of the accumulated u_e, for ease of interpretation. The first several charges produced *CSR*s of approximately 0.02 or less, resulting in very little accumulated γ_{DSS} and no $u_{e,r}$ in the case of the first two and three charges for Elements 1 and Element 2. As the charge weights and corresponding *CSR*s increased, γ_{tp} was exceeded to produce non-zero $r_{u,r}$ which accumulated steadily with each additional charge. The maximum *CSR* during the DBP was approximately 0.36 measured using TGP S14 (not shown), associated with a small charge located approximately 3 m from the center of the Sand Array, compared to 0.313 and 0.223 in TGPs S10 and S11 (Fig. 7a and b).

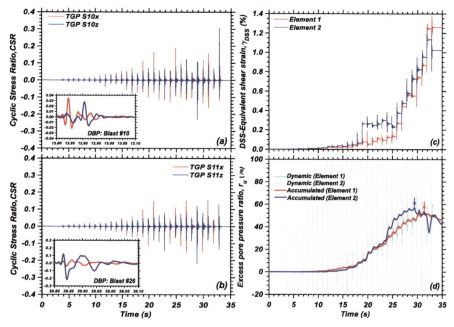

Fig. 7. Time histories of the seismic response observed within the Sand Array, including cyclic stress ratios for: (a) TGP S10, with inset showing Blast #10, (b) TGP S11, with inset showing Blast #26, (c) DSS-equivalent shear strain, and (d) excess pore pressure.

Figure 7 shows that the relationship between the non-uniform blast-induced $CSRs$ and r_u is somewhat difficult to discern. In comparison, there appears to be a direct relationship between γ_{DSS} and r_u, with increasing strains leading directly to increased accumulation of excess pore pressures. The maximum γ_{DSS}, $\gamma_{DSS,max}$, observed during the DBP was 1.371 and 1.200% for Elements 1 and 2, respectively, associated with $r_{u,max}$ and $r_{u,r}$ of 64 and 53%, and 72 and 57%, respectively. Apparent in Fig. 7d, drainage within the Sand Array initiated during Blast #26 in Element 2 and Blast #28 in Element 1 (indicated by arrows). The 3D excess pore pressure field generated by the blast program initiated hydraulic gradients that were sufficiently large to lower the u_e within the Sand Array. The partial drainage led to the development of smaller γ_{DSS} than would have been expected for a fully-undrained response, and was accompanied by a smaller reduction in the large-strain shear modulus as described in Jana and Stuedlein (2021a). Although this observation provides additional evidence for the effect of partial drainage during shaking to provide greater shearing resistance and stiffness (e.g., Adamidis et al., 2019; Ni et al., 2021), the 3D u_e field generated during the DBP differs from that anticipated under earthquake ground motions.

5.3 Comparison of *In-Situ* and Strain- and Stress-Controlled Laboratory Responses

One of the main goals of the controlled blasting test program at the Port of Portland site was to establish the *in-situ* dynamic response of natural, medium dense sands towards improving the assessment of the liquefaction hazard at the site. The main benefits of direct *in-situ* testing is that the soil response can be observed under its existing stress state within a large volume, without the detrimental effects of sample disturbance on the natural soil fabric, developed over thousands of years at this site, and without artificially-imposed boundary conditions. Side-by-side comparison of the *in-situ* and laboratory element test results serve identify similarities and differences and the role of natural soil fabric on the seismic response of liquefiable sands.

Comparison of the *in-situ* and laboratory element test results within the framework of the cyclic strain method (Dobry et al., 1982; Dobry and Abdoun, 2015) first requires pairing the $r_{u,r}$ associated with each blast pulse to the corresponding $\gamma_{DSS,max}$. The use of 1 s delays between detonations allowed for the ready identification of $r_{u,r}$, which is defined as the excess pore pressure ratio in the quiescent period following passage of any given *S-wave* and immediately prior to the arrival of the following blast pulse. Figure 8a presents the variation of $\gamma_{DSS,max}$ with $r_{u,r}$ observed during the TBP (provided here to indicate the linear- and nonlinear-elastic responses) and the DBP. The TBP and DBP indicate $r_{u,r}$ of approximately 0.1 to 0.3% for $\gamma_{DSS,max} = \gamma_{tp}$ ranging 0.008% and 0.010% during the Test and Deep Blast Programs (Fig. 8a inset), consistent with the strain-controlled DSS tests on the reconstituted specimens with $N = 30$ and the previously reported γ_{tp} summarized by Dobry and Abdoun (2015) for laboratory element, centrifuge, large-scale laboratory, and field tests with $50 \leq \sigma'_{v0} \leq 200$ kPa. *In-situ* shear strains exceeding γ_{tp} resulted in a rapid rise in the $r_{u,r}$ observed in Element 1, with a somewhat more gradual rise in Element 2 over the range of $\gamma_{tp} \leq \gamma_{DSS,max} \leq 0.3\%$. Thereafter, excess pore pressure within Element 1 may have migrated upwards into Element 2 during the remainder of the DBP and $r_{u,r}$ in Element 2 increased more rapidly with $\gamma_{DSS,max} \leq 0.7\%$. Further increases in shear strain appear to have been arrested due to the drainage established under the 3D u_e field. In contrast, the laboratory element test results indicate a similar, though more gradual, rise in $r_{u,r}$ for $\gamma_{DSS} \leq 0.8\%$; larger shear strains resulted in continued increases in $r_{u,r}$ to ~ 100% corresponding to $\gamma_{DSS} \approx 1\%$ due to the imposed constant-volume conditions. Whereas the *in-situ* and strain-controlled cyclic DSS tests on the reconstituted sand specimens consolidated to the *in-situ* σ'_{v0}, D_r, and V_s agreed well, the differences observed for larger strains could result from the effect of multi-directional loading, differences in the soil fabric, and the redistribution and upward migration of u_e, or a combination of these effects.

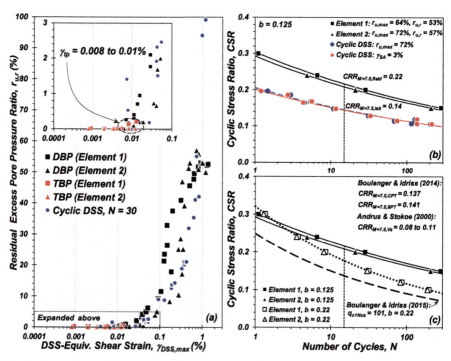

Fig. 8. Comparison of laboratory and *in-situ* test-based cyclic and dynamic responses: (a) variation of residual excess pore pressure with strain-controlled, constant-volume cyclic DSS and blast-induced DSS-equivalent maximum shear strain, (b) variation of stress-controlled, cyclic DSS and blast-induced cyclic stress ratios with the number of cycles for various cyclic performance criteria, and (c) comparison of CPT-based cyclic resistance ratios with the *in-situ*, controlled blasting-based *CRR*s for laboratory and assumed b exponents.

Figure 8b presents the comparison of the variation of *CSR* with N and N_{eq} for two liquefaction "failure" criteria and for the *in-situ* or field conditions and constant-volume, stress-controlled DSS tests at the same σ'_{v0}, D_r, and V_s. The field *CRR-N* curve shown in Fig. 8b corresponds to $b = 0.125$, equal to that determined from the stress-controlled cyclic DSS tests (Fig. 6b), for the purposes of comparison. Owing to the observed $r_{u,max} < 95$ to 100% *in-situ*, the laboratory test data was reinterpreted to compare differences in the cyclic resistance ratio, *CRR*, for two different criteria: $\gamma_{DSS} = 3\%$ and $r_{u,max} = 72\%$, the maximum observed in the Sand Array. It is recognized that liquefaction is not commonly defined using $r_{u,max} < 95$ to 100%; however, Fig. 8b shows that the differences between these two criteria are negligible, with nearly identical *CRR-N* curves over the available range in *N*. Comparison to the field-measured *in-situ CRR* with the laboratory-based cyclic resistance at $r_{u,max} = 72\%$ is therefore reasonable. Figure 8b shows that (1) the field *CRR-N* curve for Element 1 is higher than that of Element 2, owing to the lower excess pore pressures developed, and (2) the *in-situ* sand exhibits significantly larger liquefaction resistance than that of the reconstituted sand specimens at a given *N*. For example, for $N = 15$ and corresponding to $M_w = 7.5$, the *in-situ CRR* = 0.22 is ~ 50% larger than that of the laboratory specimens (*CRR* = 0.22), which has not been reduced to account for the effects of multi-directional shaking. This result agrees well with the laboratory cyclic resistances determined on frozen and cored, and unfrozen sampled sands with $D_r \approx 50\%$ reported by Yoshimi et al. (1984). Given that partial drainage occurred during the last stages of the DBP, it is likely that the *in-situ* cyclic resistance benefitted from both the effects of drainage and its natural soil fabric.

Table 1. Threshold shear strain to trigger liquefaction for $M_w = 7.5$ computed for the Sand Array within the cyclic strain framework.

Element	Blast program	Vertical effective stress $\sigma'v0$ (kPa)	*In-situ* downhole $Vs / Vs1$ (m/s)	Reference shear strain, γ_r (%)	Threshold shear strain to trigger liquefaction, γ_{cl} (%)
1	Prior to TBP	256	225 / 178	$0.042^1 / 0.066^2$	$0.139^1 / 0.071^2$
	Prior to DBP	256	192 / 151		0.087 / 0.054
2	Prior to TBP	231	218 / 177	0.040 / 0.089	0.123 / 0.053
	Prior to DBP	231	210 / 170		0.104 / 0.049

[1] γ_r derived using Darendeli (2001)
[2] γ_r derived using Menq (2003).

5.4 Comparison of the *In-Situ* Response to the Case History-Based Cyclic Strain and Stress Liquefaction Triggering Procedures

The suite of in-situ tests (SPT, CPT, V_s) and constant-volume, cyclic DSS element tests conducted on specimens prepared from split-spoon samples retrieved within the Sand Array provide the basis for comparison to the liquefaction resistance estimated using the case history-based cyclic strain and stress liquefaction procedures. The threshold shear strain to trigger liquefaction, γ_{cl}, was computed using the V_s-based cyclic strain framework updated by Dobry and Abdoun (2015) for a $M_w = 7.5$ earthquake scenario using the downhole V_s measured within the Sand Array, as summarized in Table 1. Dobry and Abdoun (2015) use the Andrus and Stokoe (2000) case history-based CRR-V_{s1} curve to link γ_{cl} to CRR. Estimation of γ_{cl} requires the use of shear modulus reduction curves, such as those proposed by Darendeli (2001) and Menq (2003), and the corresponding reference shear strain, γ_r, defined as the shear strain associated with one-half G_{max} (Table 1). Note that V_s reduced by approximately 3.5 and 10% in Elements 1 and 2, respectively, following the TBP, which was attributed to the largest magnitude of $\gamma_{DSS,max}$ imposed on Elements 1 and 2 during the TBP (Fig. 8a) and which just exceeded γ_{tp} in Element 1 (Jana and Stuedlein, 2021a). The γ_{cl} for the conditions just prior to the DBP ranges from 0.049 to 0.104% for the two elements, approximately 10 times smaller than the strains giving rise to $r_{u,max}$ (64 and 72% for Elements 1 and 2, respectively). This may be attributed to the effect of partial drainage and/or lack of correspondence of the DBP to the loading associated with a $M_w = 7.5$ earthquake.

The cyclic resistance of the medium dense sand within the Sand Array was computed using case history- and *in-situ* test-based liquefaction triggering procedures set within the cyclic stress method, including those based on the SPT and CPT (Boulanger and Idriss, 2014) and V_s (Andrus and Stokoe, 2000) for comparison to resistance determined from the DBP. The previous comparison to laboratory test results on reconstituted DSS test specimens produced the *CRR-N* exponent $b = 0.125$; however, the laboratory test results were shown to under-predict the field *CRR-N* curve as described above. Boulanger and Idriss (2004) selected $b = 0.34$ for clean sand based on the results of cyclic tests on frozen samples reported by Yoshimi et al. (1984). In reality, b can vary significantly for sands depending on the relative density, soil fabric, cementation, and other factors (Boulanger and Idriss, 2015; Verma et al., 2019; Zamani and Montoya, 2019). This prompted revision to the CPT- and SPT-based liquefaction triggering procedures (Boulanger and Idriss, 2014, Boulanger and Idriss, 2015) based on experimental data from the literature showing that the negative logarithmic slope of the *CRR-N* curve, b, tended to increase with relative density as expressed through q_{c1Ncs}. Based on the updated relationship, the exponent b corresponding the material comprising the Sand Array (with average $q_{c1Ncs} = 98$) is approximately 0.22. Figure 8c clearly shows that the implied field *CRR* depends on the assumed magnitude of b, with notable disagreement in the CRR-N curves for large N (e.g., 24% at $N = 30$). Future blast-liquefaction tests may help to provide further guidance on relationships between penetration resistance, b, and the *in-situ CRR*.

Table 2 presents *CRR*s corresponding to $b = 0.125$ (laboratory) and 0.22 (CPT) for comparison to the case history- and *in-situ* test-based *CRR*s. The case history- and *in-situ* test-based *CRR*s were computed using average penetration resistances (see Sect. 2) and method-specific overburden stress correction factors, K_σ, where applicable and for $M_w = 7.5$ (i.e., $N = 15$). Table 2 summarizes the range in *CRR*s, which indicates that the V_s-based cyclic resistances, ranging from 0.08 to 0.11, fall well below that estimated from the controlled blasting program when considering either b exponent. This is significant, as the measured $r_{u,max}$ (i.e., 72%) and $\gamma_{DSS,max}$ (1.37%) within the Sand Array was smaller than that typically attributed to liquefaction triggering within the simplified procedure (i.e., $r_{u,max} = 95$ to 100%, $\gamma_{DSS} = 3\%$). The SPT-based triggering procedure returned $CRR_{M=7.5} = 0.141$, approximately 20 and 34% smaller than the field $CRR_{M=7.5}$ for $b = 0.22$ and 0.125, respectively. Similarly, the $CRR_{M=7.5}$ computed using the CPT-based procedure returns CRRs that are 22 and 36% of that determined from the DBP and corresponding estimates of b. Differences in the available case histories used for the Andrus and Stokoe (2000) procedure and specific calibration decisions appear responsible for the differences between the V_s and penetration resistance-based CRRs summarized in Table 2. Note that the CRR calculated for the Sand Array using the CPT-based procedure implemented the mean q_{c1Ncs} corrected using the global I_c-FC correlation accompanying the Boulanger and Idriss (2014) model (n.b., SPT-based CRR used actual FC). Use of the mean FC from the split-spoon samples ($\approx 6\%$) with the CPT-based procedure returned CPT-based CRR of 0.106, closer to the V_s-based CRR and 25% lower than the comparable SPT-based procedure. Thus, consideration should be given to how FC corrections to CRR are made within the framework of the procedure-specific calibrations and what the impact could be to estimated cyclic resistance.

The comparison summarized in Table 2 underlines the observation that the *in-situ* cyclic resistance, regardless of the reasonably assumed power law parameter b, is 20% greater or more than that determined using the case history-based liquefaction triggering procedures and that *in-situ* testing can provide distinct advantages for those considering risk and mitigation of liquefaction hazards.

Table 2. Comparison of the case history-based cyclic resistance for $M_w = 7.5$ ($N = 15$) to the *in-situ* cyclic resistance for the Sand Array for the DBP.

In-Situ test method	Reference	Resistance term	Overburden stress correction, K_σ	$CRR_{M=7.5}$
SPT	Boulanger and Idriss (2014)	$N_{1,60cs} = 15$ bpf	0.90	0.141
CPT	Boulanger and Idriss (2015)	$q_{c1Ncs} = 98$[1]	0.88	0.137
V_s	Andrus and Stokoe (2000)	$V_{s1} = 151$ to 170 m/s	N/A	0.08 to 0.11
Controlled Blasting	This study	$b = 0.125$	N/A	0.215
		$b = 0.22$		0.177

[1] Using method-specific global CPT-based I_c-FC correlation; $q_{c1Ncs} = 68$ and $CRR = 0.106$ when using mean FC from split-spoon samples.

6 Concluding Remarks

The results of a blast-liquefaction test program conducted within a natural deposit of saturated, medium dense sand at a depth of 25 m are presented to demonstrate its dynamic response *in-situ* free of the effects of sample disturbance and imposed stress and drainage boundary conditions. Comparison to the response of laboratory test specimens reconstituted from samples retrieved from the same depths allow the identification of similarities and differences between the cyclic responses. The *in-situ* cyclic resistance determined through the assessment of the equivalent number of stress cycles associated with the blast-induced ground motions is compared to that computed using case history- and *in-situ* test-based liquefaction triggering procedures, allowing for direct assessment of their accuracy. The following conclusions may be drawn from this study:

1. Under the experimental conditions (e.g., charge weights, source-to-site distances) described herein, the dynamic response of the sand was controlled by *S-waves* with predominant frequencies falling within the range of earthquake ground motions. The frequency content of the *P-waves* is too large to produce appreciable displacements and strains, and therefore residual excess pore pressures, $u_{e,r}$, are controlled by low-frequency *S-waves*.
2. Dynamic shear stresses were computed for the blast-induced motions considering their largely two-dimensional nature at the experimental scale implemented, enabling quantification of the corresponding shear stress time histories. Application of the widely-used procedures to determine the equivalent number of shear stress cycles, N_{eq}, for transient earthquake ground motions were adapted to compute the corresponding N_{eq} for the blast motions.
3. Whereas the relationship between the *CSR*s and u_e was difficult to discern within the cyclic stress framework, a direct link between γ_{DSS} and $r_{u,r}$ was observed and supported previous conclusions regarding the advantages of the cyclic strain framework.
4. The multi-directional *in-situ* cyclic resistance interpreted within the cyclic stress and strain frameworks was observed to be greater than that quantified with uniaxial cyclic loading of reconstituted sand specimens consolidated to the *in-situ* σ'_{v0}, D_r, and V_s, serving to demonstrate the role of natural soil fabric and field drainage on liquefaction resistance.
5. The *in-situ* or field *CRR* depends on the assumed magnitude of the logarithmic slope of the *CRR-N* curve, b, suggesting that further refinement of the relationships between dynamic *in-situ* and cyclic laboratory test results are warranted.
6. Comparison of the field *CRR*s computed using reasonable *CRR-N* power law exponents b to those computed using case history- and *in-situ* test-based liquefaction triggering procedures indicated significant variability in their accuracy. The selected V_s-based *CRR* was up to 50% lower than that computed for the field, whereas the CPT- and SPT-based *CRR*s were 20 to 36% lower. Differences in the procedure-specific calibrations and/or available case histories appear responsible for these differences.

When coupled with the selected instrumentation scheme, controlled blasting offers an alternative method for the assessment of *in-situ* cyclic resistance unaffected by soil disturbance or imposed drainage boundary conditions, and may be readily interpreted within the cyclic stress and strain frameworks. The technique described herein can serve to further deepen the understanding of the seismic response of a wide range in geotechnical materials.

Acknowledgements. The authors gratefully acknowledge the sponsorship of this work by the Cascadia Lifelines Program (CLiP) and its members, with special thanks to sponsoring member agency Port of Portland and Tom Wharton, P.E. The authors were supported by the National Science Foundation (Grant CMMI 1663654) on this and similar work during the course of these experiments. The authors gratefully acknowledge the numerous individuals aiding in discussions and collaborative parallel work over the course of this study. The views presented herein represent solely those of the authors.

References

Adamidis, O., Madabhushi, S.P.G.: Experimental investigation of drainage during earthquake-induced liquefaction. Geotechnique **68**(8), 655–665 (2018)

Adamidis, O., Sinan, U., Anastasopoulos, I.: Effects of partial drainage on the response of Hostun sand: an experimental investigation at element level. Earthq. Geotechn. Eng. Protect. Develop. Environ. Constr. **4**, 993–1000 (2019)

Andrus, R.D., Stokoe, K.H., II.: Liquefaction resistance of soils from shear-wave velocity. J. Geotech. Geoenv. Eng. **126**(11), 1015–1025 (2000)

Bong, T., Stuedlein, A.W.: Effect of cone penetration conditioning on random field model parameters and impact of spatial variability on liquefaction-induced differential settlements. J. Geot. Geoenv. Eng. **144**(5), 04018018 (2018)

Boulanger, R.W., Idriss, I.M.: CPT and SPT based liquefaction triggering procedures. In: Report No. UCD/CGM-14/01, p. 138. UC Davis, California (2014)

Boulanger, R.W., Idriss, I.M.: Evaluating the potential for liquefaction or cyclic failure of silts and clays. In: Report No. UCD/CGM-04/01, p. 131. UC Davis, California (2004)

Boulanger, R.W., Idriss, I.M.: Magnitude scaling factors in liquefaction triggering procedures. Soil Dyn. Earthq. Eng. **79**, 296–303 (2015)

Cappa, R., Brandenberg, S.J., Lemnitzer, A.: Strains and pore pressures generated during cyclic loading of embankments on organic soil. J. Geot. Geoenv. Eng. **143**(9), 04017069 (2017)

Cox, B.R., Stokoe, K.H., II., Rathje, E.M.: An in -situ test method for evaluating the coupled pore pressure generation and nonlinear shear modulus behavior of liquefiable soils. Geotech. Test. J. **32**(1), 11–21 (2009)

Cubrinovski, M., Ishihara, K.: Empirical correlation between SPT N-value and relative density for sandy soils. Soils Found. **39**(5), 61–71 (1999)

Cubrinovski, M., Rhodes, A., Ntritsos, N., Van Ballegooy, S.: System response of liquefiable deposits. Soil Dyn. Earthq. Eng. **124**, 212–229 (2019)

Darendeli, M.B.: Development of a new family of normalized modulus reduction and material damping curves. PhD Thesis. Univ. of Texas at Austin, Austin, Texas (2001)

Dobry, R., Abdoun, T.: Cyclic shear strain needed for liquefaction triggering and assessment of overburden pressure factor K_σ. J. Geot. Geoenv. Eng. **141**(11), 04015047 (2015)

Dobry, R., Ladd, R.S., Yokel, F.Y., Chung, R.M., Powell, D.: Prediction of pore water pressure buildup and liquefaction of sands during earthquakes by the cyclic strain method. National Bureau of Standards Report 138. Gaithersburg, MD (1982)

Donaldson, A.M.: Characterization of the Small-Strain Stiffness of Soils at an In-situ Liquefaction Test Site. MS Thesis, p. 287. Oregon State University (2019)

Idriss, I.M., Boulanger, R.W.: Soil liquefaction during earthquakes. In: EERI Monograph No. 12, Earthquake Engineering Research Institute, p. 237 (2008)

Ishihara, K.: Propagation of compressional waves in a saturated soil. In: Proc. Int. Symp. Wave Prop. Dyn. Prop. Earth Mat, pp. 195–206. Univ. of New Mexico Press, Albuquerque, NM (1967)

Jana, A., Stuedlein, A.W.: Dynamic, In-situ, Nonlinear-Inelastic Response of a Deep, Medium Dense Sand Deposit. J. Geot. Geoenv. Eng. **147**(6), 04021039 (2021a)

Jana, A., Stuedlein, A.W.: Dynamic, In-situ, Nonlinear-Inelastic Response and Post-Cyclic Strength of a Plastic Silt Deposit. Can. Geot. J. **59**(1), 111–128 (2021b)

Jana, A., Donaldson, A. M., Stuedlein, A.W., Evans, T.M.: Deep, In Situ Nonlinear Dynamic Testing of Soil with Controlled Blasting: Instrumentation, Calibration, and Application to a Plastic Silt Deposit. Geotechnical Testing Journal 44(5) (2021)

Joyner, W.B., Chen, A.T.: Calculation of nonlinear ground response in earthquakes. Bull. Seis. Soc. Am. 65(5), 1315–1336 (1975)

Kayen, R., et al.: Shear-wave velocity–based probabilistic and deterministic assessment of seismic soil liquefaction potential. J. Geot. Geoenv. Eng. 139(3), 407–419 (2013)

Kramer, S.L., Sideras, S.S., Greenfield, M.W.: The timing of liquefaction and its utility in liquefaction hazard evaluation. Soil Dyn. Earthq. Eng. 91, 133–146 (2016)

Martin, G.R., Finn, W.D.L., Seed, H.B.: Fundamentals of liquefaction under cyclic loading. J. Geot. Eng. Div. 101(5), 423–438 (1975)

Mayne, P.W.: Cone penetration testing: A synthesis of highway practice. NCHRP Report, No. 368. Transportation Research Board, Washington, D.C. (2007)

Menq, F.Y.: Dynamic properties of sandy and gravelly soils. PhD Thesis. University of Texas, Austin (2003)

Mortezaie, A.R., Vucetic, M.: Effect of frequency and vertical stress on cyclic degradation and pore water pressure in clay in the NGI simple shear device. J. Geotech. Geoenv. Eng. 139(10), 1727–1737 (2013)

Ni, M., Abdoun, T., Dobry, R., El-Sekelly, W.: Effect of field drainage on seismic pore pressure buildup and kσ under high overburden pressure. J. Geot. Geoenv. Eng. 147(9), 04021088 (2021)

Rathje, E.M., Phillips, R., Chang, W.J., Stokoe, K.H. II: Evaluating Nonlinear Response In Situ, In: Proceedings of 4th Int. Conf. on Recent Adv. Geot. Earthq. Eng. Soil Dyn. San Diego, CA (2001)

Roberts, J.N., et al.: Field measurements of the variability in shear strain and pore pressure generation in Christchurch soils. In: Proc. 5th Int. Conf. on Geot. Geophys. Site Char. (2016)

Sanchez-Salinero, I., Roesset, J.M., Stokoe, K.H. II: Analytical studies of wave propagation and attenuation. In: Report, Air Force Office of Scientific Research, p. 296. Bolling AFB, Washington, D.C. (1986)

Stuedlein, A.W., Bong, T., Montgomery, J., Ching, J., Phoon, K.K.: Effect of densification on the random field model parameters of liquefiable soil and their use in estimating spatially-distributed liquefaction-induced settlement. Int. J. Geoengineering Case Histories (2021). In Press

Van Ballegooy, S., Roberts, J. N., Stokoe, K. H., Cox, B. R., Wentz, F. J., Hwang, S.: Large-scale testing of shallow ground improvements using controlled staged-loading with T-Rex. In: Proceedings of the 6th International Conference on Earthquake Geotechnical Engineering, pp. 1–4. Christchurch, New Zealand (2015)

Verma, P., Seidalinova, A., Wijewickreme, D.: Equivalent number of uniform cycles versus earthquake magnitude relationships for fine-grained soils. Can. Geot. J. 56(11), 1596–1608 (2019)

Xiao, P., Liu, H., Xiao, Y., Stuedlein, A.W., Evans, T.M., Jiang, X.: Liquefaction resistance of bio-cemented calcareous sand. Soil Dyn. Earthq. Eng. 107, 9–19 (2018)

Yoshimi, Y., Tokimatsu, K., Kaneko, O., Makihara, Y.: Undrained cyclic shear strength of a dense Niigata sand. Soils Found. 24(4), 131–145 (1984)

Youd, T.L., Idriss, I.M.: Liquefaction resistance of soils: summary report from the 1996 NCEER and 1998 NCEER/NSF workshops on evaluation of liquefaction resistance of soils. J. Geot. Geoenv. Eng. 127(4), 297–313 (2001)

Zamani, A., Montoya, B.M.: Undrained cyclic response of silty sands improved by microbial induced calcium carbonate precipitation. Soil Dyn. Earthq. Eng. 120, 436–448 (2019)

Flowslide Due to Liquefaction in Petobo During the 2018 Palu Earthquake

Mahdi Ibrahim Tanjung[1], Masyhur Irsyam[1], Andhika Sahadewa[1](✉), Susumu Iai[2],
Tetsuo Tobita[3], and Hasbullah Nawir[1]

[1] Bandung Institute of Technology, Bandung, Indonesia
sahadewa@itb.ac.id
[2] Kyoto University, Kyoto, Japan
[3] Kansai University, Osaka, Japan

Abstract. The 2018 Palu Earthquake in Indonesia triggered liquefaction that was followed by many incidents, including in massive flowsides in Petobo. In this location, flowsides with deformations reaching 1 km occurred on gentle slopes. The liquefaction mechanism causing this enormous lateral deformation is not fully understood yet. In fact, a comprehensive understanding of this phenomenon mechanism can be used to mitigate similar disasters in the future. This study describes the flow slide observations in Petobo based on geotechnical investigations, including trench, geophysical survey, CPT, boreholes, and laboratory testing. The investigation showed that the flow slide location consisted of alternating layers with high and low permeabilities. The high permeability difference during pore water pressure dissipation in the liquefied layer resulted in void redistribution in layers overlain by low permeability layer. This void ratio increase triggered a significant shear strength loss in the corresponding layer. This study also shows that differences in the soil condition and the water table depth play important roles whether a location experiences a flow slide or not. The results of this study are expected to enrich our knowledge related to liquefaction induced flow slide and can be used as a basis to evaluate flow slide potential in other locations.

Keywords: Flowslide · Liquefaction · Void redistribution · Insitu test

1 Introduction

An earthquake with moment magnitude (M_w) of 7.5 at depth of 10 km occurred in Palu, Central Sulawesi, Indonesia on September 28, 2018. The earthquake triggered large-scale liquefaction and flow slides in many areas in Palu. Baloroa, Petobo, Jono Oge, and Sibalaya are locations experienced the most disastrous flow slides (Sahadewa et al. 2019). Petobo and Balaroa are mostly residential areas, while Jono Oge and Sibalaya are dominated by agricultural lands. These four areas are located 80–110 km from the epicenter. The earthquake acceleration recorded on the surface near Balaroa, identified as a SC site class, showed peak earthquake acceleration values of 0.207 g (NS), 0.286 g (EW), and 0.341 g (UD) as reported by Okamura et al. (2019). Amongst the four areas,

© The Author(s), under exclusive license to Springer Nature Switzerland AG 2022
L. Wang et al. (Eds.): PBD-IV 2022, GGEE 52, pp. 565–579, 2022.
https://doi.org/10.1007/978-3-031-11898-2_33

Petobo experienced the largest loss of life as well as the largest flow side area of 1.43 km^2 (Mason et al. 2019). This area was a gentle slope with a shallow soil thickness prior to the earthquake.

Geographically, these 4 locations are similar, where flow slide locations were located at the end of the alluvial river descending from the hill to the valley of Palu city (Okamura et al. 2019; Mason, et al. 2019). Most of river alluvial is sandy gravel. In the location of flow slide foot, finer materials, such as clay and silt were observed between sand and gravel layers (JICA, 2019). Topographically, these 4 locations are basin where the groundwater level is shallow. In the east side of the flow slide locations, except in Balaroa, the groundwater level near the flow slide crown was also affected by the Gumbasa irrigation channel, although each location has different groundwater level characteristics.

Flow slide event due to liquefaction as large as those of Palu is very rare. The true mechanism is not clearly understood and has been an attractive subject for discussion. Several possible mechanisms include the formation of a water film due to redistribution of voids (Kokusho, 1999), shear failure due to seepage (Sento, 2004), the effect of non-cohesive content on the liquefied material that leads this material to contractive deformation (Okamura, 2019), and another mechanism is void redistributions resulting in void ratio increases in more impermeable layers that lead to significant shear strength reduction. This paper presents summary of the latest results of field investigations in Petobo and the associated laboratory tests as an effort to understand the flow slide mechanism.

2   Ground Movements

Aerial photos of Petobo before and after the 2018 Palu Earthquake were used to analyze the flow slides. A total of 49 scattered objects in the flow slide area, including houses, channels, and trees, was traced to identify the direction and magnitude of the ground surface displacement. The displacement vectors on the topography contour post-earthquake are presented in Fig. 1. The flow slide length from the crown to the toe is 2150 m. In general, the flow slide moved from the east to the west in the direction perpendicular to the contour lines.

The largest flow slide displacement is around 1000 m in the east-central part, as shown as red line in Fig. 1, about 250 m from the irrigation channel. In general, the displacement magnitude increased from the crown to the point with the greatest displacement, whereas the displacement decreased from this point to landslide foot. Petobo's deformation pattern is similar to other locations that have smaller displacements, such as one in between Petobo and Jono Oge (Bradley et al. 2019). Figure 2 shows the relationship between displacement (d) and distance to the landslide crown (s) in Petobo. This figure shows that the relationship at s values ranging from 250 m to the flow slide foot is linear this shows that the strain occurring in the compressed crust is relatively the same throughout the compression area. The strain magnitude is probably influenced by the thickness of the flow slide crust. Additionally, the point experiencing the largest movement was predicted as the initial position of the flow slide, which was followed by landslides moving to the irrigation channel.

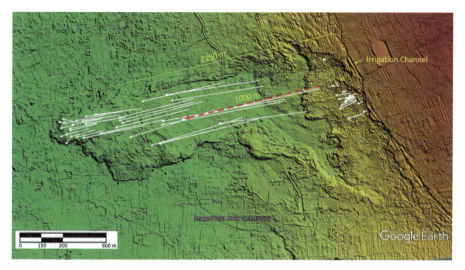

Fig. 1. Displacement vectors and contours after flowslide

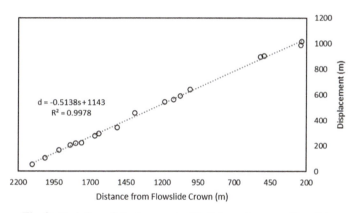

Fig. 2. Variation of displacement with distance to crown flowslide

3   Identification of Liquefied Layer

Soil profile and groundwater table depth are keys to understand landslide mechanism due to liquefaction. Therefore, 2 trenches were excavated on the east side of flow slide area in Petobo. On the trench wall, several samples were taken for laboratory testing, particularly in the layers that were considered liquefied and the impermeable layers which became the capping layers. A series of Multi-channel Analysis of Surface Wave (MASW) tests was performed nearby the trenches to evaluate the variation of shear wave velocity (V_s) with depth. In addition, secondary data from previous field investigations was collected, including 7 borehole points by JICA (2019) (Fig. 5) and 9 Cone Penetration Test (CPT) points by PT. Promisco (Figs. 6 and 7). The location of all geotechnical investigations in Petobo presented in Fig. 3.

Fig. 3. Map of the distribution of geotechnical investigation in Petobo

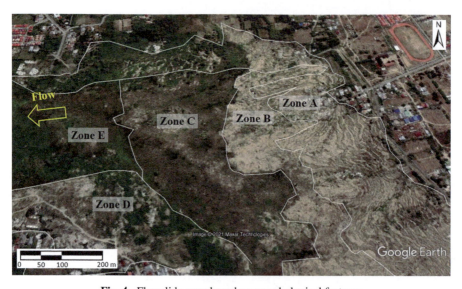

Fig. 4. Flowslide zone based on morphological features

The flow slide area generally can be divided into 5 zones based on the morphological appearance post-earthquake (Fig. 4). Zone A is an area that experienced widening and overturning. Zone B is an area that liquefaction was more pronounce resulting flatter morphology with only few soil blocks left intact. Zone C is the foundation that became the slip plane whose surface was then partially overlain by Zone B. Zone D is an area consisting of deposited materials to the south of flow slide which was highly liquefied post-earthquake. Zone E has similar deposited material to Zone D. But Zone E experienced a much larger runout distance than Zone D. Zone C and Zone E were very wet. Water started to seep at the boundary between Zone B and zone C. Subsequently, the water accumulated in Zone E which has very flat terrain.

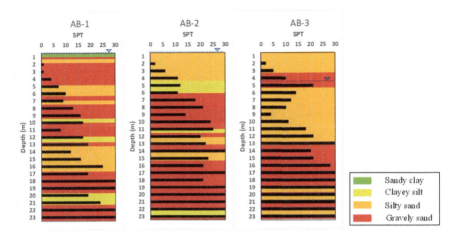

Fig. 5. Soil layer profile and SPT in the middle of the flow slide (JICA, 2019)

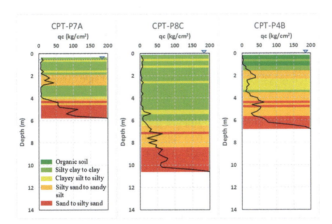

Fig. 6. Soil layer profile and cone resistance in the middle of the flowslide area

Field investigation results were studied to evaluate the soil profile in each zone. Soil layer in Zone A was characterized using Trench 1 and borehole AB-3. Layering in Zone B was evaluated using Trench 2. Soil profile in Zone C was assessed using bore hole AB-2 and CPT 4B. Subsurface condition in Zone D was identified using CPT P1. Meanwhile, soil layer in Zone E was evaluated using the borehole AB-1 and CPT P8C. Soil profile identifications were also performed in the flow slide boundary area. The toe area was evaluated using CPT P7A. The crown area was characterized using CPT-P1A.

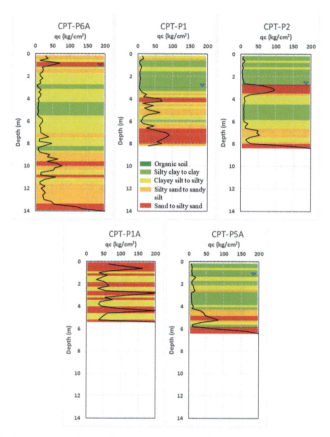

Fig. 7. Soil layer profile and cone resistance on the outside of the flowslide area

The left side of the flow slide was evaluated using points CPT P-1, CPT P-2, and CPT P5A. The right side of the flow slide was identified by points CPT P6A.

Trench 1, in Zone A, was excavated to depth of 7 m. The ground water table was found at depth of 5.85 m. This observation was relatively consistent with ground water table found at depth of 5.13 m in borehole AB-3. The layer profile in Trench 1 alternated between coarse grained materials (i.e., sand and gravelly sand) with sandy silt or silty clay (Fig. 8). This pattern was similar to that of boreholes and CPT results. In Trench 1, a relatively impermeable layer of silty clayey sand was observed at depth of 3.85 m–5.85 m interspersed with a thin layer of sand. The thin sand layer was found not continuous. This segregation may be attributed to layer stretching during flow slide.

All layers in Trench 1 had relatively low V_s values, which were less than 230 m/s. The lowest V_s value of 143 m/s was observed in the gravely sand layer containing sandy silt. The silty clay layer with alternating thin sand layers had a slightly higher V_s value of 162 m/s. This low V_s in gravely sand layer may indicate that this layer has experienced liquefaction. The gravely sand layer found at depth of 5.85 m or greater, below the silty clayey sand layer, had a V_s of 204 m/s and was prone to liquefaction. The silty

Trench 1

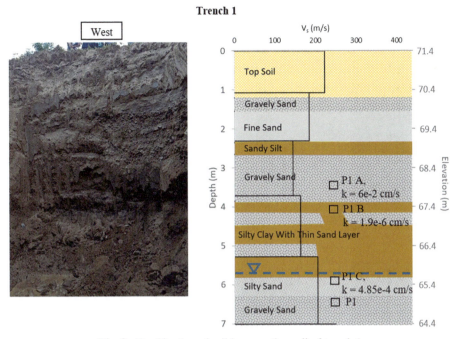

Fig. 8. Stratification of soil layer on the wall of trench 1

clayey sand layer above it was suspected as a capping layer. At depth of 7 m–8.9 m and greater, the V_s value increased with depth (Fig. 10). These layers were more resistant to liquefaction with the lowest V_s value of 248 m/s.

Trench 2, located in Zone B, was excavated to depth of 3 m (Fig. 9). The ground water level was encountered at depth of 2.85 m. The layer profile in Trench 2 wall was relatively similar to that of Trench 1. In this case, alternating sand or gravely sand with a more impermeable layer of sandy silt was observed (Fig. 9). In Trench 2, there was no impermeable layer as thick as in Trench 1. The thickness of the impermeable layer in Trench 2 was only around 15 cm–35 cm. The V_s values in Trench 2 were very small, ranging from 123 m/s–154 m/s. Due to this fact, the layer in Trench 2 wall was suspected to experience liquefaction during earthquake. Layers having quite high V_s (>300 m/s) and small liquefaction potential were observed at depth of 5.27 m or deeper (Fig. 10).

A cross-section showing the flow slide that passed through the trench test site and nearby field investigations are presented Fig. 11. Liquefaction potential is identified as the yellow layer. The green layer shows more impermeable layer. The black dotted line indicates the possible slip plane occurring along the flow slide. Above this line, materials slid to the slope foot, while materials below the line remained in place.

Based on Petobo terrain, the flow slide occurred in an area with slope of 0.5°–2.2°. Flow slide flew quite far in an area with slope of ±0.5°, forming a nearly flat area on the flow slide surface. The flow slide area in Petobo has a groundwater depth of less than 4 m.

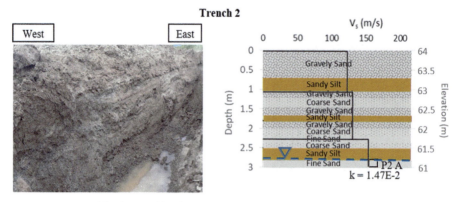

Fig. 9. Stratification of soil layer on the wall of trench 2

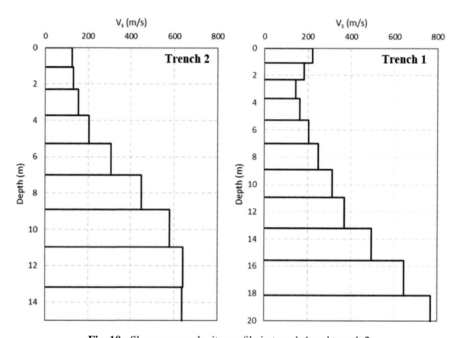

Fig. 10. Shear wave velocity profile in trench 1 and trench 2

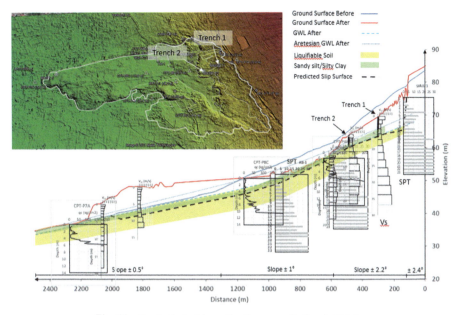

Fig. 11. Geotechnical investigation compilation in Petobo

4 Laboratory Test

A series of laboratory testing was performed on samples obtained from the trench walls, including grain size distribution, Atterberg limits, and permeability tests (Fig. 12 and Table 1). Cyclic and monotonic direct simple shear tests were conducted on samples collected from layers where liquefaction was expected. These direct shear tests allow us to assess the soil response due to cyclic loading and the effect of volume changes on residual shear strength (Fig. 13). Additionally, a liquefaction test using tubes was carried out to evaluate void redistribution process and water pressure dissipation (错误!未找到引用源。). Furthermore, ejecta sand materials on the surface were also collected.

Trench soil profile and the corresponding lab results are shown in Figs. 8 and 9, and Table 1, respectively. The gradation test showed samples from the trench wall containing fine sand and gravely sand with fine content less than 20%. It is different from the ejecta sand that was consisted of more than 20% fine grain content. This difference may be attributed to suffusion, a phenomenon where fine particle migration occurs due to seepage flow through pre-existing pores in coarse particle. In this case, during pore water pressure dissipation process, clay or silt contents were carried out upward by water flow across the voids of coarse grain soils. Field observation showed that the sand columns, where the ejecta sand escaped, consisted of coarser grains than that of materials accumulated on the surface. During pore water pressure dissipation, it appears that suffusion process was also accompanied by segregation occurred when liquefied sand material flew to the surface. This event was also observed in liquefaction tube test, where the fine grains accumulated above the liquefied layer.

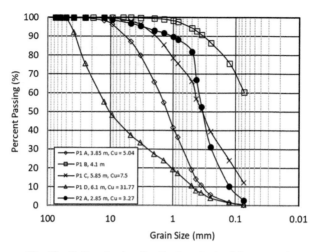

Fig. 12. Soil grain size distribution obtained from trench

Table 1. Laboratory test result obtained from trench

Parameter	P1 A	P1 B	P1 C	P1 D	P2 A
Gravel (%)	12.95	–	5.59	62.43	3.08
Sand (%)	86.68	39.76	86.79	36.98	94.18
#200 (%)	0.37	60.24	7.62	0.59	2.74
Cu	5.04	–	7.5	31.769	3.266
Cc	0.861	–	1	1.047	1.191
LL (%)	–	27.25	–	–	–
PI (%)	–	9.5	–	–	–
USCS	SP	CL	SW	GW	SP
k (cm/s)	6.04×10^{-2}	1.9×10^{-6}	4.85×10^{-4}	–	1.47×10^{-2}

The impermeable layer above the coarse grain layer contained more than 60% fines Fig. 8. Seed et al. (2003) suggests that this layer are prone to liquefaction based on the IP and LL values. Thus, in addition to serving as a capping layer due to low permeability, this layer can also serve as a slip plane.

The results of cyclic and monotonic direct simple shear tests indicate that the associated soil layer maybe liquefied during the 2018 Palu Earthquake (Fig. 13). The monotonic test showed that a layer may experience void redistribution due to temporary retention of the volume of water dissipated from the liquefied layer below it. This void redistribution leads to an increase in volume that results in decreases in effective stress and shear strength in large strains.

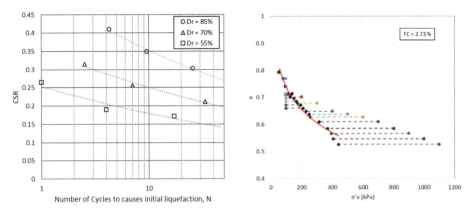

Fig. 13. Cyclic and monotonic test results from sample P2A

The liquefaction tube test facilitates observation of the distributions of sand grain movement and pore water pressure during dissipation. This test was performed by putting liquefiable soil layer in a transparent acrylic tube. Above this layer, a cap layer having lower permeability was placed. Three pore water pressure gauges were installed at 16 cm, 40 cm, and 60 cm from the bottom of the tube. A rubber mallet was used to hit the tube generating vibration. A high-resolution camcorder was used to record the event.

Soil profile post liquefaction tube test after is shown in 错误!未找到引用源。(a). A water film, indicated by blue line in 错误!未找到引用源。(a), appeared when the liquefied soil layer settled due to excess pore water pressure dissipation. The thickness of this water film increases with the magnitude of liquefaction-induced settlement. Soil settlement due to excess pore water pressure dissipation is proportional to the extent of excess pore water pressure generated during dynamic loading. Excess pore water pressure depends on the relative density of the soil. The denser the soil layer, the smaller the generated excess pore water pressure is. The smaller the generated pore water pressure, the smaller the liquefaction settlement layer is. Thus, the denser the soil, the thinner the formed water film is.

The water film was formed for some time until the pore water pressure below the cap layer was dissipated. The duration of water film occurrence depended on the permeabilities of the cap layer and the liquefied soil. The pore water pressure dissipation process length at each depth of the liquefied layer was proportional to the increase in the pore water pressure. Thus, the slope of the pore water pressure versus depth during the dissipation process was equal to that of pre-liquefaction occurrence (错误!未找到引用源。(c)). The dissipation process was affected by the cap layer. The excess pore water pressure could not be directly dissipated to the initial pore water pressure. In fact, no reduction in pore water pressure was observed for a while until it dissipated through the cap layer. If the cap layer is very impermeable or very thick, the excess pore water pressure will be trapped longer (Fig. 14).

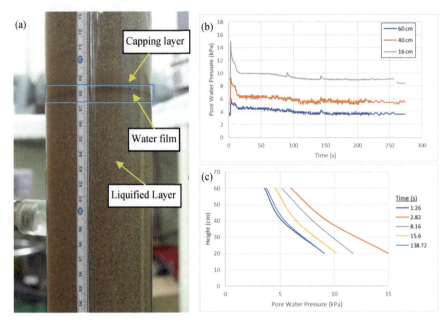

Fig. 14. (a) Post liquefaction tube test, (b) Variation of pore water pressure dissipation with time, and (c) Variation of pore water pressure with depth

The liquefaction tube test showed that there was thin layer of about 2.5 cm (Layer T) before the water film was formed (Fig. 15). Immediately after liquefaction, Layer T did not settle, whereas Layer A did. Layer T started to settle with the forming of water film layer. This is probably because the pore water pressure released by Layer A is trapped first in Layer T as it is blocked by Cap Layer (Layer C). When settlement in Layer A was overly large to be accommodated by the expansion of Layer T, ±0.5-cm water film was formed above Layer T. When the water film has formed, the pore water pressure trapped in the Layer T was then dissipated into the water film. When the pore water pressure was trapped in Layer T, this layer may experience significant strength reduction due to an increase in volume. After the water film was fully formed, the dissipation of pore water pressure below Layer C was visually identified by settlements of Layer C and layers above it. Subsequently, the pore water pressure returned to its initial pressure with the closure of water film.

Observation of strength degradation in Layer T suggested that the occurrence of this layer in the field, such as in Petobo, may trigger a slide due to liquefaction. Thus, in addition to the water film, the thin layer also contributed to the flow slide. It is also expected, even before the water film was formed, the slope has started to slip in this thin layer.

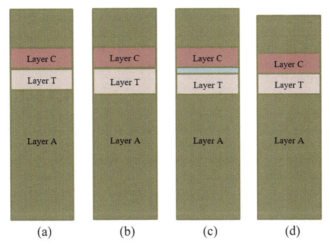

Fig. 15. Schematic of the water film formation process ((a) initial conditions, (b) pore water pressure dissipation process before the water film is formed, (c) pore water pressure dissipation process when the water film is formed, (d) finished pore water pressure dissipation

5   Flowslide Reconstruction

Petobo flow slide generally can be divided into 2 liquefied blocks based on the deformation pattern (Fig. 16). The first block, colored brown, was the fluidized block. The second block, shown in orange, was part experiencing lateral extension and back rotation. Visual observation showed that the brown block was very liquidized with a maximum displacement of up to 1 km. This block initiated displacement that eventually triggered instability of the orange block. The orange block was not as liquid as the brown block and only displaced less than 130 m. This block apparently moved as an extended solid soil block with overturning near the landslide crown.

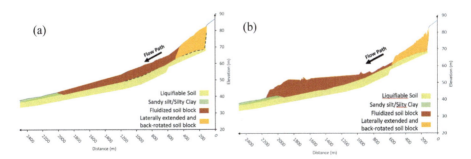

Fig. 16. Petobo flowslide schematization ((a) Before (b). After)

These orange and brown blocks contained potential liquefaction layers alternating with impermeable layers. The differences in displacement behavior between these blocks may be associated to the liquefied layer thickness and the crust layer thickness above the liquefied layer. The crust layer thickness is not only influenced by the type of soil layer but also greatly influenced by the depth of the water table. The water level in the brown block was less than 4 m, whereas that of orange block was deeper than 4 m. The liquefied layer thickness of the brown block was more than 2 m, while that of orange block was less than 2 m.

The slip plane was indicated occurring in an impermeable layer that served as a capping during liquefaction. Geotechnical investigation showed that mixtures of flow slide material were found above the liquefied layer in the brown block. Thus, the intact capping layer was difficult to identify. In the orange block, the impermeable layer that functions as capping was still intact along with the soil layer above it. The movement of the orange block did not damage the capping layer. With the movement pattern of these blocks, the liquefied soil layers did not experience much breakdown and were still under these blocks.

Kokusho (1999, 2003) described the flow slide as large soil deformation attributed to the formation of a water interlayer. In this hypothesis, a layer of water with almost zero shear strength is formed just above the liquefied layer immediately after the ground shaking. This layer is formed because the water pressure that is going to be dissipated into the layer above is covered by a more impermeable layer above it.

The flow slide mechanism cannot be comprehensively explained only based on the water film hypothesis (Okamura et al. 2019). Liquefied nature of the soil itself is also attributed to flow slide. If the sand is very contractile and shows shear strength loss, it can experience very large strains. Kokusho (2015) stated that clean sand always shows a dilative response and there is small chance of a flow slide based on undrained shear test results.

Redistribution of voids is also another phenomenon associated with flow slide due to liquefaction. Void redistribution is an important factor affecting the residual shear strength after liquefaction and is greatly affected by the permeability. Void ratio increase can occur in layers that have lower permeability. This increase eventually leads to shear strength loss of the layer. The void redistribution scheme in confined sand layers on an infinite slope is described by Malvick et al. (2003), Montgometry and Boulanger (2014), and Iai (2020).

6 Summary

An investigation of flow slide due to the 2018 Palu Earthquake in Petobo has been carried out and described in this paper. Petobo flow slide occurred in the shallow ground water table which subsequently drew the zone behind it that had deeper water elevation. The investigation showed that flow side area has layer of gravelly sand alternating with a more impermeable layers consisting of clay and silt. The liquefaction tube test showed that, after liquefaction triggering excess pore water pressure, the impermeable layer overlying the liquefied soil impeded the excess pore water pressure dissipation. In addition, during the dissipation process, the forming of a layer with increase in void ratio was observed. This void ratio increased with pore water pressure dissipation through the more impermeable layer. Layer experiencing void ratio increase was predicted to trigger instability. The results of this investigation are expected to strengthen our understanding related to flow slide due to liquefaction and can mitigate flow slide potential in other locations.

References

Bradley, K., Mallick, R., Andikagumi, H., Hubbard, J., Meilianda, E., Switzer, A., Du, N., Brocard, G., Alfian, D., Benazir, B., et al.: Earthquake-triggered 2018 Palu valley landslides enabled by wet rice cultivation. Nat. Geosci. **12**(11), 935–939 (2019)

Iai, S.: Paths forward for evaluating seismic performance of geotechnical structures. In: Kutter, B.L., Manzari, M.T., Zeghal, M. (eds.) Model Tests and Numerical Simulations of Liquefaction and Lateral Spreading, pp. 639–641. Springer, Cham (2020). https://doi.org/10.1007/978-3-030-22818-7_34

JICA: Brief explanation of "Nalodo" assessment and mitigation, presentation material in National Workshop on Joint Research, Assessment and Mitigation of Liquefaction Hazards (Lesson learned from the 2018 Palu earthquake) (2019)

Kiyota, T., Furuichi, H., Hidayat, R., Tada, N., Nawir, H.: Overview of long-distance flow-slide caused by the 2018 Sulawesi earthquake Indonesia. Soils Found. **60**(3), 722–735 (2020)

Kokusho, T.: Formation of water film in liquefied sand and its effect on lateral spread. J. Geotech. Geoenviron. Eng. ASCE **125**(10), 817–826 (1999)

Kokusho, T.: Current state of research on flow failure considering void redistribution in liquefied deposits. Soil Dyn. Earthq. Eng. **23**(7), 585–603 (2003)

Kokusho, T.: Liquefaction research by laboratory tests versus in situ behavior. In: Proceedings of the 6th International Conference on Earthquake Geotechnical Engineering, pp. 786–819. Christchurch, New Zealand (2015)

Malvick, E.J., Kutter, B.L., Boulanger, R.W., Kulasingam, R.: Shear localization due to liquefaction-induced void redistribution in a layered infinite slope. J. Geotech. Geoenviron. **132**(10), 1293–1303 (2003)

The 28 September 2018 M7.5 Palu-Donggala, Indonesia Earthquake: Version 1.0. Geotechnical Extreme Events Reconnaissance Association Report GEER-061 (2019). https://doi.org/10.18118/G63376

Montgometry, J., Boulanger, R.W.: Influence of stratigraphic interfaces on residual strength of liquefied soil. In: 34th Annual United States Society on Dams Conference, pp. 101–111. San Francisco, CA (2014)

Okamura, M., Ono, K., Arsyad, A., Minaka, U.S., Nurdin, S.: Large-scale flowslide in Sibalaya caused by the 2018 Sulawesi earthquake. Soils Found. **59**(5), 1148–1159 (2019)

Sahadewa, A., Irsyam, M., Hanifa, R., Mikhail, R., Pamumpuni, A., Nazir, R., et al.: Overview of the 2018 Palu earthquake. In: Proceedings of the 7th International Conference on Earthquake Geotechnical Engineering, pp. 857–869. Rome (2019)

Seed, R.B., et al.: Recent Advances in Soil Liquefaction Engineering: A Unified and Consistent Framework EERC-2003–06. Earthquake Engineering Research Institute, Berkeley (2003)

Sento, N., Kazama, M., Uzuoka, R., Ohmura, H., Ishimaru, M.: Possibility of postliquefaction flow failure due to seepage. J. Geotech. Geoenviron. Eng. **130**(7), 707–716 (2004)

Challenge to Geotechnical Earthquake Engineering Frontier: Consideration of Buildings Overturned by the 2011 Tohoku Earthquake Tsunami

Kohji Tokimatsu[1]([✉]), Michitaka Ishida[2], and Shusaku Inoue[3]

[1] Tokyo Soil Research Co. Ltd., Tokyo 152-0021, Japan
`tokimatsu@tokyosoil.co.jp`
[2] Shimizu Corporation, Tokyo 104-8370, Japan
`ishida.m@shimz.co.jp`
[3] Takenaka Corporation, Inzai 270-1395, Japan
`inoue.shuusaku@takenaka.co.jp`

Abstract. A simplified pseudo-static analysis is proposed to estimate the safety factors with time against instability of a building subjected to tsunami loads, based on the results of a 2D shallow water equation. The proposed analysis is used to examine the key factors affecting the performance of buildings. A 3D analysis is also conducted for one building, the performance of which has not been explained by the simplified analysis. It is shown that: (1) if the tsunami hydrodynamic and buoyant forces are the major driving forces, the proposed analysis is capable of predicting the difference in observed building performance; (2) the safety factors against overturning and sliding become minima near when the peak landward and seaward flow velocities occur, while one against uplift around when the peak inundation depth occurs; (3) the instability of building tends to occur when any of the safety factors first becomes less than one, i.e., mostly at the peak landward flow velocity; (4) the hill immediately backward of a building might have induced a lower peak landward flow velocity at a lower inundation depth and a peak seaward flow velocity at a deeper inundation depth than any other location, having affected the seaward overturning of the building; (5) the major cause of the overturning in orthogonal to the tsunami propagation direction of one building is likely due to the collision of the drifting section of other building originally located on the seaside; and (6) the interdisciplinary collaboration was enormously useful to the progress of this study.

Keywords: Tsunami load · Overturning of building · 2011 Tohoku Earthquake

1 Introduction

The tsunami caused by the 2011 Tohoku Earthquake induced catastrophic damage to various structures along the east Japanese coast. In particular, many buildings even including those supported on pile foundations were overturned and swept away in the

L. Wang et al. (Eds.): PBD-IV 2022, GGEE 52, pp. 580–596, 2022.
https://doi.org/10.1007/978-3-031-11898-2_34

town of Onagawa [7, 12–14], the location of which is shown in Fig. 1(a). The overturning direction of those buildings was mainly landward or occasionally seaward, i.e., parallel to the tsunami hydrodynamic force, or even rarely parallel to the shoreline, i.e., orthogonal to the tsunami hydrodynamic force. Sugimura et al. [12] indicated several factors that might have caused the overturning of buildings, such as hydrodynamic and buoyant forces, scour, liquefaction of underlying soil, and debris impact. It seemed important to identify the key factor controlling the problem but difficult to do so without interdisciplinary perspectives including a geotechnical viewpoint. The motivation of this study is therefore to identify the key factors differentiating the occurrence, timing, and direction of overturning based on interdisciplinary collaboration.

A simplified pseudo-static analysis is presented, for this purpose, to estimate the factors of safety with time against overturning, sliding, and uplift of a building, based on the time histories of inundation depth and flow velocity predicted by a two-dimensional shallow water analysis. The proposed analysis is then applied to examine the key factors controlling five buildings, the field performances of which are known. A three-dimensional shallow water analysis is also conducted for one building, the overturning reason of which has not been explained by the simplified pseudo-static analysis.

2   Field Performance of Target Buildings

Figure 1(b) shows a map of Onagawa town before the tsunami [2] with the area inundated by the 2011 Tohoku Earthquake tsunami [4] as well as the locations of five Buildings A to E examined in this study.

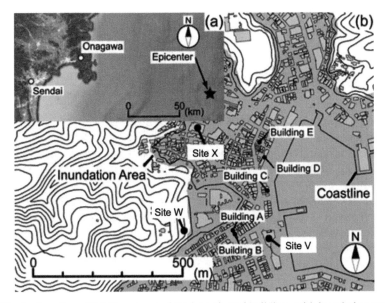

Fig. 1. Location of (a) Onagawa and (b) investigated buildings with inundation area

Figure 2 is an aerial photograph of the hardest hit area in Onagawa taken after the tsunami [2], which shows the locations of the buildings before and the after the event. Building A and D overturned landward, Building C overturned seaward, and Building E overturned almost in parallel to the coastline, while Building B remained at its original position. Digital recording open for public including the video taken from Site V [9] and photos taken from various places including Sites W and X [10, 15] has shown that Buildings A and D overturned during the first tsunami runup and Buildings C and E by the end of the first tsunami backwash.

Fig. 2. Location and overturning direction of buildings investigated (Photo [2])

Figure 3 shows Building A, a four-story RC building with pile foundation that over-turned landward during the first tsunami runup and displaced landward by 70 m from its original position. This building had been supported on 32 RC piles 4 m long and 300 mm in diameter (Figs. 3(a) and 3(b)). All piles except one pile hanging down from the building (Fig. 3(c)) were broken and cut at or near the pile head. Some of those piles were pulled up to 3 m at its original position (Fig. 3(d)), and the exposed steel reinforcement had cut remaining thinning ends. The difference in failure mode among piles suggests that the overturning of this building progressively occurred in a certain period of time.

Figures 3(a) and (b) also show Building B, a five-story RC building founded on piles that survived at its original position. The ground subsidence around the building indicated the occurrence of soil liquefaction, the influence of which was considered to be insignificant, considering that the pullout resistance of piles of Building A was almost mobilized.

Fig. 3. Photos of Buildings A and B taken before and after tsunami (Photo (a) [3])

Figure 4 shows Building C, a four-story RC building with spread foundation, which is the only one in the town having overturned seaward. The fourth floor of the building detached and was missing.

Figure 5 shows Building D, a three-story RC building with spread foundation that toppled, split into two or three sections, and drifted landward during the first tsunami runup. While the left section seen in Fig. 5(a) was found about 175 m landward from its original position (Fig. 2), the right section only about 75 m landward near Building E (Figs. 2 and 5(b)).

Fig. 4. Photos of Building C taken before and after tsunami

Fig. 5. Photos of Building D taken before and after tsunami (Photo (a) [15])

Fig. 6. Photos of Building E taken before and after the tsunami (Photo (a) [3])

Figure 6 shows Building E, a two-story RC building that toppled parallel to the shoreline, i.e., orthogonal to the tsunami propagation direction. It was supported on 14 PC piles 20 m long and 300 mm in diameter [11]. Either the pile head (Fig. 6(d)) or 0.9 m to 1.5 m below the pile head (Figs. 6(b) and 6(c)) failed in a tensile, bending, or compressive mode. This failure mode is completely different from that observed in Building A, suggesting that the failure of the piles and subsequent overturning of the building might have occurred instantly by a much larger driving force.

Table 1 summarizes the general information of the buildings, in which W is the estimated weight, H is the height, B is the width in parallel to the coastline, L is the length perpendicular to the coastline, and D_f is the depth of foundation embedment. Further information about the buildings are found elsewhere [7, 14].

Table 1. General information of buildings investigated

Building ID	W (MN)	H (m)	B (m)	L (m)	D_f (m)	Foundation Type	Pile Diam. (m)	Pile Length (m)
A	2.5	12.4	7.3	5.5	0.9	Pile	0.3	4
B	6.2	17.1	13.0	6.5	0.9	Pile	$0.3^{1)}$	$4^{1)}$
C	3.8	9.5	6.7	10.4	0.6	Spread	–	–
D	32.4	9.0	59.4	11.8	0.9	Spread	–	–
E	1.9	6.3	5.0	11.0	0.6	Pile	0.3	20

1) Assumed

3 Method to Estimate Occurrence of Overturning of Building

3.1 Simplified Method for Estimating Tsunami-Induced Overturning of Building

Figure 7(a)–(c) show simplified force diagrams for overturning, sliding, and uplift of a building subjected to tsunami-induced hydrodynamic and buoyant forces. In order to maintain the stability of the building against overturning shown in Fig. 7(a), the driving moment induced by both the tsunami hydrodynamic and buoyant forces with respect to the rotational center should be always less than the resisting moment induced either the pullout resistance and the self-weight of a building with piles or the self-weight of a building with spread foundation. The pullout resistance of a building with piles is likely either the skin friction of the piles or the tensile strength of their rebars, whichever is smaller. For the sliding shown in Fig. 7(b), the tsunami hydrodynamic force should be always less than either the shear strength of the rebar of piles and the passive earth pressure acting the embedded foundation for a building with piles or the base friction and passive earth pressure for a building with spread foundation. For the uplift shown in Fig. 7(c), the tsunami buoyant force should be always less that either the pullout resistance of a building with piles or the self-weight of a building with spread foundation.

The hydrodynamic force F_D acting on a building in the travel direction of tsunami at time t may be calculated by the following formula [1];

$$F_D(t) = \frac{1}{2}\rho C_D u(t)^2 h(t) B \tag{1}$$

where t is the time, ρ is the fluid density including sediment (1.2 t/m^3), C_D is the drag coefficient (2.0), u and h are the flow velocity and inundation depth computed by the 2D simulation for a given depth such as those shown in Fig. 8(a), and B is the building width.

The buoyant force F_B acting on a building at time t is calculated by the following formula;

$$F_B(t) = \rho g\{V(t) - V_W(t)\} \tag{2}$$

where g is gravitational acceleration (9.8 m/s^2), V is the volume of the building below the inundation depth, and V_W is the volume of water having entered the building.

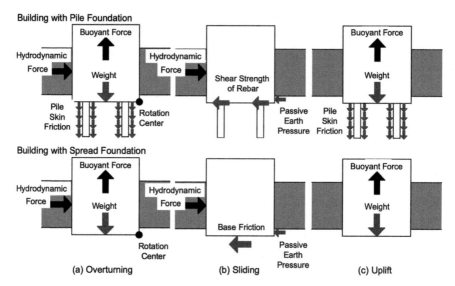

Fig. 7. Simplified free body diagrams for (a) overturning, (b) sliding, and (c) uplift mechanisms of buildings with pile or spread foundation

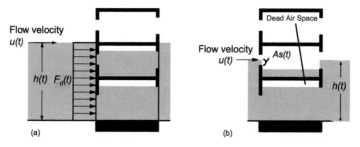

Fig. 8. Comparison of computed (a) inundation depth and (b) flow velocity with observed ones

The flow rate of the water entering the building, $\Delta V_W(t)$, during the tsunami runup was estimated by multiplying the area of the building openings between the interior and exterior inundation depths, $A_s(t)$, by the current flow velocity, $u(t)$, as shown in Fig. 8(b). The height to be filled by water in each floor was limited by either the inundation depth or the highest height of the opening of the floor that creates dead air space above, whichever is lower. $V_W(t)$ during the backwash was assumed to be equal to the volume inside the building below the inundation depth, $h(t)$, regardless of the current flow velocity, $u(t)$, and the presence of dead air space during runup. The value of V_W was also constrained by assuming that the dimension of pillar cross-section was 500 mm × 500 mm and the wall thickness was 200 mm.

The method how to evaluate other information including the base horizontal resistances of the building, the passive earth pressure acting on the embedded part of the building, and the pile pullout resistance is found elsewhere [14].

3.2 Estimation of Tsunami Hydrodynamic and Buoyant Forces by 2D Analysis

A 2D analysis of the tsunami runup was conducted with a shallow water theory [6] to estimate the variation with time of the inundation depth and flow velocity around the target buildings, which are needed in Eqs. (1) and (2). Figure 9 shows two areas used for 2D analysis, which has a dimension of either 1530 m and 570 m or 2960 m and 3660 m with a uniform mesh of 10 m by 10 m. Area A shown in Fig. 9(a) was used for Buildings A-C, while Area B for Buildings D and E, so that Site V, the video observation point, is included in both areas. The terrain model using the 2D analysis was assigned according to the digital elevation model (DEM) provided by the GSI [2] and General Ground Plan of Onagawa Port, Miyagi, available from Japan Center for Asian Historical Records [8].

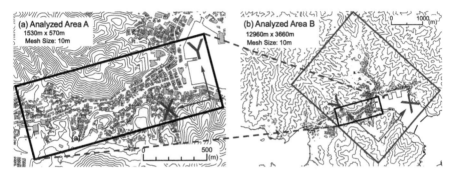

Fig. 9. Analyzed area for 2D simulation: (a) Area A and (b) Area B

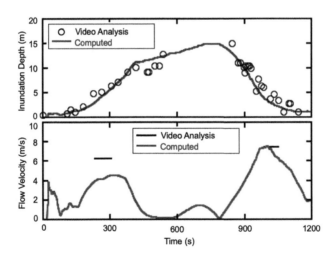

Fig. 10. Comparison of computed and observed (a) inundation depth and (b) flow velocity at Site V

The open circles in Fig. 10(a) are the inundation depth evaluated from the video recording taken at Site V in Fig. 1(b) during the first tsunami runup and backwash (Koshimura et al., [9]). The two solid black lines in Fig. 10(b) are the peak flow velocities at around 300 s during the first tsunami runup and at around 1000 s during the first tsunami backwash, estimated from the same video recording [9]. The smoothed time history of the estimated inundation depth plus the elevation of Site V was used as the time history of the sea water level prescribed at the seaside boundary of each analyzed area for the tsunami analysis.

Figures 10(a) and (b) show the computed time history of inundation depth and flow velocity at Site V from the analysis for Area A. The computed time history of inundation depth shown in Fig. 10(a) corresponds reasonably well to the observed one. The computed flow velocities shown in Fig. 9(b) also show fairly good agreements with the observed ones at the two instances. A good agreement between the observed and computed values were also found from the analysis for Area B.

Fig. 11. Computed inundation depth and flow velocity within analyzed area B at 900 s

Figure 11 shows a snap of the distribution of inundation depth and flow velocity within Area A at 900 s when the inundation depth took a peak during the first tsunami runup. The computed inundation area is consistent with that observed in the field (Fig. 1(b)). Thus, the computed time history of inundation depth and flow velocity at each building location is used to estimate the driving moment and forces with time to be imposed on each building.

4 Comparison of Field Performance with Estimated Result

4.1 Estimation of Field Performance Based on 2D Analysis

Figure 12 shows the estime histories of the driving forces/moment and resisting forces/moment for overturning, sliding, uplift for Building A. Note that the ordinate values in the overturning and sliding diagrams are normalized to the building width, while in the uplift diagram to each building weight.

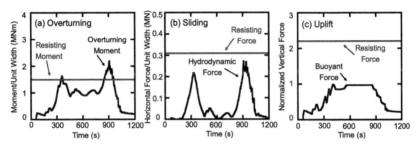

Fig. 12. Comparison between estimated driving and resisting loads to Building A with respect to (a) overturning, (b) sliding, and (c) uplift mechanisms

Figures 13, 14, 15, 16 and 17 show the time histories of safety factors against overturning, sliding, uplift for the five buildings, in which each safety factor is defined as the ratio between the corresponding driving and resisting loads for each building such as shown in Fig. 12.

The safety factors against overturning of Buildings A and D shown in Figs. 13 and 16 become less than one at around 300–500 s, suggesting that those buildings could have overturned landward about when the landward flow velocity took a peak as shown in Fig. 10(b); which is consistent with the actual field performance of those buildings.

The safety factors against overturning and sliding of Building C shown in Fig. 15 become less than one only after 900 s about when the seaward flow velocity took a peak as shown in Fig. 9(b), which is consistent with the seaward overturning of this building.

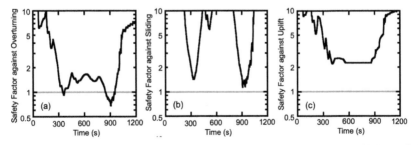

Fig. 13. Computed safety factors against (a) overturning, (b) sliding, and (c) uplift for Building A

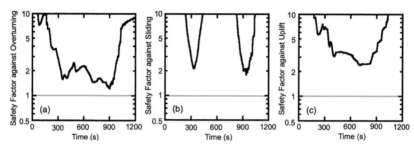

Fig. 14. Computed safety factors against (a) overturning, (b) sliding, and (c) uplift for Building B

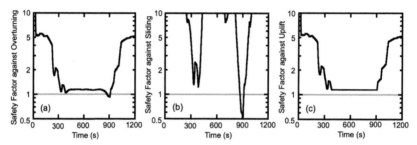

Fig. 15. Computed safety factors against (a) overturning, (b) sliding, and (c) uplift for Building C

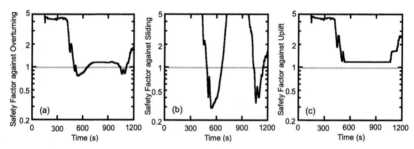

Fig. 16. Computed safety factors against (a) overturning, (b) sliding, and (c) uplift for Building D

Any of the safety factors of Building B shown in Fig. 14, in contrast, is always greater than one, confirming that this building could have survived tsunami impacts. Similarly, any of the safety factors of Building E in Fig. 17 is always less than one, suggesting its good field performance but contradicting its actual field overturning behavior. Moreover, since the overturning direction of this building was orthogonal to the direction of tsunami propagation, the actual driving moment and forces imposed in that direction would be far smaller than the computed ones, suggesting effects other than the tsunami direct forces.

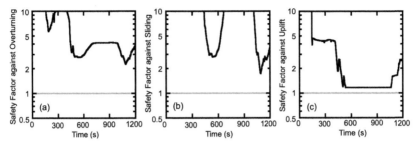

Fig. 17. Computed safety factors against (a) overturning, (b) sliding, and (c) uplift for Building E

The above discussion suggests that the proposed analysis is capable of predicting the difference in field performance of buildings on the condition that the tsunami-induced hydrodynamic and buoyant forces are the only major driving forces. Under this condition, the safety factors against overturning and sliding by the end of the first backwash, tends to take minima twice when the peak landward and seaward flow velocities occur. The overturning of building, therefore, most likely occur near when the peak landward flow velocity occurs unless the safety factor against overturning is greater than one as is the case for Building C.

In order to explore the reason for the seaward overturning of Building C, Fig. 18 compares the computed time histories of inundation depth and flow velocity at Buildings A and C. The peak flow velocities at Building C are significantly smaller and occur at a lower inundation depth during runup and at a deeper inundation depth during backwash than those at Building A, probably having led to the seaward overturning of this building. Such variation of flow velocity with time at Building C might have been caused by the presence of a hill immediately located landward of Building C.

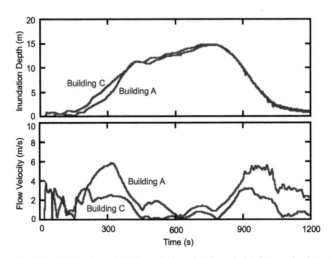

Fig. 18. Computed time histories of (a) inundation depth and (b) flow velocity at Buildings A and C

4.2 Estimation of Field Performance Based on 3D Analysis

The inconsistency between the computed result and the field performance of Building D shown in the previous section motivated us further to examine possible effects other than tsunami induced forces, i.e., the collision impact of drifting Building D on the overturning behavior of Building E.

A 3D analysis was therefore conducted with Volume of Fluid (VOF) method [5] for the restricted area encircled with lines shown in Figs. 19(a) and 19(b), which has a dimension of 176 m and 270 m and T.P. 129 m to +50 m with a uniform mesh of 1 m by 1 m. The terrain model using the 3D simulation was one used for the 2D analysis described in Sect. 3.2 plus all surviving and the target buildings (i.e., Buildings D and E). Building D was divided along its longitudinal direction into three sections that can move separately. The coefficient of friction between each section and the ground was assumed to be 0.4. The density of each section was also assumed to be 1.09 t/m^3, which corresponds to the condition where the volume of water in the building is a half of the inside volume of the building. The input boundary conditions prescribed to the seaside of the 3D area shown in Fig. 19(b) were the inundation depth and flow velocity computed by the 2D analysis, but using a more precise mesh of 4 m by 4 m plus all surviving and target two buildings as shown in Fig. 19(a). The analysis was conducted only for the tsunami runup phase (until 970 s).

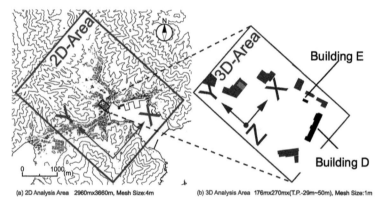

(a) 2D Analysis Area 2960mx3660m, Mesh Size:4m   (b) 3D Analysis Area 176mx270mx(T.P.-29m~50m), Mesh Size:1m

Fig. 19. Analyzed area for 3D simulation: (a) 2D area and (b) 3D area

Figure 20 shows the variation of computed inundation depth and flow velocity within the area as well as the floating behavior of Building D at times of 520 s. The building started to drift at about 480 s, and the right section of the building hit the southwest side of Building E at about 520 s and remained nearby thereafter. This is consistent with the actual field observation such as shown in Fig. 2, suggesting that the overturning of Building E in parallel to the coastline might have occurred due to the collision of the right section of Building D.

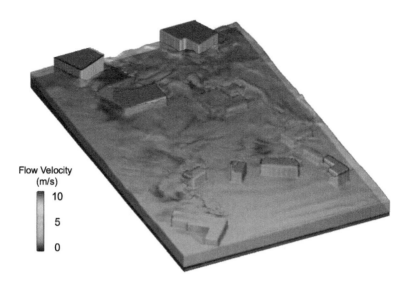

Fig. 20. Variation of computed inundation depth and flow velocity with drifting Building D at t = 520 s

Figures 21(a) and (b) show the computed time histories of hydrodynamic force acting on the shorter and longer directions (orthogonal and parallel to the tsunami propagation directions) respectively of Building E computed by the 3D simulation. Note that the collision impact force of the right section of Building D to Building E was not computed directly from the 3D analysis and thus is not included in the computed force. Figures 21(a) and (b) suggest that the horizontal force at the time of the collision is larger in the shorter direction than in the longer direction.

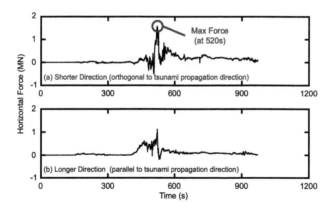

Fig. 21. Computed hydrodynamic forces acting on Building E from 3D analysis: (a) shorter direction and (b) longer direction

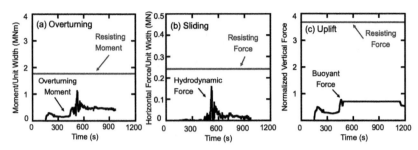

Fig. 22. Comparison between estimated driving and resisting loads to Building E with respect to (a) overturning, (b) sliding, and (c) uplift mechanisms

The proposed pseudo-static analysis was again performed for Building E along its shorter direction using the hydrodynamic force shown in Fig. 21(a). Figure 22 shows the estimated loads and resistances for overturning, sliding, and uplift of the building. The figure suggests that Building E would have overturned if the collision impact force not determined from the 3D analysis is about 2 MN.

Taking into account that the weight and velocity as well as the final location of the right section of Building D, its potential collision force to the shorter direction of Building E would be estimated to be greater than 16 MN. This is much greater than the critical one (2 MN), suggesting that the collision of the right section of Building D is the major factor having induced the overturning of Building E.

Figure 23 shows a photo of the roof of overturned Building E. The serious damage observed on its upper side, previously faced southwest, appears to demonstrate that it was caused by the collision of the right section of Building D having drifted from southwest.

Fig. 23. Serious damage observed on the roof of Building E overturned probably due to collision of Building D [3]

5 Conclusions

In order to clarify the major factors causing the difference in tsunami-induced damage among the five buildings in Onagawa, a simplified pseudo-static analysis was proposed to estimate the factors of safety in the time domain against overturning, uplift, and sliding of a building for which the hydrodynamic and buoyant forces are estimated from the 2D shallow water equations. A three-dimensional shallow water analysis was also conducted for one building, the performance of which had not been explained by the simplified analysis. It was shown that: (1) if the tsunami-induced hydrodynamic and buoyant forces are the major driving forces, the proposed analysis is capable of predicting the difference in observed building performance; (2) the safety factors against overturning and sliding become minima when the peak landward and seaward flow velocities occur, while one against uplift when the peak inundation depth occurs; (3) the overturning of building tends to occur when any of the safety factors first becomes less than one, i.e., mostly at the peak landward flow velocity; (4) the seaward overturning of one building is likely due to the topographical conditions created by the hill immediately backward, which could have made the inundation depth lower at the peak landward flow velocity and deeper at the peak seaward flow velocity than any other location, having caused the safety factor less than one for the first time during backwash; (5) the major cause of the overturning in orthogonal to the tsunami propagation direction of one building is likely due to the collision of the drifting section of other building originally located on the seaside; and (6) the interdisciplinary collaboration was enormously useful to the progress of this study.

Acknowledgements. The most part of this study was made while the first and second authors were a professor and a graduate student of Tokyo Institute of Technology. The various support provided by the university and colleagues is appreciated.

References

1. Federal Emergency Management Agency: Guideline for design of structures for vertical evacuation from tsunamis. FEMA P646 (2008)
2. Geospatial Information Authority of Japan (GSI). http://mapps.gsi.go.jp. Last Accessed 13 Jan 2022
3. Google: Memories for the future. At http://www.miraikioku.com/. Last Accessed 13 Jan 2022
4. Haraguchi, T., Iwamatsu, A.: Detailed Maps of the Impacts of the 2011 Japan Tsunami, vol. 1, pp. 167. Iwate and Miyagi prefectures, Kokon-Shoin Publishers, Aomori (2011)
5. Hirt, C.W., Nicholas, B.D.: Volume of fluid (VOF) method for the dynamics of free boundaries. J. Comput. Phys. **39**, 201–225 (1981)
6. Hirt, C.W., Richardson, J.E.: The Modeling of Shallow Flows. Flow Science, Inc., Technical Notes, FSI-99-TN48R (1999)
7. Ishida, M., Tokimatsu, K., Inoue, S.: Factors affecting overturning of building induced by tsunami run-up in the 2011 great Tohoku earthquake. In: Proc. 16th World Conf. on Earthquake Engineering, Santiago, Chile, Jan 9–13, Paper No. 3956, pp. 11 (2017)
8. Japan Center for Asian Historian Records (JACAR): Attachment, Onagawa Port, Miyagi, general ground plan, Ref. A11112289000, Relevant Documents regarding Cabinet Tohoku Bureau, File of petition documents (two), 1935 (National Archives of Japan). http://www.digital.archives.go.jp/. Last Accessed 13 Jan 2022
9. Koshimura, S., Hayashi, S.: Tsunami flow measurement using the video recorded during the 2011 Tohoku tsunami attack, pp. 6693–6696. International Geoscience and Remote Sensing Symposium, IEEE international (2012)

10. Miyagi Prefecture: Great East Japan Earthquake Archive Miyagi – Onagawa Town. Available at http://kioku.library.pref.miyagi.jp/onagawa/. Last Accessed 13 Jan 2022
11. National Institute for Land and Infrastructure Management: Ministry of Land, Infrastructure and Transport, Japan: Damage investigation report of the 2011 Tohoku-Pacific Ocean Earthquake, Technical Note of Nilim, No. 674, pp. 565 (in Japanese) (2012)
12. Sugimura, Y., Mitsuji, K.: The survey of buildings overturned and carried out by tsunami. Foundation Engineering and Equipment. **40**(12), 71–75 (in Japanese) (2012)
13. Tokimatsu, K., Tamura, S., Suzuki, H., Katsumata, K.: Buildings damage associated with geotechnical problems in the 2011 Tohoku Pacific Earthquake. Soil Found. **52**(5), 956–974 (2012)
14. Tokimatsu, K., Ishida, M., Inoue, S.: Tsunami-induced overturning of buildings in onagawa during the 2011 Tohoku earthquake. Earthq. Spectra **32**(4), 1989–2007 (2016)
15. Yahoo: East Japan Earthquake Picture Project. http://archive.shinsai.yahoo.co.jp/. Last Accessed 13 Jan (2022)

Multipoint Measurement of Microtremor and Seismic Motion of Slope Using Small Accelerometers

Lin Wang[1(✉)], Takemine Yamada[2], Kentaro Kasamatsu[2], Kazuyoshi Hashimoto[1], and Shangning Tao[1]

[1] Chuo Kaihatsu Corporation, Shinjuku-Ku, Tokyo 169-8612, Japan
wang@ckcnet.co.jp
[2] Kajima Technical Research Institute, Chofu-Shi, Tokyo, Japan

Abstract. The authors have developed a real-time monitoring system. The monitoring sensors are in small size, power-saving, low-cost. The sensor unit is embedded with a 1D micro seism accelerometer and 3D MEMS (Micro Electro Mechanical Systems) accelerometer (seismic motion/vibration) and has been verifying its field performance since 2019. In the microtremor measurement, compared the result of acceleration Fourier amplitude spectrum with speed Fourier amplitude spectrum of conventional microtremor speed sensor at the same site, almost the same results of the predominant period were obtained. The measured results of the Mj 6.9 earthquake in Miyagi prefecture (March 20, 2021) will also be introduced as a case study. The developed microseism sensor units were applied to the field test of slope over two years, the result is that it can withstand long-term measurement with an accuracy comparable to a conventional seismometer.

Keywords: Microtremor · Seismic motion · Geo-ground monitoring

1 Introduction

Many research activities have been carried out based on the methods for determining and analyzing ground types and velocity structures by utilizing microtremors and seismic motions so far. Furthermore, a method to estimate the velocity structure from the horizontal and vertical spectral ratios of seismic observation records at one point on the ground surface based on the diffuse field theory for plane waves was proposed as a past study using seismic motion (Kawase et al. 2011). In addition to the above existing methods, the multipoint measurement technology proposed by the authors can be used to evaluate the geo-ground characteristics in the local area, therefore, microtremor and seismic motion observation of the geo-ground are applied as an effective mean. The authors have developed a real-time monitoring system that is in small size, power-saving, low-cost unit embedded with a 1D accelerometer and 3D MEMS accelerometer (seismic motion/vibration) and have been verifying its field performance since 2019. In the microtremor measurement comparing the Fourier amplitude spectrum results obtained by the new type of a 1D micro seism accelerometer with the conventional microtremor sensor at the same place, almost the same results were obtained. The field observation results of some earthquakes will be introduced as case studies. It was confirmed that comparing the observation result of the embedded 3D MEMS accelerometer with the result of the conventional seismometer, it is possible and comparable to measure both the waveform and Fourier amplitude spectrum characteristics in a wide band by using the 3D MEMS accelerometer, which is comparable to the seismometer. Sensor units were set up at multiple points on the field slope, and observations were continued for more than two years. Furthermore, because of verifying the embedded accelerometer's accuracy, it was found that it can withstand long-term measurement with an accuracy comparable to that of a seismometer as described above. Based on the above results, it is considered that multi-point and reliable seismic technology and the use of the results will lead to the strengthening of disaster response capabilities when tackling ground disasters that are generally uncertain and difficult to predict.

2 Microseism and Vibration Monitoring System

2.1 Outline of Microtremor and Seismic Motion Sensors System

This system, as shown in Fig. 1, is consist of a 1D accelerometer and 3D MEMS accelerometers (see Table 1), and combines microtremor, seismic motion observation according to the purpose of microseism, earthquake, and vibration measurement while correcting the time using GNSS (Global Navigation Satellite System), real-time monitor the geo-ground, slopes, or civil engineering structures. In microtremor and seismic motion monitoring, spectrum analysis is performed at any time, the predominant period is calculated, and the calculation results are transferred to the monitoring center by wireless in LoRa (Long Range, which is a proprietary low-power wide-area network modulation technique) or FSK (Frequency-Shift Keying) mode.

Recently, some research for determining and analyzing geo-ground types and s-wave velocity structures by utilizing microtremors and seismic motions has been carried out. For example, Okimura et al. (2007) have proposed a method of observing microtremors of the foundation ground, calculating the predominant period of the ground from the results, and determining the types of ground. In addition, as a past study utilizing seismic motion, for example, Kawase et al. (2011) propose a method to estimate the velocity structure from the horizontal-vertical spectral ratio of seismic observation records at one point on the ground surface based on the diffuse wave field theory for a plane.

Considering the above methods, the evaluation using microtremor observation and seismic observation of the geo-ground are effective and useful, the multi-point monitoring system becomes necessary and important. The ground characteristics can be evaluated in planar evaluation by using such method, furthermore, the multi-point measurement technology is proposed by authors, a real-time measurement system embedded 1D accelerometer and 3D MEMS accelerometer (microtremor/seismic motion meter) have been developed. The sensor units are compact, power-saving, and the verification field test has been carried out since 2019.

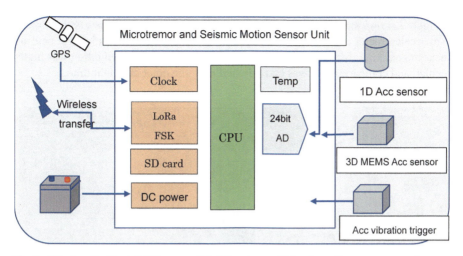

Fig. 1. Wireless (LoRa + FSK) compatible Microtremor/Seismic motion measurement sensor.

2.2 Accuracy Verification of a New Type of Microseism Accelerometer

Table 1 shows the specifications of the 1D accelerometer and 3D MEMS accelerometers embedded in the measurement system. The 1D accelerometer has high sensitivity (10 V/G) and low electrical noise level (10 Hz: 0.09 μg/$\sqrt{\text{Hz}}$), so it is used for microseism measurement. The results of comparing the 1D accelerometer with the conventional seismometer (velocity type) were shown in Figs. 3 and 4 for the purpose to explain the verification accuracy at the microtremor signal level.

Table 1. The detail of specification.

	Microtremor/Seismic acceleration	MEMS Seismic accelerometer
Acceleration range	±0.5G	±2G/±4G/±8G selection
Measurement	1D Acceleration	3D Acceleration
Sensitivity	10 V/G,0.5 mg/μ strain 24Bit A/D Converter	400 mV/G@ ± 2.5G range 24Bit A/D Converter
Electric noise	2 Hz:0.28 μg/$\sqrt{\text{Hz}}$ 10 Hz:0.09 μg/$\sqrt{\text{Hz}}$ 100 Hz:0.03 μg/$\sqrt{\text{Hz}}$	25 μg/$\sqrt{\text{Hz}}$
Frequency response	0.3–1300 Hz (±3 dB) 0.05–450 Hz (±3 dB), Selection	3 dB upper limit 1500 Hz
Note	Wireless transfer of acceleration power spectrum area ratio	
	Fourier spectrum is automatically calculated from acceleration wave	
	The power spectrum area ratio and predominant frequency of microseisms are automatically calculated	

Figure 2 shows the situation of a conventional seismometer and a new type of micro-seism accelerometer installed in the same place in Tokyo. The verification test field site is near the national road and railway, to avoid the time when the vibration was highly irregular, and comparing test was carried out during the quiet midnight time.

Fig. 2. (a) Conventional seismometer for comparison. (b) New type microseism accelerometer

Figure 3 shows the Fourier amplitude spectrum results calculated for a total of 8 waveforms by selecting the part of the recording (vertical movement component) based on the seismic speed waves of conventional seismometer with relatively little noise contamination. Figure 4 shows the Fourier amplitude spectra of the seismic acceleration wave records obtained by the new type of microseism accelerometer observed at the same location. In the range of the period of 0.1 to 1 s, a peak of the predominant period was observed in 0.3 to 0.4 s, and the comparison results of the predominant period of the conventional seismometer and the new type of microseism accelerometer were almost the same. Based on the results of this preliminary study, the new type of seismic

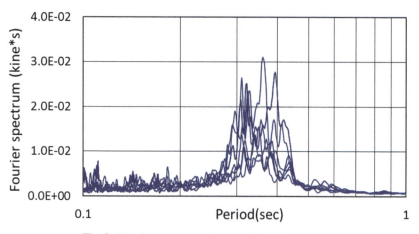

Fig. 3. Fourier spectrum of conventional seismometer

accelerometer was determined to be embedded in our sensor unit (see Table 1). For purpose of covering a wide range of measurements of acceleration, another type of accelerometer of 3D MEMS is embedded in the sensor unit, and measurements were carried out in the verification fields shown following.

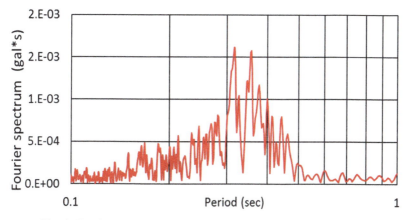

Fig. 4. Fourier spectrum of the new type of microseism accelerometer

3   Verification Field Test and Its Results

3.1   Outline of Test Slope Field

The test field site is a natural slope located at the boundary between the Ninomiya formation and the Fudoyama gravel layer in the southern part of the Oiso Hills, which is in the eastern part of Odawara city. The loam layer is deposited on the surface layer of the slope, while in the flat part near the slope, the solid gravel layer and mudstone are layered alternately under the embankment with a thickness of about 8 m that created the valley topography. Accumulation has been confirmed by a boring survey at this site. The outline of the slope to be tested is shown in Figs. 5, 6, and 7.

5 sets of 3D MEMS accelerometers and a 1D microseism accelerometer were installed on the slope, and a conventional strong-motion seismograph for comparison and verification was installed in place (nearby to No5). The verification test has been continued since 2019. The interval (horizontal distance) between adjacent sensors is 24 m to 35 m, and the angle of inclination of the slope is approximately 25 to 35 °C. After installation, more than 10 waves of seismic observation records were obtained in this test field.

A microtremor array investigation using a conventional micro seismometer (velocity) was carried out near No. 5, and it was found that the S wave velocity near the surface layer was about 450 m/s. In addition, in the horizontal-vertical spectral ratio of microtremor, a peak considered to be predominant in horizontal motion in the Rayleigh wave basic mode has been confirmed around 1 s in the period.

Fig. 5. (a) Measurement slope and sensor arrangement (plan view, the numbers in the figure are sensor numbers). (b) Measurement results of slope topography using a 3D laser scanner (looking from east to west).

Fig. 6. Outline of sensor installation status

3.2 Typical Comparing Results Based on Earthquake Measurement

Figure 8 shows the relationships of Fourier spectrum, earthquake waves of conventional seismograph, and new MEMS microseism sensor. The two types of microseism sensors almost got the same result from the monitoring of the earthquake of Miyagi Prefecture Mj6.9, Japan, 20/Mar/2021.

The graph of Fig. 8(a) shows the relationship between the Fourier spectrum and the vibration period of the two types of sensors during an earthquake, the results of the two curves almost overlap and match. Figure 8(b) shows the relationship between the seismic acceleration waveforms of the two types of sensors and the seismic motion time during an earthquake, as a result, the measured waveforms of the two types of sensors naturally overlap.

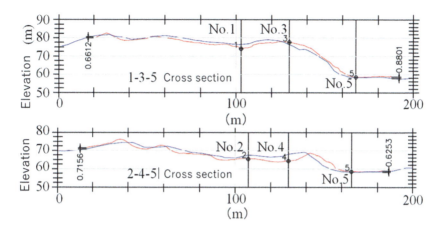

Fig. 7. Comparison of sensor installation position and measurement cross-section

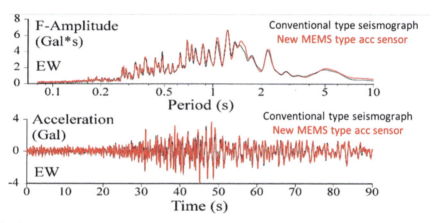

Fig. 8. Accuracy verification of earthquake measurement by using 3D MEMS acceleration sensor (Mj6.9, Earthquake of Miyagi prefecture Japan, 20/Mar/2021).

To investigate the geo-ground amplification characteristics of the slope, the observation results obtained at No. 2 installed on the flat part of the bottom of the slope, and No. 1 installed on top of the slope for the earthquake (seismic magnitude scale is Mj5.0) in central Chiba prefecture on May 6, 2020. The recorded acceleration Fourier amplitude spectrum is shown in Fig. 9. It was found that the period around 0.15 s is predominant due to the ground structure of the slope of the mountain body.

In the future, the observation records for estimating the S-wave velocity structure of the slope of the mountain body are planned to utilize.

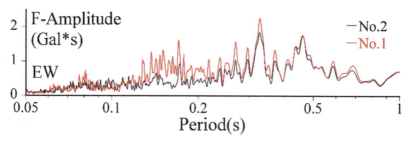

Fig. 9. Spectrum comparison (Mj5.0, Earthquake in central Chiba prefecture, Japan, 6/Jun/2020).

In Fig. 10, the acceleration Fourier amplitude spectrum of the microseisms observed by the new type of micro-seismic accelerometer is compared with the observation result by the conventional micro-seismic sensor. With a period of about 0.2 s or less, the amplitude characteristics of the two types of microseisms accelerometers are in good agreement, and it was found that the new type of accelerometer can constantly observe microseismical waves with the same accuracy as the conventional microseism sensor. To improve the observation accuracy on the long-period monitoring, authors are studying the use of another small accelerometer and the improvement of measurement method and waveform processing method.

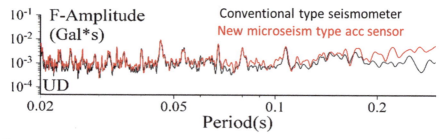

Fig. 10. Accuracy verification of new type microseism acc sensor with conventional microseism sensor

3.3 Ground Vibration Results of Slope Due to Mechanical Vibration in Multipoint Measurement

At the verification test slope field, 5 sets of sensor units were installed here for multi-point measurement of seismic response evaluation of ground vibration and seismic motion. This sensor network system includes the functions of wireless data transmission, microtremor monitoring, and earthquake monitoring, at the same time, the system supports multi-point monitoring measurement based on GPS time synchronization.

Figure 11 shows that the microseism and vibration monitoring system can sensitively monitor the micro acceleration vibrations of road roller by 5 devices installed for multi-point measurement on the slope. From left to right, the graphs show the measured acceleration vibration results of MEMS seismic sensors no.1 to no.5. From top to bottom, the graphs show three direction results of mechanical vibration waveforms, which are north-east direction, south-west direction, an up-down direction. Figure 12 shows the acceleration Fourier spectrum results. It turned out that the developed sensor sensitively measured the mechanical acceleration even in very small vibrations.

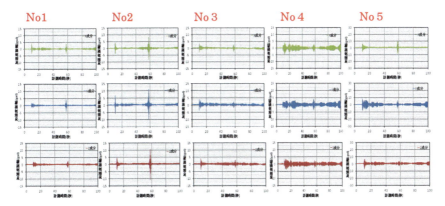

Fig. 11. Mechanical vibration waveform of slope ground by road roller vibration

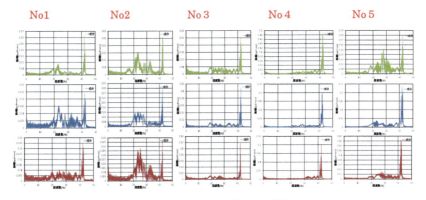

Fig. 12. Mechanical vibration Fourier spectrum (gal*sec) of 5 sensors by road roller vibration.

Based on acceleration Fourier spectrum results in Fig. 12, the predominant frequency is found to be around 42 Hz. It was presumed that the frequency of the engine vibration of the stopped road roller was picked up (Miura, 2009). Therefore, it is considered that the response dominant frequency of the slope ground due to the road roller vibration of the road roller is between 20 Hz and 30 Hz.

3.4 Seismic Slope Motion Results of Earthquakes in Multipoint Measurement

Figure 13 shows that the microseism and vibration monitoring system can sensitively monitor the earthquake (Mj5.1) that happened in Southern Ibaraki prefecture (12/Apr/2020 00:44) using five devices installed on the slope. From left to right, graphs show the observed earthquake acceleration wave results of MEMS seismic sensor no1 to no5. The graphs show the measured acceleration waveforms (gal) of NW, SW, UD directions from top to bottom. The result of K-NET Odawara station (KNG013) which is 5 km far away from the slope was also shown in right. The K-NET is a strong-motion seismograph network managed by NIED (National Research Institute for Earth Science and Disaster Resilience, Japan). The results comparing the acceleration wave of the developed seismic sensor unit with K-NET showed that the shape of the acceleration waveform is almost the same. It indirectly explained that the developed seismic sensor can measure earthquakes sensitively.

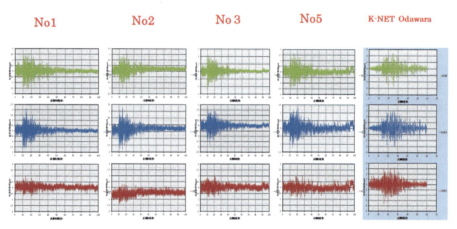

Fig. 13. Earthquake waveform of slope ground by seismic observation (12/Apr/2020 00:44, Southern Ibaraki prefecture Mj5.1, Seismic intensity 4, and K-NET (KNG013 Odawara))

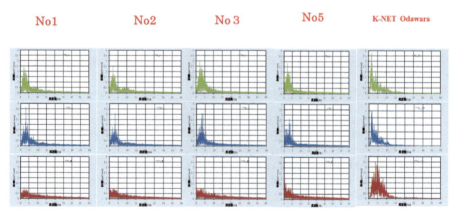

Fig. 14. The Fourier spectrum of slope ground by seismic observation (12/Apr/2020 00:44, Southern Ibaraki prefecture Mj5.1, Seismic intensity 4, and K-NET (KNG013 Odawara))

Figure 14 shows the Fourier spectrum results of measured the earthquake by seismic sensor no.1 to no.5. The result of K-NET Odawara station (KNG013) is shown in right.

Since the installation of these seismic sensors, the earthquakes that occurred in the Kanto region of Japan were mostly monitored. By long-term certification experiment for more than 2 years, the new type of microseism and vibration monitoring system developed by authors has been found to withstand long-term measurements with comparable.

4   Conclusion

Compared to conventional microtremors and seismometers, the new type of sensor developed has been found to withstand long-term measurements with comparable reliability. Based on the above results, it is considered that multi-point and reliable seismic technology and its application will lead to the strengthening of disaster response capabilities when tackling ground disasters that are generally uncertain.

When making efforts to deal with geo-ground disasters that are generally uncertain and difficult to predict, authors believe that the spread of small, inexpensive, and reliable sensing technology and the use of the results will lead to the strengthening of disaster response capabilities. For example, Table 2 shows the candidates for deploying the measurement system. If its implementation and dissemination accelerate, it is expected that it will contribute to the achievement of SDGs (Sustainable Development Goals) through the reduction of ground disasters.

Table 2. Future perspective for the usage of the new type of measurement systems.

Measurement technology	Fields of application
Microseisms/Vibration measurement	Evaluation of ground shaking
	Estimating the velocity structure of the ground and establishing a ground model
	Estimating the depth distribution of the support layer of ground
	Safety monitoring of slopes and rockfalls
	Detection of secular change of geo-ground structure, real-time damage diagnosis
	Estimating the location of the slope failure
	Verification of effects after measures such as ground improvement

References

Kawase, H., Sánchez-Sesma, F.J., Matsushima, S.: The optimal use of horizontal-to-vertical spectral ratios of earthquake motions for velocity inversions based on diffuse-field theory for plane wave. Bull. Seismol. Soc. Am. **101**(5), 2001–2014 (2011)

Ministry of Foreign Affairs: Sustainable Development Goals (SDGs) and Japan's efforts, Homepage. https://www.mofa.go.jp/mofaj/gaiko/oda/sdgs/pdf/SDGs_pamphlet.pdf

Miura, N.: Measurement method of engine oscillating force and its results. J. Jpn. Soc. Mar. Eng. **44**(2) (2009)

Okimura, T., Torii, N., Horie, H., Yogawa, C.: Application of constant tremor measurement to ground type determination. Research Center for Urban Safety and Security, Kobe University, No. 11 (March 2007)

Uchimura, T., Towhata, I., Wang, L.: Simple monitoring method for precaution of landslides watching tilting and water contents on slopes surface. Landslides **7**(3), 351–357 (2010)

Cyclic Failure Characteristics of Silty Sands with the Presence of Initial Shear Stress

Xiao Wei[1], Zhongxuan Yang[1], and Jun Yang[2(✉)]

[1] Zhejiang University, Hangzhou, Zhejiang, China
[2] The University of Hong Kong, Hong Kong, China
junyang@hku.hk

Abstract. The liquefaction of silty sands remains an outstanding issue since it triggers catastrophic hazards in recent earthquake events. The cyclic failure pattern is one of the fundamental aspects of liquefaction analysis. However, due to various influencing factors, such as packing density, confining pressure, initial shear stress, cyclic loading amplitude, etc., the failure patterns are less well understood and the underlying mechanism remains unclear. This study presents a series of laboratory testing results to identify the cyclic failure patterns of silty sands considering different soil states, fines contents, initial static shear stress, etc. The key finding is that the failure patterns are related to the states of soils and the cyclic loading characteristics, i.e., the combination of initial shear stress, density, confining stress and cyclic loading amplitude. Limitations of the existing prediction methods are discussed, and future work is also suggested.

Keywords: Liquefaction · Cyclic loading · Failure patterns · Silty sand · Initial shear

1 Introduction

Soil liquefaction remains an unsolved problem in both engineering practice and scientific research, even after decades of investigations since the 1964 Niigata earthquake. Identifying cyclic failure pattern and evaluating cyclic resistance of sand are two fundamental aspects of liquefaction assessment [1]. Previous investigations have revealed that both cyclic failure patterns and liquefaction resistance are influenced by the initial states (i.e., the packing densities and the effective confining pressures) of a given sand [2, 3]. In addition, the presence of initial static shear stress (τ_s), which is induced by the ground conditions (Fig. 1a,b), also plays an important role in altering the cyclic resistance [2, 3] and the failure pattern [4, 5], by changing the seismic stress cycles (Fig. 1c to e). It should be noted that most of the previous investigations focused on the liquefaction resistance of sands, but there have been relatively fewer investigations on the failure patterns of sands.

Based on several experimental studies using clean sands, typical failure patterns of sands are identified and categorized into flow-type failures, cyclic mobility, and plastic strain accumulation (e.g., [4]). Some researchers also suggested that there could be correlations between the monotonic and the cyclic soil responses [5, 6]. However, there is still no reliable framework to predict the cyclic failure patterns, probably because most of these studies focusing on the cyclic failure patterns of sands were based on limited testing conditions and not applicable to reveal the complicated interactions between the soil's initial states and the stress reversal condition on the cyclic failure patterns. Moreover, recent earthquake events [7, 8] revealed that liquefaction of silty sands could induce extensive damages and catastrophic hazards. This is partly because current liquefaction assessment methods are mainly based on clean sands, while the liquefaction characteristics of silty sands remain less well understood [9].

This study presents a systematic experimental program together with a detailed analysis of the failure patterns of silty sands under a variety of initial states of soil and loading conditions. Several important factors affecting the failure patterns of silty sands are identified and discussed.

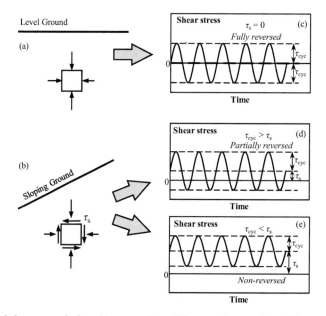

Fig. 1. Initial shear stress induced by ground conditions, and reversal conditions of stress cycles

2 Experimental Program

2.1 Materials

Toyoura sand (TS) is a uniform silica sand that consists of mainly sub-angular particles. Non-plastic crushed silica fines, with angular particle shape, were added into Toyoura sand at different fines contents (FC). The silty sands are denoted by TSS with a number indicating the fines content (e.g., TSS10 is the mixture with FC = 10%). The sand and the silt are different in particle shape, however, a recent study [10] has shown that the shape of sand particles and the particle size disparity are two major factors controlling the cyclic behavior and resistance of silty sands. The particle size distribution curves of the test materials are presented in Fig. 2 and the basic properties of these materials are summarized in Table 1.

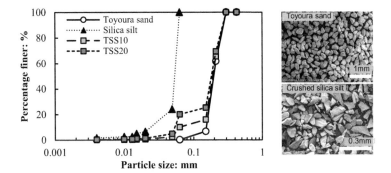

Fig. 2. Particle size distribution curves and SEM images of the tested materials

Table 1. Basic properties of the tested materials

Material	Gs	D_{50}	C_u	e_{max}	e_{min}
Toyoura sand	2.64	0.199	1.37	0.977	0.605
Crushed silica silt	2.65	0.054	2.17	–	–

2.2 Testing Procedures

Specimens (diameter $= 71.1$ mm, height $= 142.2$ mm) were reconstituted by the moist tamping method with the under-compaction technique[11]. After the percolation of CO_2 and de-aired water, the specimens were saturated by increasing the back pressure to at least 300 kPa or achieving a B value of at least 0.95.

In triaxial tests, the stress state on the maximum shear stress plane (the inclined plane with an angle of 45° to the horizontal plane) is used to simulate the stress state on the horizontal plane of a soil element that is subjected to seismic loading. For a stress state without initial static shear stress, the specimen was consolidated isotropically. For a stress state with non-zero initial static shear stress, the specimen was anisotropically consolidated by increasing the axial and the radial stresses in small increments, maintaining a constant stress ratio until the desired stress condition was reached. The initial static shear stress ratio, α, is defined by the following equation

$$\alpha = \frac{q_s}{2\sigma'_{nc}} = \frac{\sigma'_{1c} - \sigma'_{3c}}{\sigma'_{1c} + \sigma'_{3c}} \tag{1}$$

where q_s is the initial static deviatoric shear stress; $\sigma'_{nc} = (\sigma'_{1c} + \sigma'_{3c})/2$, is the effective normal stress on the maximum shear stress plane of the specimen, after consolidation; σ'_{1c} and σ'_{3c} are the axial and the radial effective stress after consolidation. After consolidation, deviatoric stress cycles with an amplitude of q_{cyc} were applied to the specimen under undrained conditions. The amplitude of the cyclic loading is characterized by the cyclic shear stress ratio, CSR, defined as follows.

$$CSR = \frac{q_{cyc}}{2\sigma'_{nc}} \qquad (2)$$

3 Failure Characteristics

3.1 Typical Failure Patterns

The typical failure patterns can be categorized into three general types, namely flow-type failure, cyclic mobility, and strain accumulation. The packing density of the specimen is one of the major factors affecting the failure patterns. For specimens at a relatively loose state, flow-type failure may take place. Figure 3 presents a typical flow-type failure, in which the specimen has a post-consolidation void ratio (e_c) of 0.906. This type of failure is characterized by a sudden and rapid development of axial strain (ε_a) without a significant pre-failure strain development, after a certain level of excess pore water pressure (Δu) development. The three plots in Fig. 3 are the stress-strain relationship (q-ε_a), stress path in the q-p' plane, and excess pore water pressure generation with the number of stress cycles (N). Figure 4 presents one more example of flow-type failure for a loose specimen ($e_c = 0.907$), but with $\alpha = 0.4$. A major difference between the failure characteristics of the two specimens is the direction of flow. For the specimen with $\alpha = 0$, the specimen failed in the triaxial extension side, but the specimen with $\alpha = 0.4$ failed in the triaxial compression side. This is due to biased cyclic stress caused by the presence of initial shear stress.

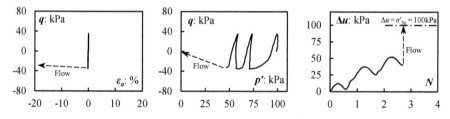

Fig. 3. Flow-type failure for TSS10, $e_c = 0.906$, $\alpha = 0$, $\sigma'_{nc} = 100$ kPa, $CSR = 0.175$

For relatively dense specimens, there are two types of failure patterns depending on the reversal conditions of the cyclic stress. Figure 5 presents a specimen loaded by reversed cyclic stress, exhibiting a typical failure pattern known as cyclic mobility. The excess pore water pressure increases while the effective stress decreases cyclically until a state called "initial liquefaction" is reached, where the effective stress transiently

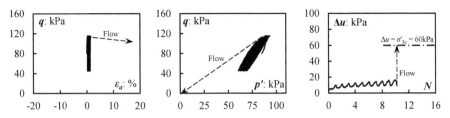

Fig. 4. Flow-type failure for TSS10, $e_c = 0.907$, $\alpha = 0.4$, $\sigma'_{nc} = 100$ kPa, $CSR = 0.175$

equals zero for the first time. The specimen undergoes two transient liquefaction states during each subsequent stress cycle when the deviatoric stress reverses its direction, and large deformation takes place when the state of the specimen is reaching the transient liquefied state. Then, the stiffness and the strength of the specimen recovered due to dilation in the following loading process. Figure 6 presents an example of plastic strain accumulation, the other type of failure pattern for dense specimens, which occurs if the cyclic stress does not reverse its direction. The major characteristic of this type of failure is the continuous accumulation of irrecoverable strain (i.e., plastic strain) in each stress cycle. The effective stress may never achieve zero even if the deviatoric stress arrived at zero in each cycle. One more characteristic of the specimen is the generation of negative excess pore water pressure due to dilation of the specimen.

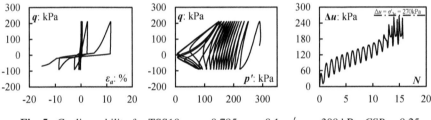

Fig. 5. Cyclic mobility for TSS10, $e_c = 0.795$, $\alpha = 0.1$, $\sigma'_{nc} = 300$ kPa, $CSR = 0.25$

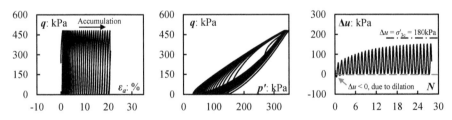

Fig. 6. Plastic strain accumulation for TSS10, $e_c = 0.793$, $\alpha = 0.4$, $\sigma'_{nc} = 300$ kPa, $CSR = 0.4$

3.2 Factors Affecting Failure Patterns

It has been shown that the failure patterns can be affected by the packing density and the reversal condition of the cyclic loading. However, other factors could alter the failure pattern of the specimens even when the packing density is the same and the reversal condition does not change.

Figure 7 presents an example that confining pressure can change the failure patterns. When the effective stress is 100 kPa, the TSS20 specimen with an initial void ratio of 0.796 exhibits plastic strain accumulation under the condition of $CSR = \alpha = 0.4$ (Fig. 7a). However, when the effective stress is 300 kPa, the TSS20 specimen with the same initial void ratio ($e_c = 0.796$) exhibit flow-type failure for $\alpha = 0.4$, while $CSR = 0.25$ (Fig. 7b). In addition, Fig. 8 presents an example that the failure pattern is affected by the loading condition. The two specimens with nearly the same void ratio ($e_c = 0.847$ and 0.849) were loaded by fully reversed stress cycles, i.e., $\alpha = 0$. For $CSR = 0.15$, cyclic mobility is observed as shown in Fig. 8a, while for $CSR = 0.2$, flow-type failure takes place as shown in Fig. 8b. The results from the two specimens indicate that the cyclic failure pattern can be affected by a small change in the amplitude of cyclic stress.

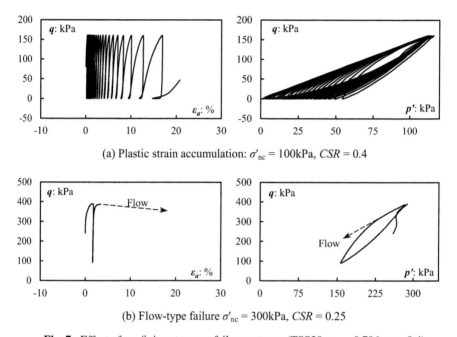

(a) Plastic strain accumulation: $\sigma'_{nc} = 100$kPa, $CSR = 0.4$

(b) Flow-type failure $\sigma'_{nc} = 300$kPa, $CSR = 0.25$

Fig. 7. Effect of confining stress on failure patterns (TSS20, $e_c = 0.796$, $\alpha = 0.4$)

4   Discussions

Mohamad and Dobry [5], as well as Chiaro et al. [6], have proposed empirical charts to predict the cyclic failure patterns of sands, by considering the reversal condition of the cyclic stress (i.e., combining initial static shear stress and the cyclic stress amplitude) and the monotonic shearing behavior of the sand. However, these proposed charts are based on limited initial states. Figure 9 presents a summary of the failure patterns that are observed for the TSS10 and TSS20, regarding the effects of void ratio and confining pressure. It is clear that the failure patterns are dependent on the initial states of sands, and even fines content. Figure 9b has shown that TSS20 failed in the pattern of flow-type but the TSS10 exhibited cyclic mobility, for specimens with nearly the same void ratio ($e_c = 0.791$–0.794) and consolidated to $\sigma'_{nc} = 300$ kPa. For this reason, a more comprehensive investigation is needed to take into account the impacts of soil states, initial static shear stress, amplitude of cyclic stress, soil properties, and so on.

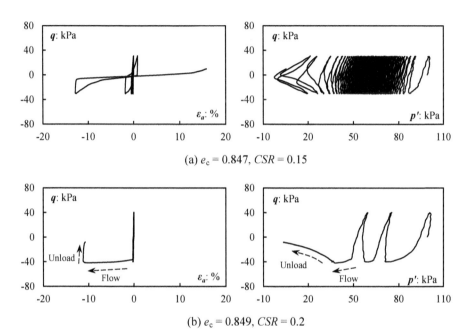

(a) $e_c = 0.847$, $CSR = 0.15$

(b) $e_c = 0.849$, $CSR = 0.2$

Fig. 8. Effect of CSR on failure patterns (TSS10, $e_c = 0.847$–0.849, $\alpha = 0$, $\sigma'_{nc} = 100$ kPa)

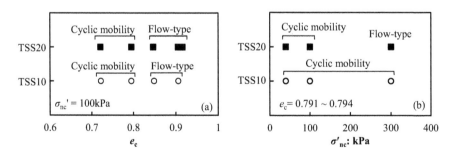

Fig. 9. Summary of effects initial states on the failure patterns of silty sands ($\alpha = 0$)

5   Conclusions

In this paper, several types of cyclic failure patterns of silty sands are presented and discussed. It is found that the failure patterns turn from flow-type failure to non-flow type (i.e., cyclic mobility and plastic strain accumulation) when the packing density of the specimens increases. For loose specimens, it appears that the presence of initial static shear stress does not affect the failure pattern. However, for relatively dense specimens, the reversal condition of the cyclic loading, which is affected integrally by the initial static shear stress and the cyclic stress amplitude, determines the failure pattern be cyclic mobility or plastic strain accumulation. In addition, the confining pressure and the amplitude of cyclic stress can also affect the failure pattern under otherwise similar conditions. The characterization of failure pattern is an important but less well-investigated issue in liquefaction evaluation and, thus, further investigations are needed.

Acknowledgement. The financial support provided by the Research Grants Council of Hong Kong (17206418), the National Natural Science Foundation of China (No. 52108351), and the Fundamental Research Funds for the Central Universities, Zhejiang University (2021QNA4021) is acknowledged.

References

1. Ishihara, K.: Liquefaction and flow failure during earthquakes. Géotechnique **43**(3), 351–451 (1993)
2. Yang, J., Sze, H.Y.: Cyclic behaviour and resistance of saturated sand under non-symmetrical loading conditions. Géotechnique **61**(1), 59–73 (2011)
3. Wei, X., Yang, J.: Cyclic behavior and liquefaction resistance of silty sands with presence of initial static shear stress. Soil Dyn. Earthq. Eng. **122**, 274–289 (2019)
4. Sze, H.Y., Yang, J.: Failure modes of sand in undrained cyclic loading: impact of sample preparation. J. Geotech. Geoenviron. Eng. **140**(1), 152–169 (2014)
5. Mohamad, R., Dobry, R.: Undrained monotonic and cyclic triaxial strength of sand. J. Geotech. Eng. **112**(10), 941–958 (1986)
6. Chiaro, G., Koseki, J., Kiyota, T.: New insights into the failure mechanisms of liquefiable sandy sloped ground during earthquakes. In: 6th International Conference on Earthquake Geotechnical Engineering, No. 200 (2015)
7. Cubrinovski, M., et al.: Geotechnical reconnaissance of the 2010 Darfield (New Zealand) earthquake. Bull. N. Z. Soc. Earthq. Eng. **43**(4), 243–320 (2010)
8. Mason, H.B., Gallant, A.P., Hutabarat, D., Montgomery, J., Reed, A.N., Wartman, J.: Geotechnical reconnaissance: the 28 September 2018 M7.5 palu-donggala, indonesia earthquake. Geotechnical Extreme Events Reconnaissance (GEER) Association, United States (2019)
9. Wei, X., Yang, J.: Characterizing the effects of fines on the liquefaction resistance of silty sands. Soils Found. **59**(6), 1800–1812 (2019)
10. Wei, X., Yang, J.: Characterising the effect of particle size disparity on liquefaction resistance of non-plastic silty sands from a critical state perspective. Géotechnique, Ahead of Print (2021)
11. Ladd, R.: Preparing test specimens using undercompaction. Geotech. Test J. **1**(1), 16–23 (1978)

Particle Fabric Imaging for Understanding the Monotonic and Cyclic Shear Response of Silts

Dharma Wijewickreme[(✉)] and Ana Maria Valverde Sancho

University of British Columbia, Vancouver, Canada
`dharmaw@civil.ubc.ca`

Abstract. The current knowledge from experimental research has shown the significant effect of particle structure (fabric) on the monotonic and cyclic shear behavior of silts, in addition to the well understood influence of void ratio and effective confining stress. Advancing the knowledge on this matter requires systematic quantification of particle fabric in a given silt matrix in terms of individual particle parameters (e.g., dimensional sizes, volumes, shapes, orientations) as well as inter-particle contact arrangements. In the research work undertaken with this background, new methodologies were developed for X-ray micro-computed tomography (μ-CT) imaging of silts with specific attention paid to: sampling and preparing of silt specimens, scanning parameters to obtain the needed image resolutions, and digital processing of images to capture individual particle data. It is shown that μ-CT imaging is able to effectively image and capture 3-dimensional fabric of silt. The particle fabric in silt specimens reconstituted via gravity deposition is illustrated using the μ-CT images produced using standard-size silica particles. The fabric(s) developed from the imaging of natural low plastic silt is also presented, and the findings are shown to be well in accord with those inferred from the mechanical laboratory element testing of the same silt. The work contributes to the accounting for fabric in understanding the macroscopic shear behavior of natural silts.

Keywords: Soil fabric/microstructure · X-ray micro-computed tomography · Low-plastic silts · Liquefaction of soils

1 Introduction

Liquefaction susceptibility of soils under seismic shaking has been studied globally for more than 50 years with much of the focus placed in the performance of saturated loose sands. Mainly as a result of the liquefaction-induced damaged observed in the 1991 Chi-Chi, 1999 Kocaeli, and 2011 Christchurch earthquakes [1–3], seismic performance of silty soils have also been receiving increased attention. With this background, the study of low-plastic silty soils in the Fraser River Delta have been a topic of extensive research at the University of British Columbia (UBC), Vancouver, BC, Canada, for over 20 years. In this research, the shear behavior of relatively undisturbed and reconstituted slurry

L. Wang et al. (Eds.): PBD-IV 2022, GGEE 52, pp. 617–630, 2022.
https://doi.org/10.1007/978-3-031-11898-2_37

deposited silts has been investigated using a variety of methods including laboratory direct simple shear (DSS) and triaxial testing. Effect of factors such as confining stress, void ratio, particle size, etc., on the monotonic and cyclic shear loading response of silts have been studied.

Soil fabric refers to the spatial arrangement of individual particles, particle groups, and pore spaces in soils [4]. Fabric focuses primarily on the characteristics of individual particles such as shape and size, and how this has led to their particulate arrangement [5, 6]; the voids in this interpretation are those that are a direct result of the physical orientation and packing of the soil particles [7]. There are two main components to fabric: (a) particle's discrete orientation; and (b) its relative position with respect to adjacent particles [8]. There is evidence to indicate the significant effect of particle fabric and microstructure on the mechanical behavior of soils [9–11]. In particular, UBC research has shown that reconstituted Fraser River silt (despite having a higher density under essentially identical initial effective consolidation stress σ'_{v0}) consistently exhibits a weaker response compared to that observed from the undisturbed specimens of the same material. These significantly different behavioral displays by the same material suggest that it is important and timely to examine and systematically account for the effect of particle structure (soil fabric) on the silt response, in addition to the traditionally well studied effects of void ratio (e) and σ'_{v0}.

Research over many years has demonstrated that a given arrangement of particles in a granular mass undergo progressive changes when subjected to shearing stresses where the concentration of contact normal tends to increase in the major principal stress direction and particles align their longitudinal axis along the minor principal stress [12, 13]. Past research in sands, comparing the behavior between soil specimens prepared using different reconstitution techniques, has shown that the macroscopic monotonic and cyclic behavior of soils is highly affected by the fabric and microstructure [9, 14–16]. Moreover, studies have demonstrated that particle reorientation of sands occurs during consolidation [17]. Likewise, particle sphericity and aspect ratio have also known to cause varying changes in sample fabric and microstructure [18]. The effect of particle fabric on the coefficient of lateral earth pressure at rest (K_0) has also been demonstrated through compression testing of sands using an instrumented oedometer [19].

With this background, a research program using X-ray μ-CT imaging technology, has been undertaken at UBC to better comprehend the macro-level mechanical shear behavior observed from geotechnical element testing of natural silts. The work undertaken at UBC so far in visualizing silt fabric has resulted in promising observations thereby providing the impetus for undertaking further research on this subject [20–23]. In particular, technology and methodology for preparing silt specimens for X-ray μ-CT imaging as well as qualitative and quantitative post-processing of images have been developed. This paper presents these approaches and demonstrates their suitability for understanding the fabric of silt-size particle matrices. Images obtained from commercially available standard-size silica granules as well as natural silt originating from the Fraser River Delta of British Columbia, Canada, have been used for the intended purpose.

2 Background

2.1 Effect of Fabric on the Behavior of Silts

Constant volume monotonic direct simple shear (DSS) testing at UBC has shown that reconstituted Fraser River Delta silt specimens prepared using slurry deposition method exhibit lower shear strength at all levels of confinement compared to those from undisturbed specimens. This behavior manifests despite reconstituted specimens having a denser arrangement compared to relatively undisturbed specimens. Moreover, the undisturbed specimens displayed a strain hardening response in contrast to the behavior observed for reconstituted specimens. The void ratio (e) and vertical effective stress (σ'_v) states after consolidation as well as after reaching relatively large shear strain levels ($\gamma \sim 15\%$) from these testing are plotted in Fig. 1. The e-σ'_v states of undisturbed Fraser River silt specimens, after-initial-consolidation and $\gamma \sim 15\%$, appear to follow straight lines. In a corresponding way, these semi-log-linear trends seem to also prevail for the after-initial-consolidation and $\gamma \sim 15\%$ states obtained for the reconstituted material. However, the lines for the reconstituted silt are at significantly different locations from that noted for the undisturbed silt; this is only explainable by the potential differences in the particle fabric and microstructure between the undisturbed and reconstituted soils.

Likewise, cyclic DSS testing has shown that reconstituted natural undisturbed silt generally exhibits a weaker response compared to that observed from the undisturbed specimens of the same material. Typical cyclic resistance ratio (CRR) observed between undisturbed specimens and reconstituted specimens in DSS loading is shown in Fig. 2. The reconstituted specimens displayed increased degree of stiffness degradation and strain accumulation potential compared to the relatively undisturbed specimen. It is evident that the decrease of void ratio and disturbance of natural fabric are two competing factors in governing the cyclic shear resistance of reconstituted material when compared to the cyclic shear resistance of undisturbed specimen. Clearly, these monotonic and cyclic DSS results highlighted the significant role played by the fabric in controlling the soil behavior and that it needs to be accounted in predicting the mechanical shear response of silts.

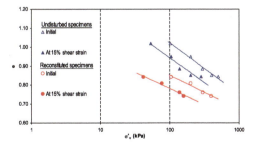

Fig. 1. Void ratio (e) versus log σ'_v relationships for undisturbed and reconstituted specimens of Fraser River silt immediately after initial consolidation compared with those after reaching 15% shear strain – extracted from [24].

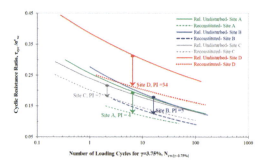

Fig. 2. Cyclic Resistance Ratio versus Number of loading cycles to reach $\gamma = 3.75\%$ curves from constant-volume cyclic DSS tests indicating the reduction of cyclic shear resistance of reconstituted specimens with respect to relatively undisturbed specimens – results for soils from three sites from the Fraser River Delta as extracted from [10, 25].

2.2 Evolution and Quantification of Fabric

The effects of soil fabric are difficult to observe and quantify in the laboratory. Several main techniques to obtain data for fabric quantification include: physical modeling using photoelasticity; imaging techniques using optical microscopy, such as scanning electron microscopy or X-ray computed tomography; numerical modeling using methods like discrete element methods (DEM). The noninvasive nature of imaging techniques represent a major advantage for conserving the soil fabric, and is increasingly being used to explore geomaterials. In this effort, methods such as scanning electron microscopy (SEM) and X-ray diffraction have led mostly to the retrieval of information on individual (or limited number of) particles; however, these methods have not been enabled to obtain a wholesome understanding of the particulate arrangements in a given soil matrix. On the other hand, by augmentation of ideas and applications emerging from the fields of medicine, computed tomography (CT) has been shown to be effective in visualizing the soil fabric. In spite of significant studies conducted using X-ray micro-computed tomography (μ-CT) with respect to sands and clays [26, 27], only a very few studies have been conducted to image fabric and microstructure of silt size particles [11, 28, 29]. As such, there is a need to further examine and systematically account for the particulate fabric in understanding the behavior of silts.

In recent years, the development of scanning technologies and high-performance computers has allowed large, three-dimensional (3D) datasets to be produced at micro- and nano-level resolutions [20, 30–32]. These advancements and refinements now present an opportunity to extend μ-CT approaches to image and capture 3-dimensional fabric of silt specimens (i.e., particles sizes ranging between 2 μm – 74 μm) and investigate their particle characteristics and influence on monotonic and cyclic shear behaviour. With this background, a comprehensive research program comprising the following multi-component objectives has been undertaken at UBC:

a) Develop an appropriate soil sampling and preparation approaches for X-ray μ-CT imaging of silts

b) Identify μ-CT devices and scanning parameters to obtain the needed resolutions for imaging of silts
c) Conduct the needed digital processing of images to capture individual "particle data" (i.e., particle length, width, breath, thickness, volume, orientation, etc.) Calibrate/verify the developed methodologies and illustrate the capacity of μ-CT imaging to assess the fabrics derived from a given silt
d) Examine the fabrics derived from the silt materials tested

This paper presents the outcomes arising from the above and demonstrates the suitability of μ-CT imaging to investigate the fabric of silt-size particle matrices and, in turn, contributes to correlating the findings with the macroscopic behavior of natural silts. For completeness, some of the findings that have been reported previously by the authors are also briefly summarized and encompassed herein.

2.3 Test Materials, Preparation of Specimens, and Imaging Devices

Description of Material Used for Imaging

The research was undertaken using two silt-size (i.e., particle sizes ranging between 2 μm to 74 μm) granular materials. The properties and index parameters of the two silts are given in Table 1. One of the materials is a commercially available pre-calibrated, standard-size silica particles manufactured by Silicycle, Quebec, Canada; the standard silica material represents the coarser range of silt particles and two shapes. The use of these standard particle sizes provided an avenue to calibrate/verify the developed procedures and methodologies, conduct "bench mark" studies, and also extending the findings to study the full range of particle sizes present in natural silts. Relatively undisturbed samples of Fraser River low plastic silt retrieved from the Lower Mainland of British Columbia in Vancouver, Canada was chosen as the second test material representing natural silts.

Table 1. Soil properties and index parameters of silts used in X-ray μ-CT imaging

Parameter	Standard-size silica particles		Fraser River silt
	Material type No.		
	I-40-63	S-45-63	
Specific gravity, Gs	2.02	1.89	2.75
Plasticity index, PI	NA	NA	7
Particle size range (μm)	40–63	45–63	2–74
Particle shape	Irregular	Spherical	Irregular

X-ray μ-CT scanning resolution and associated requirements limited the size of the specimen that could be imaged. Based on the initial studies, it was determined that specimens of 5-mm diameter or less are required to meet the resolution requirements to characterize silty material using X-ray μ-CT imaging. This required containing silt specimens in tubes made of material having a significantly lower density compared to the tested soil grains (i.e., if the outer tubing material is very dense, the X-ray source energy will be highly absorbed, thus affecting the quality of imaging of the inner material). With this in mind, metal containers were not considered. After investigating tubing made of several material types, including glass and polylactic acid (PLA) used as 3D printing material, it was found that thin-plastic tubing (having a nominal diameter D = 5.0 mm and t $\cong$ 0.14 mm thickness with tube-diameter: particle-diameter ratio > 65) produced good imaging contrast with the inner silt specimen. In addition to securely holding the silt during imaging, the chosen plastic tube could be used to obtain relatively undisturbed sub-samples from larger laboratory samples of natural soils – i.e., similar to the approach in minimizing sample disturbance by using thin-walled tubes in the field.

Imaging Devices
In seeking suitable imaging equipment, μ-CT scanners from the following university research facilities were assessed to obtain non-destructive images of the UBC silt specimens: i) Composites Research Center at UBC Okanagan Campus (UBC-O); ii) Department of Civil Engineering at Monash University, Clayton Campus, Australia (Monash); and iii) Pulp and Paper Research Center (PPRC) at UBC main campus in Vancouver (UBC-V). Based on a detailed assessment [20, 23], the ZEISS Xradia 520 Versa equipment (manufactured by Zeiss International, Oberkochen, Germany) available at the UBC-V PPRC proved to be the most suitable scanner at achieving the highest resolution, meaning the smallest voxel size, which was 0.869 μm.

Image Acquisition, Processing and Analysis
The X-ray μ-CT scans produce 2D images of greyscale intensity, this parameter relates closely to the density of the material penetrated by the X-rays. Some thousands of consecutive 2D images are later stalked and reconstructed into a 3D image (Note: images were obtained considering the central 1-mm zone of the tube sample so that the imaged zone is well away from the tube walls). A given μ-CT scan output requires further processing to perform quantitative and qualitative image analysis with the process consisting of two steps: (i) image pre-processing; and (ii) image segmentation, and these steps were performed using Avizo 9.7 software [33]. This software has been used by various researchers to successfully study particulate geomaterials [34, 35].

Reconstructed images from the scanning process were first filtered to improve the inherent noise from scanning and rectify artifacts. A given specimen should always be scanned at the highest resolution available for image sharpness; however, higher resolutions would introduce undesirably extra-noise in the resulting grayscale image [36]. In the approach developed herein at UBC, three different filters were applied: (i) 3D Median filter (MF); (ii) Recursive Exponential filter (REF); and (iii) Non-local means filter (NLMF). Typical outcomes from filtering during image-processing are given in Fig. 3. The MF is a low-pass filter that reduces the contrast and soften the edges in the image - reducing the "salt and pepper" noise induced by the scanner. The REF was used to smoothen the image. The NLMF enhanced the contrast and softened the image while preserving the edges of the particles.

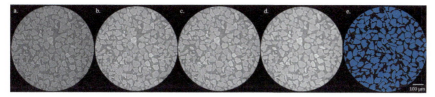

Fig. 3. Application of filters and thresholded μ-CT images: (a) raw image; (b) median filtered; (c) recursive exponential filtered; and (d) non-local means filtered; (e) final thresholded image.

Image segmentation consists of the application of a sequence of procedures to separate the in-contact solid particles from each other, and the solid particles from the void spaces. The first step to segment particles from the surrounding medium in a filtered image is typically conducted using a "thresholding" method that will produce a binary image that represents particles and its surrounding medium (usually air or water). The thresholding method by [37] is commonly used due to its simplicity and relatively fast application, and it has been widely used in sands [38] and incorporated in the Avizo software. Typical cross-sectional binary representation outcome of a thresholded image from Avizo software is illustrated in Fig. 3e.

Once data binarization is completed by thresholding methods, a segmentation algorithm separates the particles. A common procedure is to use a distance map transform and a watershed algorithm [39] that simulates water being poured over a landscape with peaks and valleys [13, 38]. Finally, the particles are individually labeled in order to extract particle data, and a quantitative label analysis is required at this stage. Commercially available software is usually capable of extracting information related to particle data. These values constitute the basis for deriving grain size distributions, rose diagrams, etc.

Fabric Quantification Approaches

The data obtained with respect to grain orientation and size can be used to quantify fabric by scalar or directional parameters. Scalar measurements include the traditional void ratio approach, while directional parameters depend on the particle orientation along the long axis or contact distributions. In a 3D study, the directionality of the long axis of a given particle (particle orientation) can be described by angles denoted by θ and φ with respect to a radial coordinate system; where θ = the direction of long axis on a horizontal plane and φ = direction of the long axis with respect to the vertical. Such particle orientation "big-data" can be expressed by statistical representations like rose diagrams – i.e., angular histogram plots which displays directional data and frequency for a determined feature – which has been used herein to demonstrate the efficacy of the μ-CT scanning.

3   Findings from Standard-Size Silica Matrices

The outcomes from investigations on standard-size silica particle matrices provided a way to calibrate/verify the suitability of μ-CT imaging technology to assess the fabric of silts and contribute to understanding the fabric of silt matrices as presented below.

3.1   Observations from Visual Inspection of Images

Figure 4(a) and (c) present a section through the raw imaging data obtained from irregular-shaped, standard-sized (40 and 63 μm size range), dry and saturated silica specimens. This greyscale intensity plots reflect the density detected by the imaging device on the plane of the section, and these values form the key input data for the image processing. Visual inspection suggests that the device resolution seem to have been sufficient to capture the full range of particle sizes above 40 μm as well as to recognize their irregular shapes. Figure 4(b) and (d) show the processed image, using Avizo software, with several filters to improve the image quality and segmentation algorithms to separate the particles, and obtain the required metrics. Again, it is notable that the segmentation algorithms have been able to identify/separate most particles in the image in spite of the limitations in digital image processing.

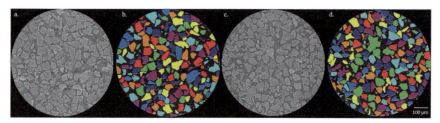

Fig. 4. Raw and processed tomography images for standard-size silica, respectively: (a) and (b) dry material; (c) and (d) saturated material.

3.2   Observations on Particle Shapes and Stratifications

A dry specimen of soil containing spherical-shaped silica zone (45 and 63 μm size range) overlying irregular-shaped silica layer (40 and 63 μm size range) was imaged. Figure 5(a) and 5(b) present a segmented section through the spherical and irregular shaped zones, respectively. Figure 5(c) presents the 3D view showing the two layers. The figures clearly demonstrate the ability of μ-CT imaging to identify/distinguish particle shapes as well as stratification/layering found in matrices of silt size particles.

3.3   Observations on Particle Size Distributions (PSDs)

The PSDs in particle assemblies form a vital component in accounting for the soil fabric. Traditionally, the PSDs are obtained by laboratory mechanical sieving and hydrometer testing as per ASTM standards [40, 41], respectively. The digital PSDs obtained from μ-CT imaging of the standard-sized silica grains were compared with those from mechanical sieving, laser diffraction analysis, as well as information available from the silica manufacturer. The details related to this work are described in detail in the companion paper in this conference by the authors [42]; as such, only the key findings are highlighted herein for brevity.

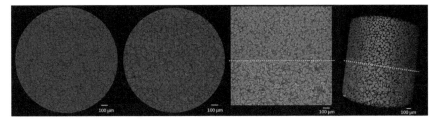

Fig. 5. Images from a subsample with two different shaped silica particles. (a) Cross-section through a spherical shaped silica zone; (b) Cross-section through an irregular-shaped silica zone; and (c) vertical section showing the layering; (d) 3D view showing the layering.

The PSDs obtained from the digital data analysis for both spherical and irregular standard-size silica particles were found to be in very good agreement with the specifications provided by the material manufacturer, and there was good alignment between the digitally derived PSDs for the saturated and dry specimens [42]. This showed that the imaging process is not significantly affected by the presence of water in the silt specimens; this is an important finding that confirms the suitability of the method to study the particle fabric in both saturated and unsaturated materials. The digital PSDs were also found to be in general accord with those from laser diffraction and mechanical analysis. Any observed differences were considered reasonable considering the limitations and differences in the techniques used. Overall, the work demonstrated the suitability of the fine-grained commercial silica as a "bench mark" research material and the potential of μ-CT for imaging silt fabric.

3.4 Evaluation of Fabric in Reconstituted Specimens

As noted earlier, the effect of particle fabric on the mechanical performance has been well recognized through tests conducted on sands. In consideration of this, two specimens of standard-sized (40 and 63 μm size range) dry silica particles were reconstituted using gravity deposition with one sample kept loose (as-deposited) and the other consolidated using vertical pressure applied using a plunger. The particle orientation data rose diagrams derived from the imaging of these specimens are presented in Fig. 6. The orientation of the principal axes of the particles in the consolidated specimen seem to align more parallel to the horizontal plane (Fig. 6b) than those of the particles in the relatively loosely deposited specimen (Fig. 6a). The results are in line with the expected possibility that the compressive stresses applied on the gravity deposited specimens would have made the longitudinal axes of the particles to assume a position relatively closer to the horizontal plane. Again, the findings demonstrate the ability of μ-CT imaging method to capture the particle fabric in silts.

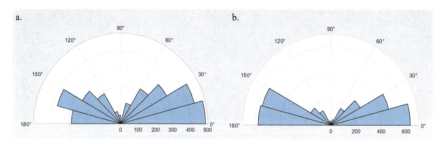

Fig. 6. Rose diagrams of particle principal axis orientation for reconstituted gravity deposited silica: (a) loose specimen; (b) consolidated specimen.

4   Findings from μ-CT Imaging of Fraser River Delta Silt

As noted earlier, 5-mm thin-walled plastic tubes sub-sampled from larger "parent" samples were required to investigate the fabric of natural silt. With this in mind, a "parent" sample was generated by one-dimensionally consolidating a reconstituted Fraser River silt by a slurry deposition to $\sigma'_{vc} = 200$ kPa in a 76-mm diameter oedometer. Three sub-samples for μ-CT imaging were obtained from the parent specimen [23].

The raw and processed images for a specific slice in each sub-sample of Fraser River silt is presented in Fig. 7 (top and bottom figures, respectively) to illustrate typical images from μ-CT technology. The 3D images, visually do not show any layering or bedding; this confirms the non-segregation (uniformity) expected by the reconstituted specimens formed using slurry deposition. The digital grain size distribution of the silt using μ-CT data for the 3 different sub-samples are presented in Fig. 8. The excellent match of the

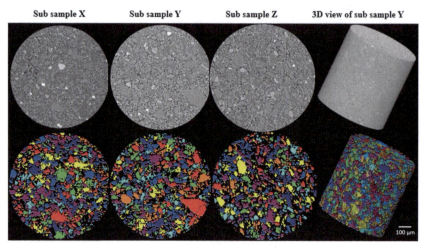

Fig. 7. Representative raw and processed images from three subsamples (X, Y, and Z) obtained from the same parent reconstituted specimen of Fraser River silt. (Note: Top Row – Raw greyscale images; Bottom Row – Segmented images).

grain size curves from these subsamples serves as evidence of very good uniformity within the parent sample from a grain size distribution point of view.

Particle orientation data derived from the 3 subsamples was used to create the rose diagrams shown in Fig. 9. Clearly, the principal axes of the particles in the subsamples seem to mainly align in directions close to the horizontal; this observation is much in accord with the previous observations related to particle orientations for gravity deposited specimens [8]. Furthermore, the similarity of the rose diagrams further supports the notion that the soil within the parent sample has a uniform constitution.

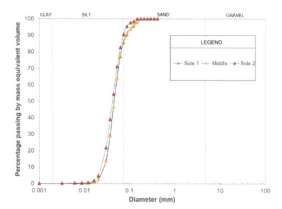

Fig. 8. Digital GSDs derived from the three subsamples (X, Y, and Z) obtained from the same parent reconstituted specimen of Fraser River silt.

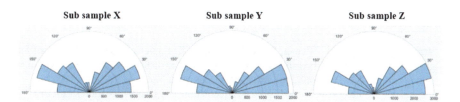

Fig. 9. Rose diagrams of particle principal axis orientation for the three subsamples (X, Y, and Z) obtained from the same parent reconstituted specimen of Fraser River silt.

5 Summary and Conclusions

Experimental evidence from laboratory shear testing clearly indicates that, in addition to the known effects of effective confining stress and void ratio, the particle fabric plays a critical role in governing the mechanical behavior of silts. Findings from the monotonic/cyclic shear testing research conducted at UBC on undisturbed and reconstituted natural low plastic silts, the void ratio and natural fabric have been realized as two important competing factors governing the shear resistance.

The research outcomes presented in this paper demonstrate the high potential of non-invasive X-ray μ-CT imaging technology to understand the soil fabric, and in turn support understanding the macro-level mechanical shear behavior of natural silts. Specially, the methodologies developed for preparing silt specimens for X-ray μ-CT imaging as well as qualitative and quantitative post-processing of such images to obtain key "particle parameters" (i.e., such as particle dimensions, volume, and orientation of axes of grains) for the expression of particle fabric are presented. Through examination of the orientation of the principal axes of particles found in specimens of standard-size silica grains, the difference in the fabric between loose and relatively dense silt is well illustrated. The validity of these observational trends is also extended to the fabric(s) derived from the imaging of natural low plastic silt originating from the Fraser River delta of British Columbia, Canada.

The methods developed so far have shown the excellent capability of X-ray μ-CT imaging to define the particle fabric for silt sizes above 40 μm. Additional work is underway to bring this resolution to image the particles below this size.

Acknowledgements. This research would not have been possible without the valuable financial support provided by the Natural Science and Engineering Research Council of Canada (NSERC). The collaborative support provided by Dr. Mark Martinez and Mr. James Drummond at the X-ray μ-CT imaging facility at the Pulp and Paper Centre at UBC main campus in Vancouver is deeply appreciated. Initial imaging support provided by the Composites Research Center at UBC Okanagan Campus (UBC-O) and the Department of Civil Engineering at Monash University, Clayton Campus, Australia (Monash) is also acknowledged. Thanks are due to Carlo Corrales for his support in the sample preparation and laboratory testing performed for this research as a UBC undergraduate research assistant.

References

1. Bray, J.D., Sancio, R.B., Riemer, M.F., Durgunoglu, T.: Liquefaction susceptibility of fine-grained soils. In: Proc. of the 11th International Conference on Soil Dynamics and Earthquake Engineering and 3d Int. Conf. on Earthquake Geotechnical Eng, pp. 655–662 (2004)
2. Idriss, I.M., Boulanger, R.W.: SPT-Based Liquefaction Triggering Procedures (2010)
3. Cubrinovski, M., Rhodes, A., Ntritsos, N., Van Ballegooy, S.: System response of liquefiable deposits. Soil Dynamics and Earthquake Engineering **124**(May 2018), 212–229 (2019)
4. Mitchell, J.K., Soga, K.: Fundamentals of Soil Behavior, 3rd edn. John Wiley & Sons, Hoboken, N.J (2005)
5. FitzPartick, E.A.: Micromorphology of Soils, 1st edn. Chapman & Hall Ltd., London (1984)
6. Santamarina, J.C.: Soil Behavior at the Microscale: Particle Forces. American Society of Civil Engineers (2001)
7. Brewer, R., Sleeman, J.R.: Soil structure and fabric: their definition and description. J. Soil Sci. **11**(1), 172–185 (1960)
8. Oda, M.: Initial fabrics and their relations to mechanical properties of granular material. Soils Found. **12**(1), 17–36 (1972)
9. Vaid, Y.P., Sivathayalan, S., Stedman, D.: Influence of specimen-reconstituting method on the undrained response of sand. Geotech. Test. J. **22**(3), 187–195 (1999)
10. Wijewickreme, D., Sanin, M.V.: Cyclic shear response of undisturbed and reconstituted low-plastic Fraser River silt. In: Proc. of the Geotechnical Earthquake Engineering and Soil Dynamics IV, pp. 1–10. ASCE, Sacramento, California (2008)

11. Wijewickreme, D., Soysa, A., Verma, P.: Response of natural fine-grained soils for seismic design practice: A collection of research findings from British Columbia. Canada. Soil Dynamics and Earthquake Engineering **124**(2019), 280–296 (2019)
12. Dabeet, A., Wijewickreme, D., Byrne, P.M.: Evaluation of Stress Strain Non-Uniformities in the Laboratory Direct Simple Shear Test Specimens Using 3D Discrete Element Analysis. Geomechanics and Geoengineering: An International Journal (2015). https://doi.org/10.1080/17486025.2014.979889
13. Fonseca, J., O'Sullivan, C., Coop, M.R., Lee, P.D.: Quantifying the evolution of soil fabric during shearing using directional parameters. Geotechnique **5 and 63**(6), 487–499 (2013)
14. Cuccovillo, T., Coop, M.: On the mechanics of structured sands. Géotechnique **49**(6), 741–760 (1999)
15. Been, K., Jefferies, M.G.: A state parameter for sands. Géotechnique **35**(2), 99–112 (1985)
16. Ibrahim, A.A., Kagawa, T.: Microscopic measurement of sand fabric from cyclic tests causing liquefaction. Geotech. Test. J. **14**(4), 371–382 (1991)
17. Paniagua, P., Fonseca, J., Gylland, A., Nordal, S.: Investigation of the change in soil fabric during cone penetration in silt using 2D measurements. Acta Geotech. **13**(1), 135–148 (2017). https://doi.org/10.1007/s11440-017-0559-8
18. Yang, H., Zhou, B., Wang, J.: Exploring the effect of 3D grain shape on the packing and mechanical behaviour of sands. Géotechnique Letters **9**(4), 1–6 (2019)
19. Northcutt, S., Wijewickreme, D.: Effect of particle fabric on the coefficient of lateral earth pressure observed during one-dimensional compression of sand. Can. Geotech. J. **50**(5), 457–466 (2013)
20. Wesolowski, M.: Application of computed tomography for visualizing three-dimensional fabric and microstructure of Fraser River Delta silt. MASc Thesis, Department of Civil Engineering. The University of British Columbia, Canada (2020)
21. Valverde, A., Wesolowski, M., Wijewickreme, D.: Towards understanding the fabric and microstructure of silt – initial findings of soil fabric from X-ray u-CT. In: proc. 17th World Conference on Earthquake Engineering. Sendai, Japan (2020)
22. Wesolowski, M., Valverde, A., Wijewickreme, D.: Towards understanding the fabric and microstructure of silt – feasibility of X-ray μ-Ct image silt structure. In: proc. 17th World Conference on Earthquake Engineering. Sendai, Japan (2020)
23. Valverde, A., Wijewickreme, D.: Towards understanding the particle fabric of silt – Assessment of laboratory specimen uniformity. In: Proc. GeoNiagara. Niagara, Canada (2021)
24. Sanin, M.V.: Cyclic shear loading response of Fraser River delta silt. Ph.D. thesis, Department of Civil Engineering. University of British Columbia, Vancouver, Canada (2010)
25. Soysa, A.: Monotonic and Cyclic Shear Loading Response of Natural Silts. M.A.Sc. thesis, Department of Civil Engineering. University of British Columbia, Vancouver, Canada (2015)
26. Hight, D.W., Leroueil, S.: Characterisation of soils for engineering purposes. In: Natural Soils Conference, p. 255 (2003)
27. Fonseca, J., Nadimi, S., Reyes-Aldasoro, C.C., O'Sullivan, C., Coop, M.R.: Image-based investigation into the primary fabric of stress-transmitting particles in sand. Soils Found. **56**(5), 818–834 (2016)
28. Zhang, M., Jivkov, A.P.: Micromechanical modelling of deformation and fracture of hydrating cement paste using X-ray computed tomography characterization. Compos. B **88**, 64–72 (2016)
29. Shen, H., Nutt, S., Hull, D.: Direct observation and measurement of fiber architecture in short fiber-polymer composite foam through micro-CT imaging. Compos. Sci. Technol. **64**, 2113–2120 (2004)
30. Ketcham, R.A., Carlson, W.D.: Acquisition, optimization and interpretation of X-ray computed tomographic imagery: applications to the geosciences **27**, 381–400 (2001)

31. Taina, I.A., Heck, R.J., Elliot, T.R.: Application of X-ray computed tomography to soil science: a literature review. Can. J. Soil Sci. **88**(1), 1–20 (2008)
32. Helliwell, J.R., et al.: Applications of X-ray computed tomography for examining biophysical interactions and structural development in soil systems: A review. Eur. J. Soil Sci. **64**(3), 279–297 (2013)
33. Thermo Fisher Scientific (TFS) (2019). Avizo 9.7
34. Fonseca, J.: The evolution of morphology and fabric of a sand during shearing. PhD. Thesis, Department of Civil and Environmental Engineering. Imperial College London, UK (2011)
35. Markussen, Ø., Dypvik, H., Hammer, E., Long, H., Hammer, Ø.: 3D characterization of porosity and authigenic cementation in Triassic conglomerates/arenites in the Edvard Grieg field using 3D micro-CT imaging. Mar. Pet. Geol. **99**, 265–281 (2019)
36. Houston, A.N., Schmidt, S., Tarquis, A.M., Otten, W., Baveye, P.C., Hapca, S.M.: Effect of scanning and image reconstruction settings in X-ray computed microtomography on quality and segmentation of 3D soil images. Geoderma **10 and 207–208**(1), 154–165 (2013)
37. Otsu, N.: A threshold selection method from gray level histograms. IEEE Trans. Systems, Man and Cybernetics **56**(4), 556–557 (1979)
38. Taylor, H.F., O'Sullivan, C., Sim, W.W.: A new method to identify void constrictions in micro-CT images of sand. Comput. Geotech. **69**, 279–290 (2015)
39. Beucher, S., Lantuejoul, C.: Use of watersheds in contour detection. Proc. of International Workshop on image processing. Rennes, France (1979)
40. ASTM D6913 / D6913M – 17: Standard Test Methods for Particle-Size Distribution (Gradation) of Soils Using Sieve Analysis. In: Annual Book of ASTM Standards. ASTM International, West Conshohocken, PA (2017)
41. ASTM D7928–21: Test Method for Particle-Size Distribution (Gradation) of Fine-Grained Soils Using the Sedimentation (Hydrometer) Analysis. In: Annual Book of ASTM Standards, ASTM International, West Conshohocken, PA (2021)
42. Valverde, A., Wijewickreme, D.: X-ray micro-computed tomography imaging to study silt fabric. In: Proc. 12TH National Conf. of Earthquake Eng. Salt Lake City, USA (2022)

Technical Framework of Performance-Based Design of Liquefaction Resistance

Jinyuan Yuan[1], Lanmin Wang[2], and Xiaoming Yuan[1(✉)]

[1] China Earthquake Administration, Institute of Engineering Mechanics, Key Laboratory of Earthquake Engineering and Engineering Vibration, China Earthquake Administration, Harbin 150080, China
yxmiem@163.com

[2] China Earthquake Administration, Lanzhou Institute of Seismology, Lanzhou 730000, China

Abstract. Seismic liquefaction hazard is always a very challenging problem in earthquake geotechnical engineering and it is necessary to develop the performance-based technology of liquefaction hazard governance. In the paper, by investigating the characteristics of seismic liquefaction hazard and combined with the existing performance-based earthquake engineering (PBEE) thought, the connotation and key points of the performance-based liquefaction hazard governance are proposed, and a technical framework of performance-based design of liquefaction resistance (PBDLR) is constructed according to the technical system construction rules. The results indicate that the liquefaction hazard has its own characteristics in the target, scale, mechanism, mode, chain effect, treatment technology and uncertainty of the influencing factors, and its technical system of hazard governance needs special consideration. The presented PBDLR, which integrates advantages of both the advanced risk management theory and the existing performance-based earthquake engineering technology system, and the elements and structure consider the sociality of the target, the integrity of the function, the hierarchy of the composition, the relevance of elements and the completeness of technology, and may provide guidance and reference for the technology development of liquefaction hazard governance in earthquake geotechnical engineering.

Keywords: Liquefaction hazard · Liquefaction resistance · Performance-based design · PBDLR

1 Introduction

Soil liquefaction under earthquakes is both a complex and interesting natural phenomenon, and meanwhile it can cause significant damage to human life and property. Soil liquefaction under earthquakes can cause the reduction or loss of foundation bearing capacity, which in turn leads to massive damage to various engineering structures and infrastructures. Notable examples include the 1964 Niigata earthquake (Ishihara and Koga, 1981), the 1976 Tangshan earthquake (Liu, 2002), the 1999 Chi-Chi earthquake (Yuan et al., 2003).

© The Author(s), under exclusive license to Springer Nature Switzerland AG 2022
L. Wang et al. (Eds.): PBD-IV 2022, GGEE 52, pp. 631–642, 2022.
https://doi.org/10.1007/978-3-031-11898-2_38

Since this century, disasters caused by soil liquefaction have continued to occur and become more serious than before, and earthquake losses due to soil liquefaction have accounted for an increasing proportion of earthquake disaster losses, such as the 2003 Xinjiang Bachu earthquake, the 2008 Wenchuan earthquake, the 2010 Maule earthquake in Chile, the 2010 Darfield earthquake in New Zealand, the 2011 New Zealand Christchurch earthquake, 2011 Tohuko earthquake in Japan, 2016 Kaikoura earthquake in New Zealand, 2016 Meno earthquake in Taiwan, China, 2018 Sulawesi earthquake in Indonesia, 2018 Songwon earthquake in China and 2019 Hokkaido earthquake in Japan, etc. Large-scale liquefaction of gravelly soils has occurred in the 2008 Wenchuan earthquake in China (Chen et al., 2009, Hou et al., 2011). The 2018 Songwon earthquake in China saw a liquefaction phenomenon of more than 200 sites stretching several kilometers, but the lowest magnitude ($M_w = 5.2$) ever recorded. The research of soil liquefaction has attracted wide attention again.

2   Characteristics of Earthquake Liquefaction Hazard

In particular, it should be noticed that the 2011 Christchurch earthquake in New Zealand (Cubrinovski et al., 2011) had a rare liquefaction damage, which became the main cause of the earthquake-induced significant losses. As shown in Fig. 1, the liquefaction caused severe damage to 16,000 residential houses and a large number of bridges, dykes, underground lifelines and other infrastructure, and eventually led to the permanent abandonment of some urban areas. Another recent severe liquefaction hazard event of concern occurred in the 2018 earthquake in Sulawesi, Indonesia (Kiyota et al., 2020). Petobo and Balaroa these two areas occurred severe soil liquefaction, where essentially all housing structures and infrastructure in the liquefaction zone were completely destroyed and liquefaction caused massive surface runoff, directly resulting in the death and disappearance of thousands of people.

Fig. 1. The liquefaction damage in 1976 Tangshan earthquake

In the past, earthquake disasters were mainly caused by the collapse of houses due to site shaking, now with the progress of human science and technology, the collapse of buildings under site shaking is becoming less and less, but the proportion of earthquake damage caused by soil liquefaction site damage is becoming noticeable increase. In modern urban construction, underground space is continuously exploited, and most of the lifeline facilities are buried underground. Modern human life and production are increasingly dependent on underground facilities, soil liquefaction which mainly destroys underground facilities, will pose a greater threat to human safety (Fig. 2).

Fig. 2. The liquefaction damage in 2011 New Zealand earthquake

Fig. 3. The liquefaction damage in 2018 Indonesian earthquake

The phenomenon of site liquefaction damage from recent major earthquakes shows that, unlike in the past, site liquefaction can cause greater economic losses and also lead to a large number of casualties, making it an extremely dangerous event that may endanger human lives. It is especially important to develop liquefaction governance technology with the concept of performance-based design.

Compared with other types of seismic hazards, site liquefaction hazards have several distinctive features as follows:

Liquefaction Disaster Targets are Comprehensive and Extensive

Liquefiable soils often exist in areas with abundant water supply, fertile land, flat surface, easy access to farming, etc., where human activities are frequent and economically developed. Seismic investigation shows that liquefaction not only causes damage to buildings, but also to infrastructure such as highways, railroads, bridges, harbors, berms, farmland and water conservancy, artificial islands, etc. Soil liquefaction is especially harmful to lifeline systems, and can cause extensive damage to both various individual engineering structures and entire cities.

The Scale of Soil Liquefaction Under Strong Earthquakes is Huge

The soil liquefaction found so far is located in the range of several tens of meters below the surface, and due to the stratified nature of geological deposition, the liquefaction under strong earthquakes will show a patchy area distribution, and large scale of liquefaction is usually occurred under strong earthquakes. In 1976 Tangshan earthquake, the liquefied zones involved an area of 24,000 km^2; the ground fracture caused by liquefaction in a village in the Wenchuan earthquake in 2008 stretched for hundreds of meters. Moreover, not all liquefaction zones have macroscopic phenomena visible on the surface such as sand and water spouts and ground fractures, the actual distribution of liquefaction zones is much larger than we can imagine.

The Mechanism of Liquefaction Causing Disaster is Special

The site is both a building support layer and a seismic wave propagation medium. Unlike the single mechanism that causes damage to engineering structures on non-liquefied sites by excessive seismic inertia forces, the site liquefaction does not directly cause engineering damage, but first causes ground failure and changes in ground shaking, and then transmits such damage and changes to engineering structures, which leads to the final damage of engineering structures. Therefore, the analysis of seismic damage by liquefaction is much more complicated than that of engineering damage on non-liquefied sites.

Liquefaction Leads to Significant Chain Effects

The site liquefaction will not only lead to the destruction of buildings, especially to the destruction of infrastructure, especially the buried lifeline system, which will lead to serious secondary disasters and cause chain reactions. After the 1906 San Francisco earthquake, the liquefaction caused the water pipes to break, which greatly hindered the delivery of water to extinguish the fires; the earthquake On March 11th, 2011 in Japan, caused many water and gas pipelines to break due to large-scale liquefaction, resulting in the breakage of water and gas pipes in the mainly reclaimed land in Urayasu City, which had a huge impact on the life of citizens; The 2016 Southern Taiwan earthquake in China, shook out the liquefaction crisis in Taiwan and caused panic in housing prices. Therefore, liquefaction many times has a chain effect and has a great impact on society.

The Liquefiable Site Treatment Technology is Complicated and Costly

Since liquefaction is the result of sudden change of physical and mechanical properties of soil under earthquake, the technology of liquefiable foundation reinforcement is special and complicated, and the cost of liquefiable foundation reinforcement is much higher than that of seismic reinforcement of buildings. This makes a great contradiction between the investment of liquefaction hazard governance and the affordability of the society. It is a difficult but necessary subject to answer how to consider both safety and economic factors, set the target in the best balance, and achieve the maximum value of liquefaction hazard governance.

The Factors Affecting Liquefaction Hazard have Significant Uncertainties

First of all, unlike the artificial materials and structures used in buildings, the carrier of liquefaction hazard is the site, which is not a product of human selection and design, but a natural geotechnical body with complex properties. Liquefaction occurs in the subsoil layer, and the uncertainty of the factors affecting the carrier itself is significant. Meanwhile, as mentioned above, the transmission process of site liquefaction to engineering structure damage and the treatment of liquefiable soils are complex and diverse, and the influencing factors also have great uncertainty. It is necessary to study the governance of liquefaction damage with significant uncertainty factors by using risk theory as a guide, but the research is also very difficult at the same time.

As it can be seen that, site liquefaction hazards have their own unique characteristics in terms of the target, scale of occurrence, mechanism, mode of occurrence, chain effect, treatment technology and uncertainty of influencing factors. Therefore, on the one hand, it is necessary to develop a technical system for governing the risk of site liquefaction hazards suitable for the development of modern society based on performance design, and on the other hand, when establishing its technical system, separate research and design are needed for the characteristics of liquefaction hazards.

3   Existing Technical Process of Soil Liquefaction Resistance

The existing technical framework of the liquefaction hazard governance in China was formed in the 1970s and 1980s, and the most typical representative is the Code for Seismic Design of Buildings, whose technical process are shown in Fig. 4. The technical system in Fig. 4 has been used until now and has played a significant role in the work of earthquake prevention and mitigation in China's engineering construction, but due to the limitations of the previous level of understanding, this system has some problems in dealing with today's complex engineering construction and needs to be improved.

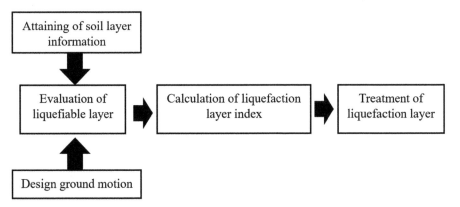

Fig. 4. The existing technical process of soil liquefaction resistance design

4   Performance-Based Earthquake Engineering and its Global Framework

In the work on performance-based seismic design led by the Pacific Earthquake Engineering Research Center (PEER), they have set the lofty goal of developing and disseminating urban earthquake risk-reduction technologies, and the performance-based earthquake engineering (PBEE) is defined as "An approach to improve decision-making about seismic risk by making the choice of performance goals and the tradeoffs to facility owners and society at large".

Several conceptual frameworks for PBEE have been developed in recent professional efforts (SEAOC Vision 2000, FEMA 273, ATC-40). They differ in details but not in concepts. Figure 5 illustrates a global framework which identifies processes, concepts, and major issues that need to be addressed in this context (Krawinkler, 1999).

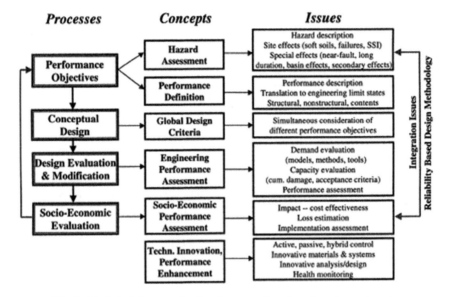

Fig. 5. Technical framework of performance-based earthquake engineering

PDEE in Fig. 5 encompass seismological, geotechnical, structural, architectural and MEP (nonstructural), and socioeconomic considerations. Each issue is associated with an extensive research agenda that covers much more than PEER can address (Krawinkler, 1999). What distinguishes this research agenda from a general shopping list is the focus on a single objective, which is the common one of providing knowledge, methods, tools, and data for development and implementation of PBEE. The challenge is to select subsets of this comprehensive research agenda that will result in substantial and measurable progress on critical aspects that bring PBEE much closer to realization (Krawinkler, 1999).

Although both serve to mitigate earthquake disasters, the PBEE technology system in Fig. 5 is significantly improved compared with the existing technical process of soil liquefaction resistance design in Fig. 3 for liquefaction treatment, which is mainly reflected in:

(1) Prioritizing performance objectives and conceptual design, which is lacking in the existing technical process of liquefaction hazard governance. Compared with the thought of performance-based seismic design, the objective of liquefaction governance system needs to be clarified.

(2) PBEE design system will be social economy as a separate element, emphasizing the cost-benefit analysis, which makes the seismic design more close to the nature of seismic work, but also reflects the seismic design can meet the requirements of personalized seismic design in a more flexible way, is a very useful idea. At present, cost-benefit analysis of liquefiable soil treatment is lacking in the soil liquefaction treatment technology system, which is particularly important for the expensive liquefiable soil treatment.

PBEE design system in Fig. 5 provides us with new ideas for reference, but from its content, it is more directly aimed at the building engineering, which is not entirely suitable for the performance-based design problem of soil liquefaction resistance. For the soil liquefaction, a separate design system of soil liquefaction resistance based on the performance concept should be established according to the characteristics of liquefaction hazard.

5   Technical Framework of Performance-Based Design of Liquefaction Resistance

5.1   Definition and Key Points of PBDLR

In this paper, the technical system of the performance-based design of soil liquefaction resistance (PBDLR) refers to a unified technical system organically linked by various technologies to eliminate or mitigate liquefaction earthquake risk under objective laws and social conditions.

According to the characteristics of site liquefaction hazard mentioned above and the risk management theory, the risk governance goal of liquefaction hazard is defined in this paper as: to establish the optimal balance between liquefaction hazard governance goal, liquefaction governance investment and social acceptable capacity, and to maximize the governance value through economic and effective countermeasures.

Taking PDEE thought as reference, in this paper, the PBDLR is defined as: an approach to improve decision-making about risk of seismic liquefaction hazard governance by selecting and balancing the performance objectives for the facility owner and the society.

In this paper, the risk governance of liquefaction hazard should be controlled by two basic factors, namely, governance objectives and performance analysis, expressed by the following:

Liquefaction hazard governance $= F$(governance objective, performance analysis)

$$(1)$$

where, the governance objective is composed of seismic ground motion fortification standard, liquefaction risk threshold, liquefaction hazard acceptability and liquefaction governance investment acceptability; The performance analysis consists of the identification, correlation, evaluation and response of liquefaction hazard.

The connotation of liquefaction hazard governance includes three main points:

(1) The liquefaction hazard due to site liquefaction risk is the main line. As shown in Formula 1, the governance of liquefaction hazard is composed of two basic elements: governance objectives and performance analysis, and liquefaction risk and its hazard play an important role in both of them. The acceptability of liquefaction hazard is one of the key factors in the governance objectives and also one of the targets of risk response in performance analysis.

(2) The principle of the establishment of governance objectives is to keep a balance between liquefaction hazard, liquefaction governance investment and social acceptability. The liquefaction damage is very serious and the governance investment is very low, so the loss risk is too large. However, the liquefaction damage caused by strong earthquakes is a rare event, so the liquefaction-resistant capacity of engineering system should be improved to a certain extent, and the excessive economic investment will exceed the social bearing capacity. It is a necessary condition to maximize the value of liquefaction hazard governance to find the optimal balance between the governance investment and the risk of liquefaction damage and the social acceptability as the principle of the establishment of the governance goal.

(3) The correlation analysis between site liquefaction risk and earthquake damage of engineering system is the key. The ultimate goal of liquefaction research is to serve the earthquake damage prevention and mitigation work of engineering system. Site liquefaction without engineering structure and infrastructure has no risk of earthquake damage, and liquefaction hazard must be related to the destruction of engineering structure and infrastructure caused by liquefaction. Sites under earthquake play two roles: seismic wave media and engineering system bearing layer, and the damage and destruction of engineering structure caused by site liquefaction are basically indirect. Therefore, only consider site liquefaction risk itself is not enough, building of the relationship between site liquefaction risk and earthquake damage risk of engineering system is also a necessary condition to maximize the value of liquefaction hazard governance.

Among the three main points of liquefaction hazard governance proposed in this paper, the program is the governance objective, and the main line is site liquefaction risk and its hazard, and the correlation between site liquefaction risk and liquefaction hazard is the key. The three are interdependent and interact with each other. Only comprehensive consideration can maximize the value of liquefaction hazard governance.

5.2 Technical Framework of PBDLR

Basic requirements for technical system construction. Technical system is a unified technical system organically linked by various technologies required to achieve specific functions under objective laws and social conditions, and should meet the following requirements:

(1) The sociality of goals. The technological system has dual attributes of nature and society and is composed in accordance with certain social purposes. It is not only the external requirements of the technological system but also the basic criteria for measuring the function of the technological system [20]. It should not only follow the natural operation law and the actual social bearing capacity, but also meet the long-term requirements of social sustainable development.
(2) Functional integrity. What the technical system pursues is not the single function of a certain technology, but the overall function formed by the combination of different technologies, and the whole is greater than the sum of its parts.
(3) The hierarchy of composition. The elements of the technical system are combined together in a certain order, and the internal elements of the technical system are the relations of mutual influence and interaction.
(4) Linkage of elements. In the technology system, all kinds of technologies are mutually prerequisite. When one kind of technology changes, related technologies should have corresponding adjustment reaction to maintain the internal balance of the technology system.
(5) Technical completeness. All kinds of technical elements are complete in the technical system. In either case, the system should operate in closed loop mode.

Technical Framework of PBDLR
According to the above requirements for technical system construction and combined with the characteristics of liquefaction damage and advantages of PBEE, this paper constructed a technical framework of performance-based design of liquefaction resistance (PBDLR), as shown in Fig. 6.

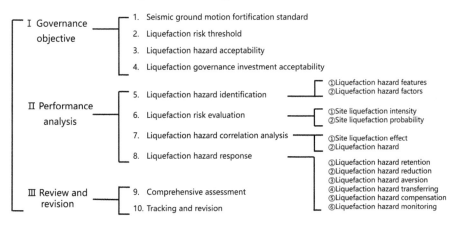

Fig. 6. Technical framework of performance-based design of liquefaction resistance (PBDLR)

Interpretation of Elements of PBDLR

In order to better understand the meanings and their relationships of the elements of PBDLR, the interpretations of the elements in Fig. 6 are given as follows:

I Governance Objective: during the action of the designed seismic ground motion, the general governance objectives of engineering system under site liquefaction are achieved, including:

1. Seismic ground motion fortification standard: the intensity and probability level of seismic ground motion for which the engineering system is located.
2. Liquefaction risk threshold: the lowest value of site liquefaction risk under possible liquefaction hazard to engineering system. If the site liquefaction risk is lower than this value the site liquefaction will not be considered.
3. Liquefaction hazard acceptability: the acceptable level and scope of earthquake damage under site liquefaction include the risk of liquefaction hazard to engineering system and the consequences of liquefaction hazard chain effect.
4. Liquefaction governance investment acceptability: the acceptable level and scope of investment in liquefaction treatment, including the economic cost of liquefiable soil treatment and the economic cost of dealing with the consequences of liquefaction hazard chain effect.

II Performance Analysis: the site liquefaction hazard and its performance after disposal are analyzed, including the identification, correlation, evaluation and response of liquefaction hazard, including:

5. Liquefaction hazard identification: identify the risk factors of site liquefaction hazard, including:

   (1) Liquefaction hazard features: integrate site liquefaction and engineering system earthquake damage events and determine their features;
   (2) Liquefaction hazard factors: determine the factors affecting liquefaction of site and earthquake damage of engineering system.

6. Liquefaction risk evaluation: determine the liquefaction risk of the site itself, including:

   (1) Site liquefaction intensity: determine the scale and severity of liquefaction at the site;
   (2) Site liquefaction probability: determine the probability of liquefaction at the site.

7. Liquefaction hazard correlation analysis: determine the liquefaction site effect associated with seismic damage of engineering system, and assess the seismic damage risk of engineering system caused by liquefaction, including:

   (1) Site liquefaction effect: The ground failure and the changes in ground motion characteristics due to liquefaction, including intensity and probability level, are considered;

(2) Liquefaction hazard: the risk assessment results of liquefaction hazard to engineering systems are given, including the intensity and probability level of economic loss, casualties, chain effect, social impact, etc.

8. Liquefaction hazard response: based on the results of the liquefaction hazard analysis, different countermeasures are taken to maximize the value of liquefaction hazard governance, including:

(1) Liquefaction hazard retention: the risk of liquefaction hazard to the engineering system is negligible if the risk of the liquefaction hazard obtained from the analysis is within the lower limit of the acceptable range.
(2) Liquefaction hazard reduction: when the risk of liquefaction hazard obtained from the analysis exceeds the lower limit of the acceptable range, measures such as soil treatment are taken to eliminate or reduce the risk of liquefaction hazard, followed by a performance reanalysis to confirm whether the objectives are met.
(3) Liquefaction hazard aversion: after the implementation of liquefaction hazard reduction measures, the analysis of the risk of liquefaction hazard still exceeds the upper limit of the acceptable range, the project can be abandoned to achieve the purpose of risk avoidance.
(4) Liquefaction hazard transferring: transferring the risk of liquefaction hazard to society through, for example, earthquake liquefaction insurance.
(5) Liquefaction hazard compensation: for liquefaction hazard bearers, a certain amount of financial compensation is given before the occurrence of seismic liquefaction disaster to reduce the pressure of liquefaction hazard bearers.
(6) Liquefaction hazard monitoring: when the established technology is difficult to achieve the desired effect or is difficult to grasp the effect, the factors with unpredictable consequences can be monitored, and according to the situation the graded early warning is put forward.

III Review and Revision: by reviewing entity performance, how well the technical system components are functioning over time is considered and in light of substantial changes, what revisions are needed, including:

9. Comprehensive assessment: the changes that may substantially affect governance objectives are identified and assessed, and the entity performance is reviewed.
10. Tracking and revision: when the established technology is difficult to achieve the expected effect or the overall operation of the technology system is problematic, the governance objectives can be adjusted or the technology system can be revised according to the actual situation.

6   Conclusions

This paper expounds the connotation and key points of the performance-based liquefaction hazard governance, and combined the characteristics of seismic liquefaction hazard with the existing PBEE, the technical framework of performance-based design of liquefaction resistance (PBDLR) is constructed. The PBDLR is specially designed for liquefaction hazard governance, and it also meets the general requirements of technical system construction, such as target sociality, functional integrity, hierarchy of composition, linkage of elements and the completeness of technology. The full implementation of PBDLR is still a long way off, but the presentation of PBDLR will provide guidance for future work of liquefaction hazard governance.

Acknowledgments. This work is jointly supported by the Scientific Research Fund of Institute of Engineering Mechanics, China Earthquake Administration (2020B07), the Key Project of National Natural Science Foundation of China (U1939209) and the Natural Science Foundation of Heilongjiang Province (ZD2019E009).

References

Chen, L.W., Yuan, X.M., Cao, Z.Z., Hou, L.Q., Sun, R.: Liquefaction macrophenomena in the great Wenchuan earthquake. Earthq Eng Eng Vib **8**(2), 219–229 (2009)

Cubrinovski, M., et al.: Soil liquefaction effects in the central business district during the February 2011 Christchurch earthquake. Seismol. Res. Lett. **82**(6), 893–904 (2011)

Hou, L.Q., Li, A.F., Qiu, Z.M.: Characteristics of gravelly soil liquefaction in Wenchuan earthquake. Appl Mech Mater **90–93**, 1498–1502 (2011)

Ishihara, K., Koga, Y.: Case studies of liquefaction in the 1964 Niigata earthquake. Soils Found **21**(3), 33–52 (1981)

Kiyota, T., Furuichi, H., Hidayat, R.F., Tada, N., Nawir, H.: Overview of long-distance flow-slide caused by the 2018 Sulawesi earthquake. Indonesia. Soils Found **60**(3), 722–735 (2020)

Krawinkler, H.: https://apps.peer.berkeley.edu/news/1999jan/advance.html (1999)

Liu, H.X.: The Great Tangshan Earthquake of 1976. Published by Earthquake Engineering Research Laboratory, California Institute of Technology, CA (2002)

Yuan, H.M., Yang, S.H., Andrus, R.D., Juang, C.H.: Liquefaction-induced ground failure: a study of the Chi-Chi earthquake cases. Eng Geol **71**(1–2), 141–155 (2003)

Deformation Mechanisms of Stone Column-Improved Liquefiable Sloping Ground Under Earthquake Loadings

Yan-Guo Zhou[(⊠)] and Kai Liu

MOE Key Laboratory of Soft Soils and Geoenvironmental Engineering, Institute of Geotechnical Engineering, Zhejiang University, Hangzhou 310058, People's Republic of China
qzking@zju.edu.cn

Abstract. Liquefaction-induced deformation in sloping ground caused heavy damage to lifelines and overlaid structures in the past earthquake events. However, there is still a great need for estimation of large deformation of sloping ground during earthquakes and the associated deformation mitigation method. In this paper, a series of element tests were conducted to investigate residual volumetric strain and shear strain of soil element with initial shear stress, corresponding to infinite sloping ground condition. The element test results indicate that residual volumetric strain estimation method proposed by Shamoto and Zhang for level ground could be able to estimate that in sloping ground with initial static stress condition, and the residual shear strain is well correlated with maximum shear strain. In addition, a series of centrifuge model tests with respect to stone column improved sloping ground were designed and conducted to explore the effect of densification and drainage of stone column on deformation in sloping ground. The mechanisms of settlement and lateral spreading mitigation by densification and drainage effect of stone columns in sloping ground were analyzed and discussed in combination with centrifuge model tests and numerical simulation results. The present study provides an effective method for evaluating post-liquefaction settlement and deep insights of deformation mitigation of stone column in sloping ground, which is of great help for developing the performance-based mitigation method for sloping ground in the future.

Keywords: Post-liquefaction deformation · Sloping ground · Stone columns · Centrifuge model test · Numerical simulation

1 Introduction

The past earthquakes have witnessed great damage to existing buildings and underground structures caused by liquefaction-induced lateral displacement in mildly sloping ground or nearly level ground with a free face (Bradley et al. 2019). Some studies have focused on this problem and several methods have been proposed to predict the lateral spreading in sloping ground (e.g., Youd et al. 2002; Zhang et al. 2004).

© The Author(s), under exclusive license to Springer Nature Switzerland AG 2022
L. Wang et al. (Eds.): PBD-IV 2022, GGEE 52, pp. 643–659, 2022.
https://doi.org/10.1007/978-3-031-11898-2_39

In order to reduce the risk of liquefaction and associated ground deformation, stone column (SC) has been used in the past over forty years, and considerable in-situ cases have proved its effectiveness. Badanagki et al. (2018) conducted a series centrifuge model tests and found that greater area replacement ratios of SC (A_r) is effective to reduce the seismic settlement and lateral deformations in gentle slopes. Zhou et al. (2021) performed three centrifuge models to individually investigate the effects of densification and drainage effect on level ground and concluded that the densification effect was the main factor and the drainage effect played the second role to reduce the settlement in a SC-improved ground. Besides, the three-dimensional numerical simulations conducted by Elgamal et al. (2009) and Asgari et al. (2013) showed that SC remediation was effective in reducing the sand stratum lateral deformation in sloping ground, the permeability of SC was a significant role in their cases. In this paper, a series of element tests with initial shear stress using Hollow Cylinder Apparatus (HCA) were conducted to investigate the residual volumetric strain and shear strain. Besides, three centrifuge model tests were performed to explore the seismic performance of SC-improved sloping ground and worked as benchmarks to validate the constitutive model and numerical method. Based on physical observations and numerical simulation results, the mechanisms of SC on deformation mitigation were preliminarily discussed.

2 Post-liquefaction Deformation with Initial Shear Stress

2.1 S-Z Method for Post-liquefaction Deformation in Level Ground

Shamoto and Zhang (1998) proposed a method for estimation of seismic settlement and horizontal displacement in level or nearly level ground (abbreviated as S-Z method in the following), based on constitutive analysis and element test evidence. According to their study, the residual volumetric strain (ε_{sr}) of a soil element could be expressed by,

$$\varepsilon_{v,r} = \frac{e_0 - e_{\min}}{1 + e_0} R_0 \cdot \gamma_{\max}^m - M_{CS.0}\gamma_{0,r} - \frac{M_{CS} - M_0}{\alpha}\gamma_{d,r} \qquad (1)$$

where e_0 and $e_{\min}$ are initial and minimum void ratio of soil, respectively; $\gamma_{\max}$, $\gamma_{0,x}$ and $\gamma_{d,r}$ are maximum shear strain, residual shear strain component independent of effective stress and residual shear strain component dependent on change in effective stress, respectively; M_{cs}, M_0 are slope of CSL (critical state line) and PTL (phase transformation line), respectively; $M_{cs,0}$ is the deviator-isotropic stress ratio at very small effective stress; R_0 and m are fitting parameters. Given that the undrained loading history is the same, ε_{vr} would reach its maximum value (ε_{vr})$_{\max}$ only when residual shear strain remains zero, namely,

$$\varepsilon_{v,r}\big|_{\gamma_r=0} = (\varepsilon_{v,r})_{\max} = \varepsilon_{vd,ir} = \frac{e_0 - e_{\min}}{1 + e_0} R_0 \cdot \gamma_{\max}^m \qquad (2)$$

2.2 The Validity of S-Z Method in Sloping Ground

In order to explore the effectiveness of above equations to sloping ground condition (i.e., initial shear stress condition), a series of undrained cyclic torsional shear tests (HCA) followed by reconsolidation were conducted, the ε_{vr} values were obtained to check the effectiveness of above expressions proposed by Shamoto and Zhang (1998).The torsional shear tests were conducted by using hollow cylinder apparatus (HCA), which details could be referred to Chen et al. (2016). The testing material is Ottawa F-65 sand with 10% Qiantang river silt, the grain size distribution and physical properties are shown in Fig. 1 and Table 1. The specimen has an outer diameter of 100 mm, an inner diameter of 60 mm, and a height of 200 mm. The specimen was prepared by placing soils in eight layers in a traxial mold using the dry deposition method. The actual relative densities of all specimens after consolidation are about 60%. After the specimens were placed in the cell, the outer and inner pressures of 20 kPa were applied, then a two-stage saturation (carbon dioxide flushing and de-aired water flushing) was carried out to ensure the Skempton's B-value of 0.95 or larger. The effective stress is 100 kPa and two initial shear stress ratio (CSR₀)were designed, where $CSR_0 = \tau_{ini}/\sigma'_c$ (σ'_c is average consolidation stress, τ_{ini} is initial shear stress).

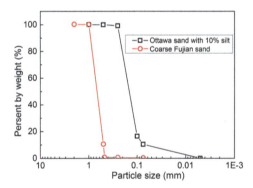

Fig. 1. Grain size distribution curves of two sands

Table 1. Physical properties of two sands.

Soil type	G_s	ρ_{max}(g/cm³)	ρ_{min}(g/cm³)	Permeability (10^{-5} m/s)
Silty sand (Ottawa sand with 10% of silt)	2.673	1.921	1.510	2.7–5.6
Coarse Fujian sand	2.644	1.713	1.489	1490–3600

The specimens were isotropically consolidated by increasing the effective stress state up to 100 kPa, then the stress state was modified by applying a drained monotonic torsional shear stress up to a specified initial shear stress. Finally, undrained constant cyclic torsional loading was applied until the shear strain reached the expected value.

Then the torsional shear stress was adjusted to the initial value under undrained condition followed by the reconsolidation process in constant shear stress ($\tau=\tau_{ini}$) with nearly constant γ_r.

For specimen without τ_{ini}, the specimens were isotropically consolidated followed by undrained constant cyclic torsional loading up to the expected shear strain, after that, the shear strain was adjusted to zero under undrained condition, and zero shear strain state was kept in the drained reconsolidation process. The above procedure is consistent to that testing procedure of Shamoto and Zhang (1998) to eliminate the effect of γ_r on ε_{vr}.

Figure 2 is a typical non-symmetrical stress–strain curve of a soil element with initial shear stress, where the maximum shear strain (γ_{max}) is usually defined as the strain increment from one stress reversal to the next. The γ_r is defined as the shear strain at shear stress (τ)=τ_{ini} for each cycle, which is consistent to Chiaro et al. (2012).

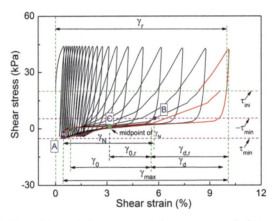

Fig. 2. Definition of shear strain components in non-symmetrical stress strain loop

Figure 3(a) shows ε_{vr} versus γ_{max} for all element tests with and without initial shear stress. It is obvious that the ε_{vr} values for tests with $CSR_0 > 0.0$ are significantly smaller than that with $CSR_0 = 0.0$, it may attribute to that the existence of γ_r reduce the value of $(\varepsilon_{vr})_{max}$, as expressed by Eqs. (1) and (2). For $CSR_0 = 0.0$, ε_{vr} could reach its maximum value of $(\varepsilon_{vr})_{max}$, because γ_r remained zero during reconsolidation process. In order to confirm the validity of Eq. (1) for cases of $CSR_0 > 0.0$, the following attempts were made to obtain the parameters and residual shear strain components ($\gamma_{0,r}$ and $\gamma_{d,r}$). The parameters of M_{cs} and M_0 were determined by traxial element tests as 1.33 and 1.14 for this kind of silty sand, respectively. Following the suggestion of Shamoto and Zhang (1998), $M_{cs.0}$ was determined as 0.142 M_{cs}, and α was taken as $1.5\sqrt{3}$. For cases of $CSR_0 > 0.0$, $(\varepsilon_{vr})_{max}$ could be obtained by,

$$\varepsilon_{vd,ir} = (\varepsilon_{v,r})_{max} = \varepsilon_{v,r} + M_{CS.0}\gamma_{0,r} + \frac{M_{CS} - M_0}{\alpha}\gamma_{d,r} \qquad (3)$$

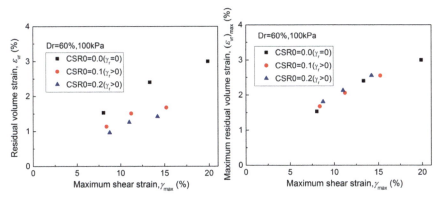

Fig. 3. The relation of maximum shear strain versus: (a) residual volumetric strain; (b) maximum residual volumetric strain

Another key issue is to determine the values of $\gamma_{0,r}$ and $\gamma_{d,r}$ in Eq. (3). It could be found in Fig. 2 that the residual shear strain started to accumulate toward to positive direction at the first cyclic loading under the effect of initial shear stress, and the stress–strain loop enlarged and moved toward positive direction with the increase of loading cycles. The movement of stress–strain loop reflects the change of element shape but not the irreversible volume strain accumulation. For this reason, the "base point" of the last stress–strain loop is depicted in Fig. 2 as point "C", it is the midpoint of shear strain increment from point "A" and "B", which are intersections of shear stress $\tau=-\tau_{ini}$ line and $\tau=\tau_{ini}$ line and the last stress–strain loop, respectively, as shown in Fig. 2. The $\gamma_{0,r}$ and $\gamma_{d,r}$ were determined following the suggestions of Shamoto and Zhang (1998). Figure 3(b) clearly shows that the revised maximum residual volumetric strain for cases of $CSR_0 > 0.0$ are well consistent to that of $CSR_0 = 0.0$. The above result indicates that the settlement estimation method proposed by Shamoto and Zhang (1998) for level ground could be applied to sloping ground condition with the premise of accurate determination of residual shear strain components (i.e., $\gamma_{0,r}$ and $\gamma_{d,r}$).

Residual shear strain has been paid more attention in the sloping ground for it closely relates to lateral spreading. Figure 2 illustrates how to determine the γ_r and γ_{max} in one cycle, and Fig. 4 shows the γ_r versus γ_{max} in this study. Also the data points for each CSR_0 are scattered, it is clearly found that the γ_r is positively correlated with γ_{max}. What's more, the γ_r value is obviously larger for larger value of CSR_0 under the same value of γ_{max}.

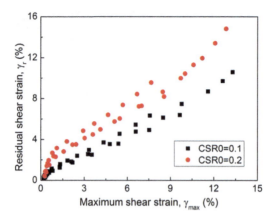

Fig. 4. The relation of maximum shear strain versus residual shear strain

3 Stone Column Improved Sloping Ground by Centrifuge Model Tests

3.1 Test Material and Model Configuration

A series of three centrifuge model tests were conducted to explore the effect of densification and drainage effect of stone column on seismic responses of sloping ground, following the same design conception from Zhou et al. (2021). Model 1 and Model 2 were designed to explore the densification effect only, both models do not include stone columns but are uniform silty sand ground. Model 1 is a loose pre-improved ground with initial void ratio of e_0, Model 2 represents the post-improved dense soil ground with void ratio of e_1. Model 2 and Model 3 were designed to explore the drainage effect of stone column, both models shared the same soil density, while stone columns were included in Model 3. The stiffness of stone column were designed to be close to that of surrounding soil to minimize the shear reinforcement.

The surrounding soil for three models was fine Ottawa F-65 sand with 10% of Qiantang river silt, and a coarse Fujian sand was chosen as stone column material. The grain size distribution curves are shown in Fig. 1 and physical properties for both materials are given in Table 1. Silicone oil with viscosity 50 times that of water was used as pore fluid.

The side and top views of model configuration of Model 3 are shown in Fig. 5, the model configuration for Model 1 and Model 2 are almost the same but both models were uniform silty sand ground. Accelerometers pore pressure transducers were placed in the center of the model ground, three pairs of bender elements were also installed. Two LVDT transducers were installed to record the surface settlement of the model ground.

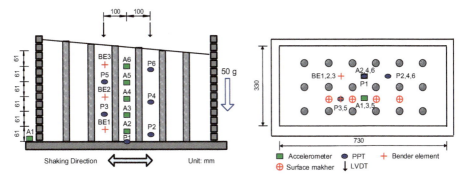

Fig. 5. Model configuration (geometry in model scale) of Model 3

3.2 Model Preparation and Test Procedure

The fine Ottawa sand and Qiantang river silt were fully mixed with a mass ratio of 9:1. Model 1 was prepared using air pluviation method with relative density (D_r) = 50%. Model 2 and Model 3 were prepared using dry tamping method with D_r = 78%.

The diameter-to-spacing (d/s) parameter in Model 3 was determined following Zhou et al. (2017) as 0.3, the diameter and spacing for SCs are 3 cm and 10 cm in model scale, and the area replacement ratio $(A_r) = \pi d^2/(4s^2) = 7\%$. The SC installation method is the same as that adopted by Zhou et al. (2021). All three models were saturated by vacuum method.

The centrifuge tests were conducted under 50 g, shear wave velocities at different depths of the model ground were measured by bender elements and then the destructive sine wave motion followed. Long enough time interval was waited for the full dissipation of excess pore water pressure (EPWP) induced by previous shaking in the model ground.

The designed shaking sequences of three models are the same, as shown in Fig. 6, to compare the differences of seismic response among three models. The input motion was a sine wave of fifteen constant cycles with dominant frequency of 1 Hz.

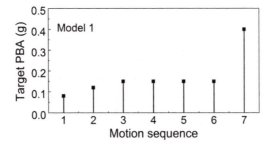

Fig. 6. Target shaking sequence in three model tests

3.3 Experimental Results and Analysis

The dynamic responses among three models were compared and analyzed to distinguish the effect of individual densification and drainage on seismic responses in sloping ground. Figure 7 shows the seismic responses between Model 1 and 2 under motion1. It could be found that densification in Model 2 slowed down the generation rate and lowered the peak of EPWP ratio at all depth. The acceleration at shallow depth in Model 1 (ie., depth = 0.5 m and 3.5 m) were de-amplified only in upslope direction but significantly amplified by high-frequency negative spikes in downslope direction. This phenomenon is only observed at shallow depth in Model 2 due to soil disturbance during modeling, the acceleration at other depths were similar to the input base motion.

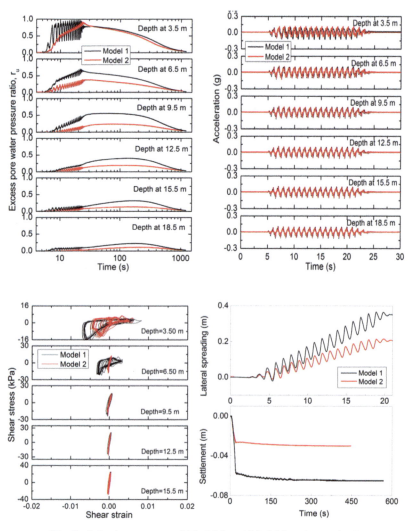

Fig. 7. Seismic responses of Model 1 and Model 2 under motion1

Figure 7(c) depicts the stress–strain loops of Model 1 and Model 2, it is obvious that densification of Model 2 reduced the shear strain at shallow depth (i.e., 3.5 m and 6.5 m), which could be attributed to the increase of shear stiffness after densification.

Figure 8 compares the seismic responses between Model 2 and Model 3 under motion3. It is found that peaks of EPWP at most depth were quite comparable between two models, while the EPWP generation rate during shaking was significantly slower at shallow depth in Model 3 due to the drainage effect. In addition, the time for fully dissipation of EPWP in Model 3 was shortened at all depth. The acceleration responses at shallow depth (i.e., depth at 3.5 m) showed significant positive spikes in Model 3

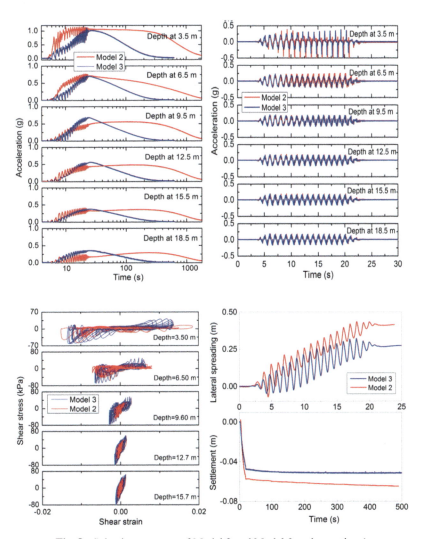

Fig. 8. Seismic responses of Model 2 and Model 3 under motion 1

compared to that of Model 2, and the acceleration at other depths were nearly the same between two models.

Figure 8(c) shows the stress–strain curves for Model 2 and Model 3. It is found that shear strain at all depth (i.e., depth above 6.5 m) were compatible between two model. The cyclic mobility phenomenon was only observed in downslope direction in Model 2 at shallow depth, while they were obvious in both downslope and upslope directions in Model 3.

4 Numerical Simulation of Centrifuge Model Test

4.1 Calibration of Constitutive Model Parameters

The unified plasticity model (CycliqCPSP), proposed by Wang et al. (2014), was adopted here to simulate stone column and silty sand materials, because it could properly reflect the dilatancy of sand and larger shear deformation under cyclic loading. This model introduces the state parameter concept thus enabling it to model sand under different states with the same set of parameters, and it has been implemented into finite difference code FLAC3D. It incorporates 14 parameters listed in Table 2, including elastic modulus constants (G_0, κ), plastic modulus parameter (h), critical state parameters (M, λ_c, e_0, ξ),state parameter constants (n_p, n_d), dilatancy parameters (dre$_{,1}$,dre$_{,2}$, dir, α, $\gamma_{d,r}$). The model parameters were determined following the methods suggested by Wang et al. (2014) and He et al. (2020).

Table 2. Model parameters for CycLiqCPSP model used in the numerical simulations.

Parameters	G_0	κ	h	M	λ_c	e_0	ξ
Sility sand	234	0.01	1.6	1.19	0.022	0.715	0.71
Stone column	353	0.01	1.2	1.28	0.029	0.788	0.70
Parameters	n_p	n_d	dre$_{,1}$	dre$_{,2}$	dir	α	$\gamma_{d,r}$
Sility sand	3.4	5.3	1.1	30	1.35	5.0	0.05
Stone column	1.4	8.0	1.1	30	1.10	30	0.05

A set of high-fidelity hollow cylinder torsional shear tests and traxial tests were conducted as benchmark element tests to calibrate the silty sand and stone column materials, respectively. The physical properties of two sands are as shown in Table 1. The model parameters are adopted as listed in Table 2, and the comparisons between tests and simulations in Fig. 9 show good agreement, especially for the post-liquefaction shear deformation.

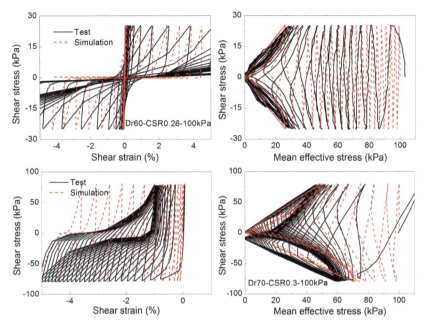

Fig. 9. Comparison between model tests and simulations: (a) Silty sand; (b) stone column

4.2 Numerical Model Configuration

A shown in Fig. 10, the plane strain numerical model was used for Model 1 and Model 2. The numerical model was constructed in prototype dimensions under 1 g gravity field. Fully coupled effective stress analysis was conducted. The initial stress state of the model was obtained using rigid box condition, then the displacement degrees of freedom for soil nodes at lateral boundaries along the shaking direction are set free, and the laminar box condition was configured to re-calculate the initial stress state. Two one-dimension columns were configured to reflect the laminar box, the nodes of which at the same elevation are tied to each other and elastic model were adopted with Young Modulus (E)= 3e5 kPa, poison ratio = 0.32. As for dynamic stage, the processed acceleration record was applied at base nodes and long enough time is needed for fully dissipation of excess pore water pressure. Small Rayleigh damping of 1% is assigned for numerical stability and to control high frequency numerical noise.

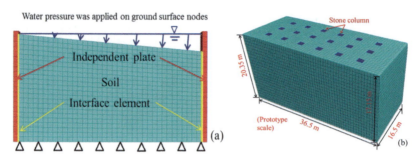

Fig. 10. Numerical model: (a) side view of boundary condition during dynamic stage; (b) 3D model mesh for Model 3

A 3D numerical model was established for Model 3 to properly reflect the 3D condition of stone column-improved ground, it has the same dimensions as the model test in prototype scale and consists of 31,900 meshes for soil materials (As shown in Fig. 10). The cylinder gravel drains are simulated with equivalent squares of the same cross section area. The permeability coefficient of the two sands are adopted the real values in Table 1. As for 3D numerical model, the displacement degree of freedom for soil nodes are free perpendicular to the shaking direction, and only soil nodes at the lateral boundary perpendicular to the direction are fix in y-direction. Other boundary conditions during static consolidation are consistent with those in 2D numerical model. The one dimensional columns in 2D model are replaced by two independent plates with one zone width to reflect the laminar box condition. The parameters of E for elastic model is adjusted as 5e4 kPa, and other parameters are the same as that in 2D model. Similarly, the nodes at two plates with the same elevation are tied each. Interface element is also used in 3D numerical simulation.

Most available constitutive models would underestimate the settlement compared to what occurs in real condition (Shahir et al. 2014). The developers of CycliqCPSP introduced a settlement improved subroutine in FLAC3D to increase the accuracy of simulated settlement, which further reduce the soil modulus with respect to the current effective stress and maximum shear strain level to simply increase post-shaking settlement. This subroutine could be switched on after shaking to significantly increase the accuracy of simulated post-shaking settlement. This subroutine was used here and the parameters were adopted as: $C_{s\_min} = 0.2$, $C_1 = 0.02$, $C_2 = 0.0$ and $C_5 = 5.0$.

4.3 Numerical Simulation Results

Figures 11 and 12 compared the simulations and test results for Model 2 and Model 3 under motion1. The simulated acceleration and EPWP reasonably match with those of model tests, especially for the generation and dissipation of EPWP. The lateral displacement time histories of model test were obtained following the method proposed by Kutter et al. (2015). The residual lateral displacement and displacement development process are well simulated with the average surface lateral spreading of model tests, but the simulated displacement fluctuation tends to be smaller during shaking period. The simulated surface settlement was also matched well with that of test when the settlement improved subroutine was adopted. The comparisons between simulations and model tests for other motions were also matched well but not shown here for space limit. The above results validate the effectiveness of CycliqCPSP and the used numerical method, which paves the road for a more in depth analysis on displacement characteristics of stone column improved ground in the following study.

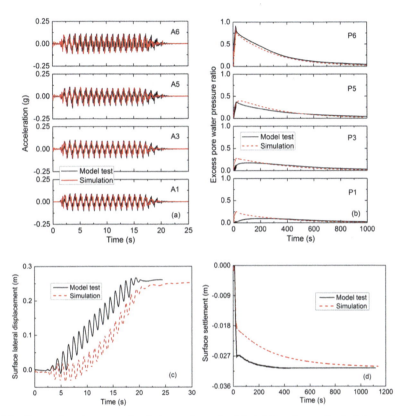

Fig. 11. Comparison between simulation and test for Model 2 under motion1: (a) acceleration time history; (b) excess pore water pressure; (c) average surface lateral spreading; (d) surface settlement

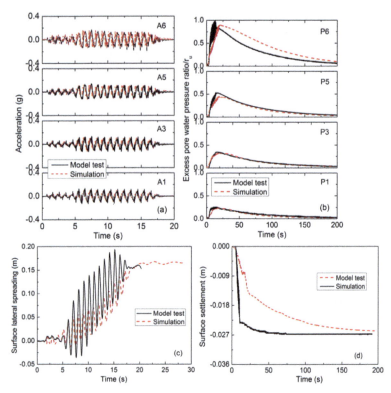

Fig. 12. Comparison between simulation and test for Model 3 under motion1: (a) acceleration time history; (b) excess pore water pressure; (c) average surface lateral spreading; (d) surface settlement

5   Deformation Mechanisms of Stone Column Improved Sloping Ground

The above model test and numerical simulation results have shown that the densification and drainage effects of stone column could reduce both surface settlement and lateral spreading in sloping ground. However, the deformation mitigation mechanisms have not been well understood, which will be preliminarily discussed in the following section.

5.1   The Effect of Densification on Settlement and Lateral Spreading

Dashti et al. (2012) identified several volumetric-induced settlement mechanisms as follows: 1) Localized volumetric strains during partially drained cyclic loading controlled by 3D transient hydraulic gradients ($\varepsilon_{\text{p-DR}}$); 2) Settlements due to sedimentation or solidification after liquefaction or soil structure break-down ($\varepsilon_{\text{p-SED}}$) and; 3) Consolidation-induced volumetric strains as excess pore pressures dissipate ($\varepsilon_{\text{p-CON}}$).

The soil skeleton after densification would become more intact with less void ratio, it will undergo less extent of soil skeleton breakdown under the same input energy, which helps to reduce the sedimentation-induced settlement.

Consolidation-induced settlement is essentially the process of soil's compression under soil own-weight over time, so it is mainly influenced by the compressibility of soil, which is conventionally expressed by compressive modulus. It is well-known compressive modulus is mainly influenced by effective stress level, it comes close to zero when soil is liquefied. In this sense, densification could significantly reduce the EPWP level (as shown in Fig. 7), thus reducing the softening extent of soil under the same shaking input motion, resulting in a less consolidation-induced settlement.

Figure 4 has validated the relation between residual shear strain and maximum shear strain. It is well understood that the shear modulus of soil will increase after densification, which will reduce the shear strain when soil undergoes the same shaking intensity, as shown in Fig. 7, so densification could reduce the residual shear strain in sloping ground. Figure 7 also shows that the soil will dilate to recover its shear modulus under smaller shear strain, which is of great help to reduce lateral spreading in sloping ground.

5.2 The Effect of Drainage on Settlement and Lateral Spreading

It could be seen from Fig. 9 that drainage effect significantly slowed down the generation rate of EPWP during shaking and delayed the time of liquefaction occurring, which would reduce the extent of soil skeleton breakdown, thus reducing the contribution of sedimentation-induced settlement ($\varepsilon_{p\text{-SED}}$) to total settlement. Besides, drainage effect expedited the dissipation of EPWP after shaking and accelerated the recovery of compressive modulus of soil, which would reduce the consolidation-induced settlement ($\varepsilon_{p\text{-CON}}$). However, the drainage effect would increase the contribution of $\varepsilon_{p\text{-DR}}$ during cyclic loading to total settlement. In this paper, the reduction of $\varepsilon_{p\text{-SED}}$ and $\varepsilon_{p\text{-CON}}$ overcame the increase of $\varepsilon_{p\text{-DR}}$, resulting in the reduction of total settlement.

Figure 8 shows that the maximum shear strain of Model 3 was smaller than that of Model 2 at shallow depth (i.e., depth = 3.5 and 6.5 m), which is attributed to drainage effect. The reduction of maximum shear strain in Model 3would result in a smaller residual shear strain according to Fig. 4. The smaller shear strain in Model 3 could be attributed to that the drainage effect reduced the extent of water absorption of soil during shear, according to Zhang et al. (1999).

In addition, the stone column material has larger permeability coefficient and liquefaction resistance compared to that surrounding soil, the "stiffer" SCs underwent smaller lateral displacement thus restricting the lateral displacement of the surrounding soil, this phenomenon was well observed in numerical simulation, as shown in Fig. 13.

Slope crest   Slope toe

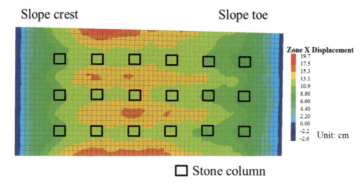

□ Stone column

Fig. 13. Top view of lateral spreading of Model 3 under motion1

6   Conclusions

A series of HCA soil element tests were conducted to identify the residual volumetric and shear strain with initial stress condition, and three centrifuge model tests and numerical simulations were performed to explore the deformation mitigation mechanisms of stone column-improved sloping ground. The following conclusions are drawn:

(1) The settlement estimation method proposed by Shamoto and Zhang (1998) for level ground could be applied to sloping ground condition given the residual shear strain component, and the maximum dynamic shear strain were found to be well correlated with residual strain;

(2) The adopted constitutive model and numerical method were validated by the centrifuge model tests with respect to acceleration, EPWP and deformation response;

(3) The densification effect could reduce sedimentation and consolidation settlement by reducing soil skeleton breakdown and softening extent of soil under the same shaking intensity respectively, in addition, it reducing the maximum shear strain thus reducing the residual shear strain in sloping condition;

(4) The drainage effect slows down the generation rate of EPWP thus reducing the sedimentation settlement and reduces consolidation settlement by expedite the dissipation process of EPWP, besides, the "stiffer column" will restrict the lateral spreading of surrounding soil.

Acknowledgements. The authors would like to acknowledge the National Natural Science Foundation of China (Nos. 51988101, 51978613 and 51778573) and the Chinese Program of Introducing Talents of Discipline to University (the 111 Project, No. B18047) for the funding support. Sincere thanks to Prof. Guoxing Chen and Dr. Qi Wu of Nanjing Tech University for their help during the soil element test, and to Mr. Qiang Ma and Mr. Zizhuang Yan of Zhejiang University for their supports in conducting the centrifuge model tests.

References

1. Asgari, A., Oliaei, M., Bagheri, M.: Numerical simulation of improvement of a liquefiable soil layer using stone column and pile-pinning techniques. Soil Dynam. Earthq. Eng. **51**, 77–96 (2013)
2. Badanagki, M., Dashti, S., Kirkwood, P.: Influence of dense granular columns on the performance of level and gently sloping liquefiable sites. J. Geotech. Geoenviron. Eng. **144**(9), 04018065 (2018)
3. Chiaro, G., Koseki, J., Sato, T.: Effects of initial static shear on liquefaction and large deformation properties of loose saturated Toyoura sand in undrained cyclic torsional shear tests. Soils Found. **52**(3), 498–510 (2012)
4. Elgamal, A., Lu, J., Forcellini, D.: Mitigation of liquefaction induced lateral deformation in a sloping stratum: three-dimensional numerical simulation. J. Geotech. Geoenviron. Eng. **135**(11), 1672–1682 (2009)
5. Wang, R., Zhang, J.M., Wang, G.: A unified plasticity model for large post-liquefaction shear deformation of sand. Comput. Geotech. **59**, 54–66 (2014)
6. Shahir, H., Mohammadi-Haji, B., Ghassemi, A.: Employing a variable permeability model in numerical simulation of saturated sand behavior under earthquake loading. Comput. Geotech. **55**, 211–223 (2014)
7. Shamoto, Y., Zhang, J.M., Tokimatsu, K.: New charts for predicting large residual post-liquefaction ground deformation. Soil Dynam. Earthq. Eng. **17**(7), 427–438 (1998)
8. Youd, T.L., Hansen, C.M., Bartlett, S.F.: Revised multilinear regression equations for prediction of lateral spread displacement. J. Geotech. Geoenviron. Eng. **128**(12), 1007–1017 (2002)
9. Zhang, G., Robertson, P.K., Brachman, R.W.I.: Estimating liquefaction-induced lateral displacements using the standard penetration test or cone penetration test. J. Geotech. Geoenviron. Eng. **130**(8), 861–871 (2004)
10. Zhang, J.M., Tokimatsu, K., Taya, Y.: Effect of water absorption in shear of post-liquefaction. Chinese J. Geotech. Eng. **21**(4), 398–404 (1999). (in Chinese)
11. Zhou, Y.G., Liu, K., Sun, Z.B., Chen, Y.M.: Liquefaction mitigation mechanisms of stone column-improved ground by dynamic centrifuge model tests. Soil Dynam. Earthq. Eng. 106660 (2021)
12. Zhou, Y.G., Sun, Z.B., Chen, J., Chen, Y.M., Chen, R.P.: Shear wave velocity-based evaluation and design of stone column improved ground for liquefaction mitigation. Earthq. Eng. Eng. Vib. **16**(2), 247–261 (2017)
13. Zou, Y.X., Zhang, J.M., Wang, G.: Seismic analysis of stone column improved liquefiable ground using a plasticity model for coarse-grained soil. Comput. Geotechnics. **125** (2020)

Liquefaction-Induced Downdrag on Piles: Insights from a Centrifuge and Numerical Modeling Program

Katerina Ziotopoulou[1]([✉]) [iD], Sumeet K. Sinha[2], and Bruce L. Kutter[1] [iD]

[1] University of California, Davis, CA 95616, USA
kziotopoulou@ucdavis.edu
[2] University of California, Berkeley, CA 94720, USA

Abstract. Earthquake-induced soil liquefaction can cause settlement around piles, which can translate to negative skin friction, resulting in the development of drag load and settlement of the piles. Despite significant research progress on the effects of liquefaction on structures and the seismic response of piles, there is still a knowledge gap in the assessment of liquefaction-induced downdrag loads on piles. The interrelationships between mechanisms affecting negative skin friction are not accounted for in current practice, typically leading to over-conservative or unsafely designed piles. A series of centrifuge model tests were performed to assess liquefaction-induced downdrag and understand the interplay of pile embedment, pile-head load, excess pore pressure generation, dissipation, reconsolidation, and ground settlement on the axial load profile during and post shaking. The two tests featured instrumented piles, loaded with a range of axial loads and embedded at a range of depths into the bearing layer, in a uniform and a layered liquefiable profile respectively. The tests showed that most of the settlement of the piles occurred during shaking. The mechanisms behind the development of liquefaction-induced drag load on piles and settlements will be described and ramifications concerning the design of piles in liquefiable soils will be discussed.

Keywords: Liquefaction · Downdrag · Centrifuge model testing

1 Introduction

Pile foundations are designed to transfer superstructure loads through positive interface shear stress (commonly called skin friction) and end-bearing resistance while meeting performance-based criteria in terms of settlements. The positive skin friction is the result of the pile settling more than the surrounding soil. In earthquake prone regions however, post-liquefaction reconsolidation soil settlements can result in soil layers settling more than the piles resulting in downward (negative) interface shear stresses, which in turn can impose additional demands on piles in terms of drag load sand settlements. The net downward force due to negative skin friction is called drag load (Q_d) and it increases the axial load on the pile beyond the pile head load (Q_f). Consequently, both the positive skin friction below the liquefied layer and the load at the pile tip increases, and the pile

L. Wang et al. (Eds.): PBD-IV 2022, GGEE 52, pp. 660–681, 2022.
https://doi.org/10.1007/978-3-031-11898-2_40

settles until enough resistance is mobilized and force equilibrium is re-established. Pile settlement caused by the drag load is known as downdrag [1]. The depth at which the relative velocity (or relative movement in time Δt) of the soil and pile is zero is known as the neutral plane [2]. Alternatively, the neutral plane is the depth at which the skin friction on the pile is zero. Above the neutral plane, the relative movement of soil is more than the pile resulting in the development of drag load; below the neutral plane, the relative movement of the pile is more than the soil resulting in the development of positive skin friction [1]. The resulting load distribution is maximized at the neutral plane. Naturally, in such cases, the estimation of the axial load distribution and pile settlement becomes an important design consideration and evaluation criterion. Most challenges related to liquefaction-induced downdrag result from the incomplete understanding of the different time- and depth-dependent mechanisms that affect drag load and pile settlement [3]. The interrelationships between mechanisms affecting negative skin friction (e.g., pore pressure generation and dissipation patterns, sequencing of settlements and reconsolidation of liquefied soils, and gapping and softening of soils around the piles) are not accounted for in current practice, typically leading to conservative designs. Figure 1 illustrates the mechanisms and terminologies of interest in this paper while [4] present a comprehensive overview of the work performed by numerous entities [5–22] on these phenomena and from multiple viewpoints.

This paper presents a comprehensive body of work performed towards resolving the mechanisms around liquefaction-induced downdrag on piles and its associated impacts on design considerations. A series of two centrifuge model tests performed to address specific questions are presented and select results are summarized, followed by the development of a numerical modeling approach the incorporates the observed mechanisms. The paper concludes by summarizing lessons learned and addressing practical implications as well as considerations for usability in field applications where centrifuge model test data are not available.

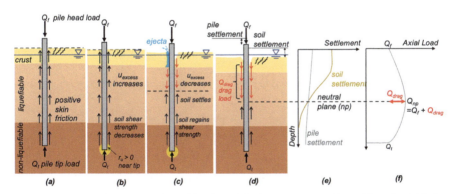

Fig. 1. Illustration of time evolution of loads and displacements experienced by pile: (a) before earthquake shaking, (b) during earthquake shaking, (c) during reconsolidation, (d) after reconsolidation and re-establishment of effective stresses in the profile, (e) depth distribution of final soil and pile settlement and definition of neutral plane, and (f) load distribution along pile peaking at the neutral plane due to the drag load.

2 Centrifuge Model Tests

2.1 Test Specifications

Two centrifuge model tests SKS02 [23] and SKS03 [24], were performed on the 9 m-radius centrifuge at the Center for Geotechnical Modeling (CGM) at the University of California Davis. The tests were performed at a centrifugal acceleration of 40 g, in a flexible shear beam container with a rigid base plate and included identical instrumented piles passing through a thick liquefiable layer (SKS02) or an interbedded liquefiable deposit (SKS03), embedded at different depths within an underlying deeper dense layer. Figure 2 shows close-up views of the two tests. The pile dimensions and the soil layers used in centrifuge tests were selected based on the parametric study of liquefaction-induced downdrag on piles by [25]. All test data are curated and publicly available through DesignSafe under PRJ-2828. All quantities presented henceforth have been converted to prototype units according to the scaling laws described by [26] unless explicitly mentioned otherwise.

Soil Layering and Instrumentation. The SKS02 model (Fig. 2 – left) consisted of a layered soil profile of a 9 m-thick loose liquefiable Ottawa F-65 sand ($D_R \approx 42$–44%, all index properties provided by [27]) layer sandwiched between 4 m of an over-consolidated kaolin clay layer ($\sigma'_p = 100$ kPa) at the top and 8 m of a dense Ottawa F-65 sand layer ($D_R \approx 86$–88%) at the bottom. Model SKS03 included six distinct soil layers. The soil profile consisted of 1 m of Monterey sand, 2 m of clay crust ($s_u \approx 35$ kPa), 4.7 m of loose liquefiable sand ($D_R \approx 40\%$), 1.3 m of clayey silt (20% Kaolin clay and 80% non-plastic silt), 4 m of medium dense sand ($D_R \approx 60\%$) and a dense sand ($D_R \approx 83\%$) beneath it. The clay crust was prepared from a lightly cemented Yolo Loam slurry with a water content of w = 50%, soil cement ratio of 3%, and cured for about two weeks underwater before the test. All sand layers consisted of Ottawa F-65 sand.

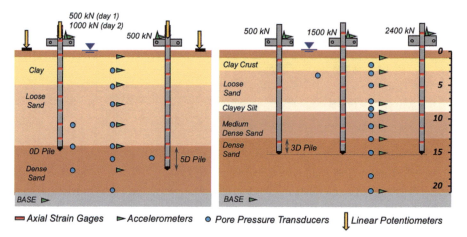

Fig. 2. Centrifuge model tests SKS02 (left) and SKS03 (right) including soil layering, instrumentation and its locations, and pile properties (embedment depths 0D, 5D, and 3D; and pile head loads defining factors of safety).

Both SKS02 and SKS03 models were instrumented with accelerometers, pore pressure transducers, linear potentiometers, and settlement markers. Linear potentiometers measured soil and pile settlements, respectively. In addition, a new non-contact displacement sensing methodology was developed using line lasers, cameras, and target markers to measure soil and pile settlement [28, 29]. To determine the state of the model at different phases of the test, cone penetration tests (CPT) and centrifuge pile penetration tests (CPPT) were performed while spinning the model at 40g, while vane shear tests (VST) were performed at 1 g after spinning down.

Pile Properties, Instrumentation, and Installation. SKS02 included two instrumented aluminum closed-ended piles with an outer diameter (D) of 635 mm and an inner diameter of 564 mm named "0DPile" and "5DPile", with the 5DPile's tip embedded five diameters into the dense sand., and the 0DPile's tip placed at the top of the dense sand. Both piles were initially loaded with a pile head mass that, after spinning up the centrifuge, represented an axial load of 500 kN. SKS03 included three identical instrumented piles with geometries identical to those of SKS02, but now denoted as 3DPileS, 3DPileM, and 3DPileL, loaded with 500kN, 1500 kN, and 2400 kN, respectively. The "3D" at the beginning of the pile names indicates that the pile tips were embedded three pile diameters into the dense sand layer.

To obtain maximum drag loads on piles, their outer surface was machined to achieve an average roughness (R_a) of 0.04 to 0.06 mm, deemed sufficient to mobilize an interface friction angle (δ) equal to the drained soil friction angle of $\phi'_{cv} = 30°$ [30]. The piles were instrumented internally with strain gages to measure the axial load distribution while not compromising the interface properties. The pile's outer and inner diameters at the model scale were approximately 15.9 mm and 14.1 mm, respectively. Nine strain gage bridges were installed with a spacing of about 5.08 cm (i.e., about 2 m at prototype

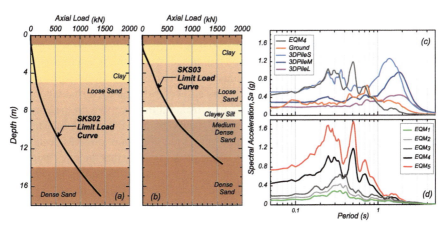

Fig. 3. Limit load curves of the piles in centrifuge model tests (a) SKS02 and (b) SKS03 for zero pile head load, and spectral accelerations recorded in SKS03: (c) input motion (EQM4) at the base of the model, the model surface, and the three piles (3DPileS, 3DPileM, and 3DPileL); and (d) all applied earthquake motions.

scale). Two-point bending moment tests on different loading axes were conducted on the piles to determine the orientation corresponding to the minimum cross-axis bending moment sensitivity which later informed the placement of the piles.

The piles were pushed before spin-up (at 1 g) to their target depths in soil layers. A split mass was clamped to the pile to produce the desired axial load with its bottom set at 1 m above the ground surface to permit large settlements of the pile head while limiting the magnitude of shaking-induced bending moments and their impacts on measuring axial loads. The total static pile capacity for the 0DPile and 5DPile in SKS02 estimated from cone penetration and centrifuge pile penetration tests was about 2700 kN and 6200 kN, respectively. As such, the SKS02 piles under the initial load of 500 kN had a static factor of safety (FS) of 5.4 and 12.4, respectively. When the load on 0DPile was increased to 1000 kN on the second day of testing, the associated static factor of safety got reduced to 2.7. In SKS03, a pile load test was used to estimate a static pile capacity of 4550 kN for all three piles, resulting in a static FS of 8, 2.6, and 1.6 for the 3DPileS, 3DPileM, and 3DPileL, respectively.

The limit load curves for the piles obtained for both the centrifuge model tests for zero pile head load are shown in Fig. 3. The limit load curve is defined as the axial load distribution of the pile corresponding to the maximum drag load the pile can develop. Assuming the pile mobilizes negative skin friction equal to its interface shear strength, the maximum drag load is estimated. Alternatively, the limit load curve for a pile can be defined as the sum of pile head load and its cumulative shaft capacity with depth. The limit load curve for SKS03 was obtained directly from the axial load distribution recorded during the pile load test, while for SKS02, it was estimated from the measured undrained shear strength in the clay layer (using the α-method of [10] with a s_u of 20 kPa) and drained shear strength in the sand layers (estimated as $\tau = K\sigma'_v \tan(\delta)$ with a $\delta = 30°$ and an average lateral stress coefficient $K = 1$).

Testing Sequence and Shaking Events. Both centrifuge models SKS02 and SKS03 were tested over the span of two days. The models were spun up to 40 g and shaken with six and five scaled Santa Cruz earthquake motions, respectively, with peak base accelerations (PBA) ranging from 0.025 to 0.4 g for SKS02, and 0.09 to 0.61 g for SKS03. EQM6 was a long-duration motion composed of one strong Santa Cruz motion followed by five small magnitudes of Santa Cruz motions. CPTs were performed in-between the shaking events. Enough time was allowed between each shaking event to completely dissipate the excess pore pressures from all the layers, with the clay crust being the controlling layer and strong shaking events (EQM3, EQM5, EQM6) needing more time for full dissipation. The predominant period of the motions was designed to be away from the first fundamental period of the piles to ensure that the piles would not undergo strong horizontal movements generating large moments, which might affect the accuracy of the axial load measurements. Shaking event EQM4 in SKS03 was a long-duration modified Santa Cruz motion [45] consisting of one large pulse followed by five small pulses, scaled to produce a PBA of 0.45 g. Figure 3 illustrates the spectra of the applied earthquake motions, as well as the spectral accelerations obtained from EQM4 at the soil surface and on the 3DPiles' masses. These spectra demonstrate that the input motions had a predominant period between 0.3–0.4 s and were indeed away from the piles' first fundamental period of 1–2 s.

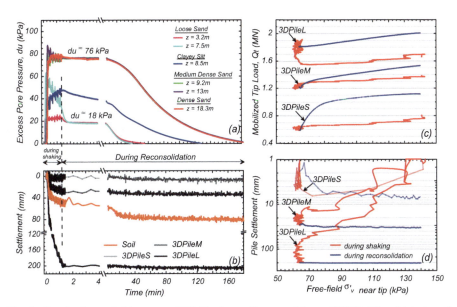

Fig. 4. Results from shaking event EQM4 of centrifuge model test SKS03. Time histories of: (a) excess pore pressure at various locations within the different layers, (b) settlement of soil and piles, (c) mobilized tip load (Q_t) for each pile versus the free-field effective stress (σ'_v) near the pile's tip, and (d) pile settlement for each pile versus the free-field effective stress (σ'_v) near the pile's tip.

2.2 Responses During and Post Shaking

The overall mechanisms observed during shaking and reconsolidation were similar across all the shaking events in the two centrifuge model tests. SKS02 is extensively described by [4, 23] while SKS03 is the focus of this paper. For SKS03, all shakings from EQM1 to EQM5 induced liquefaction in portions of the loose and medium dense sand layers, while significant excess pore pressures (du) were also measured in the dense sand layer. Figure 4a and b show the time histories of du and soil and pile settlement, respectively, for shaking event EQM4. Figure 4c-d show the mobilized tip load and pile settlement as free-field effective stress changed at the pile's tip depth (initially reduced during du generation and later recovered during dissipation and reconsolidation). Figure 5a shows select isochrones of du in the soil layers, while Fig. 5b-d show the axial load distribution in 3DPileS, 3DPileM, and 3DPileL, respectively.

Responses During Shaking. At the beginning of shaking, the piles had an initial drag load, and correspondingly an initial static neutral plane developed from the negative skin friction either from centrifuge spin-up or past shaking events (EQM1–EQM3 preceded EQM4 which is illustrated in Figs. 4 and 5). At the beginning of the shaking, some surface settlements occurred in the model (e.g., Fig. 4b). The initial surface settlement could have resulted from undrained soil movement caused by the tendency of the flat ground surface to conform to the curved g-field in the centrifuge. The generation of du during shaking in the soil layers decreased skin friction, and thus drag load and axial loads in the piles overall. In addition, the r_u build-up and strength loss throughout the model also decreased positive skin friction and the tip resistance below the initial static neutral plane. The decrease in drag load caused a decrease in axial loads in the piles (Fig. 5b-d). Typically, the decrease in drag load surpassed the decrease in shaft resistance below the neutral plane resulting in a decrease of the pile tip load (Fig. 4c). However, when large excess pore pressures developed around the heavily loaded pile 3DPileL, the loss of shaft resistance surpassed the loss of drag loads, resulting in the mobilization of larger pile tip loads (Fig. 4c).

When the soil liquefied, the drag load was reduced to zero, and high excess pore pressures were generated in the dense sand layer. At the peak of shaking, high excess pore pressures equal to du $\approx$ 80 kPa were observed even in the dense sand layer of SKS03. While the dense sand was expected to exhibit small earthquake-induced excess pore pressures, water migration from the loose sand layer to the dense layer resulted in increased excess pore pressures. The magnitude of pore pressure increase in the dense layer was found to depend on the relative thicknesses of the layers and drainage boundary conditions. Overall, the constant axial load profile of a pile demonstrated the vanishing of the drag load in the liquefied soil. This was not the case for 3DPileL, where the axial loads in the liquefied soil were not constant, later found to be erroneous as elaborated by [4]. Even though the drag load decreased the pile's tip load during shaking, all the piles settled (Fig. 4d). The pile settlement occurred because the excess pore pressures generated around the pile decreased the pile's shaft and tip capacity causing settlement until enough resistance was mobilized to achieve force equilibrium. As shown in Fig. 4d, the heavily loaded 3DPileL with a large tip load and smaller free-field effective stress near tip (σ'_v = 60 kPa) caused the pile to plunge by 200 mm in soil, while the lightly loaded piles (3DPileS and 3DPileM) suffered comparatively small settlements.

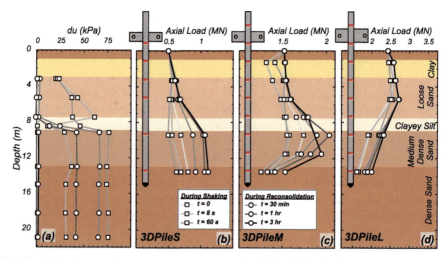

Fig. 5. (a) Isochrones of excess pore pressures, and axial load distributions in (b) 3DPileS, (c) 3DPileM, and (d) 3DPileL during shaking and reconsolidation for event EQM4 in SKS03.

Responses Post Shaking. During reconsolidation, the clay and silt layers in the centrifuge tests contributed to interesting effects such as drainage impedance and equalization of excess pore pressures, particularly in SKS02 [23]. In SKS03, the soil surface continuously settled during reconsolidation (Fig. 4b). However, the relatively low permeability of the clayey silt and the clay layer slowed down the dissipation and equalized the excess pore pressures in the soil layers beneath (Fig. 4a, Fig. 5a). The du's in the loose sand layer first equalized to 20 kPa, approximately equal to the effective stress at the bottom of the clay layer. The du's in the medium dense sand and dense sand first equalized to 76 kPa, equal to the effective stress at the bottom of the clayey silt layer. From Fig. 4b, it can be observed that reconsolidation finished in the loose sand layer within 45–60 min; however, the clayey silt layer's low permeability took more than 3 h to complete reconsolidation the layers beneath.

During reconsolidation, as du's dissipated and the soil settled, the negative skin friction and hence the drag loads increased on the piles: Fig. 5 illustrates the decrease in du and the corresponding increase in drag load and axial load in the piles at t = 30 min, 1 h, and 3 h following EQM4 of SKS03. The load at the shaft and the tip below the neutral plane increased (e.g., Fig. 4d, and b-d). Overall, after complete reconsolidation, the achieved drag load was higher than the pre-shaking initial drag load, while the neutral plane depths remained unchanged. The increase in drag load resulted from the increased mobilized negative skin friction above the neutral plane caused either 1) due to the increased soil displacement at the interface or 2) due to increased lateral stresses from densification. Still, the resulting pile settlement was smaller as the piles regained their tip capacity and stiffness (Fig. 4d) with the settlement of the piles being less than 10 mm (i.e., 1.6% of pile diameter) during reconsolidation.

2.3 Discussion on Pile Settlements and Drag Loads

Pile Settlements. Most of the settlements in the piles occurred during shaking when the excess pore pressures around the pile's shaft and tip were high. On the other hand, the pile downdrag settlement during reconsolidation was typically small ($< 2\%$ of the pile's diameter). Figure 6 summarizes the pile and soil settlement at the end of each shaking event of both the centrifuge model tests SKS02 and SKS03. The leading cause for the large settlement of the piles was the decrease in pile tip capacity and stiffness

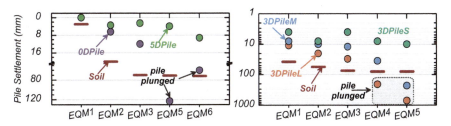

Fig. 6. Summary of the total pile and soil settlement at the end of each shaking event of centrifuge model tests: (left) SKS02 and (right) SKS03. Relative position of settlements also indicates plunging or no plunging.

from increased du's near the tip. The centrifuge tests showed that even if the bearing layer (in which the pile tip is embedded) is non-liquefiable, redistribution of excess pore pressures from the nearby liquefied layers and the generated earthquake-induced excess pore pressures can cause large du developments in the bearing layer. With the increase in the magnitude of the shaking events, the free-field du's developed near the pile's tip increased, and correspondingly the pile settlement also increased. As already discussed, during reconsolidation, while the drag loads increased, the shaft and tip capacity and stiffness also increased, producing small pile settlements. As a result, for every event (except for 5DPile and 3DPileS for small shaking events), the pile settlement during reconsolidation was much smaller than the settlement during shaking. The smaller downdrag settlement shows that drag loads do not control the pile's performance unless structurally overloaded. It is the pile settlement during shaking that governs the performance of the pile. Figure 6 also helps observe cases of plunging: for most shaking events, the settlement of the piles was smaller than the soil settlement. Only for large shaking events when the tip capacity significantly reduced and caused plunging of the piles, the pile settlement was greater than the soil. In those cases, the piles plunged when the mobilized tip load was near or exceeded the reduced pile tip capacity due to the increased neighboring free-field du (e.g., 0DPile during EQM5 and EQM6 in SKS02; 3DPileM during EQM4; and 3DPileL during EQM3 and EQM5).

[7] used load measurements from a series of centrifuge tests featuring liquefiable soils and proposed an empirical model to estimate the reduced pile tip capacity in liquefying soil $\left(Q_{t,ult}^{r_u}\right)$ as a nonlinear function of free-field excess pore pressure ratio (r_u) near the pile's tip:

$$Q_{t,ult}^{r_u} = Q_{t,ult}^0 \cdot (1 - r_u)^{\alpha_t} \tag{1}$$

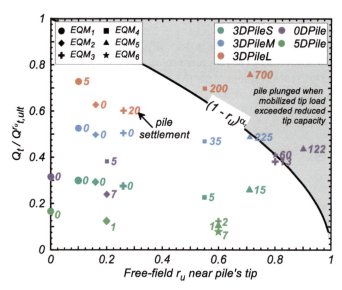

Fig. 7. Summary of pile settlement during shaking resulting from the mobilization of tip load (Q_t) and reduced tip capacity ($Q_{t,ult}^{r_u}$) caused from increased free-field excess pore pressure ratio (r_u) near the pile's tip during shaking events of centrifuge model tests: SKS02 and SKS03.

$$\alpha_t = \frac{3 - \sin \varphi'}{3(3 - \sin \varphi')} \tag{2}$$

where $Q^o_{t,ult}$ is the ultimate tip capacity when $r_u = 0$, α_t is a constant that depends only on ϕ' [7], the effective friction angle of the soil at the tip. In the centrifuge model tests, $\phi' = 30^o$ at the tip, which resulted in $\alpha_t = 0.55$. Accordingly, Fig. 7 shows the mobilized pile tip load (Q_t) and the associated settlement in piles because of the reduced pile tip capacity ($Q^o_{t,ult}$) from the increased free-field excess pore pressure ratio (r_u) near the tip for all the shaking events of centrifuge model tests SKS02 and SKS03. From Fig. 7, it can be observed that when the mobilized pile tip load (Q_t) was within the reduced pile tip capacity ($Q^o_{t,ult}$), the associated settlements in piles were small. However, when the mobilized tip load (Q_t) was near the reduced tip capacity $\left(Q^o_{t,ult}\right)$ the piles plunged and experienced significant settlements. The mapping of Fig. 7 nicely explains and summarizes the individual observations made earlier.

Drag Loads. As discussed, for each shaking event, the drag load on the piles decreased during shaking and increased during reconsolidation, eventually exceeding the pre-shaking value. Figure 8 shows axial load profiles in the piles after each shaking event in both SKS02 and SKS03. While the neutral plane depth in the piles remained almost the same across all shaking events, the drag loads kept on increasing. For example, shaking events EQM1, EQM2, and EQM3 in SKS02, increased drag loads from 250 kN to 500 kN in 0DPile, and from 700 kN to 1100 kN in the 5DPile. The increased drag loads also increased the mobilized pile tip load as seen in all depth plots of Fig. 8. This increase in drag load could be due to the gradual increase of the effective lateral stress during reconsolidation. It remains to be confirmed whether the increase in lateral stress is due to dilatancy of the soil adjacent to the pile producing increased lateral stresses locally

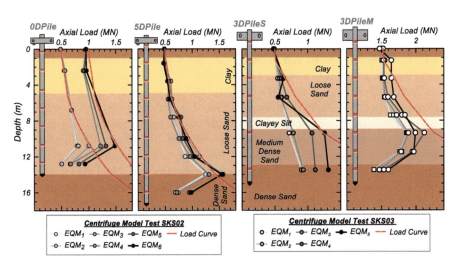

Fig. 8. Summary of axial load distribution in piles (a) 0DPile and 5DPile for centrifuge model test SKS02, and (b) 3DPileS (highest factor of safety) and 3DPileM for centrifuge model test SKS03 after reconsolidation from shaking events EQM1-EQM6.

around the pile or an artifact of the centrifuge model container flexibility. Another possible mechanism could be the increased stiffness of tip (due to soil stronger or increase in embedment from pile settlement) in each successive event resulting in small downdrag settlement and larger drag load. Shaking events EQM4, EQM5, and EQM6 in SKS02 also increased drag loads in the piles: from 150 kN to 500 kN in the 0DPile and from about 700 kN to 1100 kN in the 5DPile. However, for the 5DPile, the drag load and the axial load distribution for shaking events EQM5 and EQM6 remained almost the same. The drag load for the 5DPile may have achieved saturation at about 1100 kN, which could have resulted from the full mobilization of interface skin friction and stabilization of lateral stresses. Generally, the developed drag load was higher for the deeply embedded or lightly loaded (resulting in higher static factors of safety) piles. The deeply embedded 5DPile developed a much higher drag load than 0DPile. Among all the 3DPiles, 3DPileS with the highest static factor of safety developed the largest drag load. While estimating the exact lateral stresses developed at the interface is difficult, these results suggest that liquefaction-induced downdrag can mobilize negative skin friction equal to the interface shear strength. Thus, a conceptual limit load curve is shown in Fig. 8 to compare the axial load profile of the piles obtained after complete reconsolidation. The conceptual limit load curve was found to envelope all the piles' axial load profiles.

The development of negative skin friction requires small relative movements between the soil and pile [31]. The medium shaking events EQM2 and EQM4 of centrifuge model test SKS02, even though they did not fully liquefy the loose sand layer ($r_u \approx 50\%$), resulted in the development of negative skin friction and thus the drag loads on piles. However, the increase in the drag load was smaller in magnitude when compared to the large shaking events (EQM3, EQM5, and EQM6). The total soil settlement caused by the medium shaking events was about 15–20 mm. In SKS03, about 20–30 mm of soil settlement (during reconsolidation) achieved full drag load on piles. During the pile load test, the penetration of the pile by about 20 mm (i.e., about 3% of pile's diameter) completely removed the initial drag load on 3DPileS. [32] suggest that displacements of 0.5% to 2% of the pile's diameter are required to mobilize skin friction completely. Results from pile load tests, particularly on drilled shafts (of diameter < 2 m), show that the skin friction is fully mobilized at a relatively small displacement in the order of 10 to 30 mm, depending on the soil's D_R [33]. These results show that even when the soil layers do not fully liquefy, settlements caused by the dissipation of du, exceeding a small percentage of the pile diameter, would be enough to mobilize negative skin friction and thus drag loads in a pile. Alternatively, the 20 mm of soil settlement also corresponds to a displacement equal to 2.5 times the median grain diameter (D_{50}) of sand used in the test. [34] and [30] suggested an interface shear band thickness of 5–7 particle diameters adjacent to the interface. Assuming that the elastic settlement in the shear band with respect to the free field soil settlement is small, a displacement of 2.5 D_{50} would produce a shear strain of 50% (assuming a shear band thickness of 5 D_{50}) which would be enough to mobilize the significant negative skin friction at the pile's interface.

3 Numerical Modeling

Numerous methods have been proposed to account for the development of drag loads and estimate pile settlements in liquefiable deposits, with the neutral plane method [5, 15, 21, 22, 35] and some variations of it depending on the values considered for the negative skin friction value [36]. For example, [12] recommend taking the negative skin friction in the liquefiable layer as 50% of the positive skin friction before shaking while [10 and 37] recommend taking the negative skin friction equal to the "residual soil strength" in the liquefiable layer. The Federal Highway Administration (FHWA [14]) recommends using the neutral plane method with t-z and q-z springs calibrated from field tests. All these methods model the liquefiable layer as a consolidating layer with a defined strength without considering the effects of the event sequencing and the pattern of excess pore pressure dissipation, the soil settlement, and the evolution of soil shear strength during reconsolidation. [11] modified the neutral plane method to account for the timing of the soil settlement and dissipation of excess pore pressures in liquefiable layers but did not provide a solution for the impacts of excess pore pressure generation in the vicinity of the tip.

To address some of these gaps, a TzQzLiq numerical modeling approach in OpenSees [37] (Fig. 9) was developed and validated for modeling liquefaction-induced downdrag on piles, and is introduced herein. The model accounts for changes in the shaft (using a TzLiq material) and tip capacity (using a QzLiq material) of the pile as free-field excess pore pressures develop and dissipate in soil.

3.1 Model Setup

TzQzLiq Material. [39] developed a TzLiq material to capture the reduction of shaft capacity and stiffness as a linear function of excess pore pressure ratio $(1-r_u)$:

$$t_{ult}^{ru} = t_{ult}^{o}(1 - r_u) \tag{3}$$

where t_{ult}^{o} is the initial/ultimate shaft capacity before any excess pore pressure is generated around the shaft. The initial elastic stiffness of the material is defined as $t_{ult}^{o}/2z_{50}$, where the constant z_{50} is the displacement required to mobilize 50% of the ultimate shaft capacity (t_{ult}^{o}). The constant z_{50} is kept independent of r_u resulting in the changes in the stiffness of the TzLiq material directly proportional to the change in its capacity (t_{ult}^{ru}). [11] used the TzLiq material to study the liquefaction-induced downdrag on piles in liquefiable soils. The ultimate shaft capacity (t_{ult}^{o}) in soil layers can be obtained empirically using equations and correlations provided in [10] or can be calibrated from tests on interface shear strength and pile load tests. The z_{50} parameter essentially defines the stiffness of the shaft resistance. According to [32], displacements of 0.5% to 2% of pile diameter are required to mobilize the shaft resistance fully. [4] performed a series of centrifuge model tests and found that small displacements in the order of 1–3% of the pile's diameter were sufficient to mobilize full skin friction in soil. The nonlinear backbone curve for t-z material [39–41] is a hyperbolic curve that takes displacement equal to about four times z_{50} to mobilize > 90% of the ultimate capacity (t_{ult}^{o}). Thus, a z_{50} of 0.2% – 0.5% of the pile diameter can be assumed for modeling the stiffness of the TzLiq material.

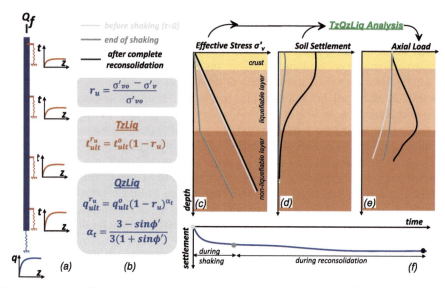

Fig. 9. Schematic illustration of the TzQzLiq numerical model for modeling liquefaction-induced downdrag on piles using (a) zero thickness interface elements with TzLiq and QzLiq materials. Model input parameters include properties of the pile, (b) TzLiq and QzLiq material properties, (c) isochrones of effective stress, and (d) soil settlement profiles. Model results include time histories of (e) axial load distribution and (f) pile settlement (after [44]).

Following the implementation of the TzLiq material by [39] and the findings of [7], a QzLiq material was developed and implemented in OpenSees [38]. The material captures the change in the pile's tip capacity and stiffness in the presence of r_u as the constitutive response of a q-z material [39, 42] scaled in proportion to r_u. The material has an ultimate load in compression related to the bearing capacity of the pile and a zero strength in tension. The properties of the material include the ultimate tip capacity (q_{ult}^o), the displacement (z_{50}) at which 50% of q_{ult}^o is mobilized, the exponent (α_t) (Eq. 2), and a time series data of free-field effective stress (σ'_v) at the pile's tip depth. The element internally evaluates r_u from the provided mean effective stress data.

Unlike shaft friction, the tip does not have an ultimate bound on its strength, especially for end-bearing piles. Displacements of the order of 10% of the pile diameter are required to significantly mobilize the pile tip capacity [32]. In the absence of pile load test data, the tip capacity, estimated from empirical correlations with soil characterization data, is taken as the mobilized resistance for tip penetration equal to 10% of the pile diameter. [43] describes the nonlinear backbone curves for tip response in dense sand; where z_{50} is taken equal to 0.125 times the displacement required to mobilize the tip capacity. A pile load test could circumvent the need to assume values for the properties (q_{ult}^o, z_{50}) of the QzLiq material.

Pile Inputs, Effective Stress and Soil Settlement Profiles. A linear elastic beam element is used to model the piles, requiring the pile's cross-section, material properties, and the superstructure dead load as inputs. Earthquake-induced cyclic axial loads can be considered by applying a time series load on the pile. If the axial loads exceed the elastic structural capacity of the pile, then nonlinear beam elements would be required without affecting the remainder of the herein presented procedure.

Time series of effective stress as a function of depth $\sigma'_v(z, t)$ and settlement $s(z, t)$ are required for the analysis. These may be obtained from a 1D or 2D site response analysis followed by a reconsolidation analysis in a finite element or a finite difference software capable of capturing the salient features of the problem. Herein, the time history of effective stress was obtained directly from the measurements of du in the centrifuge test, while the time history of the soil settlement profile was obtained from inverse analyses of measured du arrays during shaking events as described by [44] and not repeated here for brevity. The overall observation from the inverse analyses was that while most of the soil settlements occurred during reconsolidation, the soil layers suffered some immediate settlements during shaking. For shaking event EQM4 in the centrifuge model test SKS03, the immediate settlement in soil layers occurred in the first 15 s of the shaking.

3.2   Results and Validation

The TzQzLiq analysis of the piles was performed in OpenSees [38] with a mesh discretization of 0.1 m. The simulation is performed in two stages. Dead loads are applied in Stage 1, and the result of the analysis is the initial axial load distribution for Stage 2. Stage 2 is a dynamic time history analysis of the piles stepping through the time series $\sigma'_v(z, t)$ and $s(z, t)$ applied to the TzLiq and QzLiq interface elements. The limit load curves were used to obtain the ultimate capacity of the TzLiq material at different depths along the length of the pile. Backbone curves from [40] and [41] were used to model the load transfer behavior of sections of piles in sand and clay layers, respectively. The parameter z_{50} was taken as 0.3% of the pile diameter in the clay, silt, loose sand, and medium dense sand layers, and 0.15% of the pile diameter in the dense sand layer. The q-z load transfer behavior was modeled with backbone curves from [43]. Results from pile load tests were used to calibrate the properties (q^o_{ult}, z_{50}) of the QzLiq material. Figure 10a shows the satisfactory match between the centrifuge test pile load test curves and the corresponding numerical ones obtained with the selected TzLiq and QzLiq material properties. The constant (α_t) was taken as 0.55 using an effective friction angle of $\varphi' = 30°$. Tables 1 and 2 summarize the properties of the TzLiq and the QzLiq materials respectively used in the numerical analysis.

Table 1. TzLiq[a] material properties used in the analysis of piles.

Soil Layers	z_{50}(%Diameter of pile)
Clay and Silt Layers	0.31
Loose, Medium Dense Sand	0.31
Dense Sand	0.15

[a] t^o_{ult} along the length of the pile obtained from the limit load curve

Table 2. QzLiq material properties used in the analysis of piles.

Piles	z_{50}(%Diameter of pile)	q^{o}_{ult} (kN)	α_t
0DPile	7	2745	0.55
5DPile	7	7137	
3DPiles	9	4576	

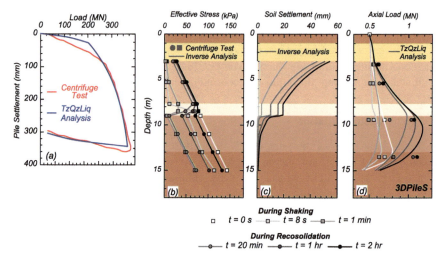

Fig. 10. TzQzLiq analysis of 3DPileS for shaking event EQM4 in centrifuge model test SKS03: (a) Calibration of QzLiq material properties against pile load test data. Profiles of (b) effective stresses obtained from centrifuge model test instrumentation, (c) soil settlement obtained from inverse analyses, and (d) axial load at selected times during shaking and reconsolidation.

Axial Load Distribution and Axial Load. Indicative results are presented for EQM4 in SKS03. Figure 10b and c illustrate the isochrones during shaking and reconsolidation of effective stress and soil settlement used in the analyses. Figure 10d shows a comparison between axial load distributions at selected times obtained from the TzQzLiq analyses for 3DPileS in the centrifuge model test SKS03. For the same pile, time histories of axial load at selected depths from the TzQzLiq analysis are compared against results from the centrifuge tests in Fig. 11. The initial axial load distribution in the TzQzLiq analysis matched the centrifuge test results quite well. The axial load in piles decreased during shaking; however, post-shaking, when the excess pore pressures dissipated and the soil settled, it again increased. For the centrifuge model test SKS03, the axial loads in 3DPileS and 3DPileM (latter not shown herein) at different depths and times matched quite well with the measured loads. This analysis is not capturing any changes in lateral stresses due to dynamic loading and this could have contributed in some discrepancies between the numerical and experimental results. Overall, the axial load distribution in the piles matched reasonably well with the centrifuge test data and this validation extended to the other piles as well.

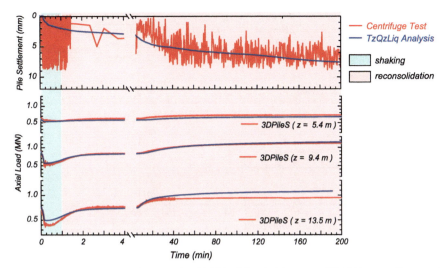

Fig. 11. Validation of TzQzLiq analysis of 3DPileS for shaking event EQM4 in centrifuge model test SKS03: time histories of settlement and axial load at three select depths.

Settlements. Figure 11 (top) shows that settlement time histories of 3DPileS from the numerical analysis matched quite well with recordings. In general, during shaking, while the axial loads in piles at all depths decreased because of the decrease in the initial drag loads from the increased du's, the loss of the shaft and tip capacity and its stiffness resulted in settlement of the piles. During reconsolidation, as effective stresses increased and the soil settled, the re-development of drag load resulted in an additional settlement of the piles. As shown by [44], piles that had a large ratio of static axial capacity to the applied head load (5DPile and 3DPileS) settled < 2 mm during shaking and < 5 mm during reconsolidation. Heavily loaded piles (3DPileM and 3DPileL) suffered large settlements during shaking (> 20 mm). Except for 3DPileL, settlement time histories of 3DPileS, 3DPileM, matched the recorded settlement quite well both during shaking and reconsolidation. For 3DPileL, the numerical model underestimated its settlement of 3DPileL during shaking by about 40% and captured it quite well during reconsolidation. While this warrants further investigations, a possible explanation could be a limitation of the constitutive model of the QzLiq material in accurately modeling the behavior of the tip. The exponent (α_t) which was assumed constant $(\alpha_t = 0.55)$ throughout the analysis may also not be a constant. Overall, the numerical model with the newly developed QzLiq material reasonably modeled the movement of piles in liquefiable soils both during shaking and reconsolidation. Results showed that the pile settlement mainly occurred during shaking when the excess pore pressures were high. During reconsolidation, the tip penetration in soil was small (< 10 mm). The piles with smaller head loads (3DPileS and 5DPile) suffered settlements less than 10 mm. Other piles settled more than about 20 mm.

Neutral Plane Depth. The validated model was used to study the time history of drag load in parallel to the evolution of the neutral plane location with depth. Figure 12

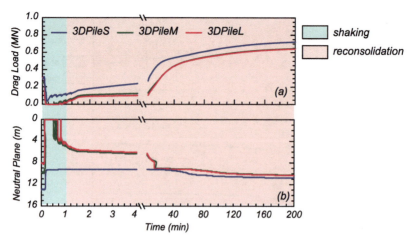

Fig. 12. Results of TzQzLiq analysis in terms of time histories of (a) drag load and (b) neutral plane (depth of maximum load Q_{np}) for 3DPileS, 3DPileM, and 3DPileL in centrifuge model test SKS03. Shaded colored areas indicate shaking and reconsolidation phases.

illustrates results for all three piles in SKS03: an increase in du during shaking generally decreased drag load and correspondingly resulted in a shallow neutral plane. As du dissipated and the soil settled, the drag load again increased, and correspondingly the neutral plane's depth increased. For the heavily loaded piles (3DPileM, 3DPileL) the drag load decreased to zero during shaking, and correspondingly the neutral plane reached the ground surface. However, changes in the neutral plane of 3DPileS were small. For all the piles, with few minutes after shaking, the neutral plane almost recovered its pre-shaking depth; however, the increase in drag load was small. During this period, the soil had settled about 10–20 mm. This shows that a small amount of soil settlement during reconsolidation is enough to recover the neutral plane. As reconsolidation progressed, drag load and depth of neutral plane both increased. After complete reconsolidation, the drag load was greater for the lightly loaded pile (3DPileS). The same was observed for the deeply embedded pile (5D) of SKS02 although not shown herein. The drag loads on 3DPileM and 3DPileL were equal, which was also confirmed by the similar axial load distribution during reconsolidation. Regardless of different head loads, all the 3DPiles of centrifuge model test SKS03 resulted in the same neutral plane depth of ~ 10.5 m after complete reconsolidation. However, the drag load on the lightly loaded pile was 15% (about 100 kN) greater than the other piles.

4   Summary and Conclusions

A series of well-instrumented centrifuge model tests were performed to address questions pertaining to the timing and magnitudes of the mechanisms surrounding the phenomenon of liquefaction-induced downdrag on piles. Results from the centrifuge model tests illuminated the mechanics of liquefaction-induced downdrag on piles through relatively simple soil profiles. This paper presented select results and observations from the centrifuge model recordings and introduced a numerical approach that was validated against the observations. Main findings are described below:

- Settlements in the order of 1 - 3% of the pile diameter in the soil relative to the pile are sufficient to mobilize significant negative skin friction.
- Full liquefaction ($r_u = 1.0$) is not a prerequisite to the development of significant drag loads. Significant drag loads were developed for shaking events that produced excess pore pressures as low as 50% of the initial effective stress. Results showed that after reconsolidation, the developed negative skin friction could equal the interface shear strength ($\tau = K \sigma'_v \tan(\delta)$) for both the non-liquefied ($r_u < 1.0$) as well as liquefied soils ($r_u = 1.0$) as opposed to findings reported by [12].
- During shaking, excess pore pressures generated in the liquefiable layer reduced the negative skin friction, decreasing the drag loads to as low as zero at full liquefaction ($u_e \approx \sigma'_v$). As pore pressures dissipated, the drag loads again increased, approaching or surpassing the drag load that existed before shaking.
- Most of the pile settlement occurred during shaking when the excess pore pressures in the soil around the pile resulted in the loss of shaft and tip capacity and their stiffness. For design, it would be recommended to check the settlement of piles for both the scenarios (1) for the generated inertial loads during shaking with the reduced shaft and tip resistance from the excess pore pressures present near the shaft and the tip even if full liquefaction does not occur (i.e., $0 < r_u < 1$); and (2) from the development of drag load following soil reconsolidation combined with applicable structural loads. Furthermore, since most of the pile settlement occurs during shaking, if feasible, piles should be embedded deep into the bearing layer to maximize their resistance and minimize their settlement during shaking. Therefore, compared to liquefaction mitigation strategies, increasing the embedment of the pile could provide a cost-effective strategy to reduce pile settlements.
- Pile settlements are generally smaller than the free-field soil settlement if the tip capacity does not significantly decrease due to increased pore pressures in its vicinity. When designing piles in liquefiable soils, pile settlement should also be compared to free field settlements. When settlements are low, a comparable settlement between piles and the free-field soil may improve post-earthquake functionality of the superstructure.

A TzQzLiq numerical analysis was performed to model liquefaction-induced downdrag on piles. Model input parameters included TzLiq and QzLiq material properties, time histories of soil settlement and effective stress profiles, and pile properties. A QzLiq material was developed and implemented in OpenSees to model the reduction of pile tip capacity and stiffness in the presence of excess pore pressures. Together, the TzLiq and the QzLiq materials account for changes in the shaft and the tip capacity of the pile as free-field excess pore pressures develop and dissipate in soil. The TzLiq and the QzLiq material properties were obtained and calibrated against the limit load curves and the pile load test results. The TzQzLiq analysis was validated against the results from the series of large centrifuge model tests presented. An inverse analysis was performed on the measured excess pore pressures arrays to obtain time histories of soil settlement profiles. Analyses results showed that the performed TzQzLiq analysis reasonably predicted the time histories of axial load distribution and settlement of piles.

The TzQzLiq numerical model presented in this paper was shown capable of modeling the response of piles during shaking and reconsolidation. The TzQzLiq analysis improves the traditional neutral plane solution method by accounting for changes in the stiffness and capacity of the pile's shaft friction and tip resistance in liquefiable soils, offers complete modeling of the liquefaction-induced downdrag phenomenon, and yields time histories of axial load distribution and settlement of piles. Analysis results on the axial load distribution, pile settlement, and drag load can aid in designing and evaluating the performance of piles in liquefiable soils.

In this paper the effective stress, soil settlement profiles, and the TzLiq and QzLiq material properties were determined from the extensive recordings of the centrifuge model tests. In practice, one would need to determine the pore pressure and free-field settlement distribution via other analysis procedures. For example, a 1D site response analysis with pore pressure models with the design earthquake followed by a reconsolidation analysis could be a reasonable approach for predicting the effective stress and soil settlement distributions with depth. Furthermore, the ultimate capacity of TzLiq and QzLiq materials can be obtained from empirical correlations based on data from soil investigation methods such as cone penetration tests. The stiffness of TzLiq material can be defined, considering that a small relative displacement of 1 - 3% of the pile's diameters is enough to mobilize the full shaft resistance in piles. Determining site-specific QzLiq material stiffness is important for accurately modeling pile settlement and drag load. Herein, pile load test data were available, and they are also advised for practical applications. In the absence of such data, a displacement equal to 10% of pile diameter can be taken as the tip penetration required to mobilize the capacity of QzLiq material [46].

Acknowledgements. This work was funded by the California Department of Transportation under Agreement 65A0688 and National Science Foundation award number CMMI 1635307. The authors would like to acknowledge Caltrans engineers and staff involved in this project for their suggestions and assistance. The support provided by the UC Davis Center for Geotechnical Modeling and staff during the conduct of the centrifuge tests is greatly appreciated.

References

1. Fellenius, B.H.: Basics of Foundation Design. Sidney, Canada: Electronic Edition (2006). available at www.fellenius.net/papers
2. Wang, R., Brandenberg, S.J.: Beam on nonlinear winkler foundation and modified neutral plane solution for calculating downdrag settlement. J. Geotechn. Geoenvironm. Eng. **139**(9), 1433–1442 (2013)
3. Rollins, K.M.: Dragload and downdrag on piles from liquefaction induced ground settlement. In: Bray, J.D., Boulanger, R.W., Cubrinovski, M., Tokimatsu, K., Kramer, S.L., O'Rourke, T., Green, R.A., Robertson, P.K., Beyzaei, C.Z. (eds.) U.S.–New Zealand–Japan International Workshop on Liquefaction-Induced Ground Movement Effects. Pacific Earthquake Engineering Research Center, University of California Berkeley, Berkeley, CA (2017)
4. Sinha, S.K., Ziotopoulou, K., Kutter, B.L.: Centrifuge model tests of liquefaction-induced downdrag on piles in uniform liquefiable deposits. Journal of Geotechnical and Geoenvironmental Engineering (2022). In Press

5. Fellenius, B.H., Siegel, T.C.: Pile drag load and downdrag in a liquefaction event. J. Geotechn. Geoenvironm. Eng. **134**(9), 1412–1416 (2008)
6. Coelho, P.A.L.F., Haigh, S.K., Madabhushi, S.P.G.: Centrifuge modelling of earthquake effects in uniform deposits of saturated sand. In: 5th International Conference on Case Histories in Geotechnical Engineering. University of Missouri Rolla, New York, NY (2004)
7. Knappett, J.A., Madabhushi, S.P.G.: Seismic bearing capacity of piles in liquefiable soils. Soils Found. **49**(4), 525–535 (2009)
8. Stringer, M.E., Madabhushi, S.P.G.: Effect of liquefaction on pile shaft friction capacity. In: 5th Int Conf Recent Advances in Geotechnical Earthquake Engineering and Soil Dynamics. Indian Society of Earthquake Technology, Bangalore, India (2010)
9. Stringer, M.E., Madabhushi, S.P.G.: Re-mobilization of pile shaft friction after an earthquake. Can. Geotech. J. **50**, 979–988 (2013)
10. AASHTO (American Association of State Highway and Transportation Officials): AASHTO LRFD Bridge Design Specifications. LRFDUS-9. AASHTO, Washington, DC (2020)
11. Boulanger, R.W., Brandenberg, S.J.: Neutral plane solution for liquefaction-induced downdrag on vertical piles. GeoTrans 470–478 (2004)
12. Rollins, K.M., Strand, S.R.: Downdrag forces due to liquefaction surrounding a pile. In: 8th U.S. National Conference on Earthquake Engineering. Earthquake Engineering Research Institute, San Francisco, CA (2006)
13. Rollins, K.M., Strand, S.R.: Liquefaction Mitigation Using Vertical Composite Drains: Full-Scale Testing for Pile Applications. Highway IDEA Project 103. Provo, UT: Transportation Research Board (2007)
14. Hannigan, P.J., Rausche, F., Likins, G.E., Robinson, B.R., Becker, M.L.: Design and Construction of Driven Pile Foundations. FHWA-NHI-16-009. Federal Highway Administration, Woodbury, MN (2016)
15. Muhunthan, B., Vijayathasan, N.V., Abbasi, B.: Liquefaction-Induced Downdrag on Drilled Shafts. WA-RD 865.1. Washington State Department of Transportation, Pullman, WA (2017)
16. Nicks, J.: Liquefaction-induced downdrag on Continuous Flight Auger (CFA) piles from full-scale tests using blast liquefaction. FHWA-HRT-17–060. Federal Highway Administration, McLean, VA (2017)
17. Rollins, K.M., Hollenbaugh, J.: Liquefaction Induced Negative Skin Friction from Blast-induced Liquefaction Tests with Auger-cast Piles. In: 6th Int Conf on Earthquake Geotechnical Engineering. International Society for Soil Mechanics and Geotechnical Engineering, Christchurch, New Zealand (2015)
18. Lusvardi, C.M.: Blast-Induced Liquefaction and Downdrag Development on a Micropile Foundation. M.S. Thesis. Department of Civil and Environmental Engineering, Brigham Young University, Provo, UT (2020)
19. Rollins, K.M., Lusvardi, C., Amoroso, S., Franceschini, M.:. Liquefaction induced downdrag on full-scale micropile foundation. In: 2nd Int Conf on Natural Hazards and Infrastructure. Chania, Greece: Innovation Center on Natural Hazards and Infrastructure (2019)
20. Elvis, I.: Liquefaction-induced dragload and / or downdrag on deep foundations within the new madrid seismic zone. Ph.D. Dissertation. Department of Civil Engineering, University of Arkansas, Fayetteville, NC (2018)
21. Vijayaruban, V.N., Muhunthan, B., Fellenius, B.H. Liquefaction-induced downdrag on piles and drilled shafts. In: 6th Int Conf on Earthquake Geotechnical Engineering. International Society for Soil Mechanics and Geotechnical Engineering, Christchurch, New Zealand (2015)
22. Fellenius, B.H., Abbasi, B., Muhunthan, B.: Liquefaction Induced Downdrag for the Juan Pablo II Bridge at the 2010 Maule Earthquake in Chile. Geotechnical Engineering Journal of the SEAGS & AGSSEA **51**(2), 1–8 (2020)

23. Sinha, S.K., Ziotopoulou, K., Kutter, B.L.: Centrifuge testing of liquefaction-induced down-drag on axially loaded piles : Data Report for SKS02. UCD/CGMDR - 21/01. Center for Geotechnical Modeling, University of California Davis, Davis, CA (2021)

24. Sinha, S.K., Ziotopoulou, K., Kutter, B.L.: Centrifuge Testing of Liquefaction-Induced Down-drag on Axially Loaded Piles : Data Report for SKS03. In: UCD/CGMDR - 21/02. Center for Geotechnical Modeling, University of California Davis, Davis, CA (2021)

25. Sinha, S.K., Ziotopoulou, K., Kutter, B.L.: Parametric study of liquefaction induced downdrag on axially loaded piles. In: 7th Int Confon Earthquake Geotechnical Engineering. International Society for Soil Mechanics and Geotechnical Engineering, Rome, Italy (2019)

26. Garnier, J., et al.: Catalogue of scaling laws and similitude questions in geotechnical centrifuge modelling. Int. J. Physi. Model. Geotechnics 7(3), 01–23 (2007)

27. Carey, T.J., Stone, N., Kutter, B.L.: Grain Size Analysis and Maximum and Minimum Dry Density of Ottawa F-65 Sand for LEAP-UCD-2017. In: Kutter, B.L., Manzari, M.T., Zeghal, M. (eds.) Model Tests and Numerical Simulations of Liquefaction and Lateral Spreading, pp. 31–44. Springer, Cham (2020)

28. Sinha, S.K., Kutter, B.L., Wilson, D.W., Carey, T., Ziotopoulou, K.: Use of Photron Cameras and TEMA Software to Measure 3D Displacements in Centrifuge Tests. In: UCD/CGM-21/01. Center for Geotechnical Modeling, University of California Davis (2021)

29. Sinha, S.K., Kutter, B.L., Ziotopoulou, K.: Measuring vertical displacement using laser lines and cameras. Int. J. Phys. Model. n Geotechn. (2021). https://doi.org/10.1680/jphmg.21.00038

30. Martinez, A., Frost, J.D.: The influence of surface roughness form on the strength of sand-structure interfaces. Géotechnique Letters 7(1), 104–111 (2017)

31. Rituraj, S.S., Rajesh, B.G.: Negative Skin Friction on Piles: State of the Art. In: Choudhary, A.K., Mondal, S., Metya, S., Babu, G.L.S. (eds.) Advances in Geo-Science and Geo-Structures, Lecture Notes in Civil Engineering, pp. 323–335. Springer, Singapore (2022)

32. Fleming, K., Weltman, A., Randolph, M., Elson, K.: Piling Engineering. CRC Press, London (2008)

33. O'Neill, M.W.: Side resistance in piles and drilled shafts. J. Geotechni. Geoenvironm. Eng. 127(1), 3–16 (2001)

34. DeJong, J.T., Westgate, Z.J.: Role of initial state, material properties, and confinement condition on local and global soil-structure interface behavior. J. Geotechni. Geoenvironm. Eng. 135(11), 1646–1660 (2009)

35. Fellenius, B.H.: Down-drag on piles in clay due to negative skin friction. Can. Geotech. J. 9(4), 321–337 (1972)

36. Fellenius, B.H.: Unified design of piled foundations with emphasis on settlement analysis. In: Current Practices and Future Trends in Deep Foundations, pp. 253–275. Contributions in Honor of George G. Gobel, ASCE, Los Angeles, CA (2004)

37. Caltrans: Liquefaction-induced downdrag. In: Caltrans Geotechnical Manual (2020). https://dot.ca.gov/programs/engineering-services/manuals/geotechnical-manual

38. McKenna, F., Scott, M.H., Fenves, G.L.: Nonlinear finite-element analysis software architecture using object composition. J. Comput. Civ. Eng. 24(1), 95–107 (2010)

39. Boulanger, R.W., Curras, C.J., Kutter, B.L., Wilson, D.W., Abghari, A.: Seismic soil-pile-structure interaction experiments and analyses. J. Geotechni. Geoenviron. Eng. 125(9), 750–759 (1999)

40. Mosher, R.L.: Load-Transfer Criteria for Numerical Analysis of Axially Loaded Piles in Sand. K-84. US Army Engineer Waterways Experiment Station, Vicksburg, MS (1984)

41. Reese, L.C., O'Neal, M.W.: Drilled Shafts: Construction Procedures and Design Methods. FHWA-IF-99–025. Federal Highway Administration, Mclean, Virginia (1987)

42. Gajan, S., Raychowdhury, P., Hutchinson, T.C., Kutter, B.L., Stewart, J.P.: Application and validation of practical tools for nonlinear soil-foundation interaction analysis. Earthq. Spectra 26(1), 111–129 (2010)

43. Vijivergiya, V.N.: Load-movement characteristics of piles. In: 4th Symposium of Waterway, Port, Coastal and Ocean Division, 2, pp. 269–284. American Society of Civil Engineering, Long Beach, CA (1977)
44. Sinha, S.K., Ziotopoulou, K., Kutter, B.L.: Numerical modeling of liquefaction-induced downdrag : validation from centrifuge tests. Journal of Geotechnical and Geoenvironmental Engineering. Under Review (2022)
45. Malvick, E.J., Kulasingam, R., Boulanger, R.W., Kutter, B.L.: Effects of Void Redistribution on Liquefaction Flow of Layered Soil – Centrifuge Data Report for EJM01. In: UCD/CGMDR-02/02. Center for Geotechnical Modeling, University of California Davis, Davis, CA (2002)
46. API (American Petroleum Institute): Recommended Practice for Planning, Designing and Constructing Fixed Offshore Platforms — Working Stress Design. API Recommended Practice 24-WSD. Washington, DC (2000)

Performance Design and Seismic Hazard Assessment

A Simplified Method to Evaluate the 2D Amplification of the Seismic Motion in Presence of Trapezoidal Alluvial Valley

Giorgio Andrea Alleanza[✉] [iD], Anna d'Onofrio[iD], and Francesco Silvestri[iD]

Department of Civil, Architectural and Environmental Engineering,
University of Napoli Federico II, Naples, Italy
giorgioandrea.alleanza@unina.it

Abstract. The seismic response of an alluvial valley is ruled by a complex combination of geometric and stratigraphic factors, which makes it hard to define a simplified amplification factor, to be adopted by the codes as it usually happens in the case of topographic and 1D stratigraphic amplification factors. This paper reports the results of an extensive parametric study aimed at analysing the influence of geometrical (e.g. shape ratio and inclination of the edges) and stratigraphic factors (e.g. impedance ratio) on the ground motion at surface of a trapezoidal valley. The soil behaviour has been assumed as linear visco-elastic. The results of the numerical analyses have been synthesised in terms of Valley Amplification Factor (VAF), defined as the ratio between the spectral ordinates obtained from 2D and 1D analysis, the latter carried out along a soil column corresponding to the centre of the valley. Based on the results of the parametric analysis, a simplified method was finally proposed, to calculate straightforward the VAF distribution along the valley.

Keywords: Seismic site response · Basin effects · Alluvial valley

1 Introduction

The buried morphology of alluvial valleys strongly affects the seismic response at surface. Physically, the complex interference phenomena among the vertical propagation of body waves upcoming from the bedrock, the focusing of inclined refracted waves towards the valley centre and the surface waves generated from the valley edges lead to a significant change in amplitude and duration of the ground motion. The key parameters which mostly influence the seismic response of these morphologies are the shape ratio of the buried valley [1, 2], the impedance ratio [3], the slope of the edges [4], the frequency content of the reference input motion [5] and the non-linear properties of the alluvial soil filling the valley [6].

In recent years, numerous studies have been carried out to quantify the amplification of the seismic motion due to the presence of an alluvial valley and to define an appropriate Valley Amplification Factor, *VAF* [7]. However, an easily calculated *VAF* capable to

© The Author(s), under exclusive license to Springer Nature Switzerland AG 2022
L. Wang et al. (Eds.): PBD-IV 2022, GGEE 52, pp. 685–692, 2022.
https://doi.org/10.1007/978-3-031-11898-2_41

adequately synthesise the effects of geometry, mechanical and non-linear soil properties on the seismic motion has not yet been found.

In order to quantify the 2D valley amplification as decoupled from the stratigraphic one, *VAF* can be defined as the ratio between peak or integral ground motion parameters computed by 2D and 1D analyses, as conventionally adopted in the evaluation of topographic amplification. Usually, in this approach the *VAF* is computed by 2D visco-elastic analyses assuming that it is not significantly affected by non-linear soil properties, that are deemed to be included in 1D stratigraphic amplification. Nevertheless, it has not yet been stated if it is possible to decouple geometric from stratigraphic effects and to quantify them separately [8].

This paper summarizes the results of an extended parametric study aimed at investigating the effects of the abovementioned parameters on the seismic response of shallow trapezoidal valleys (Fig. 1). Some charts are also proposed to be used to estimate the geometrical amplification quickly and easily along a valley profile and therefore can be considered as a starting point for the introduction of the basin effects in the building code.

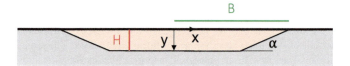

Fig. 1. Geometric scheme of the valley.

2   Geotechnical Model

An extensive set of 2D numerical seismic response analyses of symmetrical trapezoidal valleys with fixed depth, H, and variable width, $2B$ (Fig. 1), was carried out to evaluate the influence of the geometrical and mechanical parameters on the ground motion at surface. Tables 1 and 2 summarize the geometrical and mechanical properties assigned in the parametric study. Being the shape ratio H/B lower than 0.25, all the analysed valleys can be classified as 'shallow' following the criterion suggested by Bard and Bouchon [3]. For each of the analysed models, a 1D site response analysis was also carried out along the vertical at the centre of the valley. The output of this latter analysis was then compared with that obtained from 2D models with the aim of isolating the geometrical effects from the stratigraphic amplification. A Ricker wavelet accelerogram with peak amplitude of 0.45 g and mean frequency, f_m, varying between 0.05 and 11.60 Hz was used as input motion. It follows that the ratio between f_m and the fundamental frequency of the 1D column, $f_{0,1D}$, varies between 0.2 and 8, while that between the mean incident wavelength, λ_m, and the valley depth, H, varies between 20 and 0.5. Such a frequency range encompasses the occurrence of resonance phenomena at the fundamental and higher vibration modes of the reference vertical soil column.

The 2D numerical simulations were carried out with FLAC 8.0 [9] while the code STRATA [10] was adopted for 1D analyses, after verified the consistency between the predictions of both codes comparing the results of linear 1D analyses. For each different domain of analysis and value of V_S of the alluvial soil, the maximum thickness of the elements of the mesh grid was assigned through the Kuhlemeyer and Lysmer [11] relationship, considering a maximum frequency of 20 Hz.

Table 1. Geometrical properties

Thickness H (m)	Width 2B (m)	Shape ratio H/B	Slope of the edge α (°)
100	4000	0.05	90/60/45/30
	2000	0.10	
	1340	0.15	
	1000	0.20	
	800	0.25	

Table 2. Mechanical properties

Shear wave velocity (m/s)		Unit weight (kN/m^3)		Poisson ratio		Initial Damping (%)		Impedance ratio I
Bedrock $V_{S,r}$	Soil V_S	Bedrock γ_r	Soil γ	Bedrock ν_r	Soil ν	Bedrock $D_{0,r}$	Soil D_0	
800	100	22	19	0.33	0.33	0.5	5.0	9.26
	130						3.8	7.13
	180						2.8	5.15
	270						1.9	3.43
	360						1.4	2.57
	580						1.0	1.60

3   Results

1440 analyses were carried out considering 120 valley models subjected to 12 different input motions. The results of both 2D and 1D analyses were synthesized adopting the following amplification factor, A:

$$A\left(T, \frac{x}{B}\right) = \frac{S_{a,s}\left(T, \frac{x}{B}\right)}{S_{a,r}(T)} \qquad (1)$$

being $S_{a,s}\left(T, \frac{x}{B}\right)$ and $S_{a,r}(T)$ the spectral acceleration of the ground and input motion, respectively. To isolate the influence of 2D effects on the seismic response, a geometrical aggravation factor, AF, was then defined as the ratio between the 2D amplification at a generic location x/B and the 1D amplification at the valley centre:

$$AF\left(T, \frac{x}{B}\right) = \frac{A_{2D}\left(T, \frac{x}{B}\right)}{A_{1D}(T, 0)} \tag{2}$$

At each point of the mesh along the surface and for each period, the average value of AF obtained from the analyses carried out adopting 12 input motions was calculated, thus obtaining an aggravation factor independent of the input frequency, defined as:

$$\overline{AF\left(T, \frac{x}{B}\right)} = \frac{1}{m} \sum_{i=1}^{m} AF_i\left(T, \frac{x}{B}\right) \tag{3}$$

where m is the number of input signals considered, in this case 12. The factors thus obtained were further averaged within a period range between 0s and $T_{0,1D}$ (i.e. the resonance period of 1D column at the centre of the valley) to obtain a synthetic VAF, independent of period, and defined as follows:

$$VAF\left(\frac{x}{B}\right) = \frac{1}{T_{0,1D}} \int_0^{T_{0,1D}} \overline{AF\left(T, \frac{x}{B}\right)} \cdot dT \tag{4}$$

The range of periods was selected based on the observation that no significant 2D effect was observed at periods exceeding $T_{0,1D}$.

The so computed VAF along the surface depend on the shape ratio H/B and on the slope edge α of the valley, as well as on the impedance ratio I between the bedrock and the alluvial soil. In the following, VAF values greater than or equal to 1 were considered, thus neglecting any de-amplification effects that could be generated near to the edges. Figure 2a, b, c respectively show the influence of H/B, α and I on VAF. The analyses

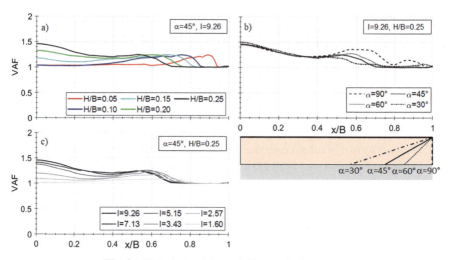

Fig. 2. VAF obtained for variable a) H/B; b) α; c) I.

results evidenced that the centre of the valleys and the zones close to the edges are characterised by the highest *VAF* values.

At the centre of the valley, the peak of the *VAF* increases with H/B and I (Figs. 2a–c) due to the constructive interference between in-phase body and surface waves, that does not depend on the slope of the edges (see Fig. 2b) [7, 8, 12]. The maximum amplification at the edge is independent of H/B (Fig. 2a) and increases with α and I (Figs. 2b-c). Its position is strongly influenced by all the factors, moving toward the valley centre as H/B and I increase, and α decreases. More details on this can be found in Alleanza [12].

4   Proposed Charts

The computed *VAF* profiles show that the presence of a buried valley induces amplification of the ground motion mainly at the centre of the valley and next to the edge. For this reason, the *VAF* profiles along the valley have been further simplified defining a piecewise linear trend, identified by five relevant points (Fig. 3). The first is the value *VAF(0)* computed at the centre of the valley ($x/B = 0$, point 0 in Fig. 3), the second (point 1 in Fig. 3) identifies the lowest value of *VAF* computed between the peaks at the centre and near the edge of the valley, the third and the fourth (points 2, 3 in Fig. 3) are used to describe the maximum value of *VAF* at the side of the valley and the last one (point 4, $x/B = 1$) is set equal to 1, therefore neglecting any attenuation at the border of the valley.

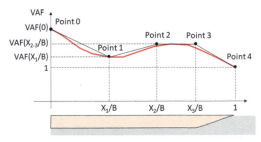

Fig. 3. Typical *VAF* profiles obtained from the numerical analysis (red line) and its simplified piecewise linear trend (black line) with the identification of 5 relevant points.

All the computed *VAF* profiles were then processed to extract the five above-described points; the relevant values were represented as contour lines in six charts as a function of the impedance ratio and shape ratio. For given H/B and I the definition of each point of the piecewise linear trend is obtained enveloping the maxima *VAF* values, in this way implicitly including the dependency on the slope angle. Figure 4 shows the charts, which can be used to describe the piecewise trend of *VAF* expected for a valley characterised by a given H/B and I.

The amplification at the valley centre, *VAF(0)*, increases with I and H/B, as shown by the chart in Fig. 4a. The magenta contour line in Fig. 4a, corresponding to *VAF(0)* = 1.05, represents the threshold below which almost no 2D amplification occurs at the centre of the valley. Particularly, $H/B = 0.1$ seems to be a threshold value of shape ratio, below which 2D effects are negligible at the valley centre. The shape ratio $H/B = 0.1$

seems to be a threshold value also for the determination of the abscissas x_1/B (Fig. 4b), x_2/B (Fig. 4d) and by x_3/B (Fig. 4f) as well as for the minimum value of VAF along the valley, $VAF(x_1/B)$ (Fig. 4c): for $H/B < 0.1$ all those values are independent of I. Instead, Fig. 4e shows that the peak value at the edges, $VAF(x_{2-3}/B)$, is independent of H/B and increases with I. In other words, in the case of very shallow valley, i.e. $H/B < 0.1$, there is almost no 2D amplification along the valley surface unless near to the edges, where the amplification is a function of the impedance ratio. Note that even for $H/B > 0.1$ there are valleys for which $VAF(0)$ is less than 1.05, but this is due to the low impedance ratio. In this case, the geometric and material damping is such that the Rayleigh waves transmitted to the valley do not have sufficient energy to significantly

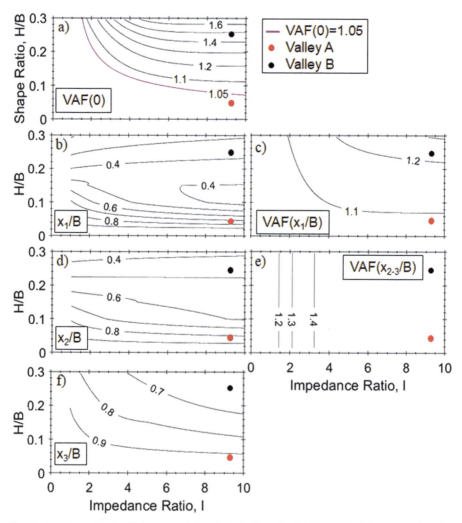

Fig. 4. Chart of: a) $VAF(0)$; b) x_1/B; c) $VAF(x_1/B)$; d) x_2/B; e) $VAF(x_{2-3}/B)$; f) x_3/B as function of H/B and I

influence the motion at the centre of the valley. Instead, for values of H/B and I such that $VAF(0) > 1.05$, the 2D effects are not negligible along the whole valley profile. Therefore, shallow valleys can be further divided into two classes, 'very shallow' if $H/B < 0.075$–0.1 and 'moderately shallow' otherwise.

The above described charts can be a useful tool for a simplified and reasonably over-conservative estimation of 2D amplification along a trapezoidal valley or with a similar shape. As an example, Fig. 5a,b shows the comparisons between the VAF obtained using the charts and those computed by numerical analyses, for two valleys with same impedance ratio ($I = 9.26$) and H/B equal to 0.05 ('very shallow' valley A) and 0.25 ('moderately shallow' valley B), respectively. The black and red points reported in the charts of Fig. 4 represent the values adopted to draw the piecewise distributions of VAF, reported in Fig. 5, for Valley A and B, respectively. From the comparison it can be concluded that the proposed method provides an acceptable estimate of the amplification in the centre of the valley, while it is over-conservative at the edge of the valley.

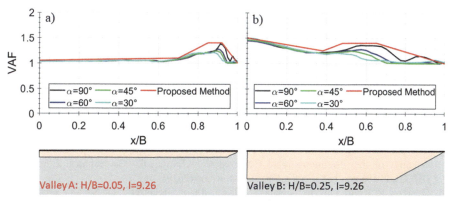

Fig. 5. Comparison between simplified and numerical VAF estimation, for all the slope of the edge, $I = 9.26$, and H/B: a) 0.05 (Valley A); b) 0.25 (Valley B)

5   Conclusion

This paper synthesises the results of extensive parametric 2D numerical analyses of the seismic response of homogeneous alluvial basins. A valley amplification factor was defined to quantify the 2D amplification along the valley: as expected, it was found to depend on the shape ratio and the edge slope of the valley, as well as on the bedrock/soil impedance ratio and on the frequency of the input motion normalised with respect to the one-dimensional resonance value. Based on these results, the paper proposes a simplified method for the evaluation of 2D amplification, adopting a set of charts allowing to compute an approximate VAF profile as a function of shape ratio and impedance ratio. The charts highlight that 'shallow valleys' can be further distinguished into two classes based on the value of the shape ratio. If $H/B < 0.1$ they can be considered as 'very shallow' basins characterised by slight 2D amplification at the centre of the valley, i.e. $VAF(0)$

(Fig. 4a) lower than 1.05–1.10 until x_1/B (Fig. 4b), gradually increasing approaching the edges. Therefore, the evaluation of site response by 1D numerical analyses can lead to a 5–10% underestimation of the spectral response along the zone between the centre of the valley and x_1/B, while it can be completely misleading in the computation of the response at the edge. If $H/B > 0.1$, instead, the 2D effects are not negligible everywhere in the valley and thus they must be taken into account and computed with simplified or numerical methods.

Acknowledgement. This work was carried out as part of WP16.1 "Seismic response analysis and liquefaction" in the framework of the research programme funded by Italian Civil Protection through the ReLUIS Consortium (DPC-ReLuis 2019–2021).

References

1. Bard, P.Y., Bouchon, M.: The seismic response of sediment-filled valleys. Part 1. The case of incident SH waves. Bull. Seismol. Soc. Am. **70**(4), 1263–1286 (1980)
2. Bard, P.Y., Bouchon, M.: The seismic response of sediment-filled valleys. Part 1. The case of incident P and SV waves. Bull. Seismol. Soc. Am. **70**(5), 1921–1941 (1980)
3. Bard, P.Y., Bouchon, M.: The two-dimensional resonance of sediment-filled valleys. Bull. Seismol. Soc. Am. **75**(2), 519–541 (1985)
4. Zhu, C., Thambiratnam, D.: Interaction of geometry and mechanical property of trapezoidal sedimentary basins with incident SH waves. Bull. Earthq. Eng. **110**, 284–299 (2016)
5. Alleanza, G.A., Chiaradonna, A., d'Onofrio, A., Silvestri, F.: Parametric study on 2D effect on the seismic response of alluvial valleys. In: Silvestri, F., Moraci, N. (eds.) Earthquake Geotechnical Engineering for Protection and Development of Environment and Constructions: Proceedings of the 7th International Conference on Earthquake Geotechnical Engineering (ICEGE 2019), pp. 1082–1089. CRC Press, Rome (2019)
6. Gelagoti, F., Kourkoulis, R., Anastasopoulos, I., Tazoh, T., Gazetas, G.: Seismic wave propagation in a very soft alluvial valley: sensitivity to ground-motion details and soil nonlinearity, and generation of a parasitic vertical component. Bull. Seismol. Soc. Am. **100**(6), 3035–3054 (2010)
7. Papadimitriou, A.G.: An engineering perspective on topography and valley effects on seismic ground motion. In: Silvestri, F., Moraci, N. (eds.) Earthquake Geotechnical Engineering for Protection and Development of Environment and Constructions: Proceedings of the 7th International Conference on Earthquake Geotechnical Engineering (ICEGE 2019), pp. 426–441. CRC Press, Rome (2019)
8. Alleanza, G.A., Allegretto, L.R., d'Onofrio, A., Silvestri, F.: Two-dimensional site effects on the seismic response of alluvial valleys: the case study of the Aterno river valley (Italy). In: Proceedings of the 20th International Conference on Soil Mechanics and Geotechnical Engineering, Sydney (2022). (Accepted paper)
9. Itasca Consulting Group, Inc.: FLAC—Fast Lagrangian Analysis of Continua, Ver. 8.0. Itasca, Minneapolis (2016)
10. Kottke, A.R., Rathje, E.M.: Technical manual for Strata. Report No.: 2008/10. Pacific Earthquake Engineering Research Center, University of California, Berkeley (2008)
11. Kuhlemeyer, R.L., Lysmer, J.: Finite element method accuracy for wave propagation problems. J. Soil Mech. Found. Div. **99**(5), 421–427 (1973)
12. Alleanza, G.A.: Study of the seismic response of alluvial valleys. Doctoral dissertation in preparation. University of Napoli Federico II, Napoli, Italy (2022)

Critical Acceleration of Shallow Strip Foundations

Orazio Casablanca [ID], Giovanni Biondi [ID], and Ernesto Cascone[(⊠)] [ID]

Department of Engineering, University of Messina, Contrada di Dio, 98166 Messina, Italy
ecascone@unime.it

Abstract. This paper focuses on the assessment of the critical acceleration of soil-foundation systems and examines the influences of various relevant parameters. Using a formula recently proposed for the evaluation of the bearing capacity of shallow strip foundations in seismic conditions, the critical acceleration of the soil-foundation system has been evaluated imposing the equilibrium condition between the limit load of the soil-foundation system and load transmitted from the foundation onto the ground surface. The proposed solution led to an iterative equation that permits the numerical evaluation of the critical acceleration.

An extensive parametric analysis allowed to examine the effects of many relevant parameters, such as the angle of shear strength of the foundation soil, the value of the surcharge acting aside the footing, the static vertical load transmitted by the foundation, the direction of the resultant of the inertial forces acting in the soil and in the superstructure.

The results obtained are provided in design charts for given sets of the relevant parameters and allow a quick evaluation of the critical acceleration of the soil-foundation system.

Keywords: Foundations · Critical acceleration · Seismic performance

1 Introduction

The seismic bearing capacity q_{ultE} of shallow strip foundations can be evaluated as:

$$q_{\text{ultE}} = c' \cdot N_{\text{cE}} + q \cdot N_{\text{qE}} + \frac{1}{2} \cdot \gamma \cdot B \cdot N_{\gamma E} \tag{1}$$

Equation (1) represents an extension of the superposition formula provided by Terzaghi (1943) for seismic condition. N_{cE}, N_{qE} and $N_{\gamma E}$ are the seismic bearing capacity factors that depend on the angle of soil shear strength φ', on the inertia forces arising in the soil involved in the plastic mechanism and on the inertia forces transmitted by the superstructure onto the foundation; γ and c' are the unit weight and the cohesion of the soil, respectively; B is the width of the foundation and q is the vertical pressure acting on the ground surface aside the foundation. Imposing the equilibrium condition between the limit load of the soil-foundation system provided by Eq. (1) and the load transmitted by the foundation onto the ground surface, it is possible to determine the critical acceleration of the system $k_{\text{h,c}}$, namely the minimum value of the horizontal seismic acceleration

L. Wang et al. (Eds.): PBD-IV 2022, GGEE 52, pp. 693–701, 2022.
https://doi.org/10.1007/978-3-031-11898-2_42

k_h in the soil that leads the soil-foundation system to failure. It is expected that $k_{h,c}$ is greatly influenced by the decrease in the seismic bearing capacity factors induced by the seismic acceleration, therefore a reliable estimate of such factors is necessary. Several solutions are available in the literature for evaluating bearing capacity factors considering both static and seismic conditions; these solutions have been obtained using different approaches such as the limit equilibrium method (e.g. [1, 15]), the lower and the upper bound theorem of limit analysis (e.g. [7, 10–12]), the method of characteristics (e.g. [2–6, 8]) and by numerical stress–strain analyses (e.g. [9, 11]). From a theoretical point of view, the procedures generally adopted to detect the pseudo-static safety factor of a shallow foundation and its critical acceleration do not differ significantly and thus, also for the latter parameter the above-mentioned methods of analysis can be reliably applied. This paper deals with the evaluation of the critical acceleration of shallow strip foundations using the seismic bearing capacity factors provided by Cascone and Casablanca [4] obtained using the method of characteristics (MC). An extensive parametric analysis has been performed in order to examine the effects of many relevant parameters, such as the angle of shear strength of the soil, the value of the surcharge acting aside the footing, the vertical load transmitted by the foundation under static conditions, the direction of the resultant of the inertial forces acting in the soil and in the superstructure. In the analysis both Prandtl and Hill mechanisms have been considered. The results are provided in design charts for given sets of the relevant parameters and allow a quick evaluation of the critical acceleration of the soil-foundation system.

2   Method of Analysis

The critical acceleration of the soil-foundation system $k_{h,c}$ was evaluated starting from the following equation representing the equilibrium condition between the limit load q_{ultE} and the load transmitted by the foundation to the soil:

$$c' \cdot e_c^s \cdot e_c^{ss} \cdot N_c + q \cdot e_q^s \cdot e_q^{ss} \cdot N_q + \frac{1}{2} \cdot B \cdot \gamma \cdot e_\gamma^s \cdot e_\gamma^{ss} \cdot N_\gamma = q_0(1 - k_{vi}) \qquad (2)$$

The left side of (2) is formally equal to (1) with the seismic bearing capacity factors N_{cE}, N_{qE} and $N_{\gamma E}$ expressed as $N_{jE} = e_j^s \cdot e_j^{ss} \cdot N_j$ (j = c, q, γ), in which the seismic corrective coefficients e_c^s, e_q^s and e_γ^s account only for the effect of soil inertia (denoted by superscript s) while, e_c^{ss}, e_q^{ss} and e_γ^{ss} account only for the effect of the inertia of the superstructure (denoted by superscript ss). Formulas of these corrective coefficients are provided in Eqs. 3–6, and the relevant coefficients are listed in Table 1:

$$e_{jE}^s = \frac{N_{jE}^s}{N_j} = \left(1 - A\frac{k_h}{1 - k_v}\cot\varphi\right)^B \sqrt{k_h^2 + (1 - k_v)^2} \qquad (3)$$

$$e_{jE}^{ss} = \frac{N_{jE}^{ss}}{N_j} = \left(1 - C\frac{k_{hi}}{1 - k_{vi}}\cot\varphi\right)^D \qquad (4)$$

$$B = b_1 \tan^2\varphi + b_2 \tan\varphi + b_3 \qquad (5)$$

$$D = d_1 \tan^2\varphi + d_2 \tan\varphi + d_3 \qquad (6)$$

In Eq. (2) q_0 is the load transmitted by the foundation to the soil under static loading conditions. Differently from most of the available solutions, the soil inertia effect, depending on the horizontal and vertical seismic acceleration coefficients k_h and k_v, and the superstructure inertia effect, depending on the seismic acceleration coefficients k_{hi} and k_{vi}, are dealt with independently. Static bearing capacity factors N_c and N_q are provided in closed algebraic form by Prandtl [13] and Reissner [14] respectively while for N_γ, the solution proposed by Cascone and Casablanca [2] has been considered (Eq. 7), in which $n = 0$ for the Hill mechanism and $n = 1$ for the Prandtl mechanism:

$$N_\gamma = (N_q - 1)\tan(1.34\varphi)\left(n + \frac{1 - n^3}{2}\right) \tag{7}$$

Table 1. Coefficients of Eqs. 5 and 6

	A	b_1	b_2	b_3		C	d_1	d_2	d_3
e_c^s	0	0	0	0	e_c^{ss}	0.4	3.89	2.33	0.02
e_q^s	0.92	0	0.51	0.12	e_q^{ss}	0.65	1.78	1.73	0.01
e_γ^s (Hill)	0.92	0.29	-0.28	0.72	$e_{\gamma E}^{ss}$ (Hill)	0.65	3.06	2.68	0.56
e_γ^s (Prandtl)	0.92	0.20	-0.01	0.53	$e_{\gamma E}^{ss}$ (Prandtl)	0.9	2.00	1.45	0.19

Introducing the nondimensional parameters $\bar{c}' = 2\,c'/\gamma B$, $\bar{q} = 2\,q/\gamma B$, $\bar{q}_0 = 2\,q_0/\gamma B$ (defined as normalized cohesion, surcharge and load transmitted to the soil) it is possible to put the equilibrium condition expressed by Eq. (2) in the following nondimensional form that is convenient for performing parametric analyses:

$$\bar{c}' \cdot e_c^s \cdot e_c^{ss} \cdot N_c + \bar{q} \cdot e_q^s \cdot e_q^{ss} \cdot N_q + e_\gamma^s \cdot e_\gamma^{ss} \cdot N_\gamma = \bar{q}_0(1 - k_{vi}) \tag{8}$$

3   Critical Acceleration

Equation (8) can be used to determine the critical acceleration coefficient $k_{h,c}$ for the soil-foundation system namely the lowest value of the horizontal seismic acceleration coefficient k_h in the soil that brings the system to failure. It is worth noting that the seismic coefficients k_h and k_v do not appear explicitly in Eq. (8) but they show up in the expressions of the seismic corrective coefficients e_c^s, e_q^s and e_γ^s; due to the non-linear relationship between k_h and k_v and the above mentioned coefficients, it is not possible to derive a closed form solution for determining $k_{h,c}$. Therefore, an iterative solution has been obtained by varying the value of k_h until Eq. (8) was satisfied. Analyses were carried out considering both Prandtl and Hill mechanisms, varying the value of $\bar{c}'$, $\bar{q}$, $\bar{q}_0$ and considering values of φ' in the range $15°$–$45°$; the values of the vertical seismic coefficient k_v acting in the soil and the horizontal coefficient k_{hi} relative to horizontal inertia forces transmitted to the foundation by the superstructure were assumed as a fraction of k_h through the coefficients $\Omega = k_v/k_h$ and $\eta = k_{hi}/k_h$, respectively, while the effect of the vertical seismic coefficient k_{vi} was neglected. Figures 1 and 2 show the values of $k_{h,c}$ obtained for the case $\Omega = \eta = 0$, for $\bar{c}'$ and $\bar{q}$ in the range 0–1 and for $\bar{q}_0$ in the range 10–50, assuming both Prandtl and Hill mechanism. The dashed line of equation $k_{h,c} = k_{h,lim} = \tan \varphi'$ plotted in the figure represents the fluidization condition defined by Richards et al. [16].

With reference to Fig. 1a for the case of Prandtl mechanism and $\bar{q}_0 = 10$ it is possible to note that for $\varphi' \leq 27.5°$ it is $k_{h,c} = 0$, denoting that the bearing capacity of the soil-foundation system is lower than the applied load. Increasing φ', $k_{h,c}$ increases until it approaches the dashed line representing the fluidization condition. Actually the solid line never overlaps the dashed line but it is very close to it; this result is consequence of the functional form of e_q^s and e_γ^s which drop to zero for $k_h = k_{h,lim}$. Therefore k_h values close to the condition $k_{h,c} = \tan \varphi'$ allow to obtain a finite value of the limit load and so values of $k_{h,c}$ close to $k_{h,lim}$. Curves depicted in Fig. 1 show also the effect of $\bar{c}'$ and $\bar{q}$ on the values of $k_{h,c}$; for a given value of φ' the larger are $\bar{c}'$ and $\bar{q}$ the larger is $k_{h,c}$. The same considerations also apply to the curves of Fig. 2; the only difference concerns the value of $k_{h,c}$ in fact, for the same values of φ', $\bar{c}'$, $\bar{q}$ and $\bar{q}_0$, the values of $k_{h,c}$ obtained considering the Hill mechanism (Fig. 2) are lower than those relative to the Prandtl one (Fig. 1). Figures 3, 4, 5 and 6 show the influence of η and Ω on $k_{h,c}$. In particular, Figs. 3 and 4 refer to the case $\bar{c}' = \bar{q} = 0$ assuming the Prandtl and the Hill mechanism respectively. It is observed that the effect of the acceleration coefficient k_{hi} relative to the superstructure plays a crucial role on the value of $k_{h,c}$. It is apparent, in fact, that particular, for a given value of $\bar{q}_0$ (Fig. 3a and b) as φ' increases $k_{h,c}$ keeps lower than $k_{h,lim}$ due to the simultaneous effect of the inertia of the superstructure that induces a reduction of the limit load. Figure 3b shows a less inclined dashed line of equation $k_{h,c} = \tan \varphi'/(1 + \Omega)$ which represents the fluidization condition in presence of vertical component of the seismic acceleration. Figures 5 and 6 show for the same cases considered in Figs. 3 and 4 the effect of $\bar{c}'$ and $\bar{q}$ on the values of $k_{h,c}$.

Plots in Figs. 1, 2, 3, 4, 5 and 6 represent simple design charts for determining $k_{h,c}$ for shallow strip foundations; they can be used starting from a known value of the static load q_0 and from the knowledge of the strength parameters φ' and c', the surchege q and the expected value of the seismic acceleration. Finally, it can be demonstrated that, depending on the values of the normalized relevant parameters, the dashed lines depicted in the plots of Figs. 1, 2, 3, 4, 5 and 6, represent the values of the horizontal component of the critical acceleration corresponding to two possible failure mechanisms of the soil-foundation system, different from bearing capacity failure and namely: horizontal sliding of the foundation, and soil fluidization according to Richards et al. [16].

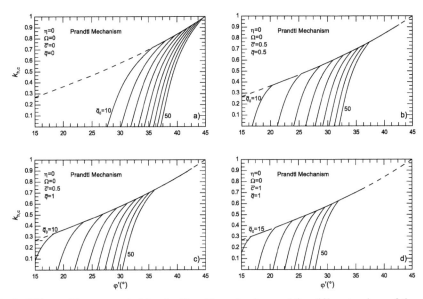

Fig. 1. Values of $k_{h,c}$ computed for the Prandtl mechanism and for different values of the normalised relevant parameters: a) $\eta = \Omega = \bar{c}' = \bar{q} = 0$; b) $\eta = \Omega = 0$; $\bar{c}' = \bar{q} = 0.5$; c) $\eta = \Omega = 0$; $\bar{c}' = 0.5$; $\bar{q} = 1$; d) $\eta = \Omega = 0$; $\bar{c}' = \bar{q} = 1$.

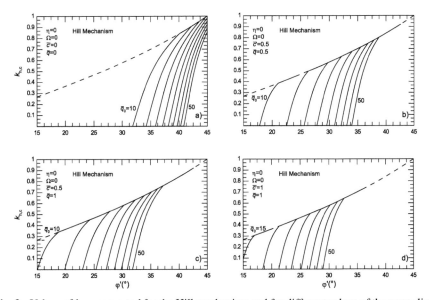

Fig. 2. Values of $k_{h,c}$ computed for the Hill mechanism and for different values of the normalised relevant parameters: a) $\eta = \Omega = \bar{c}' = \bar{q} = 0$; b) $\eta = \Omega = 0$; $\bar{c}' = \bar{q} = 0.5$; c) $\eta = \Omega = 0$; $\bar{c}' = 0.5$; $\bar{q} = 1$; d) $\eta = \Omega = 0$; $\bar{c}' = \bar{q} = 1$.

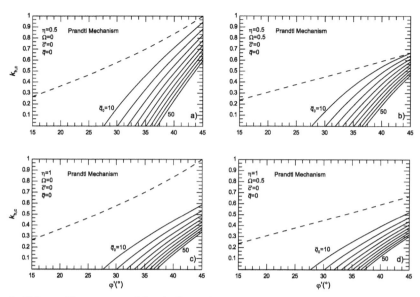

Fig. 3. Values of $k_{h,c}$ computed for the Prandtl mechanism and for different values of the normalised relevant parameters: a) $\eta = 0.5$; $\Omega = \overline{c}' = \overline{q} = 0$; b) $\eta = \Omega = 0.5$; $\overline{c}' = \overline{q} = 0$; c) $\eta = 1$; $\Omega = 0 = \overline{c}' = \overline{q} = 0$; d) $\eta = 1$; $\Omega = 0.5$; $\overline{c}' = \overline{q} = 0$.

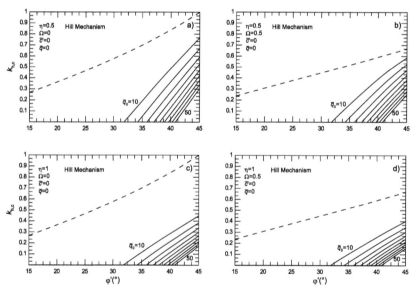

Fig. 4. Values of $k_{h,c}$ computed for the Hill mechanism and for different values of the normalised relevant parameters: a) $\eta = 0.5$; $\Omega = \overline{c}' = \overline{q} = 0$; b) $\eta = \Omega = 0.5$; $\overline{c}' = \overline{q} = 0$; c) $\eta = 1$; $\Omega = 0$; $\overline{c}' = \overline{q} = 0$; d) $\eta = 1$; $\Omega = 0.5$; $\overline{c}' = \overline{q} = 0$.

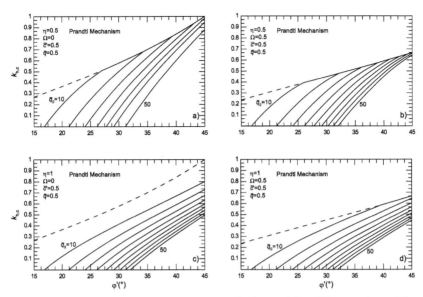

Fig. 5. Values of $k_{h,c}$ computed for the Prandtl mechanism and for different values of the normalised relevant parameters: a) $\eta = 0.5$; $\Omega = 0$; $\overline{c}' = \overline{q} = 0.5$; b) $\eta = \Omega = \overline{c}' = \overline{q} = 0.5$; c) $\eta = 1$; $\Omega = 0$; $\overline{c}' = \overline{q} = 0.5$; d) $\eta = 1$; $\Omega = \overline{c}' = \overline{q} = 0.5$.

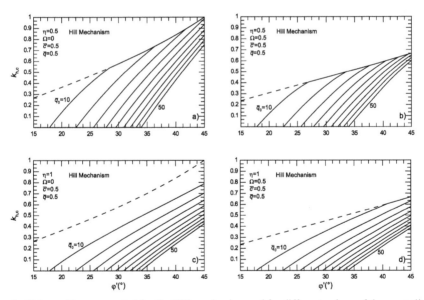

Fig. 6. Values of $k_{h,c}$ computed for the Hill mechanism and for different values of the normalised relevant parameters: a) $\eta = 0.5$; $\Omega = 0$; $\overline{c}' = \overline{q} = 0.5$; b) $\eta = \Omega = \overline{c}' = \overline{q} = 0.5$; c) $\eta = 1$; $\Omega = 0$; $\overline{c}' = \overline{q} = 0.5$; d) $\eta = 1$; $\Omega = \overline{c}' = \overline{q} = 0.5$.

4   Conclusion

This paper deals with the evaluation of the critical acceleration of shallow strip foundations using the seismic bearing capacity factors obtained by Cascone and Casablanca (2016) using the method of characteristics (MC). The proposed approach allows pointing out the parameters affecting the critical acceleration, whose values can be related to a few normalized variables $\overline{c}' = 2\ c'/\gamma B$, $\overline{q} = 2\ q/\gamma B$, $\overline{q}_0 = 2\ q_0/\gamma B$ and to the angle of soil shear strength φ'. An extensive parametric analysis has been performed in order to examine the effects of the relevant parameters. In the analysis both Prandtl and Hill mechanisms have been considered. The results are provided in design charts suitable for determining the values of $k_{h,c}$ for shallow strip foundations. The charts can be used starting from a known value of the static vertical load q_0 and from the knowledge of the soil strength parameters, the surcharge q acting aside the foundation and the expected value of the seismic acceleration.

Acknowledgements. This work is part of the research activities carried out by the University of Messina Research Unit in the framework of the work package WP16 - Contributi normativi - Geotecnica (Task 16·3: Interazione terreno-fondazionestruttura, Sub-task 16·3·2: Capacità dissipativa dell'interfaccia terreno-fondazione) of a Research Project funded by the ReLuis (Network of Seismic Engineering University Laboratories) consortium (Accordo Quadro DPC/ReLUIS 2019–2021).

References

1. Budhu, M., Al-Karni, A.: Seismic bearing capacity of soil. Géotechnique **43**, 81–187 (1993)
2. Casablanca, O., Cascone E., Biondi G.: The static and seismic bearing capacity factor Nγ for footings adjacent to slopes. Proc. Eng. **158**, 410–415 (2016). Proceedings of VI Italian Conference of Researchers in Geotechnical Engineering
3. Casablanca, O., Cascone, E., Biondi, G., Di Filippo, G.: Static and seismic bearing capacity of shallow strip foundations on slopes. Géotechnique (2021). https://doi.org/10.1680/jgeot.20.P.044
4. Cascone, E., Casablanca, O.: Static and seismic bearing capacity of shallow strip footings. Soil Dyn. Earthq. Eng. **84**, 204–223 (2016)
5. Cascone, E., Biondi, G., Casablanca, O.: Influence of earthquake-induced excess pore water pressures on seismic bearing capacity of shallow foundations. In: Earthquake Geotechnical Engineering for Protection and Development of Environment and Constructions - Proceedings of the 7th International Conference on Earthquake Geotechnical Engineering, pp. 566–581 (2019)
6. Cascone, E., Biondi, G., Casablanca, O.: Groundwater effect on bearing capacity of shallow strip footings. Comput. Geotech. **139**, 104417 (2021)
7. Dormieux, L., Pecker, A.: Seismic bearing capacity of foundation on cohesionless soil. J. Geotech. Eng. **121**, 300–303 (1995)
8. Kumar, J., Mohan Rao, V.B.K.: Seismic bearing capacity factors for spread foundations. Géotechnique **52**, 79–88 (2002)
9. Loukidis, D., Chakraborty, T., Salgado, R.: Bearing capacity of strip footings on purely frictional soil under eccentric and inclined loads. Can. Geotech. J. **45**, 768–787 (2008)

10. Mortara, G.: An exercise on the bearing capacity of strip foundations. Geotechn. Lett. **10**, 141–148 (2020)
11. Pane, V., Vecchietti, A., Cecconi, M.: A numerical study on the seismic bearing capacity of shallow foundations. Bull. Earthq. Eng. **14**, 2931–2958 (2016)
12. Paolucci, R., Pecker, A.: Seismic bearing capacity of shallow strip foundations on dry soils. Soils Found. **37**, 95–105 (1997)
13. Prandtl, L.: Über die Härte plastischer Körper. Nachr. königl. Ges. Wiss. Göttingen Math-Phys Kl **62**, 74–85 (1920)
14. Reissner, H.: Zum erddruckproblem. In: Proceedings of the 1st International Conference on Applied Mechanics, pp. 295–311. J. Waltman Jr, Delft, The Netherlands (1924)
15. Richards, R., Jr., Elms, D.G., Budhu, M.: Seismic bearing capacity and settlement of foundations. J. Geotech. Eng. **119**, 662–674 (1993)
16. Richards, R., Elms, D.G., Budhu, M.: Dynamic fluidization of soils. J. Geotech. Eng. ASCE **116**, 740–759 (1990)

Seismic Vulnerability Analysis of UHV Flat Wave Reactor Based on Probabilistic Seismic Demand Model

Jiawei Cui and Ailan Che[✉]

Shanghai Jiao Tong University, 800 Dongchuan Road, Shanghai 200240, China
alche@sjtu.edu.cn

Abstract. UHV Flat Wave Reactor is one of the important equipment of UHV converter station. Once damaged in an earthquake, it may cause the breakdown of the power system, resulting in serious economic losses and casualties. Therefore, analyzing its seismic vulnerability is significantly important. Firstly, failure modes, key vulnerable parts (weak parts), and seismic damage limit states of the structure under seismic action were analyzed by seismic damage investigation and finite element simulation. Secondly, based on the incremental dynamic analysis (IDA), the peak ground acceleration (PGA) was used as the input parameter to obtain the seismic response of key vulnerable parts. Thirdly, the seismic vulnerability curves of each key vulnerable part under different seismic damage states (complete, extensive, moderate, and slight) were generated by probabilistic seismic demand model (PSDM). Finally, through the reliability theory, the seismic vulnerability curves of UHV Flat Wave Reactor can be obtained. The results shown that this seismic vulnerability curve can give the probabilities of four seismic damage states for UHV Flat Wave Reactor under different PGA and reflects the seismic performance level of the structure, which provide reference for performance design and earthquake relief work.

Keywords: UHV Flat Wave Reactor · Seismic performance · IDA · PSDM · Seismic vulnerability

1 Introduction

Converter station is an important node of power system, which are used to transform electric energy, current and voltage [1]. Earthquake disasters are a major natural disaster that threatens the safety of converter stations. Many seismic events have proven that. Therefore, it is of great significance to study their seismic performance and seismic vulnerability to reduce the seismic risk of converter stations. The function of converter station is mainly realized by the electric facilities in the station. Therefore, to study the vulnerability of converter stations is to study the vulnerability of power facilities in the station. UHV Flat Wave Reactor is one of the important equipment of UHV DC converter station. With its low cost, light weight, convenient transportation, simple on-site installation, and low maintenance costs, it has been widely used in multiple

L. Wang et al. (Eds.): PBD-IV 2022, GGEE 52, pp. 702–710, 2022.
https://doi.org/10.1007/978-3-031-11898-2_43

transmission lines. It can be seen that Flat Wave reactor is very important to the safe operation of the converter station. Once damaged in an earthquake, it will cause serious consequences. Therefore, it is of great significance to explore its seismic vulnerability.

At present, the methods to study the seismic performance and vulnerability of power facilities mainly include seismic damage statistical analysis method [2], the mechanical calculation and experimental method [3] and the numerical and probability method [4]. The seismic damage statistics method does not have high application value due to the scarcity of seismic damage data. Due to the high cost and time consuming, it is difficult for the experimental method to obtain comprehensive analysis data through a large number of tests. However, the numerical and probability method can reduce the input of ground motion and ensure the accuracy of the result. The vulnerability analysis of electric power facilities by means of computational analysis method not only has low cost and high accuracy, but also can consider the structural performance response under various ground motions. In addition, the research on the vulnerability of tank circuit breaker is very little at present, and it has high application value to study its seismic vulnerability through computational analysis method.

Based on probabilistic seismic demand analysis, a seismic vulnerability curve of UHV Flat Wave Reactor is constructed in this paper. Firstly, the structural characteristics were analyzed, the failure mode and seismic performance evaluation index were determined, and four kinds of seismic damage states were defined. Secondly, through IDA analysis, PGA was used as the ground motion intensity index to obtain the seismic response of the key vulnerable parts. Thirdly, the seismic vulnerability curve of each key vulnerable part under different performance levels were obtained through probabilistic seismic demand analysis. Finally, the seismic vulnerability curves of UHV Flat Wave Reactor were obtained based on the theoretical calculation of reliability.

2   Method

2.1   The Basic Step of Constructing Vulnerability Curve

Seismic vulnerability curve is to establish the relationship between seismic input characteristics and structural functional damage in probabilistic way so as to predict its failure probability in future earthquakes. Probabilistic seismic demand model (PSDM) is one of the methods to construct seismic vulnerability curve of structures. Through PSDM, the structural response is determined by engineering demand parameters, which can be calculated by incremental dynamic analysis (IDA). The vulnerability of the whole structure can be determined by reliability theory according to the vulnerability of its key vulnerable parts.

2.2   Incremental Dynamic Analysis

The principle of Incremental Dynamic Analysis (IDA) is that through multiplied by a series of scaling factors, different seismic records are amplitude-modulated into a series of seismic records with the same PGA. And then, the dynamic response of the structure is calculated under the action of this set of "amplitude-modulated" seismic records. Finally, the performance of the structure is evaluated by drawing IDA curves [5]. "Amplitude modulation" can be reflected by Eq. (1) below:

$$a_\lambda = \lambda \cdot a \tag{1}$$

where, a is the recorded value of acceleration before amplitude modulation, which is the abbreviation of $a(t_i)$, $t \in \{0, t_1, \ldots, t_n\}$, λ is the adjustment coefficient, and a_λ is the recorded value of acceleration after amplitude modulation. According to the literature [6], when the earthquake intensity index is at the PGA, the amplitude levels shown in Table 1 can be used to perform the ground motion intensity amplitude modulation.

Table 1. Principle of amplitude modulation of ground motion intensity.

Characteristic period of structure T (s)	Amplitude of PGA
$T \le 0.5$ s	{0.1 g, 0.2 g, 0.3 g, …, 3.0 g}
0.5 s $< T \le 2$ s	{0.1 g, 0.2 g, 0.3 g, …, 1.5 g}
2 s $< T$	{0.1 g, 0.2 g, 0.3 g, 0.4 g, 0.5 g}

2.3 Probabilistic Seismic Demand Model

The main purpose of PSDM is to establish the probabilistic relationship between seismic intensity and seismic demand of a certain type of structure [7]. Assuming that the seismic demand (D) of the structure and seismic intensity measurement (IM) obey the lognormal distribution, the relationship is as follows:

$$\ln(D) = p \cdot \ln(IM) + \ln(m) \tag{2}$$

where, m and p are unknown parameters that must be obtained through regression analysis. Thus, an estimate of the standard deviation ($\beta_{D/IM}$) of the conditional logarithm can be further obtained:

$$\beta_{D/IM} \cong \sqrt{\frac{\sum \left(\ln(d_i) - \ln(mIM_i^p)\right)^2}{N-2}} \tag{3}$$

where, d_i is the i-th peak seismic demand; IM_i is the i-th seismic intensity peak. The failure probability of the structure is expressed as:

$$P[D \ge C/IM] = \Phi\left(\frac{\ln(IM) - \frac{\ln \overline{C} - \ln m}{p}}{\frac{\sqrt{\beta_{D/IM}^2 + \beta_C^2}}{p}}\right) \tag{4}$$

where, $\overline{C}$ is the median seismic capacity of the structure; β_c is the standard deviation of the seismic capacity of the structure. $\beta_{D/IM}$ is the logarithmic standard deviation for structural seismic demand.

2.4 Reliability Theory

Reliability theory provides the reliability analysis method of the system [7]. If the subsystems of a system are independent of each other, they are connected in series. In this case, the first-order boundary of the failure probability of the system is as below:

$$\max[P(E_i)] \leq P_{fs} \leq 1 - \prod_{i=1}^{n} [1 - P(E_i)] \tag{5}$$

where, $P(E_i)$ is the failure probability of the i-th subsystem; P_{fs} is the failure probability of the system.

3 Seismic Vulnerability Model of UHV Flat Wave Reactor

3.1 Introduction of UHV Flat Wave Reactor

UHV Flat Wave Reactor is a kind of oil-free open self-cooling structure with good inductance linearity and has been widely applied in the development of power network in China. It is mainly composed of ① Noise shield, ② Reactor body, ③ Lower hanger, ④ Upper bracket, ⑤ Miter platform, ⑥ Composite post insulator, ⑦ Insulator internode support rod, and ⑧ Subframe (Fig. 1a). The body weight is about 91 t, the total installation weight is about 135 t, and the installation height is about 20 m.

Fig. 1. The construction process of seismic vulnerability model of UHV flat wave reactor (a is the physical drawing, b is the finite element model drawing, and red points are the key vulnerability parts).

SAP2000 finite element software was used for numerical simulations, and the finite element model was constructed as shown in Fig. 1b (https://baike.baidu.com/item/SAP 2000/2016348?fr=aladdin). The structural, geometric and material parameters are shown in Table 2.

Table 2. Structural, geometric and material parameters of UHV flat wave reactor

Number	Parameter	Value
1	Height, diameter, and wall thickness of reactor body (m)	0.72, 0.96, and 0.1
2	Length, diameter and wall thickness of bushing (m)	0.4, 0.01, and 0.06
3	The elastic modulus of bushing (GPa)	100
4	Degree of inclination of bushing (°)	80
5	The type of steel	Q235

3.2 The Structural Characteristics of UHV Flat Wave Reactor

For UHV Flat Wave Reactor, the upper electrical antibody stiffness is large, rarely damage in the earthquake. According to past experience, the lower insulator bushing is brittle structure, easy to crack or break in the earthquake. Thus, failure modes and critical vulnerable locations were identified in Table 3.

Table 3. Failure modes and key vulnerability parts of UHV Flat Wave Reactor

Failure mode	Key vulnerability parts and parameters
Bending failure of bushing	The maximum bending stress of bushing root of component 1 (S_f) (red point shown in Fig. 1b)
	The maximum bending stress of bushing root of component 4 (S_a) (black point shown in Fig. 1b)

In the vulnerability analysis of power facilities, based on the principle that once earthquake damage occurs, the major impact on the function of power facilities, the recovery time, and cost, the limit state or seismic capacity of key parts refers to the quantitative description of four damage states [2]: complete, severe, moderate and slight damage. According to "Guidelines for Seismic Safety Risk Assessment of Substation (Converter Station)" [8], four performance levels of power facilities, that is four seismic damage states, was defined as complete damage, extensive damage, moderate damage, and slight damage. However, quantitative description of them was needed in vulnerability assessment. According to the structural and material properties of composite insulator bushing, it was considered that the bushing will be completely damaged when the bushing maximum tensile stress exceeds 60 MPa. According to literature [9], 0.7, 0.4 and 0.2 times of the value of complete damage were selected as the limit states of severe, moderate and slight damage, as shown in Table 4.

Table 4. Parameter values of limit states of UHV Flat Wave Reactor.

Key vulnerability parameters	Parameter values of limit states			
	Complete	Extensive	Moderate	Slight
S_f (MPa)	60	42	24	12
S_a (MPa)	60	42	24	12

3.3 IDA of UHV Flat Wave Reactor

Prior to IDA, a large number of seismic records need to be selected. The seismic record series of Pacific Earthquake Engineering Research (PEER), a strong earthquake database, was adopted. The numerical calculation shown that the characteristic period of UHV Flat Wave Reactor is 0.5735 s. Therefore, the reasonable IDA seismic record selection principle was shown in Table 5. 20 seismic records were selected from the PEER strong earthquake database.

Table 5. Selection principle of ground motion records.

Parameter	Venue category	Earthquake grouping	Earthquake influence coefficient
Attribute value	3	3	0.06

The characteristic period of UHV Flat Wave Reactor is 0.5735 s. Therefore, according to Table 1, the amplitude adjustment ratio of 20 ground motion data was 0.1 g, 0.2 g, 0.3 g, 0.4 g, 0.5 g, 0.6 g, 0.7 g, 0.8 g, 0.9 g, 1.0 g, 1.1 g, 1.2 g, 1.3 g, 1.4 g, 1.4 g, 1.5 g. In this way, a total of $20 \times 15 = 300$ ground motion records was obtained. Through IDA, the dynamic response values of two key vulnerable parts were plotted in rectangular coordinate system and logarithmic coordinate system respectively. Figure 2 shown IDA results of S_f and S_a respectively, where, the red line was the median value.

3.4 Seismic Vulnerability Curve of UHV Flat Wave Reactor

After completing IDA, the seismic vulnerability of the structure was calculated by using PSDM. Substitute the calculation results of IDA into Eqs. (1) and (2) to calculate the value of each key vulnerable parameter, as shown in Table 6.

Then, according to the calculation results in Table 6 and combining with the limit states of key vulnerable parts, the vulnerability curves can be obtained through Eq. (3). Figure 3 shown the vulnerability curves of S_f and S_a.

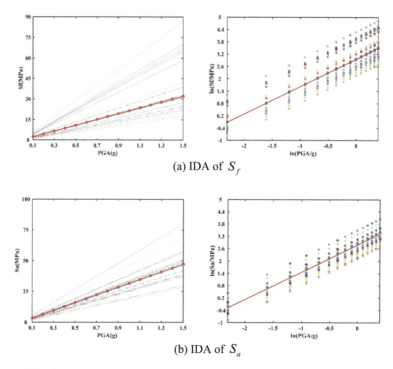

(a) IDA of S_f

(b) IDA of S_a

Fig. 2. Incremental dynamic analysis (IDA) of key vulnerable parameters.

Table 6. Logarithmic standard deviation of each key vulnerable parameter.

Parameters	Regression equation	R^2	$\beta_{D/IM}$	ρ
S_f	$\ln(S_f) = 1.23\ln(PGA) - 2.37$	0.68	0.76	0.86
S_a	$\ln(S_a) = 1.15\ln(PGA) + 1.36$	0.56	0.65	0.91

$^*R^2$: Coefficient of determination; $\beta_{D/IM}$: Logarithmic standard deviation; ρ: Correlation coefficient.

The failure of each key vulnerable part will cause the failure of UHV Flat Wave Reactor, regardless of the state of the others. Each key vulnerable part is independent from each other. Therefore, through Eq. (5), the vulnerability curve of UHV Flat Wave Reactor can be obtained, which is shown in Fig. 4.

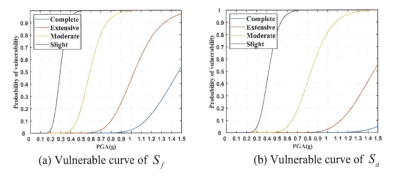

(a) Vulnerable curve of S_f (b) Vulnerable curve of S_a

Fig. 3. Seismic vulnerability curve of S_f and S_a.

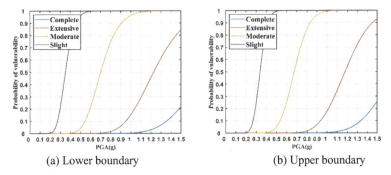

(a) Lower boundary   (b) Upper boundary

Fig. 4. Seismic vulnerability model of UHV Flat Wave Reactor.

4 Conclusion

In this study, the structural characteristics and seismic loss characteristics of UHV Flat Wave Reactor were analysed, and the geometric and material parameters of the structure are obtained and the finite element model is constructed according to the 'General Design of Power Transmission and Transformation Engineering of State Grid'. On this basis, IDA, PSDM and reliability theory are used to construct the seismic vulnerability curve. Through the seismic vulnerability curve, the probabilities of complete, severe, moderate and slight damage of this structure under different PGA can be determined, which reflect the structure seismic performance level. Therefore, the seismic design or post-seismic evaluation of UHV Flat Wave Reactor based on predetermined probability can be carried out.

Acknowledgement. This work is financially supported by the National Key R&D Program of China (2018YFC1504504).

References

1. Gjorgiev, B., Sansavini, G.: Identifying and assessing power system vulnerabilities to transmission asset outages via cascading failure analysis. Reliab. Eng. Syst. Saf. **217**, 108085 (2022)
2. Hailei, H., Jianbo, G., Qiang, X.: Vulnerability analysis of power equipment caused by earthquake disaster. Power Syst. Technol. **35**(4), 25–28 (2011). (in Chinese)
3. Paolacci, F., Giannini, R., Alessandri, S., De Felice, G.: Seismic vulnerability assessment of a high voltage disconnect switch. Soil Dyn. Earthq. Eng. **67**, 198–207 (2014)
4. Bender, J., Farid, A.: Seismic vulnerability of power transformer bushings: complex structural dynamics and seismic amplification. Eng. Struct. **162**, 1–10 (2018)
5. Vamvatsikos, D., Cornell, C.A.: Incremental dynamic analysis. Earthq. Eng. Struct. Dyn. **31**(3), 491–514 (2002)
6. Zhong, J., Zhang, J., Zhi, X., Fan, F.: Probabilistic seismic demand and capacity models and fragility curves for reticulated structures under far-field ground motions. Thin-Walled Struct. **137**, 436–447 (2019)
7. Haldar, A., Mahadevan, S.: Probability, Reliability, and Statistical Methods in Engineering Design, pp. 143–273. Wiley, New York (2000)
8. State Grid Corporation of China: Guidelines for Seismic Safety Risk Assessment of Electrical Substations (Converter Stations). China Electric Power Press, Beijing, China (2021)
9. Li, S., Lu, Z., Zhu, Z., Liu, Z., Cheng, Y., Xi, L.: Dynamic properties and seismic vulnerability of substation composite insulators. Eng. Mech. **33**(04), 91–97 (2016). (in Chinese)

Seismic Performance of the "Mediterranean Motorway" Piers Founded on Soft Soil

Filomena de Silva[1(✉)] [iD], Michele Boccardi[1], Anna d'Onofrio[1] [iD], Valeria Licata[2] [iD], Enrico Mittiga[2], and Francesco Silvestri[1] [iD]

[1] University of Naples Federico II, Naples, Italy
filomena.desilva@unina.it
[2] ANAS SpA, Rome, Italy

Abstract. The seismic performance of bridge piers is significantly influenced by stiffness and strength of the foundation soil. In this study, the effects of soil-structure interaction are investigated with reference to the piers of the viaducts of the Mediterranean Motorway, a 442 km long road connecting the southern-most part of Italy. Non-linear dynamic analyses were carried out on numerical models including soil, deep foundation and pier, under unscaled records of real earthquakes. The characteristics of the analyzed prototypes were inferred from the most recurrent features of the viaducts. The results of the analyses show that soil-structure interaction reduces the bending drift, due to significant peak and permanent foundation rotations. Hence, the overall effect of soil-structure interaction is a reduction of structural damage while, on the other hand, pier instability and tilting are enhanced. The results of the analyses were finally exploited to compute the demand hazard curve of the pier prototype, expressed in terms of mean annual rate of exceedance of the foundation rotation.

Keywords: Soil-structure interaction · Bridge pier · Deep foundation · Demand hazard curve

1 Introduction

The safe use of strategic infrastructures, such as motorways, is a crucial aspect in the management of post-earthquake emergencies. Consequently, the seismic performance of the different components of the infrastructure should be carefully evaluated. Among the infrastructural elements, bridges were found to be the most critical asset under earthquake motion (Argyroudis et al. 2019), especially when founded through direct foundations on soft soil. In such cases, the pier performance is affected by the interaction with the foundation and the surrounding soil which leads to a reduction of the bending drift and to the occurrence of foundation rocking due to the soil non-linear and plastic behavior (Anastasopoulos et al. 2014). In this study, the seismic performance of the piers and caisson foundations of the A2 - Mediterranean motorway is evaluated considering soil-structure interaction (SSI). Following the approach proposed by Bradley et al. (2010) to evaluate the seismic performance and loss of twin bridges in Christchurch, the modification to the seismic hazard and the pier fragility induced by SSI are considered by

© The Author(s), under exclusive license to Springer Nature Switzerland AG 2022
L. Wang et al. (Eds.): PBD-IV 2022, GGEE 52, pp. 711–719, 2022.
https://doi.org/10.1007/978-3-031-11898-2_44

synthesizing the results in terms of hazard demand curve expressing the mean annual rate of exceeding different values of the foundation rotation.

2 Main Features of Viaducts and Foundation Soil

The A2 - Mediterranean motorway is a 423 km long motorway, connecting three regions in the extreme southern Italy. Figure 1a shows the highway track superimposed on the simplified geological map proposed by Forte et al. (2019). Almost 50% of the Mediterranean motorway crosses the Apennines, i.e. the area with the highest seismic hazard in Italy with an expected peak ground acceleration PGA > 0.25 g with a probability of exceedance $p_{vr} = 10\%$ in 475 years.

ANAS spa, the company currently managing the motorway, counted 894 viaducts along the motorway, with the most recurrent class (425 out of 894) characterized by a span length ranging between 20 m and 40 m. Full dots in Fig. 1 indicate the localization of the viaducts belonging to such class. Most of them (99 out of 425) are founded on a coarse-grained soil mainly constituted by gravel and sand ('gs' in Fig. 1 legend).

Physical and mechanical properties of the 'gs' soil were measured through field and laboratory investigations during the renovation works of the Mediterranean Motorway. Data of 13 downhole tests DH, 20 Multichannel Analysis of Surface Waves MASW tests, 2 Extended Spatial Autocorrelation ESAC tests, 7 seismic reflection and 15 seismic refraction surveys revealed a mean bedrock depth of 23 m with a standard deviation of 15 m. The variability of the shear wave velocity with depth inferred from DH, MASW and ESAC tests is shown in Fig. 1b, together with the mean value and the standard deviation, σ. Laboratory investigations including direct shear tests provided mean values as high as 22 kN/m^3 for the unit weight, and of the order of 40° for the peak friction angle, φ.

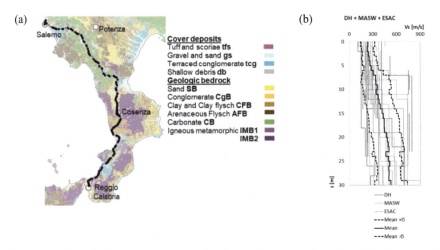

Fig. 1. Location of viaducts characterized by the maximum span length ranging between 20 m and 40 m on the geological map by Forte et al. (2019) (a) and profiles of the shear wave velocity resulting from the geophysical tests (b).

3 Description of Numerical Models

Table 1 lists the reference cases analysed as representative of the selected class of viaducts, characterized by a maximum span length between 20 m and 40 m and founded on gravelly sand. Two bedrock depths were considered, 23 m and 8 m, being the mean and the mean-σ values obtained from the interpretation of the geophysical tests. Two-dimensional models of the selected cases were generated through the finite difference code FLAC (Itasca 2011), including the soil, the foundation and the structure. The generic layout of the model is reported in Fig. 2.

Table 1. Analyzed cases.

Case	Bedrock depth H [m]	Pier height h [m]	Foundation depth D [m]	Foundation bearing capacity FoS [/]	Period T^* [s]
1	23	30	8	3.14	1.72
2	23	15	7	3.14	1.56
3	23	30	4	2.42	1.96
4	23	15	4	2.45	1.69
5	8	30	8	/	1.56
6	8	15	8	/	1.52

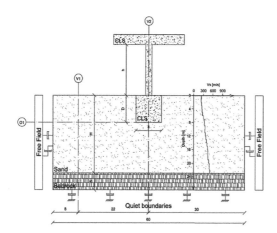

Fig. 2. Numerical model of soil, foundation and structure.

Two pier heights were considered (15 m and 30 m) to account for the pier height variability along the same viaduct related to the morphology of the crossed valleys. Also, two caisson foundation depths were assumed, 4 m and 8 m, except when the bedrock is 23 m deep, and the pier is 15 m tall. In the latter case the foundation depth was set as 7 m to obtain the same factor of safety, FoS, as case n°1 (see Table 1) in terms of vertical bearing capacity. The FoS was computed as the ratio between the tangent of the friction angle φ and a reduction of its value corresponding to the foundation failure. A perfect contact between the foundation and the surrounding soil was assumed. The infinite extension in depth of the bedrock was simulated by means of dashpots, providing viscous normal and shear stresses proportional to the compression and shear wave velocities of the bottom layer. To minimize the numerical model size, the so-called 'free-field' boundary conditions were applied to the lateral sides, simulating an ideal horizontally layered soil profile connected to the main-grid domain through viscous dashpots. Table 2 summarizes the physical and mechanical properties assigned to the soil and structural models. The bedrock was assumed elastic, while a hysteretic constitutive law with perfectly plastic Mohr-Coulomb strength criterion was adopted to model the behaviour of 'gs' soil. The small-strain shear stiffness, G_0, was assumed as variable with depth in the range reported in Table 2, following the V_S profile in Fig. 1b. The Young's modulus, E_0, was calculated from G_0 by assuming a value of the Poisson's coefficient equal to 0.27. In lack of specific laboratory data, the hysteretic soil behaviour was simulated through three-parameters sigmoid functions calibrated on the experimental curves reported by Liao et al. (2013) for gravelly sands with a grain size distribution very close to that typical for the 'gs' soil. The initial damping ratio, D_0, was introduced through the approach, assigning to the 'gs' and bedrock layers a frequency-dependent damping ratio with a minimum value $D_0 = 1\%$ and 0.5%, respectively, corresponding to the predominant frequency of the free-field 1D column of the numerical model.

Table 2. Physical and mechanical properties assigned to the soil and structural models.

	γ [KN/m^3]	E_0 [MPa]	G_0 [MPa]	D_0 [%]	φ [°]
Bedrock	20	2944	1280	0.5	/
Soil	22	310–1300	122–512	1	40
Foundation	25	30000	12500	4	/
Pier cases n°2, 4, 6	40.03	30000	12500	4	/
Pier cases n°1, 3, 5	19.77	30000	12500	4	/
Deck	97.61	30000	12500	/	/

The pier was modelled through solid elements made of a visco-elastic material with the unit weight, Young's and shear modulus reported in Table 2 and a constant damping ratio $D_0 = 4\%$. According to the typical construction technique, the horizontal cross section of the pier was assumed either circular, with a diameter of 2 m for the 15 m tall pier, or cave, with the external and internal diameters of 4.2 m and 3.7 m, respectively, for the 30 m tall pier. An equivalent unit weight was assigned to the pier material to account for the circular pier section in the 2D model. Similarly, the equivalent unit weight of the deck is calculated considering the mass of a 30 m long span.

A numerical procedure based on a random noise low-amplitude input motion (de Silva et al. 2018) was used to compute the fundamental period T^* of the soil-foundation-structure system for each case. All the results are shown in Table 1. The analyses were repeated by assigning to the soil the bedrock properties in order to derive the fixed base period, T_0. The latter resulted around 1.46 s for both structural heights, because the mass of the shortest pier is higher with respect to that of the tallest pier, due to the abovementioned differences in their horizontal cross sections. The increase of T^* due to the soil compliance is more significant for the tallest pier (cases 1 and 2 in Table 1). Such effect is accentuated by the reduction of the foundation depth (cases 3 vs 1 and 4 vs 2) and reduced when bedrock is shallow (cases 5 and 6).

Nonlinear dynamic analyses were performed by applying independently to the described numerical models the EW and NS components of 10 natural events selected from the *Selected Input Motions for Displacement-Based Assessment and Design* (SIM-BAD) database (ReLUIS 2013). The selection includes accelerations recorded at stations located on a stiff rock outcrop, i.e. with an equivalent shear wave velocity greater than 700 m/s. They were selected to cover a large range of the PGA values and to be representative of the seismic hazard of the whole road track.

4   Results of Nonlinear Dynamic Analyses

The performance of the caisson foundation under each earthquake was evaluated by computing its rotation, δ, as the difference between the settlements of its vertical sides divided by the foundation width. Figure 3 compares the maximum foundation rotation (a) and bending drift (b) of the analysed case studies. This latter, θ, was calculated from the bending displacement divided by the pier height. As expected, the peak δ is more pronounced when the foundation caisson is shorter (cases 3 and 4) while it is almost negligible for the cases of shallower bedrock (5 and 6). Conversely, the bending drift θ is higher when the foundation rocking is lower, see cases 1 vs 3, or 5 vs 1, in Fig. 3b. Numerical results highlighted that soil compliance affects the performance of piers. In particular, the common fixed base assumption, which is close to the cases 5 and 6, implies an overestimation of the bending response and neglect the foundation rocking, including the associated structural tilts.

The maximum and residual foundation rotations for the cases of rocking foundations (deep bedrock) were then correlated in a log-log plot to the spectral accelerations Sa(T*) associated with the predominant period of the model listed in Table 1.

The comparison among the equations of the power-laws fitting the data of the peak rotations shows that there is no substantial difference due to pier height between cases 1 and 2 (long caisson), while the influence of structural slenderness is slightly more marked for cases 3 and 4 (short caisson). The residual rotations are more evident for taller piers (cases 1 and 3 vs. 2 and 4) at low to moderate Sa-values. As Sa increases, they become reversed for the cases 1 and 2 and comparable for the cases 3 and 4.

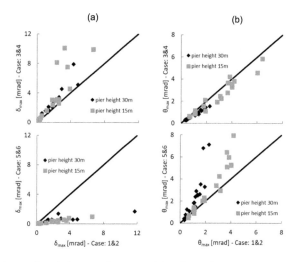

Fig. 3. Maximum foundation rotation (a) and bending drift (b) of the analyzed case studies.

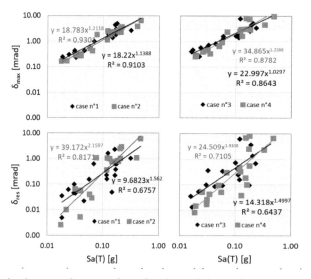

Fig. 4. Correlation between the spectral acceleration and the maximum and residual rotation of the foundation for the cases of rocking foundations.

5   Construction of the Demand Hazard Curves

The Sa-δ_{max} regressions discussed in Sect. 4 were used to estimate the expected δ_{max} demand at the two extreme seismic-prone sites, i. e. Salerno and Reggio Calabria, characterised by PGA = 0.103 g and 0.270 g at 475 yrs return period. In detail, the convolution between hazard and fragility through the well-known equation proposed by PEER (Cornell et al. 2002) was applied to compute the hazard curve of the peak foundation rotation for the analysed cases. Figure 5 shows the resulting demand hazard curves in terms of mean annual rate of exceedance of the foundation peak rotation for the analysed cases of rocking foundation. The curves for the cases 1 and 2 are practically superimposed, as well as slight differences are recognized between the curves for the cases 3 and 4 according to their close regressions shown in Fig. 4. For a given value of the peak rotation, δ_{max_t}, the mean annual rate of exceedance, $\lambda_{\delta max\_t}$, is significantly higher for the viaduct piers located in Reggio Calabria according to the higher seismic hazard of the site.

The advantage of such approach is the straightforward evaluation of the expected demand for each limit state, bypassing that of the hazard parameter. As a matter of fact, entering the graphs of Fig. 5 with the return period associated to the strategic structures for a specific limit state, it is possible to evaluate the expected peak value of piers rotation. In detail, the return periods T_R associated to the probability of exceedance, p_{VR}, defined by the Italian Code (NTC 2018) for the limit states were computed by assuming a reference life-time $V_R = 100$ years and a utilization coefficient cu $= 2$, consistently with the strategic importance of the motorway. Being $\lambda_{\delta max\_t} = 1/T_R$, the annual rate of exceeding the serviceability limit states SLD ($p_{VR} = 63\%$ and $T_R = 201$ years) and ultimate limit states SLV ($p_{VR} = 10\%$ and $T_R = 1898$ years) were calculated. The computed $\lambda_{\delta max\_t}$ values are shown in Fig. 5 through horizontal lines, highlighting that the expected peak rotation of piers located in Salerno ranges between 0.8 mrad and 1.1 mrad for SLD and between 2.9 mrad and 5.9 mrad for SLV, while the same piers settled in Reggio Calabria are likely affected by 0.9 mrad $< \delta_{max\_t} < 1.7$ mrad for SLD and 5 mrad $< \delta_{max\_t} < 10$ mrad for SLV. For both sites, the differences among the analysed cases are huger at SLV with a significant increase of the expected peak rotation when the caisson foundation is shorter (cases 3–4).

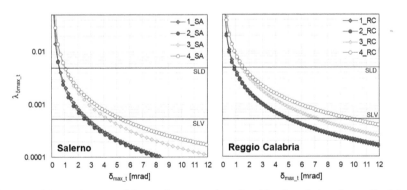

Fig. 5. Hazard demand curves of the maximum rotation of rocking foundations located in Salerno and Reggio Calabria.

According to the Italian geometric standards provided by the DM 5 11 2011, the maximum allowable inclination of the deck is 25 mrad in the case of straight stretch of the motorway and 70 mrad for circular stretch. The rocking-induced pier inclination reaches 40% of the maximum allowable value for a straight stretch at SLV in Reggio Calabria. Additionally, during earthquake the pier experiences bending drift that should be summed to that induced by the foundation rocking. According to the results of the analyses, the maximum bending drift is almost equal to the maximum foundation rotation (see Fig. 3), consequently the total drift reaches the 80% of the maximum allowable inclination of the deck.

6 Conclusions

The proposed procedure was applied for predicting the maximum and residual seismic-induced rotation of typical caisson-founded piers located along the Mediterranean motorway. The analyses were performed on the most populated class of viaducts founded on loose alluvial coarse-grained deposits with a thickness varying between 8 m and 23 m. The case of the shallowest bedrock approximates a fixed base condition of the pier, in fact negligible foundation rotations resulted from the analyses. Conversely, piers founded on the deep bedrock soil configuration are affected by significant tilting during and after the seismic event, due to the soil nonlinear and plastic behaviour. Such effect is amplified by the reduction of the foundation depth. The increase of the foundation rocking in the latter cases is associated to a general reduction of the bending drift of the piers. Hence, soil structure interaction is expected to reduce the structural damage but to enhance the pier instability and its permanent tilting. Based on the results of the nonlinear dynamic analyses, the demand hazard curves of the maximum rotation of the foundation were generated for the analysed cases. These curves enable a straightforward evaluation of the demand associated to the different limit states, bypassing that of the hazard parameter provided by the traditional approach. The expected rotation demand associated to the ultimate limit state reaches 80% of the allowable inclination of the deck stated by the Italian Code DM 5 11 2011, with even contributions of foundation rocking and pier bending motions. Anyway *ad-hoc* thresholds for the seismic performance are needed for a more accurate estimation.

References

Anastasopoulos, I., Kontoroupi, Th.: Simplified approximate method for analysis of rocking systems accounting for soil inelasticity and foundation uplifting. Soil Dyn. Earthq. Eng. **56**, 28–43 (2014)

Argyroudis, S., Mitoulis, S. A., Wint, M.G., Kayna, A.M.: Fragility of transport assets exposed to multiple hazards: state-of-the-art review toward infrastructural resilience. Reliab. Eng. Syst. Saf. **191**, 106567 (2019)

Bradley, B.A., Cubrinovski, M., Dhakal, R.P., MacRae, G.A.: Probabilistic seismic performance and loss assessment of a bridge-foundation-soil system. Soil Dyn. Earthq. Eng. **30**(5), 395–411 (2010)

Cornell, C.A., Jalayer, F., Hamburger, R.O., Foutch, D.A.: Probabilistic basis for 2000 SAC Federal Emergency Management Agency steel moment frame guidelines. J. Struct. Eng. ASCE **128**(4), 526–533 (2002)

de Silva, F., Pitilakis, D., Ceroni, F., Sica, S., Silvestri, F.: Experimental and numerical dynamic identification of Carmine bell tower in Naples (Italy). Soil Dyn. Earthq. Eng. **109**, 235–250 (2018)

Forte, G., Chioccarelli, E., De Falco, M., Cito, P., Santo, A., Iervolino, I.: Seismic soil classification of Italy based on surface geology and shear-wave velocity measurements. Soil Dyn. Earthq. Eng. **122**, 79–93 (2019)

Itasca: FLAC 7.0 – Fast Lagrangian Analysis of Continua – User's Guide. Itasca Consulting Group, Minneapolis, USA (2011)

Liao, T., Massoudi, N., Mchood, M., Stokoe, K.H., Jung, M.J., Menq, F.Y.: Normalized shear modulus of compacted gravel. In: 18th international conference on soil mechanics and geotechnical engineering Challenges and Innovations in Geotechnics, ICSMGE, 2–6 September 2013, Paris, France (2013)

NTC 2018: Norme Tecniche per le Costruzioni. DM 17/1/2018, Italian Ministry of Infrastructure and Transportation, G.U. n. 42, 20 February 2018, Rome, Italy (2018)

ReLUIS – Smerzini, C., Paolucci, R.: SIMBAD: a database with Selected Input Motions for displacement Based Assessment and Design – 3rd release by Department of Structural Engineering Politecnico di Milano, Italy. Research Project DPC (2013)

Performance-Based Estimation of Lateral Spread Displacement in the State of California: A Case Study for the Implementation of Performance-Based Design in Geotechnical Practice

Kevin W. Franke[(✉)] [iD], Clay Fullmer, Delila Lasson, Dallin Smith, Sarah McClellan, Ivy Stout, and Riley Hales

Brigham Young University, Provo, UT 84602, USA
Kevin_franke@byu.edu

Abstract. Within the last decade, many researchers have demonstrated that performance-based methods for estimating the hazard from liquefaction triggering and its effects can effectively be approximated using simplified, map-based methods. However, the development of the reference parameter maps that are necessary for the implementation of these simplified performance-based methods is major endeavor and has proven to be a significant impediment for the implementation of these methods in engineering practice. This study presents a case history of how the simplified performance-based reference parameter maps for a liquefaction-related hazard (lateral spread displacement) were recently developed for a single state known for its high seismicity (California) in the United States. Through a mentored research experience involving several undergraduate and graduate students, the development of the lateral spread reference parameter maps corresponding to three commonly used return periods (475 years, 1033 years, and 2475 years) for the State of California is summarized and presented. When combined with an existing simplified performance-based lateral spread method, the reference parameter maps described in this paper become a powerful design resource for engineers in California. Example lateral spread displacement values are computed for various parts of the state to validate the maps/approach and to demonstrate the performance-based methodology and potential uses/benefits.

Keywords: Liquefaction · Lateral spread displacement · Performance-based design

1 Introduction

1.1 Background

One of the most common hazards associated with large earthquakes is that of seismic-induced soil liquefaction and lateral spread displacement. Such displacements have been observed to cause considerable damage to infrastructure in recent events including the

L. Wang et al. (Eds.): PBD-IV 2022, GGEE 52, pp. 720–729, 2022.
https://doi.org/10.1007/978-3-031-11898-2_45

2014 Iquique, Chile; 2011 Tohoku, Japan; 2010–2011 Canterbury, New Zealand; and 2010 Concepción, Chile earthquake.

To prepare infrastructure to resist the loads induced by lateral spread displacements, geotechnical engineers routinely attempt to predict these displacements for design-level earthquakes using one or more methods. These methods include analytical, empirical, or semi-empirical prediction models. Of these methods, empirical (e.g., [1–4]) and semi-empirical (e.g., [5, 6]) are arguably the most frequently applied by engineering practitioners because of their simplicity and their basis in actual lateral spread displacements documented from the field case histories.

As with the prediction of any hazard based on field or synthetic case histories, a significant amount of uncertainty exists in the empirical and semi-empirical prediction of lateral spread displacements. Lateral spread is a complex geotechnical and geo-hydrological phenomenon that is dependent upon several factors including soil properties, site geometry, and seismic loading. While the uncertainties associated with geotechnical properties of the soil, spatial variability (i.e., soil layer continuity), and site geometry can be substantial for any particular site, the uncertainties associated with the seismic loading tend to govern all others [7]. As with other forms of seismic hazards, engineers have turned to probabilistic and performance-based methods of predicting liquefaction-induced lateral spread displacements for the purpose of quantifying risk.

While performance-based methods of predicting hazards related to liquefaction initiation [8–10] and lateral spread displacement [11] have been developed, such methods are not easily implemented in engineering practice without special computational tools and training. Rather, simplified performance-based methods [12–14] that closely approximate the results of a full performance-based analysis can prove much more useful to practitioners on typical engineering design projects. This paper demonstrates how one such simplified performance-based approach [14] can be developed for use in a high-seismicity region like the state of California in the USA. This paper will summarize the development of the lateral spread reference parameter maps for California at three return periods of interest (475, 1033, and 2475 years), and will demonstrate how those maps can be used within a publicly available GIS-based user interface using the Tethys platform to develop performance-based estimates of lateral spread for several local site conditions across numerous locations in the state.

1.2 Performance-Based Lateral Spread Displacement Prediction

To account for the uncertainties associated with seismic loading and model prediction (i.e., scatter), Franke and Kramer [11] presented a performance-based procedure for computing lateral spread displacements based on the Youd et al. [4] empirical lateral spread model. According to their procedure, the mean annual rate of exceeding some lateral spread displacement of interest (i.e., d^*) is computed as:

$$\lambda_{D_H|\mathcal{S}}\left(d^*|\mathcal{S}\right) = \sum_{i=1}^{N_{\mathcal{L}}} P\left[D_H > d^*|\mathcal{S}, \mathcal{L}_i\right] \Delta\lambda_{\mathcal{L}_i} \tag{1}$$

where

$$\overline{\log D_H} = \mathcal{L} - \mathcal{S} \tag{2}$$

$$\mathcal{L} = 1.532M - 1.406 \log R^* - 0.012R \tag{3}$$

$$S = \begin{cases} -\left(\dfrac{-16.213 + 0.338 \log S + 0.540 \log T_{15}}{+3.413 \log(100 - F_{15}) - 0.795 \log(D50_{15} + 0.1)}\right) \text{Ground Slope Geometry} \\ \\ -\left(\dfrac{-16.713 + 0.592 \log W + 0.540 \log T_{15}}{+3.413 \log(100 - F_{15}) - 0.795 \log(D50_{15} + 0.1)}\right) \text{Free - Face Geometry} \end{cases} \tag{4}$$

$$R^* = R + 10^{(0.89M - 5.64)} \tag{5}$$

$$P[D_H > d^*] = 1 - \Phi\left[\frac{\log d^* - \overline{\log D_H}}{\sigma_{\log D_H}}\right] = 1 - \Phi\left[\frac{\log d^* - \overline{\log D_H}}{0.197}\right] \tag{6}$$

In Eqs. (1) through (6), D_H is the estimated lateral spread displacement (in meters); $\mathcal{L}$ is the apparent loading parameter and represents the seismic loading affecting lateral spread displacement; S is the site parameter and represents the ability of the local site conditions to resist lateral spread displacement [ranging from about 8.0 (very high site susceptibility to lateral spread) to about 10.5 (very low site susceptibility to lateral spread)]; $\mathcal{L}$ is an increment of the hazard curve for $\mathcal{L}$ that is computed from a probabilistic seismic hazard analysis (PSHA); M is the moment magnitude; R is the shortest horizontal distance from the site to the vertical projection of a seismic source zone of fault rupture (km); R^* is a modified distance metric that reduces over-estimation from near-source seismicity; W is the free-face ratio for a free-face geometry (in percent), and is computed as the ratio of the height of the free face to the horizontal distance from site to the toe of the free-face; T_{15} is the cumulative thickness of saturated granular layers with corrected SPT resistance $(N_1)_{60}$ less than 15 blows per 0.3 m; F_{15} is the average fines content (in percent) of the soil layers comprising T_{15}; $D50_{15}$ is the average mean grain size (in millimeters) of the soil layers comprising T_{15}; $\sigma_{\log Dh}$ is the standard deviation of the Youd et al. (2002) empirical model and is equal to 0.197; and Φ is the standard normal cumulative distribution function. The valid ranges for the model parameters listed in Eqs. (3) and (4) are reported in Youd et al. (2002).

The Franke and Kramer [11] procedure produces a hazard curve for lateral spread displacement and requires the use of specialized seismic hazard analysis software such as *EZ-FRISK* [15] or *OpenSHA* [16]. The approach is akin to performing a PSHA, but using the lateral spread displacement model presented in Eq. (2) instead of a traditional ground motion prediction equation.

1.3   Simplified Performance-Based Prediction Procedure for Lateral Spread

Ekstrom and Franke [14] presented a simplified performance-based lateral spread prediction procedure that assumes a constant reference value of S (i.e., $S^{ref\ f}$) that is consistent with a site of moderately-high susceptibility to lateral spread displacement (e.g., $S^{ref} = 9.044$). The Franke and Kramer [11] procedure is then applied universally with S^{ref} to develop reference lateral spread hazard curves across a grid of geographic points. From these reference hazard curves, reference lateral spread displacements are obtained from specific hazard levels or return periods and are contoured to create lateral spread reference parameter maps. The reference lateral spread displacements in the simplified performance-based prediction procedure [mapped and contoured as $\log(D_h)^{ref}$] serve as proxies for seismic loading and account for uncertainties in the seismic loading and model scatter. However, they do not represent actual predicted ground displacements at a site because they were developed with S^{ref}.

To obtain site-specific probabilistic lateral spread displacements using a lateral spread reference parameter map and the Ekstrom and Franke [14] simplified procedure, the site-specific probabilistic lateral spread displacement at the return period of interest is computed as:

$$\log D_H = \log(D_H)^{ref} + \Delta D_H \tag{7}$$

where $\log(D_h)^{ref}$ is the value taken directly from the reference parameter map, and ΔD_h is the site-specific correction for local site conditions (i.e., S) and is computed as:

$$\Delta D_H = S^{ref} - S = 9.044 - S \tag{8}$$

In validating their simplified performance-based procedure, Ekstrom and Franke [14] observed a nearly perfect linear correlation ($R^2 = 0.997$) between their proposed procedure and the full performance-based procedure [11].

2 Development of Reference Parameter Maps for California

2.1 Grid Point Spacing and Resulting Maps

To develop the lateral spread reference parameter maps of $\log(D_h)^{ref}$ for the state of California, the commercial PSHA software *EZ-FRISK ver.8.06* [15] was used with Eq. (2), where $S = S^{ref} = 9.044$. At the time of this study, *EZ-FRISK* did not incorporate the third Uniform California Earthquake Rupture Forecast (UCERF3) [17] for the state of California, but instead incorporated the seismic source model described used in the 2008 updated of the U.S. National Seismic Hazard Maps (NSHMP) [18]. The lateral spread reference parameter maps for California developed for this study will be updated in the future once the UCERF3 seismic source model is incorporated into *EZ-FRISK*.

Grid points were laid across the state of California using the 2,475-year PGA seismic hazard maps from the U.S. NSHMP and the grid spacing recommendations developed in a previous study [19]. This grid spacing ranged from 4 km maximum spacing in areas with mapped PGA accelerations in excess to 0.64 g to 50 km maximum spacing in areas with mapped PGA accelerations less than 0.04 g.

Probabilistic values of $\log(D_h)^{ref}$ were computed at each grid point at return periods of 475, 1033, and 2475 years. These values were then mapped and contoured using Inverse Distance Weighting (IDW) interpolation [20]. The resulting lateral spread reference parameter map for California at a return period of 2,475 years is presented in Fig. 1. Due to space limitations, maps for the other two return periods are not presented in this paper, but instructions on how to access an online interactive version of the computed reference parameter values at all three return periods are provided in the next section.

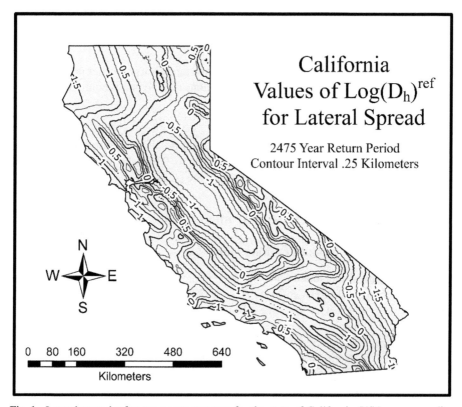

Fig. 1. Lateral spread reference parameter map for the state of California, USA, corresponding to a return period of 2475 years. For interactive online reference parameter maps at return periods of 475, 1033, and/or 2475 years, please refer to [28].

2.2 Online Interactive App for Accessing Lateral Spread Reference Parameter Values

For convenience and usability by the engineering public, the reference parameter maps that were developed for the state of California have been incorporated into an online app that is accessible to the public using an Internet browser. The Liquefaction Hazard App provides a free, graphical interface to the lateral spread reference parameters developed in these and other mapping studies [13, 14]. The app is built on the Django web framework for Python and PostgreSQL databases as implemented in the Tethys Platform [21, 22]. The Liquefaction Hazard App was built through Tethys because it targets developers of geospatial web apps and it is commonly used by other geospatial researchers [23–26]. The app currently organizes the reference parameter data by state and return period. A user chooses a state (e.g., California) and return period, and then clicks locations on an interactive map for reference parameter values that can be used in the engineer's own calculations or incorporated into a spreadsheet specifically developed for the simplified performance-based approach such as *SPLIQ* [27]. The reference parameter values for each chosen site are computed and tabulated for download in Comma Separated Values (CSV) format. Values at each selected site are computed using IDW interpolation considering only the four points nearest to the chosen location [20]. Owing to the regular spacing of the points where values were computed, the four nearest points are equivalent to using the nearest point to the northeast, northwest, southeast, and southwest of a given location. The interpolation is limited to the nearest four points because the soil properties that affect the derived reference values are best approximated by the closest available measurements. Points further away are less likely to have a similar soil composition and proximity to seismic hazards that govern the derived reference parameters. The Liquefaction Hazard App and the lateral spread reference parameter values computed for the state of California and other states in the U.S. can be accessed online [28].

3 Validation and Testing of the Reference Parameter Maps for California

3.1 Selection of Sites and Site Parameters

To validate the lateral spread reference parameter maps that were developed for the state of California, a comparative study between the full performance-based lateral spread [11] and the simplified performance-based lateral spread [14] was performed for ten different California sites at three return periods (475, 1033, and 2475 years) using two different site parameter values ($S = 9$, representing moderately high site susceptibility; and $S = 10$, representing moderately low site susceptibility). The selected sites and the corresponding mapped values of $\text{Log}(D_h)^{ref}$ are summarized in Table 1.

Using the reference lateral spread displacement parameter values presented in Table 1, lateral spread displacements were computed at each selected site, return period, and S value using the simplified performance-based approach using Equations and above. Full performance-based lateral spread displacements were then computed for each of the sites, return periods, and S values using *EZ-FRISK*. The purpose of this comparison is to observe how much error the map-based simplified performance-based approach introduces into the estimation of performance-based lateral spread displacements. A scatterplot presenting the results of the comparison is presented in Fig. 2 below.

Table 1. Selected California sites for validation study and corresponding reference lateral spread displacement values

Latitude (decimal degrees)	Longitude (decimal degrees)	City name	USGS 2,475-yr PGA (g)	Reference lateral displacement, $Log(D_h)^{ref}$		
				475yr	1033yr	2475yr
41.75	−124.407	Crescent city	1.29	1.03	1.32	1.49
40.806	−124.166	Eureka	1.49	1.17	1.41	1.56
40.591	−122.395	Redding	0.52	−0.74	−0.18	0.65
38.573	−121.511	Sacramento	0.26	−1.48	−1.28	−1.08
37.8	−122.4	San Francisco	0.72	0.62	0.79	0.92
37.353	−121.875	San Jose	0.92	0.30	0.53	0.69
36.756	−119.792	Fresno	0.28	−1.44	−1.14	−0.91
36.6	−121.888	Monterey	0.59	0.18	0.40	0.58
35.385	−119.035	Bakersfield	0.42	−0.34	−0.17	−0.02
35.264	−120.689	San Luis Obispo	0.46	−0.25	0.04	0.51
34.412	−119.691	Santa Barbara	0.98	0.71	0.93	1.07
34.011	−118.485	Santa Monica	0.83	0.11	0.46	0.70
33.984	−117.377	Riverside	0.73	0.47	0.63	0.76
32.789	−115.569	El Centro	1.03	0.18	0.30	0.41
32.713	−117.171	San Diego	0.71	−0.46	0.28	0.60

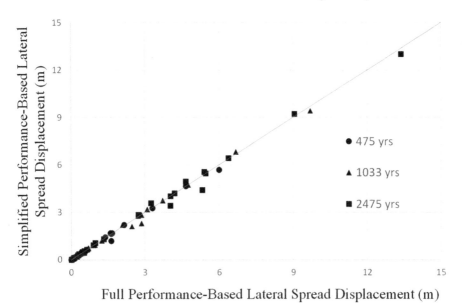

Fig. 2. Scatterplot comparisons between full performance-based lateral spread displacements and simplified performance-based lateral spread displacements for California sites

Analysis of Fig. 2 shows that the proposed simplified performance-based lateral spread prediction method [14] using the developed reference parameter maps for California provide a near-perfect approximation of the full performance-based lateral spread displacements [11] per the trendline providing a coefficient of correlation $R^2 = 0.98$ for all return periods and S values. This near-perfect correlation matches those observed in previous studies [11, 29], and is due to the linear derivation of the simplified performance-based approach. All observed scatter is due to interpolation error from the reference parameter map, but is generally less than 5%.

4 Conclusions

This study has developed, presented, and validated lateral spread reference parameter maps for the state of California at the return periods of 475, 1033, and 2475 years. These maps are to be used with the Ekstrom and Franke [14] simplified performance-based procedure for estimating lateral spread displacements. The validation study using fifteen California cities and two sets of site conditions (moderately high susceptibility, and moderately low susceptibility) demonstrates that the new reference parameter maps and corresponding simplified performance-based procedure are a practical and accurate approach to approximating probabilistic estimates of lateral spread displacement when site-specific geotechnical and topographical information is available.

References

1. Hamada, M., Towhata, I., Yasuda, S., Isoyama, R.: Study of permanent ground displacement induced by seismic liquefaction. Comput. Geotech. **4**, 197–220 (1987)

2. Bartlett, S.F., Youd, T.L.: Empirical prediction of liquefaction-induced lateral spread. J. Geotech. Eng. **121**(4), 316–329 (1995)
3. Bardet, J.-P., Tobita, T., Mace, N., Hu, J.: Regional modeling of liquefaction-induced ground deformation. Earthq. Spectra **18**(1), 19–46 (2002)
4. Youd, T.L., Hansen, C.M., Bartlett, S.F.: Revised multilinear regression equations for prediction of lateral spread displacement. J. Geotech. Geoenviron. Eng. **128**(12), 1007–1017 (2002)
5. Zhang, G., Robertson, P.K., Brachman, R.W.I.: Estimation liquefaction-induced lateral displacements using the standard penetration test or cone penetration test. J. Geotech. Geoenv. Eng. **130**(8), 861–871 (2004)
6. Faris, A.T., Seed, R.B., Kayen, R.E., Wu, J.: A semi-empirical model for the estimation of maximum horizontal displacement due to liquefaction-induced lateral spreading. In: Proceedings, 8th U.S. Nat. Conference Earthquake Eng., EERI, vol. 3, pp. 1584–1593. Oakland, CA, (2006)
7. Bazzurro, P., Cornell, C.A.: Nonlinear soil-site effects in probabilistic seismic-hazard analysis. Bull. Seismolog. Soc. America **94**(6), 2110–2130 (2004)
8. Kramer, S.L., Mayfield, R.T.: The return period of soil liquefaction. J. Geotech. Geoenviron. Eng. **133**(7), 802–813 (2007)
9. Franke, K.W., Wright, A.D.: An alternative performance-based liquefaction initiation procedure for the standard penetration test. In: Proc. Geo-Congress 2013, ASCE, pp. 846–849. Reston, VA (2013)
10. Franke, K.W., Wright, A.D., Ekstrom, L.T.: Comparative study between two performance-based liquefaction triggering models for the standard penetration test. J. Geotech. Geoenviron. Eng. **140**(5), 12 (2014)
11. Franke, K.W., Kramer, S.L.: A procedure for the empirical evaluation of lateral spread displacement hazard curves. J. Geotech. Geoenviron. Eng. **140**(1), 110–120 (2014)
12. Mayfield, R.T., Kramer, S.L., Huang, Y.M.: Simplified approximation procedure for performance-based evaluation of liquefaction potential. J. Geotech. Geoenviron. Eng. **136**(1), 140–150 (2010)
13. Ulmer, K.J., Franke, K.W.: Modified performance-based liquefaction triggering procedure using liquefaction loading parameter maps. J. Geotech. Geoenviron. Eng. **142**(3), 11p (2016)
14. Ekstrom, L.T., Franke, K.W.: Simplified procedure for the performance-based prediction of lateral spread displacements. J. Geotech. Geoenviron. Eng. **142**(7), 11p (2016)
15. EZ-FRISK homepage, Fugro USA Land, Inc. https://www.ez-frisk.com/. Accessed 27 Oct 2021
16. Field, E.H., Jordan, T.H., Cornell, C.A.: OpenSHA: a developing community-modeling environment for seismic hazard analysis. Seism. Res. Lett. **74**(4), 406–419 (2003)
17. Field, E.H., et al.: Synoptic view of the third uniform California earthquake rupture forecast (UCERF3). Seismol. Res. Lett. **88**(5), 1259–1267 (2017)
18. Petersen, M.D., et al.: Documentation for the 2008 Update of the Unites States National Seismic Hazard Maps. USGS Open File Report 2008–1128, U.S. Department of the Interior, p. 60 (2008)
19. Ulmer, K.J., Ekstrom, L.T., Franke, K.W.: Optimum grid spacing for simplified performance-based liquefaction and lateral spread displacement parameter maps. In: Proceeding of 6th International Conference on Earthquake Geotechnical Engineering, ISSMGE, London, U.K. (2015)
20. Shepard, D.: A two-dimensional interpolation function for irregularly-spaced data. In: Proceedings, 1968 ACM National Conference, pp. 517–524 (1968)
21. Khattar, R., Hales, R., Ames, D.P., Nelson, E.J., Jones, N., Williams, G.: Tethys app store: simplifying deployment of web applications for the international GEOGloWS initiative. Environ. Model. Softw. **146**, 105227 (2021)

22. Swain, N.R., et al.: A new open source platform for lowering the barrier for environmental web app development. Environ. Model. Softw. **85**, 11–26 (2016)
23. Dolder, D., Williams, G.P., Miller, A.W., Nelson, E.J., Jones, N.L., Ames, D.P.: Introducing an open-source regional water quality data viewer tool to support research data access. Hydrology **8**(2), 91 (2021)
24. Evans, S.W., Jones, N.L., Williams, G.P., Ames, D.P., Nelson, E.J.: Groundwater level mapping tool: an open source web application for assessing groundwater sustainability. Environ. Model. Softw. **131**, 104782 (2020)
25. Hales, R.C., Nelson, E.J., Williams, G.P., Jones, N., Ames, D.P., Jones, J.E.: The grids python tool for querying spatiotemporal multidimensional water data. Water **13**(15), 2066 (2021)
26. Sanchez Lozano, J., et al.: A Streamflow bias correction and performance eval-uation web application for geoglows ECMWF streamflow services. Hydrology **8**(2), 71 (2021)
27. Franke, K.W., Ulmer, K.J., Astorga, M.L., Ekstrom, L.T.: SPLiq: A new performance-based assessment tool for liquefaction triggering and its associated hazards using the SPT. In: Proceedings of Performance-Based Design in Earthquake Geotechnical Engineering III, ISSMGE, London, U.K. (2017)
28. Liquefaction Hazard App homepage. https://tethys.byu.edu/apps/lfhazard/. Accessed 10 Nov 2021
29. Franke, K.W., Youd, T.L., Ekstrom, L.T., He, J.: Probabilistic lateral spread evaluation for long, linear infrastructure using performance-based reference parameter maps. In: Proceedings of Geo-Risk 2017. ASCE, Reston, VA, USA (2017)

On a Novel Seismic Design Approach for Integral Abutment Bridges Based on Nonlinear Static Analysis

Domenico Gallese$^{(\boxtimes)}$ ⓘ, Davide Noè Gorini ⓘ, and Luigi Callisto ⓘ

Sapienza University of Rome, Rome, Italy
domenico.gallese@uniroma1.it

Abstract. This paper focuses on the seismic performance and design of a single-span integral abutment bridge (IAB), as a structural system characterised by a monolithic connection between deck and abutments. Although this is becoming a popular design solution due to its low maintenance requirements, there is still the need of developing robust design criteria for such structures under seismic conditions, mainly because of the complex soil-abutment-deck interaction. This study proposes an application of a novel design method for IABs to a reference case study inspired by a real integral bridge recently built in Italy. In the proposed method, the seismic capacity of the bridge is obtained through a nonlinear static analysis of the entire soil-structure system, in which the soil domain is perturbed by a distribution of equivalent forces aimed at reproducing the effects associated with the significant modes of the bridge. This approach is validated against the results of several dynamic analyses carried out on an advanced, full soil-structure model of the reference bridge implemented in OpenSees. Several seismic scenarios are taken into account, as well as the possibility to use an average response spectrum prescribed by technical provisions. This study demonstrates that the proposed design approach is able to reproduce quite satisfactorily the performance of the structure, in terms of maximum internal forces and displacements, with a very low computational demand.

Keywords: Integral abutment bridges · Seismic design · Nonlinear static analysis · OpenSees

1 Introduction

Integral abutment bridges (IABs) are structural systems characterised by an integral connection between deck and abutments (full transmission of forces and moments). This solution reduces to a minimum the maintenance of the bridge and the relative costs but can induce relevant internal stresses in the deck-abutment-embankment system, mainly due to thermal variations in the deck. A strategy aimed to minimise these internal stresses is to use abutment schemes characterised by a large deformability in the horizontal direction, for instance using a single row of piles as a foundation. Although numerous studies have been conducted on the static and dynamic behaviour of IABs, there is the

© The Author(s), under exclusive license to Springer Nature Switzerland AG 2022
L. Wang et al. (Eds.): PBD-IV 2022, GGEE 52, pp. 730–738, 2022.
https://doi.org/10.1007/978-3-031-11898-2_46

lack of standardized design guidelines around the world. In this regard, the present study is part of a wider research project on the dynamic response of integral abutment bridges aimed at devising simplified procedures readily applicable in seismic design. This paper illustrates a method based on the nonlinear static analysis of the entire soil-structure system, whose ability to predict the seismic performance of IABs is validated against the results of more demanding, nonlinear dynamic soil-structure analyses.

2 Reference Case Study

The reference case study is inspired by an integral single-span overpass recently built in Italy, along the A14 Adriatic highway, shown schematically in Fig. 1.a: the bridge deck is 50 m long and its cross section consists of a steel–concrete composite structure. The reinforced concrete abutments front wall presents a height of 8.0 m and width 2.2 m, and is supported by seven reinforced concrete piles, placed in a single row, with length and diameter of 20.0 m and 1.2 m, respectively.

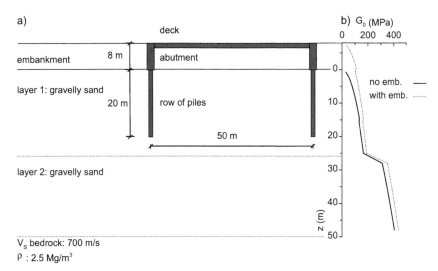

Fig. 1. (a) schematic longitudinal section of the case study; (b) profile of the small strain shear modulus G_0 with depth.

The idealised soil domain is constituted by two dry layers of gravelly sand with increasing stiffness and by a coarse-grained embankment. Figure 1.b shows the profile of the small-strain shear modulus with depth where the dashed line in the figure considers the increase in effective stresses produced by the embankment.

With the aim of investigating the dynamic response of the system, a numerical model was developed by Gallese [1] using the analysis framework OpenSees [2], while the mesh generation and the visualization of the results was performed through the pre/post-processor software GID. The focus is on the longitudinal response as the one significantly affected by dynamic soil-structure interaction [3, 4], and therefore only half of the soil-bridge system was modelled, as shown in Fig. 2.a, taking advantage of the symmetry about the vertical plane. The entire soil domain was discretized by using SSPbrick eight-noded hexahedral elements [5], with mechanical behaviour described by the Pressure Dependent Multi-Yield model developed by Yang et al. [6]. The abutment wall and the deck were modelled as assemblies of elastic beams with properties proportional to their respective influence areas. A linearly elastic behaviour was assigned to the piles, abutment walls and deck while the soil-pile and the soil-abutment contact are modelled by means of rigid links and thin layers of solid elements with reduced strength respect to the surrounding soil.

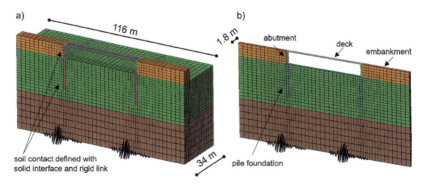

Fig. 2. (a) half of the 3D soil-bridge model used in the analysis and (b) equivalent 2D model developed in OpenSees.

An additional two-dimensional model was developed, as depicted in Fig. 2.b, where appropriate boundary conditions are employed to reproduce a plane strain deformation for the soil; plane stress conditions are instead considered for the structural elements. This 2D model considers a length of 1.8 m in the transverse, corresponding to the pile spacing, and incorporates a single pile, the abutments and the deck.

Particular attention was devoted to simulate the construction stages of the bridge. The procedure used herein, omitted for sake of brevity, follows the construction of the prototype, that was aimed at minimising the internal forces in the deck and particularly in the foundation piles under static conditions. The end of construction represents the starting point for the subsequent dynamic analyses as well as for the non-linear static analysis described in Sect. 5. The OpenSees parallel computing, performed by using the application OpenSeesSP [7], was employed to get reasonable computation times. Part of the analyses were carried out through the DesignSafe facility [8].

3 Seismic Input

The seismic hazard on stiff outcrop for the site of the bridge (Gatteo, Italy) is defined by elastic response spectra referring to soil type A, evaluated in accordance with Italian technical provisions (Italian Building Code 2018). In the present study, two seismic scenarios were investigated: Damage Limit State (DLS), with a probability of exceedance PR = 63%, and No-Collapse Earthquake (NCE), with PR = 5%.

Several seismic records were selected through the web-based PEER ground motion database (https://ngawest2.berkeley.edu/) as compatible seismic actions with the seismic demand discussed above. Figure 3.a shows the lower and upper envelopes and the average elastic response spectra of the selected records, relative to the longitudinal direction of the bridge, for the DLS and the NCE, for which 9 and 7 records were considered, respectively. Figure 3.b refers instead to the ground motion at the top of the embankment, evaluated separately through a free field site response analysis on a soil column including the embankment.

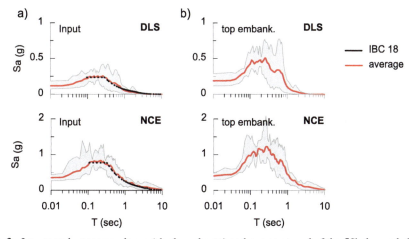

Fig. 3. Lower and upper envelopes (shadowed area) and average trend of the 5%-damped elastic spectra of the longitudinal components of the selected seismic records for the limit states DLS and NCE; (a) relative to the input motion (outcrop); (b) relative to the motion computed at the top of the embankment

4   Nonlinear Dynamic Analysis

After the construction stages, the reference soil-bridge models (Fig. 2) were subjected to the longitudinal component of the selected records. The seismic input was applied as a velocity time-history to the base of the soil domain through the interposition of viscous dampers to simulate a compliant bedrock, the latter reflecting the properties reported in Fig. 1. Periodic constraints were applied along the lateral boundaries, forcing the soil nodes placed on opposite sides at the same elevation to undergo the same displacements. Figure 4.a represents, for a selected seismic scenario belonging to the NCE, the time histories of the bending moment (for a unit deck width) in the deck-abutment node and pile head. As the 2D model was seen to provide similar results as the 3D model, it was taken as the reference for carrying out all the analyses discussed in the following, by virtue of its much higher computational efficiency.

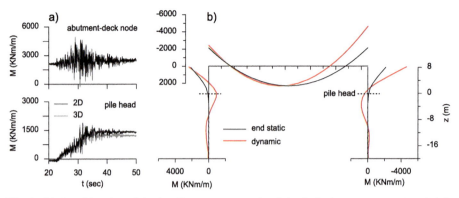

Fig. 4. (a) time histories of the bending moment at the right deck-abutment contact and right pile relative to a selected ground motion for the NCE; (b) bending moment in the structure at the end of static stage and in the instant when the maximum bending moment is reached on the right abutment.

Figure 4.b shows, for a selected ground motion belonging to the NCE, the bending moment in the structure in the instant when the maximum bending moment at the right deck-abutment contact is reached. The bending moment at the deck-abutment node elongates the top fibers of the deck and the internal fibers of the abutment. Conversely, at the top of the piles the bending moment elongates the external fibers (away from the embankment). With respect to the static conditions (black lines), increases of the bending moment are obtained at the deck-abutment contact and at the pile head, which constitute therefore the critical locations to consider for a direct assessment of the seismic performance.

5   Simplified method

As an efficient method readily applicable at the design stage, a simplified procedure is herein proposed, in which the Capacity Spectrum Method (CSM) [9] is used in conjunction with a nonlinear static analysis of the soil-bridge system to account for the nonlinear features of the mechanical behaviour. This method, commonly used for the seismic evaluation of structures, was recently extended to the case of geotechnical systems by Laguardia et al. [10]. To apply the method, two basic ingredients are needed, namely: the capacity curve of the system, and seismic demand, in the form of the acceleration-displacement elastic response spectrum (A-D response spectrum).

The capacity curve of the system is obtained by applying equivalent inertial forces to all nodes of the model. These forces are taken to be proportional to a seismic coefficient k_h, representing the ratio of the horizontal body forces to the unit weight of the soil. In the present work, the spatial distribution of k_h is aimed to reproduce the deformation patterns associated with the dominant vibration modes of the overall soil-bridge system that rule the dynamic response. To this end, a modal analysis was carried out on the 2D model referring to the small-strain elastic stiffness of the soil, that depends on the stress state. Figure 5.a shows the deformed shapes of the two dominant modes, the first and the fourth, maximising the internal stresses in the reference structural points (abutment-deck connection and the pile head). In particular, the fourth mode involves mainly the participation of the embankment (relative participating mass, M_{emb}, reported in Fig. 5.a) which has a strong influence on the structural performance as also demonstrated in some recent works [3, 4, 11].

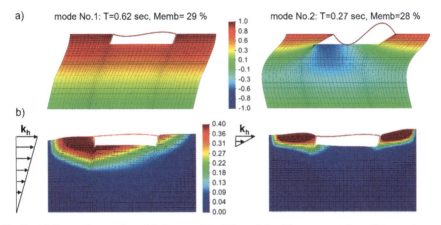

Fig. 5. (a) first and second modal shapes of the 2D model, with representation of the contours of the corresponding longitudinal displacements; (b) representation of the assumed profiles of the longitudinal seismic coefficient k_h and of the corresponding deformed shapes.

To reproduce the deformation patterns associated with these dominant modes, two linear profiles of k_h were assumed in the capacity analysis, as illustrated in Fig. 5.b: k_h varies linearly from the embankment top to a null value at the base of the soil domain and of the embankment, for the first and fourth mode, respectively. The resulting capacity curves of the soil-bridge system at hand are shown in Fig. 6.a, in which k_h and the horizontal displacement u refer to the top of the abutment.

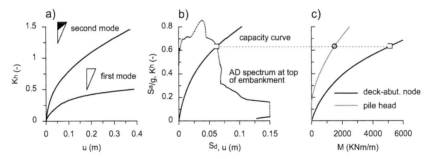

Fig. 6. (a) Capacity curves of the reference soil-structure system obtained in OpenSees; (b) evaluation of the performance point (intersection) with the proposed design method; (c) moment-seismic coefficient (k_h) curves and evaluation of the maximum forces in the structural elements.

The second ingredient of the method is the seismic demand defined by the A-D spectrum. In the proposed simplified method, a decoupled approach is employed in which the seismic demand derives from a one-dimensional site response analysis including the presence of the embankment. As in the original CSM, the equivalent damping ratio ξ of the A-D spectrum is found by an iterative procedure, i.e., by evaluating the damping ratio spectrum (using the Masing unloading–reloading rule) at the intersection of the capacity curve with the A-D, called *performance point*, and re-plotting the spectrum accordingly with the calculated damping. Once convergence on the value of ξ is attained, the actual performance point is located, as shown in Fig. 6.b–c: this provides the maximum acceleration, and the displacement of the abutment top. In turn, using the relationship between the relative bending moment and k_h obtained by the nonlinear static analysis (Fig. 6.c), the ordinate of the performance point can be used to determine the maximum internal forces in the structural elements of interest. To explore the validity of the method also when the seismic demand is provided by a technical-code spectrum, the same procedure was be applied using directly the average spectra of Fig. 3. In this case, the iteration on the damping ratio required a recalculation of the elastic response for all the ground motions, to evaluate the updated average spectrum.

6   Results and Concluding Remarks

The proposed design method is here applied to predict the seismic performance of the reference bridge of Fig. 1. To this end, Fig. 7 shows a systematic comparison between the results obtained with the proposed method (non-linear static analysis, NLSA) and with the full dynamic analysis (DYN) for the two limit states considered. Since the proposed method returns an estimate of the maximum effects induced by the seismic action, the results of the dynamic analyses refer only to the variations computed with respect to the static condition.

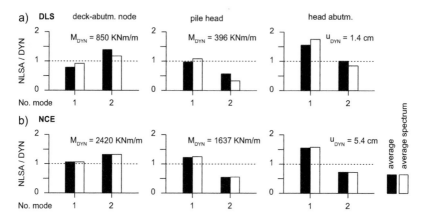

Fig. 7. Results of the proposed design method for the DLS (a) and NCE (b): ratios of the maximum bending moment at the deck-abutment node, pile head, and of the maximum displacement of the deck.

Three output quantities are considered: the maximum bending moments at deck-abutment node, and at the pile head, and the maximum deck displacement. For each output quantity the average value obtained with the dynamic analyses is reported above each plot. The application of the procedure considering the average spectrum (Fig. 3) is defined by the term "average spectrum", while the average of the results of the procedure applied to the individual ground motions is indicated as "average". It is evident that the two different ways to characterise the seismic demand do not influence significantly the average results yielded by the proposed method. Considering the simplicity and rapidity of the NLSA, the comparison with the results of the dynamic analyses is quite satisfactory. It appears that the internal forces at the deck-abutment node may be calculated using the capacity curve associated with the second mode, while the curve associated to the first mode may be used to evaluate the internal forces in the pile foundation. The first mode provides systematically greater displacements of the deck, while a limited underestimation is provided by the second mode solution.

These results demonstrate that the proposed design method may be efficiently used for a direct assessment of the seismic performance of single-span integral bridges. This requires the sole application of nonlinear static analysis methodologies, as a robust framework that, if properly validated, generalized and systematized, may be also extended for the seismic design of similar soil-structure systems.

References

1. Gallese, D.: Soil-structure interaction for the seismic design of integral abutment bridges: from advanced numerical modelling to simplified procedures. PhD thesis, Sapienza University of Rome (2022). https://doi.org/10.13140/RG.2.2.29620.32642
2. McKenna, F., Scott, M.H., Fenves, G.L.: OpenSees. nonlinear finite-element analysis software architecture using object composition. J. Comput. Civ. Eng. **24**, 95–107 (2010). https://doi.org/10.1061/(ASCE)CP.1943-5487.0000002
3. Elgamal, A., Yan, L., Yang, Z., Conte, J.P.: Three-dimensional seismic response of humboldt bay bridge-foundation-ground system. J. Struct. Eng. **134**(7), 1165–1176 (2008). https://doi.org/10.1061/(ASCE)0733-9445(2008)134:7(1165)
4. Gorini, D.N., Callisto, L., Whittle, A.J.: An inertial macroelement for bridge abutments. Géotechnique (2020). https://doi.org/10.1680/jgeot.19.P.397
5. McGann, C.R., Arduino, P., Mackenzie-Helnwein, P.: A stabilized single-point finite element formulation for three-dimensional dynamic analysis of saturated soils. Comput. Geotech. **66**, 126–141 (2015). https://doi.org/10.1016/j.compgeo.2015.01.002
6. Yang, Z., Elgamal, A., Parra, E.: Computational model for cyclic mobility and associated shear deformation. J. Geotech. Geoenviron. Eng. **129**, 1119–1127 (2003). https://doi.org/10.1061/(ASCE)1090-0241(2003)129:12(1119)
7. McKenna, F., Fenves, G.L.: Using the OpenSees Interpreter on Parallel Computers, https://opensees.berkeley.edu/ParallelProcessing.pdf. Accessed 24 Oct 2021 (2007)
8. Rathje, E., et al.: DesignSafe: a new cyberinfrastructure for natural hazards engineering. ASCE Nat. Haz. Rev. (2017). https://doi.org/10.1061/(ASCE)NH.1527-6996.0000246
9. Freeman, S.A.: Review of the development of the capacity spectrum method. ISET J. Earthq. Technol. **41**(1), 1–13 (2004)

10. Laguardia, R., Gallese, D., Gigliotti, R., Callisto, L.: A non-linear static approach for the prediction of earthquake-induced deformation of geotechnical systems. Bull. Earthq. Eng. **18**, 6607–6627 (2020). https://doi.org/10.1007/s10518-020-00949-2

11. Stefanidou, S.P., Sextos, A.G., Kotsoglou, A.N., Lesgidis, N., Kappos, A.J.: Soil-structure interaction effects in analysis of seismic fragility of bridges using an intensity-based ground motion selection procedure. Eng. Struct. **151**, 366–380 (2017). https://doi.org/10.1016/j.eng struct.2017.08.033

Influence of Seismic Displacement Models on Landslide Prediction at the Egkremnoi Coastline During the 2015 Lefkada Earthquake

Weibing Gong[1]([✉]), Dimitrios Zekkos[1], and Marin Clark[2]

[1] University of California, Berkeley, CA 94720-1710, USA
wbgong@berkeley.edu
[2] University of Michigan, Ann Arbor, MI 48109-1005, USA

Abstract. Strong earthquakes can trigger thousands of landslides in mountainous areas, and accurate prediction of landslide occurrence can reduce the risk of infrastructure and communities to earthquake-induced regional landslides. However, current methods for regional landslide prediction are still in their infancy and prediction of landslide occurrence remains a challenge. In this study, a pseudo-three-dimensional (pseudo-3D) procedure was implemented to a 1.35 km^2 area along the Egkremnoi coastline of the Lefkada island in Greece to assess the triggering of landslides and their geometry. To better understand the influence of permanent seismic displacement models on landslide prediction, seven models were employed in the prediction procedure. The adopted displacement models include four Newmark-type rigid block models and three flexible models. The results show that in this specific case and for the input parameters used in the analyses, the Bray and Macedo model achieves the best performance in landslide prediction in terms of the percentage of correctly predicted landslides and the centroid distance between mapped and predicted landslides, which are used to quantify the location accuracy of predicted landslides. The Rathje and Antonakos model predicts more landslides that overlap with mapped landslides, but also overpredicts more overall.

Keywords: Co-seismic landslides · Pseudo-3D prediction · Displacement model · Centroid distance · Correctly predicted ratio

1 Introduction

Major earthquakes can cause numerous landslides, which threaten not only infrastructure, but also people's lives and properties, and can affect communities for many years after the earthquake. Accurate prediction of co-seismic landslides within a region could assist in mitigating landslide risk. However, regional landslide prediction methods are not well developed primarily due to the significant computation costs associated with such analyses, the complexities in prediction model and the challenges in generating high-quality and reliable input data, such as topography, strength parameters, and ground

L. Wang et al. (Eds.): PBD-IV 2022, GGEE 52, pp. 739–746, 2022.
https://doi.org/10.1007/978-3-031-11898-2_47

motion intensity. Recent advances in our ability to generate high-resolution digital elevation models (DEMs) provide the needed input for regional landslide models with the goal to reliably estimate regional-scale material strength parameters, such as the cohesion and friction angle strength characteristics. Such models, properly calibrated, can be used for co-seismic landslide prediction of earthquake scenarios.

To reduce computational costs and simplify natural complexity, a Newmark-type pseudo-3D landslide prediction procedure [1, 2] was developed and implemented in this study. The major steps of the procedure include first the identification of triggering grid cells based on calculation of grid cell permanent seismic displacements due to certain earthquake excitation, and second the geomorphologic formation of landslides from the triggering grid cells. The first step is critical as it determines where predicted landslides occur. The calculated displacement of each triggering grid cell depends on the adopted seismic displacement model, which can significantly affect the landslide prediction accuracy. To assess the influence of seismic displacement models on the regional co-seismic landslide predictions, seven displacement models are adopted herein and the prediction results are compared to the mapped landslides that occurred during the 2015 earthquake in Lefkada, Greece. Comparisons among predictions are made using different metrics, i.e., number of predicted landslides, landslide area density, and accuracy in the prediction of landslide location.

2   Seismic Displacement Models

Seismic displacement models are typically divided into rigid models and flexible models. Rigid models assume that the sliding block is rigid, i.e., of infinite stiffness. Displacement prediction equations of rigid masses are simpler and require fewer parameters than flexible models since the dynamic response of the landslide mass does not need to be considered. Because the dynamic response of the landslide mass is ignored, such models are suitable for simulating shallow and stiff landslides. However, many landslides, especially deep landslides, are not infinitely stiff and the dynamic response cannot be ignored. Therefore, flexible models are better suited for landslide prediction procedure. The flexible models can be further categorized into uncoupled and fully coupled models. Uncoupled flexible models consider the dynamic response of the mass and the sliding displacement at the sliding location separately. Fully coupled flexible models consider the influence of the dynamic response of the sliding mass on the displacement along a sliding plane. Decoupled flexible models have the advantages of simpler prediction equations and fewer parameters than fully coupled flexible models that incorporate the dynamic response of the considered sliding mass. To explore the effects of displacement models on landslide prediction results, seven seismic displacement models are considered; four rigid models [3–5], two decoupled flexible models [6] and one fully coupled flexible model [7]. The specific displacement prediction equations and input parameters can be found in the relevant references [3–7]. Note that the fully coupled displacement model proposed by Bray and Macedo [7] is specifically developed for shallow crustal earthquakes, which is consistent with the 2015 Lefkada earthquake considered herein.

3 Input Data

The DEM used in the analysis is a 10-m DEM resampled from the 5-m DEM available by the Hellenic Cadastre. The peak ground acceleration (PGA) and peak ground velocity (PGV) maps are adopted from the USGS ShakeMap ATLAS 2 version [8] as this is the shaking model that also considers the recordings of the strong motion stations in the vicinity of the study region. The unit weight of the ground is 22 kN/m^3 [9]. The area of the study region is about 1.35 km^2. In addition, the whole region lies on the same geologic unit, i.e., Limestone Paxos Zone [10], and thus a pair of regional constant cohesion and friction angle, i.e., $c = 6$ kPa and $\varphi = 53°$ [1], is used as input. This estimate of strength was derived by linearly fitting the shear strength datapoints back-calculated from the mapped landslides in this region. No pore water pressures are considered as the ground was generally dry when the earthquake occurred. The triggering depth is set as 5 m, since this represents the average depth of most landslides based on the field observations [10]. Using these input parameters, landslide predictions using different seismic displacement models are made.

4 Results

4.1 Predicted Landslides

The landslides predicted by the Jibson (PGA, M$_w$) rigid (JR I) model [3], the Rathje and Antonakos (k$_{max}$, k-vel$_{max}$) uncoupled flexible (RAUC II) model [6] and the Bray and Macedo fully coupled flexible (BMFC) model [7] are used as examples for comparison with mapped landslides in Fig. 1. M$_w$ is the earthquake moment magnitude. k$_{max}$ is the maximum seismic coefficient and k-vel$_{max}$ is the maximum velocity of seismic coefficient-time history. A total of 68 co-seismic landslides were mapped along the Egkremnoi coastline. Figure 1(a) shows both the full areas and the source areas of the mapped landslides. Predicted landslides are compared to the source areas of mapped landslides. The BMFC model predicted the fewest landslides compared to the JR I model and the RAUC II model. The RAUC II model predicted the most landslides among the three displacement models. The JR I model predicted a moderate numbers of landslides, but many landslides were also predicted in the southern most part of the study area, where few landslides were observed.

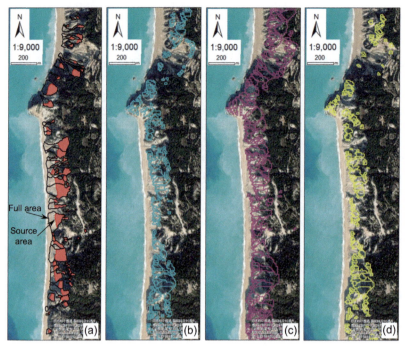

Fig. 1. (a) Mapped landslides and predicted landslides generated by (b) Jibson (PGA, M_w) model; (c) Rathje and Antonakos (k_{max}, k-vel$_{max}$) model; and (d) Bray and Macedo model

4.2 Number of Landslides and Area Density

The number of landslides and area density are two simple, but direct and important metrics to evaluate the landslide prediction procedures, as shown in Fig. 2. Note that each displacement model is listed with an abbreviated name based on the author names and the model types, as shown in Fig. 2(a). Overall, all seismic displacement models overpredict landslides. Figure 2(a) shows that the BMFC model predicts the fewest landslides, but still overpredicts the mapped landslides by a factor of 3.5. The Rathje and Antonakos (k_{max}, M_w) uncoupled flexible (RAUC I) model and the Rathje and Saygili (PGA&M_w) rigid (RSR) model have the greatest overprediction of landslides (by a factor of 5.2). Figure 2(b) illustrates the landslide area density generated by each displacement model. The area density generated by the BMFC model is closest to the mapped landslide area density and is 1.6 times the mapped landslide area. Although the RAUC I model and the RSR model generate the largest numbers of predicted landslides, the landslide area densities predicted by the RAUC II model and the Saygili and Rathje (PGA&PGV) rigid (SRR) model are highest and equal four times their mapped counterparts.

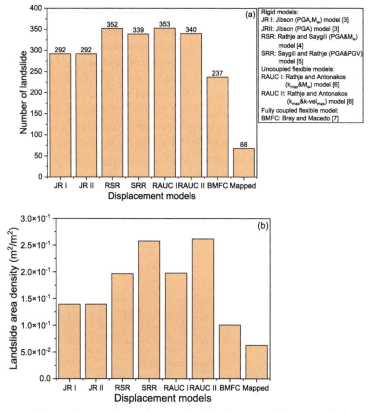

Fig. 2. Comparison of (a) number of landslide and (b) area density among different seismic displacement models and mapped landslides

4.3 Correctly Predicted Ratio and Overlapped Ratio

To further quantify the performance of the seismic displacement models, two parameters, correctly predicted ratio (CPR) and overlapped ratio (OR), are defined:

$$CPR = \frac{\text{Number of correctly predicted landslides}}{\text{Number of all predicted landslides}} \tag{1}$$

$$OR = \frac{\text{Number of mapped landslides overlapped by predicted landslides}}{\text{Number of all mapped landslides}} \tag{2}$$

In Eq. (1), we define the correctly predicted landslide number as the number of predicted landslides that overlap with mapped landslides by at least one grid cell. Thus, CPR reflects the location accuracy of predicted landslides. In Eq. (2), OR describes how many mapped landslides are correctly predicted (i.e., overlapped by predicted landslides) compared to the total population of mapped landslides. Both CPR and OR assess the landslide prediction accuracy and an ideal prediction model would result in CPR = 1 and OR = 1.

Figure 3 shows the CPRs calculated for different seismic displacement models. It is shown that BMFC model has the highest CPR, which indicates the greatest location accuracy of predicted landslides among all displacement models. The SRR model and the RAUC II model have somewhat lower CPR ratios. Figure 4 shows the ORs obtained from different seismic displacement models. It is found that the OR obtained from the BMFC model is lowest, which indicates that the BMFC model has the least overlap with mapped landslides. In comparison, the SRR model and the RAUC II model achieve the highest ORs, which is reasonable as these two models generated more predicted landslides than the BMFC model. The results indicate overall that higher ORs are accompanied by lower CPRs, indicating the challenges associated with accurate prediction of landsliding.

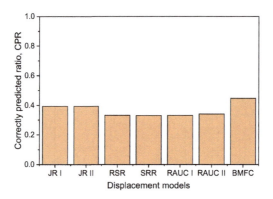

Fig. 3. Correctly predicted ratio (CPR) metric obtained from different seismic displacement models

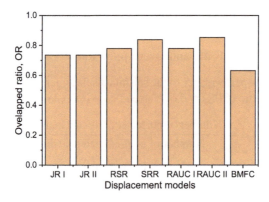

Fig. 4. Overlapped ratio (OR) metric obtained from different seismic displacement models

4.4 Centroid Distance

The centroid distance is the distance between the centroid of the predicted landslide and the centroid of the mapped landslide that is closest to the corresponding predicted landslide. Thus, the centroid distance can be used as a metric of location accuracy of predicted landslides. A large centroid distance means that the predicted landslide location accuracy is poor, while a small centroid distance indicates a high location accuracy of predicted landslides. Figure 5 shows that all models have relatively small centroid distances of about 10–20 m (median values), given that the grid cell size is 10 m, with the BMFC model having the lowest mean and median centroid distances. Note that the value of minimum centroid distance is zero for all seven displacement models.

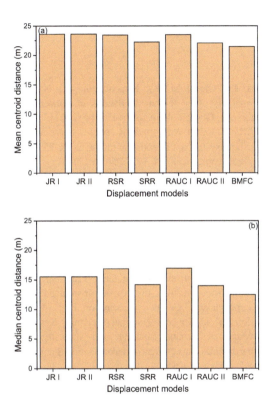

Fig. 5. Centroid distance obtained from different seismic displacement models: (a) mean and (b) median

5 Conclusions

Seven seismic displacement models, including four rigid models and three flexible models, were employed to explore their ability to predict co-seismic landslides for the Lefkada 2015 earthquake. Five metrics were used to assess the accuracy of prediction: number of landslides, area density, correctly predicted ratio, overlapped ratio and centroid distance. The results show that the Bray and Macedo model predicts the fewest landslides, and its landslide area density is closest to the mapped landslide density. The Bray and Macedo model also has the best performance in landslide prediction in terms of the correctly predicted ratio and the centroid distance. The Rathje and Antonakos (k_{max}, k-vel$_{max}$) model is best at generating predicted landslides that overlap the most with mapped landslides, which is quantified by the overlapped ratio, but also predicts more landslides overall.

Acknowledgements. The authors would like to show their sincere thanks to the financial supports provided by NASA under Grant No. 18-DISASTER18-0022 and by USGS National Earthquake Hazards Reduction Program (NEHRP) under Grant No. G17AP00088.

References

1. Gong, W., Zekkos, D., Clark, M. K., Kirschbaum, D.: Newmark-type pseudo-three-dimensional back-analysis of co-seismic landslides in Egkremnoi, Lefkada, Greece. In: Geo-Extreme Conference, Georgia, USA, pp. 405–414 (2021)
2. Gallen, S.F., Clark, M.K., Godt, J.W., Roback, K., Niemi, N.A.: Application and evaluation of a rapid response earthquake-triggered landslide model to the 25 April 2015 Mw 7.8 Gorkha earthquake, Nepal. Tectonophysics, **714**, 173–187 (2017)
3. Jibson, R.W.: Regression models for estimating coseismic landslide displacement. Eng. Geol. **91**(2–4), 209–218 (2007)
4. Rathje, E.M., Saygili, G.: Probabilistic assessment of earthquake-induced sliding displacements of natural slopes. Bull. N. Z. Soc. Earthq. Eng. **42**(1), 18–27 (2009)
5. Saygili, G., Rathje, E.M.: Empirical predictive models for earthquake-induced sliding displacements of slopes. J. Geotech. Geoenviron. Eng. **134**(6), 790–803 (2008)
6. Rathje, E.M., Antonakos, G.: A unified model for predicting earthquake-induced sliding displacements of rigid and flexible slopes. Eng. Geol. **122**(1–2), 51–60 (2011)
7. Bray, J.D., Macedo, J.: Procedure for estimating shear-induced seismic slope displacement for shallow crustal earthquakes. J.Geotech. Geoenviron. Eng. **145**(12), 04019106 (2019)
8. U.S. Geological Survey. Advanced National Seismic System (ANSS), ShakeMap, Global Region, Maps of ground shaking and intensity for event us10003ywp, GREECE. https://earthquake.usgs.gov/earthquakes/eventpage/us10003ywp/shakemap/pga?source=us&code=us10003ywp (2015)
9. Kallimogiannis, V., Saroglou, C., Zekkos, D., Manousakis, J.: 2D and 3D Back-analysis of a landslide in Egremnoi caused by the November 17 2015 Lefkada earthquake. In: 2nd International Conference on Natural Hazards & Infrastructure, Chania, Greece (2019)
10. Zekkos, D., Clark, M.: Characterization of landslides and rock mass strength leveraging the 2015 Mw 6.5 Lefkada Earthquake in Greece. Final Technical Report. U.S. Geology Survey. National Earthquake Hazards Reduction Program, Award Number G17AP00088 (2020)

Assessment of the Seismic Performance of Large Mass Ratio Tuned Mass Dampers in a Soil-Structure System

Davide Noè Gorini[1]([✉]) [ID], Guglielmo Clarizia[2], Elide Nastri[2] [ID],
Pasquale Roberto Marrazzo[2] [ID], and Rosario Montuori[2] [ID]

[1] Sapienza University of Rome, Rome, Italy
davideno.gorini@uniroma1.it
[2] University of Salerno, Fisciano, Salerno, Italy

Abstract. The use of Tuned Mass Dampers with a large mass ratio (LM-TMD) can represent an efficient means for seismic retrofitting of existing structures, intended as a superelevation of the pre-existing layout having a mass up to 20–30% of the structural mass. The mechanical properties of the LM-TMD should be designed to reproduce a proper dynamic coupling with the remaining structural system but there is still a limited knowledge about its effectiveness. Hence, this study presents a coupled numerical modelling of a soil-structure system equipped with a LM-TMD, implemented in OpenSees, as a benchmark to test the effectiveness of the non-conventional device. The full model simulates the nonlinear response of the soil-structure-TMD system and the multi-directionality of the seismic motion. The soil-structure interaction effects are quantified by comparing the results of the full model with those obtained in the case of structure with fixed-base. The results highlight the significant improvement of the structural performance when the LM-TMD is used, and the influence of soil-structure interaction on its effectiveness, as an additional source of energy dissipation not affected by the presence of the anti-seismic device.

Keywords: Soil-structure interaction · Nonlinear soil-structure-TMD model · Non-conventional TMD · Vibration control · OpenSees

1 Introduction

Tuned Mass Dampers (TMDs) are passive devices able to control structural vibration. In their basic configuration, a TMD is constituted of a mass, ranging between 1–5% of the structural mass, connected to the main structure through a linear spring and damper aimed at mitigating structural deformations under seismic and wind excitation. This allows for an effective seismic enhancement of existing structures and of the design of new ones. Through the years, several variations of the original solution have been proposed to increase the effectiveness and robustness of TMDs, such as large mass ratio TMDs [1], pendulum TMDs with friction [2], tuned inerter dampers [3] and steel frames with aseismic floors [4].

L. Wang et al. (Eds.): PBD-IV 2022, GGEE 52, pp. 747–754, 2022.
https://doi.org/10.1007/978-3-031-11898-2_48

Optimal tuning of a TMD is generally related to the significant natural frequencies of the structure. A widely used criterion was proposed by Den Hartog [5], which provided the optimal analytical solutions for the properties of a TMD used in a single-degree-of-freedom (SDOF) system subjected to harmonic excitation, neglecting the interaction with soil. Subsequently, several studies showed that the effectiveness of a Den Hartog-type TMD is no longer optimal when the linear SDOF approximation is not acceptable. This occurs, for instance, when soil-structure interaction cannot be neglected, for which the available solutions may lead to detuning the TMD from the dynamic response of the coupled soil-structure system. This was recognised for conventional TMDs through some analytical studies [9–12] and verified through experimental evidence [13, 14]. It was found that the relative soil-structure stiffness is a crucial factor affecting the effectiveness of conventional TMDs. More recently, Gorini and Chisari [15] showed that the TMD effectiveness is moreover controlled by other factors related to the foundation aspect ratio, the structural slenderness and the distribution of mass in the structural layout.

This work proposes an insight into the seismic performance of large mass ratio TMDs, LM-TMD, considering the effects associated with soil-structure interaction through advanced numerical modelling. A LM-TMD can indeed represent a valuable solution to control structural vibration. Recent studies carried out on fixed-base layouts showed that a LM-TMD can improve remarkably the structural performance compared to conventional TMDs [1] (mass ratio usually not greater than 3%). In the present study a three-dimensional, fully coupled soil-structure model equipped with a LM-TMD is analysed to point out the salient features controlling the seismic performance of this novel technology.

2   Case Study and Numerical Modelling in OpenSees

A moment resisting, reinforced concrete building resting on a homogeneous coarse-grained soil is taken as a reference in this study. The eight-floors structure has a constant inter-floor height of 3.3 m and represents an existing building dated down to the '70s, whose structural members were designed only for vertical loads according to an old non-seismic Italian Code (Italian Building Code, 1976). The LM-TMD consists of a superelevation of the existing structural layout, which has the same area in plan and develops for other two floors with the same height, reaching an additional mass equal to 20% of the initial structural mass.

A coupled numerical representation of the soil-structure-TMD system at hand was implemented in the analysis framework OpenSees [16]. The model, shown in Fig. 1, consists of 55627 finite elements, of which 53552 are solid elements representing the soil domain, while 1221 and 854 elements model the superstructure and the shallow foundations supporting the columns, respectively.

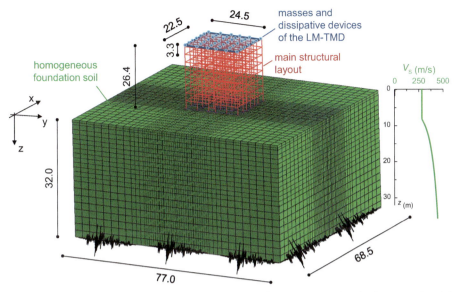

Fig. 1. Full soil-structure-TMD model implemented in OpenSees (lengths in meters), with representation of the profile of the shear wave velocity of soil, V_S, with depth.

2.1 Structural System

The structural members were modelled by using fibre-section force-based beam-column elements, reproducing the effective geometry and the highly nonlinear behaviour of the reinforced-concrete cross sections. The mass distribution of the LM-TMD was simply modelled as additional masses placed in correspondence of the top of the columns of the last floor. These masses are connected to the main structural system using a proper combination of elastomeric bearing elastic-plastic elements and flat slider bearing elements, available in the OpenSees library, as illustrated in Fig. 2. The properties of the dissipative devices were calibrated on the basis of the optimum vibration period of the tuning system according to the design abacuses reported in a previous work [17], relative to fixed-base structural systems.

The superstructure is supported by shallow foundations, which are connected by reinforced-concrete beams, reproduced by using shell elements and force-based beam-column elements, respectively, with elastic behaviour.

2.2 Subsoil

The reference building is assumed located in the high seismicity Pantano region (Messina Strait, Italy). The soil domain is inspired by the subsoil conditions of the region, studied in [18]: it is dry and composed of a homogeneous coarse-grained soil, reflecting the typical features of Messina Gravels, down to the deconvolution depth of the seismic input placed at 32 m of depth (see Sect. 2.3). The soil was discretised through SSPbrick hexahedral elements, available in the OpenSees library, whose behaviour was described by the Pressure Dependent, Multi-Yield hardening model (PDMY) developed by Yang et al. [19]. The model calibration refers to the one carried out in [18]. In particular, the stiffness properties are calibrated to reproduce the experimental profile of the small strain shear wave velocity shown in Fig. 1, whereas the friction angle and the cohesion are equal to 38° and 0 kPa, respectively.

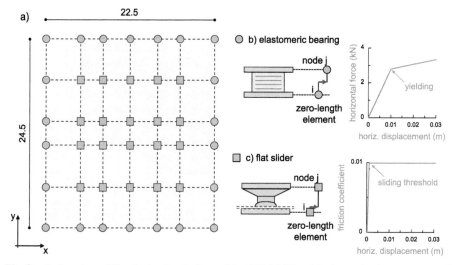

Fig. 2. a) plan view of the dissipative devices of the LM-TMD and b,c) constitutive responses of the elastomeric bearings and flat sliders, respectively.

2.3 Seismic Input

The seismic input for the soil-structure-TMD model was defined according to the seismic demand for the Pantano region [18]. This paper focuses on a Life Safety Limit State (return period of 475 years), whose design spectrum on stiff outcrop is represented in Fig. 3. Eight natural seismic records were selected through the PEER ground motion database (https://ngawest2.berkeley.edu/) as spectrum compatible seismic action. The average response spectrum of the horizontal components of the selected records is shown in Fig. 3. In the following, only the response relative to the Northridge Vasquez Rock Park (NVRP) record is analysed (blue line in Fig. 3).

For the site at hand, the bedrock is located at a depth of 475 m and a deconvolution procedure was carried out, analogous to the one performed in [18], to reduce the vertical extension of the soil-structure-TMD model. This depth resulted from a sensitivity study on a nonlinear, free field soil column, aimed at identifying an effective depth beyond which the soil response can be regarded as linear. In this manner, the propagation of the seismic waves from the bedrock up to the effective depth was studied through a free field site response analysis; then, the motion at the effective depth was applied to the nodes along the lower boundary of the coupled model. The spectra of the deconvoluted seismic motion are shown in Fig. 3. The propagated motion is generally de-amplified, expect for the peaks around 0.4 s and 1.0 s belonging to the spectral region of interest for the dynamic response of the superstructure (first and eighth vibration periods equal to 1.5 s and 0.3 s, respectively).

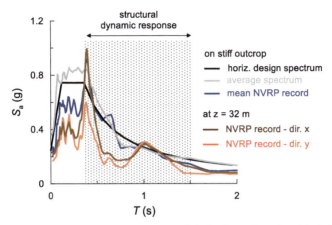

Fig. 3. Comparison between the design spectrum of the horizontal motion (black line), the 5%-damped elastic response spectra of the horizontal components of the reference NVRP record on stiff outcrop and at the effective depth of 32 m.

2.4   Analysis Procedure

A staged analysis procedure was adopted in which the construction sequence of the reference building was progressively simulated for a realistic reproduction of the stress state in the soil. In the dynamic stage, the longitudinal and transverse lateral boundaries were constrained to undergo the same motion since these are located far enough from the structure to ensure the free field response. The seismic input was applied to the base of the model as displacement time histories along directions x and y (see Fig. 1). The highly demanding dynamic analyses on the full model were carried out through the OpenSeesSP interpreter [20] for parallel calculation and by using the super-computing resources provided by the DesignSafe facility [21].

3   Seismic Performance and Concluding Remarks

The seismic performance of the LM-TMD was investigated through dynamic analyses in the time domain, considering the presence or not of the LM-TMD and comparing the results of the rigorous direct approach, described in Sect. 2, with the case in which soil-structure interaction is neglected. The reference models are: coupled soil-structure-TMD model with no TMD, CM-noTMD; coupled soil-structure-TMD model with the 20% mass ratio TMD, CM-TMD20; structural model with no soil-structure interaction effects and with no TMD, FB-noTMD; structural model with no soil-structure interaction effects and with the 20% mass ratio TMD, FB-TMD20. In the FB-models, the seismic input is constituted by the free field ground motion computed at the foundation level.

The LM-TMD effectiveness is expressed as the reduction of the maximum interstorey drift, MID, with respect to the case without TMD. Figure 4 shows the profiles of the MID along the elevation of the building for the four reference models subjected to the bi-component NVRP record. The MID tends to increase with the elevation, expect for the case of FB-noTMD for which the MID of floor 1 is considerable because of the lack of the soil compliance and of the mitigating effect of the LM-TMD. The MID is extremely high at floor 6 in all cases because of the activation of a plastic plane mechanism during the seismic event, as an expected consequence of the use of non-seismic technical provisions in the design of the existing building.

The LM-TMD is able to reduce the MID up to about 35% in the models with fixed base, with a comparable effect in the two horizontal directions. This affects the number of damaged elements in the structure, whose calculation is omitted for brevity, which reduces to about 35% in the systems equipped with the LM-TMD. The interaction with soil reduces the LM-TMD effectiveness of about 37%, because of the elastic-plastic behaviour of the foundation soils that alters the dynamic response of the structural system with a consequent partial detuning of the dissipative device. On the other hand, apart from floor 6 (plastic plane mechanism), soil-structure interaction concurs in mitigating structural vibration with a more pronounced effect in direction y, probably because of the lower severity of the y-component of the NVRP record causing more limited plastic strains in the structural members.

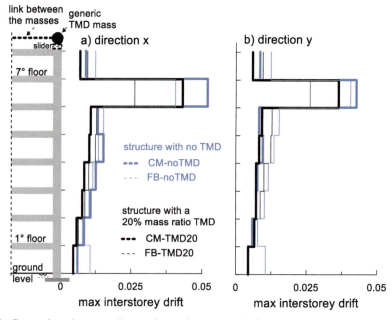

Fig. 4. Comparison between the maximum interstorey drifts in the x and y directions, a,b) respectively, obtained with the four reference models.

This mitigating effect of soil-structure interaction is partly emphasised by the non-linear response of the soil, that is not negligible at this limit state, as demonstrated by the post-earthquake configuration of the foundations in the coupled models described in Table 1. At the end of the seismic event, the structure undergoes an additional permanent translation in the horizontal plane, of about 0.045 m, impressed by the foundations (negligible rotation of the whole foundation system), as well as a uniform increase of the settlements of a factor equal to 2.6 (settlement under static conditions = 0.031 m), produced by the directional coupling of the plastic response of the soil. Therefore, no substantial permanent distortions of the superstructure are induced by the soil. These permanent effects are moreover not significantly altered by the presence of the LM-TMD (noTMD vs LM-TMD in Table 1).

Table 1. Permanent displacements u and rotation δ of the foundation system in the case with and without LM-TMD.

Quantity	no TMD	LM-TMD
$u_{1,x}$ (m)	0.036	0.035
$u_{1,y}$ (m)	0.026	0.026
$u_{1,z}$ (m)	0.081	0.084
δ_{xy} (°)	0.003	0.003
$u_{21,z}$ (m)	0.002	0.001
$u_{41,z}$ (m)	0.003	0.003

plan view   vertical section

In conclusion, these preliminary results point out some significant features of the seismic performance of LM-TMDs, as a favourable solution to mitigate structural vibration of existing buildings. In the case at hand, soil-structure interaction has a noticeable negative effect on the LM-TMD effectiveness, but, by contrast, reduces moderately the maximum structural deformations, aspects that should be considered in design. Vice versa, the response of the LM-TMD does not alter the seismic performance of the foundations. This seems to make valid the available frameworks for seismic assessment of conventional TMDs also to LM-TMDs, which will constitute the next step of this ongoing research towards the development of an optimised design criterion.

References

1. De Angelis, M., Perno, S., Reggio, A.: Dynamic response and optimal design of structures with large mass ratio TMD. Earthq. Eng. Struct. Dyn. **41**, 41–60 (2012). https://doi.org/10.1002/eqe.1117
2. Matta, E.: A novel bidirectional pendulum tuned mass damper using variable homogeneous friction to achieve amplitude-independent control. Earthq. Eng. Struct Dyn. **48**, 653–677 (2019). https://doi.org/10.1002/eqe.3153
3. Deastra, P., Wagg, D., Sims, N., Akbar, M.: Tuned inerter dampers with linear hysteretic damping. Earthq. Eng. Struct Dyn. **49**, 1216–1235 (2020). https://doi.org/10.1002/eqe.3287

4. Xiang, Y., Koetaka, Y., Nishira, K.: Steel frame with aseismic floor: from the viscoelastic decoupler model to the elastic structural response. Earthquake Engng. Struct. Dyn. **50**, 1651–1670 (2021). https://doi.org/10.1002/eqe.3419
5. Den Hartog, J.P.: Mechanical vibrations. In: Mc-Graw Hill (eds), New York (1956)
6. Mohebbi, M., Joghataie, A.: Designing optimal tuned mass dampers for nonlinear frames by distributed genetic algorithms. Struct. Des. Tall Special Build. **21**(1), 57–76 (2012)
7. Pourzeynali, S., Salimi, S., Kalesar, H.: Robust multi-objective optimization design of TMD control device to reduce tall building responses against earthquake excitations using genetic algorithms. Sci. Iranica **20**(2), 207–221 (2013)
8. Poh'sie, G.H., Chisari, C., Rinaldin, G., Fragiacomo, M., Amadio, C.: Optimal design of tuned mass dampers for a multi-storey cross laminated timber building against seismic loads. Earthq. Eng. Struct. Dyn. **45**(12), 1977–1995 (2016)
9. Farshidianfar, A., Soheili, S.: Ant colony optimization of tuned mass dampers for earthquake oscillations of high-rise structures including soil-structure interaction. Soil Dyn. Earthq. Eng. **51**(1), 14–22 (2013)
10. Bekdas, G., Nigdeli, S.M.: Metaheuristic based optimization of tuned mass dampers under earthquake excitation by considering soil-structure interaction. Soil Dyn. Earthq. Eng. **92**, 443–461 (2017)
11. Salvi, J., Pioldi, F., Rizzi, E.: Optimum tuned mass dampers under seismic soil-structure interaction. Soil Dyn. Earthq. Eng. **114**, 576–597 (2018)
12. Gorini, D.N., Chisari, C.: Effect of soil-structure interaction on seismic performance of tunes mass dampers in buildings. In: Earthquake Geotechnical Engineering for Protection and Development of Environment and Constructions, pp. 2690–2697 (2019). https://doi.org/10.1201/9780429031274
13. Jabary, R.N., Madabhushi, S.P.G.: Tuned mass damper effects on the response of multi-storied structures observed in geotechnical centrifuge tests. Soil Dyn. Earthq. Eng. **77**, 373–380 (2015)
14. Jabary, R.N., Madabhushi, S.P.G.: Tuned mass damper positioning effects on the seismic response of a soil-MDOF-structure system. J. Earthq. Eng. **22**(2), 281–302 (2017)
15. Gorini, D.N., Chisari, C.: Impact of soil-structure interaction on the effectiveness of Tuned Mass Dampers. Earthquake Eng. Struct. Dynam. **51**(6), 1501–1521 (2022). https://doi.org/10.1002/eqe.3625
16. McKenna, F., Scott, M.H., Fenves, G.L.: OpenSees. nonlinear finite-element analysis software architecture using object composition. J. Comput. Civ. Eng. **24**, 95–107 (2010). https://doi.org/10.1061/(ASCE)CP.1943-5487.0000002
17. Marrazzo, P.R.: Tuned Mass Damper for the Risk Mitigation of Existing R.C. Buildings, MSc Thesis, Department of Civil Engineering, University of Salerno (2020)
18. Gorini, D.N.: Soil-structure interaction for bridge abutments: two complementary macro-elements, PhD thesis. Rome, Italy: Sapienza University of Rome, https://iris.uniroma1.it/handle/11573/1260972 (2019)
19. Yang, Z., Elgamal, A., Parra, E.: Computational model for cyclic mobility and associated shear deformation. J. Geotech. Geoenviron. Eng. **129**, 1119–1127 (2003)
20. McKenna, F., Fenves, G.L.: Using the OpenSees Interpreter on Parallel Computers. https://opensees.berkeley.edu/ParallelProcessing.pdf (2007)
21. Rathje, E.M., et al.: Designsafe: a new cyber infrastructure for natural hazards engineering. Nat. Haz. Rev. **18**(3), 06017001-06017001 https://doi.org/10.1061/(ASCE)NH.1527-6996.0000246 (2017)

Required Strength of Geosynthetics in Seismic Reinforced Soil Retaining Wall in Multi-tiered Configuration

Shilin Jia[1,2], Fei Zhang[1,2($\boxtimes$)], Xiaoyi Lu[1,2], and Yuming Zhu[1,2]

[1] Key Laboratory of Ministry of Education for Geomechanics and Embankment Engineering, Hohai University, Nanjing 210098, China
feizhang@hhu.edu.cn

[2] Jiangsu Province's Geotechnical Research Center, Nanjing 210098, China

Abstract. Based on the limit equilibrium method, seismic stability analyses of the reinforced multi-tiered wall are conducted to calculate the required strength of geosynthetics. The available reinforcement tensile capacity against the front and rear-end pullout is used to determine the distribution of the required strength and the strength of connection to the facing for each layer. The calculated solutions are compared with the results obtained from the experiments and numerical simulations for the verification of the presented method. A series of parametric study is carried out to investigate seismic effects on the stability of reinforced multi-tiered wall and its failure mechanism. The positions of maximum value of the required tensile force are closer to the end rear of the reinforcements resulting in the pullout failures at the reinforcements of the uppermost wall. Increasing offset distance and tier number can yield less reinforcement strength required for seismic stability. The height of the uppermost wall can be selected as 0.6–0.7 times of the total height for a two-tiered wall seismic design.

Keywords: Reinforced soil · Retaining wall · Stability analyses · Limit equilibrium · Seismic design · Tiered configuration

1 Introduction

Geosynthetic-reinforced soil (GRS) retaining walls are widely used in the construction of mountainous highways, railroads, airports and other high-fill projects because of its superior performance, green ecology, convenient construction and area saving. As the wall height increases, more strength of the reinforcement is required to maintain the stability of GRS walls, which could make the construction more difficult and cost [1, 2]. In order to solve this problem, adopting the multi-tiered configuration and the load shedding effect brought by steps to ensure the stability of the high retaining wall becomes a feasible solution.

The post-earthquake investigations [3–6] of several GRS retaining walls showed that the seismic performance of GRS walls is satisfactory. Several shaking table tests [7–11] on GRS wall without tiered were performed to investigate their dynamic response and

© The Author(s), under exclusive license to Springer Nature Switzerland AG 2022
L. Wang et al. (Eds.): PBD-IV 2022, GGEE 52, pp. 755–761, 2022.
https://doi.org/10.1007/978-3-031-11898-2_49

behavior. Safaee et al. [12] conducted shaking table tests on multi-tiered GRS wall, and the results showed that multi-tiered configuration is good for seismic performance of GRS walls. Similar results had been obtained from the numerical method by Liu et al. [13]. This study extended the analytical method of Han & Leshchinsky [14] from the simple GRS wall to the multi-tiered configuration. The seismic effects are involved in the stability analyses to determine the required strength of reinforcement.

2 Stability Analysis Method for Multi-tiered GRS Retaining Wall

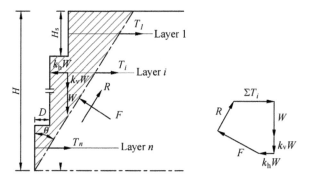

Fig. 1. Notation and forces of a multi-tiered GRS wall

A baseline model of multi-tiered GRS wall is used here, as shown in Fig. 1. The height of the wall denotes H and the number of the tier denotes N. To formulate the presented problem, the same assumptions are made as Han & Leshchinsky [14]. Assuming that the angle between failure surface and the vertical wall surface is θ, the forces on the shaded wedge-shaped sliding body include: the self-weight of the sliding mass W, normal force F, the shearing resistance R, horizontal reinforcement tensile load T_i (i indicates the i-th layer of reinforcement), horizontal seismic load $k_h W$, vertical seismic load $k_v W$. k_h and k_v are horizontal and vertical seismic acceleration coefficients respectively. Establishing the equilibrium equations along the horizontal and vertical directions can obtain the expression of the total reinforcement tensile load required for stability, as

$$\sum T_i = \left(\frac{1}{2}H^2 \tan\theta - DH_s\frac{N(N-1)}{2}\right)\gamma\left(\frac{k_h}{R_c} + \frac{1 + k_v}{R_c}\frac{\cos\theta - \sin\theta\tan\varphi}{\sin\theta + \cos\theta\tan\varphi}\right) \quad (1)$$

where R_c = coverage ratio of reinforcement (varying between 1.0 for full coverage to 0 for no reinforcement, here $R_c = 1.0$); D = offset distance; H_s = height of the single tier; $= \gamma$ unit weight of the backfill; φ = friction angle of the backfill.

In order to determine the distribution of the reinforcement tensile load along the laying length, the available front and rear-end pullout resistance should be considered, and calculated following Han and Leshchinsky [14], as

$$T_{(f)-i} = T_{o-i} + 2x_i\sigma_i C_i R_c \tan\varphi \quad (2)$$

$$T_{(e)-i} = 2(L_i - x_i)\gamma z_i C_i R_c \tan \varphi \qquad (3)$$

where T_{o-i} = available connection strength between reinforcement layer i and wall facing; x_i = distance to a slip surface from the wall facing along reinforcement layer i; z_i = height of overburden soil; C_i = the interaction coefficient between reinforcement and reinforced fill, $C_i = 0.8$; L_i = total length of the reinforcement layer i.

3 Comparisons

Safaee et al. [12] conducted shaking table tests using multiple seismic waves to compare the seismic performance of single-tiered walls with that of multi-tiered walls. The peak ground acceleration (PGA) equals 0.9 g at the time of failure damage of the wall. According to the suggestion on relation of the horizontal acceleration coefficient k_h and the peak ground acceleration (PGA) from FHWA [15], the k_h approximately equal to 0.45. For the same model, using the present method can obtain the position of maximum value of the required tensile force (T_{max}) for each layer, as shown in Fig. 2. The failure surface derived by the earth pressure theory of FHWA [15] is also given. The T_{max} position indicate a composite failure mechanism occurred in the GRS wall. The predicated failure surface from the T_{max} position is in good agreement with the failure surface obtained by Safaee et al. [12], but deeper than that of FHWA method.

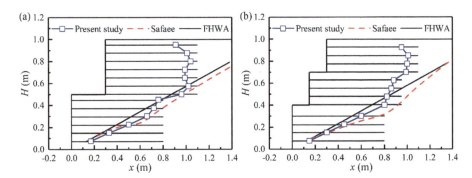

Fig. 2. Comparisons of the potential failure surfaces: (a) $N = 1$; (b) $N = 2$

4 Results and Discussions

Using the presented method, a baseline case is given here as an example: a two-tiered GRS wall with reinforcement space (distance between adjacent reinforcements) $S_v = 0.4$ m, backfill friction angle $\phi = 34°$, backfill unit weight $\gamma = 20$ kN/m³, upper wall height $H_1 = 6$ m, lower wall height $H_2 = 6$ m, offset distance $D = 2$ m, reinforcement length $L = 0.8\ H$, horizontal seismic acceleration coefficient $k_h = 0.2$. A parametric study (e.g., horizontal acceleration coefficient, tier number and upper wall height) is carried out to investigate their influences on the seismic behavior of the multi-tiered configuration at a limit state.

4.1 Horizontal Acceleration Coefficient

Figure 3a shows the effects of the seismic acceleration on T_o and T_{max} for each reinforcement layer of tiered GRS wall. The facing connection strength T_o at each layer is very small but increases abruptly at the step. For $k_h = 0.1$, the required strength T_{max} is uniform. As expected, the required T_{max} increases with the seismic acceleration k_h. However, the T_{max} of the lower reinforcements largely increase. As shown in Fig. 3b, the position of T_{max} is closer to the rear end of the reinforcement and then yields more strength T_{max} required at the lower layers. As k_h increases, the critical slip surface changes from the internal failures to the composite failures.

4.2 Tier Number

Figure 4a shows the effects of the tier number on T_o and T_{max} for each reinforcement layer of tiered GRS wall under seismic condition. The load shedding effect due to the tier provide more contributions on seismic stability. As the tier number N increases, the required T_{max} reduces. Similar to the case of $N = 2$, the value of T_o increases abruptly at each step for $N = 3$. Figures 4b–d illustrate the positions of T_{max} for GRS wall with tier number $N = 1$, 2 and 3, respectively. Increasing the tier number could yield the seismic wall more stable.

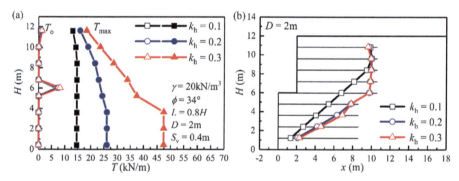

Fig. 3. Influence of horizon acceleration coefficient on T_o and T_{max}: (a) Values of T_o and T_{max}; (b) Position of T_{max}

4.3 Height of the Uppermost Tiered Wall

Selecting various heights of the uppermost tiered wall calculates the required reinforcement strength T_{max}. Figures 5a and 5b illustrate the variation of H_1/H on the maximum T_{max} for each layer. The value of $\max(T_{max})$ reduces gradually with increasing ratio of H_1/H and then suddenly decreases. It indicates that, there is a suitable height H_1 selected for seismic stability, especially for large magnitude of the seismic acceleration. As shown in Fig. 5b, increase of the offset width could yield less $\max(T_{max})$ and larger abrupt decrease. Figure 6 illustrates the distribution of T_{max} for the uppermost tiered

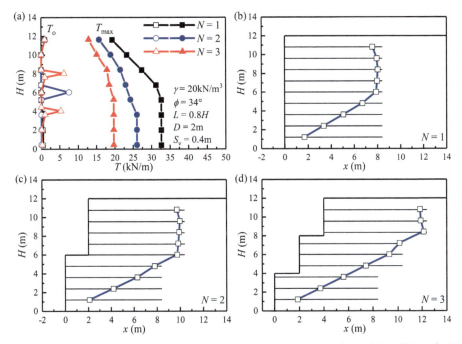

Fig. 4. Influence of tier number on seismic stability: (a) T_o and T_{max}; (b) position of T_{max} for $N = 1$; (c) position of T_{max} for $N = 2$; (d) position of T_{max} for $N = 3$

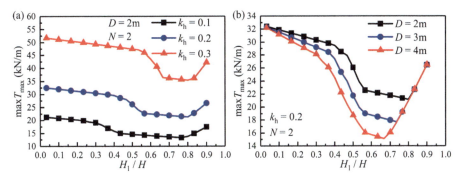

Fig. 5. Variation of max (T_{max}) with H_1/H: (a) Horizontal seismic coefficients; (b) Tier number

wall at $H_1/H = 0.3$, 0.5, 0.6 and 0.8. For $H_1/H = 0.3$ and 0.5, the positions of T_{max} at the uppermost tier are closer to the end rear of the reinforcement, which could yield pullout failures. As the uppermost wall height increases, more tension resistances of the reinforcements at the upper layers are mobilized. It is recommended to select the two-tiered GRS wall with unequal height of the upper and lower walls, and the height ratio of the upper wall prefers 0.6–0.7.

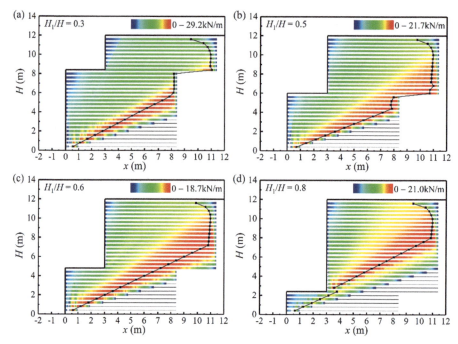

Fig. 6. Reinforcement load distribution: (a) $H_1/H = 0.3$; (b) $H_1/H = 0.5$; (c) $H_1/H = 0.6$; (d) $H_1/H = 0.8$

5 Conclusions

This paper proposed an analytical method for calculating the reinforcement tension strength required for the seismic stability of multi-tiered GRS retaining walls. Combining with the available front and rear-end pullout resistances, the distribution of tension strength and facing connect strength are determined at each layer. The following conclusions are made based on the results, as

1. As the seismic acceleration increases, more reinforcement strength are required for stability of tiered walls. The positions of T_{max} are closer to the end rear of the reinforcements, which results in the pullout failures at the reinforcements of the uppermost wall.
2. Increasing the tier number of wall and offset width yield less reinforcement strength required for seismic stability.
3. The height of the uppermost wall has significant influences on the seismic stability of tiered GRS walls, especially for large magnitude of the earthquakes. For a two-tiered GRS wall, the height of the uppermost wall can be selected as 0.6–0.7 times of the total height for seismic design.

References

1. Koerner, R.M.: Design with Geosynthetics. Fifthed. Prentice-Hall Inc, Englewood Cliffs, New Jersey (2010)
2. Yoo, C., Song, A.R.: Effect of foundation yielding on performance of two-tier geosynthetic reinforced segmental retaining walls: a numerical investigation. Geosynth. Int. **20**(3), 110–120 (2007). https://doi.org/10.1680/gein.2006.13.5.181
3. Sandri, D.: A performance summary of reinforced soil structures in the Greater Los Angeles area after the Northridge earthquake. Geotext. Geomembr. **15**(4–6), 235–253 (1997). https://doi.org/10.1016/S0266-1144(97)10006-1
4. Huang, C.C., Chou, L.H., Tatsuoka, F.: Seismic displacement of geosynthetic reinforced soil modular block walls. Geosynth. Int. **10**(1), 2–23 (2003). https://doi.org/10.1680/gein.10.1.2.37160
5. De-Pei, Z., Jian-Jing, Z., Zhang, Y.: Seismic damage analysis of road slopes in Wenchuan earthquake. Chin. J. Rock Mechan. Eng. **29**(3), 565–576 (2010). https://doi.org/10.1016/S1876-3804(11)60004-9
6. Chen, Q., Yang, C.W., Jian-jing, Z., Zhou, X.W.: Study on failure mechanism of high-large reinforced earth retaining wall during 5.12 Wenchuan earthquake. Railway Eng. (in Chinese) **9**, 73–77 (2010). https://doi.org/10.3969/j.issn.1003-1995.2010.09.023
7. Xu, P., Hatami, K., Jiang, G.: Study on seismic stability and performance of reinforced soil walls using shaking table tests. Geotext. Geomembr. **48**(1), 82–97 (2020). https://doi.org/10.1016/j.geotexmem.2019.103507
8. Panah, A.K., Yazdi, M., Ghalandarzadeh, A.: Shaking table tests on soil retaining walls reinforced by polymeric strips. Geotext. Geomembr. **43**(2), 148–161 (2015). https://doi.org/10.1016/j.geotexmem.2015.01.001
9. Yazdandoust, M.: Investigation on the seismic performance of steel-strip reinforced-soil retaining walls using shaking table test. Soil Dyn. Earthq. Eng. **97**, 216–232 (2017). https://doi.org/10.1016/j.soildyn.2017.03.011
10. El-Emam, M.M., Bathurst, R.J.: Influence of reinforcement parameters on the seismic response of reduced-scale reinforced soil retaining walls. Geotext. Geomembr. **25**(1), 33–49 (2007). https://doi.org/10.1016/j.geotexmem.2006.09.001
11. Yazdandoust, M.: Experimental study on seismic response of soil-nailed walls with permanent facing. Soil Dyn. Earthq. Eng. **98**, 101–119 (2017). https://doi.org/10.1016/j.soildyn.2017.04.009
12. Safaee, A.M., Mahboubi, A., Noorzad, A.: Experimental investigation on the performance of multi-tiered geogrid mechanically stabilized earth (MSE) walls with wrap-around facing subjected to earthquake loading. Geotext. Geomembr. **49**(1), 130–145 (2021). https://doi.org/10.1016/j.geotexmem.2020.08.008
13. Liu, H., Yang, G., Ling, H.I.: Seismic response of multi-tiered reinforced soil retaining walls. Soil Dyn. Earthq. Eng. **61**, 1–12 (2014). https://doi.org/10.1016/j.soildyn.2014.01.012
14. Han, J., Leshchinsky, D.: General analytical framework for design of flexible reinforced earth structures. J. Geotech. Geoenviron. Mental Eng. **132**(11), 1427–1435 (2006). https://doi.org/10.1061/(ASCE)1090-0241(2006)132:11(1427)
15. Berg, R.R., Christopher, B.R., Samtani, N.C.: Mechanically Stabilized Earth Walls and Reinforced Soil Slopes Design and Construction Guidelines. Federal Highway Admin-istration, Washington, D.C. (2009)

Study on Design Method of Vertical Bearing Capacity of Rock-Socketed Piles Based on Reliability

Zhongwei Li, Hanxuan Wang, Guoliang Dai[✉], and Fan Huang

Southeast University, Nanjing, China
daigl@seu.edu.cn

Abstract. The traditional vertical bearing capacity design method of rock-socketed piles uses the constant safety factor K to assess the reliability of pile foundations, and this method leads to the high cost or the insufficient safety reserve of some pile foundations. The Bayesian optimization estimation was used to process the complete data of 144 rock-socketed piles, 65 sets of skin resistance data and 31 sets of end resistance data in rock-socketed segment. The analysis of the probability limit state design method was carried out for the five partial resistance coefficients: the whole pile, the skin resistance of soil section, the resistance of rock section, the skin resistance of the rock section and the end resistance of the rock section. The relationship between the partial coefficients, embedment ratio and uniaxial saturation compressive strength of rock was established, and the suggested partial coefficients was proposed. Finally, the partial coefficients were introduced into the calculation formula of the vertical bearing capacity of rock-socketed piles in the *Code for Design of Ground and Foundation of Highway Bridge and Culvert*. This formula can provide reference for further revision and improvement of existing normative practices.

Keywords: Rock-socketed pile · Reliability · Design method · Bayesian optimization

1 Introduction

The traditional method of designing the vertical bearing capacity of rock-socketed piles is mainly the constant value design method, using the safety factor K to assess the reliability of the capacity of piles. However, this generalized and fixed safety factor determined by experience does not guarantee that each pile has the same reliability level, and that causes vagueness of rock-socketed pile failure probability, uncertainty of safety factor meaning, ambiguity of each parameter value and too high or too low design safety reserve. Therefore, it is very necessary to introduce the reliability theory into the design of rock-socketed piles in combination with the traditional method.

In the reliability analysis of pile foundations, Ochiai et al. [1] used the number of standard penetrations to determine pile side friction coefficients, pile end resistance and to analyze the reliability of vertical bearing capacity of bored friction piles. Hansen

L. Wang et al. (Eds.): PBD-IV 2022, GGEE 52, pp. 762–772, 2022.
https://doi.org/10.1007/978-3-031-11898-2_50

et al. [2] analyzed the design program of reliability. Along with the foundation and development of the above reliability theory, the foundation design method based on reliability analysis has gradually become popular all over the world, such as the LRFD the American highway foundation code compiled by Barker et al. [3] and Canadian national building code by Becker [4]. These codes have further applied the reliability analysis theory to pile foundation design and made the industry more standardized. In China, Zheng [5] conducted a series of studies based on reliability theory, pile foundation testing, Bayesian estimation of resistance coefficient and vertical bearing capacity analysis. However, researches in the design method of vertical bearing capacity of rock-socketed piles based on reliability are rare, and the codes still take the safety factor method as the design basis.

In this study, the probability limit state design method was used as the basis, the selected target reliability was integrated with Chinese and American codes, the statistical variables were processed using Bayesian optimization, and the JC method and recursive method were used to calculate each partial coefficients considering the uniaxial saturated compressive strength of the rock mass and the embedded ratio. Thereby, the design method was further optimized, and the new expression of bearing capacity design was proposed by combining the empirical parameters and design habits of traditional design methods.

2 Engineering Data Processing Based on Bayesian Optimization Estimation

The mean and standard deviation of multiple random variables need to be used in the process of calculating the reliability index, while the engineering data collected in this study are selected from different projects and cover various types of geological conditions. Therefore, the resistance factor [6, 7] is introduced to normalize the test results and the calculated. The resistance factor is shown in Eq. (1).

$$\lambda_R = \frac{Q}{Q_c} \tag{1}$$

where λ_R is the resistance factor, Q is the limit value of pile bearing capacity (kN); Q_c is the limit value of monopiles bearing capacity calculated by the code's method. Through many scholars' research [8] it is can be known that the resistance factor is a random variable, and its probability distribution obeys lognormal distribution. Because in actual engineering, some test data have large deviations, which have a large impact on the calculation results, so Zheng et al. [9] proposed to use Bayesian optimization to process the sample data, eliminate the bad data and use the other data to find out the mean and variance of the target statistical variables, and first construct the deviation factor according to Eq. (2).

$$\zeta_i = \frac{|\lambda_{Ri} - \mu_{\lambda R}|}{\mu_{\lambda R}}, \ 1 \le i \le n \tag{2}$$

where λ_{Ri} is the ith resistance factor, ζ_i is the deviation factor, and $\mu_{\lambda R}$ is the mean value of the resistance factor. The statistics of many resistance factors are then classified according to ζ_i: (1) when $\zeta_i < 0.25$, the measured data are close to the actual value and are defined as "good data"; (2) when $0.25 \leq \zeta_i < 0.5$, they are defined as "general data"; (3) when $\zeta_i \geq 0.5$. defined as "bad data". After classifying the resistance factors, the mean and variance of the resistance factors can be calculated according to Eq. (3).

$$\mu_{up} = \exp(\mu_u + 0.5\sigma_u^2)\,\sigma_{up}^2 = \mu_{up}^2[\exp(\sigma_u^2) - 1]\,\mu_u = \frac{\mu_P\sigma_L^2 + \mu_L\sigma_P^2}{\sigma_L^2 + \sigma_P^2}\,\sigma_u^2 = \frac{\sigma_L^2\sigma_P^2}{\sigma_L^2 + \sigma_P^2} \tag{3}$$

where μ_{uP}, μ_u, μ_P, μ_L are the lognormal, posterior, prior, and likelihood distribution means respectively. σ_{uP}, σ_u, σ_P, σ_L are the lognormal, posterior, prior, and likelihood distribution standard deviations. The original test data is known to be divided into "good data", "general data" and "bad data". The "bad data" were discarded. The remaining "good data" is taken as the likelihood distribution data, and the "general data" is taken as the a priori distribution data, which is brought into the Eq. (3) to calculate the mean and variance of the resistance factor. The mean of resistance factor $\mu_{\lambda R}$ is 1.372 and the variance $\sigma^2_{\lambda R}$ is 0.068.

3   Reliability Calculation Method Based on JC Method

The JC method is a practical method for approximating the reliability index, which uses the Rackwitz-Fiessler method [10] for standard normalization of the non-normal random variables in the limit state equation, and then simplifies the solution process by using Taylor's formula, and finally approximates the reliability index of the structure by assuming check point and iterative calculations.

3.1   Calculation Process

According to the definition, the reliability index of the structure is the ratio of the mean value of the function to the standard deviation i.e., $\beta = \mu_z/\sigma_z$. In the calculation of the reliability index of the monopile bearing capacity, the following limit state equation (Eq. (4)) is constructed.

$$g = R - G - Q = 0 \tag{4}$$

where R is the monopile resistance, G is the permanent load and Q is the variable load. Since R, G and Q are random variables and have many influencing factors. It is inconvenient to obtain statistical parameters, so introduce the Eq. (5) into Eq. (4) to obtain Eq. (6)[11].

$$\lambda_R = \frac{R}{R_k}\,\lambda_G = \frac{G}{G_k}\,\lambda_Q = \frac{Q}{Q_k}\,\rho = \frac{Q_k}{G_k}\,\alpha_R = \frac{1}{\gamma_R}R_k\alpha_R = G_k\gamma_G + Q_k\gamma_Q \tag{5}$$

$$g = (\lambda_R/\alpha_R)\,(\gamma_G/\rho + \gamma_Q) - (\lambda_G/\rho + \gamma_Q) = 0 \tag{6}$$

$$\begin{cases} \mu_{\lambda R} = 1.372 \; \mu_{\lambda G} = 1.08 \; \mu_{\lambda Q} = 1.115 \\ \delta_R = 0.068 \quad \delta_G = 0.13 \quad \delta_Q = 0.18 \\ \gamma_R = 2.5 \qquad \gamma_G = 1 \qquad \gamma_Q = 1 \end{cases} \tag{7}$$

where λ_R, λ_G, λ_Q are the resistance factor, constant load factor, variable load factor respectively; R_k, G_k, Q_k are the standard values of resistance, constant load and variable load used in the code calculation; γ_R, γ_G, γ_Q are the partial coefficients of resistance, constant load, and variable load.

The JC method performs a normal approximation in the end of a non-normally distributed random variable. The equivalent normalization condition for the JC method is that X_i and $X_i{}'$ comply with Eq. (8) at the test point $x_i{}^*$, and then Eq. (9) can be obtained.

$$f_{X_i'}(x_i^*) = \frac{1}{\sigma_{X_i'}} \varphi\left(\frac{x_i^* - \mu_{X_i'}}{\sigma_{X_i'}}\right) = f_{X_i}(x_i^*) \; F_{X_i'}(x_i^*) = \Phi\left(\frac{x_i^* - \mu_{X_i'}}{\sigma_{X_i'}}\right) = F_{X_i}(x_i^*) \tag{8}$$

$$\mu_{X_i'} = x_i^* - \Phi^{-1}[F_{X_i}(x_i^*)]\sigma_{X'} \quad \sigma_{X'} = \frac{\varphi\{\Phi^{-1}[F_{X_i}(x_i^*)]\}}{f_{X_i}(x_i^*)} \tag{9}$$

where $\varphi(\cdot)$ is the standard normal distribution probability density function and $\Phi^{-1}(\cdot)$ is the inverse function of the standard normal cumulative distribution. By using Eq. (8) the mean and variance of the λ_R, λ_G, λ_Q equivalent normalized quantities can be accessed.

The reliability index can be found by iterating through the determined means and standard deviations of the basic variables and the iterative process is as follows: (1) Assume the initial test point $\boldsymbol{x}^*$, generally set $x_i{}^* = \mu_i$; (2) Calculate the sensitivity coefficient α_i, as shown in Eq. (9); (3) Solve the reliability index β; (4) Calculate the new test point $\boldsymbol{x}^*$, using Eq. (11). (5) If the iteration termination condition is met (the new $\boldsymbol{x}^*$ is close to the pre-iteration $\boldsymbol{x}^*$), terminate the iteration, otherwise repeat steps (2)–(4) with the new $\boldsymbol{x}^*$.

$$\alpha_i = \frac{\left.\frac{\partial g}{\partial x_i}\right|_{x^*} \sigma_{xi}}{\sqrt{\sum_{i=1}^{n} \left(\sigma_{xi} \left.\frac{\partial g}{\partial x_i}\right|_{x^*}\right)^2}} \tag{10}$$

$$\beta = \frac{\mu_Z}{\sigma_Z} = \frac{\sum_{i=1}^{n} (\mu_{xi} - x_i^*) \left.\frac{\partial g}{\partial x_i}\right|_{x^*}}{\sum_{i=1}^{n} \alpha_i \left.\frac{\partial g}{\partial x_i}\right|_{x^*} \sigma_{xi}} \tag{11}$$

$$x_i^* = \mu_{xi} - \alpha_i \beta \sigma_{xi} \quad i = (1, 2, \cdots, n) \tag{12}$$

3.2 Calculation Results

According to the above data, the reliability index is 5.213 by using the JC method and the calculation formula in the Chinese code. Considering the safety level corresponding to the target reliability index, $\beta = 4.2$ and $\beta = 4.7$ will be subsequently selected as the target reliability for the design of rock-socketed piles, their partial coefficients will be calculated and compared, and finally each coefficient will be determined under the requirements of safety and economy.

4 Design Method of Probabilistic Limit State for Rock-Socketed Piles

4.1 Partial Coefficients for Total Resistance

In generalized total resistance partial coefficient analysis of the rock-socketed piles, the calculation result is equivalent to the safety factor K of the original allowable stress method, which does not fully reflect the advantages of the probabilistic limit design method, and the formula of the vertical bearing capacity of the embedded rock section is closely related to the uniaxial saturation compressive strength of the rock in the codes. Therefore, the rock-socketed piles are classified according to the uniaxial saturation compressive strength at the different pile ends. The total resistance coefficients for each type of piles are calculated separately. Due to the limitation of the statistical data, the resistance factors were classified by $\sigma_c < 5$ MPa, $5 \leq \sigma_c < 5$ MPa, and 15 MPa $\leq \sigma_c$, which can basically distribute the data evenly in each category and ensure the adequacy of the sample data. Bayesian optimization statistics were performed on the data of each category, and then the total resistance partial coefficient was calculated using the JC method, and the results are shown in Table 1.

Table 1. Calculated total resistance partial coefficient γ_R for each type of piles

Type of codes	Load partial coefficients	$\sigma_c < 5$ MPa		$5 \leq \sigma_c < 15$ MPa		15 MPa $\leq \sigma_c$	
		$\beta = 4.2$	$\beta = 4.7$	$\beta = 4.2$	$\beta = 4.7$	$\beta = 4.2$	$\beta = 4.7$
AASHTO	$\gamma_G = 1, \gamma_Q = 1$	1.72	1.89	1.85	2.03	2.20	2.43
	$\gamma_G = 1.2, \gamma_Q = 1.4$	1.38	1.52	1.49	1.63	1.77	1.96
Chinese highway code	$\gamma_G = 1, \gamma_Q = 1$	1.41	1.54	1.51	1.63	1.82	1.98
	$\gamma_G = 1.2, \gamma_Q = 1.4$	1.14	1.24	1.21	1.31	1.46	1.60

The load partial coefficients $\gamma_G = 1.2$, $\gamma_Q = 1.4$ and $\gamma_G = 1$, $\gamma_Q = 1$, respectively, are used as the basis for Fig. (1) to compare and analyze the differences in the values of the resistance partial coefficients.

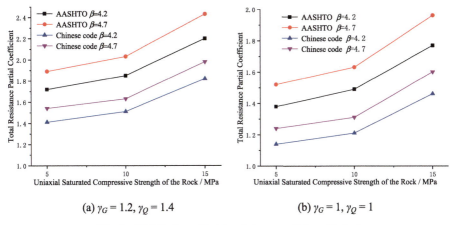

(a) $\gamma_G = 1.2, \gamma_Q = 1.4$ (b) $\gamma_G = 1, \gamma_Q = 1$

Fig. 1. Distribution of total resistance partial coefficient

Comparing the calculation results of AASHTO and the Chinese highway code, it can be seen that the resistance partial coefficient obtained from AASHTO are greater than those obtained from the Chinese code. The fundamental reason is that the mean and variation coefficients of the load effects of AASHTO are larger than those of the Chinese code, and the number of samples taken by AASHTO in the United States is larger than that Chinese code and its range and variability are wider, so it is appropriate to take the value according to the result of AASHTO calculation, i.e., according to row 2 of Table 1.

4.2 Partial Coefficients of Resistance of Soil Section and Rock-Socketed Section

The vertical bearing capacity of the rock-socketed piles is generally divided into the upper soil friction resistance and the embedded section resistance, and the composition and bearing mechanism of the two sections of resistance are quite different. Therefore, it is in accordance with the design law to calculate the partial coefficients of two parts respectively. In the previous part, the total partial coefficient γ_R of the rock-socketed pile in different uniaxial saturation compressive strength intervals has been obtained, and this part will use the recursive method (η method) to solve the soil resistance partial coefficient γ_s and rock resistance partial coefficient γ_r.

Let the total resistance of the pile be provided by n parts, let $\mu_{ki} = \mu_i$ then to obtain the ith layer resistance partial coefficient as shown in Eq. (13), then let each part of the resistance ratio η_i, bring Eq. (14) into Eq. (13) to obtain Eq. (15).

$$\gamma_i^{-1} = 1 - \beta \delta_i \frac{\sigma_i}{\sigma_z} \tag{13}$$

$$R = \sum_{i=1}^{n} R_i \, \eta_i = \frac{m_i}{m_R} \sum \gamma_i^{-1} \eta_i = \gamma_R^{-1} \sum \eta_i = 1 \tag{14}$$

$$\gamma_i^{-1} = 1 - \frac{1 - \gamma_R^{-1}}{\eta_i} \frac{(\delta_i \eta_i)^2}{\sum (\delta_i \eta_i)^2} \tag{15}$$

where m_i is the mean value of the ith resistance and m_R is the mean value of the total resistance; σ_i, σ_R are the standard deviation of the ith resistance and the standard deviation of the total resistance, respectively. It is known that $\mu_i/\mu_{ki} = \overline{\lambda}_i$, since $\mu_{ki} = \mu_i$ is assumed, the Eq. (15) is corrected by dividing γ_i^{-1} by the $\overline{\lambda}_i$. The results of the calculations of the resistance partial coefficients for the upper soil and embedded rock sections are shown in Table 2.

Table 2. Statistics of subfactors for embedded piles

Applicable codes	Lode partial coefficients	Reliability index	Resistance partial coefficients	$\sigma_c < 5$ MPa	$5 \leq \sigma_c < 15$ MPa	15 MPa $\leq$ σ_c
AASHTO	$\gamma_G = 1.2$ $\gamma_Q = 1.4$	$\beta = 4.2$	γ_s	1.15	1.22	1.28
			γ_r	1.66	1.71	1.95
		$\beta = 4.7$	γ_s	1.19	1.27	1.33
			γ_r	1.97	1.95	2.21
Chinese highway code	$\gamma_G = 1$ $\gamma_Q = 1$	$\beta = 4.2$	γ_s	1.16	1.23	1.30
			γ_r	1.72	1.74	2.01
		$\beta = 4.7$	γ_s	1.20	1.27	1.34
			γ_r	2.02	1.95	2.24

According to the data in Table 2, the results are compared between the two codes for load partial coefficients $\gamma_G = 1.2$, $\gamma_Q = 1.4$ and $\gamma_G = 1$, $\gamma_Q = 1$. The two calculation results under the same target reliability index are close to each other, so the AASHTO calculation form with $\gamma_G = 1.2$, $\gamma_Q = 1.4$ are taken for the sake of unification with the superstructure design and for the sake of unification with the advanced experience of international pile foundation reliability design. At the same time, the resistance partial coefficient of rock embedded section is generally larger than the upper soil, which indicates that the value of rock embedded section resistance partial coefficient is more discrete and has a large deviation from the measured value, because the rock embedded section resistance is related to rock embedded ratio (h_r/d, the ratio between the depth of pile embedded in the rock and the pile diameter), so the next part will discuss in detail on the calculation of rock embedded section resistance partial coefficient on the basis of considering the rock embedded ratio.

4.3 Partial Coefficients of Friction Resistance and End Resistance at Rock-Socketed Section

The embedded rock section resistance is composed of two parts: the friction resistance and the end resistance, and the Eq. (15) is still can be used to calculate the partial coefficients of each part when the resistance ratio of each part is known, while the resistance ratio of each part is directly related to the embedded rock ratio, so this section first fits the equation of the relationship between the embedded rock ratio h_r/d and the resistance ratio (η_{rs}/η_{rp}, the ratio of the friction resistance to the end resistance at the embedded rock section) according to the statistical data. The fitted curves are shown as Fig. (2).

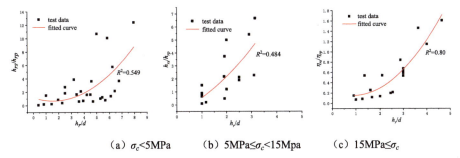

(a) $\sigma_c < 5$MPa   (b) 5MPa$\leq\sigma_c<$15Mpa   (c) 15MPa$\leq\sigma_c$

Fig. 2. Scatter and fit plots of the correlation between η_{rs}/η_{rp} and hr/d

After determining h_r/d, the η_i can be calculated according to the fitting function, and the coefficients of each component are obtained through Eq. (15), and the final proposed values of the partial coefficients are shown in Table 3.

Table 3. Partial coefficients of friction resistance and end resistance for embedded sections

Reliability index	Partial coefficients in rock section	The uniaxial saturation compressive strength σ_c	Embedded rock ratio									
			1	2	3	4	5	6	7	8	9	10
$\beta = 4.2$	γ_{rs}	$\sigma_c < 5$ MPa	0.85	0.85	0.85	1	1.1	1.2	1.3	1.3		
		$5 \leq \sigma_c < 15$ MPa	1.1	1.1	1.1	1.3	2					
		15 MPa $\leq \sigma_c$	1.3	1.5	2.4	3	3					
	γ_{rp}	$\sigma_c < 5$ MPa	1.2	1.2	1.3	1.3	1	0.8	0.7	0.7		
		$5 \leq \sigma_c < 15$ MPa	1.5	1.5	1.5	1.3	1.2					
		15 MPa $\leq \sigma_c$	2.7	2.7	2.2	1.7	1.4					
$\beta = 4.7$	γ_{rs}	$\sigma_c < 5$ MPa	0.9	0.9	0.9	1	1.3	1.4	1.5	1.6		
		$5 \leq \sigma_c < 15$ MPa	1.3	1.3	1.3	1.6	3.3					

(continued)

Table 3. (*continued*)

Reliability index	Partial coefficients in rock section	The uniaxial saturation compressive strength σ_c	Embedded rock ratio									
			1	2	3	4	5	6	7	8	9	10
		$15\ \text{MPa} \le \sigma_c$	1.3	1.6	2.8	3.9	4					
	γ_{rp}	$\sigma_c < 5\ \text{MPa}$	2.0	2.0	2.3	2.5	1.6	1	0.8	0.7		
		$5 \le \sigma_c < 15$ MPa	1.8	1.8	1.8	1.6	1.4					
		$15\ \text{MPa} \le \sigma_c$	3.1	3.1	2.5	1.8	1.5					

5   Proposed Design Formula for Rock-Socketed Piles

Through the above discussion, the total resistance partial coefficient, soil friction resistance partial coefficient, rock section end and friction resistance partial coefficient of rock-socketed piles are obtained respectively. For these piles in highway bridges, culverts and other large projects, the vertical bearing capacity R_a of a monopile based on the reliability theory should be calculated according to the following formula.

$$R_a = \frac{Q_{sk}}{\gamma_s} + \frac{Q_{rsk}}{\gamma_{rs}} + \frac{Q_{rpk}}{\gamma_{rp}} \ge \gamma_G S_{G_k} + \gamma_Q S_{Q_k} \qquad (16)$$

$$Q_{sk} = \zeta_s u \sum_{i=1}^{n} l_i q_{ik} \quad Q_{rpk} = 2 \times c_1 A_p f_{rk} \quad Q_{rsk} = 2 \times u \sum_{i=1}^{n} c_{2i} h_i f_{rki} \qquad (17)$$

where R_a is the design value of monopile vertical bearing capacity (kN); Q_{sk}, Q_{rsk}, Q_{rpk} are the total ultimate friction resistance of soil, the total ultimate friction resistance of embedded rock section, the total ultimate end resistance of embedded rock section (kN); S_{Gk}, S_{Qk} are the standard value of constant load, the standard value of variable load (kN); γ_s, γ_{rs}, γ_{rp}, γ_G, γ_Q are the friction resistance partial coefficient of soil, the friction resistance partial coefficient of embedded rock section, the end resistance partial coefficient of embedded rock section, the constant load partial coefficient and the variable load partial coefficient. If the influence of the embedded rock ratio is considered, the value of each partial coefficient is referred to Table 3, and if the influence of the embedded rock ratio is not considered, the value of each partial coefficient is referred to Table 4. The rest of the calculation parameters are the same as those in the *Code for Design of Ground and Foundation of Highway Bridge and Culvert*.

Table 4. Partial coefficients disregarding the embedded rock ratio

Reliability index	Load partial coefficients	The uniaxial saturation compressive strength σ_c	γ_s	γ_{rs}	γ_{rp}
$\beta = 4.2$	$\gamma_G = 1.2$ $\gamma_Q = 1.4$	$\sigma_c < 5$ MPa	1.2	1.0	1.3
		$5 \leq \sigma_c < 15$ MPa	1.2	2.0	1.3
		15 MPa $\leq \sigma_c$	1.3	3.0	2.1
$\beta = 4.7$	$\gamma_G = 1.2$ $\gamma_Q = 1.4$	$\sigma_c < 5$ MPa	1.2	1.1	2.6
		$5 \leq \sigma_c < 15$ MPa	1.3	3.3	1.4
		15 MPa $\leq \sigma_c$	1.3	4.0	2.3

6 Conclusion

In this study, the formula in *Code for Design of Ground and Foundation of Highway Bridge and Culvert* is selected and the collected engineering data of a large number of embedded piles are processed using Bayesian optimization estimation. Analytical studies based on the probabilistic limit state design method were carried out for the partial coefficients of total resistance, soil friction resistance, embedded rock section resistance, friction resistance of embedded rock section and end resistance of embedded rock section for the rock-socketed piles, respectively. The following conclusions were obtained.

(1) By calibrating the reliability of the current code's formula and referring to the relevant codes in the United States, it is determined that the target reliability index is appropriate to choose $\beta = 4.2$ and $\beta = 4.7$.
(2) The influence of the statistical parameters taken from the AASHTO code and the Chinese highway code on the target reliability index was compared and analyzed, and it was determined that the AASHTO code parameters were used and the load partial coefficients were determined to be $\gamma_G = 1.2$ and $\gamma_Q = 1.4$, respectively.
(3) The partial coefficients satisfying the target reliability index are discussed and be introduced into the bearing capacity calculation formula for design method considering the rock uniaxial compressive strength and rock embedded ratio. The research data in this study are slightly insufficient, and its applicability needs to be verified by many projects.

References

1. Ochiai, H.F., Otani, J.S., Matsui, K.T.: Performance factor for bearing resistance of bored friction piles. Struct. J. **31**,12–16 (1994)
2. Hansen, P.F.F., Madsen. H.O.S., Tjelta, T.I.T.: Reliability analysis of a pile design. Reliab. Marine Struct. **8**(2), 171–198 (1995)
3. Barker, R.M.F., Duncan, J.M.S., Rojiani, K.B.T.: Manuals for the Design of Bridge Foundations. National Research Council, Washington D.C. (1991)

4. Becker, D.E.F.: Limit state design for foundations—Part II: development for national building code of Canada. Geotech **336**, 984–1007 (1996)
5. Zhang L, F., Einstein H, S.: End bearing capacity of drilled shafts in rock. Geotech Geoenviron Eng. 124 (7): 574–584 (1998)
6. Phoon, K.K.F.: Reliability-based Design in Geotechnical Engineering. Computions and Applications. 1st edn. CRC Press, New York (2008)
7. Dithide, M.F., Phoon, K.K.S., Wet, D.M.T.: Characterization of model uncertainty in the static pile design formula. J. Geotech. Geoenviron. Eng. **137**(1), 70–85 (2010)
8. Rowe, R.K.F., Armitage, H.H.S.: A design method for drilled piers in soft rock. Can Geotech. **24** (1), 126–142 (1987)
9. Zheng, J.J.F., Xu, Z.J.S., Liu, Y.T.: Bayesian optimization estimation of pile resistance coefficient. Chinese J. Geotech. Eng. **34**(09), 1716–1721 (2012)
10. Zhou, S.T.F., Li, H.G.S.: Generalized from based on the generalized rackwitz-fiessler method. J. Mech. Eng. **50**(16), 6–12 (2014)
11. Bian, X.Y.F., Zheng, J.J.S., Xu, Z.J.T.: Reliability analysis of pile bearing capacity considering pile foundation design method. Chin. J. Geotech. Eng. **35**(S2), 1099–1102(2013)

Performance-Based Probabilistic Assessment of Liquefaction Induced Building Settlement in Earthquake Engineering

Chenying Liu[1], Jorge Macedo[1]([✉]), and Gabriel Candia[2]

[1] School of Civil and Environmental Engineering, Georgia Institute of Technology, Atlanta, GA 30332, USA
Jorge.macedo@ce.gatech.edu
[2] Universidad del Desarrollo, Las Condes, Chile

Abstract. The Current engineering procedures for estimating liquefaction-induced building settlements (LIBS) utilize deterministic or pseudoprobabilistic approaches that separate the estimation of ground motion intensity measures (IMs) hazard from the estimation of LIBS hazard. In contrast, in a performance-based probabilistic approach, the estimation of the IM hazard is coupled with the estimation of the LIBS hazard. As a result, engineers can directly obtain LIBS estimates corresponding to a selected design hazard level (or return period), which is more consistent with performance-based engineering design. In this study, we make new developments for the performance-based probabilistic assessment of LIBS hazard, including 1) the performance-based assessment of LIBS considering the hazard from a single IM in terms of scalar probabilistic seismic hazard assessment (PSHA), 2) the performance-based assessment of LIBS considering the hazard from multiple IMs in terms of vector PSHA, 3) deaggregation of earthquake scenarios from LIBS hazard curves, and 4) treatment of aleatory variability and epistemic uncertainties. We implement the developments in a computational platform named "LIBS" to facilitate their use in engineering practice. Finally, we share the insights from the comparison between the performance-based and pseudoprobabilistic-based estimates of LIBS hazard.

Keywords: Liquefaction-induced building settlement · Probabilistic assessment · Performance-based earthquake engineering

1 Introduction

Liquefaction-induced ground failure contributes significantly to earthquake damage, in which ground that was solid before shaking can transform temporarily into a fully softened state with little strength and stiffness. Liquefaction can induce damage in shallow founded buildings by producing significant building settlement and damage in previously stable buildings, which can tilt, deform, and collapse. Figure 1 shows some examples of damage caused by liquefaction-induced building settlements (LIBS) in urban centers after recent earthquakes. LIBS has historically caused substantial damage in cities,

L. Wang et al. (Eds.): PBD-IV 2022, GGEE 52, pp. 773–782, 2022.
https://doi.org/10.1007/978-3-031-11898-2_51

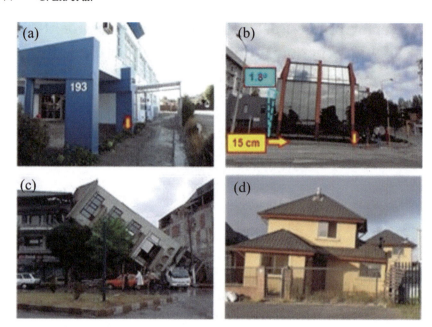

Fig. 1. Examples of liquefaction-induced building settlement damage after recent earthquakes in (a) and (b) New Zealand, (c) Turkey, and (d) Chile. Adapted from Zupan (2014), Bray et al. (2014), Sancio (2003), and Bertalot et al. (2013).

notably Adapazarı in 1999, Christchurch in 2010–2011, Urayasu in 2011, and Palu in 2018.

As discussed in Macedo et al. (2018), there are different frameworks for estimating LIBS when analytical LIBS models are available. These frameworks are categorized as deterministic, pseudoprobabilistic, and performance-based probabilistic. In a deterministic approach, an earthquake design scenario is first identified, often using the deaggregation results from a probabilistic seismic hazard assessment (PSHA). The characteristics of the identified earthquake scenario are used as inputs into a ground motion model (GMM) to evaluate the ground motion intensity measure (IM) of interest. The estimated IM, along with soil and building properties, are used as inputs into a LIBS model to estimate LIBS. Often the uncertainty in the estimated LIBS is introduced through the standard deviation of the analytical LIBS model. In a pseudoprobabilistic framework, which currently dominates in engineering practice, the IM of interest is estimated from a PSHA study that considers all the potential earthquake scenarios (not only one or few as in the deterministic framework). The PSHA study provides the annual rate of exceedance for different IM thresholds, also known as the IM hazard curve, which is used to select the IM of interest given a hazard design level. Once the IM is selected, the next steps are similar to those described before for the deterministic framework. In a performance-based probabilistic framework, a convolution between the entire IM hazard curve obtained from a PSHA study and a LIBS analytical model is performed. Hence, the uncertainties in the ground motion and properties of a geotechnical system can be considered. This is in contrast with the pseudoprobabilistic framework, where only one IM level is selected. One of the results of a performance-based assessment is a LIBS hazard curve, which provides the annual rate of exceedance for different LIBS thresholds. Thus, engineers can directly estimate the LIBS associated with a selected design hazard level, which is more consistent with performance-based engineering concepts as the design hazard is associated with the EDP (Engineering Demand Parameter) of interest and not the IM. Performance-based methods do not assume a consistency between selected IM and EDP hazard design levels, which is implicit in pseudoprobabilistic frameworks (see Macedo et al. (2018) and Rathje and Saygili (2011) for more details). Instead, the entire IM hazard is used to estimate the EDP hazard.

Despite the advantages of performance-based approaches, the current estimation of LIBS, however, relies on deterministic or pseudoprobabilistic approaches, due to the simplicity in their implementation. In this study, we present new developments for the performance-based probabilistic assessment of LIBS, including 1) performance-based assessment of LIBS hazard curves for models that consider a single IM in the context of scalar PSHA, 2) performance-based assessments of LIBS for models that consider multiple IMs using vector PSHA, 3) deaggregation of earthquake scenarios directly from LIBS hazard curves, and 4) integration of aleatory variability and epistemic uncertainty through a logic tree approach. We have implemented the new developments on a computational platform (Candia et al. 2019) that facilitates the straightforward performance-based evaluation of LIBS in engineering practice. Finally, we compare performance-based and pseudoprobabilistic approaches to estimate LIBS and share the insights from these comparisons.

2 Previous Studies on Estimating LIBS

The initial research efforts to understand the mechanisms associated with liquefaction-induced damage used shaking tables and centrifuge tests, considering saturated clean sand (loose to medium dense) deposits that supported rigid shallow foundations (e.g., Dashti 2009; Liu and Dobry 1997; Yoshimi and Tokimatsu 1977). These studies allowed researchers to broadly categorize the mechanisms associated with LIBS as shearing-induced, volumetric-induced, or ejecta-induced (e.g., Bray and Dashti 2014; Dashti 2014; Bray et al. 2017). The different mechanisms associated with LIBS are schematically presented in Fig. 2.

Other researchers have also used advanced constitutive models in nonlinear dynamic soil-structure-interaction (SSI) effective stress analyses to understand liquefaction-induced building damage mechanisms better. These analyses have been generally used as benchmarks for the performance of buildings over liquefiable soils observed in centrifuge tests and well-documented case histories. (e.g., Luque and Bray 2015; Karimi and Dashti 2016; Dashti and Bray 2013; Karamitros et al. 2013). These additional research efforts have shown that nonlinear dynamic SSI effective stress analyses can capture many of the key aspects of the soil-structure interaction of buildings over liquefiable soils. Hence, nonlinear dynamic SSI effective stress analyses have also been used to formulate few analytical models to estimate LIBS (e.g., Karamitros et al. 2013; Bray

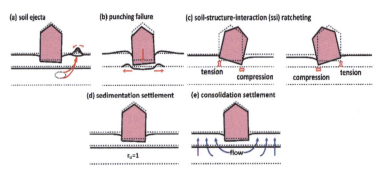

Fig. 2. Mechanisms associated with LIBS: settlements due to ground loss associated with (a) soil ejecta; shear-induced settlement from (b) loss of bearing capacity (e.g., punching failure), or (c) soil-structure-interaction (SSI) ratcheting; and volumetric-induced settlement from (d) sedimentation, or (e) post-liquefaction and reconsolidation (Bray et al. 2017).

and Macedo 2017; Bullock et al. 2018). In this study, we use the Bray and Macedo (2017) and Bullock et al. (2018) models, henceforth referred to as BM2017 and B2018, respectively.

The BM2017 and B2018 models are robust analytical LIBS models. They have been developed considering a large number of realistic ground motions, several soil profiles and building configurations, soil-structure interaction with flexible buildings, and quantification of the uncertainty in the LIBS estimate, and validation against case histories. The BM2017 model is based on a large set of numerical analyses by Macedo and Bray (2018) using the PM4Sand constitutive model (Boulanger and Ziotopoulou 2015), and the B2018 model is based on a large set of numerical analyses by Karimi et al. (2017) using the PDMY02 model (Elgamal et al. 2002; Yang et al. 2003). Both the BM2017 and B2018 models are used as inputs for the performance-based developments performed in this study.

3   Performance-Based Assessment of LIBS

The LIBS hazard can be obtained by convoluting the seismic hazard and a LIBS model. For instance, consider a tectonic setting with Ns seismic sources and a LIBS model formulated in terms of a single IM. The settlement hazard curve can be estimated using Eq. 1.

$$\lambda_{EDP}^{k}(s) = -\sum_{i=1}^{N_s} \sum_{j=1}^{N_m} \int_{IM} P\left(EDP > s|im, m_j, \beta^k, \theta^k\right) P_M\left(m_j|im\right) \Delta\lambda_{IM}^{i,k} d(im)$$

(1)

where k denotes the k-th realization of epistemic uncertainty in the form of alternative: (i) seismic hazard curves, (ii) building parameters, and (iii) soil properties, $\lambda_{EDP}^{k}(s)$ is the annual rate of exceedance of EDP at threshold s for the k-th realization of epistemic uncertainty, im is a realization of IM, $\Delta\lambda_{IM}^{i,k} = \frac{d\lambda_{IM}^{i,k}}{d(im)}$ is the derivative of the IM hazard curve in the i-th source, β^k is the k-th realization of building parameters, and θ^k is the k-th realization of soil properties. m_j is M_w (earthquake magnitude) value representing the j-th magnitude bin, and the conditional probability $P_M\left(m_j|im\right)$ can be estimated from the IM hazard deaggregation. Similarly, the settlement hazard for models defined in terms of two or more IMs, say IM_1, IM_2, ... IM_n, can be computed from the n-dimensional integral in Eq. 2; in this case, $IM = [IM_1\ IM_2 \ldots IM_n]^T$, $im = [im_1 im_2 \ldots im_n]^T$ represents a realization of IM, and $\Delta\lambda_{IM}^{i,k} = \partial\lambda_{IM}^{i,k}/\partial(im)$ is the joint rate of occurrence for IM obtained from a vector PSHA.

$$\lambda_{EDP}^k(s) = -\sum_{i=1}^{N_s}\sum_{j=1}^{N_m}\int_{IM} P\left(EDP > s|im, m_j, \beta^k, \theta^k\right)P_M\left(m_j|im\right)\Delta\lambda_{IM}^{i,k}d^n im \quad (2)$$

Equation 1 to 2 provide the LIBS hazard for a single realization of the epistemic uncertainty. In a logic tree with N_k realizations to account for epistemic uncertainties (i.e., N_k branches and weighting factors w_k), the mean hazard is computed directly as the weighted sum of hazard curves λ_{EDP}^k, as shown in Eq. 3.

$$\lambda_{EDP}(s) = \sum_{k=1}^{N_k}\lambda_{EDP}^k(s)w_k \quad (3)$$

As previously mentioned, LIBS are estimated using the BM2017 and B2018 models in this study. The BM2017 model provides LIBS estimates as the sum of the settlements associated with shearing (D_s), volumetric (D_v), and ejecta mechanisms (D_e), which can be calculated according to Bray and Macedo (2017), Juang et al. (2013), and Hutabarat (2020), respectively. The B2018 model estimates LIBS using a soil layering and soil properties inferred from standard penetration test (SPT) or CPTu (cone penetration test) data, building parameters, and CAV (cumulative absolute velocity) as the ground motion IM.

4 Deaggregation of Earthquake Scenarios from LIBS Hazard Curves

Suppose the contributions from earthquake scenarios and ground motion scenarios to an IM hazard level in Eq. 1 are binned in nM values of M_w, nR values of R_{rup} (rupture distance), and $n\varepsilon$ values of epsilon (ε), where ε is the number of standard deviations above the median IM in the generation of IM realizations in Eq. 1. This results in a total $nScen = nM \cdot nR \cdot n\varepsilon$ ground motion scenarios. Let's consider nIM threshold levels for IM, and $nlibs$ threshold levels for LIBS in Eq. 1. Under these considerations, the contributions to the IM hazard can be stored in a matrix $[\lambda_{IM}]$ with $nScen$ rows and nIM columns, and the annual rate of occurrence $\Delta\lambda_{IM}$ in Eq. 1 can be approximated using Eq. 4:

$$\Delta\lambda_{IM} = -\frac{d\lambda_{IM}}{d(IM)} \approx -\frac{\lambda_{IM_j} - \lambda_{IM_{j+1}}}{IM_j - IM_{j+1}} = \frac{RO_{IM}}{IM_j - IM_{j+1}} \quad (4)$$

where $RO_{IM} = \lambda_{IM_j} - \lambda_{IM_{j+1}}$, is the rate of occurrence of the ground motion intensity IM_j, and the operations are performed with components of the matrix $[\lambda_{IM}]$. Now, considering a fixed LIBS level, and fixed β^k, θ^k parameters, we can evaluate Eq. 1 for each ground motion scenario and at each IM level. This will result in a matrix $[\lambda_{LIBS_p}]$ with *nScen* rows and *nIM* columns that contain partial contributions to the LIBS rates of occurrences. Finally, the contribution to the LIBS hazard from all IM levels is estimated by summing up the columns of $[\lambda_{LIBS_p}]$, which will result in a vector of length *nScen* denoted as $[\lambda_{LIBS}]$ that can be used as a proxy to perform the deaggregation of the LIBS hazard curve as in Eq. 5:

$$[\lambda_{LIBS}] = \left[\lambda_{LIBS_1}, \lambda_{LIBS_2}, \ldots \lambda_{LIBSnScen}\right]^T \quad Deagg_{LIBS_i} = \lambda_{LIBS_i} / \sum_{p=1}^{nscen} \lambda_{LIBSp} \quad (5)$$

where $Deagg_{LIBS_i}$ represents the LIBS hazard deaggregation for the *i*-th earthquake scenario. Because the earthquake scenarios are evaluated from the LIBS hazard and not the IM hazard, this procedure is more consistent with performance-based engineering.

5   Illustrative Implementation of Performance-Based Approaches

In this illustrative example, we consider the subsurface soil information and building properties described in Liu et al. (2021). We consider a single seismic source and two magnitude recurrence scenarios: MR1 and MR2, which differ in the source's activity rate and the slope of the magnitude recurrence relationship (more details of the source models are in Liu et al. (2021)). In our implementation of the BM2017 model, the seismic hazard needs to account for jointly occurring PGA (peak ground acceleration), Sa1 (spectral acceleration at 1 s), and CAV_{dp} (modified version of *CAV*, check Campbell and Bozorgnia (2011)) using vector-PSHA. For the B2018 model, only the CAV scalar hazard is required. These IMs are estimated using the GMMs developed in Liu et al. (2021), Campbell and Bozorgnia (2014), and Campbell and Bozorgnia (2011). Figure 3(a) and Fig. 3(b) show the LIBS hazard curves for the two scenarios considered. Table 1 presents the LIBS contribution from different mechanisms considering the scenarios MR1 and MR2 and return periods of 475 and 2475 years. The D_s estimates using BM2017 are approximately 1.35 times (on average) larger than the D_s estimates from the B2018 for both MR1 and MR2 scenarios, and this ratio is approximately stable for the two return periods. In general, the BM2017 procedure provides larger estimates than the B2018 model, but the estimates from these two procedures indicate a similar performance. For example, the estimates for the MR1 scenario vary from 210 to 400 for 475 years of return period and from 560 to 650 for 2475 years of return period. These estimates indicate that the building performance will likely be poor in the two cases. Thus, both the BM2017 and B2018 procedures are useful in estimating LIBS as a performance index. Figure 1(c) and Fig. 1(d) show the deaggregation of earthquake scenarios obtained from B2018-based LIBS hazard curves for the MR1 and MR2 scenarios.

We also estimate settlements using a pseudoprobabilistic approach. The difference between the performance-based and pseudoprobabilistic estimates is quantified using the parameter $\Delta = Ln(LIBS_{PB}) - Ln(LIBS_{PP})$, where $LIBS_{PB}$ and $LIBS_{PP}$ are the LIBS estimates from performance-based and pseudoprobabilistic approaches, respectively. In terms of the B2018 model, the LIBS estimates from the performance-based assessment are higher than those from the pseudoprobabilistic assessment. Furthermore, Δ is higher in the MR1 scenario compared to the MR2 scenario because of the higher

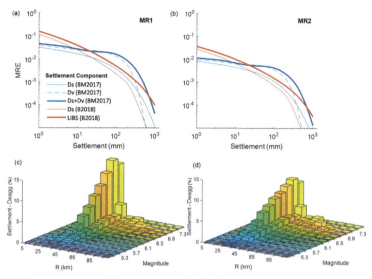

Fig. 3. LIBS hazard curves and deaggregation obtained using the performance-based approach. The hazard deaggregation considers the 475 yr return period for earthquake scenarios: (c) MR1 (LIBS$_{475}$ = 205 mm), and (d) MR2, (LIBS$_{475}$ = 70 mm)

Table 1. LIBS estimates considering performance-based and pseudoprobabilistic approach

Return Period (yr)	Scenario	Performance Based Settlement (mm)								Pseudo-Prob. Settlement (mm)								Δ	
		BM2017				B2018				BM 2017				B2018				BM 2017	B2018
		D_s	D_v	D_g	LIBS	D_s	$D_v + D_g$	LIBS		D_s	D_v	D_g	LIBS	D_s	$D_v + D_g$	LIBS			
475	MR1	152	268	59	400	104	156	210		173	233	59	481	87	56	142		-0.18	0.39
	MR2	48	160	59	220	37	43	70		63	144	59	273	32	21	53		-0.21	0.27
2475	MR1	355	370	59	650	255	305	560		402	242	59	727	211	135	346		-0.11	0.48
	MR2	203	294	59	450	155	138	292		217	239	59	533	125	80	205		-0.17	0.35

seismic activity in the MR1 scenario, which derives in larger $\Delta\lambda_{IM}$ values in Eq. 1, causing more contribution to the annual rate of exceedance for a given LIBS value. Thus, the LIBS estimates in a performance-based approach depend on the entire CAV hazard curve instead of a single CAV value. Finally, Δ tends to increase as the return period increases. In addition, the LIBS estimates from a pseudoprobabilistic approach are not directly related to a return period. For example, consider the MR1 scenario, the current practice would assume that CAV at a return period of 475 years would lead to a LIBS estimate consistent with 475 years. However, the LIBS estimate from the pseudoprobabilistic approach (142 mm) is actually associated with a return period of approximately 300 years. In terms of the results using the BM2017 model, the LIBS estimates from performance-based assessments are comparable but lower than those from pseudoprobabilistic assessments. The differences between the BM2017 and B2018 models in terms of the results obtained from performance-based and pseudoprobabilistic approach may result from the epistemic uncertainty in the estimation of LIBS (e.g., the two models have different functional forms and use different number of IMs, etc.).

6 Conclusions

We present new developments for the performance-based probabilistic assessment of LIBS. Performance-based approaches should be preferred in engineering practice because they better incorporate the uncertainties in the ground motion hazard, soil properties, and building properties. Furthermore, the performance-based assessments enable a hazard-consistent estimate of LIBS, which is more rational as engineers should design for the expected LIBS given a hazard design level and not for the expected IM. We use an illustrative example to implement the performance-based approach considering the BM2017 and B2018 LIBS models. The performance-based estimates of D_s using the BM2017 and B2018 procedures are generally comparable. Differences can be observed for the two models in the final LIBS estimates which come from the differences in the contribution of non-shearing mechanisms to LIBS. However, both models indicate consistent performance levels in the examples considered for this study. Hence, we recommend using the two procedures equally weighted in engineering practice to account for the epistemic uncertainty in LIBS. In addition, we recommend performing a deaggregation from LIBS hazard curves, which is more consistent with performance-based engineering concepts. Comparing performance-based with pseudoprobabilistic procedures, the B2018 model provides larger LIBS estimates in the performance-based case, with the differences increasing as the return period increases and depending on the activity rate in the seismic sources being considered. Thus, the differences between performance-based and pseudoprobabilistic procedures are expected to depend on the activity of seismic sources and the hazard design level under consideration. In the case of the BM2017 procedure, the performance-based LIBS estimates are more comparable with the pseudoprobabilistic estimates in some scenarios.

References

Candia, G., Macedo, J., Jaimes, M.A., Magna-Verdugo, C.: A new state-of-the-art platform for probabilistic and deterministic seismic hazard assessment. Seismol. Res. Lett. **90**(6), 2262–2275 (2019)

Bray, J., Cubrinovski, M., Zupan, J., Taylor, M.: Liquefaction effects on buildings in the central business district of Christchurch. Earthq. Spectra **30**(1), 85–109 (2014)

Campbell, K.W., Bozorgnia, Y.: NGA-West2 ground motion model for the average horizontal components of PGA, PGV, and 5% damped linear acceleration response spectra. Earthq. Spectra **30**(3), 1087–1115 (2014)

Zupan, J.D.: Seismic Performance of Buildings Subjected to Soil Liquefaction. University of California, Berkeley (2014)

Sancio, R.B.: Ground Failure and Building Performance in Adapazari. Turkey. University of California, Berkeley (2003)

Bertalot, D., Brennan, A.J., Villalobos, F.A.: Influence of bearing pressure on liquefaction-induced settlement of shallow foundations. Géotechnique. **63**(5), 391–399 (2013)

Liu, C., Macedo, J., Candia, G.: Performance-based probabilistic assessment of liquefaction-induced building settlements. Soil Dyn. Earthq. Eng. **151**, 106955 (2021)

Hutabarat, D.: Effective Stress Analysis of Liquefaction Sites and Evaluation of Sediment Ejecta Potential. University of California of Berkeley (2020)

Macedo, J., Bray, J., Abrahamson, N., Travasarou, T.: Performance-based probabilistic seismic slope displacement procedure. Earthq. Spectra **34**(2), 673–695 (2018)

Rathje, E.M., Saygili, G.: Estimating fully probabilistic seismic sliding displacements of slopes from a pseudoprobabilistic approach. J. Geotech. Geoenviron. Eng. **137**(3), 208–217 (2011)

Dashti, S.: Toward Developing an Engineering Procedure for Evaluating Building Performance on Softened Ground. University of California, Berkeley (2009)

Liu, L., Dobry, R.: Seismic response of shallow foundation on liquefiable sand. J. Geotech. Geoenviron. Eng. **123**(6), 557–567 (1997)

Yoshimi, Y., Tokimatsu, K.: Settlement of buildings on saturated sand during earthquakes. Soils Found. **17**(1), 23–38 (1977)

Bray, J.D., Dashti, S.: Liquefaction-induced building movements. Bull. Earthq. Eng. **12**(3), 1129–1156 (2014)

Dashti, S., Bray, J.D., Pestana, J.M., Riemer, M., Wilson, D.: Mechanisms of seismically induced settlement of buildings with shallow foundations on liquefiable soil. J. Geotech. Geoenviron. Eng. **136**(1), 151–164 (2010)

Macedo, J., Bray, J.D.: Key trends in liquefaction-induced building settlement. J. Geotech. Geoenviron. Eng. **144**(11), 04018076 (2017)

Luque, R., Bray, J.: Dynamic analysis of a shallow-founded building in Christchurch during the Canterbury earthquake sequence. In: 6th International Conference on Earthquake Geotechnical Engineering, pp. 1–4 (2015)

Karimi, Z., Dashti, S.: Seismic performance of shallow founded structures on liquefiable ground: validation of numerical simulations using centrifuge experiments. J. Geotech. Geoenviron. Eng. **142**(6), 04016011 (2016)

Dashti, S., Bray, J.D.: Numerical simulation of building response on liquefiable sand. J. Geotech. Geoenviron. Eng. **139**(8), 1235–1249 (2013)

Karamitros, D.K., Bouckovalas, G.D., Chaloulos, Y.K.: Seismic settlements of shallow foundations on liquefiable soil with a clay crust. Soil Dyn. Earthq. Eng. **46**, 64–76 (2013)

Bray, J.D., Macedo, J.: 6th Ishihara lecture: simplified procedure for estimating liquefaction-induced building settlement. Soil Dyn. Earthq. Eng. **102**, 215–231 (2017)

Bullock, Z., Karimi, Z., Dashti, S., Liel, A., Porter, K.: Key parameters for predicting residual tilt of shallow-founded structures due to liquefaction. In: Geotechnical Earthquake Engineering and Soil Dynamics V: Numerical Modeling and Soil Structure Interaction (2018)

Macedo, J., Bray, J.D.: Key trends in liquefaction-induced building settlement. J. Geotech. Geoenviron. Eng. **144**(11), 04018076 (2018)

Boulanger, R.W., Ziotopoulou, K.: PM4Sand (Version 3): A sand plasticity model for earthquake engineering applications. Center for Geotechnical Modeling Report No. UCD/CGM-15/01, Department of Civil and Environmental Engineering, University of California, Davis, Calif., 2015.

Karimi, Z., Dashti, S., Bullock, Z., Porter, K., Liel, A.: Key predictors of structure settlement on liquefiable ground: a numerical parametric study. Soil Dyn. Earthq. Eng. **113**, 286–308 (2017)

Elgamal, A., Yang, Z., Parra, E.: Computational modeling of cyclic mobility and post-liquefaction site response. Soil Dyn. Earthq. Eng. **22**(4), 259–271 (2002)

Yang, Z., Elgamal, A., Parra, E.: Computational model for cyclic mobility and associated shear deformation. J. Geotech. Geoenviron. Eng. **12**, 1119–1127 (2003)

Campbell, K.W., Bozorgnia, Y.: Predictive equations for the horizontal component of standardized cumulative absolute velocity as adapted for use in the shutdown of US nuclear power plants. Nucl. Eng. Des. **241**(7), 2558–2569 (2011)

Juang, C.H., Ching, J., Wang, L., Khoshnevisan, S., Ku, C.S.: Simplified procedure for estimation of liquefaction-induced settlement and site-specific probabilistic settlement exceedance curve using cone penetration test (CPT). Can. Geotech. J. **50**(10), 1055–1066 (2013)

Seismic Hazard Analysis, Geotechnical, and Structural Evaluations for a Research Reactor Building in the Philippines

Roy Anthony C. Luna[1]([✉]), Patrick Adrian Y. Selda[1], Rodgie Ello B. Cabungcal[1], Luis Ariel B. Morillo[1], Stanley Brian R. Sayson[1], and Alvie J. Asuncion-Astronomo[2]

[1] AMH Philippines, Inc., Quezon City, Philippines
racluna@amhphil.com
[2] Philippine Nuclear Research Institute, Quezon City, Philippines

Abstract. The Philippine Research Reactor-1 (PRR-1) Building, which is operated by Philippine Nuclear Research Institute (PNRI) but has been shut down since 1988, is planned to be used as housing for a subcritical assembly. Considering that the building is old and not in use, there is a need to assess and evaluate its overall integrity.

This paper presents the seismic hazard analysis, geotechnical, and structural evaluations done for the Reactor-1 Building. The geotechnical evaluation includes the subsurface characterization and derivation of shear strength parameters from field and laboratory testing. Probabilistic Seismic Hazard Analysis (PSHA) was performed to quantify the seismic hazard on site for different hazard levels based on recurrence interval. Upon performing PSHA, site specific response spectra were developed for different return periods and damping ratios.

Structural evaluation, which includes structural idealization, analytical procedures, and calculations, was also conducted to determine the building integrity. Retrofitting is recommended for structural members shown to be inadequate to achieve the performance objective of life safety subjected to a 475-year return period earthquake. Under classification of Hazard Category 4 according to IAEA guidelines, PRR-1 facilities may proceed with operation of a subcritical assembly below 0.1 MW provided retrofit measures are undertaken to meet the objective.

Keywords: Geotechnical evaluation · Probabilistic seismic hazard analysis · Response spectra · Structural evaluation · Retrofitting

1 Introduction

The Philippine Research Reactor-1 (PRR-1) Building, housing the first and only nuclear reactor in the Philippines, is planned to be used to establish a subcritical assembly. In line with this, the structure was analyzed to check its structural integrity.

This paper presents the seismic hazard analysis and geotechnical evaluation done for the structural analysis of the Philippine Research Reactor-1 Building. Its purpose is to produce response spectra at different levels of ground motions and subsequently use these for seismic and retrofitting design of the building and the subcritical assembly.

© The Author(s), under exclusive license to Springer Nature Switzerland AG 2022
L. Wang et al. (Eds.): PBD-IV 2022, GGEE 52, pp. 783–791, 2022.
https://doi.org/10.1007/978-3-031-11898-2_52

2 Seismicity and Geology

The tectonic framework of the Philippines may be subdivided into the Philippine Mobile Belt and Palawan Seismic Zones 4 and 2. The Philippine Mobile Belt is the portion of the archipelago bounded in the west by the Manila Trench, Negros Trench, and Cotabato Trench; and in the east by the East Luzon Trough and Philippine Trench.

The project site is in a region that is tectonically and seismically active as it is located just outside the Philippine Mobile Belt. The project area is generally underlain by the Guadalupe Tuff Formation (GTF) – the regional bedrock of Metro Manila, a horizontally bedded rock commonly referred to as "adobe". The nearest fault trace, West Valley Fault (WVF), which is part of the active Valley Fault System that cuts through the Greater Metro Manila, is found approximately 3.4 km from the site according to the Philippine Institute of Volcanology and Seismology (PHIVOLCS) fault map.

3 Site Description

3.1 Structure

The PRR-1 facility (Fig. 1) is an open-pool type nuclear research reactor obtained under the Atoms for Peace Program. Its operation began in 1963 at a power of 1 MW. Unreliable instrumentation, however, were encountered due to aging in the late 1970s leading to the complete replacement of its instrumentation in 1980 (Philippine Decommissioning Plan). The reoperation of the reactor building is planned with a subcritical assembly for training and basic research in nuclear science and technology.

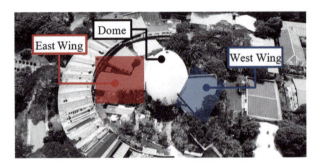

Fig. 1. Aerial shot of east and west wings of reactor-1 building

3.2 Subsurface Conditions

The geotechnical investigation program included the drilling of one (1) borehole of 11.22 m. The results of the field and laboratory testing show that the site subsoil consists of very dense sand soil underlain by siltstone layers. Based on these results, geotechnical strength parameters were determined using correlations developed by Bowles, which are necessary for the analysis and design of foundations and substructures (Table 1).

Table 1. Idealized subsurface profile and equivalent shear strength parameters

Depth, m	Soil classification (USCS)	SPT N-value	Relative condition/Consistency	Geotechnical parameters		
				γ (kN/m^3)	c (kPa)	ϕ (°)
0.00 – 0.72	SM	'refusal'	Very Dense	20	0	35
0.72 – 11.22	Siltstone	[coring]	RQD = 10% to 17% q_u = 0.86 to 1.54 MPa	20	15	38

The regional map (Fig. 2) provided by PHIVOLCS suggests that the expected V_{s30} of the project site should lie within the 760 to 1,500 m/s range, which is categorized as rock site class (S_B) under the National Structural Code of the Philippines (NSCP) 2015. This is consistent with the regional geology and the results of the geotechnical investigation; however, a representative V_{s30} of 760.0 m/s is prudently adopted given the lack of in-situ geophysical data (Fig. 3).

Fig. 2. V_{s30} site model map of metro Manila (PHIVOLCS, 2017)

The recommended allowable bearing capacity is 900 kPa at a depth of 1.5 m from the existing grade line assuming an isolated foundation system. Terzaghi's bearing capacity theory was used in the computation of the ultimate bearing capacity. A factor of safety of 3.0 was adopted to obtain the allowable bearing pressure. Deformation criteria, settlements, were considered in coming up with the bearing capacity.

4   Probabilistic Seismic Hazard Analysis

Probabilistic Seismic Hazard Analysis (PSHA) is the more widely adopted method for describing seismic hazard of an area in terms of potential shaking. By considering all possible earthquake events and resulting ground motions, with their corresponding probabilities of occurrence, a full distribution of levels of ground shaking intensity and their associated rates of exceedance can be obtained. PSHA aims to quantify and combine these uncertainties to produce an explicit distribution of probable future ground shaking that may occur at a particular site [4]. To plot response spectra, PSHA models account for these uncertainties by using the Total Probability Theorem. The PSHA model is analyzed using the OpenQuake engine by the Global Earthquake Model (GEM) Foundation.

4.1 Modification Factors to the Seismic Hazard

Directionality effects reflect the amplification of ground motion when it acts at the optimum orientation that causes the maximum response of the structure. Ground motions considering directionality effects are represented as either the median response computed over all possible rotation angles (RotD50 component) or as the maximum response computed over all possible rotation angles (RotD100 component). In this study, the RotD100 ground motion is derived from the median response based on the models developed by Shahi and Baker [14]. Moreover, directivity effects should be considered when an active seismic source is within 10 km of the project site [2]. It is associated with the orientation of the wave propagation relative to the site. It is necessary to create distinct response spectra for both Fault-Normal (FN) and Fault-Parallel (FP) components to accurately capture Near-Source Effects (NSE) [15].

In this study, the FN and FP components are decomposed from the RotD100 ground motion using the model developed by Bayless and Somerville in 2013. When an active source is within 5 km of the site, the FN component is closely aligned with the RotD100 component, which is used in this study.

The predictive model developed by Razaeian [13], which utilizes damping scaling factors (DSF), was adopted to convert spectral acceleration ordinates, based on a 5% reference damping ratio, to other damping ratios of interest ($\beta\%$). In addition to the damping ratio of 5%, 2% and 0.5%, which are common for steel structures and tank design, are adopted for several hazard levels.

4.2 Logic Tree

A logic tree is employed to formally specify the epistemic uncertainties related to the GMPE's for the corresponding tectonic regions, and to source models for the respective seismogenic parameters.

NGA-West2 GMPE's: Abrahamson, Silva, and Kamai (ASK14); Boore, Stewart, Seyhan, and Atkinson (BSSA14); Campbell and Bozorgnia (CB14); and Chiou and Youngs (CY14) are adopted in PSHA calculations for shallow crustal events. Meanwhile, for subduction interface events, the GMPE's by Youngs et al. (Y97); Zhao et al. (Z16); Abrahamson et al. (BCH15); and Atkinson and Boore (AB03). Finally, for subduction intraslab events, the same set of GMPE's is used, except that Y97 is replaced by the GMPE developed by Garcia et al. (G05).

The branching and weights of the logic tree is adopted for all source models. It must be noted that each branching level will apply to all items in the previous branch. The hypocentral depth distributions are treated as aleatory uncertainties. To address the uncertainties related to utilizing the GR equation, adjustments/standard deviations in the GR parameters are incorporated into the logic tree.

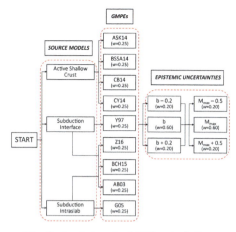

Fig. 3. Logic tree for PSHA calculations

5 Development of Response Spectra

Site-specific response spectra were developed probabilistically for seven hazard levels. The Service Level Earthquake (SLE) response spectrum corresponds to the hazard brought about by relatively weak, but frequent ground motions. In the design of PRR-1, the SLE is associated with the ground motion with a 100-year return period in accordance with International Atomic Energy Agency (IAEA) guidelines [9].

NSE's are neglected at service level due to the intensity associated with the hazard's short return period. In addition to NSE's becoming negligible, the behavior of structural elements is not expected to go beyond the linear-elastic range.

Depending on the type of structure and its design criteria, different seismic hazard levels are provided in the FN and FP directions. Figure 4 presents the SLE response spectrum and the 5%-damped horizontal design ground motions in the form of response spectra for recurrence intervals of 475, 1,000, 2,475, 4,975, 10,000, and 100,000 years.

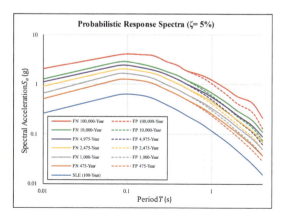

Fig. 4. 5%-damped uniform hazard response spectra

DSFs were generated for damping ratios of 2.0% and 0.5% considering the characteristics of WVF and the hazard brought about by a M7.2 earthquake (Table 2).

Table 2. Damping scaling factors for 2% and 0.5% damping

T (s)	DSF, 2.0%	DSF, 0.5%	T (s)	DSF, 2.0%	DSF, 0.5%
0 (PGA)	1.00	1.00	0.50	1.29	1.65
0.075	1.19	1.57	0.75	1.29	1.60
0.10	1.22	1.64	1.00	1.26	1.51
0.15	1.27	1.75	1.50	1.25	1.49
0.20	1.30	1.80	2.00	1.23	1.44
0.25	1.31	1.79	3.00	1.21	1.39
0.30	1.30	1.73	4.00	1.19	1.35
0.40	1.30	1.69			

6   Structural Evaluation Methodology

A mathematical model composed of beam, and plate elements was created to study the behavior of the structure based on the as-built plans of the reactor building. This idealized model of the structure was subjected to the prescribed loads and a structural finite element analysis is carried out through an elastic approach (Fig. 5). The analysis was conducted using STAAD.Pro by Bentley Systems.

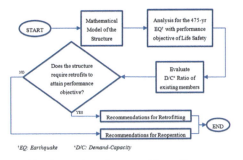

¹EQ: Earthquake   ²D/C: Demand-Capacity

Fig. 5. Methodology workflow diagram

7   Structural Evaluation

The analysis of the reactor building shows that several structural members prove inadequate when subjected to a 475-year return period earthquake based on the provisions of the NSCP 2015.

The walls are grouped based on location, thickness, and existing reinforcing bars (Fig. 6). Generally, the existing rebars are sufficient to provide required tensile strength to the walls except for a Wall B portion at PBS-3 basement level, which exhibits deficiency in flexural capacity when lateral loads are applied to the structure.

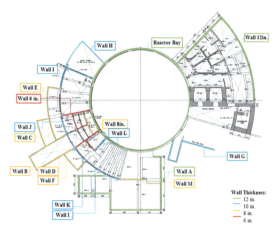

Fig. 6. Wall evaluation key map

The dome of the building is separated into segments which were identified based on existing reinforcement bars. At the bottom-most segment, allowable tensile capacity at the outer face of the dome is exceeded by a 1% margin. Additionally, it is determined that existing reinforcing bars at the dome are sufficient.

The column capacities prove adequate in resisting the loads imposed on the structure. However, the current dimensions are below the minimum dimensions set by NSCP 2015 Sect. 418.7.2 and below the minimum flexural capacities for columns set by NSCP 2015 Sect. 418.7.3. The columns that fail in these two provisions can mostly be found in the second and third level wherein majority of the columns in the east wing and west wing are not braced by walls.

About 60% of the beams in the structure do not satisfy the provisions set in Chapter 4 of NSCP 2015 for shear, mostly on the provision on maximum spacing for shear reinforcements. Among the beams that did not satisfy the shear provisions, only approximately 2% have failed due to load requirements. For flexure, about 25% of the beams do not satisfy the provisions. Majority of these beams do not satisfy requirements for minimum required flexural reinforcements for moment capacity design of beams. Only a few beams have inadequate flexural capacity due to load requirements.

8 Recommendation for Retrofitting and Reoperation

It is recommended to perform the retrofit of the structural members shown to be inadequate based on the analysis to ensure that the performance objective of life safety when subjected to a 475-year return period earthquake is achieved.

It is advised to perform concrete jacketing for the first and second floor columns of the east and west wing to meet the minimum requirements set for these members. This increases the capacity and stiffness of the structural members thereby enabling the existing columns to achieve the minimum requirements set by NSCP 2015.

Basement level of Wall B must be provided with additional reinforcing bars and wall thickness through shotcrete. Meanwhile, to provide additional tensile strength to critical portion of the dome, carbon fiber-reinforced polymers are to be placed at the outer face of the dome at its northwest and southeast portions from elevations of 9.5 m to 12 m.

Steel jacketing is also advised on the beams of the reactor building to strengthen these without significant increase in the cross-sectional area of the beams. Additionally, exposed beams at the roof deck level are to be retrofitted through concrete jacketing where additional flexural and shear reinforcements are required.

For the reoperation of the reactor building, since inadequacies in the structural members are already present when subjecting the structure to an earthquake with a return period of 475 years, it is not advised to operate the reactor building as nuclear research facility under Hazard Category 3 or higher. This is due to the recommendation of IAEA that research facilities operating within populated areas and above the power range of 0.1 MW would escalate the category of the research facility from Hazard Category 3 to Hazard Category 2 as specified in the guidelines of IAEA [9]. Without appropriate retrofit designs to meet the safety requirements for a Hazard Category 2 structure, namely a design to ensure safety for hazards with return period of 10,000 years, it is concluded that operating the reactor building again as a research facility is not safe.

However, since the current operation of PRR-1 can be classified as Hazard Category 4 as its subcritical assembly operates below 0.1 MW, continued operation of the subcritical assembly below 0.1 MW is feasible provided that the recommended retrofit measures are undertaken to meet the life-safety performance objective.

9   Conclusion

Based on the results of the geotechnical investigation program, the upper subsoil layer is generally composed of very dense, which is underlain by rock formation (siltstone). The bearing capacity was also found to be 900 kPa for isolated foundation systems.

PSHA was performed to quantify the seismic hazard (horizontal response spectra) that can be experienced on-site. Seismic hazards associated with ground motions with return periods or recurrence intervals of 100, 475, 1,000, 2,475, 4,975, 10,000, and 100,000 years were developed. All response spectra peaked at a period of 0.10 s. This suggests that resonance may occur for very rigid structures. Based on the 2,475-year disaggregation plots, near-source seismic events within 10 km, mainly attributed to the Valley Fault System, are the major contributors to the overall hazard.

Retrofitting is recommended for structural members shown to be inadequate to achieve the performance objective of life safety subjected to a 475-year return period earthquake. Under classification of Hazard Category 4 according to IAEA guidelines, the facilities of PRR-1 may proceed with operation of a subcritical assembly below 0.1 MW provided retrofit measures are undertaken to meet the life-safety performance objective.

References

1. American Society of Civil Engineers (ASCE): ASCE Structural Engineering Standards 7–16: Minimum Design Loads for Buildings and Other Structures. American Society of Civil Engineers Inc., Virginia, USA (2016)
2. Association of Structural Engineers of the Philippines (ASEP) National Structural Code of the Philippines Volume 1: Buildings, Towers, and Other Vertical Structures. Quezon City: Association of Structural Engineers of the Philippines, Inc (2015)
3. Asuncion-Astronomo, A., Olivares, R.U., Romallosa, K.M., Marquez, J.M.: Utilizing the Philippine research reactor-1 TRIGA fuel in a subcritical assembly. In: Buenos Aires: International Conference on Research Reactor (2019)
4. Baker, J.W.: An Introduction to Probabilistic Seismic Hazard Analysis (PSHA) White Paper (2015). https://web.stanford.edu/~bakerjw/Publications/Baker_(2015)_Intro_to_PSHA.pdf
5. Bayless, J., Somerville, P.: Bayless-Somerville Directivity Model. PEER Report 2013/09, Pacific Earthquake Engineering Research Center, University of California, Berkeley, USA (2013)
6. Dimas, V-A., Kaynia, A.M., Gibson, G.: Modelling Southeast Asian subduction zones for seismic hazard assessments using global tectonics and GIS techniques. In: International Commission on Large Dams (2015)
7. Elnashai, A., El-Khoury, R.: Earthquake Hazard in Lebanon. Imperial College Press (2004)
8. Geoportal Philippines Municipal and City Boundaries from NSO. nso_2010_mun_hhpop_webmercator shapefiles. http://www.geoportal.gov.ph/. Accessed 19 Dec 2019
9. International Atomic Energy Agency (IAEA) Safety Standards Series Safety Guide No. NS-G-1.6: Seismic Design and Qualification for Nuclear Power Plants (2003)
10. Koo, R., Mote, T., Manlapig, R.V., Zamora, C.: Probabilistic seismic hazard assessment for central Manila in Philippines. In: Australian Earthquake Engineering Society (AEES) Poster Presentation (2009)
11. OpenStreetMap Contributors (2020). https://planet.openstreetmap.org/. Accessed 09 Jan 2020
12. PHIVOLCS (n.d.) FaultFinder. Retrieved from faultfinder.phivolcs.dost.gov.ph
13. Rezaeian, S., Bozorgnia, Y., Idriss, I.M., Abrahamson, N., Campbell, K., Silva, W.: Damping Scaling of Response Spectra for Shallow Crustal Earthquakes in Active Tectonic Regions. PEER Report 2012/01, Pacific Earthquake Engineering Research Center, University of California, Berkeley, USA (2012)
14. Shahi, S.K., Baker, J.W.: An empirically calibrated framework for including the effects of near-fault directivity in probabilistic seismic hazard analysis. Bull. Seismol. Soc. Am. **101**(2), 742–755 (2011). https://doi.org/10.1785/0120100090
15. Somerville, P.G., Smith, N.F., Graves, R.W., Abrahamson, N.A.: Modification of empirical strong ground motion attenuation relations to include the amplitude and duration effects of rupture directivity. Seismol. Res. Lett. **68**(1), 199–222 (1997)
16. Thenhaus, P.C., et al.: Estimates of the regional ground motion hazard of the Philippines. National Disaster Mitig. Philippines, 45–60 (1994)
17. Torregosa, R.F., Sugito, M., Nojima, N.: Strong motion simulation for the Philippines based on seismic hazard assessment. J. Nat. Dis. Sci. **23**(1), 35–51 (2001)
18. van Stiphout, T., Zhuang, J., Marsan, D.: Seismicity declustering, community online resource for statistical seismicity analysis(2012). https://doi.org/10.5078/corssa52382934
19. Wang, Y.J., Chan, C.H., Lee, Y.T., Ma, K.F., Shyu, J.H., Rau, R.J., Cheng, C.T.: Probabilistic seismic hazard assessment for Taiwan. Terr. Atmos. Oceanic Sci. **27**(3), 325–340 (2016). https://doi.org/10.3319/TAO.2016.05.03.01(TEM)

Integrating Local Site Response Evaluations in Seismic Hazard Assessments

Roy Anthony C. Luna, Ramon D. Quebral, Patrick Adrian Y. Selda,
Francis Jenner T. Bernales[✉], and Stanley Brian R. Sayson

AMH Philippines Inc., Quezon, Philippines
francis.bernales@amhphil.com

Abstract. In recent years, site-specific seismic hazard studies have steadily been incorporated in large infrastructure projects as performance-based earthquake assessment is continuously being assimilated into engineering practice. The first component of this framework is the characterization of (engineering) bedrock or "seismic base" accelerations, often using probabilistic seismic hazard analysis (PSHA) with ergodic (global) ground motion models. A major part of the analysis effort is the accounting of local site effects, which have been known to alter surface ground motions significantly. Moreover, for sites with soft/loose deposits, the ground shaking intensity is expected to be high, and thus, the onset of liquefaction may also be a concern. Liquefaction effects are often evaluated using simplified methods via triggering curves, but there is an increasing trend towards using nonlinear effective stress analysis to better simulate surface ground motions and obtain a more appropriate structural response.

In this paper, a case study involving a site situated on soft sedimentary deposits is discussed to show the integration of seismic hazard and site response analyses. This study highlights the importance of advancing the state-of-practice to integrate local site response evaluations in seismic hazard assessment.

Keywords: Site response · Seismic hazard analysis · Ergodic models

1 Introduction

In recent years, site-specific seismic hazard analysis/assessment (SHA) studies have been incorporated steadily in large infrastructure projects as performance-based earthquake assessment is continuously being assimilated into engineering practice. Modern versions of various codes and guidelines usually adopted as design references in infrastructure projects have been supporting the increasing use of the performance-based methodology by providing dedicated sections on site-specific studies in characterizing seismic demands.

This paper discusses the application of site-specific seismic hazard analysis in an example site underlain by soft sedimentary deposits. In the following sections, information regarding the subsurface characterization, as well as details on the site-specific seismic hazard analysis and its integration with site response analysis will be discussed.

© The Author(s), under exclusive license to Springer Nature Switzerland AG 2022
L. Wang et al. (Eds.): PBD-IV 2022, GGEE 52, pp. 792–800, 2022.
https://doi.org/10.1007/978-3-031-11898-2_53

2 Subsurface Characterization

Geophysical surveys using Seismic Refraction Survey (SRS) and Multi-Channel Analysis of Surface Waves (MASW) were conducted to obtain in-situ seismic wave velocities. From the survey lines, seismic velocity profiles can be generated—presenting a tomography of Compressional Wave (P-wave) and Shear Wave (S-wave) velocities (V_p and V_s, respectively).

The site is generally underlain by soft sedimentary deposits. MASW V_s readings generally do not exceed 200 m/s. For sites where the average representative V_s is less than 200 m/s, Site Response Analysis (SRA) is typically warranted to emulate ground shaking and soil response more properly, and in effect, obtain a more accurate representation of the seismic demand at the surface. For this purpose, the stratigraphic profile was idealized to generate best-estimate shear wave velocity and shear strength profiles.

3 Seismic Hazard Analysis

3.1 Connection to Performance-Based Earthquake Engineering

SHA can be defined as the process of quantifying the plausibility of the overall effects of earthquakes on an area primarily in terms of potential shaking. SHA can be thought of as a means to mitigate other seismic effects (e.g., landslides, liquefaction, tsunami incursion). The main purpose of this study is to provide designers with different levels of ground motions where seismic design parameters can be extracted from.

For evaluating the seismic performance (e.g., strains, displacements, damage states) of structures at different levels, ground motions associated to varying return periods (i.e., hazard levels) are warranted. In this study, the life safety and collapse prevention seismic performance objectives are to be assessed using the Design Basis Earthquake (DBE) and Maximum Considered Earthquake (MCE), respectively. In line with this, Probabilistic Seismic Hazard Analysis (PSHA) is performed to obtain ground motions with the required recurrence intervals listed in Table 1.

Table 1. Seismic performance objectives

Level of earthquake	Performance objective
DBE: 10% Probability of exceedance in 50 years (475-year return period)	Code level/Life safety: Moderate structural damage; extensive repairs may be required
MCE: 2% Probability of exceedance in 50 years (2,475-year return period)	Collapse prevention: extensive structural damage; repairs are required and may not be economically feasible

3.2 Uniform Hazard Spectra

Fundamentally, PSHA considers all possible earthquake events and resulting ground motions, with their corresponding probabilities of occurrence, to obtain a full distribution of levels of ground shaking intensity and their associated rates of exceedance. The plot of the intensity, which may be represented by the spectral acceleration, against the rate of exceedance is called a hazard curve. Spectral acceleration ordinates at varying periods from different hazard curves at a target rate of exceedance can be combined to form the uniform hazard spectrum.

PSHA models account for the propagation of uncertainties associated with ground motions by using the Total Probability Theorem, which states that the overall probability of a system can be obtained by summing the individual probabilities of events that consider a particular uncertainty. Mathematically, this can be written as:

$$\lambda_a = \sum_{i=1}^{N_s} v_i \sum_{j=1}^{N_m} \sum_{k=1}^{N_r} P\big(S_a > a|m_j, r_k\big)P\big(M = m_j\big)P(R = r_k) \tag{1}$$

where S_a is the spectral acceleration, a is the ground acceleration level of interest, λ_a is the mean annual rate of exceedance of any ground acceleration, v indicates seismic sources, and $P(S_a > a \mid m_j, r_k)$ is the probability of exceeding any ground acceleration level given any combination of magnitude (m) and distance (r).

Taking temporal uncertainties into account, PSHA models the Probability of Exceedance (PoE) of an earthquake's occurrence within a time frame or window, t, such that it follows the Poisson distribution:

$$P(S_a > a) = 1 - e^{-\lambda_a t} \tag{2}$$

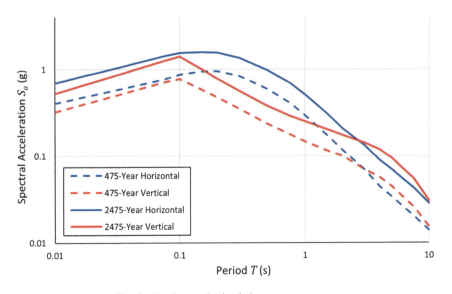

Fig. 1. 5%-Damped seismic base response spectra

Site-specific UHS at the seismic base layer were generated using the PSHA approach for the two (2) return periods considered in this study. Figure 1 provide the horizontal and vertical seismic base ($V_s = 374.93$ m/s) response spectra for a damping ratio of 5%.

4   Site Respose Analysis

4.1   Base Motions

Careful consideration is applied in selecting empirical seed motion time-histories. Seismological features of the earthquake event and local site conditions at the recording station shall be similar with the earthquake scenario being considered. The choice of a feasible scenario event must be based on a disaggregation analysis.

Based on the relative contribution of shallow crustal earthquake ruptures in the area, seven (7) records in the ground motion suite representing the shallow tectonic regime is deemed to be reasonable (Stewart et al. 2014). The suitable seed acceleration time-histories were downloaded from the PEER NGA-West2 Ground Motion Database.

To properly carry out SRA, the response spectra of the input base time-history records should be compatible with the seismic base target spectrum. Stewart et al. (2014) explains that unlike for structural applications, input signals used for developing nonlinear site amplification functions need only be compatible with the seismic base target spectrum at a select period (e.g., site period) instead of enforcing compatibility across a very wide range of frequencies. The representative V_{s30} for the site is 155.99 m/s—corresponding to a fundamental period of 0.54 s. In consideration of the homogenized subsurface condition, the period range of concern should at least encompass 0.50 to 0.70 s.

Utilizing the 475-year spectrum as the reference hazard level, Amplitude Scaling was carried out by multiplying the seismic base seed acceleration time-histories with a constant scaling factor. The scaling factors used in this study were obtained using PEER's algorithm based on the least resulting mean squared error between the scaled motion's response spectrum and the seismic base target spectrum.

4.2   Soil Constitutive Model

The nonlinear, strain-dependent behavior of soils under cyclic loading—such as earthquakes—is captured in part by Modulus Reduction and Damping curves (MRD Curves). The MRD curves, including shear strengths, of each layer in the 1D soil column are derived from the results of the geotechnical and geophysical program.

The MRD Curves, combined with models for unloading-reloading or hysteretic behavior, give the basic unit of a constitutive model for cyclic response of soils. The primary constitutive model adopted in this study is the General Quadratic/Hyperbolic (GQ/H) model. The GQ/H model can capture the cyclic behavior of soils quite well at large (approaching strength level) and small strain levels. In other words, the nonlinear and dissipative characteristics of soils can be simulated in the time domain when the numerical model is subjected to cyclic or seismic loading.

4.3 Site Response Modeling

When performing SRA, the input ground motion time-histories are applied at the base of the 1D soil column—herein referred to as "seismic base layer" (SBL). Seismic waves then propagate from the SBL, through the stratification, and up to the surface. As this occurs, seismic waves are either reflected back or are allowed to refract into the strata. This phenomenon is primarily controlled by the interplay of forcing frequency content and the mechanical properties of the subsurface profile.

In reality, reflected waves continuously propagate downwards throughout the strata, and refracted waves continuously propagate laterally in the free field. Thus, appropriate numerical boundary conditions must be defined to prevent seismic waves from getting trapped inside the model—propagating indefinitely until completely dissipated.

In DEEPSOIL, each layer in the soil profile is discretized into a lumped-mass model represented by a corresponding mass, nonlinear spring, and a dashpot for small-strain equivalent viscous damping. For the SBL, definition of an "elastic half-space" allows for only the partial reflection of downward waves back up through the profile. This allows for part of the elastic wave energy to be dissipated into the bedrock (Hashash et al. 2010). As input base motions used herein are recorded at an equivalent rock outcrop, an elastic half-space is used.

4.4 Site Response Simulation Results

Simulations of site response were carried out for seven (7) pairs of horizontal ground motions considering fifty-one (51) soil profiles, randomized from the best-estimate V_s profile using the Toro (1995) algorithm to account for uncertainties. The fit-and-reduction process in calibrating the constitutive model is done automatically in DEEPSOIL for each randomized profile or "realization". Consequently, a total of 714 SRA simulations were carried out in this study.

The surface response spectrum determined for each SRA calculation run, per realization (grey dashed line), is shown in Fig. 2. The mean response spectrum for the base layer (black solid line) is also included in the figure for comparison. Evidently, the short-period intensities for ($T < 0.3$ s) are either damped or attenuated. Meanwhile, for long-period accelerations, the intensities are amplified—peaking near the site period (0.5 to 0.6 s).

4.5 Merging Site Response with Seismic Hazard Analysis

The randomization scheme, in account of uncertainties in site properties, returns a range of surface response values, particularly surface S_a. A statistical distribution can then be chosen to represent these value ranges. Typically, the surface S_a is often normalized with respect to the base S_a to obtain a site amplification factor or ratio. The implementation of site amplification factors in PSHA applications usually takes a log-normal distribution, wherein the mean (in natural log units) $\mu_{\ln Y}$ is defined using the following nonlinear expression (Stewart et al. 2017):

$$\mu_{\ln Y} = f_1 + f_2 \ln \left(\frac{x_{IM,ref} + f_3}{f_3} \right) \tag{3}$$

where f_1, f_2, f_3 are model regression parameters, $x_{IM,ref}$ is the amplitude of shaking for the reference site condition (taken as $V_{s30} = 760$ m/s).

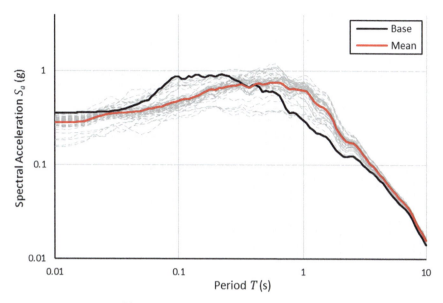

Fig. 2. 5%-Damped SRA response spectra

Site amplification factors are obtained per SRA calculation for each realization, taking note of its corresponding $x_{IM,ref}$. Any ground motion intensity measure can be taken for x_{IMref}, but in this study, it was taken as the PGA at $V_{s30} = 760$ m/s.

Site amplification factors are then compiled for each desired oscillator period. The equation for $\mu_{\ln Y}$, presented above, is utilized to fit through the points provided by the SRA prediction, given their corresponding $x_{IM,ref}$. This process is repeated for a suite of desired period values.

The mean amplification ratios can then be calculated and compared with the Seyhan and Stewart (SS14) ergodic model. The recommended mean amplification typically transitions to the semi-empirical ergodic model for periods beyond the site period.

The recommended site amplification spectra are utilized to modify the hazard curves from PSHA for the reference (SBL) condition. Two methods were used to perform the modification of "rock" hazard curves to incorporate local site effects from SRA—hybrid method (Cramer 2003) and convolution method (Bazzurro and Cornell 2004).

The hybrid method simply multiplies the ordinate from the rock hazard curve at a given exceedance probability by the mean amplification conditional on the same ordinate. A common drawback to this method is that the standard deviation used in the formulation is for the rock condition; in effect, the change in ground motion variability due to local site effects is not captured. In contrast, the convolution method incorporates the uncertainties in the site amplification spectra, in addition to the mean estimates.

Figure 3 shows the 2,475-year UHS from the two methods just discussed. For comparison, the UHS from ergodic site terms in PSHA, the surface response spectra from SRA, as well as the reference condition UHS are also plotted.

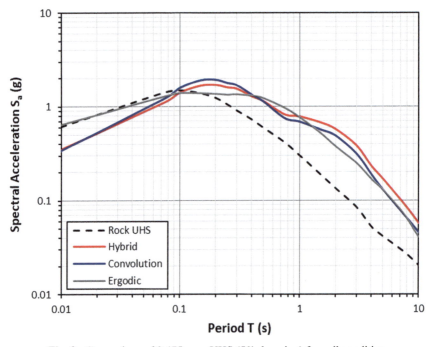

Fig. 3. Comparison of 2,475-year UHS (5% damping) for soil condition

5 Surface Response

5.1 Surface Ground Motion Suite

A ground motion suite shall also be developed for dynamic analysis at the surface level. In contrast to Amplitude Scaling, the ground motions used for the evaluation of structural performance shall need to be amplitude-compatible with the target hazard (response spectrum) across a wide range of frequencies associated to the modes of vibration of the structures within the site.

The reasoning behind this approach is that recorded ground motions should be able to better reflect the earthquake phenomena than the simulated motions from SRA. Even though the variabilities in site geology and ground motion characteristics are incorporated by simulating across several profile and input motion realizations, other aspects of the site response physics are still left out in one-dimensional SRA. Effects of surface waves, basin geometry, and site topography are still important features of site response that are readily present in recorded time-histories.

A similar rigorous process of selecting input motions at the SBL applies to the surface ground motion suite. However, as this development will be utilized for dynamic analyses of structures, there is a slight difference in the code requirements. In ASCE 7–16, the minimum ground motions are set to 11 time-histories that are assessed based on their maximum-rotated (i.e., RotD100) components after applying the modification or matching procedure.

5.2 Spectral Matching of Surface Motions

Spectral matching is the process of modifying the amplitude and/or frequency content of a certain ground motion such that its response spectrum will match a given target response spectrum. In doing so, the modified record can be surmised to be an event that may happen on-site. Still, due to the wide uncertainty associated to earthquakes, codes and standards require using a practical number of matched ground motions to capture the mean response of a system subjected to different kinds of earthquakes.

The spectral matching procedure utilizes the wavelet algorithm of Al Atik and Abrahamson (2010). The quality of the matching is assessed by looking at how well-matched the modified ground motions compare with the time-histories (velocity and displacement) of the original record. The primary criterion is to check if the non-stationary parameters of the original ground motion is well-maintained even after matching.

6 Conclusions

This paper discusses the application of site-specific seismic hazard analysis in an example site underlain by soft sedimentary deposits. The procedures for site-specific seismic hazard analysis and the integration of local site effects based on SRA are demonstrated. The resulting surface response spectra, in effect, is consistent with the desired hazard level defined by its exceedance probability or return period. However, key issues remain that should be considered to advance the state-of-practice in integrated site-specific studies as part of the performance-based earthquake engineering framework.

One of the issues involves the development of the site amplification spectra required for the modification of the rock hazard curve to incorporate site effects. As mentioned in Stewart et al. (2014), multiple suites of base input motions may be necessary to properly constrain the parameters in the site amplification function. The multiple suites are then scaled to target response spectra from different hazard levels to cover a wider range of intensity levels. In this study, only a single hazard level (10% PoE in 50 years) is considered, making the simulated surface motions to cluster around a narrow range of intensity levels (here 0.3–0.4 g only). Since this approach will affect the mean amplification spectra, then changes in the UHS at the soil condition are expected as well.

Liquefaction effects may need to be incorporated by accounting for pore pressure generation as well. This can be carried out by performing nonlinear effective stress SRA. Moving forward, further studies should extend the framework implemented herein for a fully probabilistic seismic liquefaction hazard analysis.

References

Al Atik, L., Abrahamson, N.: An improved method for nonstationary spectral matching. Earthq. Spectra **26**(3), 601–617 (2010). https://doi.org/10.1193/1.3459159

American Society of Civil Engineers Minimum Design Loads and Associated Criteria for Buildings and Other Structures Reston, Virginia: American Soc. of Civil Engineers (2016)

Bazzurro, P., Cornell, C.A.: Nonlinear soil-site effects in probabilistic seismic-hazard analysis. Bull. Seismol. Soc. Am. **94**, 2110–2123 (2004)

Cramer, C.H.: Site-specific seismic-hazard analysis that is completely probabilistic. Bull. Seismol. Soc. Am. **93**, 1841–1846 (2003)

Hashash, Y.M.A., Phillips, C., Groholski, D.R.: Recent advances in non-linear site response analysis. In: 5th International Conference on Recent Advances in Geotechnical Earthquake Engineering and Soil Dynamics San Diego, California, 24–29 May 2010 (2010)

Pacific Earthquake Engineering Research Center PEER NGA-West2 Ground Motion Database (2012). https://www.ngawest2berkeleyedu

Seyhan, E., Stewart, J.P.: Semi-empirical nonlinear site amplification from NGA-West2 data and simulations. Earthq. Spectra **30**(3), 1241–1256 (2014). https://doi.org/10.1193/063013EQS 181M

Stewart, J.P., Afshari, K., Hashash, Y.M.A.: Guidelines for Performing Hazard-Consistent One-Dimensional Ground Response Analysis for Ground Motion Prediction, PEER Report No 2014/16, Pacific Earthquake Engineering Research Center. Univ. of California, Berkeley (2014)

Stewart, J.P., Afshari, K., Goulet, C.: Non-ergodic site response in seismic hazard analysis. Earthq. Spectra **33**(4), 1385–1414 (2017). https://doi.org/10.1193/081716eqs135m

Toro, G.R.: Probabilistic models of site velocity profiles for generic and site-specific ground-motion amplification studies, Technical Report No 779574, Brookhaven National Laboratory, Upton, NY (1995)

Performance-Based Design Review of a Reinforced Earth Retaining Wall for a Road Embankment Project in the Philippines

Roy Anthony C. Luna, Jenna Carmela C. Pallarca$^{(\boxtimes)}$, Patrick Adian Y. Selda, Rodgie Ello B. Cabungcal, Marvin Renzo B. Malonzo, and Helli-Mar T. Trilles

AMH Philippines, Inc., Quezon City, Philippines
jenna.pallarca@amhphil.com

Abstract. Several infrastructure projects such as light rail transit system, national roads, expressways, and underground rapid transit lines are currently being developed in the Philippines to address the huge gap in transportation infrastructure. One of the major infrastructure projects is an elevated toll expressway that connects several business districts in the southern areas of Metro Manila. Located near the coastal area of Manila Bay, the expressway alignment is underlain by soft clay and loose sand layers where long-term settlements and liquefaction are expected. In this context, a need to transition from conventional design methods to Performance-Based Design is essential in providing a cost-effective solution.

This paper presents the methodology utilized in evaluating stability and deformation of road embankment protection designs involving Mechanically Stabilized Earth walls. The assessment of the proposed ground improvement by Soil-Cement Columns to minimize settlements and to mitigate the onset of geohazards such as liquefaction are discussed. This paper also shows the instabilities observed by performing Slope Stability Analysis, and their corresponding deformations evaluated by Finite Element Analysis. Recommendations for optimization and design improvements based on Client's risk tolerance, and further studies to be undertaken are identified.

Keywords: Numerical modelling · Slope stability · Liquefaction · Deformation · MSE wall

1 Introduction

Several infrastructure projects are being built in the Philippines. These include light rail transit system, national roads, expressways, and underground rapid transit line. Therefore, a need for a more advanced design methodology, especially in the field of geotechnical engineering design is necessary. This paper focuses on an elevated expressway that aims to decongest city traffic, connect major business districts in the Southern Metro Manila area, and to provide faster and efficient regional movement of people and goods.

The project site rests generally on loose sand and soft clay materials. To address potential occurrence of geohazards, the contractor of the project proposed to improve the ground by means of Soil-Cement Columns (SCC). The resulting soil-binder composite material is expected to have higher soil properties relative to the properties of the native soil. This paper presents the general design methodology adopted in assessing the adequacy of the design of the resulting composite soil to increase bearing capacity, and to mitigate or minimize possible geohazards such as settlement, liquefaction, and slope instability.

Figure 1 shows the geometry of the critical section focused on this paper. The embankment height at this section is 6.5 m, and the road width is 11.2 m. The Soil Cement Columns spaced at 1.46 m have a diameter of 900 mm and length of 20 m.

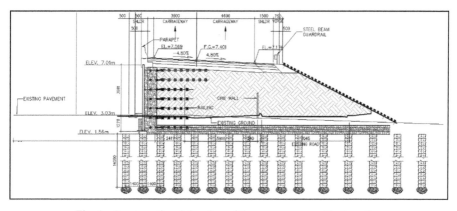

Fig. 1. Cross-sectional drawing of the expressway critical section

2   Design Review Methodology

Boreholes were drilled along the alignment of the project site to determine the local subsurface conditions. The materials encountered are composed of very loose to medium dense sand with some very soft to very stiff clay with varying thicknesses ranging from 20 to 25 m underlain by the bedrock. Geotechnical parameters of the native or untreated soil were then evaluated from correlating the SPT N-values with several published studies such as Bowles (1997) [1]. The soil-cement columns and surrounding native soil were characterized as a composite soil block. The equivalent geotechnical parameters of the composite soil are functions of the area replacement ratio and the geotechnical parameters of the native soil.

New Generation Attenuation Relationships was adopted in estimating the seismic demand on-site. The NGA-West attenuation models for active fault systems in 2008 and 2014 developed by the Pacific Earthquake Engineering Research Center (PEER) [2] was used in determining the Peak Ground Acceleration (PGA), needed for the analyses.

In the Philippines, it is common practice to adopt a k_h value taken as half of the surface PGA in accordance with the works of Kavazanjian et al. in 1997 [3]. On the other hand, k_v usually falls within one-fourth (25%) to two-thirds (66.67%) of its horizontal counterpart. For simplicity, the k_v value was taken as one-half (50%) of k_h.

With the presence of loose materials and shallow ground water table at the site, it is imperative to conduct liquefaction analysis. For this analysis, the Boulanger and Idriss 2014 [4] method was adopted; and, the liquefaction-induced settlement was estimated using the method Pradel developed in 1998 [5].

It is crucial that the strength properties of the composite soil would be able to resist liquefaction; hence, the composite soil materials were also evaluated for liquefaction susceptibility. The analysis was an iterative process wherein the SPT N-values of each liquefiable soil layers were refined until adequate safety factors against liquefaction are achieved.

The expressway project will be composed of high embankments for the ramps and elevated roads. Therefore, it is warranted to estimate the settlement of the subsurface after the placement of the embankment. Settlement analysis of both native and composite soil was carried out with the aid of Settle 3D software. Short and long-term settlements after the construction duration for the embankment of 12 months, and post-construction settlement were compared.

Slope stability analysis using Limit Equilibrium Method was conducted to determine the global stability of the MSE wall. The GLE/Morgenstern-Price Method was used for both circular and non-circular analysis of the slope. The table below presents the different loading cases used in the analysis.

Table 1. Global slope stability analysis scenarios

Case	Pore pressure ratio, r_u	Seismic coefficient		Minimum allowable FS
		k_h	k_v	
Normal and static	0.0	0.0	0.0	1.5
High pore water pressure and static	0.4	0.0	0.0	1.3
Moderate pore water pressure with strong earthquake	0.2	0.25	±0.125	1.1

Considering the results of the SSA, deformation analysis by way of Finite Element Method (FEM) using the numerical analysis tool, PLAXIS, was carried out. This is aimed at establishing the earthquake-induced localized deformations, and assessing if these will be tolerable. The primary criterion for designing slope protection systems is the serviceability performance in terms of both vertical deformation (e.g. settlement, heaving) and lateral deflection. As such, deformation analysis was performed using Finite Element Method (FEM) to establish both static and seismic performance of the slope protection system. The main advantage of FEM is its ability to estimate the deformations leading up to the failure or time point of interest. These deformations correspond to the movement of the soil mass showing how it will behave upon application of loads.

Section 11.10.4 of AASHTO LRFD 2012 [6] details the maximum deformation criteria for MSE walls. For MSE walls with full height precast concrete facing panels, total settlement should be limited to 2.0 in. (50 mm) for the pseudo-static condition. Since lower settlements are expected in normal condition, an allowable settlement of 1.0 in. (25 mm) is set for static conditions. The limiting differential settlement should be 1/500.

The maximum allowable lateral displacement of MSE wall, as prescribed by BS 8006 Code of Practice for Strengthened/Reinforced Soils and Other Fills [7], may be taken as 0.5% of the height of the supported embankment for the static condition. For the pseudo-static case, twice the allowable deflection for the static case or 1% of the height of the supported embankment may be set.

3 Results of Analysis

3.1 Subsurface Conditions

The succeeding table presents the geotechnical parameters of both native and composite soil evaluated from the SPT N-values obtained from field testing.

Table 2. Subsurface condition and geotechnical parameters of critical section

Depth, m	Soil classification	SPT N-value	Geotechnical parameters Native soil; Composite soil			
			γ (kN/m^3)	c (kPa)	$\phi(^{\circ})$	E (MPa)
0.0−7.5	SM/SP-SM	2–8	17	0; 115	28; 29	5; 80
7.5−9.0	GM/SM	19	18	0; 115	32	2; 70
9.0−13.5	SC/CL/CH	0–2	12	6; 115	0; 9	5; 80
13.5−16.5	GM/SM	9–13	18	0; 115	32	10; 80
16.5−19.5	CH	7–10	17	40; 140	0; 9	30
19.5−23.0	CH/CL	26–36	20	200; 245	0; 9	75
23.0−25.5	Siltstone	'coring'	20	200	30	80

3.2 Seismic Parameters

It can be surmised that the most influential seismic source for the project area is the West Valley fault (WVF) [8] which could produce a magnitude 7.2 earthquake. Based on the proximity of the alignment to the WVF, the expected surface PGA from the attenuation models is estimated to be 0.45 g. For prudence, PGA was rounded up to 0.50 g to be adopted in the analyses.

3.3 Liquefaction Analysis

The upper 19.5 m depth of the native subsurface is liquefiable, and the liquefaction-induced settlements is approximately 670 mm. However, once N-value of at least 25 is attained by the SCCs, liquefiable layers will be limited only at depths below the proposed SCC lengths. This is presented in Fig. 2.

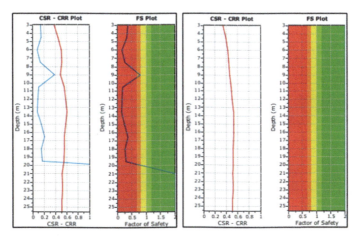

Fig. 2. CSR-CRR and FS plots (Left: Native soil; Right: composite soil)

3.4 Settlement Analysis

The results suggest that without SCC, settlement of 485 mm will take place upon application of load, and long-term post-construction consolidation settlement of 300 mm is expected because of the presence of clay layers. Upon application of embankment and traffic loadings at the ground with SCCs, the calculated settlements significantly decreased compared to the values considering native soil condition. Long term settlement of 55 mm is still expected to occur over a period of 4 years. This is presented in Fig. 3.

3.5 Slope Stability Analysis

Generally, for each section, there are several slip circles passing through the native soil having FS less than the allowable. Therefore, the embankment requires ground improvement of the underlying native soil to prevent slope instabilities under the loading conditions presented.

Based on the results of SSA (Fig. 4), ground improvement by SCC will be able to intercept the potential slip circles passing through the loose/soft soil layers as evident in the higher FS presented below. However, the FS is still lower than the established criteria in Table 1. In order to determine its performance once this condition is reached, FEM was conducted.

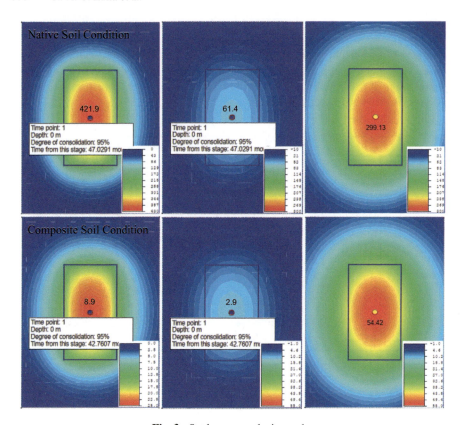

Fig. 3. Settlement analysis results

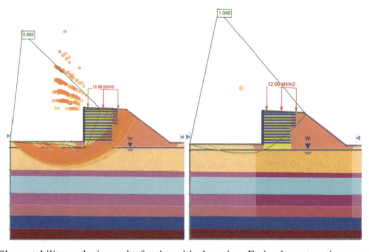

Fig. 4. Slope stability analysis results for the critical section. Embankment resting on native soil (Left); embankment resting on improved ground (Right)

3.6 Deformation Analysis

For all the FEM runs of the native soil with pseudo-static condition, the soil model collapsed due to excessive deformations. To illustrate, the deformation contours in Fig. 5 show the presence of a failure envelope cutting through the whole embankment and the upper soft layers of the existing ground.

At normal conditions, the fill settlements are generally unsatisfactory. Wall deflections exceeds the tolerable limits. In the event of an earthquake, excessive deformations are experienced. The tensile strength of the geostraps in all stations are adequate for the case of the unimproved soil. For static conditions, the maximum reinforcement tension experienced by the geostraps in each wall are all less than their pullout strengths.

In order to reduce the deformations to tolerable conditions, improvement of the soft and loose soils under the embankment with the use of soil cement columns is necessary.

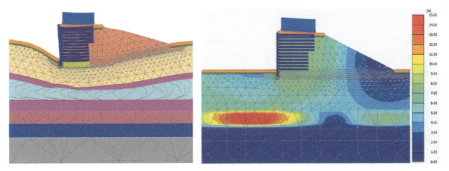

Fig. 5. Resulting exaggerated deformed mesh and defomation contour for the native ground (pseudo-static)

At normal conditions, the wall deflections and fill settlements resting on the improved ground are less than 10 mm and are within tolerable limits. This signifies the reduction of deformations upon improving the underlying soft and loose soil. However, for the pseudo-static cases, the computed deflections and settlements remain excessive (100–235 mm) at the upper portions of the wall, as shown in Fig. 6.

Geostraps have adequate strength against tensile failure for both the static condition and pseudo-static conditions. The pullout strength of the geostraps are generally adequate. However, the maximum reinforcement tension experienced by the geostraps in some of the upper layers at each location exceeds the pullout strength for the pseudo-static condition.

Fig. 6. Resulting deformation contour for the improved ground (pseudo-static)

4   Conclusions and Way Forward

Based on the results of the analyses conducted for the structure, Soil-Cement Columns will minimize liquefaction-induced settlements and layers with excessive settlements. The results of SSA for pseudo-static loading condition yielded inadequate factors of safety for global stability. Therefore, the deformation was analyzed using FEM to determine the performance of the embankment during seismic conditions.

In the context of performance-based design approach, improvement in the design may still be done by using the site-specific seismic parameters. Refinement of the established failure criteria may be done to optimize the design of both SCC and MSE wall.

References

1. Bowles, J.: Foundation Analysis and Design, 5th edn. McGraw-Hill, USA (1997)
2. Spudich, P., et al.: Final Report of the NGA-West2 Directivity Working Group. Pacific Earthquake Engineering Research Center, University of California, Berkeley (2013)
3. Kavazanjian, E., Matasovic, N., Hadj-Hamou, T., Sabatini, P.J.: Geotechnical Engineering Circular No. 3: Design Guidance: Geotechnical Earthquake Engineering For Highways (1997)
4. Boulanger, R., Idriss, I.: CPT and SPT-Based Liquefaction Triggering Procedures. Center for Geotechnical Modelling. Department of Civil and Environmental Engineering University of California. Davis, California (2014)
5. Pradel, D.: Procedure to evaluate earthquake-induced settlements in dry sandy soils. J. Geotech. Geoenviron. Eng. **124**(4), 364–368 (1998)
6. American Association of State Highway and Transportation Officials. AASHTO LRFD Bridge Design Specifications (2012)
7. Britist Standard. Code of practice for strengthened/reinforced soils and other fills (BS 8006–1 (2010)
8. Philippine Institute of Volcanology and Seismology. Distribution of Active Faults and Trenches in the Philippines (2019)

Geotechnical and Seismic Design Considerations for Coastal Protection and Retaining Structures in Reclaimed Lands in Manila Bay

Gian Paulo D. Reyes[✉], Roy Anthony C. Luna, John Michael I. Tanap, Marvin Renzo B. Malonzo, and Helli-mar T. Trilles

AMH Philippines, Inc., University of the Philippines Diliman, Guezon, Philippines
gian.reyes@amhphil.com

Abstract. In recent years, more and more reclamation projects along the coastlines of the Philippines' capital, Manila, are being envisioned and realized. With the commonly known poor ground conditions in coastal areas, characterized by soft and loose sedimentary soils, coupled with the combination of hazards that endanger the country, both seismic and climate related, the need for performance based, advanced engineering analyses has never been more important in the local design scene. This paper presents the geotechnical and seismic considerations in the long-term design of coastal protection and retaining structures of land reclamation projects in Manila Bay, aimed at addressing the multiple hazards inherent on the geographical location of the country. State-of-the-art technologies and approaches in geotechnical characterization and seismic analysis are discussed, including the seismic tests using seismic velocity logging (SVL) to obtain relevant dynamic properties of the soil, probabilistic seismic hazard analysis (PSHA) to quantify the overall seismic hazard of the area and obtain appropriate ground motions, and nonlinear time-history analysis (NLTHA) by finite element method (FEM) to simulate the response of the land mass and structures during earthquake events.

A case study on a massive land reclamation development along the shorelines of Manila Bay is presented.

Keywords: Coastal protection · Reclamation · Seismic velocity logging · Seismic hazard analysis · Nonlinear time-history analysis

1 Introduction

In recent years, the demand for more property development on reclaimed areas in the Philippines has escalated as real estate and infrastructure projects within the nation's capital, Manila, become more and more essential in boosting the country's economy and its drive to become a 'developed' nation. To date, there are more than twenty (20) proposed and on-going reclamation projects along Manila Bay comprising of various types of developments and infrastructures. One that will be discussed in this paper is a proposed 2,000+ hectare major infrastructure development located northwest of Manila Bay.

As with all land reclamation projects, one key component is the design and engineering of the land platform, dikes, and coastal protection structures that will ensure that the superstructures can be built safely on top of the reclaimed area. Additional design considerations are made to account for the engineering consequences of the country's geographical location. Being situated along the Pacific Ring of Fire, the Philippines accounts for 3.2% of the world's seismicity, posing threats of ground shaking, landslides, and liquefaction all over the country, and has more than twenty (20) typhoons passing over it every year. This paper focuses on the geotechnical engineering and seismic hazard assessment of such reclamation and coastal structures. A design approach is discussed which is specifically tailor-made to design against the unique hazards present in the country due to its geographical location using state-of-the-art analyses and technologies in geotechnical and earthquake engineering such as seismic velocity logging (SVL) geophysical test, probabilistic seismic hazard analysis (PSHA), and nonlinear dynamic time-history analysis (NLTHA) using finite element numerical method (FEM). Details of which are discussed using the case study project.

2 Project Background

The proposed 2,000+ hectare land reclamation project in Manila Bay is one of the latest flagship infrastructure projects of the country that aims to spearhead the economic growth of a particular group of provinces adjacent to Metro Manila. Project information shall not be disclosed at this point in time. Phase 1 of the project is the reclamation of the land where the infrastructures will be built on; this will be the main focus of this paper. The structures requiring engineering design and solutions are the land platform and the coastal protection comprising of armor rock slopes that will confine the entire area. Given the very soft subsurface conditions at the project area, a ground improvement scheme of preloading with prefabricated vertical drains or PVDs was deemed to be the most economical solution. The general construction methodology is to backfill the entire area until the pre-determined PVD installation levels, install PVDs, backfill until required surcharge level including settlement compensation heights, allow consolidation to occur, then finally construct the armor rock protection bund. About 8m to 10m high fill embankments are going to be placed to reclaim the overall land footprint and attain a specified target final ground level. After consideration of ground settlements, the final height of the armor rock slope protection ranges from 4 m to 6 m.

3 Geotechnical Conditions

In order to characterize the geotechnical subsurface conditions at the project area, various geotechnical investigations were carried out comprising of more than 160 boreholes with standard penetration tests (SPT), triaxial strength tests (UU, CU, CD), field vane shear tests, one-dimensional consolidation tests, and routine geotechnical laboratory tests; more than 60 cone penetration tests with pore pressure (CPTu); and a few number of seismic velocity logging (SVL) tests. One particular test that this paper would want to highlight is the SVL. Seismic velocity logging is an intrusive non-destructive method used to determine the physical properties of the underlying soil or rock surrounding a borehole and the speed with which seismic waves propagate through the strata. The test is conducted by lowering a PS suspension logger probe that has an acoustic wave source and geophone receivers into the borehole. The source, which is located bottom of probe, will apply waves that will then travel through the soil/rock material and the geophones, located top of probe, will receive and record the signal. The resulting outputs are the mechanical and dynamic properties of the soil such as shear wave velocity, modulus of elasticity, shear modulus of elasticity, and Poisson's ratio which are important parameters for the subsequent seismic hazard analysis (SHA) and nonlinear time-history analysis (NLTHA).

From the results of the field and laboratory tests, it was observed that the existing subsurface conditions at the project area consist of thick layers of very soft clays, extending from 5 m to 30 m below the existing ground level. These layers are underlain by stiff to very stiff clays and pockets of very dense sands. Subsurface idealization was performed, and parameters were subsequently derived for the geotechnical analyses conducted. Additionally, dynamic properties such as E50ref are derived from references provided by PLAXIS and are also checked with established literatures (i.e., Bowles) and previous project experiences.

Table 1 to Table 3 present the general summary of soil parameters.: Summary of settlement properties of soil (Table 2).

Table 1. Summary of settlement properties of soil

Soil layer	Compression index, Cc [−]	Creep index, Cα [−]	Coefficient of consolidation, Cv [m²/yr]
Very soft clay	−0.3–0.8	−0.020–0.045	−2.0
Stiff clay	−0.4–0.6	−0.015–0.020	−2.0

Table 2. Summary of strength properties of soil

Soil layer	Unit weight [kN/m³]	Undrained shear strength, Su [kPa]
Very soft clay	−5	−0–30
Stiff clay	−16	−30–70
Very stiff clay	−19	−80–120
Hard clay	−20	−150–250

4   Seismic Hazard Analysis

Seismic hazard analysis (SHA) is the process of quantifying the overall seismic hazard of an area in terms of acceleration. The probabilistic approach (PSHA) in performing

Table 3. Summary of dynamic properties of soil

Soil layer	Shear wave velocity, vs [m/s]	Shear modulus, G_0 [MPa]	Triaxial loading stiffness, E_{50}^{ref} [MPa]
Very soft clay	−90–140	−20–30	−5–10
Stiff clay	−140–200	−30–60	−10–20
Very stiff clay	−230–280	−60–90	−30–40
Hard clay	−300–430	−140–370	>−40

SHA quantifies seismic hazard at different levels of risk depending on the recurrence interval or return period of the design ground motion. PSHA also considers multiple seismic sources simultaneously and accounts for uncertainties related to distance, time, recurrence, and size (magnitude).

In performing SHA, empirically-formulated attenuation models are utilized to determine the expected surface acceleration by estimating how seismic waves propagate and travel from source to site. Attenuation models are commonly referred to as Ground Motion Prediction Equations (GMPE), and these equations were formulated using globally-acquired earthquake information (e.g. epicenter location, depth, and magnitude). The New Generation Attenuation West2 (NGA-West2) GMPE's developed by the Pacific Earthquake Engineering Research (PEER) Center were used for fault systems, and the BC Hydro GMPE (Abrahamson et al., 2016, 2018) was used for subduction zone sources.

For the case study project, the required earthquake return periods are 475-year or 10% probability of exceedance in 50 years, also known as Design Basis Earthquake (DBE), and 2,475-year or 2% probability of exceedance in 50 years, also known as Maximum Credible Earthquake (MCE). The output target response spectra at rock outcrop for each return period at 5%-damping are presented in Fig. 1.

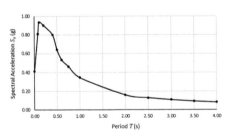 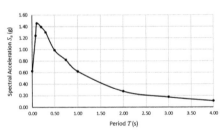

Fig. 1. (Left) 5%-damped DBE response spectra for rock outcrop, (Right) 5%-damped MCE response spectra for rock outcrop.

5   Spectral Matching and Ground Motion Selection

To carry out a nonlinear time-history analysis (NLTHA) for the armor rock slopes, a ground motion suite of records of at least three (3) ground motion records should be developed. The ground motion records selected must have similar characteristics to the seismic source with the most contribution to the overall seismic hazard at the project area and shall be spectrally matched with the target response spectra discussed in the previous section. Spectral matching is the process of modifying the amplitude and/or frequency content of a certain ground motion record such that the original record's response spectrum matches the response spectrum obtained from the SHA. By doing so, the modified record can be surmised to be an event that may happen on-site given the nature of the potential earthquake generators, their respective recurrence parameters, and their rupture/focal mechanisms.

Spectral matching was done using SeismoMatch 2018 by Seismosoft. SeismoMatch performs spectral matching using the wavelet algorithm of Al Atik and Abrahamson (2010), which is an update of the original algorithm proposed by Abrahamson (1992). This wavelet algorithm utilizes an improved tapered cosine adjustment function that prevents drift in the modified velocity and displacement time series without baseline correction. The quality of the matching is assessed by looking at how well-matched ground motions compare with the original records' time-histories (velocity and displacement). The primary criterion would be to check if the non-stationary parameters of the original ground motion is well-maintained even after matching. The matched response spectra (dashed plots) of the 3 ground motions are shown in Fig. 2.

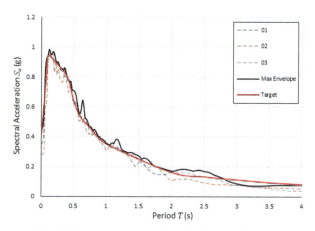

Fig. 2. Response spectra of matched ground motions.

6   Nonlinear Time-History Analysis

Nonlinear time-history analysis (NLTHA) measures the response of a soil-mass over the period of during and after the application of a full ground motion time-history record. With NLTHA, the stress state of materials is allowed to exceed the linear-elastic region and behave with nonlinear material properties (plastic or elastoplastic behavior). As such, moving past the limitation of being within the linear-elastic region may approximate or realize a more realistic behavior of the soil-mass (e.g. plastic hinge manifestation, stiffness degradation, hysteretic damping). NLTHA utilizes earthquake records that must capture the expected hazard on-site as described in the previous section. In geotechnical analyses, NLTHA is performed via numerical modeling, where the process is often referred to as "Site Response Analysis" (SRA). SRA is a process that measures accelerations and deformations by means of deconvolution-convolution cycles. Deconvolution and Convolution refer to the attenuation and amplification of ground accelerations, respectively, as seismic waves propagate across the subsurface. The deconvolution-convolution cycle is generally performed in a multi-step manner. SRA was carried out using a finite element method (FEM) numerical analysis tool, PLAXIS 2D. FEM is a numerical technique used for finding approximate solutions by continuum-based methods wherein the governing equations describing the state of stress of the soil/rock mass are derived on the principles of conservation of mass, momentum, and energy. These equations, together with the prescribed boundary conditions, result in a nonlinear boundary value problem. It is noted that the nonlinear analysis is carried out in the time domain.

NLTHA was performed for the armor rock protection slopes using the 3 spectrally matched ground motions discussed in the previous section. See Fig. 4 for the acceleration time-history record of one of the selected ground motions. In PLAXIS 2D, the Hardening Soil Small-strain (HSS) constitutive model (Benz 2007) was used to model the soil materials which involves dynamic parameters such as the shear wave velocity, Vs, shear modulus of elasticity, G0, and three moduli of elasticity, E50, Eoed, and Eur. The HSS model is an advanced model that simulates more realistic soil behavior by using three different input stiffnesses that accounts for non-linearity, stress-dependency, and inelasticity. It also has the ability to capture hysteretic behavior of soils at large and small strain levels, which is effective in generating realistic soil response for such non-linear dynamic analysis. More details on the HSS model can be found in the Benz (2007) and PLAXIS references. A compliant base boundary condition was set to account for the outcrop ground motions and a factor of 0.5 was applied to the time-histories to consider only the upward travelling seismic waves. This was done in reference to the guidelines presented by PLAXIS and Mejia (2006) where only half the magnitude of the earthquake time history should be applied when using outcrop motions. More details on this are discussed in the said references. Moreover, the finite element model was extended such that there will be no 'confining effects' that will impact the results, i.e., the model is extended in both the horizontal and vertical directions to ensure that the stress distribution and amplification/attenuation of seismic waves are not limited and concentrated at boundaries of the model.

A general summary of the results of the NLTHA is presented in Table 4. The magnitudes of each time-history record are also indicated. Figure 4 shows a typical deformed mesh and deformation contour outputs of the NLTHA in PLAXIS 2D.

The resulting deformations are found to be within the allowable limits set for the project. It is noted that the horizontal deformations, ux, were assessed in terms of relative displacements. Time-history record no. 1 (TH-1) incurred the most deformation at the crest and toe of the armor rock slope, having the highest magnitude among the three ground records. From the deformation contour diagram in Fig. 3, it can be observed

Table 4. Summary of NLTHA results. ux and uy are horizontal and vertical deformations

Ground motion	Maximum calculated deformation (mm)			
	Crest		Toe	
	ux	uy	ux	uy
TH-1 [Mw = 7.37]	−40	−220	−120	−8
TH-2 [Mw = 7.13]	−30	−215	−100	−1
TH-3 [Mw = 6.19]	−40	−210	−100	−1

that the concentration of the displacements (warmer colors) is within the armor rock section with it being the confining structure for the entire reclamation area. Traces of potential slip failure planes can also be seen. Figure 4 shows the deformed mesh where it can be observed that maximum vertical displacements are located right below the crest of the armor rock where the most volume of fill is placed, and maximum horizontal displacements occurred at the upper soft soil layers with lower shear strength.

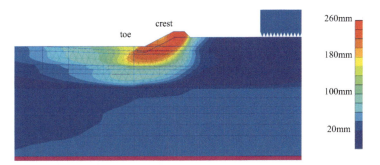

Fig. 3. Snapshot of the deformation contours of the armor rock slope. Full extent of model not shown. Colors represent the deformation contours with red (warmer colors) as the highest in magnitude and blue (colder colors) as the lowest. Magenta line at the bottom represents the ground motion application. Blue vertical arrows represent the loads applied on top of the land platform.

In terms of the surface accelerations that resulted from the site response analysis, there has been an observed reduction in the peak ground acceleration (PGA), i.e., acceleration at period equals zero, at the land platform surface when comparing it to the PGA value at rock site. However, acceleration amplifications were also observed in the mid-to-long period range. See Fig. 5. This is expected as the seismic waves travel from the rock up to the soft-to-stiff soil layers. These are important values in the design of the superstructures which is part of the next phase of the development.

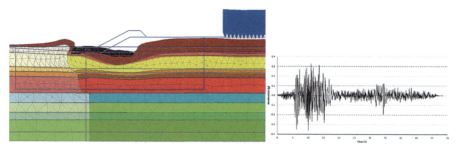

Fig. 4. (Right) Snapshot of the deformed mesh of armor rock slope. Figure is scaled up 20.0 times; full extent of model not shown. Lighter shaded soil materials represent the unimproved conditions while the darker shaded soil materials represent the improved conditions. (Left) Spectrally-matched acceleration time-history of one of the selected ground motions (TH-1).

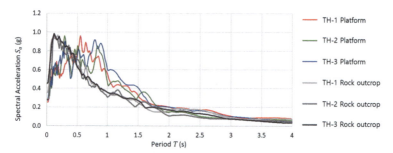

Fig. 5. Comparison of calculated response spectra at platform level vs rock site level for 3 ground motions.

7   Conclusion

The design of dikes and coastal protection structures for reclamation projects in the Philippines has been undergoing continuous improvement and development throughout the years as more and more reclamations projects are being envisioned and realized in the country. The constant rise of state-of-the-art geophysical and geotechnical engineering tests and analyses has led to the advancement in data gathering and design analysis of such structures. A combination of classical soil mechanics theories and advanced nonlinear numerical analyses has helped in providing engineering solutions that are more accurate, elaborate, realistic, and at the same time cost-effective. However, it is extremely important to remember that all of these analyses only provide the best estimates and simulations of the behavior of the structures and its surrounding. It is still best practice to conduct validation or proof tests, monitoring measurements, and recalibration analysis to verify all theoretical results with the actual observations.

References

Abrahamson, N., Silva, W., Kamai, R.: Summary of the ASK14 ground motion relation for active crustal regions. Earthq. Spectra **30**(3), 1025–1055 (2014)

Baker J.W. 2015. Introduction to Probabilistic Seismic Hazard Analysis

Benz, T.: Small-Strain Stiffness of Soils and its Numerical Consequences. Universität Stuttgart (2007)

Bowles, J.E.: Foundation Analysis and Design. 5th Edition (1997)

Lambe, W.T., Soil, W.R.V., Mechanics.: Massachusetts Institute of Technology. John Wiley & Sons, New York (1969)

Mejia, J., Dawson, E.: Earthquake deconvolution for FALC. In: 4th International FLAC Symposium on Numerical Modeling in Geomechanics. Madrid, Spain (2006)

Philippine Institute of Volcanology and Seismology. Distribution of Active Faults and Trenches in the Philippines (2019)

PLAXIS. PLAXIS Material Models Connect Edition V20

PLAXIS. PLAXIS 2D Dynamics Manual

Mote, R.T., Manlapig, R., Zamora, C.: Probabilistic Seismic Hazard Assessment for Central Manila in Philippines. AEES (2009)

A Framework for Real-Time Seismic Performance Assessment of Pile-Supported Wharf Structures Incorporating a Long Short-Term Memory Neural Network

Liang Tang[1,2(✉)], Yi Zhang[1,2], Zheng Zhang[1,2], Wanting Zhang[1,2], and Xianzhang Ling[1,2]

[1] School of Civil Engineering, Harbin Institute of Technology, Harbin 150090, China
hit_tl@163.com

[2] Heilongjiang Research Center for Rail Transit Engineering in Cold Regions, Harbin 150090, China

Abstract. Past earthquake disasters have shown that pile-supported wharf structures are susceptible to severe damage during earthquakes, and thus it is important to assess the seismic performance of pile-supported wharf structures. Effective post-earthquake performance assessment requires a prompt and accurate assessment. However, existing performance assessment methods cannot simultaneously meet these requirements. In recent years, with the progress of software and hardware technology, machine learning has surpassed the traditional methods in many fields such as computer vision and natural language processing. The time history of ground motion acceleration is a typical time series data, and most of the traditional machine learning methods are not good at dealing with such high dimensional time series data. Recurrent Neural Network (RNN), as a technology suitable for extracting and learning features of high-dimensional time-series data, has attracted more and more attention. A graphical user interface is developed to combine nonlinear dynamic time history analysis of pile-supported wharf structures with the implementation of Recurrent Neural Network (RNN). A framework for real-time seismic performance assessment of pile-supported wharf structures incorporating a Long Short-Term Memory (LSTM) neural network is proposed. The framework is built around a workflow that establishes mapping rules between ground motions and structural performance via pile-supported wharf structures models. In this paper, the analysis framework and the main components of the graphical user interface are presented.

Keywords: Numerical framework · Seismic performance assessment · Long short-term memory neural network · Pile-supported wharf structures

1 Introduction

Pile-supported wharf structures are a critical component of the construction of port facilities due to its many merits. But such a pile-supported wharf structure is susceptible

L. Wang et al. (Eds.): PBD-IV 2022, GGEE 52, pp. 818–825, 2022.
https://doi.org/10.1007/978-3-031-11898-2_56

to severe damage during major earthquake events. The seismic damage to pile-supported wharf structure has been frequently reported in many recent seismic cases [1–3]. To mitigate the potential seismic-induced damage of a pile-supported wharf structure, it is crucial to assess its seismic performance during earthquakes.

The finite element analysis has drawn considerable attention of engineers to assess its seismic performance of pile-supported wharf structure owing to its low cost and modeling generality [4–6]. However, the earthquake occurs very quickly. Although the finite element analysis method is accurate, it has a large amount of calculation, and has a real time defect. In recent years, with the progress of software and hardware technology, machine learning has surpassed the traditional methods in many fields. Several prior studies in earthquake engineering utilized Recurrent Neural Network (RNN) or Long Short-Term Memory (LSTM) for capturing the features of the time history of ground motion data [7–9]. As a result, nonlinear dynamic time history analysis of pile-supported wharf structures is combined with the realization of the LSTM, developed based on LSTM of pile-supported wharf structures seismic response analysis of finite element analysis software (Wharf LSTM).

In this paper, the elements of analysis framework and the main components of the graphical user interface are presented to highlight the underlying analysis framework capabilities and range of potential applications.

2   The Wharf LSTM GUI

The open-source computational platform OpenSees (http://opensees.berkeley.edu) [10] provides the possibility for three-dimensional numerical analysis of pile-supported wharf structures. In this open-source computational platform, a wide range of linear and nonlinear soil and structural elements is available [11, 12]. Based on the finite element model modeling foundation and the superiority of the machine learning method, the team developed this three-dimensional finite element software (Wharf LSTM). The Wharf LSTM user interface allows for: establish pile-supported wharf structures finite element model, carry on Pushover analysis and LSTM analysis, complete quickly the seismic performance assessment of pile-supported wharf structures.

3   Wharf LSTM Analysis Framework

Based on a large number of historical earthquake ground motion and accurate nonlinear dynamic time history analysis method to obtain damage state of pile-supported wharf structures, the mapping relationship between the acceleration time history of ground motion and the damage state of pile-supported wharf structures was learned by using the LSTM neural network, and the training model of a LSTM neural network was obtained, which was embedded in Wharf LSTM software. Wharf LSTM software can be used to establish a refined finite element model of pile-supported wharf structures. The damage state of pile-supported wharf structures can be determined by Pushover analysis. The ground motion data can be obtained based on the open-access ground motion databases (e.g. the PEER-NGA West/West2/East database in the United States and the K-NET database in Japan) or actual ground motion. A reasonable method was used for selection and pretreatment to obtain the required acceleration time history, input the preloaded LSTM model, and compare the damage criterion to quickly complete the seismic performance assessment of pile-supported wharf structures.

4   Elements of the Wharf LSTM GUI

Among the main Wharf LSTM GUI component are the finite element model generation of pile-supported wharf structures, Pushover analysis, definition of damage state of pile-supported wharf structures, selection and pretreatment of ground motion, LSTM model training and analysis, and seismic performance assessment of pile-supported wharf structures. These elements of the Wharf LSTM GUI are briefly addressed below. Figure 1 shows the main user interface of Wharf LSTM.

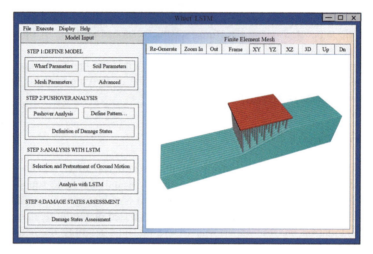

Fig. 1. The main user interface of Wharf LSTM.

4.1   Finite Element Model of Pile-Supported Wharf Structures

In Wharf LSTM computing software, the finite element model of pile-supported wharf structures needs to be defined. The piles are modeled as linear beam-column elements or nonlinear beam-column elements with fiber cross sections. The deck is also modeled using linear beam-column element. And the deck is assumed to be capacity designed so that it responds in the elastic range. Different constitutive models and unit descriptions of soil models, soil-pile interaction, boundary conditions and other modeling details are based on earlier research [6, 13–16].

4.2   Pushover Analysis

Wharf LSTM has two methods available for Pushover analysis, include: force-based and displacement-based. The Pushover analysis window is shown in Fig. 2. If Force-Based Method is chosen, please enter the parameters of force increment (per step);

If Displacement-Based Method is chosen, please enter the displacement increment parameters (per step). The pushover load/displacement linearly increases with step in a monotonic pushover mode. The load/displacement is applied at the deck.

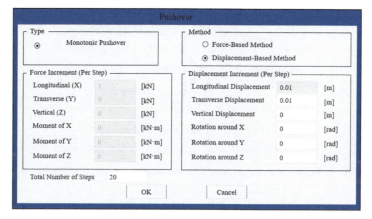

Fig. 2. The Pushover analysis window.

The output of the Pushover analysis include: 1) Response time histories and profiles for pile; 2) Response relationships (force–displacement as well as moment–curvature) for pile; 3) Deformed mesh, contour fill, and animations.

4.3   Definition of Damage State of Pile-Supported Wharf Structures

According to the results of moment curvature analysis and the initial axial force of pile, the pile yield moment distribution diagram of pile-supported wharf structures can be obtained. By comparing the pile internal force obtained in Pushover analysis with the pile yield moment at the corresponding position, the development position of the pile plastic zone can be determined, as shown in Fig. 3. Taking deck displacement as monitoring index, the damage evolution process of pile-supported wharf structures can be captured according to the above damage state location method and the data obtained from Pushover analysis.

In this study, definition of damage state of pile-supported wharf structures according to the International Navigation Association (PIANC) (2001), as shown in the Table 1. PIANC proposed qualitative criteria for determining the grade of damage of a pile-supported wharf structure. However, PIANC did not provide a clear quantitative criterion.

In order to better apply the concepts to subsequent seismic performance assesment, combined with the damage evolution process of pile-supported wharf structures captured, the damage criteria for each degree are determined by taking deck displacement as monitoring index. The specific displacement limits are de-tailed as follows:

Limit value 1 (boundary between degree I and degree II): When deck displacement reaches a certain value, plastic hinge appears at a pile top, and the overall bearing capacity of pile-supported wharf structures decreases to a certain extent. As the plastic

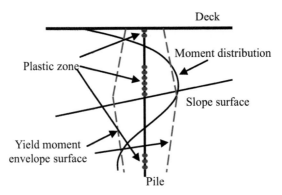

Fig. 3. Location diagram of pile plastic zone [17].

Table 1. Proposed damage criteria for pile-supported wharf structures [18].

Level of damage	Degree I	Degree II	Degree III	Degree IV
Pile (peak response)	Essentially elastic response with minor or no residual deformation	Controlled limited inelastic ductile response and residual deformation intending to keep the structure repairable	Ductile response near collapse (double plastic hinges may occur at one or limited number of piles)	Beyond the state of Degree III

hinge located at the pile top is convenient for maintenance and reinforcement after the earthquake, it belongs to the repairable category, and the deck displacement at this time is defined as limit value 1.

Limit value 2 (boundary between degree II and degree III): When deck displacement reaches a certain value, the piles produce new plastic hinges at a certain depth below the slope surface, and the overall bearing capacity of pile-supported wharf structures will further decline. Because the plastic hinge in the soil is inconvenient to repair and reinforce after the earthquake, belongs to the category of irreparable. Based on the limited bearing capacity of the plastic hinge itself and the possibility of the plastic hinge at the pile top being repaired in time after the earthquake, the pile-supported wharf structure still belongs to the category of single plastic hinge structure, and the deck displacement at this time is defined as limit value 2.

Limit value 3 (boundary between degree III and degree IV): When deck displacement reaches a certain value, another new plastic hinge in soil is generated at the bearing layer of pile, which also belongs to the category of irreparable. At this point, even if the piles undergo effective earthquake repair, the pile-supported wharf structure will also belong to the category of double plastic hinges structure. Therefore, it can be judged that the whole pile-supported wharf structure has lost its bearing capacity at this time. The deck displacement at this time is defined as limit value 3.

Therefore, based on Pushover analysis, the damage state of pile-supported wharf structures at each degree is quantitatively determined by taking deck displacement as an index.

4.4   Selection and Pretreatment of Ground Motion

For real-life implementations, it is desirable if ground motion selection and pretreatment is carried out by considering the characteristics of the seismic setting of the target region and matching with the input characteristics of LSTM neural network. Good selection and pretreatment of ground motion can improve the performance of the training model of a LSTM neural network and improve computational efficiency. The selection and pretreatment of ground motion in this software include three steps of amplitude adjustment, frequency adjustment and duration adjustment [19]. The processing flow is shown in Fig. 4.

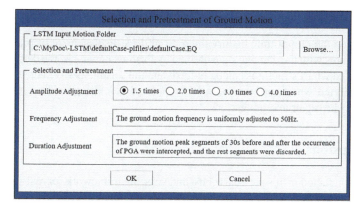

Fig. 4. The selection and pretreatment of ground motion.

Amplitude adjustment: Due to the damage of pile-supported wharf structures caused by ground motion in the ground motion database is mostly slight damage, in order to enrich the ground motion data and comprehensive analysis of pile-supported wharf structures, therefore, amplitude adjustment of acceleration time histories obtained in database is carried out in different proportions, and acceleration time histories are multiplied by amplification coefficients of 1.0, 2.0, 3.0 or 4.0.

Frequency adjustment: The obtained ground motion data is composed of waveforms recorded at different sampling frequencies. The input of ground motion data at different sampling frequencies into the LSTM model will lead to the degradation of the performance of the LSTM training model. Therefore, the sampling frequency of ground motion data should be adjusted so that the sampling frequency of all records is the same, and the sampling frequency is uniformly adjusted to 50 Hz.

Duration adjustment: LSTM model is not limited by the duration of ground motion data, but different ground motion data require different training space due to different duration, and different space placement will cause some negative effects. Therefore, the ground motion peak segments of 30 s before and after the occurrence of PGA were intercepted, and the rest segments were discarded.

4.5 LSTM Model Training and Analysis

The LSTM model embedded in the Wharf LSTM software learns the mapping relationship between the acceleration time history of ground motion and deck displacement through the LSTM neural network. It is worth noting that the machine learning model continuously learns the relationship between the two with the help of a large amount of ground motion data to achieve better accuracy. The user can input the time-history of the studied ground motion acceleration, through the above processing method, load the LSTM model, conduct LSTM analysis, and output the time history diagram of the deck displacement of the pile-supported wharf structures.

4.6 Seismic Performance Assessment of Pile-Supported Wharf Structures

By comparing the output time history diagram of deck displacement of pile-supported wharf structures with the damage state defined by Pushover analysis, the damage level of pile-supported wharf structures can be judged, and the seismic performance assessment of pile-supported wharf structures can be completed quickly, providing reference for post-earthquake repair and other work.

5 Conclusions

The study conducted in this paper, a robust and versatile framework for real-time seismic performance assessment of pile-supported wharf structures is given by combining the neural network of Long Short-Term Memory neural network and finite element method, and the performance indexes and limits of pile-supported wharf structures in different stages during earthquake are determined. It is important to note that the reported outcomes are a direct consequence of modelling assumptions and applying Long Short-Term Memory neural network, so the research in this paper should adapt the framework to more accurately and effectively represent the specific practice scenarios of interest, as needed.

Acknowledgements. This work was supported by the National Key R&D Program of China (Grant No. 2016YFE0205100), the National Natural Science Foundation of China (Grant Nos. 51578195 and 41902287), and the Technology Research and Development Plan Program of Heilongjiang Province (Grant No. GA19A501).

References

1. Werner, S.D., Dickenson, S.E., Taylor, C.E.: Seismic risk reduction at ports: case studies and acceptable risk evaluation. J Waterw Port Coast **123**(6), 337–346 (2000)
2. Mylonakis, G., Syngros, C., Gazetas, G., Tazoh, T.: The role of soil in the collapse of 18 piers of Hanshin Expressway in the Kobe earthquake. Earthq Eng Struct D **35**(5), 547–575 (2010)
3. Green, R.A., Olson, S.M., Cox, B.R., Rix, G.J., Rathje, E., Bachhuber, J., French, J., Lasley, S., Martin, N.: Geotechnical Aspects of Failures at Port-au-Prince Seaport during the 12 January 2010 Haiti Earthquake. Earthq Spectra **27**(S1), S43–S65 (2011)
4. Shafieezadeh, A., Desroches, R., Rix, G.J., Werner, S.D.: Seismic performance of pile-supported wharf structures considering soil-structure interaction in liquefied soil. Earthq Spectra **28**(2), 729–757 (2012)
5. Su, L., Lu, J., Elgamal, A., Arulmoli, A.K.: Seismic performance of a pile-supported wharf: Three-dimensional finite element simulation. Soil Dyn Earthq Eng **95**, 167–179 (2017)
6. Su, L., Wan, H.P., Dong, Y., Frangopol, D.M., Ling, X.Z.: Efficient uncertainty quantification of wharf structures under seismic scenarios using gaussian process surrogate model. J Earthq Eng. (3), 1–22 (2018)
7. Panakkat, A., Adeli, H.: Recurrent neural network for approximate earthquake time and location prediction using multiple seismicity indicators. Comput-Aided Civ Inf **24**(4), 280–292 (2010)
8. Bhandarkar, T., Vardaan, K., Satish, N., Sridhar, S., Ghosh, S.: Earthquake trend prediction using long short-term memory RNN. Int. J. Electrical Computer Eng. **9**(2), 1304 (2019)
9. Kim, T., Song, J., Kwon, O.S.: Pre- and post-earthquake regional loss assessment using deep learning. Earthq Eng Struct D. (3) (2020)
10. Mckenna, F., Fenves, G.L.: Open system for earthquake engineering simulation (OpenSees). Pacific Earthquake Engineering Research Center. University of California (2013)
11. Elgamal, A., Lu, J., Forcellini, D.: Mitigation of liquefaction-induced lateral deformation in a sloping stratum: three-dimensional numerical simulation. J. Geotechnical Geoenvironmental Eng. **135**(11), 1672–1682 (2009)
12. Spacone, E., Filippou, F.C., Taucer, F.F.: Fibre beam-column model for non-linear analysis of R/C frames: part 1. Formulation. Earthq Eng Struct D **25**(7), 711–725 (2015)
13. Zhang, X., Tang, L., Ling, X., Chan, A., Jinchi, L.: Using peak ground velocity to characterize the response of soil-pile system in liquefying ground. Eng Geol. S217649360 (2018)
14. Su, L., Wan, H.P., Dong, Y., Frangopol, D.M., Ling, X.Z.: Seismic fragility assessment of large-scale pile-supported wharf structures considering soil-pile interaction. Eng Struct. **186**(MAY 1), 270–281 (2019)
15. Su, L., et al.: Seismic fragility analysis of pile-supported wharves with the influence of soil permeability. Soil Dyn Earthq Eng. **122**(JUL.), 211–227 (2019)
16. Zhang, X., Tang, L., Ling, X., Chan, A.: Critical buckling load of pile in liquefied soil. Soil Dyn Earthq Eng **135**, 106197 (2020)
17. Chiou, J.S., Chiang, C.H., Yang, H.H., Hsu, S.Y.: Developing fragility curves for a pile-supported wharf. Soil Dyn. Earthq. Eng. **31**(5–6), 830–840 (2011)
18. Bernal, A., Blazquez, R., Burcharth, H.F., Dickenson, S.E., Sugano, T.: Seismic Design Guidelines for Port Structures (2001)
19. Xu, Y., Lu, X., Cetiner, B., Taciroglu, E.: Real-time regional seismic damage assessment framework based on long short-term memory neural network. Comput-Aided Civ Inf (2020)

Evaluation of the Liquefaction Hazard for Sites and Embankments Improved with Dense Granular Columns

Juan Carlos Tiznado[1]($\boxtimes$) (ID), Shideh Dashti[2] (ID), and Christian Ledezma[1] (ID)

[1] Pontificia Universidad Católica de Chile, Santiago, Chile
jctiznad@ing.puc.cl
[2] University of Colorado at Boulder, Boulder, CO, USA

Abstract. Dense granular columns (DGC) have become a common soil improvement strategy for critical embankment structures founded on potentially liquefiable deposits. The state-of-practice for the design of DGCs is limited to simplified methods that consider, separately, the three-primary liquefaction-mitigation mechanisms provided by these columns: (i) installation-induced densification; (ii) enhanced drainage; and (iii) shear reinforcement. Critical aspects, such as the effects of soil-column-embankment interaction, site characteristics and layer-to-layer interaction, ground motion characteristics beyond the peak ground acceleration, and the total uncertainty, are not included in current engineering design procedures. In this work, we present results from a numerical parametric study, previously validated with dynamic centrifuge test results, to evaluate the liquefaction hazard in layered profiles improved with DGCs. The criteria for various degrees of liquefaction are based on peak excess pore pressure ratios and shear strains observed within each layer. Our study includes different properties and geometries for both soil and DGCs, various confining pressures induced by an overlying embankment, as well as a large collection of ground motions from shallow crustal and subduction earthquakes. We performed a total of 30,000 3D, fully coupled, nonlinear, dynamic finite-element (FE) simulations in OpenSees using a state-of-the-art soil constitutive model (PDMY02), whose properties were calibrated based on both element level laboratory tests and a free-field boundary-value problem modeled in the centrifuge. The results from this parametric study are used to develop a probabilistic predictive model for the triggering of liquefaction in embankment sites treated with DGCs.

Keywords: Liquefaction triggering · Dense granular columns · Finite-element modeling · Centrifuge modeling · Probabilistic models

1 Introduction

Dense granular columns (DGC) have become a common soil improvement strategy for critical geo-structures founded on potentially liquefiable deposits. Although proven effective in case histories from past earthquakes [1] as well as previous experimental and numerical studies (e.g., [2, 3]), the influence of various mitigation mechanisms provided

© The Author(s), under exclusive license to Springer Nature Switzerland AG 2022
L. Wang et al. (Eds.): PBD-IV 2022, GGEE 52, pp. 826–833, 2022.
https://doi.org/10.1007/978-3-031-11898-2_57

by DGCs and their seismic interactions with site and ground motion characteristics is not yet well understood in the context of liquefaction triggering and softening in the treated ground.

The state-of-practice for the design of DGCs is currently limited to simplified approaches that consider, in a de-coupled manner, their primary liquefaction-mitigation mechanisms: (i) installation-induced densification; (ii) enhanced drainage; and (iii) shear reinforcement. Critical aspects, such as the effects of soil-column-embankment interaction, site characteristics and layer-to-layer interaction, ground motion characteristics beyond the peak ground acceleration, and the total uncertainty, are not included in current engineering design procedures.

In this work, we present results from a numerical parametric study, previously validated against dynamic centrifuge test results, to evaluate the liquefaction hazard in layered profiles improved with DGCs. The criteria for various degrees of liquefaction are based on excess pore pressure ratios and shear strains observed within each layer. Our study includes different properties and geometries for both soil and DGCs, various confining pressures induced by an overlying embankment, as well as a large collection of ground motions from shallow crustal and subduction earthquakes. We performed a total of 30,000 3D, fully-coupled, nonlinear, dynamic finite-element (FE) simulations in OpenSees using the PDMY02 soil constitutive model, whose properties were calibrated based on both element level laboratory tests and a free-field boundary-value problem modeled in the centrifuge. The results from this parametric study are used to propose a probabilistic predictive model for the triggering of liquefaction in embankment sites treated with DGCs.

2 Numerical Modeling

The numerical simulations considered a fully saturated, layered profile of granular soils with variations in relative density (D_r) and hydraulic conductivity (k), treated with DGCs of varying area replacement ratio (A_r), shear stiffness ratio (G_r), and k. The base model was adopted from the soil profile tested in centrifuge by [4], as shown in Fig. 1a.

The pressure-dependent, multi-yield surface, plasticity-based constitutive model (PDMY02) [5] was chosen to simulate the nonlinear response of saturated granular soils during dynamic loading. The PDMY02 model parameters used for Ottawa sand of different relative densities were adopted based on calibrations against drained and undrained, cyclic and monotonic triaxial tests [6] and a centrifuge free-field test performed at University of Colorado Boulder (CU) involving the same soil layers of interest in this study [7, 8] as well as the number of cycles to trigger liquefaction based on empirical observations of liquefaction surface manifestation [9]. Calibration of soil parameters for the dense layer of Monterey sand and the silt capping layer was only based on the number of cycles to trigger liquefaction from previous cyclic simple shear tests in the literature [10]. Finally, properties for DGCs were selected based on recommendations of [2] and strength and permeability tests performed at CU on the same soil type representing DGCs in the centrifuge [4, 11].

Figure 1a shows the model setup and instrumentation in the single-drain dynamic centrifuge test (with a prototype to model scale of 70) performed by [4]. Figure 1b shows

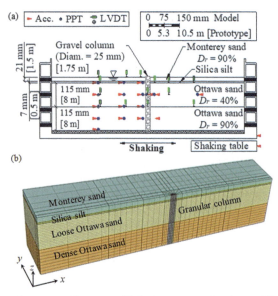

Fig. 1. (a) Single-column centrifuge test with its instrumentation layout [4]; (b) Finite-element mesh configuration of the centrifuge experiment.

the OpenSees model simulating the Badanagki et al. centrifuge experiment [4]. Only half of the container was modeled in the direction perpendicular to shaking. Nodes at the left and right edges of the model were tied together according to their elevation in the x- and z-directions. Nodes on both the inner symmetry plane and the opposite boundary were fixed against out-of-plane displacements. The DGC and the surrounding soil element nodes were assumed to be tied together at their nodes. Finally, nodes at the base of the model were fully fixed, and those at the soil surface were modeled with a prescribed zero-pore-pressure condition. 3-D BrickUP 8-node elements were used to model the soil layers. The maximum allowable element size was determined based on the estimated small-strain shear wave velocity of the soil profile and the maximum frequency content of the input motion in the centrifuge. In general, the numerical simulations could successfully capture: (i) the peak magnitude and rate of excess pore pressure generation at different depths and locations with respect to the DGC up to the point of liquefaction, and (ii) the peak magnitude of transient shear strain within different layers; which provided confidence in the ability of the simulations to produce reasonable predictions of liquefaction triggering in layered liquefiable deposits treated with DGCs. The models for liquefaction triggering detailed subsequently rely on both the magnitude of peak excess pore pressure ratio and shear strain developed within each soil layer.

3 Numerical Parametric Study

A comprehensive numerical parametric study was performed to: (i) evaluate the influence of different soil, DGC, and ground motion input parameters (IPs) on the patterns of excess pore water pressure (EPWP) generation and shear strain accumulation, and (ii) produce a dataset for developing a probabilistic predictive model for liquefaction triggering in layered sites improved with DGCs.

In these simulations, the soil-mitigation system was modeled as a unit cell, representative of a regular, repeating grid of DGCs. All models considered a DGC with a typical diameter $(d) = 1$ m. Therefore, column spacing was automatically accounted for through changes in A_r. The bedrock was modeled as an elastic medium by considering dashpots at the base and the input motions were applied to the base nodes as shear force–time histories. By using a quasi-Monte Carlo (QMC) sampling of cases, a total of nearly 30,000 simulations was performed with the aid of the Summit supercomputer at CU and the parallel version of OpenSees. This study considers 11 soil model geometries, as shown in Fig. 2. These profile geometries were designed to account for layer-specific characteristics, the presence of multiple liquefiable or critical layers, spatial variations in permeability through including a thin low-permeability silt cap, the geometry and characteristics of DGCs, and the effects of soil-mitigation and layer-to-layer interaction.

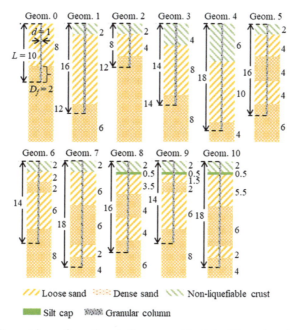

Fig. 2. Geometric configurations in the parametric study (dimensions in meters).

Table 1 summarizes the soil and granular column input parameters (IPs) that were varied in the numerical parametric study for each soil geometry (0 to 10). The variables considered were relative density of the critical or looser Ottawa sand layer(s) (D_{rc}), area replacement ratio (A_r) of DGC, ratio of maximum shear modulus of the DGC to that of the surrounding critical soil layer (G_r), ratio of hydraulic conductivity of the DGC to that of the surrounding critical soil layer (k_r), hydraulic conductivity of the critical layer itself (k_c), and the surcharge load applied at the ground surface (q) by a hypothetical embankment or foundation.

Table 1. Input Parameters (IP) considered for the sampling of cases in the parametric study

IP	D_{rc} (%)	A_r (%)	k_r	G_r	k_c (m/s)	q (kPa)
Range	30–90	0–20	0–100	2–8	Silty to Clean Sand	0–200
Values	30; 40; 60; 70; 90	0; 5; 10; 20	0; 1; 100	2; 5; 8	10^{-6}; 10^{-5}; 10^{-4}	0;50; 100; 150; 200

A suite of 151 outcropping rock earthquake motions was selected from the Bullock et al. database [12] and used as input to the numerical parametric study. It included 74 records from normal, reverse, and strike-slip shallow crustal earthquakes and 77 records from interface and intraslab subduction earthquakes. These motions covered a wide of range parameters in terms of earthquake magnitude, closest distance to rupture, shaking intensity, frequency, duration, and rate of energy buildup.

4   Influence of IPs on Peak EPWP and Shear Strain

Figure 3 shows the tornado diagrams summarizing the relative sensitivity of $r_{u,max}$ and γ_{max} in a given soil layer to various input parameters (IPs) and ground motion intensity measures (IMs), based on the results of the comprehensive numerical parametric study. Overall, the cumulative absolute velocity of the outcropping rock motion, CAV_{OR}, the post-treatment (i.e., after column installation) relative density (D_r), and hydraulic conductivity (k_{soil}) were the most influential parameters on both of $r_{u,max}$ and γ_{max} within a given layer. These IPs were followed by the layer's confining stress (represented by its depth, D), thickness (H), and embankment load (q). A_r played a more significant role in determining the extent of γ_{max} compared to $r_{u,max}$.

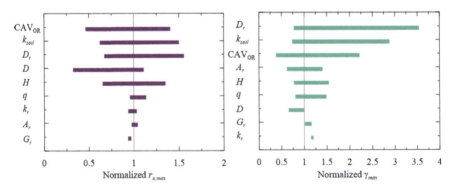

Fig. 3. Tornado diagrams for maximum excess pore pressure ratio and peak shear strain. X-axes are normalized by the median values of $r_{u,max}$ and γ_{max}, respectively.

5   Probabilistic Model for Liquefaction Triggering

Using the numerically generated database, the risk of liquefaction was evaluated based on the triggering criteria proposed by Olson et al. [13], which depend on the peak values of excess pore pressure ratio ($r_{u,max}$), maximum shear strain (γ_{max}), and a limit shear strain value (γ_{limit}) for each layer. In this study, we considered the values of γ_{limit} for relatively loose ($D_r \sim 30\%-50\%$), medium dense ($D_r \sim 50\%-70\%$), and dense ($D_r > \sim 70\%$) clean sands proposed by Olson et al. [13] (i.e., $\gamma_{limit} = 2\%$, 1.5%, and 1.2%, respectively). Therefore, the following definitions regarding liquefaction triggering were adopted in the predictive model that follows: (1) no liquefaction or NL (when $r_{u,max} < 0.8$); (2) marginal liquefaction or ML (when $0.8 \leq r_{u,max} \leq 0.9$ and $\gamma_{max} < \gamma_{limit}$); and (3) full liquefaction or FL (when $0.8 \leq r_{u,max} \leq 0.9$ and $\gamma_{max} \geq \gamma_{limit}$ or $r_{u,max} \geq 0.9$, whichever happens first in the numerical analyses).

After processing the results from the numerical parametric study, multinomial logistic regression was employed to develop the functional forms of a probabilistic predictive model for liquefaction triggering in layered sites improved with DGCs. We selected predictors for the model using the LASSO or least absolute shrinkage and selection operator [14]. In the multinomial logistic framework, a baseline or reference level needs to be defined to estimate the logit functions, L_j (in this case, $j = 1$ or $j = 2$), for all other categories relative to the baseline assuming a linear trend. Taking FL as the reference level, the logit functions can be evaluated as:

$$L_j = \beta_{0,j} + \beta_{1,j} ln\left(\frac{q}{G_{r-i}^2}\right) + \beta_{2,j}\left(\frac{A_r G_{r-i} k_{r,flag-i}}{100}\right) + \beta_{3,j}\sqrt{D_{r-i}} + \beta_{4,j}k_{flag-i}$$
$$+\beta_{5,j}\left(\frac{H_i}{D_i}\right) + \beta_{6,j}ln(CAV_{OR}) + \beta_{7,j}Env \tag{1}$$

where $\beta_{0,j}$, $\beta_{1,j}$, ... $\beta_{7,j}$, are regression coefficients determined via penalized maximum likelihood and summarized in Table 2.

Table 2. Coefficients for the model predicting liquefaction triggering.

Parameter	Value	Parameter	Value
$\beta_{0,1}$	0.905	$\beta_{0,2}$	-1.890
$\beta_{1,1}$	0.256	$\beta_{1,2}$	0.268
$\beta_{2,1}$	0.405	$\beta_{2,2}$	0.181
$\beta_{3,1}$	0.986	$\beta_{3,2}$	0.496
$\beta_{4,1}$	1.229	$\beta_{4,2}$	0.852
$\beta_{5,1}$	-0.794	$\beta_{5,2}$	-0.593
$\beta_{6,1}$	-2.289	$\beta_{6,2}$	-1.070
$\beta_{7,1}$	2.128	$\beta_{7,2}$	0.928

In these equations, q is the surcharge load applied at the ground surface in kilopascals (kPa). $G_{r\text{-}i}$, is defined as the ratio of equivalent DGC shear modulus, $G_{eq\text{-}DGC,i}$, to the surrounding soil layer shear modulus, $G_{soil,i}$:

$$G_{r-i} = \frac{G_{eq-DGC,i}}{G_{soil,i}}; \text{ where } G_{eq-DGC,i} = G_{DGC,i} \cdot \frac{H_{DGC,i}}{H_i} + G_{soil,i} \cdot \left(1 - \frac{H_{DGC,i}}{H_i}\right) \quad (2)$$

$G_{DGC,i}$ is the DGC maximum shear modulus averaged within the i-th layer, $H_{DGC,i}$ is its corresponding length, and H_i is the layer thickness, which is always equal or greater than $H_{DGC,i}$. The column-to-soil hydraulic conductivity ratio for the i-th layer, $k_{rflag\text{-}i}$, is a flag defined as: 1 for inhibited drainage; 2 for no change in drainage capacity; and 3 for enhanced drainage. The term $D_{r\text{-}i}$ (in percentage) refers to the relative density of the i-th layer. The flag for soil hydraulic conductivity, $k_{flag\text{-}i}$, in a given layer was defined as 1 if $k_{soil,i} \leq 10^{-6}$ m/s; 2 if $10^{-6} < k_{soil,i} \leq 10^{-5}$ m/s; 3 if $10^{-5} < k_{soil,i} \leq 10^{-4}$; and 4 if $k_{soil,i} > 10^{-4}$ m/s. D_i is the depth to the middle of the i-th layer measured from the ground surface. Finally, CAV_{OR} is the cumulative absolute velocity of the outcropping rock input ground motion in cm/s. The parameter Env was set as 1 for shallow crustal and 2 for subduction earthquakes, respectively.

6 Concluding Remarks

The results from a comprehensive numerical parametric study were used to develop the functional forms of a probabilistic model for evaluating liquefaction triggering in liquefaction-susceptible granular soil profiles treated with DGCs. The presented model allows users to estimate probabilities of full, marginal, and no liquefaction within each critical layer in the presence of DGCs, following the definitions proposed by Olson et al. [13]. The model is intended for use as an initial design tool for the mitigation of soil liquefaction with DGCs. It is expected that, after validation with case histories of sites improved with DGCs, new and improved versions of the model include other important aspects such as the influence of soil variability, multidirectional shaking, and the presence of intermediate soils (e.g., silts, silty sands, and low-plasticity clays) in the profile.

References

1. Hausler, EA.: Influence of ground improvement on settlement and liquefaction: A study based on field case history evidence and dynamic geotechnical centrifuge tests. Ph.D. thesis, Dept. of Civil and Environ. Eng., Univ. of California, Berkeley (2002)
2. Rayamajhi, D., Ashford, S.A., Boulanger, R.W., Elgamal, A.: Dense granular columns in liquefiable ground. I: Shear reinforcement and cyclic stress ratio reduction. J. Geotech. Geoenviron. Eng. **142**(7), 4016023 (2016)
3. Tiznado, J.C., Dashti, S., Ledezma, C., Wham, B.P., Badanagki, M.: Performance of embankments on liquefiable soils improved with dense granular columns: observations from case histories and centrifuge experiments. J. Geotech. Geoenviron. Eng. **146**(9), 04020073 (2020)
4. Badanagki, M., Dashti, S., Kirkwood, P.: Influence of dense granular columns on the performance of level and gently sloping liquefiable sites. J. Geotech. Geoenviron. Eng. **144**(9), 04018065 (2018)

5. Elgamal, W.A., Yang, Z., Parra, E., Ragheb, A.: Modeling of cyclic mobility in saturated cohesionless soils. Int. J. Plast. **19**(6), 883–905 (2003)
6. Badanagki, M.: Centrifuge modeling of dense granular columns in layered liquefiable soils with varying stratigraphy and overlying structures. Doctoral dissertation, Dept. of Civil, Environ., and Architect. Eng., Univ. of Colorado at Boulder (2019)
7. Ramirez, J., Barrero, A.R., Chen, L., Dashti, S., Ghofrani, A., Taiebat, M., Arduino, P.: Site response in a layered liquefiable deposit: evaluation of different numerical tools and methodologies with centrifuge experimental results. J. Geotech. Geoenviron. Eng. **142**(10), 04018073 (2018)
8. Hwang, Y.W., Ramirez, J., Dashti, S., Kirkwood, P., Liel, A.B., Camata, G., Petracca, M.: Seismic interaction of adjacent structures on liquefiable soils: insight from centrifuge and numerical modeling. J. Geotech. Geoenviron. Eng. **147**(8), 04021063 (2021)
9. NCEER: Proceedings of the NCEER workshop on evaluation of liquefaction resistance of soils. In: Youd, T.L., Idriss, I.M.: (Eds.), Technical Report No. NCEER-97–0022 (1997)
10. Karimi, Z., Dashti, S.: Seismic performance of shallow founded structures on liquefiable ground: validation of numerical simulations using centrifuge experiments. J. Geotech Geoenviron Eng. **142**(6), 04016011 (2016)
11. Li, P., Dashti, S., Badanagki, M., Kirkwood, P.: Evaluating 2D numerical simulations of granular columns in level and gently sloping liquefiable sites using centrifuge experiments. Soil Dyn. Earthquake Eng. **110**, 232–243 (2018)
12. Bullock, Z., Dashti, S., Liel, A., Porter, K., Karimi, Z., Bradley, B.: Ground-motion prediction equations for Arias intensity, cumulative absolute velocity, and peak incremental ground velocity for rock sites in different tectonic environments. Bull. Seismol. Soc. of America **107**(5), 2293–2309 (2017)
13. Olson, S.M., Mei, X., Hashash, Y.M.A.: Nonlinear site response analysis with pore-water pressure generation for liquefaction triggering evaluation. J. Geotech. Geoenviron. Eng., **146**(2), 04019128 (2020)
14. Tibshirani, R.: Regression shrinkage and selection via the lasso: a retrospective. J. R. Stat. Soc. Ser. B (Methodol.) **73**(3), 267–288 (1996)

History of Liquefaction Hazard Map Development and a New Method for Creating Hazard Maps for Low-Rise Houses

Susumu Yasuda(⊠) ⓘ

Tokyo Denki University, Hatoyama, Saitama, Japan
yasuda@g.dendai.ac.jp

Abstract. This paper introduces the development history, current situation and problems concerning hazard maps for soil liquefaction. A zonation manual produced by the TC 4 of ISSMFE in 1993, is introduced first. Then, the development of hazard maps for liquefaction in Japan is reviewed. A hazard map for liquefaction was first created by Ishihara and Ogawa in 1978. Currently, hazard maps exist for all administrative divisions of 47 administrative zones. However, they are not being fully used. In order to be fully used, hazard maps must be upgraded, mainly improving the following three points: i) accuracy and reliability, ii) maps specifically for low-rise housing, and iii) maps showing timelines of emergency risks immediately after an earthquake. In addition, there is a lack of risk communication between the government, home builders and residents. Therefore, in order to solve this problem, Japan's Ministry of Land, Infrastructure, Transport and Tourism organized a technical committee to propose a method for creating liquefaction hazard maps, and published the "Liquefaction Hazard Map Guide for Risk Communication" in February 2021. It is expected that such risk communication will promote measures against liquefaction.

Keywords: Soil liquefaction · Hazard map · Earthquake · Risk communication

1 Introduction

Liquefaction was recognized worldwide for the first time following the Niigata Earthquake and the Alaska Earthquake, which both occurred in 1964. Since these earthquakes, many methods to predict liquefaction have been developed. There are two kinds of prediction methods: those based on geomorphological conditions and those based on geotechnical data, such as SPT N-value. Therefore, hazard maps for liquefaction in many areas of the world have been created based on one or a combination of these prediction methods. In 1993, a zonation manual entitled "Manual for Zonation on Seismic Geotechnical Hazards" was produced by the TC 4 [1]. Several methods to create hazard maps for soil liquefaction and examples of hazard maps developed around the world by 1993 were introduced in the manual.

© The Author(s), under exclusive license to Springer Nature Switzerland AG 2022
L. Wang et al. (Eds.): PBD-IV 2022, GGEE 52, pp. 834–844, 2022.
https://doi.org/10.1007/978-3-031-11898-2_58

Recently, hazard maps for liquefaction have been made in many countries, including Japan, the USA and Turkey. It has been especially urgent to prepare hazard maps in Japan, because liquefaction-induced damage occurs there every one-to-two years. Liquefaction hazard maps should be used to alleviate the damage caused by liquefaction to houses, schools, public facilities, industrial facilities, lifelines, roads, etc. However, they are not being fully used. In order to be used, hazard maps must be upgraded.

2 Development of Hazard Maps in Japan

Soon after the 1964 Niigata Earthquake, geomorphological conditions at liquefied and unliquefied zones were investigated, and it was concluded that present river beds, old river beds and reclaimed lands were most likely to be liquefied. Kuribayashi and Tatsuoka [2] surveyed the literature to find information on liquefied zones during 44 past Japanese earthquakes. Based on these studies, geomorphology-based criteria were established to grossly classify the liquefaction potential of a given area into one of three groups, as shown in Table 1 (Iwasaki et al. [3]). The Public Works Research Institute then classified the liquefaction potential of all land in Japan according to this microzoning technique.

Table 1. A zonation procedure based upon geomorphological information (Iwasaki et al. [3]).

Rank	Geomorphological units	Liquefaction potential
A	Present river bed, Old river bed, Swamp, Reclaimed land, Interdune lowland	Liquefaction likely
B	Fan, Natural levee, Sand dune, Flood plain, Beach, Other plains	Liquefaction possible
C	Terrace, Hill, Mountain	Liquefaction not likely

The soil condition in Niigata City was investigated just after the 1964 Niigata Earthquake based on existing boring data and on data obtained from newly conducted borings. Then, a simple liquefaction prediction method based on critical SPT N-value was derived by several researchers. In 1970, liquefaction was first taken into consideration in a design code, the code for harbor facilities. Over the next four years, the consideration of liquefaction was also introduced in the design codes for highway bridges, railways and buildings. Prediction methods in this early stage were based on critical SPT N-values, as mentioned above. In 1971, Seed and Idriss [4], in the US, proposed another simple prediction method based on the safety factor (resistance factor) against liquefaction, F_L, which is the ratio of the liquefaction strength (cyclic resistance) of soils, R (or CRR) to earthquake-induced cyclic stress, L (or CSR). In Japan, several modified F_L methods have been proposed based on undrained cyclic triaxial tests and seismic response analyses. In 1972, Ishihara and Ogawa [5] made a microzoning map on liquefaction of a limited area of downtown Tokyo based on existing boring data.

In 1980, the design code for highway bridges was revised to reflect the liquefaction potential predicted by the safety factor against liquefaction F_L, or so called "F_L method," based on the studies of laboratory soil tests conducted by Iwasaki et al. [6, 7]. Since then, several F_L methods have been developed and introduced in almost all seismic design codes when the codes were revised. However, the formulae to evaluate R and L differ slightly by code. In particular, the formula for estimating R from the SPT N value differs for highway bridges, building foundations, port and harbor facilities, railway structures, and tailings dams. Please refer to the book edited by the Japanese Geotechnical Society [8]. However, the method of measuring the N value is specified in JIS A 1219, and the test is conducted by a unified method in Japan. Of these formulas, the formulae used in the code for highway bridges are mostly used for the zoning of liquefaction potential in each administrative division. Iwasaki et al. [6] further quantified the severity of possible liquefaction at any site by introducing a factor called the liquefaction potential index, P_L, described in Fig. 1. Most liquefaction hazard maps in Japan have been displayed in this P_L. Liquefaction hazard maps existed in 13 municipalities as of 1991, but by 2018 they were present in all 47 administrative division.

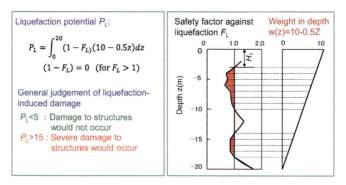

Fig. 1. Definition of liquefaction potential, P_L.

3   Problems with Hazard Maps for Soil Liquefaction

3.1   Problems Preventing the Wide Use of Hazard Maps in Society

Liquefaction hazard maps should be used to alleviate the damage caused by liquefaction to houses, schools, public facilities, industrial facilities, lifelines, roads, etc. However, they are not being fully used. In order to be fully used, hazard maps must be upgraded, mainly improving the following three points: i) accuracy and reliability, ii) maps specifically for low-rise housing, and iii) maps showing timelines of emergency risks immediately after an earthquake.

3.2   Accuracy and Reliability

Liquefaction occurs almost every one-two years during major earthquakes in Japan, so the accuracy of a hazard map before an earthquake can be verified by comparing it with the actual liquefied area. For example, during the 2011 Great East Japan (Tohoku) Earthquake, liquefaction occurred in several cities for which liquefaction hazard maps had been created. In Abiko City, liquefaction occurred only in a limited, narrow area, although the estimated liquefiable area was very wide. In general, local governments tend to overestimate the liquefiable area so that residents will not complain when future earthquakes occur. Also, the cost of creating a hazard map is generally low because these maps use only some of the existing boring data and a liquefaction-estimation method proposed for other cities or countries. No effort has been made to add new borings or to develop a unique estimation method.

So, accurate and reliable hazard maps must be made, especially by considering local soil characteristics. In general, liquefaction strength R for hazard maps is estimated from SPT N-value and fines content. However, hazard maps would be more accurate if they were based on liquefaction strength that also considered local soil characteristics, as determined by soil samplings and laboratory liquefaction tests. In addition, the reliability of hazard maps must be confirmed by comparing actual liquefaction-induced damage in future earthquakes to damage caused by past earthquakes.

I present an example in which this problem was overcome for Fukuoka City (Yasuda et al. [9]). As shown in Fig. 2 (1), the geomorphological conditions of Fukuoka City can be divided into four types: alluvial plain, sand dune, artificially reclaimed land, and mountain or hill. And there are three sand layers that must be examined for liquefaction: i) an alluvial sand layer which exists in most of the alluvial plain, ii) a sand dune layer which exists along the natural coast, and iii) an artificially reclaimed sand layer. As the first step, as shown in Fig. 2 (1), undisturbed samples were taken at the five sites where the above three types of sand were deposited. At those sites, as shown in Fig. 2 (2), excavation was performed to the depth where the sand layer was deposited, and undisturbed samples were taken by the block sampling method using thin walled tubes. Then, undrained cyclic triaxial tests were conducted to measure the R. Test results showed that the R of the alluvial sand was higher than the R of the reclaimed sand, dune sand and reconstituted alluvial sand. Therefore, comparing the cyclic triaxial test results with the formula used in the specification for highway bridges, a modified formula was derived to be applied to soil in Fukuoka City. There is no guideline for such a modification method in Japan, but since the author was involved in the development of the method for the specification for highway bridges [6, 7], I revised the formula based on that experience. Then, assuming a maximum surface acceleration of 200 cm/s^2, liquefiable layers were predicted at 13 soil cross sections. In the next step, the possible damage was judged based on the relationship between the thickness of the liquefiable layer, H_2, and the thickness of the upper non-liquefiable layer, H_1, shown in Fig. 3 (1) (Ishihara [10]). Figure 3 (2) shows the microzoning map for liquefaction thus determined.

The Fukuoka-ken Seiho-oki Earthquake, with a magnitude of 7.0 (Mj), occurred in 2005. Liquefaction occurred in many reclaimed lands and caused damage to quay walls, tanks, and sheds. The closed circles plotted in Fig. 3 (2) show liquefied sites. Almost all liquefied sites are located in the areas where structures had been deemed susceptible to damage. Therefore, the microzonation for liquefaction was fairly valid.

(1) Geomorphological map and sampling sites   (2) Undisturbed soil sampling

Fig. 2. Geomorphological map and sampling sites of Fukuoka City [9]

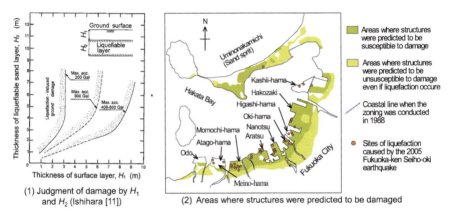

(1) Judgment of damage by H_1
and H_2 (Ishihara [11])

(2) Areas where structures were predicted to be damaged

Fig. 3. Hazard map and liquefied sites by the Fukuoka-ken Seiho-oki Earthquake [9].

3.3 Hazard Maps Specifically for Low-Rise Housings

Soil liquefaction causes various damage to structures, such as the settlement of buildings with raft foundations, the collapse of bridges with pile foundations, the uplift of manholes, the tilting of quay walls, the sliding of embankments, and the ground flow of gentle slopes. Therefore, it is important for hazard maps to show not only the liquefiable area but also the probable, quantitative, liquefaction-induced damage to all structures. However, current hazard maps use the P_L, or other indices only to indicate the probability and severity of liquefaction, and it is impossible to estimate the specific damage. It is necessary to create hazard maps that can quantitatively predict the damage to each structure.

In Japan, almost all new structures have been designed to prevent damage due to liquefaction. However, many small structures, such as low-rise housing and flat roads, are still designed without consideration of the effect of liquefaction. Therefore, the most useful hazard maps in Japan would be for low-rise housing and flat roads.

Several studies on the effect of the depth of the water table or the thickness of the unliquefied surface layer, H_1 on the damage to wooden houses were conducted for several cities damaged by the 2011 Tohoku Earthquake. Based on these studies, the Ministry of Land, Infrastructure, Transport and Tourism (MLIT) proposed a new criterion to estimate the liquefaction-induced damage to wooden houses in 2013 [11]. Figure 4 shows the ground classifications based on this new criterion, in which the possibility of damage can be estimated by P_L and the H_1. Using this method, hazard maps for liquefaction-induced damage to low-rise housing can be created.

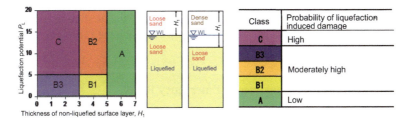

Fig. 4. A new method to estimate the liquefaction-induced damage to houses (MLIT [11]).

I present an example of a hazard map for general structures and low-rise housing in Hiroshima City prepared by the author and his colleagues, not by the Hiroshima City Government, partially quoting from Yasuda and Ishikawa [12]. Figure 5 (1) shows a series of representative soil profile models from upstream to downstream of the Ohta River along the A-A' line in Fig. 5 (2). A loose sand layer with SPT N-value of 5 to 10 is deposited from the ground surface to several meters below the surface in the whole area. A comparatively dense sand layer is deposited under the loose sand layer in the north (upstream) zone, and a soft clayey layer is deposited under the loose sand layer in the south (downstream) zone. The water table is 3 to 5 m deep in the upstream zone and rises to a depth of 1 to 2 m in the southern part of the city. Figure 5 (2) show hazard maps for damage to general structures based on P_L and for liquefaction-induced damage to low-rise houses based on P_L and H_1. The estimated P_L in the north zone is larger than that in the south zone because the loose sandy layer is thick in the north zone. However, the distribution of estimated damage to low-rise houses is the opposite. In

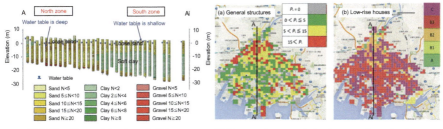

(1) Soil profile models along A-Aí line   (2) Damage to general structures and wooden houses

Fig. 5. Hazard maps in Hiroshima City. (Yasuda and Ishikawa [12]).

the north zone, severe damage to low-rise houses is not predicted because the ground water table is deeper than 3 m in many meshes. So, the depth of the water table must be carefully considered in the preparation of a hazard map for liquefaction-induced damage to wooden houses.

Groundwater levels fluctuate seasonally. According to the results of measuring the groundwater level for two years from June 2012 in Chiba City, where liquefaction occurred widely due to the 2011 Tohoku Earthquake, there were fluctuations of GL-1.2 m to GL-1.6 m during that period [13]. However, it rose to GL-0.6 m in the short term during heavy rains such as typhoons. When creating hazard maps, existing boring data is used, so the effects of seasonal fluctuations contained in them are unknown. Therefore, it is necessary to compare the groundwater level of the mesh to be examined with the groundwater level of the surrounding meshes, to examine the validity of the groundwater level setting.

3.4   Hazard Maps Showing Timelines of Emergency Risks

In urban areas, liquefaction damages many structures besides houses. Serious damage to flat roads occurred and interrupted traffic in residential areas along Tokyo Bay during the 2011 Tohoku Earthquake as shown in Fig. 6 (1) and 1 (2). Many sewage manholes were uplifted due to liquefaction during major earthquakes after the 1993 Kushio-oki Earthquake as shown in Fig. 6 (3). If a fire occurs in a residential area, inhabitants must evacuate quickly. In coastal areas, people have to escape from tsunamis that may occur

Fig. 6. Examples of damage to flat roads due to liquefaction.

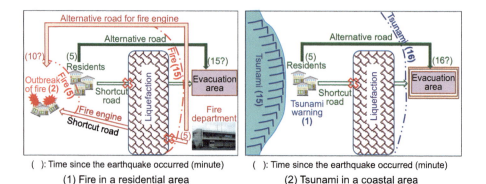

Fig. 7. Ideas for hazard maps that show the timelines of emergency risks.

soon after earthquakes. Therefore, hazard maps should show not only the probability of liquefaction, but also timelines of such temporary risks immediately after an earthquake as shown in Fig. 7.

4 New Guide by the Japanese MLIT

In Japan, measures against liquefaction are being taken for the foundations of large structures, such as middle-rise buildings, bridges and storage tanks. In addition, as mentioned above, liquefaction hazard maps have been created for many areas. Nevertheless, for residential areas, not many measures against liquefaction for the soil beneath low-rise houses, roads, and lifelines have not been taken. This is due to a lack of risk communication between the government, home builders and residents. Therefore, in order to solve this problem, MLIT organized a technical committee (chaired by the author) for three years to propose a method for creating liquefaction hazard maps that facilitate risk communication between the government, home builders, and residents. The MLIT published the "Liquefaction Hazard Map Guide for Risk Communication" in February 2021 [14]. In Japan, geomorphological conditions are classified by 250 m meshes, so the committee also created a "Schematic Map of the Possibility of Liquefaction" using the geomorphological classification and published it on MLIT's website in 2020.

Figure 8 shows the procedure for creating hazard maps that facilitate risk communication specified in this guide. First, a plan to create a detailed hazard map is made by a local government by referring to the "Schematic Map of the Possibility of Liquefaction." In creating the "Map of Liquefaction Susceptibility based on detailed Geomorphological Conditions", existing geomorphological classification maps, aerial photographs, topographic maps, digital elevation models and maps showing artificial terrain, such as reclaimed land, are collected. Then, detailed geomorphological classification is performed not by mesh display but by area display, and a "Map of Liquefaction Susceptibility" is created based on Table 2.

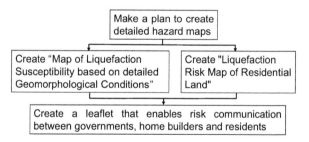

Fig. 8. Procedure for creating hazard maps for risk communication (MLIT [14]).

Table 2. Classification of liquefaction susceptibility by geomorphological information [14].

Liquefaction susceptibility		Natural terrain and artificial terrain	
High ↑ Low ↓	Natural terrain	Old river channel, Dune edge, Lowlands between dunes, Inter-sandbar lowland
	Artificial terrain	Reclaimed land, Excavated and backfilled land, Land filled in a wetland
	Natural terrain	Delta, Coastal lowland, Natural levee, Depressions on sandbar and dune
	Artificial terrain	Polder, Land filled in a depression, Land filled in a valley
	Natural terrain	Reef, Gravel island, Flooded lowland, Backswamp
	Natural terrain	Dune (excluding edge and lowland), Fan, Valley bottom lowland
	Natural terrain	Mountain, Hill, Foothills sedimentary area, Plateau

In creating the "Liquefaction Risk Map of Residential Land", the depth distribution of the F_L is estimated by collecting past boring data. Standard seismic motion is a medium motion with a maximum ground acceleration of 200 cm/s² and a magnitude of 7.5. Based on the P_L and the H_1, the possibility of damage to a low-rise house due to liquefaction is determined by the method shown in Fig. 4. Then, the risk level of low-rise housing against liquefaction is displayed on the map in ranks A to D, as shown in Fig. 5 (2) (b). However, this alone makes it difficult for residents and home builders to specifically imagine the risks of liquefaction. Therefore, the guide also provides a simple method to evaluate the amount of settlement and tilt angle of a house. In this method, the formula proposed by Steinbrenner to estimate the amount of settlement of a structure on an elastic body is used. However, the elastic modulus decreases due to liquefaction. Therefore, the relationship among the F_L, the R and the shear modulus ratio, G_1/σ'_c used in the analysis program ALID [15], shown in Fig. 9, is used as the rate of decrease in elastic modulus due to liquefaction. However, the limit of the reduction rate of elastic modulus is set to 1/300.

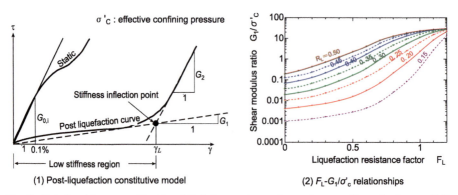

Fig. 9. Relationship among F_L, R and G_1/σ'_c to estimate the settlement of low-rise houses [14].

Finally, by adding the following information to the above two maps, a leaflet that enables risk communication between governments, home builders and residents is created by the local government. It is expected that such risk communication will promote measures against liquefaction.

(1) Adverse effects of road traffic obstruction due to liquefaction on evacuation behavior immediately after an earthquake
(2) Adverse effects of liquefaction damage on life after an earthquake and guidelines for this period
(3) Measures for the whole residential area and for individual houses

5 Conclusions

Though many hazard maps have been prepared in the world, current maps are not being fully used. They must be upgraded based on more studies in order to be fully used:

(1) Accurate and reliable hazard maps must be made, especially by considering local soil characteristics.

(2) It is important for hazard maps to show not only the liquefiable area but also lique faction-induced damage to low-rise housing.

References

1. Technical Committee for Earthquake Geotechnical Engineering, TC4, ISSMFE.: Manual for zonation on seismic geotechnical hazards. The Japanese Geotechnical Society, 209 (1999)
2. Kuribayashi, E., Tatsuoka, F.: Brief review of soil liquefaction during earthquakes in Japan. Soils Found. 15(4), 81–92 (1975)
3. Iwasaki, T., Tokida, K., Tatsuoka, F., Watanabe, S., Yasuda, S., Sato H.: Microzonation for soil liquefaction potential using simplified methods, Proc., 3rd International Conference on Microzonation 3, 1319–1330 (1982)
4. Seed, H.B., Idriss, I.M.: Simplified procedure for evaluating soil liquefaction potential. J. SMFD, ASCE 97(SM9), 1249–1273 (1971)
5. Ishihara, K., Ogawa, K.: Liquefaction susceptibility map of downtown Tokyo, Proc., 2nd International Conference on Microzonation 2, 897–910 (1978)
6. Iwasaki, T., Tatsuoka, F., Tokida, K., Yasuda, S.: A practical method for assessing soil lique-faction potential based on case studies at various sites in Japan, Proc. of the 2nd International Conference on Microzonation 2, 885–896 (1978)
7. Tatsuoka, F., Iwasaki, T., Tokida, K., Yasuda, S., Hirose, M., Imai, T., Kon-no, M.: A method for estimating undrained cyclic strength of sandy soils using standard penetration resistances. Soils Found. 18(3), 43–58 (1978)
8. The Japanese Geotechnical Society: Remedial measures against soil liquefaction, Balkema, 439 (1998)
9. Yasuda, S., Nagase, H., Tanoue, Y.: Microzonation for seismic geotechnical hazards and actual damage during the 2005 Fukuoka-ken Seiho-oki Earthquake. Soils Found. 46(6), 885–894 (2011)
10. Ishihara, K.: Stability of natural deposits during earthquakes. In: Proceedings, 11th International Conference on Soil Mechanics Foundation Eng. l, pp. 321–376 (1985)
11. MLIT: Technical guidelines for determining the possibility of liquefaction damage to residential land (2013, in Japanese)
12. Yasuda, S., Ishikawa, K.: Liquefaction-induced damage to wooden houses in Hiroshima and Tokyo during future earthquakes. In: Proc., 16th European Conference on Earthquake Engineering, Paper No.1014 (2018)

13. Yasuda, S., Hashimoto, T: New project to prevent liquefaction-induced damage in a wide existing residential area by lowering the ground water table. Soil Dynamics Earthquake Eng. **91**, 246–259 (2016)
14. MLIT: Liquefaction Hazard Map Guide for Risk Communication (in Japanese). https://www.mlit.go.jp/toshi/toshi_tobou_tk_000044.html. Accessed on 31 Oct 2021 (2021)
15. Yasuda, S., Yoshida, N., Adachi, K., Kiku, H., Ishikawa, K.: Simplified evaluation method of liquefaction-induced residual displacement. J. Japan Association Earthquake Eng. **17**(6), 1–20 (2017)

Dynamic Response Analysis of Slope Based on 3D Mesh Model Reconstruction and Electrical Resistivity Topography

Hanxu Zhou and Ailan Che[✉]

School of Naval Architecture, Ocean and Civil Engineering, Shanghai Jiao Tong University, 800 Dongchuan Road, Shanghai 200240, China
alche@sjtu.edu.cn

Abstract. Seismic motion is one of the important factors triggering slope instability and failure. In areas with frequent earthquakes, such as the southwestern region of China, the motion of earthquakes with small magnitude could induce the reactivation of slope, resulting in destruction to the surrounding public transportation infrastructure. Along the Dayong expressway in Yunnan province, China, a slope is situated in Damieju village, Chenghai town, Yongsheng county, having the trend developing into geological disaster. In order to investigate the potential risk of Damieju slope under earthquakes, electrical resistivity measurement containing 3 measuring lines were conducted in the field. Based on the 3D mesh model reconstruction method, the numerical model of Damieju slope is built using resistivity data, considering the spatial distribution characteristic of silt clay and mudstone. And the dynamic response of Damieju slope model is simulated using the commercial software Abaqus. The seismic wave record under the 2019 Yongsheng 4.9 earthquake is adopted as the input wave. The amplification coefficient of PGA reaches the maximum value of 4.18. The maximum value of displacement appears at the area with approximate height of 60 m, where could be the most unstable area of Damieju slope under the influence of seismic motion.

Keywords: Electrical resistivity topography · Mesh model reconstruction · Numerical simulation · Dynamic response

1 Introduction

Seismic motion is one of the principal causes of landslides around the world (Djerbal et al. 2020). Seismic landslides have characteristics of large number, vast scale and strong destructiveness. Therefore, the dynamic response analysis of unstable slope under earthquake is of great significance for landslide risk assessment and disaster prevention and reduction. With the improvement of computing power in recent years, numerical simulation has been a common method to analyze the seismic performance of slopes (Bhandari et al. 2016). The numerical simulation method can establish simplified numerical model according to the actual situation, and it also can simulate the behavior of slope

© The Author(s), under exclusive license to Springer Nature Switzerland AG 2022
L. Wang et al. (Eds.): PBD-IV 2022, GGEE 52, pp. 845–852, 2022.
https://doi.org/10.1007/978-3-031-11898-2_59

based on the different constitutive equations. However, in traditional mesh model building process, the model of slope is commonly simplified due to the complicated spatial distribution of geotechnical materials. The simplifications of slope model might affect the accuracy of dynamic response analysis of slopes.

In the 1980s, 3D mesh model reconstruction was gradually applied in industrial product design and product performance analysis (Agathos and Azariadis, 2018). At present, 3D mesh model reconstruction is widely used in medical simulation, composite performance research and dynamic analysis of material microstructure. In common mesh model reconstruction, the mesh model containing detailed spatial shape characteristics is established based on the images of computed tomography (CT). The reconstruction model of medical components like bone or viscera can be built from CT images for mechanical behavior analysis or process simulation (Agathos and Azariadis, 2018). For compound material, mesh model reconstruction method is adopted to analyze the effect of microstructure on macro properties (Huang and Li, 2013). In geotechnical engineering, mesh model reconstruction is adopted for standard specimen to study the influence of crack or pore on macro-mechanical properties (Song et al. 2020). It can be found that the mesh model reconstruction method is constrained by the scale of CT images in the previous studies. It results in the fact that it can be hard to reconstruct a mesh model for slopes due to its large size.

With the development of field test means in geotechnical engineering, electrical resistivity measurement has been extensively adopted to investigate risky slopes (Zhou and Che, 2021). Similar to the CT images, electrical resistivity measurement could determine the internal structure of slope using resistivity parameters. In the presented study, focusing on the Damieju slope along Dayong highway which is considered as a potential hazard under seismic motion. Electrical resistivity measurement is conducted for Damieju slope and the geotechnical materials of the slope are determined based on resistivity values. The mesh model of the slope is reconstructed using the results of electrical resistivity measurement. And the dynamic response of Damieju slope is calculated and analyzed with finite element software Abaqus.

2   Field Investigation of the Slope in Damieju Village

2.1   Overview of the Slope

The study area is located in the Damieju village, Chenghai town, Yongsheng county, Lijiang city, Yunnan province, China, as shown in Fig. 1. The position of Yunnan province is at the junction of Asia-Europe plate and Indian Ocean plate. The special tectonic conditions result in the frequent seismic activities around the study area. Figure 1 shows the epicenter of historical earthquakes with magnitude beyond 4.5 in Yunnan province from 1949 to 2018. Under the influence of frequent seismic motion, the stability of slopes decreases and has the possibility of failure sliding.

Dayong highway is a national way in Yunnan province, connecting Dali city and Yongsheng county, which is also one of the important and convenient passages for exit in Southwest China. During the field investigation, an ancient landslide was found adjacent to Dayong highway and defined as a potential geological hazard, which could pose a threat to the traffic safety of highway under earthquake, as shown in Fig. 1.

The recent earthquake with higher magnitude around the study area was the Ms4.9 Yongsheng earthquake occurred on July 21, 2019, whose epicenter was approximately 35 km away from the Damieju village (Fig. 1). The deposits of the slope can be easily activated by seismic motion. Therefore, it is necessary to analyze the response of the slope under earthquakes to avoid the effect of slope failure on traffic safety.

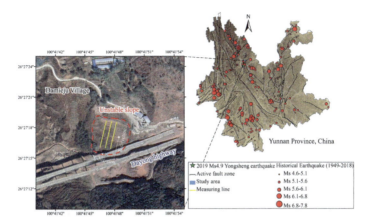

Fig. 1. Study area and location of Damieju slope

2.2 Electrical Resistivity Measurement

Electrical resistivity measurement was conducted to clarify the geological condition of Damieju slope. The Wenner (AMNB) acquirement method was adopted. Three measuring lines containing linear electrode arrays were arranged with a spacing of 40 m, as shown in Fig. 1. The length of each measuring line was about 480 m, containing 240 electrodes with a spacing of 2 m.

The 3D resistivity result is presented in Fig. 2, and the color corresponds to the value of resistivity. The apparent resistivity data were filtered to remove the outlier. Good convergence between measured data and resistivity results was achieved. It can be seen that the investigation depth was approximately 30 m, and the maximum value of resistivity reached 1600 $\Omega \cdot$m. The distribution of resistivity values within the slope was basically layered, and the value of resistivity increased with the depth. The drill hole information reveals that the Damieju slope is a typical soil slope consisted of overburden clay layer mixed with gravels and underlying mudstone formation. Based on the difference of resistivity values between clay and mudstone, the geotechnical material with relatively lower resistivity in the upper layer can be defined as the clay layer corresponding to the green color, and the material with higher resistivity values at bottom can be determined as the mudstone corresponding to the yellow or other colors.

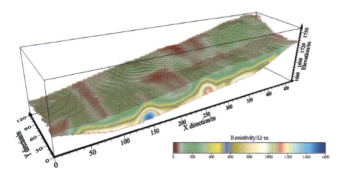

Fig. 2. Resistivity result of Damieju slope

2.3 Geotechnical Material Recognition

The physical and mechanical behaviors have significant differences for soil and rock. Therefore, it needs to establish different sets for various kinds of geotechnical materials to distinguish the property of soil and rock. The electrical resistivity values are adopted as the reference to recognize and segment the material categories quantitively.

The frequency of resistivity values is counted and the statistics result is shown in Fig. 3. It can be noticed that there are two principal peaks in the statistics results. One peak corresponds to the resistivity range of 0–300 Ω·m and the other peak corresponds to the range of 400–800 Ω·m. The frequency can be considered as the proportion of geotechnical material within certain resistivity range in the whole study area. There are two main kinds of materials including clay and mudstone distributed in the study area based on the drill hole information. Therefore, it can be deduced that the two peaks correspond to the component of clay and mudstone respectively. The minimum value (425 Ω·m) of the valley between two peaks is determined as the resistivity threshold for geotechnical material segmentation. The material with resistivity value lower than 425 Ω·m is recognized as clay layer, and the one with resistivity higher than 425 Ω·m is recognized as mudstone.

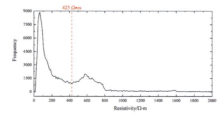

Fig. 3. Frequency statistic analysis of resistivity values

3   Mesh Model Reconstruction

3.1   Methodology of Model Reconstruction

The result of electrical resistivity measurement is the 3D discrete point cloud data with equal spacing. Mesh model reconstruction is a method to build mesh model for finite element simulation on the basis of discrete point cloud data. The methodology of mesh model reconstruction is based on the storage pattern of finite element mesh model in computer (Wang and Che, 2019). The reconstruction model is composed of eight-node hexahedral elements. And the size of each element in reconstruction model is the same as the spacing of discrete points in resistivity result.

The reconstruction mesh model mainly includes two parts of data to record the model information. One part is the number and the spatial position of all nodes. The nodes of reconstruction mesh model are corresponding to the discrete points of resistivity results. The other part is the number of element and the number of eight nodes within the element. This part of data describes the information of element property, and different sets of elements like clay and mudstone are recorded separately.

For the discrete point cloud data of resistivity results, each point is given a number, as show in Fig. 4. Then the first part data containing the number and the spatial position of all nodes is recorded. Every adjacent eight nodes are established as an element. Excel software is used for inquiry of the number of eight nodes within each element based on spatial position. And Hypermesh software is adopted to integrate the reconstruction mesh model.

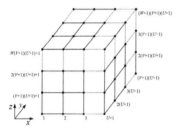

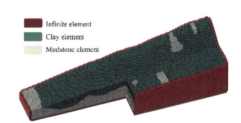

Fig. 4. Node number of 3D mesh model   **Fig. 5.** Reconstruction model of Damieju slope

3.2   Mech Model of Damieju Slope

The reconstruction mesh model of Damieju slope is presented in Fig. 5. The size of element shall not be greater than 1/10 of the seismic wave length. And the spacing of discrete points in resistivity results (2.0 m) is much smaller than 1/10 of the seismic wave length, so the size of element is set to 2.0 m × 2.0 m × 2.0 m. The scale of whole reconstruction model is 414 m × 120 m × 100 m in x, y, and z direction, respectively. The mesh model totally contains 383,437 nodes and 360,222 elements. The green elements in Fig. 5 represents the surface clay sets and the gray elements indicates the mudstone sets. The infinite elements which can reduce the reflection of seismic wave, are added to the lateral and bottom sides of reconstruction model, as shown as the red sets in Fig. 5.

4 Numerical Simulation of Damieju Slope Under Seismic Motion

4.1 Input Motion and Physical Parameters

The commercial software Abaqus is used to carry out the numerical simulation of Damieju slope. The acceleration wave record of Chenghai observation station in 2019 Ms4.9 Yongsheng earthquake is selected as the input motion of dynamic response simulation. The location of Chenghai observation station is only 1.5 km away from the Damieju village. However, the magnitude of Yongsheng earthquake was not high, and the amplitude of acceleration record wave was too small to reveal the dynamic response characteristics of Damieju slope. Therefore, the acceleration waveform is amplified to the design value of peak ground acceleration with maximum of 3.0 m/s^2, based on the local seismic fortification intensity. The amplified seismic wave and its frequency spectrum are shown in Fig. 6. The duration of acceleration wave is 30 s. The loading direction is horizontal and the loading position is the bottom nodes of mudstone elements. Dynamic implicit algorithm is adopted in the calculation process.

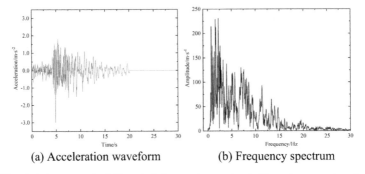

(a) Acceleration waveform   (b) Frequency spectrum

Fig. 6. Wave record of Chenghai observation station in Yongsheng earthquake

The parameters of clay and mudstone are derived from the laboratory geotechnical test, as shown in Table 1. In order to reduce the computing time of numerical simulation, the clay and mudstone material are assumed as the elastic material, and the plastic behavior of materials is not considered in simulation. The initial in-situ stress balance is carried out for the reconstruction model and then the seismic dynamic calculation steps are conducted.

Table. 1. Material parameters

Material	Density (kg/m^3)	Modulus (MPa)	Poisson ratio
Clay	2100	50	0.35
Mudstone	2480	1000	0.3

4.2   Acceleration Response

The acceleration and displacement calculation results of various positions of Damieju slope is summarized to analyze the dynamic response characteristics. Figure 7 shows the acceleration time history record of different height on slope surface. In the curve with height of 0 m, the peak value appears at the time of 5 s, being consistent with the input waveform. Except for the slight difference in acceleration value and phase, the variation trend between curves with different height and input seismic wave is basically the same.

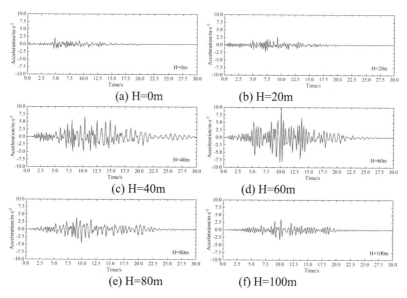

(a) H=0m

(b) H=20m

(c) H=40m

(d) H=60m

(e) H=80m

(f) H=100m

Fig. 7. Acceleration time history curve

The peak acceleration of input motion is 3.0 m/s^2, and it can be seen from Fig. 7 that Damieju slope has obvious amplification effect on seismic motion. Figure 8 shows the distribution of the maximum value of acceleration. The maximum value of acceleration reaches 12.55 m/s^2, which indicates that the amplification coefficient of Damieju slope reaches 4.18. And the maximum value of acceleration appears at the position with approximate height of 60 m. The amplification effect is not only related to the topography features, but also related to the thickness of surface soil. The results reveal that the reconstruction mesh model could quantitively consider the spatial distribution characteristics of geotechnical materials.

4.3   Displacement Response

Figure 9 shows the distribution of the maximum value of displacement. It can be found that the maximum value of displacement reaches 20.4 cm and it appears at the position with approximate height of 60 m. It indicates that this area may slide first under the influence of seismic motion.

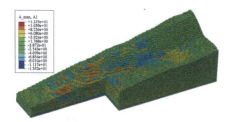

Fig. 8. Distribution of maximum value of acceleration

Fig. 9. Distribution of maximum value of displacement

5 Conclusion

This study presents a numerical analysis of Damieju slope based on 3D mesh model reconstruction using electrical resistivity. The electrical resistivity measurement is conducted in the field to investigate the geological conditions of the slope. And the clay layer and mudstone basement are determined based on the frequency statistics results and resistivity values. Based on the 3D mesh model reconstruction method, the numerical mesh model of Damieju slope considering detailed geological conditions is established. The recorded acceleration wave of 2019 Yongsheng earthquake is selected as the input motion of numerical calculation. The dynamic response analysis of Damieju slope reveals that the slope has obvious amplification effect on seismic motion, and the amplification coefficient of PGA reaches the maximum value of 4.18. The maximum value of displacement appears at the area with approximate height of 60 m, where could be the most unstable area of Damieju slope under the influence of seismic motion.

Acknowledgements. This work is financially supported by the National Key R&D Program of China (2018YFC1504504).

References

Agathos, A., Azariadis, P.: 3D reconstruction of skeletal mesh models and human foot biomodel generation using semantic parametric-based deformation. Int. J. Comput. Appl. **42**, 127–140 (2018)

Bhandari, T., et al.: Numerical modelling of seismic slope failure using MPM. Comput. Geotech. **75**, 126–134 (2016)

Djerbal, L., et al.: Assessment and mapping of earthquake-induced landslides in Tigzirt City. Algeria. Natural Hazards **87**, 1859–1879 (2020)

Huang, M., Li, Y.: X-ray tomography image-based reconstruction of microstructural finite element mesh models for heterogeneous materials. Comput. Mater. Sci. **67**, 63–72 (2013)

Song, R., et al.: Effects of pore structure on sandstone mechanical properties based on micro-CT reconstruction model. Advances in Civil Eng. 1–21 (2020)

Wang, Q., Che, A.: Reconstruction method of 3D element mesh models for unfavorable geology based on CT technology..Chin. J. Rock Mechan. Eng. **38**(6), 1222–1232 (2019)

Zhou, H., Che, A.: Geomaterial segmentation method using multidimensional frequency analysis based on electrical resistivity tomography. Eng. Geol. **284**, 105925 (2021)

Printed by Printforce, the Netherlands